Engineering of Smart Agriculture

Engineering of Smart Agriculture

Editors

Paweł Kiełbasa
Tadeusz Juliszewski
Sławomir Kurpaska

MDPI

Basel • Beijing • Wuhan • Barcelona • Belgrade • Novi Sad • Cluj • Manchester

Editors
Paweł Kiełbasa
University of Agriculture in Krakow
Kraków, Poland

Tadeusz Juliszewski
University of Agriculture in Krakow
Kraków, Poland

Sławomir Kurpaska
University of Agriculture in Krakow
Kraków, Poland

Editorial Office
MDPI
St. Alban-Anlage 66
4052 Basel, Switzerland

This is a reprint of articles from the Special Issue published online in the open access journal *Applied Sciences* (ISSN 2076-3417) (available at: https://www.mdpi.com/journal/applsci/special_issues/Engineering_Smart_Agriculture).

For citation purposes, cite each article independently as indicated on the article page online and as indicated below:

Lastname, A.A.; Lastname, B.B. Article Title. *Journal Name* **Year**, *Volume Number*, Page Range.

ISBN 978-3-0365-9496-5 (Hbk)
ISBN 978-3-0365-9497-2 (PDF)
doi.org/10.3390/books978-3-0365-9497-2

Cover image courtesy of Paweł Kiełbasa

Contents

Paweł Kiełbasa, Tadeusz Juliszewski and Sławomir Kurpaska
Special Issue on the Engineering of Smart Agriculture
Reprinted from: *Appl. Sci.* **2023**, *13*, 8523, doi:10.3390/app13148523 1

Amani Abdulrahman Albraikan, Mohammed Aljebreen, Jaber S. Alzahrani, Mahmoud Othman, Gouse Pasha Mohammed and Mohamed Ibrahim Alsaid
Modified Barnacles Mating Optimization with Deep Learning Based Weed Detection Model for Smart Agriculture
Reprinted from: *Appl. Sci.* **2022**, *12*, 12828, doi:10.3390/app122412828 5

Jiedong Feng, Yaqin Sun, Kefei Zhang, Yindi Zhao, Yi Ren, Yu Chen, et al.
Autonomous Detection of *Spodoptera frugiperda* by Feeding Symptoms Directly from UAV RGB Imagery
Reprinted from: *Appl. Sci.* **2022**, *12*, 2592, doi:10.3390/app12052592 21

Praveen Kumar Jayapal, Eunsoo Park, Mohammad Akbar Faqeerzada, Yun-Soo Kim, Hanki Kim, Insuck Baek, et al.
Analysis of RGB Plant Images to Identify Root Rot Disease in Korean Ginseng Plants Using Deep Learning
Reprinted from: *Appl. Sci.* **2022**, *12*, 2489, doi:10.3390/app12052489 37

Selami Kesler, Abdil Karakan and Yüksel Oğuz
Real-Time Strawberry Plant Classification and Efficiency Increase with Hybrid System Deep Learning: Microcontroller and Mobile Application
Reprinted from: *Appl. Sci.* **2022**, *12*, 8860, doi:10.3390/app12178860 53

Fangyi Li, Xumeng Li, Huang Huang, Hao Xiang, Chunyun Guan and Mei Guan
An Image Processing Method for Measuring the Surface Area of Rapeseed Pods
Reprinted from: *Appl. Sci.* **2023**, *13*, 5129, doi:10.3390/app13085129 71

Zihao Meng, Lixin Zhang, He Li, Runmeng Zhou, Haoran Bu, Yongchao Shan, et al.
Design and Application of Liquid Fertilizer pH Regulation Controller Based on BP-PID-Smith Predictive Compensation Algorithm
Reprinted from: *Appl. Sci.* **2022**, *12*, 6162, doi:10.3390/app12126162 91

Wilmer Quimbita, Edison Toapaxi and Jacqueline Llanos
Smart Irrigation System Considering Optimal Energy Management Based on Model Predictive Control (MPC)
Reprinted from: *Appl. Sci.* **2022**, *12*, 4235, doi:10.3390/app12094235 111

Dorijan Radočaj, Ivan Plaščak, Goran Heffer and Mladen Jurišić
A Low-Cost Global Navigation Satellite System Positioning Accuracy Assessment Method for Agricultural Machinery
Reprinted from: *Appl. Sci.* **2022**, *12*, 693, doi:10.3390/app12020693 129

Nawab Khan, Ram L. Ray, Hazem S. Kassem and Shemei Zhang
Mobile Internet Technology Adoption for Sustainable Agriculture: Evidence from Wheat Farmers
Reprinted from: *Appl. Sci.* **2022**, *12*, 4902, doi:10.3390/app12104902 143

Volodymyr Bulgakov, Aivars Aboltins, Semjons Ivanovs, Hristo Beloev, Volodymyr Nadykto, Yevhen Ihnatiev and Jüri Olt
Theory of Movement of Machine-Tractor Unit with Trailer Haulm Harvester Machine
Reprinted from: *Appl. Sci.* **2022**, *12*, 3901, doi:10.3390/app12083901 163

Lexing Deng, Tianyu Liu, Ping Jiang, Fangping Xie, Junchi Zhou, Wenhan Yang and Aolin Qi
Design of an Adaptive Algorithm for Feeding Volume–Traveling Speed Coupling Systems of Rice Harvesters in Southern China
Reprinted from: *Appl. Sci.* **2023**, *13*, 4876, doi:10.3390/app13084876 . . . . **177**

Leonardo Vita and Davide Gattamelata
Analytical Method for Assessing Stability of a Counterbalanced Forklift Truck Assembled with Interchangeable Equipment
Reprinted from: *Appl. Sci.* **2023**, *13*, 1206, doi:10.3390/app13021206 . . . . **197**

Yuan Gao, Guozhong Zhang, Hongchang Wang, Abouelnadar Salem, Jianwei Fu and Yong Zhou
Measuring System Design and Experiment for Ground Pressure on Seeding Skateboard of Rice Direct Seeding Machine
Reprinted from: *Appl. Sci.* **2021**, *11*, 10024, doi:10.3390/app112110024 . . . . **209**

Bongki Lee, Donghwan Kam, Yongjin Cho, Dae-Cheol Kim and Dong-Hoon Lee
Comparing Performances of CNN, BP, and SVM Algorithms for Differentiating Sweet Pepper Parts for Harvest Automation
Reprinted from: *Appl. Sci.* **2021**, *11*, 9583, doi:10.3390/app11209583 . . . . **227**

Seo-Yong Shin, Myoung-Ho Kim, Yongjin Cho and Dae-Cheol Kim
CFD Analysis and Validation of a Foreign Material Winnowing Machine for Pepper Harvester
Reprinted from: *Appl. Sci.* **2022**, *12*, 6134, doi:10.3390/app12126134 . . . . **247**

Jesus David Chaux, David Sanchez-Londono and Giacomo Barbieri
A Digital Twin Architecture to Optimize Productivity within Controlled Environment Agriculture
Reprinted from: *Appl. Sci.* **2021**, *11*, 8875, doi:10.3390/app11198875 . . . . **263**

Fudi Chen, Yishuai Du, Tianlong Qiu, Zhe Xu, Li Zhou, Jianping Xu, et al.
Design of an Intelligent Variable-Flow Recirculating Aquaculture System Based on Machine Learning Methods
Reprinted from: *Appl. Sci.* **2021**, *11*, 6546, doi:10.3390/app11146546 . . . . **275**

Ousmane Wane, Luis F. Zarzalejo, Francisco Ferrera-Cobos, Ana A. Navarro, Alberto Rodríguez-López and Rita X. Valenzuela
Generation of Typical Meteorological Sequences to Simulate Growth and Production of Biological Systems
Reprinted from: *Appl. Sci.* **2023**, *13*, 4826, doi:10.3390/app13084826 . . . . **291**

Iris Ramaj, Steffen Schock, Shkelqim Karaj and Joachim Müller
Influence of Self-Compaction on the Airflow Resistance of Aerated Wheat Bulks (*Triticum aestivum* L., cv. 'Pionier')
Reprinted from: *Appl. Sci.* **2022**, *12*, 8909, doi:10.3390/app12178909 . . . . **307**

Ali Mousavi, Ebrahim Asadi Aghbolaghi, Ali Khorramifar, Marek Gancarz, Yousef Darvishi, Mateusz Stasiak, et al.
Life Cycle Assessment for Environmental Impact Reduction and Evaluation of the Energy Indices in Lettuce Production
Reprinted from: *Appl. Sci.* **2022**, *12*, 10348, doi:10.3390/app122010348 . . . . **327**

Khulekani Cyprian Mbatha, Charmaine Ntokozo Mchunu, Sydney Mavengahama and Nontuthuko Rosemary Ntuli
Effect of Poultry and Goat Manures on the Nutrient Content of *Sesamum alatum* Leafy Vegetables
Reprinted from: *Appl. Sci.* **2021**, *11*, 11933, doi:10.3390/app112411933 . . . . **339**

Chaosai Liu, Yang Zhou, Guixiang Chen, Deqian Zheng and Longfei Yue
Compression and Fungal Heat Production in Maize Bulk Considering Kernel Breakage
Reprinted from: *Appl. Sci.* **2022**, *12*, 4870, doi:10.3390/app12104870 . . . . **353**

Alba Cruz Coronel, Conrado Parraguirre Lezama, Yesenia Pacheco Hernández, Olga Santiago Trinidad, Antonio Rivera Tapia and Omar Romero-Arenas
Efficacy of Four In Vitro Fungicides for Control of Wilting of Strawberry Crops in Puebla-Mexico
Reprinted from: *Appl. Sci.* **2022**, *12*, 3213, doi:10.3390/app12073213 . . . . **367**

Arkadiusz Stepień, Katarzyna Wojtkowiak and Ewelina Kolankowska
Effect of Commercial Microbial Preparations Containing *Paenibacillus azotofixans, Bacillus megaterium* and *Bacillus subtilis* on the Yield and Photosynthesis of Winter Wheat and the Nitrogen and Phosphorus Content in the Soil
Reprinted from: *Appl. Sci.* **2022**, *12*, 12541, doi:10.3390/app122412541 . . . . **383**

Felipe Augusto Moretti Ferreira Pinto, Victor Biazzotto Correia Porto, Rafaela Ara újo Guimarães, Carolina da Silva Siqueira, Mirian Rabelo de Faria, José da Cruz Machado, et al.
Detection and Factors That Induce *Stenocarpella* spp. Survival in Maize Stubble and Soil Suppressiveness under Tropical Conditions
Reprinted from: *Appl. Sci.* **2022**, *12*, 4974, doi:10.3390/app12104974 . . . . **401**

Bogdan Saletnik, Grzegorz Zaguła, Aneta Saletnik, Marcin Bajcar, Ewelina Słysz and Czesław Puchalski
Method for Prolonging the Shelf Life of Apples after Storage
Reprinted from: *Appl. Sci.* **2022**, *12*, 3975, doi:10.3390/app12083975 . . . . **413**

Bogdan Saletnik, Grzegorz Zaguła, Aneta Saletnik, Marcin Bajcar, Ewelina Słysz and Czesław Puchalski
Effect of Magnetic and Electrical Fields on Yield, Shelf Life and Quality of Fruits
Reprinted from: *Appl. Sci.* **2022**, *12*, 3183, doi:10.3390/app12063183 . . . . **429**

Addis Lemessa, Ernest Popardowski, Tomasz Hebda and Tomasz Jakubowski
The Effect of UV-C Irradiation on the Mechanical and Physiological Properties of Potato Tuber and Different Products
Reprinted from: *Appl. Sci.* **2022**, *12*, 5907, doi:10.3390/app12125907 . . . . **451**

Grzegorz Zaguła, Bogdan Saletnik, Marcin Bajcar, Aneta Saletnik and Czesław Puchalski
Preliminary Research on the Influence of a Pulsed Magnetic Field on the Cationic Profile of Sunflower, Cress, and Radish Sprouts and on Their Germination Rate
Reprinted from: *Appl. Sci.* **2021**, *11*, 9678, doi:10.3390/app11209678 . . . . **471**

Ernest Popardowski and Paweł Kiełbasa
Influence of Broadleaved Wood Conditioning by Pulsed Electric Field on Its Combustion Heat Characteristics
Reprinted from: *Appl. Sci.* **2022**, *12*, 5048, doi:10.3390/app12105048 . . . . **485**

applied sciences

MDPI

Editorial

Special Issue on the Engineering of Smart Agriculture

Paweł Kiełbasa *, Tadeusz Juliszewski and Sławomir Kurpaska

Faculty of Production and Power Engineering, University of Agriculture in Krakow, Balicka Av. 116B, 30-149 Krakow, Poland; tadeusz.juliszewski@urk.edu.pl (T.J.); slawomir.kurpaska@urk.edu.pl (S.K.)
* Correspondence: pawel.kielbasa@urk.edu.pl

1. Introduction

The monograph presents an extract from the reality of smart agriculture, where the combination of modern technologies, innovative solutions, and sustainable approaches to food production classifies this part of science as highly interdisciplinary, multifaceted, and technologically advanced. Furthermore, innovative methods that are shaping the future of the food system are presented.

The need to increase productivity, optimize natural resources, and minimize environmental impact requires new approaches. In this context, smart agriculture is emerging as a solution that combines technology, data, and science to achieve sustainable, efficient, and innovative food production. This issue introduces the field of smart farming, which encompasses a range of advanced technologies such as the Internet of Things (IoT), Artificial Intelligence (AI), robotics, automation, drones, and precision agriculture.

The use of these tools allows the monitoring and optimization of crop conditions, precise fertilization, minimization of water and energy usage, and improvement of crop quality and quantity. In addition, plant monitoring systems are described, which, by means of sensors and data analysis, provide farmers with valuable information about plant health, soil moisture, temperature, and other factors affecting crop growth. A significant part of this monograph deals with the automation of agricultural processes, where robots and machines undertake tasks with high precision and accuracy, contributing to the farmer's efficiency. It outlines how smart farming can contribute to reducing greenhouse gas emissions, minimizing water usage, reducing waste, and protecting biodiversity. Practical examples from different regions of the world where smart farming has already been successfully implemented, benefiting both farmers and consumers, are also highlighted. This monograph aims not only to provide an understanding of smart agriculture, but also to inspire the reader to think about the future of agriculture and the ways in which modern food production methods can be improved. By understanding the technological potential and being aware of the growing challenges, we are able to move towards a more sustainable, efficient, and resilient food system.

Citation: Kiełbasa, P.; Juliszewski, T.; Kurpaska, S. Special Issue on the Engineering of Smart Agriculture. *Appl. Sci.* **2023**, *13*, 8523. https://doi.org/10.3390/app13148523

Received: 13 July 2023
Accepted: 19 July 2023
Published: 24 July 2023

2. Contemporary Systems for Intelligent Farming

This monograph delves into various aspects of intelligent agriculture research, covering studies of crop diseases, pest detection, plant nutrition, precision farming technologies, and post-harvest techniques using machine learning-based artificial intelligence algorithms. The diverse range of matters covered in the monograph emphasizes the multidisciplinary nature of modern agricultural research and ongoing efforts to improve yield, quality, and sustainability. This monograph provides a comprehensive overview of the latest research and technological advances in the field of agriculture, offering valuable insights and practical solutions from image processing methods for precise area measurement [1] to the design of adaptive algorithms for efficient rice harvesting [2]. It also analyzes the generation of meteorological sequences for simulating the growth of biological systems [3] and analytical methods for assessing forklift stability [4]. In addition, the resource use of deep learning models for weed detection in smart agriculture [5] and the impact of microbial preparations on yields and soil nutrients [6] are discussed. In addition, it examines the energy consumption and environmental impact of lettuce production [7] and the resistance to airflow in aerated wheat masses [8]. Finally, it investigates real-time plant classification using deep learning techniques [9] and the design of a controller for pH regulation of liquid fertilizers [10]. It includes research on the analysis and validation of a foreign material shaking machine for pepper harvesters [11], the effect of pulsed electric field conditioning on leaf-wood burning characteristics [12], and the detection and factors that induce *Stenocarpella* spp. survival in maize stubble and soil suppression [13]. Moreover, this monograph investigates the adoption and use of mobile internet technology in sustainable agriculture among wheat farmers [14], compression and heat generation by fungi in bulk maize with consideration of kernel cracking [15], and the implementation of a smart irrigation system with consideration of optimal energy management [16]. In addition, a method for extending the shelf life of apples after storage using low magnetic fields [17] and the theory of motion of a machine-tractor-trailer combination for harvesting beet tops [18] have been investigated. These studies provide valuable insights into innovative technologies, management practices, and theoretical frameworks that contribute to the development of agricultural practices, crop quality, and resource efficiency. One of the areas of interest in this monograph is the control of strawberry wilt disease caused by *Fusarium solani*, which is an economic challenge for strawberry producers in Mexico [19]. The in vitro efficacy of four fungicides against *F. solani* is being evaluated to identify effective means of controlling this devastating disease [19]. Similarly, the use of autonomous drones equipped with deep learning algorithms to detect *Spodoptera frugiperda*, a destructive pest of maize, based on foraging symptoms observed in RGB images, is being investigated [20]. This application of digital technologies has the potential to revolutionize pest detection in precision agriculture. The use of image processing and deep learning algorithms to identify root rot disease in Korean ginseng plants from RGB plant images has been proposed [21]. An inexpensive method for assessing the positioning accuracy of a global navigation satellite system was presented, providing a flexible framework for assessing the precision of agricultural machinery [22]. The efficiency of selective fertilization of leafy vegetables with manure was also investigated [23]. Other research has developed a device to measure the ground pressure of paddy fields using internet of things technology and wireless data transmission [24]. In addition, the effect of pulsed magnetic fields on the cation profile of sunflower, cress, and radish sprouts was investigated, revealing potential benefits for seed germination and nutrient content [25]. This monograph also discusses advances in agricultural machinery. Image recognition algorithms have been developed and compared, which play a key role in effectively distinguishing between different parts of the sweet pepper plant for automation purposes. Using techniques such as normalized difference vegetation index (NDVI) and local feature analysis, this study compares the performance of different algorithms. The results demonstrate the capabilities of these algorithms, with significant successes achieved by the convolutional neural network (CNN) approach [26]. An architecture potentially capable of optimizing productivity has been proposed, as it uses simulation software to optimize (i) climate control strategies related to crop microclimate control and (ii) crop management treatments [27]. The design of an intelligent recirculating variable flow aquaculture system, based on machine learning methods, is presented to optimize productivity while maintaining a clean and stable aquatic environment [28]. Furthermore, this monograph deals with the effects of UV-C radiation on the mechanical and physiological properties of potato tubers and various agro-products. The review highlights the dose-dependent nature of UV-C treatment and its potential to extend shelf life and improve quality [29]. Finally, this monograph looks at the effects of magnetic and electric fields on fruit yield, shelf life, and quality. The literature on the use of magnetic and electric fields in agricultural production is reviewed, highlighting their potential to improve plant growth, firmness, ripening, and nutrient content [30]. Overall, this monograph presents a comprehensive body of research that contributes to the development of agricultural practices, disease management, crop quality, and resource efficiency.

3. The Future of Smart Farming

The degree of technical sophistication of modern agriculture and the variety of information, mechatronic, and satellite systems used allow us to conclude that this will be one of those disciplines whose development will be progressive. The multi-disciplinary nature of the research allows for the concentration of experts from different fields in one place and the creation of joint scientific projects that complement each other. Therefore, this is the first of many monograph editions related to smart agriculture.

Author Contributions: Conceptualization, P.K.; methodology, P.K.; formal analysis, P.K., T.J. and S.K.; supervision, P.K., T.J. and S.K. All authors have read and agreed to the published version of the manuscript.

Acknowledgments: Publication of this monograph would not be possible without the contributions of various talented authors, hardworking and professional reviewers, and the dedicated editorial team of Applied Sciences. Congratulations to all authors—regardless of the final decisions of the submitted manuscripts, the feedback, comments, and suggestions from the reviewers and editors helped the authors improve their papers. We would like to take this opportunity to express our sincere gratitude to all the reviewers. Finally, we would like to express our gratitude to the editorial team of *Applied Sciences*, with special thanks to the managing editor from the Beijing MDPI Branch Office.

Conflicts of Interest: The authors declare no conflict of interest.

References

1. Li, F.; Li, X.; Huang, H.; Xiang, H.; Guan, C.; Guan, M. An Image Processing Method for Measuring the Surface Area of Rapeseed Pods. *Appl. Sci.* **2023**, *13*, 5129. [CrossRef]
2. Deng, L.; Liu, T.; Jiang, P.; Xie, F.; Zhou, J.; Yang, W.; Qi, A. Design of an Adaptive Algorithm for Feeding Volume–Traveling Speed Coupling Systems of Rice Harvesters in Southern China. *Appl. Sci.* **2023**, *13*, 4876. [CrossRef]
3. Wane, O.; Zarzalejo, L.F.; Ferrera-Cobos, F.; Navarro, A.A.; Rodríguez-López, A.; Valenzuela, R.X. Generation of Typical Meteorological Sequences to Simulate Growth and Production of Biological Systems. *Appl. Sci.* **2023**, *13*, 4826. [CrossRef]
4. Vita, L.; Gattamelata, D. Analytical Method for Assessing Stability of a Counterbalanced Forklift Truck Assembled with Interchangeable Equipment. *Appl. Sci.* **2023**, *13*, 1206. [CrossRef]
5. Albraikan, A.A.; Aljebreen, M.; Alzahrani, J.S.; Othman, M.; Mohammed, G.P.; Ibrahim Alsaid, M. Modified Barnacles Mating Optimization with Deep Learning Based Weed Detection Model for Smart Agriculture. *Appl. Sci.* **2022**, *12*, 12828. [CrossRef]
6. Stępień, A.; Wojtkowiak, K.; Kolankowska, E. Effect of Commercial Microbial Preparations Containing *Paenibacillus azotofixans*, *Bacillus megaterium* and *Bacillus subtilis* on the Yield and Photosynthesis of Winter Wheat and the Nitrogen and Phosphorus Content in the Soil. *Appl. Sci.* **2022**, *12*, 12541. [CrossRef]
7. Mousavi, A.; Aghbolaghi, E.A.; Khorramifar, A.; Gancarz, M.; Darvishi, Y.; Stasiak, M.; Miernik, A.; Karami, H. Life Cycle Assessment for Environmental Impact Reduction and Evaluation of the Energy Indices in Lettuce Production. *Appl. Sci.* **2022**, *12*, 10348. [CrossRef]
8. Ramaj, I.; Schock, S.; Karaj, S.; Müller, J. Influence of Self-Compaction on the Airflow Resistance of Aerated Wheat Bulks (*Triticum aestivum* L., cv. 'Pionier'). *Appl. Sci.* **2022**, *12*, 8909. [CrossRef]
9. Kesler, S.; Karakan, A.; Oğuz, Y. Real-Time Strawberry Plant Classification and Efficiency Increase with Hybrid System Deep Learning: Microcontroller and Mobile Application. *Appl. Sci.* **2022**, *12*, 8860. [CrossRef]
10. Meng, Z.; Zhang, L.; Li, H.; Zhou, R.; Bu, H.; Shan, Y.; Ma, X.; Ma, R. Design and Application of Liquid Fertilizer pH Regulation Controller Based on BP-PID-Smith Predictive Compensation Algorithm. *Appl. Sci.* **2022**, *12*, 6162. [CrossRef]
11. Shin, S.-Y.; Kim, M.-H.; Cho, Y.; Kim, D.-C. CFD Analysis and Validation of a Foreign Material Winnowing Machine for Pepper Harvester. *Appl. Sci.* **2022**, *12*, 6134. [CrossRef]
12. Popardowski, E.; Kiełbasa, P. Influence of Broadleaved Wood Conditioning by Pulsed Electric Field on Its Combustion Heat Characteristics. *Appl. Sci.* **2022**, *12*, 5048. [CrossRef]
13. Pinto, F.A.M.F.; Porto, V.B.C.; Guimarães, R.A.; Siqueira, C.d.S.; Faria, M.R.d.; Machado, J.d.C.; Medeiros, H.N.; Silva, D.D.d.; Santos Neto, H.; Pozza, E.A.; et al. Detection and Factors That Induce *Stenocarpella* spp. Survival in Maize Stubble and Soil Suppressiveness under Tropical Conditions. *Appl. Sci.* **2022**, *12*, 4974. [CrossRef]
14. Khan, N.; Ray, R.L.; Kassem, H.S.; Zhang, S. Mobile Internet Technology Adoption for Sustainable Agriculture: Evidence from Wheat Farmers. *Appl. Sci.* **2022**, *12*, 4902. [CrossRef]
15. Liu, C.; Zhou, Y.; Chen, G.; Zheng, D.; Yue, L. Compression and Fungal Heat Production in Maize Bulk Considering Kernel Breakage. *Appl. Sci.* **2022**, *12*, 4870. [CrossRef]

16. Quimbita, W.; Toapaxi, E.; Llanos, J. Smart Irrigation System Considering Optimal Energy Management Based on Model Predictive Control (MPC). *Appl. Sci.* **2022**, *12*, 4235. [CrossRef]
17. Saletnik, B.; Zaguła, G.; Saletnik, A.; Bajcar, M.; Słysz, E.; Puchalski, C. Method for Prolonging the Shelf Life of Apples after Storage. *Appl. Sci.* **2022**, *12*, 3975. [CrossRef]
18. Bulgakov, V.; Aboltins, A.; Ivanovs, S.; Beloev, H.; Nadykto, V.; Ihnatiev, Y.; Olt, J. Theory of Movement of Machine-Tractor Unit with Trailer Haulm Harvester Machine. *Appl. Sci.* **2022**, *12*, 3901. [CrossRef]
19. Coronel, A.C.; Parraguirre Lezama, C.; Pacheco Hernández, Y.; Santiago Trinidad, O.; Rivera Tapia, A.; Romero-Arenas, O. Efficacy of Four In Vitro Fungicides for Control of Wilting of Strawberry Crops in Puebla-Mexico. *Appl. Sci.* **2022**, *12*, 3213. [CrossRef]
20. Feng, J.; Sun, Y.; Zhang, K.; Zhao, Y.; Ren, Y.; Chen, Y.; Zhuang, H.; Chen, S. Autonomous Detection of *Spodoptera frugiperda* by Feeding Symptoms Directly from UAV RGB Imagery. *Appl. Sci.* **2022**, *12*, 2592. [CrossRef]
21. Jayapal, P.K.; Park, E.; Faqeerzada, M.A.; Kim, Y.-S.; Kim, H.; Baek, I.; Kim, M.S.; Sandanam, D.; Cho, B.-K. Analysis of RGB Plant Images to Identify Root Rot Disease in Korean Ginseng Plants Using Deep Learning. *Appl. Sci.* **2022**, *12*, 2489. [CrossRef]
22. Radočaj, D.; Plaščak, I.; Heffer, G.; Jurišić, M. A Low-Cost Global Navigation Satellite System Positioning Accuracy Assessment Method for Agricultural Machinery. *Appl. Sci.* **2022**, *12*, 693. [CrossRef]
23. Mbatha, K.C.; Mchunu, C.N.; Mavengahama, S.; Ntuli, N.R. Effect of Poultry and Goat Manures on the Nutrient Content of *Sesamum alatum* Leafy Vegetables. *Appl. Sci.* **2021**, *11*, 11933. [CrossRef]
24. Gao, Y.; Zhang, G.; Wang, H.; Salem, A.; Fu, J.; Zhou, Y. Measuring System Design and Experiment for Ground Pressure on Seeding Skateboard of Rice Direct Seeding Machine. *Appl. Sci.* **2021**, *11*, 10024. [CrossRef]
25. Zaguła, G.; Saletnik, B.; Bajcar, M.; Saletnik, A.; Puchalski, C. Preliminary Research on the Influence of a Pulsed Magnetic Field on the Cationic Profile of Sunflower, Cress, and Radish Sprouts and on Their Germination Rate. *Appl. Sci.* **2021**, *11*, 9678. [CrossRef]
26. Lee, B.; Kam, D.; Cho, Y.; Kim, D.-C.; Lee, D.-H. Comparing Performances of CNN, BP, and SVM Algorithms for Differentiating Sweet Pepper Parts for Harvest Automation. *Appl. Sci.* **2021**, *11*, 9583. [CrossRef]
27. Chaux, J.D.; Sanchez-Londono, D.; Barbieri, G. A Digital Twin Architecture to Optimize Productivity within Controlled Environment Agriculture. *Appl. Sci.* **2021**, *11*, 8875. [CrossRef]
28. Chen, F.; Du, Y.; Qiu, T.; Xu, Z.; Zhou, L.; Xu, J.; Sun, M.; Li, Y.; Sun, J. Design of an Intelligent Variable-Flow Recirculating Aquaculture System Based on Machine Learning Methods. *Appl. Sci.* **2021**, *11*, 6546. [CrossRef]
29. Lemessa, A.; Popardowski, E.; Hebda, T.; Jakubowski, T. The Effect of UV-C Irradiation on the Mechanical and Physiological Properties of Potato Tuber and Different Products. *Appl. Sci.* **2022**, *12*, 5907. [CrossRef]
30. Saletnik, B.; Zaguła, G.; Saletnik, A.; Bajcar, M.; Słysz, E.; Puchalski, C. Effect of Magnetic and Electrical Fields on Yield, Shelf Life and Quality of Fruits. *Appl. Sci.* **2022**, *12*, 3183. [CrossRef]

applied sciences

MDPI

Article

Modified Barnacles Mating Optimization with Deep Learning Based Weed Detection Model for Smart Agriculture

Amani Abdulrahman Albraikan [1], Mohammed Aljebreen [2], Jaber S. Alzahrani [3], Mahmoud Othman [4], Gouse Pasha Mohammed [5,*] and Mohamed Ibrahim Alsaid [5]

1 Department of Computer Sciences, College of Computer and Information Sciences, Princess Nourah Bint Abdulrahman University, Riyadh 11671, Saudi Arabia
2 Department of Computer Science, Community College, King Saud University, Riyadh 11437, Saudi Arabia
3 Department of Industrial Engineering, College of Engineering at Alqunfudah, Umm Al-Qura University, Makkah 24382, Saudi Arabia
4 Department of Computer Science, Faculty of Computers and Information Technology, Future University in Egypt New Cairo, Cairo 11835, Egypt
5 Department of Computer and Self Development, Preparatory Year Deanship, Prince Sattam Bin Abdulaziz University, Al-Kharj 16278, Saudi Arabia
* Correspondence: g.mohammed@psau.edu.sa

Abstract: Weed control is a significant means to enhance crop production. Weeds are accountable for 45% of the agriculture sector's crop losses, which primarily occur because of competition with crops. Accurate and rapid weed detection in agricultural fields was a difficult task because of the presence of a wide range of weed species at various densities and growth phases. Presently, several smart agriculture tasks, such as weed detection, plant disease detection, species identification, water and soil conservation, and crop yield prediction, can be realized by using technology. In this article, we propose a Modified Barnacles Mating Optimization with Deep Learning based weed detection (MBMODL-WD) technique. The MBMODL-WD technique aims to automatically identify the weeds in the agricultural field. Primarily, the presented MBMODL-WD technique uses the Gabor filtering (GF) technique for the noise removal process. For automated weed detection, the presented MBMODL-WD technique uses the DenseNet-121 model as feature extraction with the MBMO algorithm as hyperparameter optimization. The design of the MBMO algorithm involves the integration of self-population-based initialization with the standard BMO algorithm. At last, the Elman Neural Network (ENN) method was applied for the weed classification process. To demonstrate the enhanced performance of the MBMODL-WD approach, a series of simulation analyses were performed. A comprehensive set of simulations highlighted the enhanced performance of the presented MBMODL-WD methodology over other DL models with a maximum accuracy of 98.99%.

Keywords: smart agriculture; weed management; crop productivity; computer vision; deep learning

Citation: Albraikan, A.A.; Aljebreen, M.; Alzahrani, J.S.; Othman, M.; Mohammed, G.P.; Ibrahim Alsaid, M. Modified Barnacles Mating Optimization with Deep Learning Based Weed Detection Model for Smart Agriculture. *Appl. Sci.* **2022**, *12*, 12828. https://doi.org/10.3390/app122412828

Academic Editors: Paweł Kiełbasa, Tadeusz Juliszewski and Sławomir Kurpaska

Received: 27 October 2022
Accepted: 16 November 2022
Published: 14 December 2022

1. Introduction

Agriculture is confronting enormous difficulties, which include threats from diseases, weeds, and pests, varying climate, severe scarcity of water resources and arable lands, etc. [1]. Many efforts were exerted to weed control over the years by farmers and researchers to solve the difficulties brought by weeds [2]. Weeds appear everywhere arbitrarily in the domain and compete with crops for sunlight, water, and nutrients, which causes harmful effects on crop quality and yields if not controlled properly. Many studies have illustrated a strong correlation between weed competition and crop yield loss [3]. There are different weed varieties that are detrimental to crop productivity and should be identified in the initial stage of growth. The weed's growth in the crop would affect fundamental resources such as minerals, sunlight, water, fresh air, soil, etc., which are the fundamental necessities of the crop [4]. It has been found that 35% of crops are ruined because of the growth of various weeds in the agricultural field. This study aimed to learn various techniques and tools employed by the researchers to classify and detect weeds, which are essential for assessing the growth of weeds [5]. Numerous computer-related approaches, such as wireless sensor networks (WSN), artificial intelligence (AI), and some other approaches, are used that enhance agriculture.

Weed detection in plants is now becoming a great challenge since there is not much effort put into weed detection. Techniques to realize field weed identification through computer vision (CV) technology primarily use conventional deep learning (DL) and image processing [6]. If weed identification has been carried out by conventional image-processing technology, then feature extraction, such as shape, color, and texture of the image, and integration with conventional ML classifiers, such as the Support Vector Machine (SVM) or random forest (RF) algorithm, for weed detection are essential [7]. Such techniques must devise features manually and highly depend on pre-processing methods, quality of feature extraction, and image acquisition algorithms. Classical algorithms for detecting agricultural weeds had a focus on direct identification of the weed; yet, there were substantial variances in weed species [8]. With the advances in computational power and growth in data volume, DL methodologies are capable of extracting multidimensional and multiscale spatial semantic feature information of weeds by using the Convolutional Neural Network (CNN) because of its enhanced data expression abilities for images, evading demerits of conventional extracted approaches [9,10]. Hence, CNN models have gained higher attention in the research community.

This manuscript introduces a Modified Barnacles Mating Optimization with Deep Learning based weed detection (MBMODL-WD) technique. The presented MBMODL-WD technique initially pre-processes the input images via the Gabor filtering (GF) technique to eradicate the noise. For automated weed detection, the presented MBMODL-WD technique uses the DenseNet-121 model as feature extractor. Moreover, the MBMO algorithm is used for the hyperparameter optimization process. At last, the Elman Neural Network (ENN) method was applied for the classification of images into plants and weeds. To demonstrate the enhanced performance of the MBMODL-WD approach, a series of simulation analyses were performed.

2. Literature Review

Sodjinou et al. [11] introduce a segmentation approach relevant to the compilation of K-means and semantic segmentation methods for segmenting weeds and crops in color images. The two distinct databases of agronomic images are employed for segmenting methods. Everything except the plants has been eliminated from the images with the use of the threshold method. Then, utilizing U-net and the K-means subtractive approach for the segmentation of weeds and crops, semantic segmentation has been implemented. A U-net for the segmentation of weeds and wheat on images was given in [12]. An image classifier task has been employed for choosing the backbone network for the encoding part. The abovementioned task on a similar dataset will be exploited for pretraining and choosing the decoding network. The training process implemented TL. Sa et al. [13] formulates a weed segmentation and mapping structure that processes multispectral images received from drones by utilizing deep neural networks (DNNs). Many researchers are increasingly making an effort in weed or crop semantic segmentation just by considering single images for classification and processing.

A weed species and density evaluation approach rely upon an image semantic segmentation NN that has been modeled in [14]. To train the network, an amalgamation of fine-tuning and pretraining training approaches has been employed. The pretraining datasets are the images that just have a single weed species in a single image. The weeds will be labeled mechanically by an image segmentation approach related to the minimum error threshold and Excess Green (ExG). The fine-tuning data are real imageries comprising many crops and weeds and are manually labeled. In [15], an AI-oriented method has been modeled to find weeds' unintentional growth on agricultural fields and rise the control rate, concerning technological advances. This presented technique will train the dataset by utilizing pretrained DL techniques and has been optimized through metaheuristic optimization approaches through the selection of the species-related activations that can be sent to Softmax, which was in the final layer of DL techniques.

In [16], a technique related to DL has been modeled for weed segmentation from the images. This approach may segment weeds from crops and soil in the images. This semantic segmentation approach has been advanced using a simplified U-net. An image augmentation method has been devised because of the difficulty of image labeling for the semantic segmentation of weeds. The semantic segmentation network is trained by a two-stage training technique made up of fine-tuning and pretraining. Abdalla et al. [17] intended to assess three TL approaches with the use of a VGG16-oriented encoder net for segmenting the oilseed rape images in a field including high-density weeds. Three TL methods utilizing a VGG16-based encoder method have been modeled, and their performances will be compared with a VGG19-related encoder net.

3. The Proposed Model

In this article, we have introduced a new MBMODL-WD technique for the automated identification of weeds using computer vision techniques and improved crop productivity. It encompasses a series of processes: GF-based image pre-processing, DenseNet-121 feature extractor, MBMO-based hyperparameter optimizer, and ENN-based classification. Figure 1 demonstrates the block diagram of the MBMODL-WD system.

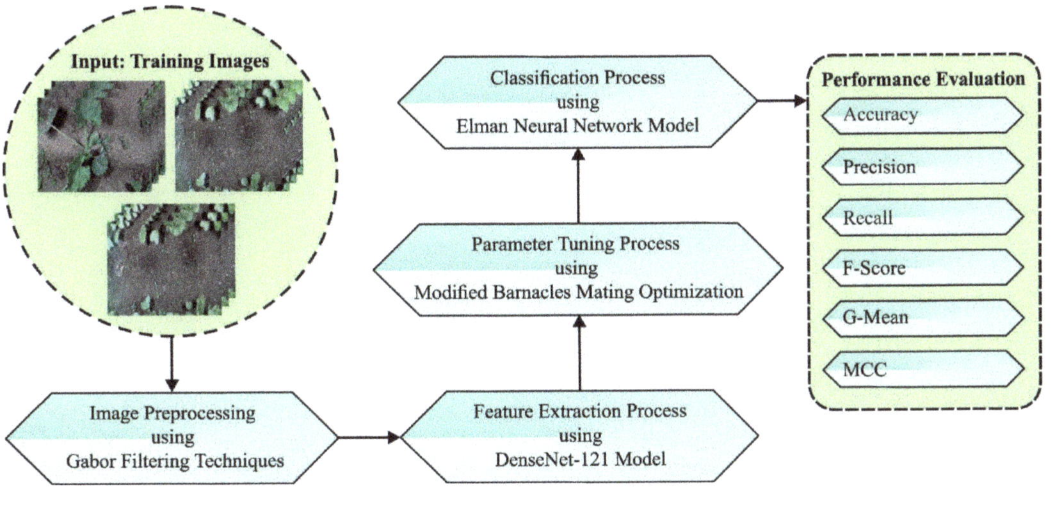

Figure 1. Block diagram of MBMODL-WD system.

3.1. Image Pre-Processing

Primarily, the presented MBMODL-WD technique removes noise using the GF technique. Gabor filter has been employed in the presented system for enhancing the ridges and relaxing the valleys through the enforcement of short-term Fourier transformation including Gaussian window in the spatial domain [18]. It helps to gain deviations in characteristics and textures in the fingerprint images for distinct scales and orientations. Such statistical features generated image features that can be significantly accentuated by utilizing the frequency information and orientations in fingerprint images by fine-tuning a

Gabor filter. A set of Gabor filters was employed on image $I(x, y)$ in distinct frequencies having distinct orientations, utilizing the Gabor function $g(x, y)$, as defined by Equation (1).

$$g(x, y) = \exp\left(-\frac{x'^2 + \gamma^2 y'^2}{2\sigma^2}\right)\cos\left(2\pi\frac{X'}{l} + \phi\right) \quad (1)$$

Here $\chi' = xcox\theta + ys\dot{m}\theta$ and $y' = ycox\theta - xs\dot{m}\theta$.

This Gabor transform will be applied in the Gaussian envelope σ, in addition to the χ and y directions.

3.2. Feature Extraction

For automated weed detection, the DenseNet-121 model produces feature vectors. A CNN accomplishes higher performance in the domain of image classification [19]. However, training a CNN from scratch is not easier because classifier accuracy based on hyperparameter tuning, namely learning rate, initial weight, number of epochs, optimizers, and dropout, needs a wide-ranging amount of labeled training datasets and higher computation power. This problem can be leveraged by means of the TL method. In the TL model, the training time can be decreased by the weight attained from the pretrained mechanism that is applied as an initial weight for training the novel problem. This process of reusing the pretrained weight leads to lower generalization errors. The DenseNet framework is popular since the DenseNet module motivates feature reuse, attenuates the gradient vanishing problems, decreases the parameter count, and enhances feature propagation. In DCNN, all the layers are interconnected to other layers as a feed forward pattern. Every layer in DenseNet accepts the feature map of each previous layer as other inputs and passes on its own feature map to every succeeding layer. Therefore, the n-th layer has n inputs of each previous layer.

In general, a CNN changes the size of feature maps via the down-sampling of layers. However, DenseNet facilitates feature concatenation and down-sampling by separating the network into multiple densely connected dense blocks. The feature map size in the block is unchanged. Inside the dense block, the size of the feature maps is similar which assists in performing concatenation while outside the dense block, while pooling and convolution operations were performed to down-sample. At the end of every dense layer, a transition block or layer was added. The transition layer comprises a batch normalization layer, a 1×1 convolutional, and 2×2 average pooling layers. The transition layer changes the size of the feature map. Therefore, the DenseNet comprises 1 classification, 117 Conv, and 3 transitions, making the size of the layer 121 [19].

Here, the MBMO algorithm as a hyperparameter optimizer is applied. Barnacles are certain kinds of arthropods which constitute an infraclass Cirripedia based on crabs and lobsters [20]. A barnacle is a marine animal that lives in shallow and tidal waters. They can be found all around the seawater and are raised on hard surfaces in seawater. After they hatch eggs, barnacle larvae were disseminated in the water to find and stick to hard surfaces. Indeed, hard surfaces cover the bodies of barnacles and enhance shell plates. They must seek a balance between managing ever-longer penises and accomplishing more mates in a turbulent flow. A novel optimization technique, named the barnacles mating optimizer (BMO) technique, based on these behaviors has been introduced. The balance behavior can be devised depending upon the Hardy–Weinberg equilibrium as follows:

The initial population of barnacles for the solution can be determined by:

$$X = \begin{Bmatrix} \chi_1^1 & \chi_1^N \\ \vdots & \vdots \\ \chi_n^1 & \chi_n^N \end{Bmatrix} \quad (2)$$

where n defines candidate number, and N denotes count of decision parameters according to lower and higher bounds:

$$l_b = \left[l_b^1, \ldots, l_b^i\right] \quad (3)$$

$$u_b = \left[u_b^1, \ldots, u_b^i\right] \quad (4)$$

From the expression, u_b and l_b indicate the upper and lower bounds of variable i. By assessing the objective function for each candidate, better outcomes to worst outcomes are stored and arranged at the initial iteration.

The presented method involves exploration and exploitation. The offspring generation can be performed by the sperm cast method as an exploration term:

$$b_D = rand(n) \quad (5)$$

$$b_M = rand(n) \quad (6)$$

where b_D and b_M denote the mated parents.

Depend upon the Hardy–Weinberg theory, the BMO method considers the parents' genotype frequencies or inheritance features in the offspring generation to model the reproduction process:

$$X_i^{N_{new}} = pX_{b_D}^N + qX_{b_M}^N \quad (7)$$

Now, $X_{b_M}^N$ and $X_{b_D}^N$ characterize the variables of Mum and Dad candidates, correspondingly, and p defines a pseudo-random value disseminated between 0 and 1, $= (1-p)$.

When the candidate choice for mating exceling pl amount is taken into account initially, then the exploration term can take place:

$$X_i^{N_{new}} = rand \times X_{b_M}^n \quad (8)$$

In Equation (8), $rand$ defines a random integer between zero and one. The recently produced offspring for exploration can be created by Mum's candidate.

The offspring will be added and analyzed to the parents to extend the solution matrices from the candidate dimension. Hence, to choose fifty percent top solutions, a technique was utilized for arranging individual dimension, and the inappropriate solution was removed. The MBMO algorithm involves the integration of self-population-based initialization with the standard BMO algorithm. Similar to other metaheuristic models, BMO has a population-related optimization method that begins with random initialization. This implies that it needs a control variable to determine the population size. Nevertheless, it is worth noting that the selection of population size to resolve case problems becomes difficult. The self-adaptive population will regulate the population size at the iteration. Now, the initial population size in the initial iteration can be accomplished by a self-adaptive population:

$$PopSize = 10 \times d \quad (9)$$

where d signifies the problem dimension and it can be defined as follows:

$$PopSize_{new} = \max\left(d,\ round(PopSize + r \times PopSize)\right) \quad (10)$$

where r defines a random number between -0.5 and 0.5.

Algorithm 1: Pseudocode of BMO

```
Initializing the population of barnacles X_i
Compute the fitness of all the barnacles
Arrange for locating an optimum outcome at the top of populations (T = the optimum solution)
while (I < Maximal iterations)
        Fixed the value of pl
        if selective of Dad and Mum = pl
                for all the variables
                        Off spring generation:
                end for
        else if selective of Dad and Mum > pl
                for all the variables
                        Off spring generation:
        end for end if
        Bring the present barnacle back once it moves outside the boundaries
        Compute the fitness of all the barnacles
        Arranging and upgrading T if there is an optimum solution
        l = l + 1
end while
Return T
```

3.3. *Weed Detection and Classification*

For the weed recognition process, the MBMODL-WD technique exploited the ENN model. The ENN model is a common example of a dynamic recurrent network, and its architecture comprises an input layer with specific context nodes, an output, and hidden layers [21]. The key advantage of ENN is that the context node might be employed for remembering the prior hidden node activation, which makes it appropriate in the fields of dynamic system identification and prediction control. Consider the external input of the network as u. The output is y, and the output of the hidden layer is χ. Hence:

$$x(k) = f\left[w_k^1 x_c(k) + w_k^2 u(k-1)\right] \tag{11}$$

$$x_c(k) = x(k-1) \tag{12}$$

$$y(k) = g\left[w_k^3 x(k)\right] \tag{13}$$

where $w_k^1 w_k^2$, and w_k^3 indicate the weight connection matrices from context to implicit layers, the input to hidden layers, and hidden to output layers, correspondingly. Now, f and g represent the transfer function of implicit and output layers.

$$x_c(k) = x(k-1) = f\left[x_{k-1}^1 x_c(k-1) + x_{k-1}^2 u(k-2)\right] \tag{14}$$

Then,

$$x_c(k-1) = x(k-2) \tag{15}$$

where $x_c(k)$ depends on w_{k-1}^1, w_{k-2}^2 at distinct moments; hence, $x_c(k)$ represents the dynamic recursive method. Consequently, the BP model applied to Elman regression NN training is the dynamic BP learning mechanism.

4. Experimental Validation

The proposed model is simulated using the Python 3.6.5 tool. The proposed model is experimented on PC i5-8600k, GeForce 1050Ti 4 GB, 16 GB RAM, 250 GB SSD, and 1 TB HDD. The parameter settings are given as follows: learning rate: 0.01, dropout: 0.5, batch size: 5, epoch count: 50, and activation: ReLU.

The weed detection results of the MBMODL-WD model are tested using a dataset, compassing 3000 images. The dataset holds 287 crop images and 2713 weed images, as defined in Table 1. Figure 2 illustrates some sample images.

Table 1. Dataset details.

Class	No. of Images
Crop	287
Weed	2713
Total Number of Images	**3000**

Figure 2. Sample weed images.

The confusion matrices of the MBMODL-WD model on the weed detection process are demonstrated in Figure 3. The outcomes displaying the MBMODL-WD model properly recognized the crop and weed images under all aspects.

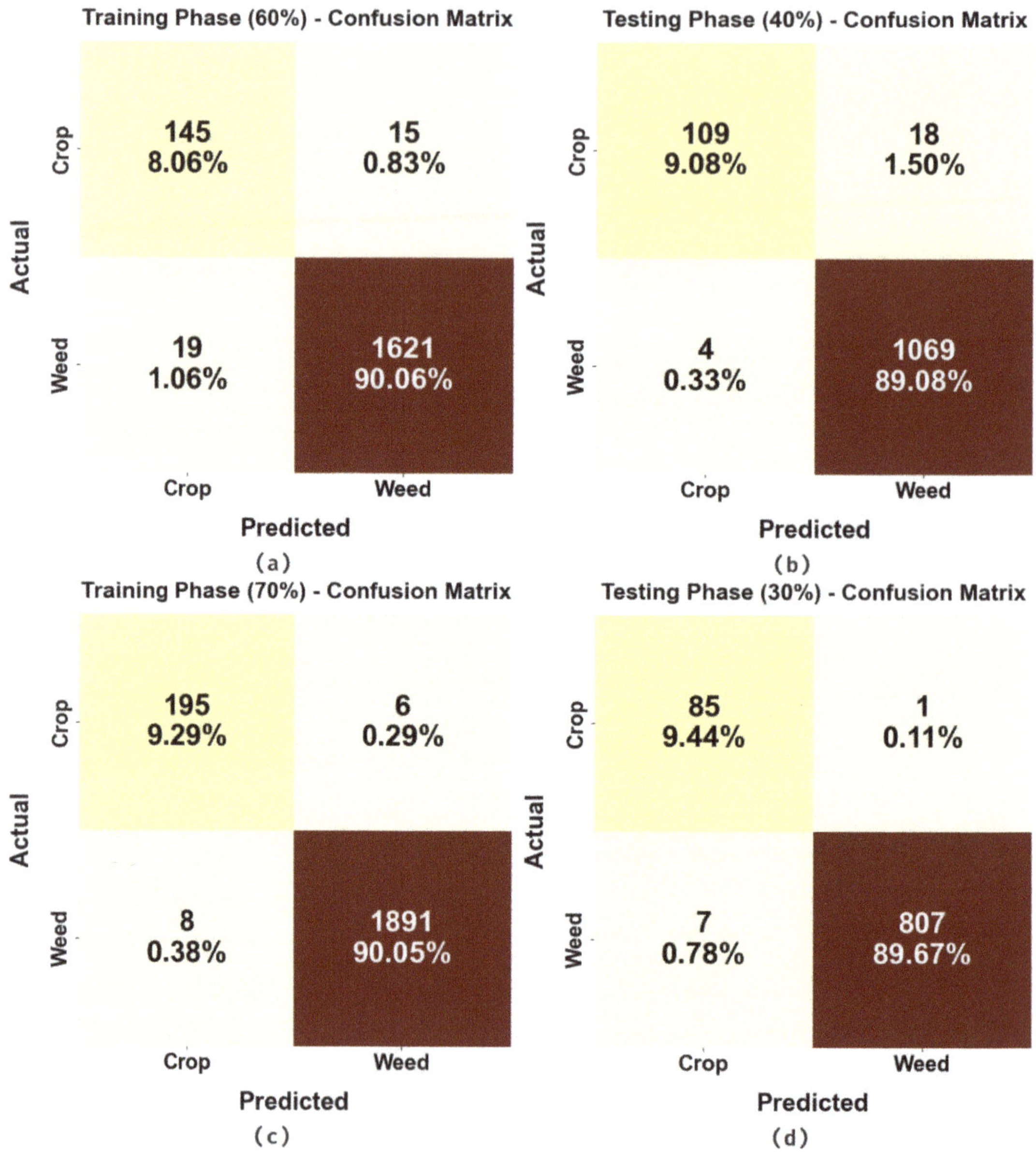

Figure 3. Confusion matrices of MBMODL-WD system, (**a**,**b**) TR and TS databases of 60:40, and (**c**,**d**) TR and TS databases of 70:30.

Table 2 and Figure 4 report the weed detection results of the MBMODL-WD method on 60% of TR and 40% of TS databases. The simulation values revealed the MBMODL-WD method properly classified the crop and weed images. On 60% of TR database, the MBMODL-WD model reached an average $accu_{bal}$ of 94.73%, $prec_n$ of 93.75%, $reca_l$ of 94.73%, F_{score} of 94.23%, MCC of 88.48%, and G_{mean} of 94.64%. Concurrently, on 40% of TS database, the MBMODL-WD approach gained an average $accu_{bal}$ of 92.73%, $prec_n$ of 97.40%, $reca_l$ of 92.73%, F_{score} of 94.91%, MCC of 90.01%, and G_{mean} of 92.47%.

Table 2. Weed detection outcome of MBMODL-WD system under 60:40 of TR/TS databases.

Class	$Accu_{bal}$	$Prec_n$	$Reca_l$	F_{score}	MCC	G_{mean}
Training Phase (60%)						
Crop	90.62	88.41	90.62	89.51	88.48	94.64
Weed	98.84	99.08	98.84	98.96	88.48	94.64
Average	**94.73**	**93.75**	**94.73**	**94.23**	**88.48**	**94.64**
Testing Phase (40%)						
Crop	85.83	96.46	85.83	90.83	90.01	92.47
Weed	99.63	98.34	99.63	98.98	90.01	92.47
Average	**92.73**	**97.40**	**92.73**	**94.91**	**90.01**	**92.47**

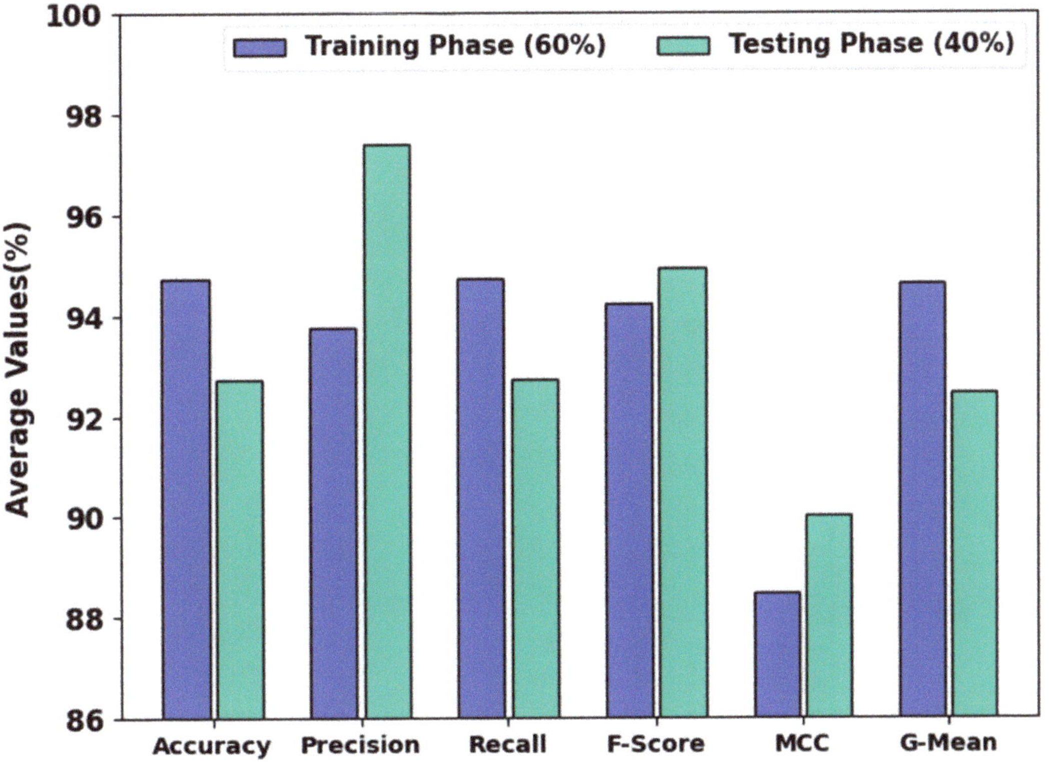

Figure 4. Average weed classification analysis of MBMODL-WD system under 60:40 of TR/TS databases.

Table 3 and Figure 5 depict the weed detection results of the MBMODL-WD system on 70% of TR and 30% of TS databases. The simulation outcomes stated that the MBMODL-WD system properly classified the crop and weed images. On 70% of TR database, the MBMODL-WD system reached an average $accu_{bal}$ of 98.30%, $prec_n$ of 97.87%, $reca_l$ of 98.30%, F_{score} of 98.08%, MCC of 96.17%, and G_{mean} of 98.29%. Simultaneously, on 30% of TS database, the MBMODL-WD algorithm attained an average $accu_{bal}$ of 98.99%, $prec_n$ of 96.13%, $reca_l$ of 98.99%, F_{score} of 97.51%, MCC of 95.08%, and G_{mean} of 98.99%.

Table 3. Weed detection outcome of MBMODL-WD system under 70:30 of TR/TS databases.

Class	$Accu_{bal}$	$Prec_n$	$Reca_l$	F_{score}	MCC	G_{mean}
Training Phase (70%)						
Crop	97.01	96.06	97.01	96.53	96.17	98.29
Weed	99.58	99.68	99.58	99.63	96.17	98.29
Average	**98.30**	**97.87**	**98.30**	**98.08**	**96.17**	**98.29**
Testing Phase (30%)						
Crop	98.84	92.39	98.84	95.51	95.08	98.99
Weed	99.14	99.88	99.14	99.51	95.08	98.99
Average	**98.99**	**96.13**	**98.99**	**97.51**	**95.08**	**98.99**

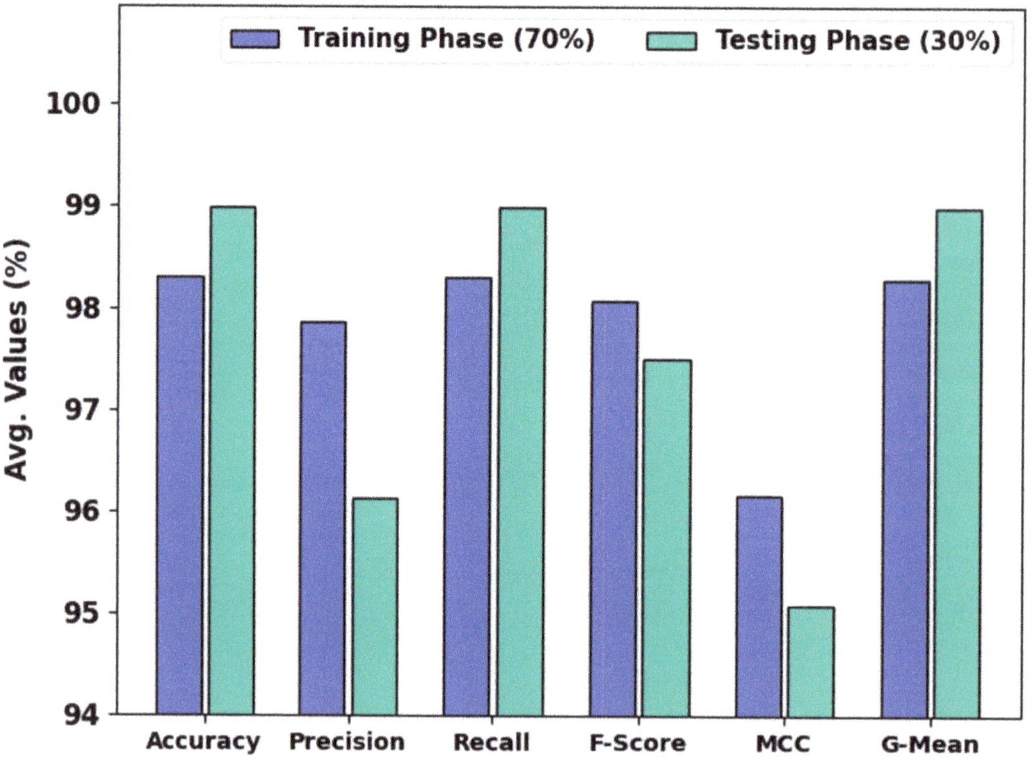

Figure 5. Average weed classification analysis of MBMODL-WD system under 70:30 of TR/TS databases.

The TACC and VACC of the MBMODL-WD approach are investigated on weed detection performance in Figure 6. The figure states that the MBMODL-WD methodology revealed an improved performance with higher values of TACC and VACC. It is evident that the MBMODL-WD system has attained superior TACC outcomes.

Figure 6. TACC and VACC analyses of MBMODL-WD system on weed classification.

The TLS and VLS of the MBMODL-WD methodology are tested on weed detection performance in Figure 7. The figure points out that the MBMODL-WD approach has exposed optimum performance with decreased values of TLS and VLS. It is noticeable that the MBMODL-WD method has resulted in lesser VLS outcomes.

Figure 7. TLS and VLS analyses of MBMODL-WD system on weed classification.

A noticeable precision–recall study of the MBMODL-WD system under test database is represented in Figure 8. The figure pointed out that the MBMODL-WD algorithm has enhanced values of precision–recall values in various classes.

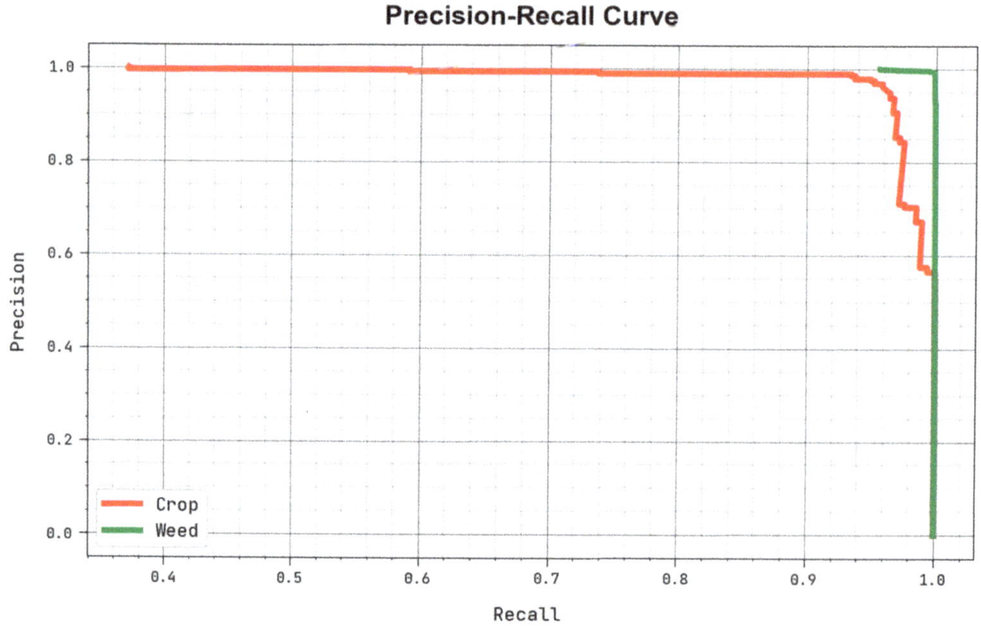

Figure 8. Precision–recall analysis of MBMODL-WD system.

A comprehensive ROC investigation of the MBMODL-WD methodology under test database is depicted in Figure 9. The results denoted the MBMODL-WD algorithm has displayed its capability in categorizing in various classes.

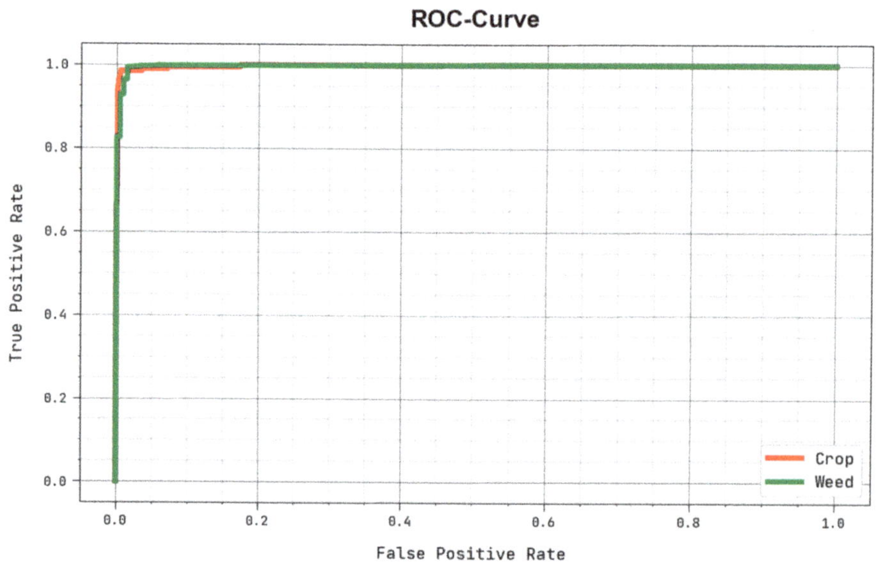

Figure 9. ROC curve analysis of MBMODL-WD system.

Table 4 reports a detailed weed detection result of the MBMODL-WD model and existing models [22]. Figure 10 examines the $prec_n$ and $reca_l$ results of the MBMODL-WD method and recent methods. The results demonstrated the MBMODL-WD method reached enhanced values of $prec_n$ and $reca_l$. For instance, based on $prec_n$, the MBMODL-WD model has shown a higher $prec_n$ value of 96.13%. In contrast, the RF, KNN, SVM, ResNet-101, VGG-16, SVM-Pixel-based, and HLBODL-WDSA models have shown reduced $prec_n$ of 95.24%, 62.43%, 91.32%, 93.15%, 93.96%, 85.41%, and 95.84%, respectively. In addition, in terms of $reca_l$, the MBMODL-WD approach has exposed a maximally $reca_l$ value of 98.99%. In contrast, the RF, KNN, SVM, ResNet-101, VGG-16, SVM-Pixel-based, and HLBODL-WDSA models have shown lesser $reca_l$ values of 93.93%, 61.85%, 91.24%, 94.47%, 92.76%, 85.79%, and 95.18%, correspondingly.

Table 4. Comparative analysis of MBMODL-WD system with other algorithms.

Methods	$Accu_{bal}$	$Prec_n$	$Reca_l$	MCC	G_{mean}
MBMODL-WD	98.99	96.13	98.99	95.08	98.99
RF Model	95.53	95.24	93.93	88.63	89.87
KNN Model	62.81	62.43	61.85	35.83	93.29
SVM Model	94.39	91.32	91.24	84.00	90.34
ResNet-101 Model	93.52	93.15	94.47	92.90	92.80
VGG-16 Model	93.29	93.96	92.76	92.93	89.07
SVM-Pixel-based	85.84	85.41	85.79	86.81	93.62
HLBODL-WDSA	98.96	95.84	95.18	93.34	95.16

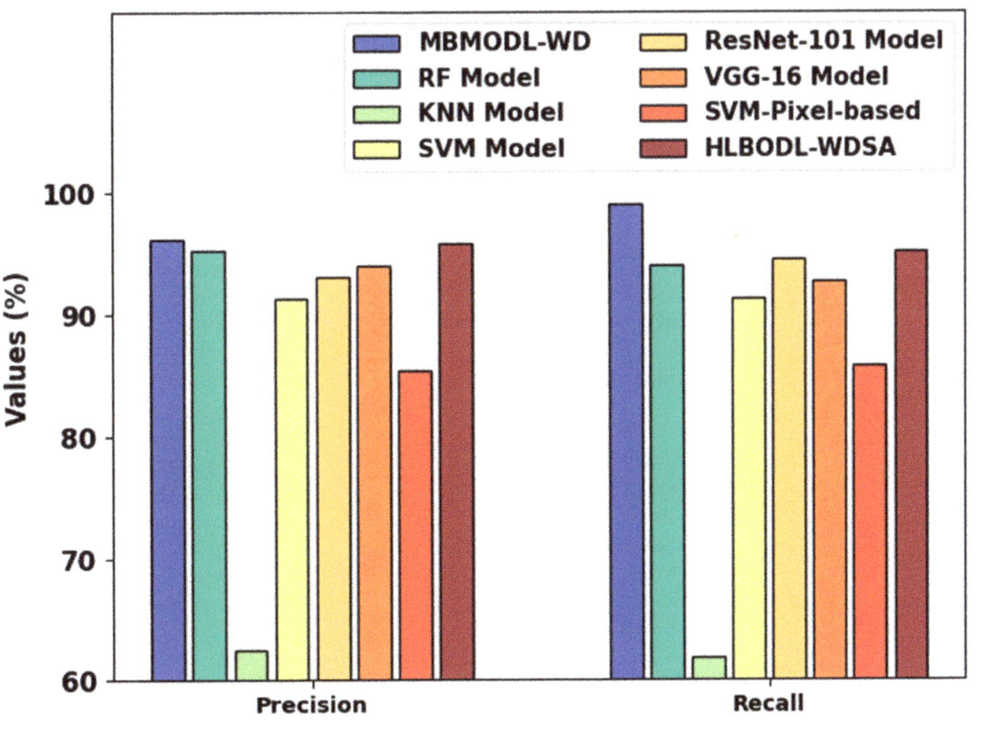

Figure 10. $Prec_n$ and $reca_l$ analyses of MBMODL-WD system with other weed classification algorithms.

Figure 11 observes the MCC and G_{mean} results of the MBMODL-WD algorithm and recent approaches. The outcomes exhibited that the MBMODL-WD system gained improved values of MCC and G_{mean}. For instance, with respect to MCC, the MBMODL-WD model outperformed a maximal MCC value of 95.08%. In contrast, the RF, KNN, SVM, ResNet-101, VGG-16, SVM-Pixel-based, and HLBODL-WDSA systems outperformed lower MCC values of 88.63%, 35.83%, 84%, 92.90%, 92.93%, 86.81%, and 93.34%, correspondingly.

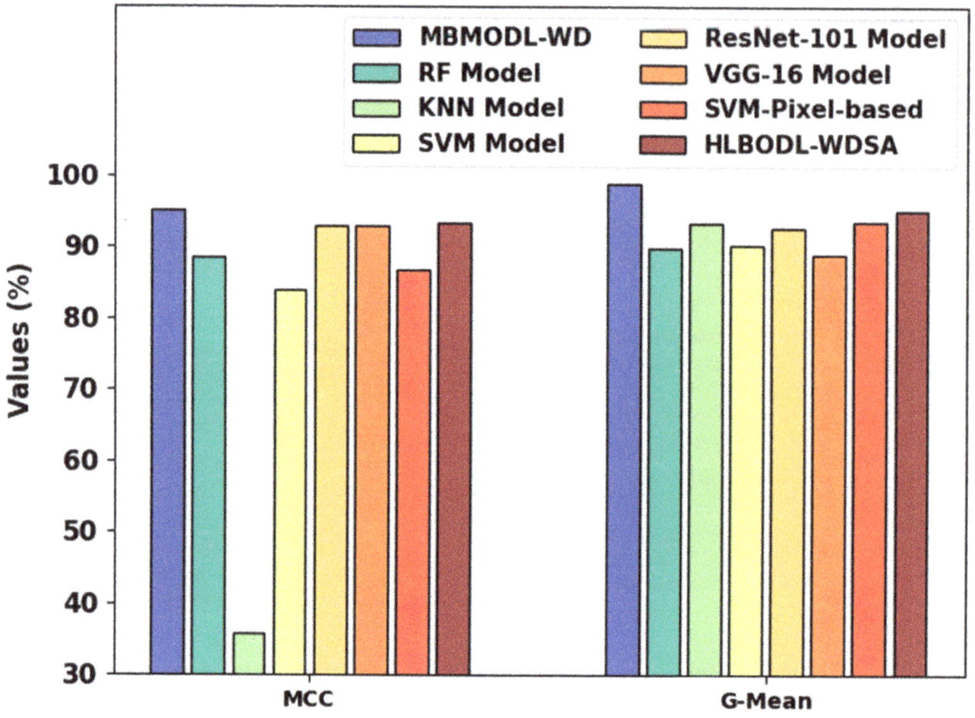

Figure 11. MCC and G_{mean} analyses of MBMODL-WD system with other weed classification algorithms.

Moreover, in terms of G_{mean}, the MBMODL-WD system has depicted a superior G_{mean} value of 98.99%. In contrast, the RF, KNN, SVM, ResNet-101, VGG-16, SVM-Pixel-based, and HLBODL-WDSA algorithms have demonstrated minimal G_{mean} values of 89.87%, 93.29%, 90.34%, 92.80%, 89.07%, 93.62%, and 95.15%, correspondingly. These results reassured the superior outcomes of the MBMODL-WD model.

5. Conclusions

In this article, we have modeled a new MBMODL-WD technique for the automated identification of weeds using computer vision techniques and improving crop productivity. Primarily, the presented MBMODL-WD technique removes noise using the GF technique. For automated weed detection, the DenseNet-121 model produces feature vectors with the MBMO algorithm as hyperparameter optimizer. At last, the ENN method was applied for the classification of images into plants and weeds. To demonstrate the enhanced performance of the MBMODL-WD approach, a series of simulation analyses were performed. A comprehensive set of simulations highlighted the enhanced performance of the presented MBMODL-WD methodology over other DL models, with a maximum accuracy of 98.99%. In the future, the performance of the proposed model can be improved using hybrid DL classification models. In addition, the computational complexity of the proposed model needs to be examined in the future.

Author Contributions: Conceptualization, A.A.A.; Data curation, M.I.A.; Funding acquisition, A.A.A.; Methodology, J.S.A. and M.A.; Project administration, G.P.M.; Supervision, A.A.A.; Validation, M.O.; Visualization, M.I.A.; Writing—original draft, A.A.A., J.S.A. and M.A.; Writing—review & editing, M.O. All authors have read and agreed to the published version of the manuscript.

Funding: Princess Nourah bint Abdulrahman University Researchers Supporting Project number (PNURSP2022R191), Princess Nourah bint Abdulrahman University, Riyadh, Saudi Arabia. The authors would like to thank the Deanship of Scientific Research at Umm Al-Qura University for supporting this work by Grant Code: 22UQU4340237DSR55. Research Supporting Project number (RSP2022R459), King Saud University, Riyadh, Saudi Arabia.

Institutional Review Board Statement: Not applicable.

Informed Consent Statement: Not applicable.

Data Availability Statement: Data sharing is not applicable to this article as no datasets were generated during the current study.

Acknowledgments: Princess Nourah bint Abdulrahman University Researchers Supporting Project number (PNURSP2022R191), Princess Nourah bint Abdulrahman University, Riyadh, Saudi Arabia. Research Supporting Project number (RSP2022R459), King Saud University, Riyadh, Saudi Arabia.

Conflicts of Interest: The authors declare that they have no conflict of interest. The manuscript was written through contributions of all authors. All authors have given approval to the final version of the manuscript.

References

1. Peteinatos, G.G.; Reichel, P.; Karouta, J.; Andújar, D.; Gerhards, R. Weed identification in maize, sunflower, and potatoes with the aid of convolutional neural networks. *Remote Sens.* **2020**, *12*, 4185. [CrossRef]
2. Yang, J.; Bagavathiannan, M.; Wang, Y.; Chen, Y.; Yu, J. A comparative evaluation of convolutional neural networks, training image sizes, and deep learning optimizers for weed detection in Alfalfa. *Weed Technol.* **2022**, *36*, 1–11. [CrossRef]
3. Sabzi, S.; Abbaspour-Gilandeh, Y.; Garcia-Mateos, G. A fast and accurate expert system for weed identification in potato crops using metaheuristic algorithms. *Comput. Ind.* **2018**, *98*, 80–89. [CrossRef]
4. Wang, Q.; Cheng, M.; Xiao, X.; Yuan, H.; Zhu, J.; Fan, C.; Zhang, J. An image segmentation method based on deep learning for damage assessment of the invasive weed Solanum rostratum Dunal. *Comput. Electron. Agric.* **2021**, *188*, 106320. [CrossRef]
5. Dadashzadeh, M.; Abbaspour-Gilandeh, Y.; Mesri-Gundoshmian, T.; Sabzi, S.; Hernández-Hernández, J.L.; Hernández-Hernández, M.; Arribas, J.I. Weed classification for site-specific weed management using an automated stereo computer-vision machine-learning system in rice fields. *Plants* **2020**, *9*, 559. [CrossRef]
6. Wang, S.; Han, Y.; Chen, J.; Zhang, K.; Zhang, Z.; Liu, X. Weed Density Extraction based on Few-shot Learning through UAV Remote Sensing RGB and Multi-spectral Images in Ecological Irrigation Area. *Front. Plant Sci.* **2022**, *12*, 3456. [CrossRef]
7. Wang, Y.; Zhang, X.; Ma, G.; Du, X.; Shaheen, N.; Mao, H. Recognition of weeds at asparagus fields using multi-feature fusion and backpropagation neural network. *Int. J. Agric. Biol. Eng.* **2021**, *14*, 190–198. [CrossRef]
8. Roy, A.M.; Bhaduri, J. Real-time growth stage detection model for high degree of occultation using DenseNet-fused YOLOv4. *Comput. Electron. Agric.* **2022**, *193*, 106694. [CrossRef]
9. Khan, W.; Raj, K.; Kumar, T.; Roy, A.M.; Luo, B. Introducing Urdu Digits Dataset with Demonstration of an Efficient and Robust Noisy Decoder-Based Pseudo Example Generator. *Symmetry* **2022**, *14*, 1976. [CrossRef]
10. Yaacob, M.E.; Nobily, F.B.; Lu, L.; Che Ya, N.N.; Aziz, A.A.; Dupraz, C.; Yahya, M.S.; Hassan, S.N.S.; Mamun, M.A.A. Tropical Weed Identification in Large Scale Solar Photovoltaic Infrastructures: Potential Impacts on Field Operation. *Available at SSRN 4075575*. [CrossRef]
11. Sodjinou, S.G.; Mohammadi, V.; Mahama, A.T.S.; Gouton, P. A deep semantic segmentation-based algorithm to segment crops and weeds in agronomic color images. *Inf. Process. Agric.* **2022**, *9*, 355–364. [CrossRef]
12. Zou, K.; Liao, Q.; Zhang, F.; Che, X.; Zhang, C. A segmentation network for smart weed management in wheat fields. *Comput. Electron. Agric.* **2022**, *202*, 107303. [CrossRef]
13. Sa, I.; Popović, M.; Khanna, R.; Chen, Z.; Lottes, P.; Liebisch, F.; Nieto, J.; Stachniss, C.; Walter, A.; Siegwart, R. WeedMap: A large-scale semantic weed mapping framework using aerial multispectral imaging and deep neural network for precision farming. *Remote Sens.* **2018**, *10*, 1423. [CrossRef]
14. Zou, K.; Wang, H.; Yuan, T.; Zhang, C. Multi-species weed density assessment based on semantic segmentation neural network. *Precis. Agric.* **2022**, 1–24. [CrossRef]
15. Toğaçar, M. Using DarkNet models and metaheuristic optimization methods together to detect weeds growing along with seedlings. *Ecol. Inform.* **2022**, *68*, 101519. [CrossRef]
16. Zou, K.; Chen, X.; Wang, Y.; Zhang, C.; Zhang, F. A modified U-Net with a specific data argumentation method for semantic segmentation of weed images in the field. *Comput. Electron. Agric.* **2021**, *187*, 106242. [CrossRef]

17. Abdalla, A.; Cen, H.; Wan, L.; Rashid, R.; Weng, H.; Zhou, W.; He, Y. Fine-tuning convolutional neural network with transfer learning for semantic segmentation of ground-level oilseed rape images in a field with high weed pressure. *Comput. Electron. Agric.* **2019**, *167*, 105091. [CrossRef]
18. Ahsan, M.; Based, M.A.; Haider, J.; Kowalski, M. An intelligent system for automatic fingerprint identification using feature fusion by Gabor filter and deep learning. *Comput. Electr. Eng.* **2021**, *95*, 107387.
19. Chhabra, M.; Kumar, R. A Smart Healthcare System Based on Classifier DenseNet 121 Model to Detect Multiple Diseases. In *Mobile Radio Communications and 5G Networks*; Springer: Singapore, 2022; pp. 297–312.
20. Norouzi, A.; Shayeghi, H.; Olamaei, J. Multi-objective allocation of switching devices in distribution networks using the Modified Barnacles Mating Optimization algorithm. *Energy Rep.* **2022**, *8*, 12618–12627. [CrossRef]
21. Fan, Q.; Zhang, Z.; Huang, X. Parameter conjugate gradient with secant equation based elman neural network and its convergence analysis. *Adv. Theory Simul.* **2022**, *5*, 2200047. [CrossRef]
22. Alrowais, F.; Asiri, M.M.; Alabdan, R.; Marzouk, R.; Hilal, A.M.; Gupta, D. Hybrid leader based optimization with deep learning driven weed detection on internet of things enabled smart agriculture environment. *Comput. Electr. Eng.* **2022**, *104*, 108411. [CrossRef]

applied sciences

MDPI

Article

Autonomous Detection of *Spodoptera frugiperda* by Feeding Symptoms Directly from UAV RGB Imagery

Jiedong Feng [1], Yaqin Sun [1,2,*], Kefei Zhang [1,2], Yindi Zhao [1,2], Yi Ren [1], Yu Chen [1,2], Huifu Zhuang [1,2] and Shuo Chen [1]

1 School of Environment and Spatial Informatics, China University of Mining and Technology, Xuzhou 221116, China; stdfengjiedong@cumt.edu.cn (J.F.); profkzhang@cumt.edu.cn (K.Z.); zhaoyd@cumt.edu.cn (Y.Z.); 07182463@cumt.edu.cn (Y.R.); chenyu@cumt.edu.cn (Y.C.); huifuzhuang@cumt.edu.cn (H.Z.); stdchensh@cumt.edu.cn (S.C.)

2 Key Laboratory of Land Environment and Disaster Monitoring MNR, China University of Mining and Technology, Xuzhou 221116, China

* Correspondence: syqin@cumt.edu.cn

Abstract: The use of digital technologies to detect, position, and quantify pests quickly and accurately is very important in precision agriculture. Imagery acquisition using air-borne drones in combination with the deep learning technique is a new and viable solution to replace human labor such as visual interpretation, which consumes a lot of time and effort. In this study, we developed a method for automatic detecting an important maize pest—*Spodoptera frugiperda*—by its gnawing holes on maize leaves based on convolution neural network. We validated the split-attention mechanism in the classical network structure ResNet50, which improves the accuracy and robustness, and verified the feasibility of two kinds of gnawing holes as the identification features of *Spodoptera frugiperda* invasion and the degree. In order to verify the robustness of this detection method against plant morphological changes, images at the jointing stage and heading stage were used for training and testing, respectively. The performance of the models trained with the jointing stage images has been achieved the validation accuracy of ResNeSt50, ResNet50, EfficientNet, and RegNet at 98.77%, 97.59%, 97.89%, and 98.07%, with a heading stage test accuracy of 89.39%, 81.88%, 86.21%, and 84.21%.

Keywords: *Spodoptera frugiperda*; deep learning; convolutional neural network; corn insect

Citation: Feng, J.; Sun, Y.; Zhang, K.; Zhao, Y.; Ren, Y.; Chen, Y.; Zhuang, H.; Chen, S. Autonomous Detection of *Spodoptera frugiperda* by Feeding Symptoms Directly from UAV RGB Imagery. *Appl. Sci.* **2022**, *12*, 2592. https://doi.org/10.3390/app12052592

Academic Editors: Paweł Kiełbasa and Manuel Armada

Received: 5 January 2022
Accepted: 25 February 2022
Published: 2 March 2022

1. Introduction

Spodoptera frugiperda, originating from the American continent, has invaded Europe, Asia, and Africa [1]. As a migratory pest, *Spodoptera frugiperda* has a strong survival ability and a rapid reproduction rate, colonizing the above continents in a short time and causing great damage to corn, rice, and other main food crops [2–5].

At present, the main standard method to control this pest is pesticides, including (a) detecting the occurrence and status of pests by field sampling investigation, which relies on agronomists or trained surveyors [3], and (b) spraying pesticides evenly in the corresponding area [6,7]. It is simple and easy to indiscriminatingly spray, but the process of obtaining the information is time consuming and laborious, which depends on the subjectivity of surveyors. Uniform spraying would cause pesticide waste and environmental pollution [8,9]. In this context, there is an urgent need for a low-cost, high-efficiency, and high-precision method to quickly and effectively obtain field information, including the occurrence location, extent, and overall distribution of insect pests [10].

There have been several research studies focusing on the identification of pests and diseases affecting plant leaves. Most of the image data come from ground-based sensors such as mobile phones and digital cameras [11–13], and a small part of this is collected by unmanned aerial vehicles (UAV), which belong to remote sensing (RS) technology [14–16]. RS has been frequently adopted as a rapid, non-destructive, and cost-effective means for plant disease and pest detection that can be adapted to different scenarios and different objects [17–19]. All of the abbreviations in the introduction are found in Table 1.

Table 1. Abbreviations.

CNN	Convolutional Neural Networks	UAV	Unmanned Aerial Vehicle
RS	Remote Sensing	RGB	R (red), G (green), B (blue)

Compared to satellite remote sensing and aerial remote sensing, UAVs have great advantages in terms of cost, operation, carrying, etc. [19], and they have been widely used in crop classification, growth monitoring, yield estimation, and other aspects, especially for large fields [20].

On the other hand, deep learning—originating from machine learning—has gradually gained popularity because of its ability to automatically extract representative features from a large number of input images [21,22]. Konstantinos et al. developed CNN models to perform plant disease detection and diagnosis using simple leaves images of healthy and diseased plants through deep learning methodologies [23]. Chen et al. used the UNet-based BLSNet to automatic identify and segment the diseased region of Rice bacterial leaf streak from the camera photos [24]. The appearance of the attention mechanism also further improves the performance of the network [22,25].

The following are studies based on UAV imagery combined with machine learning or deep learning: Tetila et al. detected soybean foliar diseases subjected to biological stress based on the simple linear iterative clustering segmentation method through foliar physical properties using RGB imagery captured by the low-cost unmanned aerial vehicle model DJI Phantom 3 [26]. Harvey et al. used an unmanned aerial vehicle (UAV) to acquire high-resolution images in the field, and they built an automated, high-throughput system based on a convolutional neural network (CNN) for the detection of northern leaf blight of maize plants [27]. Jin et al. proposed a computerized system based on CNN to process images captured by UAVs at low altitudes, which can detect *Fusarium* wilt of radish with high accuracy [28]. Ryo et al. used CNN to implement a detection method of virus-infected plants in a potato seed production field, with UAV RGB images being captured at an altitude of 5–10 m from the ground [7].

Compared with disease studies, insect pests are more flexible. There are two primary approaches to insect identification [29]: (i) direct, focusing on the ontology of the insects, and (ii) indirect, which focus on the damage caused by the insects [6]. For example, Liu et al. used a field insect light trap to obtain images and combined the CNN and attention mechanism to construct a direct classification model for insect identification [30]. Zhang focused on the significant change in the plant's leaf area index caused by *Spodoptera frugiperda* to indirectly monitor the infestation [31]. On the other hand, using the camera to closely capture pest images is also a widely used method, such as Li et al. integrating Convolutional Neural Network (CNN) and non-maximum inhibition for positioning and counting aphids in rice images obtained by a close view camera, achieving 0.93 accuracy and 0.885 mAP by optimizing key parameters and feature extraction network [13].

The above methods may have defects in accuracy or cannot be applied to large area practice. Thanks to the development of UAV technology, the pest identification based on UAV images is worth further research [32]. Ana et al. carry small aircraft RGB camera drones to obtain the vineyard plant image, and the application of geometric vision and computer vision technology, combined with landform factors on the influence of pests on the vineyard of the quantitative analysis for the farm digital management provides accurate low-cost information, which helps in the implementation and improvement of farm management and decision-making processes [33]. Farian et al. also used the corn leaves damaged by *Spodoptera frugiperda* and applied VGG16 and InceptionV3 to detect the infected corn leaves captured by the UAV (UAV) remote sensing technology while using the angular detection method in computer vision to strengthen the feature representation and improve the detection accuracy [34].

This paper presents a CNN-based deep learning system for the automatic detection of maize leaves infected by *Spodoptera frugiperda*; from RGB UAV remote sensing images at high spatial resolution. UAV remote sensing images have excellent potential for agricultural data acquisition, while deep learning has agricultural data processing potential. Through the combination, this study is based on a multi-stage pest detection classification model applied to actual maize production environmental characteristics based on the ResNest model. The model has the following capabilities: (1) Collecting corn images from the actual field agricultural production conditions for the automatic detection of leaves infected by *Spodoptera frugiperda*; (2) According to the feeding characteristics of corn grass, Accurately and quickly determining the pest stage of the infected leaves, providing a reliable reference for the formulation and implementation of prevention and control measures; (3) The potential and generalization ability of indirect pest detection based on UAV remote sensing images are verified. This provides a reference for the automated detection of pest invasion status in the field. The remainder of this paper is organized as follows: In Section 2, we describe the study area, data collection, and methods. Section 3 presents results, and Section 4 provides a discussion. Finally, Section 5 summarizes this work and highlights future works.

2. Materials and Methods

2.1. Study Area

The UAV RGB imagery of the maize pest *Spodoptera frugiperda* was captured at the Experimental Station (117.552616, 34.309942) of the China University of Mining and Technology, Xuzhou city, Jiangsu Province, China. Image are shown in Figure 1. The experimental site was invaded by the grass moth because of a later maize planting cycle than the surrounding fields. When we started the data collection, the larvae in the field were in a transition phase from low to medium age.

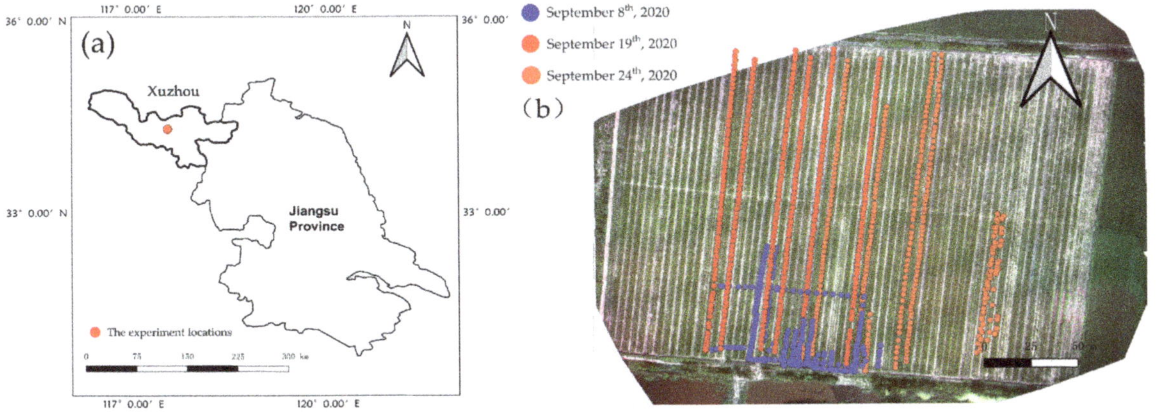

Figure 1. (**a**) The study area located in Xuzhou and the experimental field; (**b**) images of the experimental field obtained on 8, 9, and 24 September 2020 using a low-altitude UAV equipped with an RGB sensor. Each of the colors represents a different experiment date. Each drone image has its own coordinate.

2.2. UAV Image Collection

The image acquisition device was a DJI Mavic air2 equipped with a half-inch CMOS sensor, which can achieve an effective pixel rate of 48 million. What is more, it is an ultra-small drone, weighing just 570 g, which is capable of capturing high-resolution images (standard red–green–blue or RGB photos) of corn at ultra-low altitudes.

The data were collected three times in September 2020 during the critical growth period of maize across the jointing stage to the heading stage, and the specific time and resolution are shown in Table 2. In the jointing stage, corn was collected for the first time, and in the heading stage, it was collected for the second and third time. Between these two periods, maize grows rapidly, and its morphology changes greatly, especially the appearance of stamen changes the overall morphology of the maize plant to a great extent, which can be applied to the generalized type test of model. That is the reason we set the interval. The specific differences between the two stages are shown in Figure 2Part A below, and maize leaf categories are shown in the Figure 2Part B.

Table 2. Date and quantity of data collection.

Date	#Images	Stage	Image Resolution
8 September 2020	295	Jointing	4000 × 3000
19 September 2020	249	Heading	6000 × 8000
24 September 2020	84 272	Heading	6000 × 8000 4000 × 3000

Figure 2. (**Part A**) The picture on the left shows corn in the jointing stage, the right picture shows corn in the heading stage; (**Part B**) The red box shows severely infected corn leaves, the blue box shows slightly infected corn leaves, and the purple boxes show corn leaves in healthy condition.

The flight speed was controlled at 1.5 m/s and the flight altitude was from 2 to 5 m from the ground, which was very close to the corn canopy. Moreover, the shooting angle of the flight path along the ridges of the field was 90°, with more efficient harvesting of corn canopy information. The specific information of the images is shown in Table 2, including the date, number, and image resolution.

2.3. Image Preprocessing

The data processing included two main steps: Cropping and Classing.

Cropping: Due to the size of the images, we tailored them from the sizes in Table 2 to 200 × 200 to speed up training and to reduce the pressure on the graphics memory. We used the OpenCV-Python tool in Python language to read and crop images in batches. For visual effect, Figure 3 shows the conversion process of an image from 1 × 3000 × 4000 to 25 × 200 × 200.

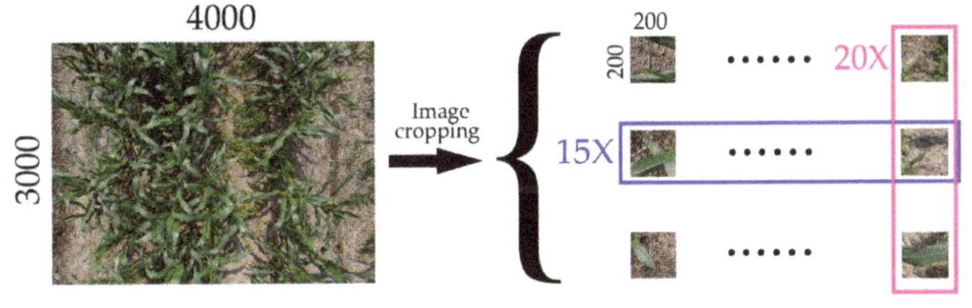

Figure 3. An image cropped from 1 × 3000 × 4000 to 300 × 200 × 200.

Classing: After cropping, the main body of the image is basically composed of corn leaves, and part of the image is land and weeds. By combining the edge detection tool in Opencv2-Python library and RGB channel calculation, the image containing only land and weeds is removed. According to the habits of *Spodoptera frugiperda* and specific representation on the image, we divided the rest of the images into 3 categories by visual interpretation, as shown in Figure 4 below.

(Condition 1): Healthy leaves: Green healthy leaves, complete without damage.

(Condition 2): Translucent silver window (TSW), the first instar and part of the second instar of *Spodoptera frugiperda* only feed on one side of corn leaves, creating a translucent silver windowpane.

(Condition 3): Irregular wormhole (IW), the rest of the instars of *Spodoptera frugiperda* cause significant irregular pores, and some of the leaves infected in the undeveloped period show symmetry in the holes.

(Condition 4): Other objects, the picture contains only land and weeds (drop).

For pest control, the earlier the intervention, the fewer losses and pesticides used, so the first translucent silver windowpane is the most important object. However, in the image, most of the leaves were presented as healthy. In order to balance the number of positive and negative samples, we selected the number of healthy leaves in condition 1 according to the number of infected leaves in conditions 2 and 3, and we removed the leaves in condition 4 at the same time.

After the above processing, we finally obtained more than 5000 maize images in the joint stage, including 2043 healthy leaves, 1866 condition 1 images, and 1430 condition 2 images. The quantities are shown in Table 3. The above data are used for training and validation of the models. During the training, the ratio of training to verification was 9:1.

Figure 4. Classing: Four categories, two stages.

Table 3. Number of processed images.

Stage	Health	TSW	IW	Sum
Jointing	2043	1904	1744	5691
Heading	532	417	596	1545

To test the robustness of the model detection ability, 1545 images in the heading stage were used, including 532 in condition 1 and 417 in condition 2. This part of the image does not participate in training and verification at all, and it was used as an independent test set for testing the model after training.

2.4. Augmentation

The training of a deep learning model requires a very large amount of data, so we used data augmentation to amplify the data [35]. Data enhancement technology can enhance images and reduce over-fitting by flipping, mirroring, and contrast transformation without changing the original form of an image [36]. In the classification task, the image geometric transformation, color space enhancement, random erasure, and feature space enhancement operation can change the image status without changing the image category so as to improve the quantity and quality of data, play the effect of reducing the distance between the training data and the test data, and reduce the overfitting in the model training process [37,38].

For example, the contrast change can change the brightness of the image to a certain extent and enhance the sensitivity of the image to the illumination change. In this study, we used a variety of image enhancement methods, as shown in Figure 5 below.

Figure 5. Augmentation: (**a**) The original image; (**b**,**c**) Horizontal flips and rotations of 90°, respectively; (**d**) Shift scale rotate; (**e**) Random grid transformation; (**f**,**g**) Contrast transformation; (**h**) Random Contrast transformation; (**i**) IAASharpen; (**j**) IAAEmboss; (**k**) CLAHE; (**l**) Shift cutout.

2.5. Convolutional Neural Network

After a long period of development since AlexNet [39], convolutional neural networks, composed of a convolution layer, a pooling layer, and a fully connected layer, have evolved into series of models that can automate the extraction of features through training iterations [40,41]. ResNet [42], through the application of residual blocks, has solved the problem of network degradation and parameter disappearance with a continuous increase in neural network layers, making an indelible contribution to the progress of deep learning, called Deep Convolutional Neural Network (DCNN). DCNN can automatically extract the features of convolutional check images with different specifications to obtain higher data classification accuracy, and it has become the most common identification method [43].

Based on ResNet, the integration of different methods leads to the development of various network structures, such as grouping convolution [44], self-attention mechanism [45], and selective attention mechanism [46]. Therefore, in this study, the feasibility of the feeding symptoms method based on maize *Spodoptera frugiperda* was verified by using several kinds of ResNet related networks, including ResNet, ResNeS [47], SE-Net, and SK-Net. Although the residual structure has been widely applied with its simple structure and convenient modular design, its performance is not satisfactory in downstream applications, which is affected by factors such as receptive field size and channel interaction. Recently, the successful application of the channel and attention mechanism has introduced new possibilities for its improvement. ResNext first introduced the idea of grouping convolution. The SE-Net introduces a channel-attention mechanism for feature construction by adaptively recalibrating the channel feature response. SK-Net extracted the channel information from the feature map through the construction of the grouped channels. Therefore, according to the idea of taking the channel as the operation unit and dividing the input data into more fine-grained weighted subgroups or subchannels based on the global context, it is able to build a channel-based split attention structure. During training, each subgroup is able to perform different mapping abstractions on the input channel data of its own part so as to build different feature representations. In the model, the module is named a distraction block. Thanks to the simple and modular structure, the distraction blocks can perform multiple reuse and stacking and then construct the universal structure bodies similar to the same residue model. Therefore, the block can be simply described as replacing the original residual part of the attention operations with channels as units and thus giving the corresponding weight to the identity.

At the same time, to increase the comparison, other classic network models such as EfficientNet [48] and RegNet [49] are also selected. The networks used in this study were consistent with the architecture of ResNet50, and the original block was replaced with a split-attention block, as shown in the Figure 6 below.

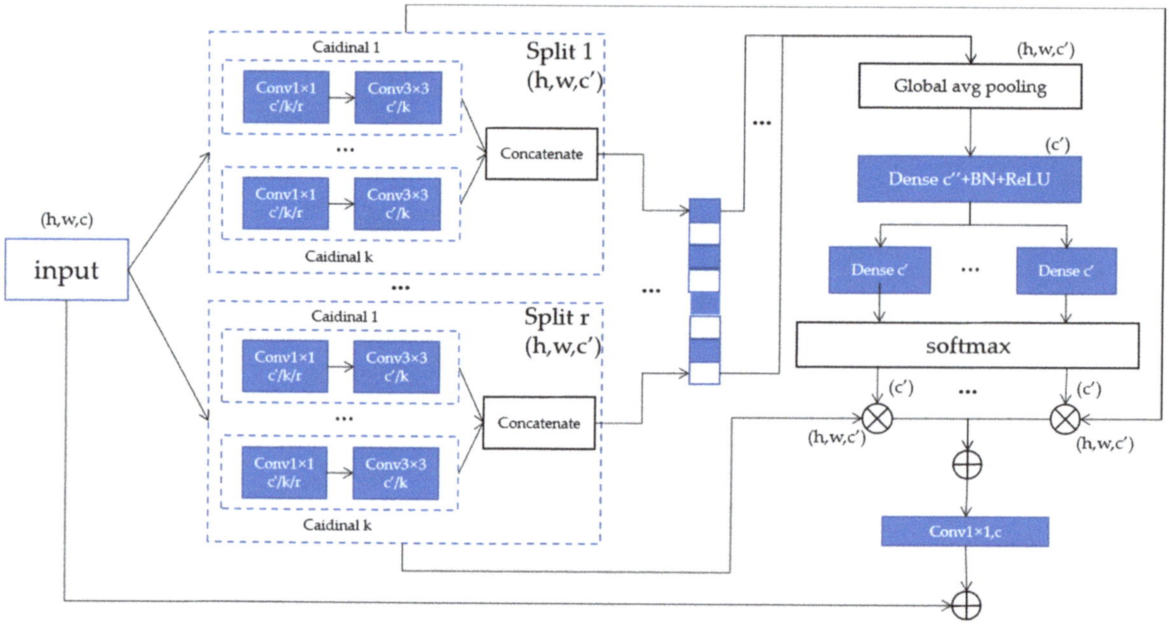

Figure 6. ResNest block.

2.6. Transfer Learning

Transfer learning can transplants the weights obtained through the pre-training of large data sets to the network. Fine tuning based on these weights can accelerate the network training speed and reduce the amount of data required for training [50,51]. In this paper, the ImageNet data set [52] was used as the source data to pre-train the model.

3. Results

3.1. Experimental Setup

In this experiment, we used Pytorch 1.4 as the framework, which is an open-source package in deep learning based on the python programming language. The selected optimizer was Stochastic Gradient Descent (SGD) with the momentum of 0.9, the initial learning rate was 0.002, which decreases with loss, the batch size was 32, and the loss function was CrossEntropy. The training was performed on a machine with the graphics processor of NVIDIA GTX2080s and 32 GB of memory. We trained and tested the models with the data set consisting of the corn leaf in the jointing stage, and to valid the robust ability, we tested the models with the heading stage data set. Figure 7 illustrates the processes involved in obtaining the images used for the experiments.

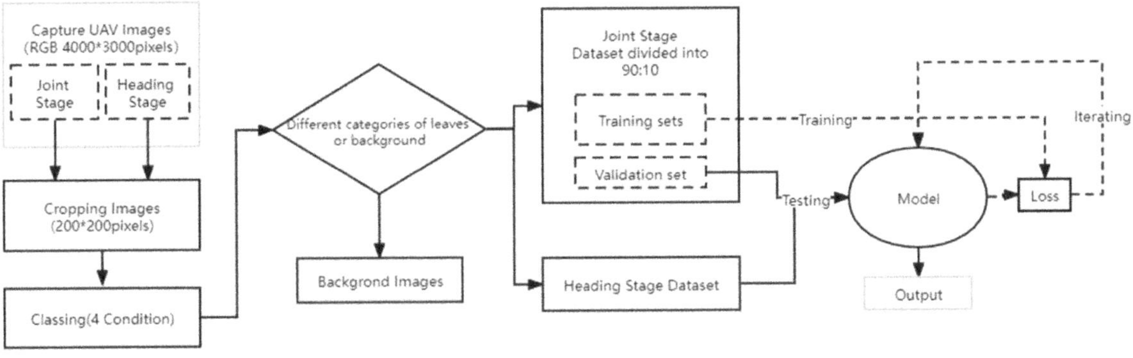

Figure 7. Illustration of the process used in this study.

3.2. Evaluation Parameters

Model performance was assessed using six parameters: Accuracy, Sensitivity, Specificity, Precision, F1 Score, and Kappa.

Accuracy is used as the main method to calculate the accuracy of a model, and the classification ability of the model is represented by the proportion of the correct number of samples in the total number of samples. The specific formula is as follows, and in the formula, T (True) and F (False) represent whether the prediction is correct, and P (Positive) and N (Negative) represent the category of the model prediction; TP and TN is the sum of True Predictions, and FP and FN are the opposite:

$$Accuracy = \frac{TP + TN}{TP + TN + FP + FN} \tag{1}$$

Sensitivity represents the model's recognition ability for positive samples, consisting of TP and FN:

$$Sensitivity = \frac{TP}{TP + FN} \tag{2}$$

Specificity is defined to show true negative assessment ability, consisting of TN and FP:

$$Specificity = \frac{TN}{TN + FP} \tag{3}$$

Precision shows the accuracy of all model-identified positive samples, consisting of TN and FP:

$$\text{Precision} = \frac{TP}{TP + FP} \tag{4}$$

F1 Score is an aggregative indicator based on the harmonic mean of precision and recall.

$$\text{F1 score} = \frac{2 * \text{Precision} * Sensitivity}{\text{Precision} + Sensitivity} \tag{5}$$

Kappa coefficient is a consistency test confusion matrix-based indicator with values between −1 and 1; closer to 1 indicates the overall effect of the classification. In the formula, a_i and b_i represent the true number and the predicted number of the i category, respectively, and the sum means the number of all data.

$$\text{Kappa} = \frac{p_0 - p_1}{1 - p_1}$$
$$p_0 = Accuracy \tag{6}$$
$$p_1 = \frac{a_1 \times b_1 + a_3 \times b_3 + a_3 \times b_3}{sum^2}$$

3.3. Experimental Results

In this study, we compared the performance of four models in the data set, and the results of each model are shown in Figure 8 below. The accuracy for ResNeSt50, ResNet50, EfficientNet, and RegNet is 98.77%, 97.59%, 97.89%, and 98.07%. It can be seen from the data that all the four network structures can obtain high reconnaissance accuracy in this classification problem. Among them, the accuracy of ResNet50 with split attention is the highest. At the same time, with the addition of transfer learning, all networks basically reach the steady state in about 20 epochs, which means in the production environment, we can complete the training and validation of the model in a relatively short time.

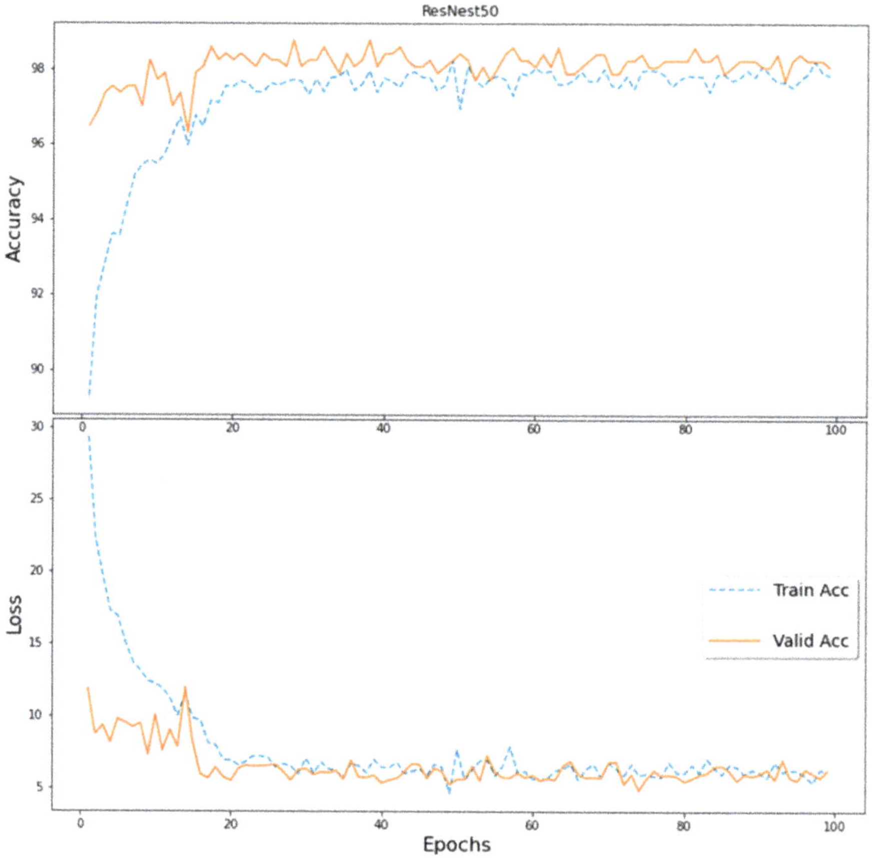

Figure 8. Accuracy and loss of the ResNest50 during training and validation.

An image was randomly selected from the test data set (clipping completed), and the model operation was carried out. The infected image blocks in the calculation results were given different colors according to their severity, and then, they were spliced together for display. A blue box represents a slight silver window, while a red box represents an irregular wormhole (see Figure 9).

Figure 9. Visualization of model's outputs. The upper part of the figure is the original image, the lower part is the image after adding the infected color block, and the left and right yellow image cells are the enlarged image of the infected area.

4. Discussion

In this study, we proposed the deep learning model to detect the invasion of *Spodoptera frugiperda* by the features of the damaged leaves. At the same time, four different neural network structures are used to verify the feasibility of the proposed method. In addition, in order to verify the ability of this feature against maize morphological change, the four models were trained on the images at the jointing stage and tested on the images at the heading stage. The appearance of stamen and the fall of stamen in corn leaves images at the heading stage has a certain influence on the overall structure and color of the image. However, the neural network based on features of the damaged leaves still has good accuracy. Accuracy, Sensitivity, Specificity, Precision, F1 score, and Kappa were used to demonstrate the recognition ability of maize leaves with holes (see Table 4) (TSW: Translucent silver window, IW: Irregular wormhole).

Table 4. Test results—based on images at the heading stage.

Model	Class	Health	TSW	IW	Accuracy	Sensitivity	Specificity	Precision	F1 Score	Kappa
	Health	473	47	12		0.89	0.97	0.94	0.91	
ResNeSt50	TSW	7	398	12	89.39%	0.950.86	0.90	0.97	0.86	0.84
	IW	22	64	510			0.97	0.96	0.90	
	Health	390	37	105		0.73	0.97	0.94	0.82	
ResNet50	TSW	14	349	54	81.88%	0.84	0.91	0.79	0.81	0.72
	IW	13	57	526		0.88	0.82	0.77	0.82	
	Health	456	39	37		0.86	0.95	0.91	0.88	
EfficientNet	TSW	18	377	22	86.21%	0.90	0.90	0.77	0.83	0.79
	IW	25	72	499		0.84	0.93	0.89	0.86	
	Health	419	30	83		0.79	0.97	0.94	0.86	
RegNet	TSW	15	342	60	84.21%	0.82	0.93	0.82	0.82	0.76
	IW	10	46	540		0.91	0.84	0.79	0.84	

It can be seen from the table that the models based on four different network structures all have a good ability to identify the infected leaves from the corn images at the heading stage. However, compared with the original valid accuracy, the current accuracy has a degree of decline, respectively 89.39%, 81.88%, 86.21%, and 84.21%. The split-attention models outperformed the origin ResNet50 structures and the classical network model on the performance in terms of Accuracy, Precision, etc. ReNest50 also achieves the best results on the Kappa coefficients and F1 Score. What is more, we can see the differences between these networks more clearly in the CAM (Class Activation Map based on the average gradient) in Figure 10 below. Compared with the results of other network structures, the split-attention network can identify the target more accurately and closely.

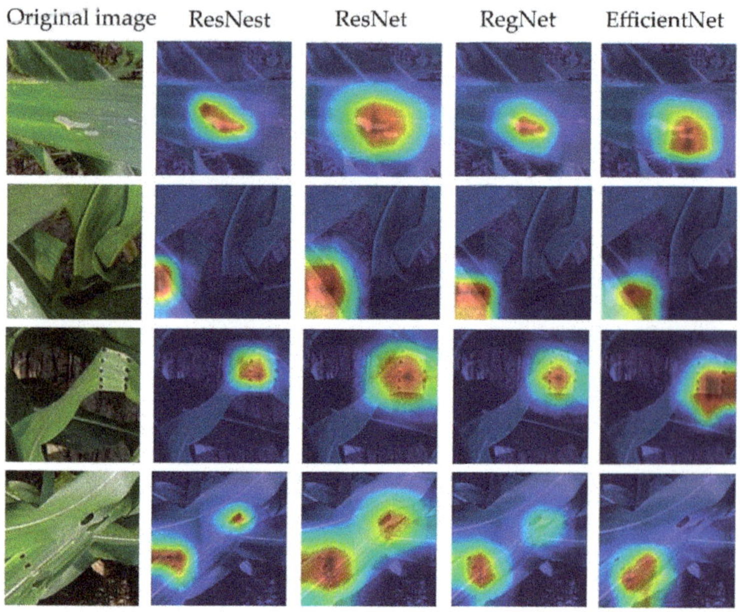

Figure 10. Attention map base on grad CAM.

5. Conclusions and Future Directions

This study aimed to detect maize images that included leaves infected by *Spodoptera frugiperda* in the early stages. Four different models including ResNeSt50, ResNet50, EfficientNet, and RegNet were used to verify the feasibility of using the above features for recognition and explore the split-attention mechanism to improve the accuracy and robustness of the model. The ResNeSt50 network achieved a high accuracy of 98.77% in the validation data set based on the jointing stage and of 89.39% in the test data set based on the heading stage. The model demonstrated its ability to identify infected maize leaves at various stages and allowed to classify them according to the degree of infection. In the process of model construction, methods such as data enhancement and transfer learning are adopted to speed up model construction, reduce overfitting, and improve robustness. Accurate treatment can carry out according to an image's coordinates the grade and distribution of infected leaves, which can significantly reduce the use of pesticides and assist in the implementation of biological control.

Although the model can accurately and quickly identify and judge the maize leaves present in the image for insect pests, the following problems still need to be further studied and explored in practical application: (1) under the condition of positive projections, so that some pest leaves may be ignored due to occlusion; (2) the image acquisition parameters such as height, angle, and resolution and actual field planting conditions; (3) the overall statistical analysis of field pest distribution and subsequent application should be further explored with agronomic knowledge. In future research, we will use the model of *Spodoptera frugiperda* based on more accurate network architecture for real-time field corn image recognition. In addition, according to the optimal resolution combination obtained, we will conduct a new round of data collection in this year's maize planting period to further verify the method. At the same time, we will collect more data to build a model that can identify more pests and diseases faster and more accurately.

Author Contributions: Conceptualization, Y.S. and J.F.; methodology, J.F.; software, J.F.; validation, Y.Z., J.F. and Y.R.; formal analysis, J.F.; investigation, S.C. and J.F.; resources, S.C.; data curation, S.C. and J.F.; writing—original draft preparation, J.F.; writing—review and editing, Y.S., K.Z., Y.C., H.Z. and Y.Z.; visualization, J.F. and Y.R.; supervision, K.Z.; project administration, Y.S. and K.Z.; funding acquisition, K.Z. and Y.S. All authors have read and agreed to the published version of the manuscript.

Funding: This research was funded by the Fundamental Research Funds for the Central Universities (Grant No. 2017XKQY019). This work was supported by a project founded by the Priority Academic Program Development (PAPD) of Jiangsu Higher Education Institutions.

Institutional Review Board Statement: Not applicable.

Informed Consent Statement: Not applicable.

Data Availability Statement: The data presented in this study are available within the article.

Conflicts of Interest: The authors declare no conflict of interest.

References

1. Spark, A.N. A Review of the Biology of the Fall Armyworm. *Florida Entomol.* **1979**, *62*, 82–87. [CrossRef]
2. Food and Agriculture Organization of the United Nations. Map of the Worldwide Spread of Fall Armyworm since 2016. 2022. Available online: http://www.fao.org/fall-armyworm/monitoring-tools/faw-map/en/ (accessed on 24 February 2022).
3. Sarkowi, F.N.; Mokhtar, A.S. The fall armyworm (Faw) spodoptera frugiperda: A review on biology, life history, invasion, dispersion and control. *Outlooks Pest Manag.* **2021**, *32*, 27–32. [CrossRef]
4. Poisot, A.; Hruska, A.; Fredrix, M. Integrated Management of the Fall Armyworm on Maize. 2018. Available online: https://www.preventionweb.net/publication/integrated-management-fall-armyworm-maize-guide-farmer-field-schools-africa (accessed on 24 February 2022).
5. Ministry of Agriculture and Rural Affairs. The General Office of the Ministry of Agriculture and Rural Affairs on Continuously Strengthening the Prevention and Control of Grass Moths. 2019. Available online: http://www.zzys.moa.gov.cn/tzgg/201907/t20190731_6321854.htm (accessed on 24 February 2022). (In Chinese)
6. Bieganowski, A.; Dammer, K.; Siedliska, A.; Bzowska-Bakalarz, M.; Bereś, P.K.; Dąbrowska-Zielińska, K.; Pflanz, M.; Schirrmann, M.; Garz, A. Sensor-based outdoor monitoring of insects in arable crops for their precise control. *Pest Manag. Sci.* **2020**, *77*, 1109–1114. [CrossRef] [PubMed]
7. Sugiura, R.; Tsuda, S.; Tamiya, S.; Itoh, A.; Nishiwaki, K.; Murakami, N.; Shibuya, Y.; Hirafuji, M.; Nuske, S. Field phenotyping system for the assessment of potato late blight resistance using RGB imagery from an unmanned aerial vehicle. *Biosyst. Eng.* **2016**, *148*, 1–10. [CrossRef]
8. Dammer, K.; Adamek, R. Sensor-Based Insecticide Spraying to Control Cereal Aphids and Preserve Lady Beetles. *Agron. J.* **2012**, *104*, 1694–1701. [CrossRef]
9. Karimzadeh, R.; Hejazi, M.J.; Helali, H.; Iranipour, S.; Mohammadi, S.A. Assessing the impact of site-specific spraying on control of *Eurygaster integriceps* (Hemiptera: Scutelleridae) damage and natural enemies. *Precis. Agric.* **2011**, *12*, 576–593. [CrossRef]
10. Bock, C.H.; Poole, G.H.; Parker, P.E.; Gottwald, T.R. Plant Disease Severity Estimated Visually, by Digital Photography and Image Analysis, and by Hyperspectral Imaging. *Crit. Rev. Plant Sci.* **2010**, *29*, 59–107. [CrossRef]
11. Nanni, L.; Maguolo, G.; Pancino, F. Research on insect pest image detection and recognition based on bio-inspired methods. *arXiv* **2019**, arXiv:1910.00296.
12. Thenmozhi, K.; Reddy, U.S. Crop pest classification based on deep convolutional neural network and transfer learning. *Comput. Electron. Agric.* **2019**, *164*, 104906. [CrossRef]
13. Li, W.; Chen, P.; Wang, B.; Xie, C. Automatic Localization and Count of Agricultural Crop Pests Based on an Improved Deep Learning Pipeline. *Sci. Rep.* **2019**, *9*, 1–11. [CrossRef]
14. Zhang, X.; Yue, Q.; Meng, F.; Fan, C.; Zhang, M. Identification of maize leaf diseases using improved deep convolutional neural networks. *IEEE Access* **2018**, *6*, 30370–30377. [CrossRef]
15. Alvaro, F.; Sook, Y.; Sang, K.; Dong, P. A robust deep-learning-based detector for real-time tomato plant diseases and pests recognition. *Sensors* **2017**, *17*, 2022. [CrossRef]
16. Selvaraj, M.G.; Vergara, A.; Ruiz, H.; Safari, N.; Blomme, G. AI-powered banana diseases and pest detection. *Plant Methods* **2019**, *15*, 1–11. [CrossRef]
17. Salamí, E.; Barrado, C.; Pastor, E. UAV Flight Experiments Applied to the Remote Sensing of Vegetated Areas. *Remote Sens.* **2014**, *6*, 11051–11081. [CrossRef]

18. Zhang, C.; Kovacs, J.M. The application of small unmanned aerial systems for precision agriculture: A review. *Precis. Agric.* **2012**, *13*, 693–712. [CrossRef]
19. Matese, A.; Toscano, P.; di Gennaro, S.F.; Genesio, L.; Vaccari, F.P.; Primicerio, J.; Belli, C.; Zaldei, A.; Bianconi, R.; Gioli, B. Intercomparison of UAV, Aircraft and Satellite Remote Sensing Platforms for Precision Viticulture. *Remote Sens.* **2015**, *7*, 2971–2990. [CrossRef]
20. Zhang, X.; Zhang, K.; Sun, Y.; Zhao, Y.; Zhuang, H.; Ban, W.; Chen, Y.; Fu, E.; Chen, S.; Liu, J.; et al. Combining Spectral and Texture Features of UAS-Based Multispectral Images for Maize Leaf Area Index Estimation. *Remote Sens.* **2022**, *14*, 331. [CrossRef]
21. Toda, Y.; Okura, F. How Convolutional Neural Networks Diagnose Plant Disease. *Plant Phenomics* **2019**, *2019*, 9237136. [CrossRef]
22. Zhao, S.; Peng, Y.; Liu, J.; Wu, S. Tomato Leaf Disease Diagnosis Based on Improved Convolution Neural Network by Attention Module. *Agriculture* **2021**, *11*, 651. [CrossRef]
23. Ferentinos, K.P. Deep learning models for plant disease detection and diagnosis. *Comput. Electron. Agric.* **2017**, *145*, 311–318. [CrossRef]
24. Chen, S.; Zhang, K.; Zhao, Y.; Sun, Y.; Ban, W.; Chen, Y.; Zhuang, H.; Zhang, X.; Liu, J.; Yang, T. An Approach for Rice Bacterial Leaf Streak Disease Segmentation and Disease Severity Estimation. *Agriculture* **2021**, *11*, 420. [CrossRef]
25. Liu, J.; Zhang, K.; Wu, S.; Shi, H.; Zhao, Y.; Sun, Y.; Zhuang, H.; Fu, E. An Investigation of a Multidimensional CNN Combined with an Attention Mechanism Model to Resolve Small-Sample Problems in Hyperspectral Image Classification. *Remote Sens.* **2022**, *14*, 785. [CrossRef]
26. Tetila, E.C.; Machado, B.B.; Belete, N.; Guimaraes, D.A.; Pistori, H. Identification of Soybean Foliar Diseases Using Unmanned Aerial Vehicle Images. *IEEE Geosci. Remote Sens. Lett.* **2017**, *1*, 5. [CrossRef]
27. Wu, H.; Wiesner-Hanks, T.; Stewart, E.L.; DeChant, C.; Kaczmar, N.; Gore, M.A.; Nelson, R.J.; Lipson, H. Autonomous Detection of Plant Disease Symptoms Directly from Aerial Imagery. *Plant Phenome J.* **2019**, *2*, 1–9. [CrossRef]
28. Ha, J.G.; Moon, H.; Kwak, J.T.; Hassan, S.I.; Dang, M.; Lee, O.N.; Park, H.Y. Deep convolutional neural network for classifying Fusarium wilt of radish from unmanned aerial vehicles. *J. Appl. Remote Sens.* **2017**, *11*, 1. [CrossRef]
29. Lima, M.C.F.; Leandro, M.E.D.D.A.; Valero, C.; Coronel, L.C.P.; Bazzo, C.O.G. Automatic Detection and Monitoring of Insect Pests—A Review. *Agriculture* **2020**, *10*, 161. [CrossRef]
30. Liu, L.; Wang, R.; Xie, C.; Yang, P.; Wang, F.; Sudirman, S.; Liu, W. PestNet: An End-to-End Deep Learning Approach for Large-Scale Multi-Class Pest Detection and Classification. *IEEE Access* **2019**, *7*, 45301–45312. [CrossRef]
31. Zhang, J.; Huang, Y.; Yuan, L.; Yang, G.; Jingcheng, Z.; Zhao, C. Using satellite multispectral imagery for damage mapping of armyworm (*Spodoptera frugiperda*) in maize at a regional scale. *Pest Manag. Sci.* **2016**, *72*, 335–348. [CrossRef]
32. Suwa, K.; Cap, Q.H.; Kotani, R.; Uga, H.; Kagiwada, S.; Iyatomi, H. A comparable study: Intrinsic difficulties of practical plant diagnosis from wide-angle images. In Proceedings of the IEEE International Conference on Big Data, Los Angeles, CA, USA, 9–12 December 2019; pp. 5195–5201. [CrossRef]
33. Del-Campo-Sanchez, A.; Ballesteros, R.; Hernandez-Lopez, D.; Ortega, J.F.; Moreno, M.A.; on behalf of Agroforestry and Cartography Precision Research Group. Quantifying the effect of Jacobiasca lybica pest on vineyards with UAVs by combining geometric and computer vision techniques. *PLoS ONE* **2019**, *14*, e0215521. [CrossRef]
34. Ishengoma, F.S.; Rai, I.A.; Said, R.N. Identification of maize leaves infected by fall armyworms using UAV-based imagery and convolutional neural networks. *Comput. Electron. Agric.* **2021**, *184*, 106124. [CrossRef]
35. Blok, P.M.; van Evert, F.K.; Tielen, A.P.M.; van Henten, E.J.; Kootstra, G. The effect of data augmentation and network simplification on the image-based detection of broccoli heads with Mask R-CNN. *J. Field Robot.* **2021**, *38*, 85–104. [CrossRef]
36. Kuznichov, D.; Zvirin, A.; Honen, Y.; Kimmel, R. Data Augmentation for Leaf Segmentation and Counting Tasks in Rosette Plants. In Proceedings of the IEEE/CVF Conference on Computer Vision and Pattern Recognition Workshops, Long Beach, CA, USA, 16–17 June 2019; pp. 2580–2589. [CrossRef]
37. El-Kenawy, E.-S.M.; Ibrahim, A.; Mirjalili, S.; Eid, M.M.; Hussein, S.E. Novel Feature Selection and Voting Classifier Algorithms for COVID-19 Classification in CT Images. *IEEE Access* **2020**, *8*, 179317–179335. [CrossRef]
38. Shorten, C.; Khoshgoftaar, T.M. A survey on Image Data Augmentation for Deep Learning. *J. Big Data* **2019**, *6*, 60. [CrossRef]
39. Gonzalez, T.F. *Handbook of Approximation Algorithms and Metaheuristics*; Chapman and Hall/CRC: New York, NY, USA, 2007; 1432p. [CrossRef]
40. LeCun, Y.; Bengio, Y.; Hinton, G. Deep learning. *Nature* **2015**, *521*, 436–444. [CrossRef] [PubMed]
41. Girshick, R.; Donahue, J.; Darrell, T.; Malik, J. Rich feature hierarchies for accurate object detection and semantic segmentation. In Proceedings of the 2014 IEEE Conference on Computer Vision and Pattern Recognition, Columbus, OH, USA, 23–28 June 2014; pp. 580–587. [CrossRef]
42. He, K.; Zhang, X.; Ren, S.; Sun, J. Deep residual learning for image recognition. In Proceedings of the 2016 IEEE Conference on Computer Vision and Pattern Recognition (CVPR), Las Vegas, NV, USA, 27–30 June 2016.
43. Zhang, X.; Han, L.; Dong, Y.; Shi, Y.; Huang, W.; Han, L.; González-Moreno, P.; Ma, H.; Ye, H.; Sobeih, T. A Deep Learning-Based Approach for Automated Yellow Rust Disease Detection from High-Resolution Hyperspectral UAV Images. *Remote Sens.* **2019**, *11*, 1554. [CrossRef]
44. Xie, S.; Girshick, R.; Dollar, P.; Tu, Z.; He, K. Aggregated residual transformations for deep neural networks. In Proceedings of the IEEE Conference on Computer Vision and Pattern Recognition (CVPR), Honolulu, HI, USA, 21–26 July 2017; pp. 5987–5995. [CrossRef]

45. Hu, J.; Shen, L.; Albanie, S.; Sun, G.; Wu, E. Squeeze-and-Excitation Networks. *IEEE Trans. Pattern Anal. Mach. Intell.* **2020**, *42*, 2011–2023. [CrossRef]
46. Li, X.; Wang, W.; Hu, X.; Yang, J. Selective kernel networks. In Proceedings of the 2019 IEEE/CVF Conference on Computer Vision and Pattern Recognition (CVPR), Long Beach, CA, USA, 15–20 June 2019; pp. 510–519.
47. Zhang, H.L.H.; Wu, C.; Zhang, Z.; Zhu, Y. ResNeSt: Split-Attention Networks. *arXiv* **2020**, arXiv:2004.08955.
48. Tan, M.; Le, Q.V. EfficientNet: Rethinking model scaling for convolutional neural networks. In Proceedings of the Thirty-sixth International Conference on Machine Learning, ICML, Long Beach, CA, USA, 9–15 June 2019; pp. 10691–10700.
49. Radosavovic, I.; Kosaraju, R.P.; Girshick, R.; He, K.; Dollar, P. Designing network design spaces. In Proceedings of the 2020 IEEE/CVF Conference on Computer Vision and Pattern Recognition (CVPR), Seattle, WA, USA, 13–19 June 2020; pp. 10425–10433.
50. Afifi, A.; Alhumam, A.; Abdelwahab, A. Convolutional Neural Network for Automatic Identification of Plant Diseases with Limited Data. *Plants* **2020**, *10*, 28. [CrossRef]
51. Hu, R.; Zhang, S.; Wang, P.; Xu, G.; Wang, D.; Qian, Y. The identification of corn leaf diseases based on transfer learning and data augmentation. *ACM Int. Conf. Proceeding Ser.* **2020**, 58–65. [CrossRef]
52. Deng, J.; Dong, W.; Socher, R.; Li, L.-J.; Li, K.; Fei-Fei, L. ImageNet: A large-scale hierarchical image database. In Proceedings of the 2009 IEEE Computer Society Conference on Computer Vision and Pattern Recognition, Miami, FL, USA, 20–25 June 2009; pp. 248–255.

applied sciences

MDPI

Article

Analysis of RGB Plant Images to Identify Root Rot Disease in Korean Ginseng Plants Using Deep Learning

Praveen Kumar Jayapal [1], Eunsoo Park [1], Mohammad Akbar Faqeerzada [1], Yun-Soo Kim [2], Hanki Kim [1], Insuck Baek [3], Moon S. Kim [3], Domnic Sandanam [4] and Byoung-Kwan Cho [1,5,*]

1 Nondestructive Bio-Sensing Laboratory, Department of Biosystems Machinery Engineering, College of Agriculture and Life Science, Chungnam National University, 99 Daehak-ro, Yuseong-gu, Daejeon 34134, Korea; jpraveenkumar5288@gmail.com (P.K.J.); besoo0407@gmail.com (E.P.); akbar.faqeerzada@gmail.com (M.A.F.); zxcvkhk@gmail.com (H.K.)

2 R&D Headquarters, Korea Ginseng Corporation, 30 Gajeong-ro, Yuseong, Daejeon 34128, Korea; gintechkim@kgc.co.kr

3 Environmental Microbial and Food Safety Laboratory, Agricultural Research Service, United States Department of Agriculture, Powder Mill Road, BARC-East, Bldg 303, BARC-East, Beltsville, MD 20705, USA; insuck.baek@usda.gov (I.B.); moon.kim@usda.gov (M.S.K.)

4 Department of Computer Applications, National Institute of Technology, Tiruchirappalli 620015, India; domnic@nitt.edu

5 Department of Smart Agriculture System, College of Agricultural and Life Science, Chungnam National University, 99 Daehak-ro, Yuseong-gu, Daejeon 34134, Korea

* Correspondence: chobk@cnu.ac.kr

Abstract: Ginseng is an important medicinal plant in Korea. The roots of the ginseng plant have medicinal properties; thus, it is very important to maintain the quality of ginseng roots. Root rot disease is a major disease that affects the quality of ginseng roots. It is important to predict this disease before it causes severe damage to the plants. Hence, there is a need for a non-destructive method to identify root rot disease in ginseng plants. In this paper, a method to identify the root rot disease by analyzing the RGB plant images using image processing and deep learning is proposed. Initially, plant segmentation is performed, and then the noise regions are removed in the plant images. These images are given as input to the proposed linear deep learning model to identify root rot disease in ginseng plants. Transfer learning models are also applied to these images. The performance of the proposed method is promising in identifying root rot disease.

Keywords: Korean ginseng; root-rot-disease; plant segmentation; deep learning

Citation: Jayapal, P.K.; Park, E.; Faqeerzada, M.A.; Kim, Y.-S.; Kim, H.; Baek, I.; Kim, M.S.; Sandanam, D.; Cho, B.-K. Analysis of RGB Plant Images to Identify Root Rot Disease in Korean Ginseng Plants Using Deep Learning. *Appl. Sci.* **2022**, *12*, 2489. https://doi.org/10.3390/app12052489

Academic Editors: Paweł Kiełbasa, Tadeusz Juliszewski and Sławomir Kurpaska

Received: 24 January 2022
Accepted: 23 February 2022
Published: 27 February 2022

1. Introduction

Plants play a vital role in the life of human beings and animals. They are major sources of food, medicine, shelter, etc. Plant phenotyping is becoming essential with the increase in food demand globally. It deals with the quantitative measurement of the structural and functional properties of plants—that is, the process of determining plant traits such as chlorophyll content, water content, leaf surface area, and leaf count, and disease identification. The conventional methods involve the manual measurement of the key plant traits. These methods depend on the knowledge of plant breeding experts and farmers. The drawbacks of the conventional methods are that they are expensive, inaccurate, and time-consuming, and many of them are destructive in nature. To overcome these difficulties and limitations, modern researchers are using computer vision techniques [1] and deep learning models in plant phenotyping.

Recently, global agriculture and modern research programs have faced difficulties in plant breeding [2]. Many plant phenotyping communities have been developed in different countries and are involved in solving problems in plant phenotyping. Many researchers have worked on various plant phenotyping methods to identify plant traits from plant images. Wu and Nevatia [3] developed an occluded object segmentation method for plant phenotyping. Praveen Kumar and Domnic [4,5] developed various plant segmentation and leaf counting models for plant phenotyping. Dellen et al. [6] identified the growth signatures of rosette plants from time-lapse video. Grand-Brochier et al. [7] studied various methods to extract the tree leaves from natural plant images. Reeve Legendre et al. [8] developed a low-Cost chlorophyll fluorescence imaging to detect the stress in Catharanthus roseus plants. Raghav Khanna et al. [9] developed a spatio-temporal spectral framework for plant stress phenotyping. Martínez-Ferri et al. [10] conducted a study in an environmental condition where only white root rot disease stress was applied to avocado roots and the other conditions favored plant growth. This study was conducted to provide information on physiological change that occurs during the initial stages of R. necatrix infection on avocado roots. It showed the effect of root rot disease on leaf chlorophyll content.

Biotic and abiotic plant stresses act as important factors for crop yield. In order to protect the plants from these stresses and to prevent a reduction in crop yield, plant breeders depend on plant phenotyping methods and genetic tools to accurately identify plant traits.

Korean ginseng (Panax ginseng Meyer) is a famous herbal plant that is sensitive to biotic stress. It is a shade-loving plant and useful for strengthening human immunity. It is highly sensitive to heat stress [11,12]. The roots of Korean ginseng have high pharmacological efficacy [13]. The size, appearance, and shape of the roots determine the quality and value of the roots. Generally, Ginseng plants are cultivated for several years, and the roots of the plants are harvested for sale between the fourth and sixth years. Ginseng roots are at risk of soil-borne diseases caused by nematodes, fungi, and bacteria in these long cultivation periods [14]. Among these fungi are the most common pathogens causing various diseases such as anthracnose, alternaria blight, root rot, gray mold, botrytis blight, etc. [15,16].

The fungus Cylindrocarpon destructans is one of the most harmful pathogens, causing root rot and rusty root disease. Root rot disease can significantly reduce ginseng production [17–20]. Cylindrocarpon destructans also gives rise to thick-walled resting spores known as chlamydospores that can survive for more than 10 years in the soil. This can lead to the development of root rot disease at any point of time while the spores remain in the soil [21].

The root rot disease in grapevine plants cause symptoms in aboveground plant region [22–24]. These include stunted shoots, wilting leaves, low fruit production and dwarf leaves. Sometimes, the plants do not show any symptoms of disease (asymptomatic) in the aboveground plant region [23,25]. The roots cannot support the aboveground plant regions and they start to die. Hence, there will be less photosyntetic activities in the plants and the plant slowly starts to die. A study [26] on symptomatic and asymptomatic effects of this disease has been conducted in 15 plant species.

There are many existing methods [27,28] to identify the presence of Cylindrocarpon destructans, but these methods are destructive in nature. The root rot disease has been identified from the plant leaves using hyperspectral leaf images and machine learning techniques [25]. The root images are analyzed using feature extraction methods [29] and deep learning techniques [30] to find the root rot disease in Lentil. In order to obtain the root images, the plants are removed from the pots and the root images are captured.

Hence, non-destructive identification of root rot disease is necessary to prevent a reduction in crop yield. In order to develop a low-cost, high-throughput, non-destructive, image-based analysis model, RGB images can be used. It is easy to collect numerous RGB images at a low cost and in a short time. Furthermore, various analyses are possible with these numerous data. The visible information from the collected RGB images helps in identifying the state and morphological changes in the plants through RGB-image-based plant phenotyping.

Nowadays, deep learning models are widely used to solve various complex real-time problems. Furthermore, their usage is increasing in solving agriculture-related problems. Among them are deep-learning-based plant phenotyping methods. The deep learning models can be applied to the collected data to perform various analyses.

In the proposed method, various deep learning models are applied to identify biotically stressed (root rot) Korean ginseng plants based on RGB plant images. Furthermore, a new simple linear deep learning model is proposed. The proposed method involves three steps: (i) a region growing method for plant region segmentation, (ii) noise removal, and (iii) deep learning models for identifying biotically stressed (root rot) ginseng plants. In the first phase, the plant region is segmented from the raw RGB images using the seeded region growing method. Then, in the second step, the noise regions are removed from the segmented plant region. Finally, these noise-free segmented plant images are given as input to the proposed deep learning model to identify root-rot-diseased ginseng plants. The main advantages of the proposed method are that it (i) is non-destructive, (ii) is high throughput, (iii) does not require ground-truth images, (iv) is robust, and (v) is cost effective.

The structure of the remaining portion of this paper is as follows: Section 2 describes the materials and methods. Section 3 explains the results and provides a discussion, and finally, conclusions are given in Section 4.

2. Materials and Methods

2.1. Dataset Preparation

Dormant Panax ginseng roots were obtained from the Ginseng National Research Center in Daejeon, South Korea. After storage for 1 month at 4 degrees Celsius to break dormancy, the roots were planted in small pots. The soil used in our experiment was collected from the field and was contaminated with the soil-borne pathogen (Cylindrocarpon destructans) that causes root rot disease in ginseng plants. The collected soil was divided into two halves. One-half of the soil was sterilized using an autoclave [31]. The soil was sterilized at a temperature of 121 degrees Celsius for 20 min. The sterilization was repeated three times under similar conditions to ensure the healthiness of the soil. The plants were then grown in two different soil conditions: (i) pots containing the healthy soil and (ii) pots containing the soil mixed with soil-borne pathogens (Cylindrocarpon destructans). To grow the plants, the growing conditions were set to a temperature of 20 degrees Celsius with relative humidity of 60–70%, manual watering once a week, and continuous light intensity of 15,000 Lx. In our experiment, biotic stress was the only stress applied to the ginseng plants. Among the diseased plant images, the asymptotic diseased plant images [25] exist and did not show any symptoms in leaves but in roots. Figure 1 shows the plant growth conditions in the chamber.

Figure 1. Plant growth chamber.

RGB images of the ginseng plants were captured one day every week between the 10th week and 16th week using a Canon EOS 700D camera. The imaging device is shown in Figure 2. These images were stored in a computer separately for further analysis. The entire dataset of images was divided into training and testing images. There were

2112 healthy plant images and 1991 diseased plant images in the training dataset. The testing dataset contained 760 healthy plant images and 770 diseased plant images. Image augmentation [32,33] was adopted in our experiment. The variations such as Rotation in the range of ±180 degrees; Height and/or width shifting of ±0.1 of image dimensions; Contrast variation with the factor between 0.75 and 0.95; Illuminance variations with a factor of 0.5 and 1.5 are randomly applied. Figure 3 shows a few augmented images.

Figure 2. Imaging device.

Figure 3. Data augmentation. (**a**) Original plant image; (**b**) Image with Contrast variation; (**c**) Image with Illuminance variation; (**d**) Rotated image; (**e**) Height and/or width shifted image.

Of the training dataset, 10% were used for validation, and the remaining data in the training dataset were used for training the model. The images were carefully divided for training, validation, and testing such that plants in different stages were distributed evenly among the datasets in order to ensure that the proposed method is capable of identifying diseased plants at different stages. Once the images were collected during the 16th week, the plants were harvested to check the ginseng roots and root images were also captured in order to cross-verify the proposed model. The size of the plant and root images was 3456 × 5184.

2.2. Proposed Method

In a study conducted by Martínez-Ferri et al. [10], it was found that root rot disease has effects on the physiology of leaves. The authors in [25] studied the early identification of root rot disease in Grapevine. They identified the root rot disease by analyzing the hyperspectral images of the Grapevine leaves and matched the leaves symptoms with root symptoms. These studies motivated us to develop a high-throughput, non-destructive model that can be used to identify root rot disease from physiological changes in ginseng leaves. The proposed method consists of three steps. In the first step, the plant region is segmented based on the region growing algorithm. The purpose of plant segmentation is to improve the accuracy and robustness of the proposed method. In the second step, the noise regions are removed from the segmented plant regions. Finally, root rot disease is identified by the proposed deep learning model. The workflow for the proposed method is presented in Figure 4.

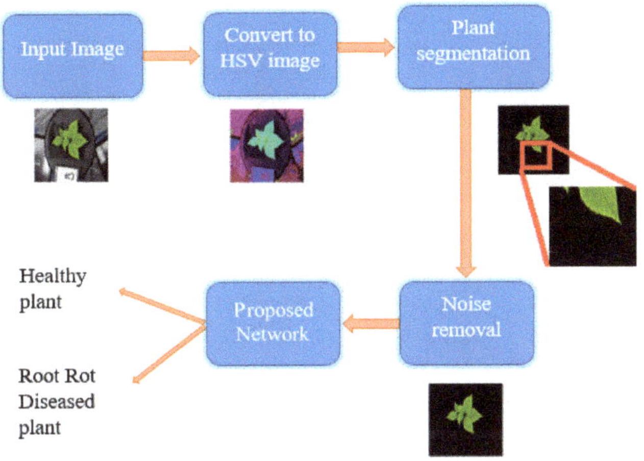

Figure 4. Workflow of the proposed method.

2.2.1. Plant Segmentation

In this step, the plant regions are segmented from the input plant images using the region growing algorithm. To segment the plant images, initially, the RGB plant images are resized into 227 × 227 × 3 and converted to HSV images. Then, the plant region is segmented based on Algorithm 1.

The selection criterion (G_i) is given in the following equation:

$$G_i = \{0.1 < H(v_i) > 0.4 \text{ and } S(v_i) > 0.1\}, \quad (1)$$

where $H(v_i)$ denotes the ith pixel value of the H plane and $S(v_i)$ denotes the ith pixel value of the S plane in the HSV plant image.

A pixel is considered a seed pixel (s) when it is the first occurring pixel that satisfies selection criterion G_i in the HSV plant image. The similarity criterion plays a major role in the region growing process. This has to be chosen such that the plant region can be segmented from the plant image. In the proposed method, the similarity criterion ($G_{i,j}$) was chosen as given in Equation (2):

$$G_{i,j} = \begin{cases} 1, & \text{if } G_i \text{ and } G_j \text{ are true} \\ 0, & \text{otherwise} \end{cases} \quad (2)$$

where i and j are neighboring pixels of the plant image.

Equations (1) and (2) are used in the region growing algorithm (Algorithm 1) to segment the plant region.

Algorithm 1 Region growing method for plant segmentation

INPUT (I): HSV Plant image
OUTPUT (P): Segmented plant region

B = Background region
s = seed pixel
$G_{i,j}$ = similarity criterion between the ith pixel and its non-visited neighboring jth pixel.
$G_{s,i}$ = similarity criterion between the s and ith pixel.
Initialize: P = { }, B = { } and $G_{s,i}$=0.
Select the seed pixel (s) as discussed in Section 2.2.1
for each unassigned neighboring pixel of s **do**
    $G_{i,j} = G_{i,j} + G_{s,i}$
    if $G_{i,j}$ remains_unchanged **then**
        Add that pixel to B
        Search for a new seed pixel from the nearby unassigned pixels of the image.
    else
        Add that pixel to P
        Update the current pixel as seed pixel.
    end if
end for
Repeat until all pixels are assigned to either P or B

2.2.2. Noise Removal

The second step is noise removal from the segmented image. The segmented plant region contains certain background regions like the reflection of light, background pixels similar to the diseased portion, etc., as noise regions. These noise regions affect the root rot disease identification. Hence, it is very important to remove these noise regions. Utmost care should be taken in the noise removal stage because infected leaf regions may be removed along with the noise regions present in the segmented plant region. In the proposed method, the noise removal is done in the following way: (i) Initially, the connected components (similar pixels connected with each other) are identified in the segmented image. (ii) Then, the area of each component is calculated. (iii) Finally, the small objects are removed by thresholding. In the proposed method, based on experimentation, the threshold value was chosen to be 3500. After removing the noise region from the segmented region (P), it is mapped with the input HSV plane. There will be pixels in the HSV plane that do not map with P which are considered as background pixels. These background pixels are removed in HSV plane and the remaining plant region is used as the mask over the original RGB plant image to obtain the plant region.

Since the aim of the proposed method is to identify the root-rot-diseased plants (without ground truth images, so that it can be applied in various environments where the creation of ground truth images is tedious) and not plant segmentation, the segmentation accuracy was not calculated as this calculation requires ground truth images.

2.2.3. Root Rot Disease Identification

Once the plant regions are segmented from the plant images, they are given as the input to the proposed deep learning model to identify whether the plants are infected with root rot disease. Image augmentation techniques such as random rotation and random rescaling were used during the model training. The architecture of proposed model is shown in Figure 5, and the details are presented in Table 1. Furthermore, transfer learning models were used to identify the plants infected by root rot disease. In transfer learning, deep learning models are trained to learn the features in a domain and the models are optimized to learn more features in a specific domain. The various transfer learning models used in our experimentation were AlexNet, VGG19, SqueezeNet, DarkNet19, ResNet18, and ResNet101. In the proposed model, there are four convolutional layers. Three of these convolutional layers are followed by the maxpooling layer. All these layers are followed by two fully connected layers. The second fully connected layer is followed by a softmax layer and a classification layer. The numbers of neurons in the first and second fully connected layers are 1024 and 2, respectively.

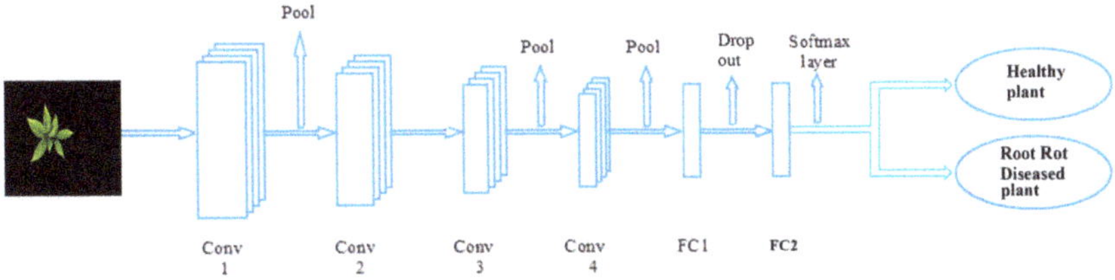

Figure 5. The proposed model Architecture.

The following is a brief discussion about the functions of the different layers in the deep learning models. The convolutional layer identifies the local feature maps from the input image. The size of the features can be calculated by the following equation:

$$F = [(W - K + 2P)/S] + 1 \tag{3}$$

where F denotes the feature map size, W is the input image size, K is the filter size, P denotes the padding, and S denotes the stride.

The activation layer used in the proposed model is a ReLU layer. The ReLU function gives output the same as the input if the input is greater than 0. It is given by

$$ReLU(x) = max(0, x) \tag{4}$$

where x is the input to the ReLU function.

Maxpooling is used to down-sample the obtained feature maps from the convolutional layer. The size of the output after maxpooling is given by the following equation:

$$O = floor((I_y - L)/S) \tag{5}$$

where O is the output size of the maxpooling function, I_y is the input shape, L is the pooling window size, and S is the stride.

The fully connected layers generate the final features by integrating the output of previous layers, and these final features are used for classification or regression. The softmax layer assigns the probabilities for each class in the problem. It is given by the following equation:

$$M(x)_i = \frac{exp(x_i)}{\sum_{j=1}^{n} exp(x_j)} \tag{6}$$

where M is the softmax, x is the input vector, and $exp(x_i)$ and $exp(x_j)$ are the exponential function for input and output vectors, respectively.

The classification layer calculates the loss and performs the classification tasks. The loss function used in the proposed model is cross-entropy. In our experiment, the loss function is given by the following equation:

$$L_C = -\sum_{i=1}^{2} log(m_i),\ for\ 2\ classes \tag{7}$$

where L_C is the cross-entropy loss; t_i is the truth value, having 0 or 1 as its value; and m_i is the ith class softmax probability.

Table 1. Details of the proposed model Architecture.

Layer	Details
INPUT	227 × 227 × 3
CONV1	16 3 × 3 filters with stride [1 1] and padding 'same'
Maxpool1	3 × 3 max pooling with stride [2 2] and padding 'same'
CONV2	32 3 × 3 filters with stride [1 1] and padding 'same'
CONV3	32 3 × 3 filters with stride [1 1] and padding 'same'
Maxpool2	3 × 3 max pooling with stride [2 2] and padding 'same'
CONV4	8 3 × 3 filters with stride [1 1] and padding 'same'
Maxpool3	3 × 3 max pooling with stride [2 2] and padding 'same'
FC1	1 × 1 × 1024
Drop out	0.5
FC2	1 × 1 × 2
Softmax	1 × 1 × 2

3. Results and Discussion

The proposed method was implemented in Matlab (R2020b) on a system with a 64-bit Windows operating system, 16 GB memory, and a 2.9 GHz Intel i5 processor. Initially, the plant region was segmented from the plant images using Algorithm 1. Once the noise region was removed from the plant region, these images were given as input to the transfer learning models and the proposed model.

3.1. Performance Evaluation Measures

The measures used for evaluating the proposed model are as follows: (i) Precision (P) is the ratio of the number of correctly identified root-rot-diseased plants to the total number of identified root-rot-diseased plants. It is given in Equation (8). (ii) Recall (R) is the ratio of the number of correctly identified root-rot-diseased plants to the total number of root-rot-diseased plants. It is given in Equation (9). (iii) The F_1 score is the weighted average of P and R. It is given in Equation (10). (iv) Accuracy (Acc) is the proportion of correctly identified plants and is given by Equation (11).

$$Precision(P) = \frac{TP}{TP+FP} \tag{8}$$

$$Recall(R) = \frac{TP}{TP+FN} \tag{9}$$

$$F_1\ score = \frac{2 \times P \times R}{P+R} \tag{10}$$

$$Accuracy = \frac{TP+FN}{TP+FN+FP+TN} \tag{11}$$

Here, TP is True Positive, which denotes the number of root-rot-diseased plants correctly identified; FP is False Positive, which denotes the number of plants falsely identified as root rot diseased; FN is False Negative, which denotes the number of root-rot-diseased plants that were not identified as such; and TN denotes the number of healthy plants correctly identified.

3.2. Manual Design of Model: Effects of Layers and Hyperparameters

The proposed deep learning model was trained with the SGDM (Stochastic Gradient Descent with Momentum) solver. The values of the initial learning rate, number of epochs, batch size, and momentum were chosen as 0.001, 100, 64, and 0.9, respectively. These parameter values were chosen such that the model has high performance. The manual selection involved in designing the architecture is explained as follows.

3.2.1. Layers

In designing the proposed model, the appropriate layers were selected in such a way that the model exhibited better classification accuracy. The proposed model was also tested with fewer layers and more layers. The model did not learn the necessary root rot disease features when designed with fewer layers, resulting in lower accuracy, whereas the model learned more features but showed overfitting when it was designed with more layers. In our design, combinations such as (Conv2, Pool2), (Conv3, Pool2), (Conv4, Pool2), (Conv4, Pool3), (Conv5, Pool2), (Conv5, Pool3), (Conv6, Pool2) and (Conv6, Pool3) were evaluated. The high accuracy and low error was achieved when the number of convolution layers was increased from two to four and the number of pooling layers used was three. However, when the number of layers was increased, the accuracy was reduced. Hence, a further increase in the number of layers did not have a significant effect on the accuracy. Thenmozhi and Srinivasulu Reddy [34] studied the effects of the number of layers in the performance of their model. They investigated the layer depth of the CNN model and analyzed the effect of the number of layers in their model for pest classification by varying the number of convolution layers from three to seven. They observed that the accuracy decreased with an increase in the number of layers, and lower error was observed for six convolution and five pooling layers, avoiding the occurrence of degradation. It can be observed from Table 2 that the proposed model performed better for four convolution and three pooling layers and effectively reduced overfitting.

Table 2. Performance evaluation of proposed model for different numbers of layers.

Layer Combinations	Accuracy (%)
Conv2Pool2	78.76
Conv3Pool2	83.59
Conv4Pool2	88.17
Conv4Pool3	90.73
Conv5Pool2	86.86
Conv5Pool3	85.49
Conv6Pool2	84.64
Conv6Pool3	83.26

3.2.2. Number of Epochs

The proposed model was trained for up to 100 epochs. The initial learning rate and mini-batch size were set to 0.001 and 64, respectively. The accuracy and loss curves during model training is shown in Figure 6. It can be observed from both the curves that the loss values are stable, and after the 75th epoch the accuracy results are closer to 100% with less fluctuation for training and closer to 90% for validation and obtaining 90.73% at 100th epoch.

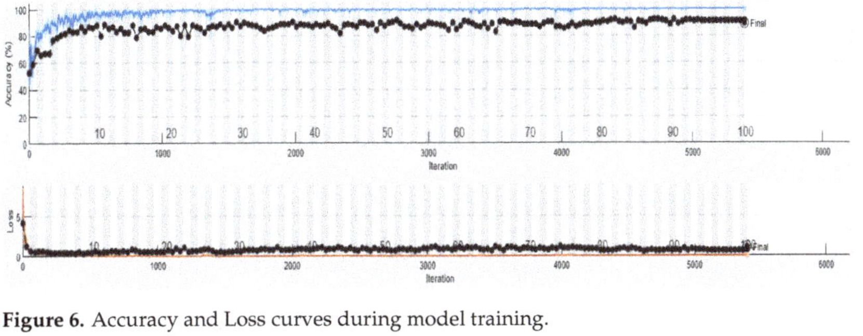

Figure 6. Accuracy and Loss curves during model training.

3.2.3. Initial Learning Rate

The initial learning rate is one of the key factors that determines model performance. The learning process speed is high when the model is trained with a higher learning rate, but there will be an increase in the loss function. The loss function can be decreased with a lower learning rate. Hence, it is essential to set the appropriate learning rate when training a model. Xia et al. [35] performed a study on pest classification and varied the initial learning rate between 0.0006 and 0.0014; the best performance was obtained at an initial learning rate of 0.0001. In our experiment, the proposed model was evaluated with various initial learning rate values, such as 0.000001, 0.00001, 0.0001, 0.001, and 0.01. The number of epochs and size of mini-batches were set to 100 and 64, respectively, during the evaluation of the model. The performance evaluation is shown in Table 3. It can be observed from Table 3 that the proposed model performed better when the initial learning rate was set to 0.001.

Table 3. Performance evaluation of proposed model for various initial learning rates.

Learning Rate	Accuracy (%)
0.000001	85.16
0.00001	87.06
0.0001	88.04
0.001	90.73
0.01	78.76

3.2.4. Mini-Batch Size

The mini-batch size is an important parameter when training a deep learning model. The model takes a long time and huge memory to run if the batch size is larger. Hence, it is essential to select the appropriate mini-batch size for better performance of deep learning models [36]. In our experiment, the proposed model was evaluated with various sizes of mini-batches, such as 1, 16, 32, 64, 128, and 256. The performance of the proposed model is high with the mini-batch size of 64. This is shown in Table 4. The proposed model was evaluated for 100 epochs with a learning rate of 0.001. Based on the experiment, a mini-batch size of 64 was chosen to train proposed model such that the convergence precision of proposed model is increased. It can be observed from Table 4 that a further increase in mini-batch size did not improve the model performance. Hence, the proposed model was trained with a mini-batch size of 64.

Table 4. Performance evaluation of proposed model for various mini-batch sizes.

Mini-Batch Size	Accuracy (%)
1	85.16
16	85.95
32	88.04
64	90.73
128	87.12
256	86.47

3.3. *Machine Learning Based Optimization in Designing the Model*

The manual hyperparameter designing in a model has a few limitations such as the dependency of knowledge and experience of the users, time consuming to find the optimal hyperparameters, and difficulty in handling huge data when the number of parameters and their range of values increase. The users require the capacity to identify the relations between the hyperparameters and the results obtained through visualization tools [37]. In order to overcome these limitations, machine learning based hyperparameter optimization has been used in recent researches. One among them is bayesian optimization [38]. This was used in our experiment and the initial learning rate and batch size were set to 0.0006 and 64, respectively. The number of epoch was set to 82.

The performance evaluation of the model with manual selection and machine learning based optimization is given in Table 5. It is observed from Table 5 the performance of the model is increased with machine learning based optimization. Hence, this result is used to compare with the other models.

Table 5. Performance evaluation of proposed model: Manual selection Vs Machine learning based optimization.

Method for Model Design	Accuracy (%)-Validation Data	Accuracy (%)-Test Data
Manual selection	90.73	87.78
Machine learning based optimization	93.86	89.02

3.4. *Overall Performance of Proposed Model for Root Rot Disease Identification in Ginseng Plants*

The plant regions were initially segmented from the healthy and diseased plant images as shown in Figures 7 and 8, respectively. The time taken for the proposed segmentation method was less than 5 s per image. After segmenting the plant region, the root rot disease identification was performed by transfer learning models and proposed model. Sample plant images that were correctly identified by proposed model are shown in Figures 9 and 10. These figures show that the leaves and roots of the healthy and diseased plants have a major difference in their appearance. It can be observed from these figures that the diseased plants have shorter primary roots, whereas the length of the primary roots in the healthy plants is greater than that in the diseased plants. Furthermore, it can be observed that there are many secondary and tertiary roots in healthy plants, but there are no such roots in the diseased plants. The healthy and root-rot-diseased plants were identified in a promising manner by proposed model. This shows the performance of the proposed model in identifying root rot disease in ginseng plants.

Figure 7. Plant segmentation from healthy plant images. The top row presents raw plant images; the bottom row presents segmented plant images.

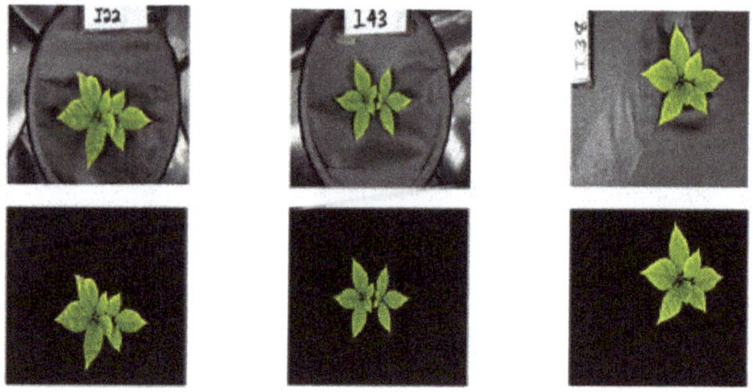

Figure 8. Plant segmentation from diseased plant images. The top row presents raw plant images; the bottom row presents segmented plant images.

(a) (b) (c)

Figure 9. Correct identification of healthy plants by proposed model. (**a**) Raw plant; (**b**) Segmented plant region; (**c**) Plant root.

(a) (b) (c)

Figure 10. Correct identification of diseased plants by proposed model. (**a**) Raw plant; (**b**) Segmented plant region; (**c**) Plant root.

The Figure 11 shows sample images on which proposed model failed. There exist a few asymptotic plants in the dataset. They do not show any symptoms in leaf but infected by the pathogens. The proposed model fails to identify these plants as shown in Figure 12. The times taken for training the transfer learning models and the proposed model are given in Table 6. In order to analyze the improvement in model performance due to plant segmentation, the deep learning models were also tested with raw plant images as input. The performance of the deep learning models is given in Tables 7 and 8.

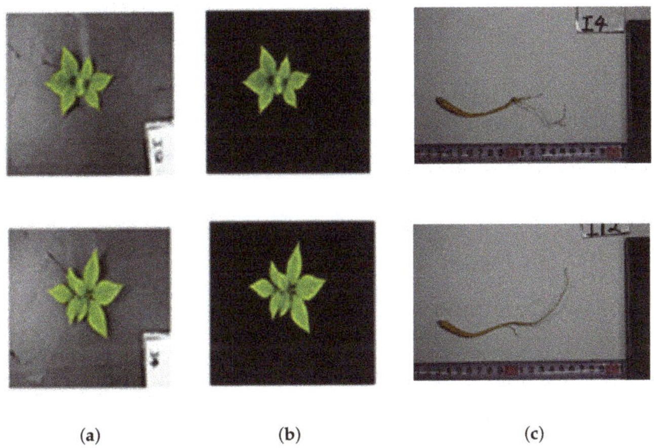

Figure 11. Incorrect identification of diseased plants by proposed model. (**a**) Raw plant; (**b**) Segmented plant region; (**c**) Plant root.

Figure 12. Incorrect identification of diseased plant images. The top row presents raw images; the bottom row presents segmented images.

Tables 7 and 8 show that the deep learning models performed well when the segmented plant images were given as input. It is observed from Table 6 that the proposed model required less training time when compared with a few transfer learning models except Resnet-18 and Squeezenet. Furthermore, the inference time of proposed model is lower than the other models as shown in Table 9. It is observed from Table 10 that the number of parameters of proposed model are less when compared with other models. These show that the proposed model can be practically applicable in a high-throughput system.

Table 6. Training times of various deep learning models.

Deep Learning Model	Training Time
Alexnet	25 min 46 s
VGG19	54 min 32 s
Squeezenet	7 min 16 s
Darknet19	50 min 57 s
Resnet18	12 min 51 s
Resnet101	94 min 35 s
Proposed model	18 min 11 s

Table 7. Performance comparison of various deep learning models with raw plant images (from testing dataset) as input.

Deep Learning Model	Precision	Recall	F_1 Score	Accuracy
Alexnet	0.60	0.87	0.71	0.64
VGG19	0.64	0.91	0.75	0.69
Squeezenet	0.64	0.93	0.76	0.71
Darknet19	0.64	0.93	0.76	0.71
Resnet18	0.66	0.95	0.78	0.73
Resnet101	0.63	0.89	0.74	0.68
Proposed model	0.64	0.92	0.76	0.70

Table 8. Performance comparison of various deep learning models with segmented plant regions (from testing dataset) as input images.

Deep Learning Model	Precision	Recall	F_1 Score	Accuracy
Alexnet	0.89	0.82	0.85	0.86
VGG19	0.87	0.84	0.85	0.85
Squeezenet	0.89	0.77	0.83	0.84
Darknet19	0.89	0.77	0.83	0.84
Resnet18	0.88	0.84	0.86	0.86
Resnet101	0.88	0.78	0.83	0.84
Proposed model	0.95	0.82	0.88	0.89

Table 9. Inference times (for testing dataset) of various deep learning models.

Deep Learning Model	Inference Time
Alexnet	13.15 s
VGG19	21.63 s
Squeezenet	14.41 s
Darknet19	19.45 s
Resnet18	17.39 s
Resnet101	35.93 s
Proposed model	8.97 s

Table 10. Parameters of various deep learning models.

Deep Learning Model	Parameters
Alexnet	61 M
VGG19	138 M
Squeezenet	12 M
Darknet19	22 M
Resnet18	11 M
Resnet101	44 M
Proposed model	6 M

The proposed model achieved a high precision value of 0.95 and a recall value of 0.82. Additionally, it achieved a high F_1 score of 0.88 and accuracy of 0.89, which are higher than those of the other transfer learning models. These imply that the proposed proposed can accurately predict 89% of diseased plants in the created dataset. Furthermore, the training time and inference time of proposed model are comparatively lower, increasing the practicality of this model. Overall, the proposed model has several advantages in real-time deployment.

4. Conclusions

In this paper, a new deep learning model was proposed to identify root rot disease in ginseng plants. Initially, the plant regions are segmented using the region growing method, and then the noise in the segmented plant regions is removed. Finally, the root-rot-diseased plants are identified by the proposed model. The performance of proposed model was compared with that of transfer learning models. The proposed model achieved an F_1 score of 0.88 and accuracy of 0.89 in less inference time when compared with the transfer learning models. Overall, the proposed model has several advantages, such as high throughput, cross-platform applicability, and robustness. However, a further improvement in performance is required, as well as the early detection of disease. This can be achieved by using various imaging sensors such as hyperspectral, fluorescence, etc., and corresponding image analysis methods. Additionally, this experiment can be improved by applying various plant stresses along with root rot disease.

Author Contributions: Conceptualization, P.K.J. and B.-K.C.; data curation, E.P., M.A.F. and Y.-S.K.; data analysis, P.K.J., H.K., I.B., M.S.K. and B.-K.C.; model development, P.K.J., H.K., D.S. and B.-K.C.; writing—original draft, P.K.J. and B.-K.C.; writing—review and editing, B.-K.C.; supervision, B.-K.C. All authors have read and agreed to the published version of the manuscript.

Funding: This research was supported by a grant from the Korean Society of Ginseng Funded by Korean Ginseng Corporation.

Institutional Review Board Statement: Not applicable.

Informed Consent Statement: Not applicable.

Data Availability Statement: Not applicable.

Conflicts of Interest: The authors declare no conflict of interest.

References

1. Minervini, M.; Scharr, H.; Tsaftaris, S. Image analysis: The new bottleneck in plant phenotyping [applications corner]. *IEEE Signal Process. Mag.* **2015**, *32*, 126–131. [CrossRef]
2. Furbank, R.T.; Tester, M. Phenomics—Technologies to relieve the phenotyping bottleneck. *Trends Plant Sci.* **2011**, *16*, 635–644. [CrossRef]
3. Wu, B.; Nevatia, R. Detection and segmentation of multiple, partially occluded objects by grouping, merging, assigning part detection responses. *Int. J. Comput. Vis.* **2009**, *82*, 185–204. [CrossRef]
4. Kumar, P.; Domnic, S. Computer Vision for Green Plant Segmentation and Leaf Count. In *Modern Techniques for Agricultural Disease Management and Crop Yield Prediction*; IGI Global: Hershey, PA, USA, 2020; pp. 89–110.
5. Kumar, J.P.; Domnic, S. Rosette plant segmentation with leaf count using orthogonal transform and deep convolutional neural network. *Mach. Vis. Appl.* **2020**, *31*, 1–14.
6. Dellen, B.; Scharr, H.; Torras, C. Growth signatures of rosette plants from time-lapse video. *IEEE/ACM Trans. Comput. Biol. Bioinform.* **2015**, *12*, 1470–1478. [CrossRef] [PubMed]
7. Grand-Brochier, M.; Vacavant, A.; Cerutti, G.; Kurtz, C.; Weber, J.; Tougne, L. Tree leaves extraction in natural images: Comparative study of preprocessing tools and segmentation methods. *IEEE Trans. Image Process.* **2015**, *24*, 1549–1560. [CrossRef]
8. Legendre, R.; Basinger, N.T.; van Iersel, M.W. Low-Cost Chlorophyll Fluorescence Imaging for Stress Detection. *Sensors* **2021**, *21*, 2055. [CrossRef] [PubMed]
9. Khanna, R.; Schmid, L.; Walter, A.; Nieto, J.; Siegwart, R.; Liebisch, F. A spatio temporal spectral framework for plant stress phenotyping. *Plant Methods* **2019**, *15*, 1–18. [CrossRef]
10. Martínez-Ferri, E.; Zumaquero, A.; Ariza, M.; Barceló, A.; Pliego, C. Nondestructive detection of white root rot disease in avocado rootstocks by leaf chlorophyll fluorescence. *Plant Dis.* **2016**, *100*, 49–58. [CrossRef]

11. Jayakodi, M.; Lee, S.C.; Yang, T.J. Comparative transcriptome analysis of heat stress responsiveness between two contrasting ginseng cultivars. *J. Ginseng Res.* **2019**, *43*, 572–579. [CrossRef]
12. Lee, J.S.; Lee, J.H.; Ahn, I.O. Characteristics of resistant lines to high-temperature injury in ginseng (Panax ginseng CA Meyer). *J. Ginseng Res.* **2010**, *34*, 274–281. [CrossRef]
13. Lee, S.M.; Bae, B.S.; Park, H.W.; Ahn, N.G.; Cho, B.G.; Cho, Y.L.; Kwak, Y.S. Characterization of Korean Red Ginseng (Panax ginseng Meyer): History, preparation method, and chemical composition. *J. Ginseng Res.* **2015**, *39*, 384–391. [CrossRef]
14. Yu, Y.; Ohh, S. Research on ginseng diseases in Korea. *Korean J. Ginseng Sci.* **1993**, *17*, 61–68.
15. Farh, M.E.A.; Kim, Y.J.; Kim, Y.J.; Yang, D.C. Cylindrocarpon destructans/Ilyonectria radicicola-species complex: Causative agent of ginseng root-rot disease and rusty symptoms. *J. Ginseng Res.* **2018**, *42*, 9–15. [CrossRef]
16. Park, Y.H.; Kim, Y.C.; Park, S.U.; Lim, H.S.; Kim, J.B.; Cho, B.K.; Bae, H. Age-dependent distribution of fungal endophytes in Panax ginseng roots cultivated in Korea. *J. Ginseng Res.* **2012**, *36*, 327. [CrossRef]
17. Reeleder, R.; Brammall, R. Pathogenicity of Pythium species, Cylindrocarpon destructans, and Rhizoctonia solani to ginseng seedlings in Ontario. *Can. J. Plant Pathol.* **1994**, *16*, 311–316. [CrossRef]
18. Kang, Y.; Kim, M.R.; Kim, K.H.; Lee, J.; Lee, S.H. Chlamydospore induction from conidia of Cylindrocarpon destructans isolated from ginseng in Korea. *Mycobiology* **2016**, *44*, 63–65. [CrossRef]
19. Chung, H.S. Studies on Cylindrocarpon destructans (Zins.) Scholten causing root rot of ginseng. *Rep. Tottori Mycol. Inst.* **1975**, *12*, 127–138.
20. Cho, D.H.; Park, K.J.; Yu, Y.H.; Ohh, S.; Lee, H. Root-rot development of 2-year old ginseng (Panax ginseng CA Meyer) caused by Cylindrocarpon destructans (Zinssm.) Scholten in the continuous cultivation field. *Korean J. Ginseng Sci.* **1995**, *19*, 175–180.
21. Cho, D.H.; Yu, Y.H.; Kim, Y.H. Morphological characteristics of chlamydospores of Cylindrocarpon destructans causing root-rot of Panax ginseng. *J. Ginseng Res.* **2003**, *27*, 195–201.
22. Ricciolini, M.; Rizzo, D. *Avversità della Vite e Strategie di Difesa Integrata in Toscana*; Press Service srl: Sesto Fiorentino, Italy, 2007.
23. Baumgartner, K.; Rizzo, D.M. Spread of Armillaria root disease in a California vineyard. *Am. J. Enol. Vitic.* **2002**, *53*, 197–203.
24. Daniele; Prodorutti, F.; de Luca, A.; Pellegrini, I.; Pertot. *I Marciumi Radicali Della Vite*; Safe Crop: San Michele all'Adige, Italy, 2007.
25. Calamita, F.; Imran, H.A.; Vescovo, L.; Mekhalfi, M.L.; La Porta, N. Early Identification of Root Rot Disease by Using Hyperspectral Reflectance: The Case of Pathosystem Grapevine/Armillaria. *Remote Sens.* **2021**, *13*, 2436. [CrossRef]
26. Kolander, T.; Bienapfl, J.; Kurle, J.; Malvick, D. Symptomatic and asymptomatic host range of Fusarium virguliforme, the causal agent of soybean sudden death syndrome. *Plant Dis.* **2012**, *96*, 1148–1153. [CrossRef] [PubMed]
27. Jang, C.S.; Lim, J.H.; Seo, M.W.; Song, J.Y.; Kim, H.G. Direct Detection of Cylmdrocarpon destructans, Root Rot Pathogen of Ginseng by Nested PCR from Soil Samples. *Mycobiology* **2010**, *38*, 33–38. [CrossRef]
28. Seifert, K.; McMullen, C.; Yee, D.; Reeleder, R.; Dobinson, K. Molecular differentiation and detection of ginseng-adapted isolates of the root rot fungus Cylindrocarpon destructans. *Phytopathology* **2003**, *93*, 1533–1542. [CrossRef] [PubMed]
29. Marzougui, A.; Ma, Y.; Zhang, C.; McGee, R.J.; Coyne, C.J.; Main, D.; Sankaran, S. Advanced imaging for quantitative evaluation of Aphanomyces root rot resistance in lentil. *Front. Plant Sci.* **2019**, *10*, 383. [CrossRef]
30. Marzougui, A.; Ma, Y.; McGee, R.J.; Khot, L.R.; Sankaran, S. Generalized Linear Model with Elastic Net Regularization and Convolutional Neural Network for Evaluating Aphanomyces Root Rot Severity in Lentil. *Plant Phenomics* **2020**, *2020*, 2393062. [CrossRef]
31. Werner, H.; Kindt, R.; Borneff, J. Testing the sterilisation effect of autoclaves by means of biological indicators (author's transl). *Zentralblatt Fur Bakteriol. Parasitenkd. Infekt. Hyg. Erste Abt. Orig. Reihe B Hyg. Prav. Med.* **1975**, *160*, 458–472.
32. Kutlugün, M.A.; Sirin, Y.; Karakaya, M. The effects of augmented training dataset on performance of convolutional neural networks in face recognition system. In Proceedings of the 2019 Federated Conference on Computer Science and Information Systems (FedCSIS), Leipzig, Germany, 1–4 September 2019; pp. 929–932.
33. Naranjo-Torres, J.; Mora, M.; Hernández-García, R.; Barrientos, R.J.; Fredes, C.; Valenzuela, A. A review of convolutional neural network applied to fruit image processing. *Appl. Sci.* **2020**, *10*, 3443. [CrossRef]
34. Thenmozhi, K.; Reddy, U.S. Crop pest classification based on deep convolutional neural network and transfer learning. *Comput. Electron. Agric.* **2019**, *164*, 104906. [CrossRef]
35. Xia, D.; Chen, P.; Wang, B.; Zhang, J.; Xie, C. Insect detection and classification based on an improved convolutional neural network. *Sensors* **2018**, *18*, 4169. [CrossRef] [PubMed]
36. Lee, M.; Xing, S. A study of tangerine pest recognition using advanced deep learning methods. *Preprints* **2018**, 2018110161. [CrossRef]
37. Jain, A. Complete Guide to Parameter Tuning in Gradient Boosting (Gbm) in Python. Analyticsvidhya.com. 2016. Available online: https://www.analyticsvidhya.com/blog/2016/02/complete-guide-parametertuning-gradient-boosting-gbm-python (accessed on 14 January 2022).
38. Wu, J.; Chen, X.Y.; Zhang, H.; Xiong, L.D.; Lei, H.; Deng, S.H. Hyperparameter optimization for machine learning models based on Bayesian optimization. *J. Electron. Sci. Technol.* **2019**, *17*, 26–40.

applied sciences

MDPI

Article

Real-Time Strawberry Plant Classification and Efficiency Increase with Hybrid System Deep Learning: Microcontroller and Mobile Application

Selami Kesler [1], Abdil Karakan [2,*] and Yüksel Oğuz [3]

[1] Electrical and Electronics Engineering, Faculty of Engineering, Pamukkale University, 20160 Denizli, Turkey
[2] Electrical Department, Dazkırı Vocational School, Afyon Kocatepe University, 03204 Afyonkarahisar, Turkey
[3] Electrical and Electronics Engineering, Faculty of Technology, Afyon Kocatepe University, 03204 Afyonkarahisar, Turkey
* Correspondence: akarakan@aku.edu.tr; Tel.: +90-507-118-12-48

Abstract: The strawberry plant has three life stages: seedling, blooming, and crop. It needs different acclimatization conditions in these life stages. A dataset consisting of 10,000 photographs of the strawberry plant was prepared. Using this dataset, classification in convolutional neural networks was performed in Matrix Laboratory (MATLAB). Nine different algorithms were used in this process. They were realized in ResNet101 architecture, and the highest accuracy rate was 99.8%. A low-resolution camera was used while growing strawberry plants in the application greenhouse. Every day at 10:00, a picture of the strawberry plant was taken. The captured image was processed in ResNet101 architecture. The result of the detection process appeared on the computer screen and was sent to the microcontroller via a USB connection. The microcontroller adjusted air-conditioning in the greenhouse according to the state of the strawberry plant. For this, it decided based on the data received from the temperature, humidity, wind direction, and wind speed sensors outside the greenhouse and the temperature, humidity, and soil moisture sensors inside the greenhouse. In addition, all data from the sensors and the life stage of the plant were displayed with a mobile application. The mobile application also provided the possibility for manual control. In the study, the greenhouse was divided into two. Strawberries were grown with the hybrid system on one side of the greenhouse and a normal system on the other side of the greenhouse. This study achieved 9.75% more crop, had a 4.75% earlier crop yield, and required 8.59% less irrigation in strawberry plants grown using the hybrid system.

Keywords: deep learning; convolutional neural networks; MATLAB; hybrid system; mobile application; productivity

Citation: Kesler, S.; Karakan, A.; Oğuz, Y. Real-Time Strawberry Plant Classification and Efficiency Increase with Hybrid System Deep Learning: Microcontroller and Mobile Application. *Appl. Sci.* **2022**, *12*, 8860. https://doi.org/10.3390/app12178860

Academic Editors: Paweł Kiełbasa, Tadeusz Juliszewski and Sławomir Kurpaska

Received: 2 July 2022
Accepted: 31 August 2022
Published: 3 September 2022

1. Introduction

Strawberries are a fruit species grown in a wide variety of ecological conditions across the world, and strawberry production is increasing every year. Strawberry cultivation is generally done with traditional agriculture in soil. Production with traditional agriculture causes many diseases and production efficiency is not very high. The yield is very high when production utilizes soilless agriculture in greenhouses. Strawberry plants grow very quickly when grown in greenhouses. This means that the needs of strawberry plants change very quickly. The faster the response to the situation, the higher the yield of the strawberry will be. Using deep learning, fast and very highly accurate detection can be achieved. There are many studies with deep learning in the literature.

Kawasaki et al. studied cucumber leaves. They tried to detect leaf diseases in cucumbers. A total of 7520 samples were tested in their study. Half of the photographs used in their sampling were taken in poor conditions. As a result of this study, they obtained an accuracy rate of 82.3% with convolutional neural networks [1]. Sladojevic et al. worked on disease detection by using leaf images. They identified 13 different classifications in their study. In this study, they used more than 30,000 samples for convolutional neural networks. They reached an accuracy rate of 91.11% in the experimental study [2]. Mohanty et al. used a publicly available dataset. In that study, the dataset consisted of 54,306 images and detected 26 diseases. Convolutional neural networks were also used in that study. As a result of their study, they reached an accuracy rate of 99.35% in an extended test set of [3]. Nachtigall et al. worked on the detection of 6 diseases in apple trees. They used a dataset of 2539 images in their study. As a result of their study, they achieved an accuracy rate of 97.3% at the best performance [4]. DeChant et al. worked on the detection of northern leaf blight disease in maize plants. In that study, they used a dataset consisting of 1796 images. They reached an accuracy rate of 96.7% with the dataset that they used [5]. Lu et al. studied rice diseases. Their work was on the detection of 10 common rice diseases. The dataset used in their study consisted of 500 images. They reached an accuracy rate of 95.48% with a CNN-based model [6].

Brahimi et al. worked on the detection of general diseases in plants. They used AlexNet and GoogleNet architectures in convolutional neural networks in their studies. They used a dataset consisting of 14,828 images in their study. In their studies, they reached an accuracy rate of 97.35% with AlexNet and an accuracy rate of 97.71% with GoogleNet [7]. Rangarajan and his friends worked on the detection of diseases in tomatoes. In their study, the detection of six diseases was performed and images of one healthy tomato were used. There were 13,262 images in the dataset that they used. VGG16 and AlexNet architectures were used for the detection of diseases. In that study, 97.29% accuracy was obtained with VGG16 and 97.49% accuracy was achieved with AlexNet [8]. Khandelwal and Raman studied general diseases in plants. In their study, ILSVRC 2012 architecture was used in convolutional neural networks on a dataset consisting of 86,198 images. The results of their study reported that an accuracy rate of 99.37% was reached [9]. Waheed et al. worked on corn plants. They tried to detect diseases that occur in maize plants with convolutional neural networks. DenseNet architecture was used in convolutional neural networks in that study. The results of their study reported that they reached an accuracy rate of 98.06% [10]. Epinso et al. performed work on solar panels. In their study, they tried to identify two classifications of solar panels: defective and intact. Convolutional neural networks were used in their classification. They used 345 images in their study. The results of their study report that they reached an accuracy rate of 70% [11]. Tang et al. worked on the detection of diseases in grapes. They used convolutional neural networks in their work. A total of 4062 grape leaf pictures were used as the dataset in their studies. They created four classifications: one healthy and three disease. They used a ShuffleNet architecture. An accuracy rate of 99.14% was found in the best-trained learning model [12]. Wang et al., while working on the detection diseases in rice, classified rice grains into four classifications: healthy rice and three groups of diseased rice. Their sample consisted of 503 grains of healthy rice, 523 with brown spots, 779 with leaf blast, and 563 with rice hispa damage. They used DenseNet121, VGG16, ResNet50, and MobileNet architectures in convolutional neural networks. The results of that study reported that MobileNet had an accuracy rate of 90.2% [13].

When the literature is examined, the studies are mostly on classification. Photographs of the plant were taken for classification. A dataset was created with these photographs. With this dataset, classification was performed in convolutional neural networks in MATLAB. In order to increase the accuracy rate, the photos of the plants were taken in a laboratory environment, not in real living areas. The classification process was carried out only in one phase of the plant. The most important feature of this study is that the classification process is not only carried out at one stage of the strawberry plant, but over the whole lifetime. The classification process is carried out from the planting of the strawberry plant as a seedling to the end of the crop yield. In this study, a high accuracy rate of 99.8% was achieved. The most important reason for this was a trial planting of the strawberry plants. The dataset consists of photographs of strawberry plants in this trial planting. Cocopeat was used in the trial planting and subsequent real planting. Thus, the similarity rate was increased in both cultivations. In this study, the real planting was performed in a greenhouse. For the classification process, a picture was taken every day at 10:00 in the morning with a low-resolution camera while the strawberry plant was growing in the greenhouse. This picture was tested with the ReNet101 architecture, which previously gave the highest accuracy rate of four different algorithms. The result of the process was transferred to the microcontroller via USB. The microcontroller adjusted the air conditioning in the greenhouse according to the state of the strawberry plant. To make this decision, it used the data received from the temperature, humidity, wind direction, and wind speed sensors outside the greenhouse and the temperature, humidity, and soil moisture sensors inside the greenhouse. In addition, all data from the sensors and the life stage of the plant were displayed with a mobile application. This mobile application also provided the ability for manual control. In this study, the greenhouse was divided into two. The hybrid system was grown on one side of the greenhouse and normal strawberries were grown on the other side. Thus, the comparison was made.

2. Materials and Methods

The study was carried out in three stages. Classification of three different growth stages of strawberry plant—seedling, flowering, and crop—was carried out. Secondly, this classification was applied in the real greenhouse. For this, the strawberries grown in the greenhouse were photographed every day at 10:00 in the morning. This photo was tested on the ResNet101 architecture. The result of this operation was sent to the Arduino microcontroller. The microcontroller adjusted air conditioning in the greenhouse according to the state of the strawberry plant. To make this decision, it used data from the temperature, humidity, wind direction, and wind speed sensors outside the greenhouse and the temperature, humidity, and soil moisture sensors inside the greenhouse. Finally, using a mobile application, all data from the sensors and the life stage of the plant were displayed. The mobile application also enabled manual control. Figure 1 shows the general structure of the system.

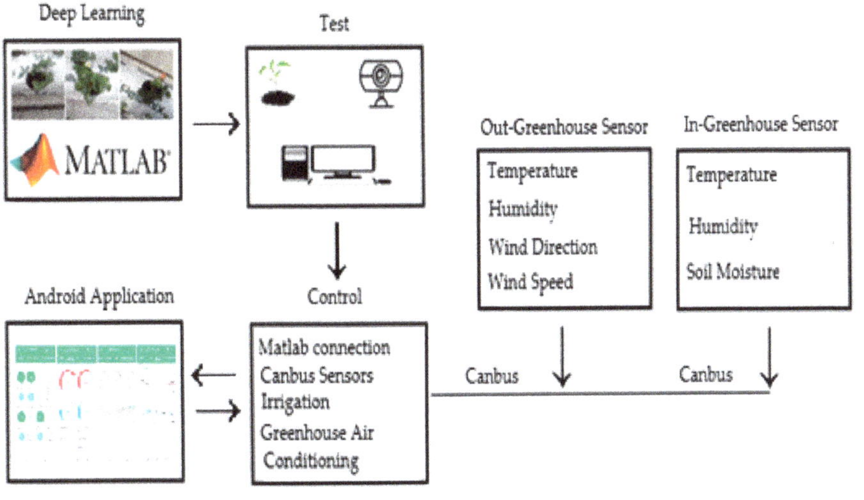

Figure 1. General structure of the system.

2.1. Dataset

A dataset consisting of 10,000 photographs was used in the study. Trial cultivation was done for this dataset. The dataset consists of photographs obtained from the trial planting and photographs obtained from the outside greenhouses. The dataset includes five different diseased strawberry groups and three different growth stages of strawberry plants. Figure 2 shows three healthy photographs of the strawberry plant used in the dataset.

Figure 2. Three healthy photos (seedling, flowering, and crop) of the strawberry plant used in the dataset.

In this study, the process was controlled. For this, three healthy states of strawberry plant were determined. Trial sowing was performed to determine the status of the strawberry plants. Thus, the similarity of the photographs to be taken later in the greenhouse to the photographs that form the dataset was increased. No disease was detected in the strawberry plants during the trial planting. In this study, photos of diseased strawberries from the surrounding greenhouses and the internet were used. Figure 3 shows the diseased strawberry plant photos used in the dataset.

Figure 3. Strawberry diseases (Botrytis cinerea, Sphaerotheca macularis fsp. fragariae, Mycosphaerella fragariae, Tetranychus urticae, Frankliniella occidentalis) used in the dataset.

2.2. Deep Learning

Deep learning is a field of study that covers artificial neural network algorithms with many hidden layers. The most well-known of the deep learning algorithms is the convolutional neural network algorithm. It is one of the most frequently used algorithms in image classification problems. The main structure of convolutional neural networks is based on computer vision and deep learning. CNNs obtain a different layered output by applying the specified filter to the image step by step. The CNN in this study was implemented in MATLAB to create the necessary layers of the CNN architecture. Figure 4 shows the architecture of a convolutional neural network.

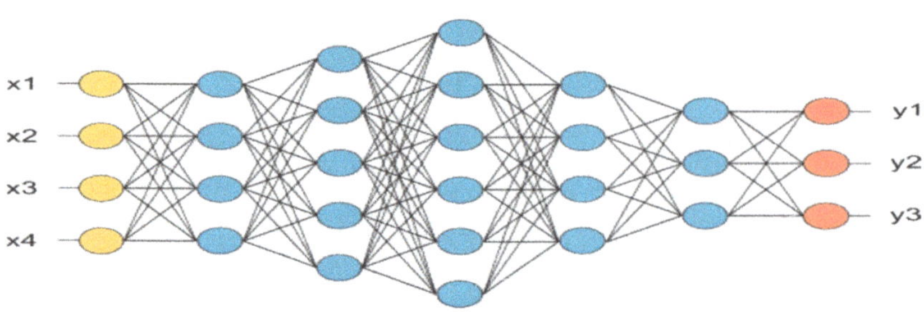

Figure 4. Architecture of a convolutional neural network.

2.3. *AlexNet*

AlexNet was created by Alex Krizhevsky [14]. The AlexNet model won the ImageNet competition held in 2012 [15]. The input to AlexNet was a 256×256 RGB image. This means that all images and all test images in the training set must be 256×256. If the input image was grayscale, it was converted to an RGB image by multiplying the single channel to get a three-channel RGB image. It had 60 million parameters and 650,000 neurons. It took five to six days to train on two GTX 580 3GB GPUs. With this architecture, the computerized object identification error rate was reduced from 26.2% to 15.4%. The architecture was designed to classify 1000 objects. The filters were 11×11 in size and the number of step shifts was 4. The architecture given in Figure 5 consists of 5 convolution layers, a pooling layer, and 3 fully connected layers.

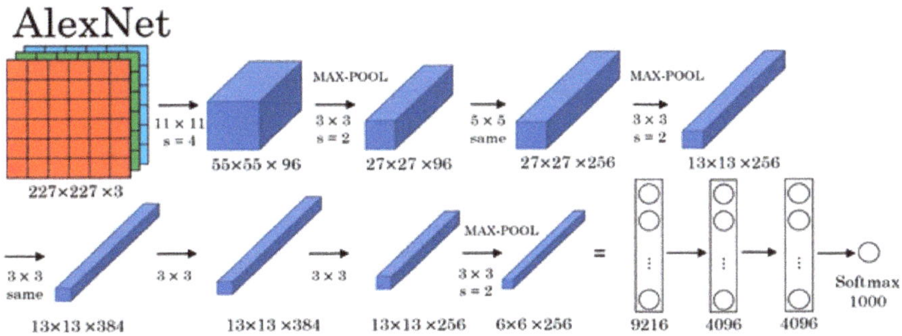

Figure 5. AlexNet architecture.

2.4. *VGGNet*

VGG Net was designed by Matthew Zeiler and Rob Fergus. They won the ImageNet competition held in 2013 [16]. With the designed VGG Net model, the error rate in object recognition was reduced to 11.2%. This architecture was an enhanced version of the AlexNet architecture and is shown in Figure 6.

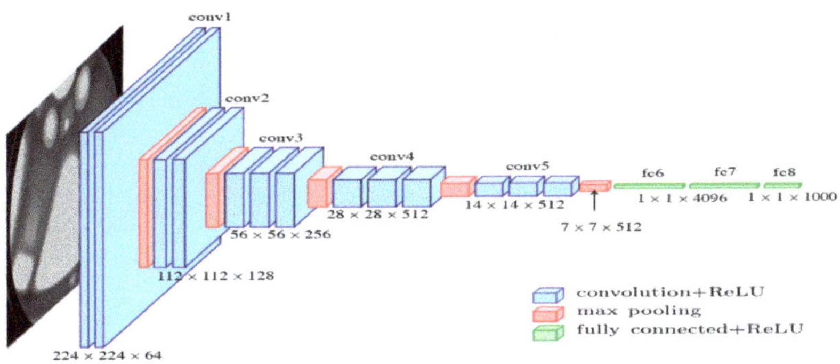

Figure 6. VGGNet architecture.

2.5. *GoogleNet*

GoogleNet was developed by Szegedy [17], who won the ImageNet competition held in 2015 [16]. GoogleNet has a complex architecture due to the incorporation of Inception modules. GoogleNet has 22 layers and a 5.7% error rate. The GoogleNet architecture was the first CNN architecture to move away from stacking convolution and pooling layers in a sequential structure [18]. Additionally, this new model represented an important advancement in memory and power usage, as stacking all the layers and adding many filters adds computational and memory costs and increases the probability of memorization.

GoogleNet used modules connected in parallel to overcome this problem. The GoogleNet network architecture is shown in Figure 7.

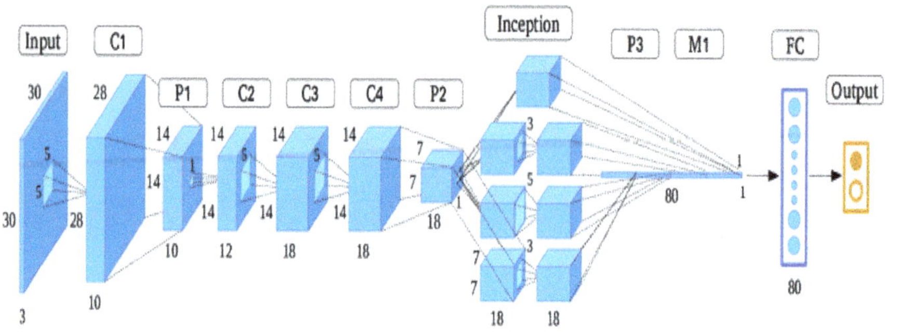

Figure 7. GoogleNet architecture.

2.6. ResNet

ResNet was developed by He [19]. It was an architecture more deeply designed than all other architectures. It consisted of 152 layers and was also the winner of the ImageNet 2016 competition, with an error rate of 3.6%. Depending on their skills and expertise, humans generally had an error rate of 5–10%. ResNet's first 34-layer network architecture is shown in Figure 8.

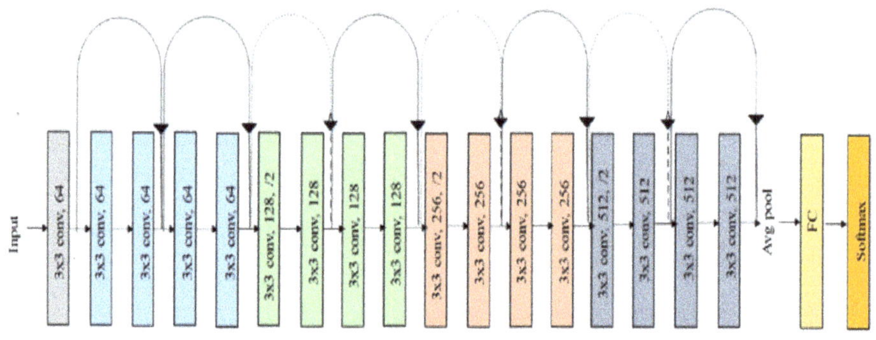

Figure 8. ResNet architecture.

The ResNet architecture consisted of residual blocks. In the residual block, it gave an F(x) result after the convolution-ReLu-convolution series of input x. This result was then added to the original input x and expressed as H(x) = F(x) + x (Figure 9).

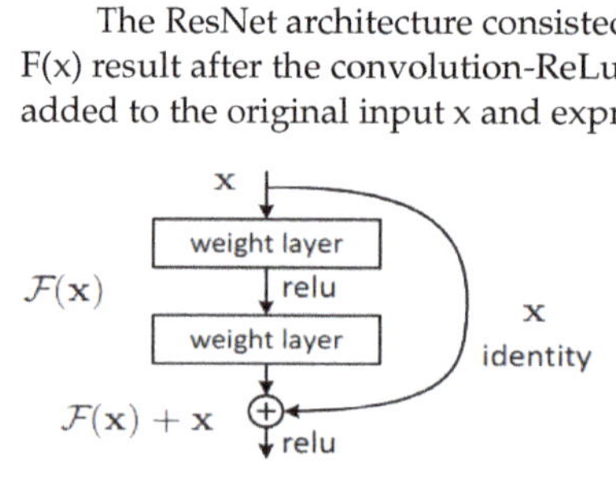

Figure 9. ReLu-convolution series.

When the literature was examined, it was understood that the lowest error rate belongs to the ResNet101 architecture. With the increase in the convolution layer, the error rate decreased to a rate of 3.6%. Although the error rate of humans varied, it varied between 5–10% on average. This showed that ResNet101 architecture gives better results than humans.

2.7. Applied System

In this study, the diseases that may occur in the life process of the strawberry plant were determined. In doing this, deep learning was used. During this process, a webcam was fixated on the plant grown in the greenhouse. The webcam took pictures of the strawberry plant in the greenhouse at 10:00 every morning. This photograph was tested in the deep learning algorithm which gave the highest accuracy rate. This process took place in MATLAB. The result of the process was reported by communicating with the Arduino microcontroller. According to the treatment of the disease detected by the microcontroller, a chemical or chemical mixture was applied to the plant. According to the condition of the plant, the indoor climate is adjusted. In this way, both crop productivity and irrigation efficiency were increased with timely delivery of the required amount of irrigation and humidification. Figure 10 shows the realized system.

Figure 10. (**a**) Exterior view of the application greenhouse. (**b**) Internal view of the control board. (**c**) View of the strawberry plants, including irrigation system, webcam, and spraying sprinklers. (**d**) Four different spraying tanks, mixing tank, and control system. (**e**) General view of the application system with the humidification system included. (**f**) System controlled by deep learning and normal system view.

2.8. Mobile Application

In this study, the system was monitored and controlled remotely. For this, an android application interface was designed. With this interface, all data produced by the system were displayed. In addition, remote control ability was also provided. Figure 11 shows the interface designed for the remote monitoring and control of the system.

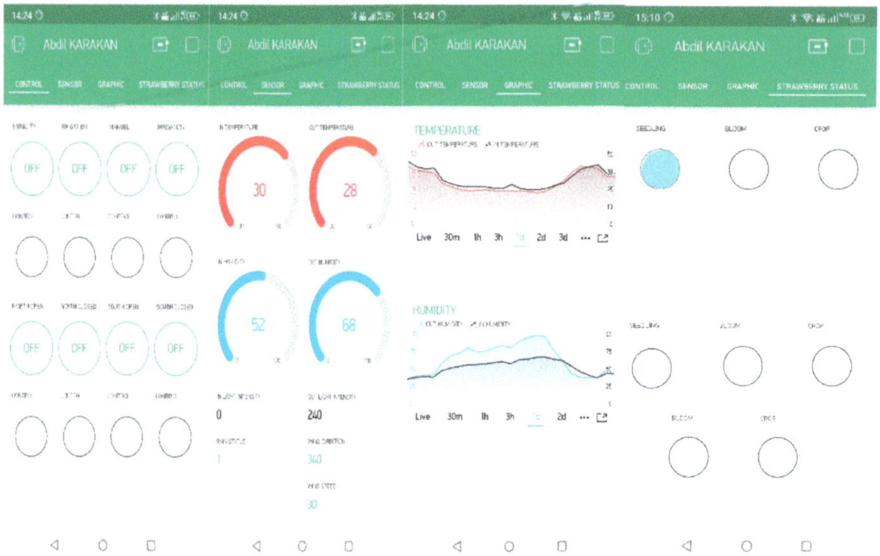

Figure 11. Control, sensor, graphic, and strawberry status interface display of mobile application.

The designed interface consists of four parts: control, sensor, graphic, and strawberry status. In the control section, the system can be controlled either manually or automatically. In this study, when manual control was selected, the north and south ventilation could be opened and closed, the irrigation and humidification of the strawberries grown with deep learning could be controlled, and the irrigation of the normal strawberry plants could be controlled. The control part of the interface provides feedback. When any button in the control part is pressed, the button is first activated and its color changes. The control light at the bottom changes according to the output of the microcontroller. The light is on when the output is active. Thanks to this feedback, it is possible to check whether the system is working or not. In the sensor section, data from sensors inside and outside the greenhouse are displayed. These are temperature, humidity, wind speed, wind direction, rain information, and luminous intensity. In the graphic section, the temperature and humidity values are displayed as instantaneous, 30 min, 1 h, 3 h, 1-day, and 2-day data. In the strawberry status pane, the life stage of the strawberry is shown. The fuzzy logic control values required for humidification are shown. The value from the soil moisture sensor is also displayed.

3. Results

The study which is carried out here covers the complete production process. This process is 45 days from planting the strawberry plant to the crop time. In this study, 70% of the data were used for training and 30% of the data were used for testing. Figure 12 shows the training process of the AlexNet architecture used in the system.

AlexNet is one of the first architectures used in convolutional neural networks. In the AlexNet architecture, a photo with dimensions of 227 × 227 is used. The depth parameter is 6 million. The rate of error in the AlexNet architecture varies between 15.4% and 26.2%. The main reason for the high level of change in this rate is the dataset. The more similar the photos which made make up the dataset are, the lower the error rate is. In this study, a trial plant was created to minimize the rate of error. Using photographs from the planting seedling of the strawberry plant planting to the end of the crop stage in the trial planting, a dataset was created. Because of this, the accuracy rate was 95.74% and the rate of error was 4.26%. The detection error rate of a normal person varies between 5 and 10%; the AlexNet architecture reached a lower error rate than a normal person. Figure 13 shows the training process of the VGGNet architecture used in the system.

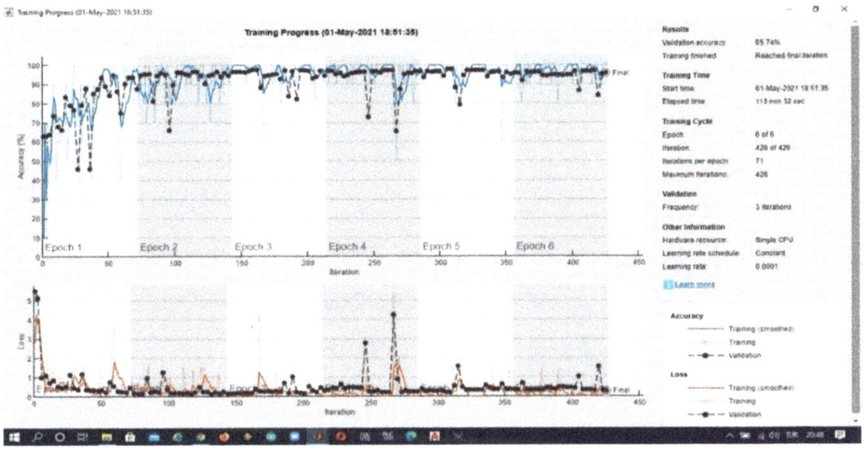

Figure 12. AlexNet training progress.

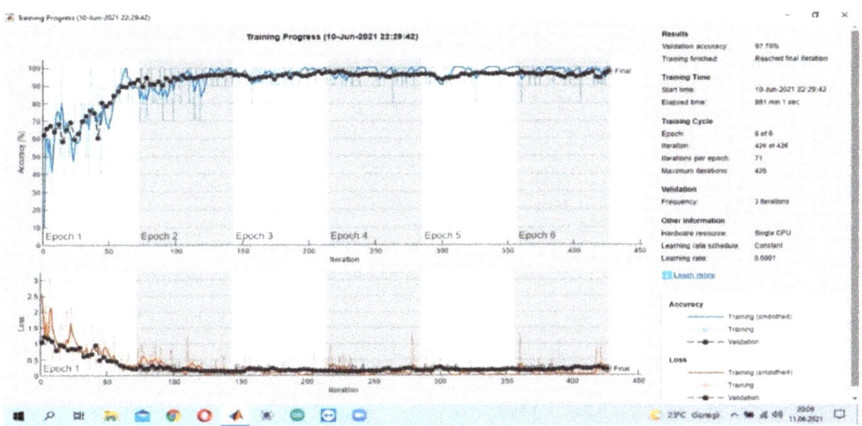

Figure 13. VGGNet training progress.

In VGGNet, the depth is 19 and there are 144 million parameters. VGGNet is greater than AlexNet in both depth and parameters. This is reflected in the rate of accuracy. The accuracy rate of VGGNet was 97.70%, approximately 2% higher than the accuracy rate of AlexNet. The training time in VGGNet was 961 min. Buddha was the longest training period. Figure 14 shows the training process of the GoogleNet architecture used in the system.

GoogleNet took had a training period of 24 min. The accuracy rate was the lowest, with 68.74%. It had a much higher error rate than normal people. Figure 15 shows the training process of the ResNet architecture used in the system.

Nine different architectures were used in this study. Of these architectures, the learning outcomes of the AlexNet, GoogLeNet, VGGNet, and ResNet architectures are shown. The other architectures shown are versions with increased accuracy rates. The ROC curves of the four different architectures used have been calculated. Figure 16 shows ROC graphs belonging to four different architectures.

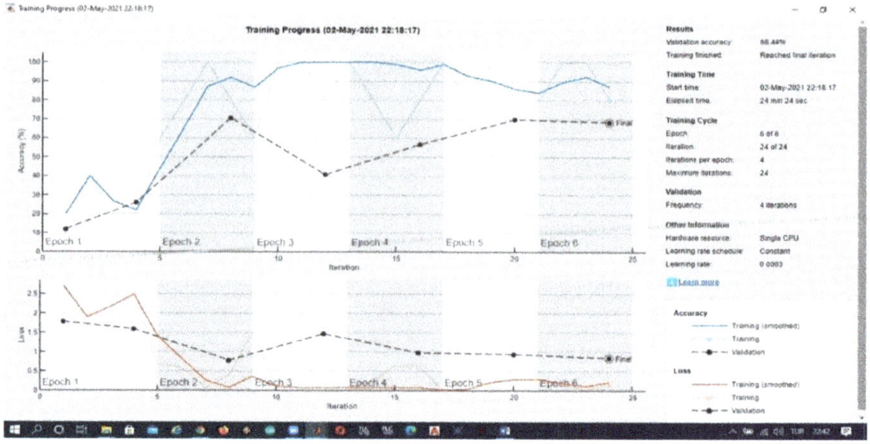

Figure 14. GoogleNet training progress.

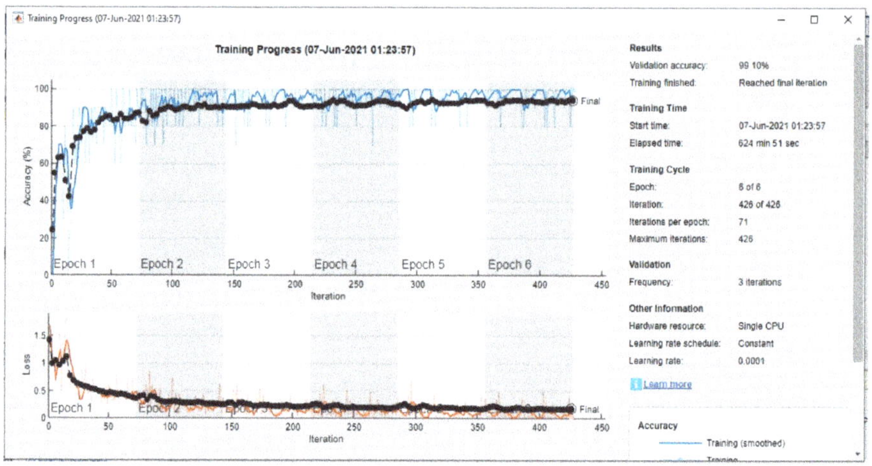

Figure 15. ResNet101 training progress.

This study was performed in the program MATLAB. A picture was taken with the webcam and processed at 10:00 every morning. It worked with ResNet101 architecture in MATLAB for every process. Training was done with ResNet101 architecture. Testing was done with training files. The test program and program output are shown in Figure 17.

This study was carried out in the real-life environment of the plant. The photo of the plant was assessed at 10:00 every morning. The size of the captured image was first reduced. ResNet101 architecture, which gave the highest accuracy rate in the study, uses 224 × 224 photos as an input. After the size of the captured photo was reduced, the test process was carried out in the ResNet101 architecture. After the test process was finished, the result was obtained and the captured picture was shown on the computer screen. There was a USB connection between MATLAB and the microcontroller. The code for the microcontroller was written in MATLAB. After the test process was finished, the result was sent to the microcontroller. Figure 18 shows the test algorithm.

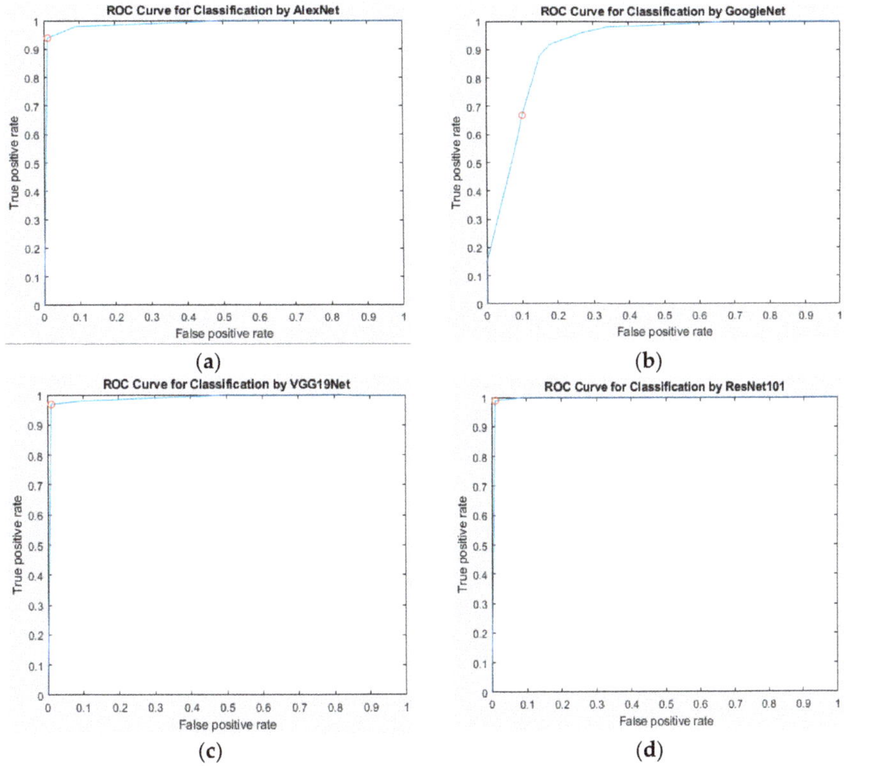

Figure 16. (**a**) ROC graph of AlexNet architecture. (**b**) ROC graph of GoogleNet architecture. (**c**) ROC graph of VGG19Net architecture. (**d**) ROC graph of the ResNet101 architecture.

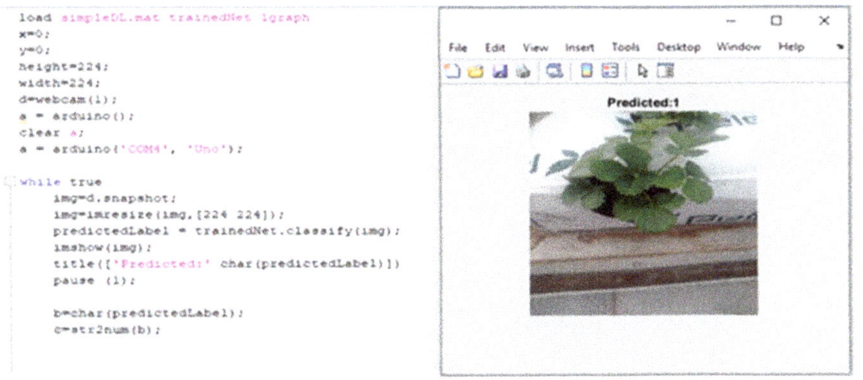

Figure 17. Test program and program output.

In this study, Canbus communication protocol was used for the communication of internal sensor, external sensor, irrigation, humidification, air conditioning, and the control part of the interface. Thanks to Canbus, communication was provided over long distances with less cables. All systems were connected via a single line with Canbus. Each system was given an ID number. Thus, the required system ID number can be easily found. In this study, 50 Kbit/sec was the preferred communication speed. In this way, data transmission was provided up to 1 km away. Figure 19 shows the general algorithm of the system.

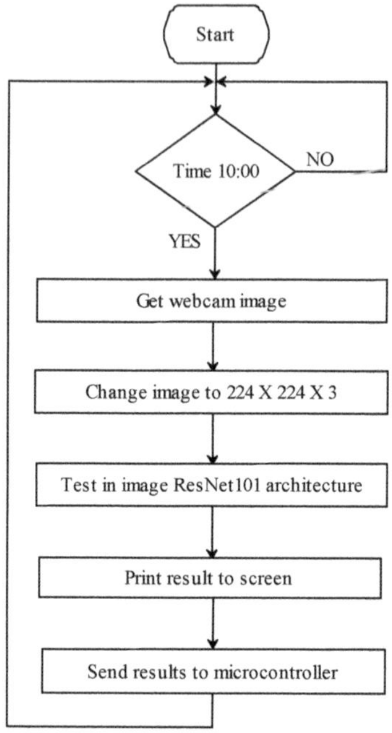

Figure 18. Deep learning test algorithm.

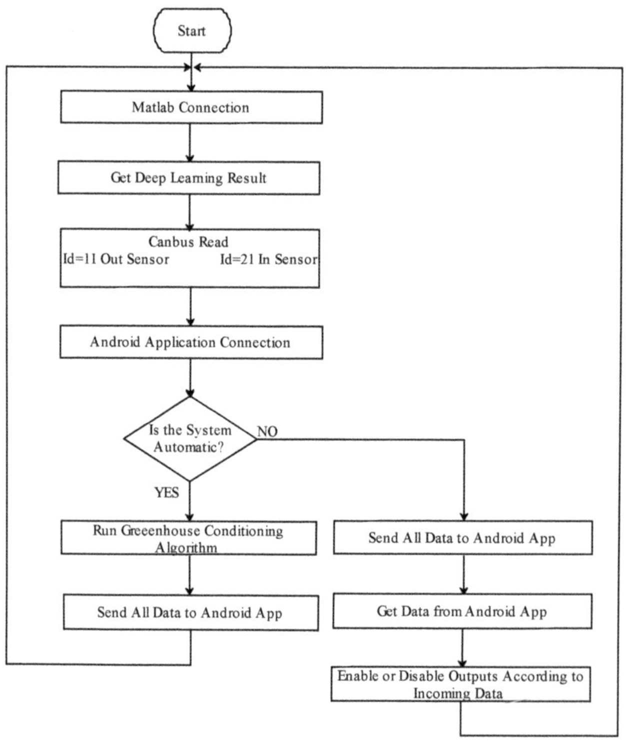

Figure 19. General algorithm of the system.

More than one algorithm was used in the study. Especially in automatic control, this was very important for the system to function by itself. The greenhouse air conditioning algorithm was designed to provide the ideal living environment for the plant. Figure 20 shows the indoor air conditioning algorithm.

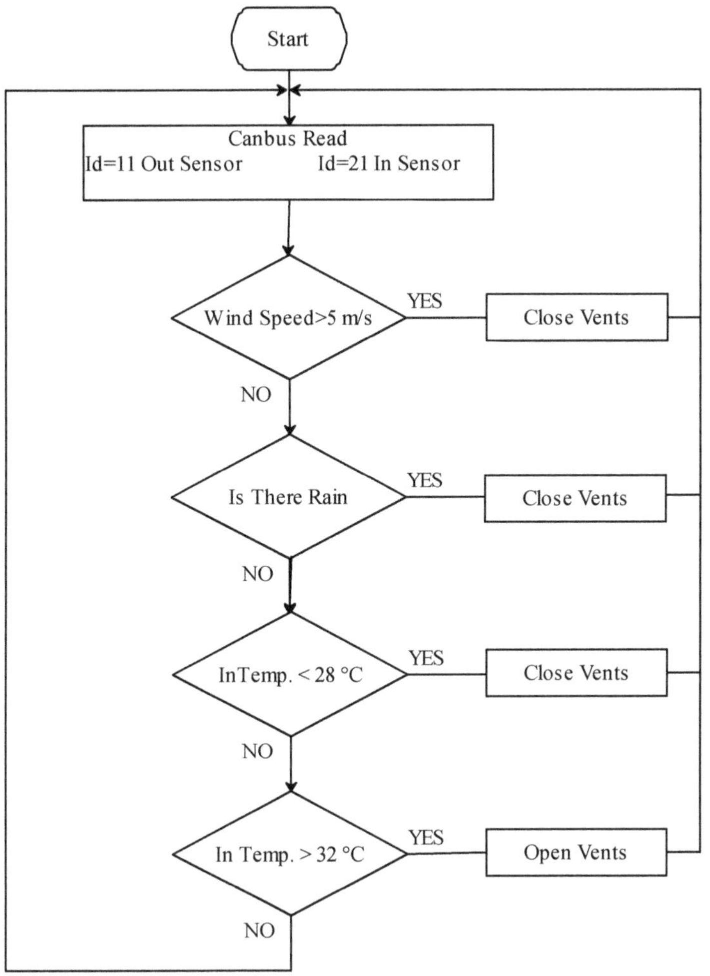

Figure 20. Automatic greenhouse air conditioning algorithm.

The work carried out covers the entire process from the planting of the strawberry seedling to the end of the crop yield. During this process, the detection and treatment of diseases that may occur in the strawberry plants were carried out. For this reason, diseases that originated from five different strawberry plants were detected. At the same time, the determination of the three different life stages of the strawberry plant was carried out. This study was carried out while growing strawberry plants in the greenhouse.

One of the most important differences between this study and other studies in the literature is that this study was carried out in a real greenhouse, not in a laboratory environment. All procedures were performed while the strawberry plant was growing. For this, strawberry planting was carried out in a 16 m^2 greenhouse. Following this planting, every stage of the strawberry plant was photographed. The dataset was created from these photographs. This dataset was used by nine different deep learning architectures. The same dataset was used by each architecture. The architectures used, accuracy rates, and other data are shown in Table 1.

Table 1. Algorithms used in the study.

Architecture Name	Image Input Size	Depth	Parameters	Training Time	Test Time	Accuracy Rate %
AlexNet	227 × 227	8	61 million	113 min	13.39 s	95.74
GoogleNet	224 × 224	22	7.0 million	24 min	7.56 s	68.44
ResNetInceptionV2	299 × 299	164	55.9 million	78 min	10.08 s	87.21
ResNetInceptionV3	299 × 299	48	23.9 million	151 min	8.53 s	93.77
ResNet18	224 × 224	18	11.7 million	426 min	7.55 s	97.05
ResNet50	224 × 224	50	25.6 million	553 min	8.54 s	94.75
ResNet101	224 × 224	101	44.6 million	878 min	9.57 s	99.80
VGG16Net	224 × 224	16	138 million	426 min	12.33 s	95.25
VGG19Net	224 × 224	19	144 million	113 min	13.58 s	97.70

Nine different algorithms were used in the study. Of the algorithms used, the ResNet101 architecture gave the highest accuracy rate. ResNet101 architecture had 99.80% accuracy, while a normal person had an error rate of 5–10%. An error rate that is less than human error was determined in this study. All algorithms were loaded with the same database. Although the photos in the database were the same, their sizes differed. Some algorithms used 224 × 224 photos, while others required 227 × 227 or 229 × 229 photos. For the training times, the highest training time was the SqueezeNet architecture, with 981 min. The lowest training time was GoogleNet, with 24 min. GoogleNet had the lowest training time but the highest error rate. The same computer was used for all training periods of the algorithms. Thus, the error rate arising from the system was minimized. Accuracy rates were not directly proportional to the training times. Thus, the parameters and depth of the algorithm were effective. Although SqueezeNet had the highest training time, the accuracy rate remained at 91.31%. Its 1.24 million parameters and 18 depths were far below those of the ResNet101 architecture.

This study was carried out to increase the productivity of strawberry plants. With this in mind, a disease that can occur in the growth phase of the strawberry plant was determined. With this determination, the necessary chemical or chemical mixture was given to the plant and recovery was achieved as soon as possible. At the same time, three life stages of the plant—seedling, blooming, and crop—were determined. Between these stages of the strawberry plant, the climate needs to be adjusted. The sooner that the indoor air conditioning becomes suitable for the stage of the plant, the more that the efficiency will increase. In Figure 21, the efficiency of the system irrigated with the normal and the hybrid system is compared.

In addition to the outdoor temperature and humidity values, soil moisture was also measured in the irrigation system. Since the soil moisture was very low in the first seedling planting, the water yield was less. Afterwards, the water yield increased a lot.

In this study, the strawberry plant grew very quickly. A total of 5652 g of crop was obtained in the plants grown with deep learning. A total of 5150 g of strawberry crop was obtained in the plants grown normally. This equates to 9.75% more product when using deep learning. Figure 22 shows a comparison of the strawberry production.

The strawberry plants grown with deep learning yielded more crops and were grown in a shorter time. In the plants that grew with deep learning, time to flowering was on average 2 days shorter. Crop giving was started 2 days earlier on average. Figure 23 comparison of the growth times of the strawberry plant.

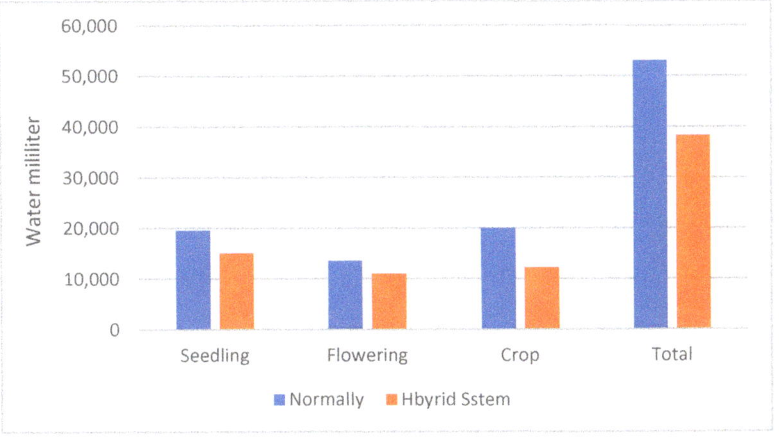

Figure 21. Comparison of water consumption using irrigation.

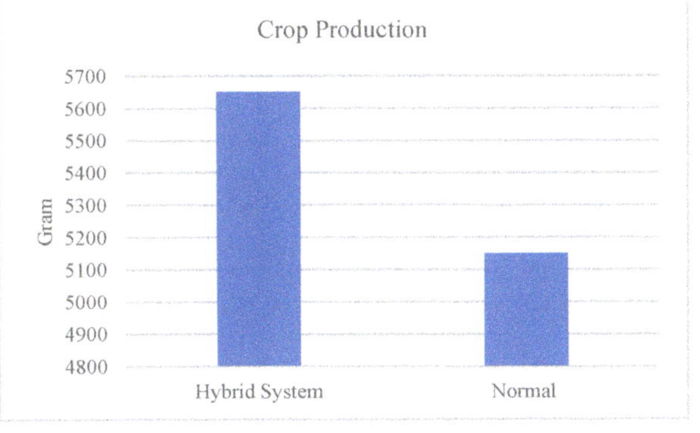

Figure 22. Comparison of strawberry productions.

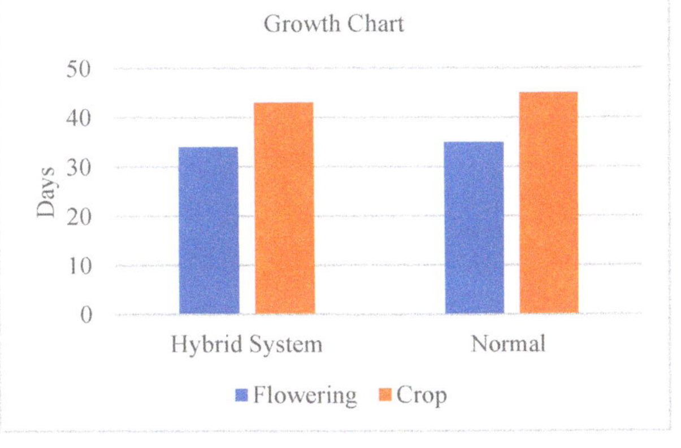

Figure 23. Comparison of the growth times of the strawberry plant.

4. Discussion

When the literature was examined, many studies have been carried out on disease detection with deep learning. These studies covered the leaf of the diseased plant. Then, all pictures were taken in a laboratory environment so that they were the same at the bottom.

The datasets consisted of the taken photographs. These photographs were not taken in the real-life environment of the plant and did not cover the entire life cycle of the plant. Table 2 shows the results of the studies in the literature.

Table 2. Literature's studies.

Authors	Architecture Name	Accuracy Rate %
Brahimi et al. [7]	AlexNet, GoogleNet	97.71
Rangarajan et al. [8]	AlexNet, VGG16	97.49
Khandelwal et al. [9]	GoogleNet	99.37
Waheed et al. [10]	Denset	9.,06
Espinosa et al. [11]	GoogleNet	70
Tang et al. [12]	ShuffleNet	99.14
Wang et al. [13]	Denset121, VGG16, ResNet50, MobileNet	90.02
Vasconeza et al. [20]	MobileNet, Inpection V2	93
Sajad et al. [21]	GoogleNet	9.05
Daobilige et al. [22]	Inception V3	95.48
Samyak et al. [23]	MobileNet	95.70
Yibin et al. [13]	Denset121, VGG16, ResNet50, MobileNet	90.02
Alkan et al. [24]	AlexNet, ResNet, GoogleNet	99.06
Ashwinkumar et al. [25]	MobileNet	98.50
Valeri et al. [26]	GoogleNet	99.72
Ben et al. [27]	VGG16	90
Cen et al. [28]	1D-CNN	9969
Xioa et al. [29]	MobileNet	98.,1
Paymode et al. [30]	VGG16	95.71
Anh et al. [31]	VGG16	91.8
Yuana et al. [32]	VGG16	98.15
Weng et al. [33]	SqueezeNet	99.71

When we looked at the literature, most studies used one or more architectures, but no more than five architectures were used. Ten different architectures were used in the study. Thus, a comparison was performed between ten architectures.

When the accuracy rates in the literature were examined, they varied according to the study. The most important factor in this study was the photos in the dataset. The closer the photos in the dataset were to the photo to be tested, the higher the accuracy. In order to increase the accuracy rate in this study, a trial sowing was conducted. Soilless agriculture cocopeat was used in this trial planting. The dataset was comprised of the photos taken during this trial planting. As a result, the later rate increased since the same cocopeat was in the photo in the recognition process of the disease. This study had the highest accuracy rate among the literature.

The studies in the literature have only taken one phase of the plant into account. This study covered the entire life cycle of the plant, the time elapsed from the plant's seedling stage to its yielding.

This study was conducted to increase the productivity of strawberry plant. In this pursuit, a disease that may occur in the growth phase of the strawberry plant was determined. With this determination, the necessary chemical or chemical mixture was given to the plant and recovery was achieved as soon as possible. At the same time, three life stages of the plant—seedling time, blooming time, and crop time—were determined. Between these stages of the strawberry plant, the climate needs to be adjusted. The sooner the greenhouse air conditioning becomes suitable for the stage of the plant, the more that efficiency will increase.

5. Conclusions

In this study, five diseases and three healthy stages of strawberry plant were determined. This determination was conducted not only in the laboratory environment, but over the entire plant growing process in the greenhouse. Convolutional neural networks were used to perform the detection. In this study, the detection accuracy results were as follows: GoogleNet, 68.44%; InceptionResNetV2, 87.21%; InceptionV3, 93.77%; ResNet50, 94.75%; ResNet18, 97.05%; VGG16, 95.25%; AlexNet, 95.74%; VGG19, 97.70%; and ResNet101, 99.80%. In this study, fuzzy logic control was used to increase efficiency in water and humidity. As a result of the application of the deep learning system, 9.75% more crops were obtained. At the same time, the harvesting time was shortened by 4.75%. An efficiency increase of 8.51% was achieved in the irrigation system.

Author Contributions: Supervision, S.K.; Writing—original draft, A.K.; Writing—review & editing, Y.O. All authors have read and agreed to the published version of the manuscript.

Funding: This research received no external funding.

Institutional Review Board Statement: This study not involving humans or animals, so this research do not applicable.

Informed Consent Statement: For studies not involving humans.

Data Availability Statement: The data presented in this study are available on request from the corresponding author.

Conflicts of Interest: The authors declare no conflict of interest.

References

1. Kawasaki, Y.; Fujita, E.; Uga, H.; Kagiwada, S.; Iyatomi, H. Basic investigation on a robust and practical plant diagnostic system. In Proceedings of the 2016 15th IEEE International Conference on Machine Learning and Applications (ICMLA), Anaheim, CA, USA, 18–20 December 2016; pp. 989–992.
2. Sladojevic, S.; Arsenovic, M.; Anderla, A.; Culibrk, D.; Stefanovic, D. Deep neural networks-based recognition of plant diseases by leaf image classification. *Comput. Intell. Neurosci.* **2016**, *2016*, 3289801. [CrossRef] [PubMed]
3. Mohanty, S.P.; Hughes, D.P.; Salathé, M. Using deep learning for image-based plant disease detection. *Front. Plant Sci.* **2016**, *7*, 1419. [CrossRef] [PubMed]
4. Nachtigall, L.G.; Araujo, R.M.; Nachtigall, G.R. Classification of apple tree disorders using convolutional neural networks. In Proceedings of the 2018 IEEE 28th International Conference on Tools with Artificial Intelligence, San Jose, CA, USA, 6–8 November 2016; pp. 472–476.
5. DeChant, C.; Wiesner-Hanks, T.; Chen, S.; Stewart, E.L.; Yosinski, J.; Gore, M.A.; Nelson, R.J.; Lipson, H. Automated identification of northern leaf blight-infected maize plants from field imagery using deep learning. *Phytopathology* **2017**, *107*, 1426–1432. [CrossRef] [PubMed]
6. Lu, Y.; Yi, S.; Zeng, N.; Liu, Y.; Zhang, Y. Identification of rice diseases using deep convolutional neural networks. *Neurocomputing* **2017**, *267*, 378–384. [CrossRef]
7. Brahimi, M.; Arsenovic, M.; Laraba, S.; Sladojevic, S.; Boukhalfa, K.; Moussaoui, A. Deep Learning for Plant Diseases: Detection and Saliency Map Visualisation. In *Human and Machine Learning*; Springer: Cham, Switzerland, 2018; pp. 93–117.
8. Rangarajan, A.K.; Purushothaman, R.; Ramesh, A. Tomato crop disease classification using pre-trained deep learning algorithm. *Procedia Comput. Sci.* **2019**, *133*, 1040–1047. [CrossRef]
9. Khandelwal, I.; Raman, S. Analysis of Transfer and Residual Learning for Detecting Plant Diseases Using Images of Leaves. *Comput. Intell. Theor. Appl. Future Dir.* **2019**, *II*, 295–306.
10. Waheed, A.; Goyal, M.; Gupta, D.; Khanna, A.; Hassanien, A.; Pandey, H. An optimized dense convolutional neural network model for disease recognition and classification in corn leaf. *Comput. Electron. Agric.* **2020**, *175*, 105456. [CrossRef]
11. Espinosa, A.R.; Bressan, M.; Giraldo, L.F. Failure signature classification in solar photovoltaic plants using RGB images and convolutional neural networks. *Renew. Energy* **2020**, *162*, 249–256. [CrossRef]
12. Tang, Z.; Yanga, J.; Lia, Z.; Grape, F.Q. Grape disease image classification based on lightweight convolution neural networks and channel wise attention. *Comput. Electron. Agric.* **2020**, *178*, 105735. [CrossRef]
13. Wang, Y.; Wang, H.; Peng, Z. Rice diseases detection and classification using attention based neural network and bayesian optimization. *Expert Syst. Appl.* **2021**, *178*, 114770. [CrossRef]
14. Krizhevsky, A.I.; Sutskever, G.E. Imagenet classification with deep convolutional neural networks. *Adv. Neural Inf. Process. Syst.* **2012**, 1097–1105. [CrossRef]
15. Russakovsky, O.; Deng, J.; Su, H.; Krause, J.; Satheesh, S.; Ma, S.; Huang, Z.; Karpathy, A. ImageNet Large Scale Visual Recognition Challenge. *International Journal of Computer Vision.* **2015**, *115*, 211–252. [CrossRef]
16. Zeiler, M.D.; Fergus, R. Visualizing and Understanding Convolutional Networks. In Proceedings of the European Conference on Computer Vision, Zurich, Switzerland, 6–12 September 2014; pp. 818–833.

17. Szegedy, C.; Liu, W.; Jia, Y.; Sermanet, P.; Reed, S.; Anguelov, D.; Erhan, D.; Vanhoucke, V.; Rabinovich, A. Going deeper with convolutions. In Proceedings of the IEEE Conference on Computer Vision and Pattern Recognition, Boston, MA, USA, 7–12 June 2015; pp. 1–9.
18. Ghazi, M.M.; Yanikoglu, B.E. Plant identification using deep neural networks via optimization of transfer learning parameters. *Neurocomputing* **2017**, *235*, 228–235. [CrossRef]
19. He, K.M.; Zhang, X.Y.; Ren, S.Q.; Sun, J. Deep Residual Learning for Image Recognition. In Proceedings of the 2016 IEEE Conference on Computer Vision and Pattern Recognition, Las Vegas, NV, USA, 27–30 June 2016; pp. 770–778.
20. Vasconeza, J.P.; Delpianob, J.S.; Vougioukasc, F.; Auat, C. Comparison of convolutional neural networks in fruit detection and counting: A comprehensive evaluation. *Comput. Electron. Agric.* **2020**, *173*, 105348. [CrossRef]
21. Sajad, S.; Razieh, P.; Mohammad, H.R.; Gines, G.; Juan, I.A. Estimation of nitrogen content in cucumber plant (*Cucumis sativus* L.) leaves using hyperspectral imaging data with neural network and partial least squares regressions. *Chemom. Intell. Lab. Syst.* **2021**, *217*, 104404.
22. Daobilige, S.; Yongliang, Q.; He, K.; Salah, S. Real time detection of inter-row yegrass in wheat farms using deep learning. *Biosyst. Eng.* **2021**, *204*, 198–211.
23. Samyak, S. PlantifyAI: A Novel Convolutional Neural Network Based Mobile Application for Efficient Crop Disease Detection and Treatment. *Procedia Comput. Sci.* **2021**, *191*, 469–474.
24. Alkan, A.; Abdullah, M.Ü.; Abdullah, H.O.; Assaf, M.; Zhou, H. A smart agricultural application: Automated detection of diseases in vine leaves using hybrid deep learning. *Turk. J. Agric. For.* **2021**, *45*, 717–729. [CrossRef]
25. Ashwinkumar, S.; Rajagopal, S.; Manimaran, V.; Jegajothi, B. Automated plant leaf disease detection and classification using optimal MobileNet based convolutional neural networks. *Mater. Today Proc.* **2022**, *51*, 480–487. [CrossRef]
26. Valeri, M.; Carlos, E.; Laura, A.; Jose, M.; Jorge, I.; Hamurabi, G.; Huizilopoztli, L.; Rafael, M.; Carlos, A. Comparison of Convolutional Neural Network Architectures for Calssification of Tomato Plant Diseases. *Appl. Sci.* **2020**, *10*, 1245.
27. Ben, M.; Rivas Casado, M. JellyNet: The convolutional neural network jellyfish bloom detector. *Int. J. Appl. Earth Obs. Geoinf.* **2021**, *97*, 102279.
28. Chen, S.; Du, X.; Zhao, W.; Guo, P.; Chen, H.; Jiang, Y. Olive oil classification with Laser-induced fluorescence (LIF) spectra using 1-dimensional convolutional neural network and dual convolution structure model. *Spectrochim. Acta Part A Mol. Biomol. Spectrosc.* **2022**, *21*, 121418. [CrossRef] [PubMed]
29. Xiao, K.; Zhou, L.; Yang, H.; Yang, L. Phalaenopsis growth phase classification using convolutional neural network. *Smart Agric. Technol.* **2022**, *2*, 100060. [CrossRef]
30. Paymode, A.S.; Malode, V.B. Transfer Learning for Multi-Crop Leaf Disease Image Classification using Convolutional Neural Network VGG. *Artif. Intell. Agric.* **2022**, *6*, 23–33. [CrossRef]
31. Anh, P.T.Q.; Thuyet, D.Q.; Kobayashi, Y. Image classification of root-trimmed garlic using multi-label and multi-class classification with deep convolutional neural network. *Postharvest Biol. Technol.* **2022**, *190*, 11956. [CrossRef]
32. Yuana, P.; Qiana, S.; Zhaia, Z.; Martínezb, J.F.; Xua, H. Study of chrysanthemum image phenotype on-line classification based on transfer learning and bilinear convolutional neural network. *Comput. Electron. Agric.* **2022**, *194*, 106679. [CrossRef]
33. Weng, Z.; Meng, F.; Liu, S.; Zhanga, Y.; Zheng, Z.; Gong, C. Cattle face recognition based on a Two-Branch convolutional neural network. *Comput. Electron. Agric.* **2022**, *194*, 1066871. [CrossRef]

applied sciences

MDPI

Article

An Image Processing Method for Measuring the Surface Area of Rapeseed Pods

Fangyi Li [1], Xumeng Li [1], Huang Huang [1], Hao Xiang [2], Chunyun Guan [1,3,*] and Mei Guan [1,3,*]

1 College of Agronomy, Hunan Agricultural University, Changsha 410128, China
2 College of Information and Intelligence, Hunan Agricultural University, Changsha 410128, China
3 National Oilseed Crops Improvement Center in Hunan, Changsha 410128, China
* Correspondence: guancy2011@aliyun.com (C.G.); gm7142005@hunau.edu.cn (M.G.)

Abstract: An image processing method that considers pods to be irregular cylinders composed of several oblique cylinder slices with different diameters was proposed to achieve the "highly accurate, highly efficient and large-scale" target of measuring the surface area of rapeseed pods. The total side area of all the oblique cylinder slices, specifically the pod surface area, was calculated. A high-precision 3-dimensional method was used to measure and correct the actual area of the silique for the first time. The results of the measurement accuracy analysis showed that the image processing method could accurately measure the surface area of rapeseed pods. The average measurement error was 2.46%, and the root-mean-square error (RMSE) was 0.92 cm^2. To prove the superiority of this method, we measured the same test samples using four other methods: the Clark formula, the Leng formula, flattening scanning, and quasi-cylinder side area methods. The accuracy and efficiency of the image processing method were much higher than the other four measurement methods. The surface area of multiple pods from 83 rape plants was measured using the image processing method; the results were consistent with the expectations of the experimental design. The 3D measurement and image processing technology were compared and analyzed, and the latter was preliminarily designed for future rape pod seed testing. Thus, this method can provide technical support to measure the surface area of numerous rapeseed pods.

Keywords: rapeseed pod; surface area measurement; image processing; 3-D measurement; side area of oblique cylinders; rapeseed pod seed testing machine

Citation: Li, F.; Li, X.; Huang, H.; Xiang, H.; Guan, C.; Guan, M. An Image Processing Method for Measuring the Surface Area of Rapeseed Pods. *Appl. Sci.* **2023**, *13*, 5129. https://doi.org/10.3390/app13085129

Academic Editor: Joachim Müller

Received: 4 October 2022
Revised: 2 March 2023
Accepted: 25 March 2023
Published: 20 April 2023

1. Introduction

Unlike many other crops, the primary photosynthetic organs of rapeseed include branches, stems, and pods in addition to the leaves [1]. Once the plants enter the flowering stage, the pods rapidly proliferate. The surface area of pods reaches its maximum and tends to become constant after approximately 25 days of flowering. Simultaneously, the leaves wither gradually and their area decreases sharply; therefore, the pods gradually overtake the role of leaves as the primary photosynthetic organ in rape [1]. As an essential source of photosynthesis, pods function as the "source" [1–3] and "sink" for grain formation, significantly influencing the yield and quality of rape grain [4–8].

The surface area of rapeseed pods is an important index widely used to measure the photosynthetic capacity of pods. It is generally believed that cultivars that produce large pods with substantial surface areas are more efficient at photosynthesis and produce larger amounts of grain. Rape pods are irregularly shaped, with varying lengths and quantities of long-bead chains. Thus, measuring the surface area of rape pods is difficult [9]. The surface area of rape pods refers to the measured surface area of the functional part of the pods. The measurement of pod surface area can be divided into two dimensions: the single-pod surface area and the multi-pod surface area (Figure 1). These areas will be referred to as "single-pod measurement" and "multi-pod measurement", with the former serving as the basis of the latter. Currently, the methods for single pod measurement are relatively well-established and include the Clark formula [10], Leng formula [9,11], flattening scanning [12,13], and quasi-cylinder side area methods [14]. The Clark and Leng formula methods have a similar principle of estimating the unknown pod surface area by establishing a regression equation between the pod length/width and the actual surface area. The principle and operation of these two methods are relatively simple and widely used [15–21]. The results of these methods are obtained indirectly through mathematical models; thus, the "actual area" of the pods used for their correlation analysis might not be accurate owing to the limitations of the measurement techniques and conditions at that time, i.e., the measurements may not be highly accurate. Flattening scanning obtains the area value by directly measuring the flattened pod peel. These results are more accurate than the two methods described earlier as the pod peel is directly measured. However, owing to the complex operation and inefficient measurements that result from these models, they are only used to measure the "actual area" when there are few pods [22]. In addition, it is difficult to completely flatten the pod peels because of the shape of their chain beads; this results in a measured value that is theoretically less than the "actual area". The quasi-cylinder and the quasi-cone side area methods treat the pod as an approximate cylinder or cone. The pod length and average width are measured and inserted into the formula to calculate the cylinder or cone side area to obtain the measured value [23]. Although the operation of these methods is relatively simple, they exhibit calculation errors because the pods are not actually standard cylinders or cones.

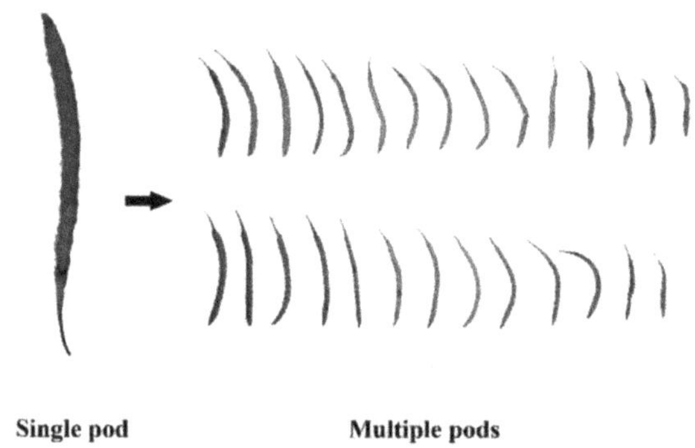

Figure 1. Two dimensions of the surface area measurement of rapeseed pods.

In the past, the common method used to measure the surface area of rapeseed pods based on the dimensions of a single pod was inaccurate, primarily because there was no method at that time that could directly measure the surface area of irregularly shaped pods. Therefore, approximate measurement results were obtained by calculating these parameters based on formulas or flattening the skin shape. The measurement of the surface area of irregularly shaped objects became accurate when three-dimensional (3-D) measurement technology became commonplace; 3-D laser technology was recognized as one of the most favorable methods for the highly accurate measurements of objects [24]. 3-D measurement technology uses a computer vision method to reconstruct a 3-D model of the object [25]. It is necessary to use a laser scanning system to obtain the image and use image processing, vision technology, and point-cloud data processing technology [26] to perform in-depth mining and extract the features of the 3-D data of the measured object [27,28] to realize the 3-D digitization of the relevant measurement indices of the measured object. 3-D technology has been widely used in recent years to measure the related terrestrial space position and size because it can restore the actual shape and characteristics of the measured object truly. The relevant measurement indicators [29], such as distance and object surface area and volume, can be directly calculated. There are many application examples of the 3-D measurement technology in measuring the surface area and volume of objects, including the surface area and volume of fish [30], the surface area of coral [31], cranial volume [32], and the surface area of human feet [33] and female breasts [34]. In the field of agriculture, there are reports of the non-destructive measurement of the surface area and volume of sweet potato (*Ipomoea batatas*) [35] and storage roots [36] using an inexpensive 3-D scanner. Therefore, to our knowledge, a high-precision 3-D laser scanning system, which ensures the accuracy of measurement results, was used to measure the pod surface area for the first time. The results were compared and corrected with the measurement results of the image processing technology proposed in this experiment.

The multi-pod measurement is a process of the batch measurement of multiple rapeseed pods to obtain the total area value. It is widely used and has more practical significance than measuring a single pod. The primary methods used for this measurement include accumulating a single pod area and fresh-weight measurement methods. The accumulation of the area of a single pod was used to measure the surface area of all the pods; then, the measured value that had accumulated was obtained. The method used to measure the single pod measurement described above is still required to measure a single pod. The accumulation measurement of a single pod area can be challenging because of the highly complex operations and the heavy workload involved. Therefore, this technique has been briefly discussed in the experimental methods section of a few studies, and some systematic studies have reported it. The fresh-weight method is based on the area and mass of a few samples determined by single pod measurement, i.e., scanning flattening. The approximate area of the sampled pods can be obtained by data transformation on the ratio of the mass of the sampled pods to the mass of the pods with a known area. Although such methods are simple and efficient, they produce considerable errors owing to the uneven texture of rapeseed pods [14]. In summary, single pod measurement is currently the mainstay for measuring rapeseed pod surface areas because most methods have difficulty being both "highly accurate" and "highly efficient".

This study is a subproject of the "Rapeseed Ideal Plant Type Breeding Project" of the Hunan Branch of the National Oilseed Crops Improvement Center (Changsha, China). This study aimed to screen ideal rapeseed plant types with high grain yields using multiple pod areas of a single plant dimension as important indices. Based on the results obtained by our group and focusing on the measurement goal of "highly accurate, highly efficient and large-scale", we established a correction model using the measurement principle of oblique-cylinder side area calculus and the results with a high-precision 3-dimensional (3-D) laser scanner. We proposed a new method to determine the areas of rapeseed pod surfaces on image processing and to conduct precise and efficient measurements of multiple pod surface areas.

2. Materials and Methods

2.1. Experimental Materials and Equipment

2.1.1. Materials

The single rape plant samples used in this experiment were from the rapeseed experimental base of Hunan Agricultural University, Changsha, China (conducted from October 2017 to May 2018). The cultivars included the early maturing cultivar 420 (winter *Brassica napus* L.A1, compact variety), medium maturing cultivar 1035 (winter *B. napus* L.A2, standard variety), and hybrid cultivar 991 (winter *B. napus* L.A3, tall variety). Three fertility treatments (compound fertilizer with total nutrient $\geq$ 45%, each containing 15% of N, P_2O_5, and K_2O) were applied, specifically 0.053 kg/m^2 (B1), 0.068 kg/m^2 (B2), and 0.083 kg/m^2 (B3). Each treatment was repeated three times. A total area of 10.8 m^2 for each plot was planted with a density of 15 plants/m^2. The rape plants were sown in October 2017, and inter-seedling and transplantation were conducted at the seedling stage. Different cultivars and fertilization levels were used to create single plant samples of different sizes and shapes. On 29 April 2018 (the pod maturity period), the pod surface area of 3~4 representative plants in each plot was measured in the dimension of a single plant, yielding a total of

83 plants that were measured. In this study, the pod surface area of a single plant refers to the total surface area of all pods of the plant, including the areas of pod beaks.

2.1.2. Equipment and Software

Equipment: ASUS integrated computer (Processor: CORE i5-8250U; Memory: 8 G/1 T; Display: 23.8 in LED; Graphics card: NVIDIA GeForce 930MX; Operating system: Windows 10 Home), an Apple iPhone 6 (Memory: 64 G; Display screen: 4.7 in, 326 ppi; Chip: A8; Camera: 8 million pixels, automatic focusing, f/2.2 aperture), an ultra-high precision and high-speed 3-D laser scanning system LDI-SurveyorZS (LDI-SLP250) [37], Li-3000C Portable Area Meter (LI-COR, Lincoln, NE, USA), Bluetooth remoter, white cardboard, Vernier caliper, and ruler.

Software: Image Processing System for measuring the surface area of rapeseed pods [38] (developed by our group: invention patent; referred to as Image Processing System from this point onward); DPSV7.05 statistical analysis software [39]; Adobe Photoshop CS (San Jose, CA, USA); Microsoft Office 2016 (Redmond, WA, USA).

2.2. Measurement Method and Technical Roadmap

A correction model to measure the surface area of a single pod using the image processing method needed to be established before the batch measurements of multiple pod surface areas. Several pods of different sizes were collected from different cultivars and plants and sent to a dimension measurement and Data Services Company for single pod measurement using a high-precision 3-D laser scanning system (referred to as the "3-D method" from this point on). The measured value was then treated as the "actual surface area" (referred to as the "3-D area"). A portion of the pods sampled above were used as modeling samples, and the single pods were measured using the image processing method. The correlation between the measurement results of the image processing method and the corresponding 3-D area was analyzed, enabling the establishment of the correction model. The remaining sampled pods were used as test samples, for which the image processing method was also used to measure the surface area of single pods. The corrected area was obtained by correcting the measurement results of the image processing method based on the correction model. The accuracy was then tested by analyzing the relative error and RMSE of the corrected area and the corresponding 3-D area. Moreover, the single-pod measurement of the test samples was performed using the Clark formula, Leng formula, flattening scanning, and quasi-cylinder side area methods. Their accuracy was assessed by analyzing their relative error, RMSE, and the corresponding 3-D areas. We then compared the measurement accuracy of the image processing method and the other four methods; this enabled us to determine whether the image processing method had a relative advantage. In addition, the measurement efficiency of the image processing method and the other four methods was tested by the batch measurement of multiple pod surface areas using the image processing method. The results of measurements were corrected using the correction model, and the rationality of the measurement results was also analyzed. A technical flowchart of this study is shown in Figure 2.

2.3. Measurement Principle and Operation

2.3.1. Image Processing Method

The application of the image processing method depends on the integrated use of image processing technology and data analysis. The basic principle is that the pod is considered an irregular column composed of several circular slices with different diameters. The length of the column is the length of the pod, and each circular slice can be regarded as an oblique cylinder with equal height and different diameters. The diameter of each circular slice represents the width of the corresponding position of the pod. The side area of the irregular column of the pod is the sum of the side areas of all the circular slices, which can be estimated using calculus [40], and is considered to be the approximate surface area of the pod. The process of image processing involves calculating the value of pod length and the width at different positions transformed from the image of pods; these are required for the use of calculus through image processing. The correlation between the estimated pod surface area and the corresponding pod "actual area" was then analyzed. A correction model was established to correct the measurement results of the other pod images. The specific procedure and methods were as follows:

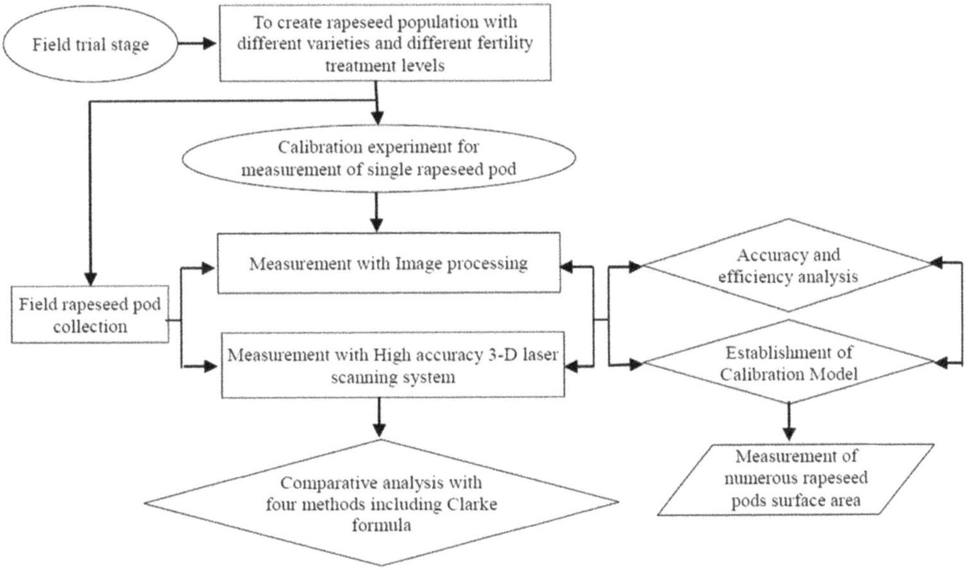

Figure 2. Research flowchart of the proposed methodology for measuring the rapeseed pod surface area.

(1) Image acquisition device. We configured an image acquisition device, as shown in Figure 3, which used consistent parameters to obtain highly defined and high-quality images. The device was primarily composed of five parts: a long-strip desktop, a self-use iPhone, a tripod for fixing the mobile phone, white background cardboard for placing the pods, and a remote Bluetooth device that can control the photography of the mobile phone. Using a long strip desktop, we established the batch placement of pods and the assembly line operation and improved the efficiency of imaging acquisition. The tripod fixed the shooting distance, unified the shooting background, and provided an identity guarantee for later image processing and comparison analysis. The selection of white background cardboard helped to separate the pod and background during later image segmentation. Bluetooth was used to prevent the mobile phone from shaking during shooting, which affects the image quality.

(2) The pixel transformation equation was generated. The image acquisition device was used to capture an object space frame of reference (Figure 4) and calculate the transformation equation of pixel coordinates and actual coordinates. The transformation equation of pixel position (i, j) and actual coordinates (x, y) used the binary quadratic functions Equation (1) [41], and the parameters were obtained by least-square fitting. The goal was to ensure that the distance between the adjacent points on the transformed coordinates of the transformation map was closer to their actual distance.

$$\begin{cases} x = a_1 \cdot i^2 + b_1 \cdot i \cdot j + c_1 \cdot j^2 + d_1 \cdot i + e_1 \cdot j + f_1 \\ y = a_2 \cdot i^2 + b_2 \cdot i \cdot j + c_2 \cdot j^2 + d_2 \cdot i + e_2 \cdot j + f_2 \end{cases} \quad (1)$$

Figure 3. The image acquisition scheme.

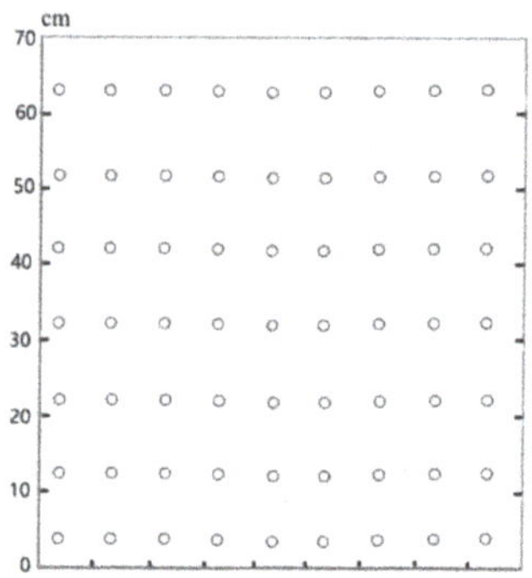

Figure 4. Diagram of coordinate transformation points.

After calculation, the transformation equation between the position of the pixel (i, j) and the actual coordinate (x, y) applied the binary quadratic functions as follows:

$$\begin{cases} x = -7.3314 \times 10^{-6} \cdot i^2 + 6.7344 \times 10^{-7} \cdot i \cdot j - 2.2592 \times 10^{-7} \cdot j^2 + 0.090528 \cdot i - 0.0089604 \cdot j \\ y = -1.4521 \times 10^{-6} \cdot i^2 - 7.2268 \times 10^{-6} \cdot i \cdot j + 7.7158 \times 10^{-8} \cdot j^2 + 0.014176 \cdot i + 0.09199 \cdot j \end{cases} \quad (2)$$

(3) Image acquisition. The images can be acquired using multiple people (Figure 3). One person was designated to take photographs and the others were responsible for placing, numbering, recycling, and recording the pods. The pods should be arranged in parallel on the white cardboard, and their numbers on each whiteboard should be the same as that in principle (e.g., 40), which is convenient for later analysis. The flash function implemented in the mobile phone was used during the shooting process to acquire the image with consistent exposure.

(4) Image processing. First, the rapeseed pod photos were preprocessed through image binarization transformation and denoising. The primary purpose was to facilitate the segmentation of the later image. For comparison, gray-scale and super-green transformation can be selected as image transformation, and median filtering can be selected as denoising. Second, the images were segmented to separate the background and pods. The image obtained this time was a white background, and the green of the pods and white of background were significantly different in gray scale. Therefore, the threshold method based on the gray scale was used for segmentation. Repeated comparisons indicated that the Otsu threshold segmentation and the maximum inter-class and intra-class distance

ratio methods were the best. The intra-class variance (ICV) provided by Nobuyuki Otsu was defined as follows:

$$ICV = PA \times (MA - M)^2 + PB \times (MB - M)^2 \tag{3}$$

where M is the average of the image gray value. If any gray value t is taken, the histogram can be divided into the foreground color A and background color B. The average values of the two parts were MA and MB, respectively. The proportion of pixels in A and B to the total number of pixels was recorded as PA and PB, respectively. This optimal threshold t was the value that maximized the ICV.

The positions of each pod pixel set and the boundary pixel set were extracted from the pod batch. Using the transformation equation of the position of the pixel points and the actual coordinates described in step 2, the position sets of the pod and the boundary pixel points were transformed into their actual coordinates. The image processing process is shown in Figure 5.

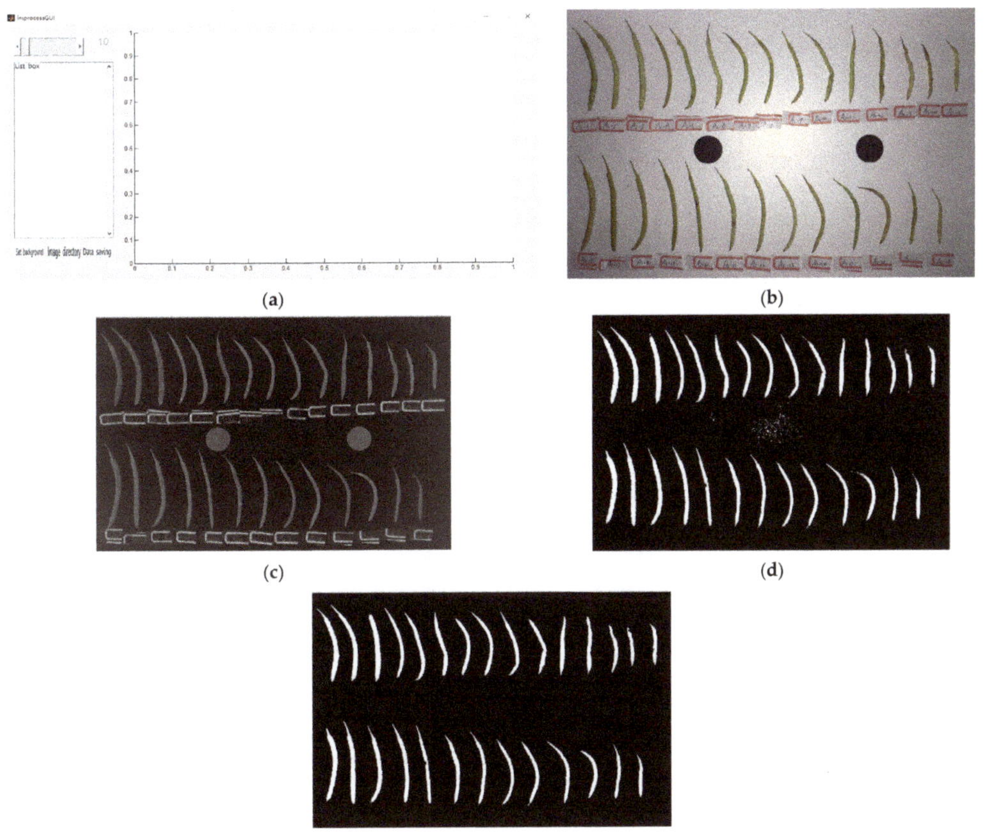

Figure 5. Image processing process: (**a**) Image processing system interface; (**b**) Loading the artwork; (**c**) Image preprocessing; (**d**) Image segmentation; (**e**) Filling and removing impurities.

(5) Surface area calculation. The surface area of each pod was estimated separately, and the steps were as follows:

A. The actual coordinates of the pod pixel set were fitted using a linear equation.

B. Coordinate rotation. The actual coordinates of the pod pixel set and the boundary pixel set were rotated based on the inclination angle of the fitting line, so that the fitting line of the new coordinates of the pod pixel set was parallel to the horizontal line. The new coordinates of the pod pixel set and the boundary pixel set were then obtained.

C. Fitting the boundary curve of the pods. The polynomial function was used to fit the pod boundary pixels (Figure 6) and the two pod boundary curves are shown below as follows:

$$\begin{cases} elu(x) = \sum_{i=1}^{n} \alpha_i x^i \\ eld(x) = \sum_{i=1}^{n} \beta_i x^i \end{cases} x \in [a, b] \quad (4)$$

where, a, b is the value range of x, α_i, β_i is the corresponding i-order coefficient, and n is the maximum number of pod convex parts. The value here is 15. α_i, β_i is solved by least squares fitting.

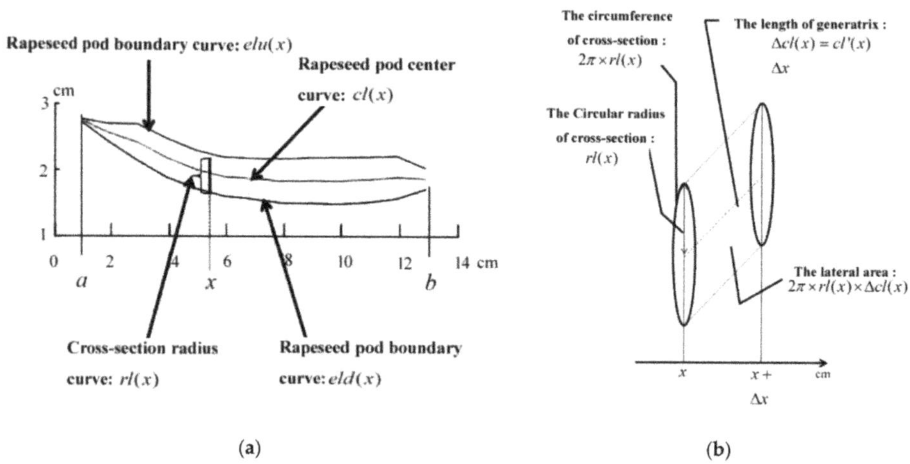

Figure 6. The indices related to rapeseed pods and their surface area calculation. (**a**) Rapeseed pod curve of calculations of the boundary, center, and cross-section radius curves; (**b**) Calculation of the side area of the cylinder of rape pod slice.

D. Calculation of the central curve of pods $cl(x)$. The center curve of pods was calculated according to the two fitted pod boundary curves (Figure 6). The equation used to calculate the pod center curve was as Equation (5).

$$cl(x) = (elu(x) + eld(x))/2 \quad (5)$$

E. Calculation of the cross-section radius curve of the pods. The cross-section radius curve of the pods was calculated based on the two fitted pod boundary curves, whereby the radius of the x-point was half the difference between the corresponding upper boundary and the lower boundary of the pod. The equation of calculation was as follows:

$$rl(x) = |elu(x) - eld(x)|/2 \quad (6)$$

F. Estimation of the surface area S of the pods. The surface area of the pods was estimated based on the obtained center curve $cl(x)$ and the cross-section radius curve $rl(x)$ of the pods (Figure 6). The calculation algorithm used for the surface area of the pod was as follows:

$$S = 2\pi \int_a^b rl(x) dcl(x) \quad (7)$$

where $rl(x)$ is the cross-section radius curve of the pods, $cl(x)$ is the center curve of the pods, $dcl(x)$ is the height of the cylinder for fitting the pod segments, and a and b are the range of x values.

(6) Correction of the pod surface area. The cross-section of the actual pod is not a regular circle but an ellipse. When the radius of a circle is the same as the long axis of the ellipse, the circumference of the circle becomes longer than that of the ellipse. Similarly, the pod surface area measured by the image processing method is theoretically larger than that of the actual pod, and therefore, it is necessary to correct the pod surface area measured by the image processing method. Consequently, pods from different cultivars with varying sizes and shapes were pooled separately for measurement. After the image sampling, the actual area was measured using a high-precision 3-D laser scanning system. The correlation between the measured value of the image processing method and the 3-D area value was analyzed and a correction model was then established to correct the surface area of pods sampled from other images.

2.3.2. High-Precision 3-D Laser Scanner Measurement Method

The instrument used to measure the actual area of the pods was an LD-SurveyorZS ultra-high precision 3-D laser scanning system (LDI; Logistic Dynamics, Inc., Seattle, WA, USA) (Figure 7). The system configuration included a 3-D laser probe (sensor) system, a probe movement and positioning system, a 3-D laser scanning auxiliary fixture, scanning control software, and data processing software. The laser scanning probe used in this measurement was LDI-SLP250, which is accurate at scanning up to 0.01 mm. It can be used to detect the vast majority of industrial product sizes. The 3-D laser probe was installed on the probe movement and positioning system (mainframe) and calibrated so that the data detected could be converted to the coordinate system of the measuring machine. The 3-D laser probe system was the sensing device used for acquiring 3-D data. It works with a triangular measurement method comprising a high-precision laser transmitter located in the middle of the scanning probe and a CMOS image sensor with a certain angle with the laser projection installed on both sides of the probe. Upon projecting the laser onto the surface of an object, the laser line is misaligned by the different 3-D surface structures. The misalignments are then read and analyzed by the CMOS image sensor. This process, combined with the movement of the probe and the function of its positioning system, enables the 3-D structural data of the measured object surface to be obtained.

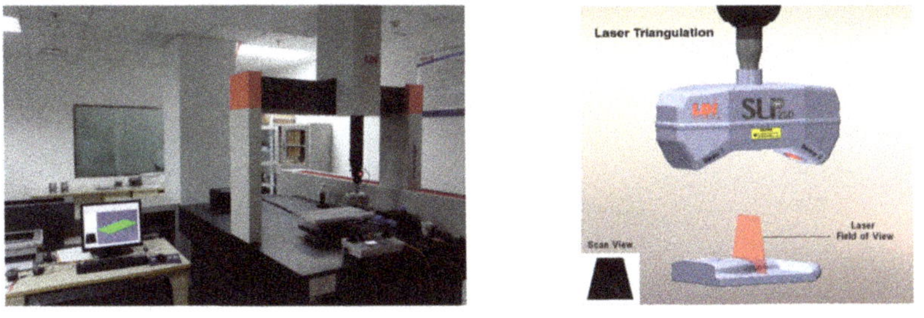

Figure 7. LDI-Surveyor ZS/SLP ultra-high accuracy high-speed 3-D laser scanning system and the structure of scanning probe.

No pretreatment was required when the 3-D laser scanning system (described in Figure 7) was used to scan and measure the pods as long as the pods were arranged and fixed on the scanning fixture in sequence (Figure 8). Two fixtures can be used for clamping to improve the efficiency of laser scanning, and a remarkable scanning fixture can hold 20–30 pods at a time. The scanned data are encapsulated by Geomagic Qualify (ControlX; 3-D Systems, Cary, NC, USA). The geometric attributes, such as volume and surface area, can be calculated directly after encapsulating the point cloud data (Figure 8).

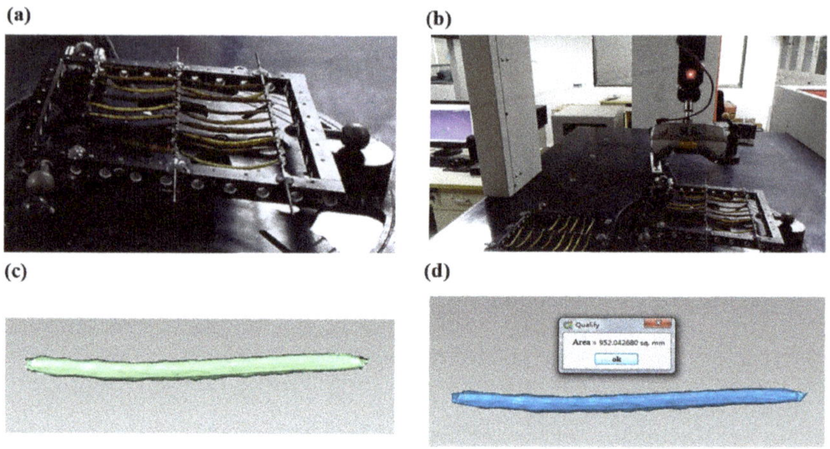

Figure 8. Scanning of rapeseed pods using a 3-D laser scanning system and the data processing process: (**a**) Pod scanning fixture; (**b**) Process for the 3-D laser scanning system to scan pods; (**c**) Pod selection; (**d**) Calculation of the pod surface area.

2.3.3. The Other Four Measurement Methods Used for Comparison

(1) The Clark formula method [42] is shown below:

$$\begin{cases} S = \pi \bar{d} h_1 + \frac{1}{3}\pi \bar{d} h_2 \\ h_1 = 0.8H, h_2 = 0.2H \end{cases} \tag{8}$$

where S is the surface area of the pods, H is the length of the pods, and $\bar{d}$ is the average width of the pods. The length and average width of the pods were measured using a ruler and Vernier caliper, respectively.

(2) The Leng formula method [43] is shown below:

$$\begin{cases} S = -0.6 + 2.4S_m \\ S_m = LB_m \end{cases} \tag{9}$$

where S is the surface area of the pods, L is the length of the pods, and B_m is the maximum width of the pods. The method used for measurement was the same as that of the Clark formula method.

(3) Flattening scanning method: Pods were cut from the middle line, then the two peels were fully flattened and then wrapped with plastic. The surface area of each pod was scanned and measured using a leaf area meter (Li-3000C Portable Area Meter).

(4) Quasi-cylinder side area method [14]. The pods were considered regular cylinders, and the cylinder side area obtained represented the pod surface area as follows:

$$S = \pi H \overline{D} \tag{10}$$

where S is the surface area of the pods, H is the maximum length of the pods, and $\overline{D}$ is the average pod width. The measurement method was the same as that of the Clark formula method.

3. Results and Analysis

3.1. Analysis of the Accuracy of Measurement

Three rape plants were randomly selected from nine plots (three cultivars and three fertility treatments), and three pods of different sizes were picked from each plant, resulting in 81 pods. Among these pods, 54 pods of various sizes were collected from two randomly selected plants from each plot and used as modeling samples to establish a correction model using the image processing method. The remaining 27 pods of different sizes (collected from another plant in each plot) were used as samples to analyze and compare the accuracy and efficiency of measurement between the image processing method and the other four methods. The modeling and test samples described above were randomly selected and were in triplicate.

3.1.1. Comparison of the Measurement Results between the Image Processing and 3-D Methods

The measurement results of the modeling samples obtained by the image processing and the 3-D methods were significantly and linearly correlated; the average correlation coefficient R of the three replicates was 0.92 ** (** $p < 0.01$), which was the same as those used in the following contents. Before the correction, the average area of the modeled sample measured using the image processing method was 1.23-fold larger than that of the 3-D method; this is consistent with the previous proposition that it is essential to correct the measurement results of the image processing method. A univariate linear regression model established using the measurement results of the modeling sample (which utilized the image processing and 3-D methods) was used to correct the measurement results of samples measured using the image processing method. The average relative error of the corrected value compared with that of the 3-D method data was 2.46%, while the average RMSE was 0.92 cm^2. An analysis of the accuracy of the image processing methods is shown in Table 1.

3.1.2. Comparison of the Results of the Other Four Measurement Methods with the 3-D Method

Single pods of the test samples described above were measured using the Clark formula, Leng formula, flattening scanning, and quasi-cylinder side area methods. A comparison with the measurement results of the corresponding 3-D method indicated significant measurement errors in the four methods. The flattening scanning method was the most accurate, followed by the Leng formula, Clark formula, and quasi-cylinder side area methods, with an average relative error of 16.53%, 19.51%, 22.10%, and 40.89%, respectively, and an average RMSE of 1.64 cm^2, 2.27 cm^2, 2.39 cm^2, and 3.90 cm^2, respectively. An analysis of the accuracy of the other four measurement methods is shown in Table 2. The flattening scanning method was relatively accurate because it involved the direct measurement of fully flattened pods. However, the pods were difficult to flatten completely because of the irregular bulge of the epidermis. In addition, the beak easily fell off during the unfolding of the pod epidermis, resulting in the loss of some portions of the epidermal area. Therefore, the value of the measured result was smaller than that of the corresponding 3-D area. The Leng formula method is similar to that of the Clark formula method, whose principal goal was to indirectly measure the seed pods by establishing the regression model between the pod length, width, and surface areas. The large relative error was because the cultivars and morphological characteristics of the pods used for modeling with the two methods were substantially different from those used in this experiment. For example, the rape cultivars used in the Leng formula method were spring *B. napus* and spring *B. campestris*, while the cultivar used in this experiment was winter *B. napus*. Therefore, when the surface area of the pods was measured using similar coefficient regression methods, the corresponding regression model should be established for the pod group from a particular cultivar and with a certain shape range before measurement. This is the so-called "one measurement and one coefficient" principle. In addition, the "actual surface area of pods" referred to in the regression model established using the Leng and Clark formula methods could be inaccurate owing to the limitation of measurement conditions at that time. The largest error in measurement of the quasi-cylinder side area method could be explained by the highly slender beak of the pods and the position close to the pod handle. The side area is smaller than the cylinder with the average fruit width. Moreover, the cross-sections of the pods were oval or irregular and not a regular circle, resulting in a measured value larger than the 3-D area. Therefore, the four methods could be used directly to accurately measure the pods in this experiment.

Table 1. Analysis of the accuracy of results measured by the image processing method.

Experiment	The Model Established with the Modeling Samples				Accuracy Analysis of the Testing Samples				
	Number of Samples	The Average Measurement Value of the 3-D Method (cm^2)	The Average Measurement Value of the Image Processing Method (cm^2)	Correlation Coefficient	Number of Samples	The Average Measurement Value of the 3-D Method (cm^2)	The Average Measurement Value of the Image Processing Method (cm^2)	Relative Error	RMSE (cm^2)
Replicate 1	54	8.44	10.25	0.92 **	27	8.42	8.74	3.78%	0.95
Replicate 2	54	8.46	10.47	0.91 **	27	8.39	8.25	1.67%	0.89
Replicate 3	54	8.40	10.41	0.91 **	27	8.49	8.33	1.93%	0.93
Average	54	8.43	10.38	0.92 **	27	8.43	8.44	2.46%	0.92

Note: The model established with the modeling samples: A total of 54 pods from different sizes of two rapeseed plants were randomly selected from each plot as modeling samples. The surface area of pods was measured by 3-D and image processing methods, respectively. The values measured by the two methods were analyzed by regression analysis and a correction model was established. An analysis of the accuracy of the testing samples was performed as follows: A total of 27 pods with different sizes of another rapeseed plant in each plot were used as test samples. The surface area of pods was measured using a 3-D method and image processing, respectively, and then the correction model was used to correct the measured value of the image processing method. The corrected value was compared with the measured value of the 3-D method, i.e., an accuracy analysis. "**" means significant test $p < 0.01$.

Table 2. Analysis of the accuracy of the measurement results using the other four methods.

Experiment	Number of Samples	The Average Measurement Value of the 3-D Method (cm^2)	Clark Formula Method			Leng Formula Method			Flattening Scanning Method			Quasi-Cylinder Side Area Method		
			The Average Measurement Value (cm^2)	Relative Error	RMSE (cm^2)	The Average Measurement Value (cm^2)	Relative Error	RMSE (cm^2)	The Average Measurement Value (cm^2)	Relative Error	RMSE (cm^2)	The Average Measurement Value (cm^2)	Relative Error	RMSE (cm^2)
Replicate 1	27	8.42	10.63	26.25%	2.64	10.43	23.80%	2.52	7.21	14.43%	1.66	12.27	45.67%	4.24
Replicate 2	27	8.39	10.08	20.15%	2.12	9.85	17.44%	1.99	6.78	19.10%	1.78	11.63	38.63%	3.61
Replicate 3	27	8.49	10.19	19.92%	2.40	9.96	17.30%	2.32	7.13	16.05%	1.49	11.75	38.37%	3.86
Average	27	8.43	10.30	22.10%	2.39	10.08	19.51%	2.27	7.04	16.53%	1.64	11.88	40.89%	3.90

Note: The 27 test samples used in the accuracy analysis of these image processing methods were also used to measure the pod surface area with the 3-D method and the other four methods shown above. The values measured by these four methods were compared with the 3-D method.

3.1.3. Comparison of the Results of Measurement Obtained by the Image Processing Method with Those of the Other Four Methods of Measurement

The same test samples were measured using image processing and the other four measurement methods described above, and the results were analyzed comparatively. The corresponding values of the relative error and the RMSE of measurement (reflecting the degree of dispersion) of the image processing method were lower than those of the other four methods. They were also approximately 14.07% and 0.72 cm^2 lower than those of the flattening scanning method, which was the most accurate of the four measurement methods. Therefore, the accuracy was much higher when the image processing method was used when compared to the other four measurement methods. The primary reasons for the increased accuracy of the image processing method were as follows: (a) the image processing method regarded the pod as an irregular cylindrical body composed of several inclined cylinder slices with different diameters, which objectively restored the characteristics of the irregular long-chain bead shape for the pod; and (b) the models of measured values were corrected using image processing, and the 3-D methods were established, which substantially improved the accuracy of measurement. The use of a high-precision 3-D method to measure the actual area of pods had not been previously reported. The high accuracy of such a 3-D method directly determines the high accuracy of our image processing method.

Based on this analysis, we suggest that correcting the measurement results by establishing a correction model is necessary. We introduced the correction step into the image processing method to determine whether more accurate measurements could be obtained by correcting other measurement methods. The comparative analysis was conducted using the same modeling sample and detection sample, considering that the calculations of the Clark formula, Leng formula, and quasi-cylinder side area methods require different coefficients configured with pod length and pod width because modeling and correction using the same samples would lead to the same results. Thus, only the Clark formula method was selected as a representative for comparison. After the calculations, a significant linear correlation was observed between the measurement results of the modeling samples produced by the Clark formula, flattening scanning, and the 3-D method. The results of modeling and accuracy analysis are shown in Table 3. After correction, the measurement accuracy of the Clark formula and flattening scanning methods were significantly improved. The average relative error was 2.72% and 2.74%, respectively, and the average RMSE was 1.15 cm^2 and 0.87 cm^2, respectively. Comparatively, the modeling correlation coefficient of the image processing and flattening scanning methods was higher than that of the Clark formula method. The smallest average value of relative error was obtained using the image processing method. The flattening scanning method generated the smallest average value of the RMSE, followed by the image processing method. The Clark formula method yielded the largest value. A comprehensive analysis suggests that even if the other measurement methods are corrected, the image processing method still has comparative advantages in the correlation with the actual value, relative error, and degree of dispersion.

Table 3. Comparison of the accuracy of measurement of the correction results between the image processing method and the other two methods.

Experiment	Number of Samples	The Average Measurement Value of the 3-D Method (cm^2)	Image Processing Method				Clark Formula Method				Flattening Scanning Method			
			Correlation Coefficient of Modelling	The Average Measurement Value (cm^2)	Relative Error	RMSE (cm^2)	Correlation Coefficient of Modelling	The Average Measurement Value (cm^2)	Relative Error	RMSE (cm^2)	Correlation Coefficient of Modelling	The Average Measurement Value (cm^2)	Relative Error	RMSE (cm^2)
Replicate 1	27	8.42	0.92 **	8.74	3.78%	0.95	0.87 **	8.78	4.21%	1.09	0.96 **	8.71	3.47%	1.18
Replicate 2	27	8.39	0.91 **	8.25	1.67%	0.89	0.85 **	8.25	1.59%	1.09	0.92 **	8.04	4.15%	0.86
Replicate 3	27	8.49	0.91 **	8.33	1.93%	0.93	0.88 **	8.29	2.37%	1.26	0.90 **	8.54	0.59%	0.58
Average	27	8.43	0.92 **	8.44	2.46%	0.92	0.87 **	8.44	2.72%	1.15	0.92 **	8.43	2.74%	0.87

Note: The correction method of the image processing, Clark formula, and the flattening scanning methods were modeled, and the values measured were corrected with the same modeling and testing samples. The correction values measured by the two methods were compared with those measured by the image processing method. "**" means significant test $p < 0.01$.

3.2. Analysis of Measurement Efficiency

The efficiency of measurement was tested using modeling and detection samples (81 pods) during the precision analysis of the experiment. Using the Leng formula and quasi-cylinder side area methods, we performed the calculations by only measuring the pod length and width with similar efficiency. The Clark formula was the only method that was selected as the representative for testing. The 81 pods were measured using image processing, flattening scanning, and Clark formula methods. Each method was conducted by the same person and timed with a stopwatch. Using the image processing method, the time for placing and shooting was 172 s. This value excluded the preparation time, such as preliminary numbering, which was the same as those used in the following methods. The time for subsequent image processing was 162 s, and the measurement time was 334 s. For the flattening scanning method, the common time of flattening and plastic wrapping, scanning, and measurement were 1620 s, 405 s, and 2025 s, respectively. The length and width measurement of pods using the Clark formula method was 567 s. Since data post-processing with these three methods could be performed in batches, the post-processing time was not recorded, and the operation time was the total measurement time. The analysis of the "efficiency test" used above found that the image processing method was the most efficient, followed by the Clark formula method. The flattening scanning method was the least efficient.

3.3. Analysis of the Results of Measurement for Multiple Pod Surface Areas

Multi-pod measurements were conducted on 83 rape plant samples in the pod stage using the image processing method. The results were corrected by the correction model established during replicate 2 of the accuracy test experiment (Figure 9). Five people were selected to take photos of the pods during the early period of the experiment, and four were responsible for image processing during the later period. A total of 1203 pod photos were processed; this took approximately 20 h. The measurement of 39,501 pods from 83 rape plant samples provided a total surface area of 330,189 cm^2; the average surface area of pod per plant was 3978 cm^2. The average surface area of the pods from each plant (A1, A2, A3, B1, B2, and B3) were 3403 cm^2, 3943 cm^2, and 4567 cm^2, and 3435 cm^2, 4127 cm^2, and 4451 cm^2, respectively. The measurement results of the pod surface areas of the samples from different cultivars and fertility treatments were significantly different, consistent with our experimental design concept.

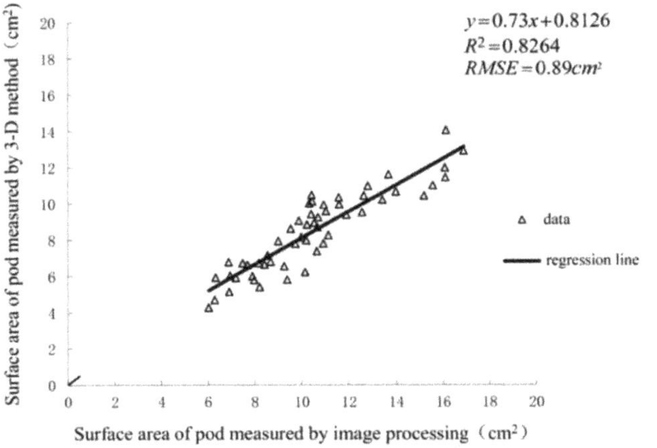

Figure 9. Correction model established by image processing method and the 3-D method. The range of the pod surface area was measured as 6.03–16.89 cm^2.

4. Discussion

4.1. 3-D Measurement Method and a Comparison of Image Processing Methods

In this experiment, the "high accuracy, high efficiency, and large scale" measurement of rapeseed pods was realized using the image processing method. The key link was to use the 3-D measurement method to measure the "actual surface area" of a small number of rapeseed pods and use this value as the basis for correcting the measured value of the image processing method. A key question is: why not directly use the 3-D measurement method to measure the surface area of rapeseed pods and develop this image processing method? This is primarily because of the limitations and shortcomings of the 3-D measurement method itself. First, although the 3-D measurement method is highly accurate, many preparations are necessary before the values can be measured, and the operation is more complicated. It requires special measurement equipment and software, which has high requirements for the surveyor and is difficult for general staff to master. Secondly, the 3-D measurement is relatively inefficient. According to the 3-D measurement reported in this experiment as an example, it is necessary to pack the pods, scan with the 3-D scanning camera, and perform calculations with platform-specific software. Among them, only the scanning link takes a long time. For example, this experiment takes 7 or 8 min to scan 12 pods at a time, so it is difficult to scan thousands of pods in a short time. In addition, freshly harvested pods generally stay fresh for approximately 3 days, which does not allow adequate time for their measurement. Third, the 3-D measurement equipment and platform are expensive, and the construction needed to use them is very expensive [44]. The ultra-precision and high-speed 3-D laser scanning system LDI-SurveyorZS (LDI-SLP250) used in this experiment is worth more than one million RMB. The image processing method is typically more convenient to operate. The only equipment needed is a digital camera, tripod, and other conventional equipment. It is also highly efficient at measuring. Because the correction link for the measurement result is designed, the measurement accuracy can also meet the needs of crop science research. Therefore, combining the image processing method and the 3-D measurement method for the "actual area" measurement of a few rape pods establishes a correction model of the image processing method, which is easier to operate, more efficient, and economical for measuring the surface area of several rapeseed pods.

4.2. The Application of Image Processing Technology in a Future Rapeseed Pod Testing Machine

With the continuous increase in crop phenomics, researchers are committed to developing various phenomics analytical platforms [45]. The combination of these platforms with various "omics" studies will trigger a new technological revolution, and the "high-throughput phenomics era" [46] will profoundly affect the development of various fields of crop science. As an important index of rapeseed plant type, the technology for measuring the surface areas of rapeseed pods will be reconstructed from theory and technology by integrating robot technology, sensor technology, optics, cloud computing, and artificial intelligence. The technology will then be developed to ensure automation, high efficiency and accuracy, mass, and simple and easy operation. The image processing measurement technology of the rapeseed pod surface area proposed in this study will provide a design principle for the future design of a rapeseed pod testing machine with high-throughput characteristics. The basic framework and design ideas are as follows:

(1) Conceptual design. The rapeseed pod testing machine is roughly composed of three parts: a pod transmission device, an image acquisition device, and a data processing software platform. The pod transmission device primarily automates pod placement, eliminating the involvement of people in the placement process. The pods are automatically and efficiently evenly placed through the transmission device. After the image acquisition device takes photos, it automatically places and transmits the next set of pods. The image acquisition device closely cooperates with the pod transmission device and rhythmically takes photos that conform to the later image segmentation and data processing. The data processing software platform is primarily used to process the pod images and calculate the related phenotypic indices.

(2) Overall design. The conceptual design described above indicates that the rapeseed pod test machine primarily comprises three parts, as shown in Figure 10. The pod conveying device primarily comprises a feeder, blower, flow controller, conveyor belt, rapeseed pods, and other components. The blower evenly disperses many accumulated pods, which fall evenly on the conveyor belt without overlapping and blocking. The flow controller includes a valve to control the density of the pod by controlling the size of the airflow from the blower. The conveyor belt is a pod conveyance device and a background for pod shooting made of white plastic material. The image acquisition device is permanently installed on the conveyor belt and made as a dark box to prevent external light interference. The top surface of the interior is equipped with a mobile phone or a digital camera bracket, and an LED flat lamp is installed to ensure that the shooting light source is stable and sufficient. The data processing software platform primarily comprises computer and special data processing software. For the convenience of operation, the data processing platform can also operate and control the machine.

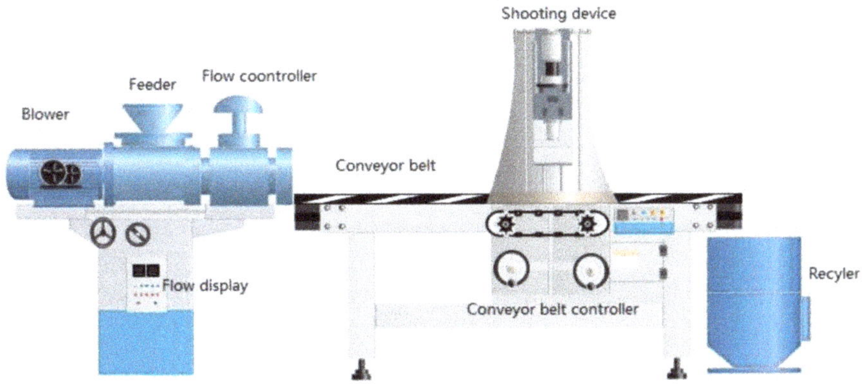

Figure 10. Design model of rapeseed pod testing machine.

(3) Operational process. The device is started after manual feeding, and the accumulated pods are evenly blown onto the conveyor belt under the combined action of gravity and the blower. The scattered pods keep a certain rhythm and speed in the dark box. The system software controls the mobile phone or digital camera to shoot the scattered pods and then performs the subsequent image processing and data calculation.

5. Conclusions

In this study, we proposed an image processing method in which the pod was regarded as an irregular cylinder composed of several inclined cylinder slices with different diameters. We showed that the surface area of the pods could be calculated as the sum of the side areas of all the inclined cylinder slices by calculus, and the corresponding empirical study was also conducted. The application of this method allows for "highly accurate, highly efficient and large-scale" measurement of numerous rapeseed pods. "Highly accurate" stems from the fact that this method uses image processing technology and calculus principles to transform the pod image into a mathematical model that can be calculated by a computer, which restores the basic characteristics of rapeseed pods. The high-precision 3-D method was first used to measure and correct the actual area of pods, which is also a vital premise to ensure "high accuracy." Our experiment proved that the image processing method is more accurate than the four commonly used methods, such as the Clark formula. "Highly efficient" can be explained by the fact that no manual or instrumental measurement is needed for the pods, which only need an orderly arrangement in a suitable background and position for photography. This process is completed by multiple people, which substantially improves measurement efficiency. Currently, pods are arranged and photographed manually during the measurement process. Thus, the measurement efficiency will be significantly improved if automatic batch operation by machines can be achieved in the future. "Large scale" primarily lies in the achievement of measuring multiple pods. Since an assembly line operation process has been established for placing and photographing the pods, the corresponding photos serve as records, facilitating subsequent inspection, thus avoiding repeated and missed measurements or errors. This study measured the surface area of nearly 40,000 pods from three different rape cultivars subjected to three different fertility treatments. The process did not involve many people, and neither was it time-consuming. The measurement results met the expectations of the experimental design and achieved the measurement aim of "highly accurate, highly efficient, and large-scale" for the surface area of rapeseed pods. Therefore, this method provides technical support for the surface area measurement of numerous rapeseed pods.

Author Contributions: Conceptualization, C.G., M.G., H.H. and F.L.; methodology, X.L. and F.L.; software, X.L. and H.X.; data analysis, F.L.; investigation, F.L. and X.L.; writing—original draft preparation, F.L.; writing—review, M.G. and H.H.; funding acquisition, M.G., F.L. and X.L.; All authors have read and agreed to the published version of the manuscript.

Funding: This research was funded by the China Agriculture Research System of MOF and MARA (CARS-13); Hunan Agriculture Research System of DARA, China; the Project from the Department of Education of Hunan Province, China (20C0970); and the Key research projects of Science and Technology Department of Hunan Province, China (2022NK2047).

Institutional Review Board Statement: Not applicable.

Informed Consent Statement: Not applicable.

Data Availability Statement: Not applicable.

Conflicts of Interest: The authors declare no conflict of interest.

References

1. Guan, C.Y. *Fruiting Organs and Yield Formation in Rapeseed*; Science Press: Beijing, China, 2017.
2. Ling, Q.H. *Crop Population Quality*; Shanghai Scientific & Technical Publishers: Shanghai, China, 2000.
3. Leng, S.H.; Tang, Y.; Li, Q.L.; Zuo, Q.S.; Yang, P. Studies on source and sink of rapeseed I. Regulation of pod size on source and sink in rapeseed after flowering. *Chin. J. Oil Crop Sci.* **2005**, *27*, 37–40.
4. Guan, C.Y. *Physiological Ecology and Modern Cultivation Techniques of High-Quality Rapeseed*; China Agriculture Press: Beijing, China, 2013.
5. Inanaga, S.; Kumura, N.; Murata, Y. Research on material production on rapeseed- characteristics of photosynthesis and material production in pod layer. *Chin. J. Oil Crop Sci.* **1981**, 3, 84–87.
6. Inanaga, S.; Kumura, N.; Murata, Y. Research on material production on rapeseed-photosynthesis, respiration and carbon metabolism of pod. *Chin. J. Oil Crop Sci.* **1981**, 3, 74–78, 81.
7. Chen, T. *Influence of Leaf and Silique Photosynthesis on Seeds Yield and Seed Soil Quality of Oilseed Rape*; Northwest A&F University: Yangling, China, 2016.
8. Li, N.; Song, D.J.; Peng, W.; Zhan, J.P.; Shi, J.Q.; Wang, X.F.; Liu, G.H.; Wang, H.Z. Maternal control of seed weight in rapeseed (*Brassica napus* L.): The causal link between the size of pod (mother, source) and seed (offspring, sink). *Plant Biotechnol. J.* **2019**, *17*, 736–749. [CrossRef]
9. Leng, S.H.; Zhu, G.R. Calculation Method of Rapeseed pod Surface Area. *Chin. J. Oil Crop Sci.* **1991**, *13*, 76–77.
10. Liu, H.L. *Cultivation of Rape*; Shanghai Scientific & Technical Publishers: Shanghai, China, 1987.
11. Wang, H.X. Effects of Seeding Quantity and Fertilizer Treatments on the Structural Characters of Rape Silique Layer. *J. Anhui Agric. Sci.* **2017**, *45*, 33–36.
12. Liu, R.F.; Huang, S.Y.; Nie, Y.P.; Xu, S.Y. Automated detection research for number and key phenotypic parameters of rapeseed silique. *Chin. J. Oil Crop Sci.* **2020**, *42*, 71–77.
13. Wang, C.L.; Hai, J.B.; Yang, J.L.; Tian, J.H.; Chen, W.J.; Chen, T.; Luo, H.B.; Wang, H. Influence of leaf and silique photosynthesis on seeds yield and seeds oil quality of oilseed rape (*Brassica napus* L.). *Eur. J. Agron.* **2016**, *74*, 112–118. [CrossRef]
14. Chen, X.C.; Wang, H.H.; Gu, Z.Q. Study on the Measurement of Rape Pod Surface Area. *J. Hubei Agric. Coll.* **2004**, *24*, 170–173.
15. Zhang, Y.W.; Wang, Z.Y.; Li, D.R.; Zhang, X.; Chen, W.J.; Tian, J.H. Study on Diurnal Changes of Photosynthetic Characteristics in Pods of Brassica napus. *Acta Agric. Boreali-Occident. Sin.* **2008**, *17*, 174–180.
16. Leng, S.H.; Yang, G.; Chen, X.L.; Chen, W.D.; Ye, J.A. Effects of N application on pod characteristics among different layer of pod canopy in Ningza No. 1 (*Brassica napus* L.). *Chin. J. Oil Crop Sci.* **2002**, *24*, 25–28.

17. Gao, J.Q.; Pu, H.M.; Long, W.H.; Chen, X.J. Changes in dry matter and oil accumulation during pod development of rapeseed. *Jiangsu Agric. Sci.* **2007**, *35*, 50–52.
18. Ding, X.Q. Study on characters of silique and seed in spring rape (*B. campestris* L.). *Chin. J. Oil Crop Sci.* **1996**, *18*, 28–30.
19. Bai, Y.Z. The research for the growing course of three kinds rapeseed and rapefruit. *J. Xinjiang Norm. Univ. (Nat. Sci. Ed.)* **1993**, *12*, 54–57.
20. Pu, H.M.; Qi, C.K.; Fu, S.Z. The growth characteristics and source-sink effect of pods in rapeseed. *Jiangsu Agric. Sci.* **1993**, *21*, 22–25.
21. Zhao, H.; Sun, C.C. Study on the Relationship between the morphological differences of pod and seeds in Brassica napus. *Shanghai Agric. Sci. Technol.* **1990**, *20*, 17–19.
22. Wang, C.L.; Wang, Z.L.; Chen, T.; Yang, J.L.; Chen, W.J.; Mu, J.X.; Tian, J.H.; Zhao, X.G. Relationship between Yield and Photosynthesis of Leaf and Silique of Different *Brassica napus* L. Varieties during Reproduction Period. *Acta Bot. Boreali-Occident. Sin.* **2016**, *36*, 1417–1426.
23. Li, C.F.; Wang, C.F. Effects of nitrogen nutrition on pods development in rape (*B. napus*). *Acta Agron. Sin.* **1988**, *14*, 329–335.
24. Mao, F.R.; Wang, L. Measurement Technology of 3D Laser Scanning. *J. Astronaut. Metrol. Meas.* **2005**, *25*, 1–6.
25. Tong, S.; Xu, X.G.; Yi, C.T.; Shao, C.Y. Overview on vision-based 3D reconstruction. *Appl. Res. Comput.* **2011**, *28*, 2411–2417.
26. Yang, B.S.; Liang, F.S.; Huang, R.G. Progress, Challenges and Perspectives of 3D LiDAR Point Cloud Processing. *Acta Geod. Cartogr. Sin.* **2017**, *46*, 1509–1516.
27. Sun, Y.C.; Ge, B.Z.; Zhang, Y.M. Review for the 3D Information Measuring Technology. *J. Optoelectron.·Laser* **2004**, *15*, 248–254.
28. Nguyen, H.; Nguyen, D.; Wang, Z.Y.; Kieu, H.; Le, M. Real-time, high-accuracy 3D imaging and shape measurement. *Appl. Opt.* **2015**, *54*, A9–A17. [CrossRef] [PubMed]
29. Ma, L.G. Classification and Application of Terrestrial Laser Scanners. *Geospat. Inf.* **2005**, *3*, 60–62.
30. Rantung, J.; Oh, J.M.; Kim, H.K.; Oh, S.J.; Kim, S.B. Real-Time Image Segmentation and Determination of 3D Coordinates for Fish Surface Area and Volume Measurement based on Stereo Vision. *J. Inst. Control. Robot. Syst.* **2018**, *24*, 141–148. [CrossRef]
31. Reichert, J.; Schellenberg, J.; Schubert, P.; Wilke, T. 3D scanning as a highly precise, reproducible, and minimally invasive method for surface area and volume measurements of scleractinian corals. *Limnol. Oceanogr.-Methods* **2016**, *14*, 518–526. [CrossRef]
32. Sholts, S.B.; Warmlander, S.K.T.S.; Flores, L.M.; Miller, K.W.P.; Walker, P.L. Variation in the Measurement of Cranial Volume and Surface Area Using 3D Laser Scanning Technology. *J. Forensic Sci.* **2010**, *55*, 871–876. [CrossRef]
33. Telfer, S.; Woodburn, J. The use of 3D surface scanning for the measurement and assessment of the human foot. *J. Foot Ankle Res.* **2010**, *3*, 19. [CrossRef]
34. Lee, H.Y.; Hong, K.; Kim, E.A. Measurement protocol of women's nude breasts using a 3D scanning technique. *Appl. Ergon.* **2004**, *35*, 353–359. [CrossRef]
35. Villordon, A.; Gregorie, J.C.; LaBonte, D. Direct Measurement of Sweetpotato Surface Area and Volume Using a Low-cost 3D Scanner for Identification of Shape Features Related to Processing Product Recovery. *Hortscience* **2020**, *55*, 722–728. [CrossRef]
36. Villordon, A.Q.; LaBonte, D. Using a Low-cost 3D Scanner for Non-destructive Measurement of Surface Area and Volume of Storage Roots in Postharvest Research. *Hortscience* **2017**, *52*, S65.
37. SGKS. LDI Ultra-High Precision 3D Laser Scanning Measurement System. 2022. Available online: http://www.sgkslabs.com/resources-v.asp?id=2 (accessed on 20 August 2022).
38. Li, X.M.; Deng, Y.P.; Wang, X.H.; Li, F.Y.; Guan, C.Y. A New Method for Rapid Estimation of Rapeseed Pod Surface Area. China CN201911170476, 21 February 2020.
39. Tang, Q.Y.; Zhang, C.X. Data Processing System (DPS) software with experimental design, statistical analysis and data mining developed for use in entomological research. *Insect Sci.* **2013**, *20*, 254–260. [CrossRef]
40. James, S. *Calculus*; World Publishing Corporation: Beijing, China, 2004.
41. Li, X.M.; Wang, X.H.; Wei, H.L.; Zhu, X.G.; Peng, Y.L.; Li, M.; Li, T.; Huang, H. A technique system for the measurement, reconstruction and character extraction of rice plant architecture. *PLoS ONE* **2017**, *12*, e0177205. [CrossRef] [PubMed]
42. Li, C.; Chen, J.H.; Shang, W.Y. Study on the processes of the development of pods and seeds in *Brassica napus* L. *Chin. J. Oil Crop Sci.* **1988**, *10*, 23–26.
43. Leng, S.H.; Hui, F.H.; Zuo, Q.S.; Tang, Y. Regulations of N application on pod qualities of different branches in rapeseed. *Chin. J. Oil Crop Sci.* **2003**, *25*, 60–63.
44. Zhang, Q.F.; Sun, X.S. Measuring Principle and Developmental Prospect of 3D Laser Scanner. *Beijing Surv. Mapp.* **2011**, *25*, 39–42.
45. Duan, L.F.; Yang, W.N. Research advances and futurescenarios of rice phenomics. *Chin. Bull. LifeSciences* **2016**, *28*, 1129–1137.
46. Mu, J.H.; Chen, Y.Z.; Feng, H.; Li, W.J.; Zhou, L.B. A new revolutionin crop breeding: The era of high-throughput phenomics. *Plant Sci. J.* **2016**, *34*, 962–971.

applied sciences

MDPI

Article

Design and Application of Liquid Fertilizer pH Regulation Controller Based on BP-PID-Smith Predictive Compensation Algorithm

Zihao Meng, Lixin Zhang *, He Li, Runmeng Zhou, Haoran Bu, Yongchao Shan, Xiao Ma and Ruihao Ma

School of Mechanical and Electrical Engineering, Shihezi University, Shihezi 832003, China; mzh9823@163.com (Z.M.); lihe@stu.edu.cn (H.L.); zhourunmeng@stu.edu.cn (R.Z.); buhaoran@stu.edu.cn (H.B.); shanyongchao@stu.edu.cn (Y.S.); maxiao@stu.edu.cn (X.M.); maruihao@stu.edu.cn (R.M.)

* Correspondence: zhlx2001329@163.com

Abstract: The pH value of liquid fertilizer is a key factor affecting crop growth, so it is necessary to regulate its pH value. However, the pH regulation system has the characteristics of nonlinearity and time lag, which makes it difficult for the conventional controller to achieve accurate pH control. By analyzing the regulation process, this paper designs a BP-PID-Smith prediction compensator, which compensates for the error between the actual model and the theoretical model and improves the control accuracy. The pH regulation system with STM32F103ZET6 as the control core was also developed, and the performance tests were carried out under different flow rates to compare with the regulation system of PID-Smith and Smith algorithms. The experimental results showed that the maximum overshoot of the BP-PID-Smith prediction compensator was 0.27% on average, and the average adjustment time for pH value reduction from 7.5 to 6.8 was 71.39 s, which had good practicality and robustness to meet the actual control demand.

Keywords: water and fertilizer integration; pH adjustment; BP-PID-Smith algorithm; estimated compensation

Citation: Meng, Z.; Zhang, L.; Li, H.; Zhou, R.; Bu, H.; Shan, Y.; Ma, X.; Ma, R. Design and Application of Liquid Fertilizer pH Regulation Controller Based on BP-PID-Smith Predictive Compensation Algorithm. *Appl. Sci.* **2022**, *12*, 6162. https://doi.org/10.3390/app12126162

Academic Editor: Rafael López Núñez

Received: 11 May 2022
Accepted: 16 June 2022
Published: 17 June 2022

1. Introduction

In China, the use of traditional methods to irrigate crops is prone to the problem of water abuse and China's fertilizer use rate ranks first in the world all the year round, but the utilization rate of chemical fertilizers is not high. The integrated water and fertilizer technology is a new technology that combines irrigation and fertilization. According to the nutritional needs of crops, the pH value of liquid fertilizers is precisely regulated so that the roots of crops can fully absorb nutrients [1]. This technology can effectively reduce the pollution of fertilizers to soil and groundwater, and protect the ecological environment [2].

However, there is often a time lag in the process of pH adjustment, and the change in pH value of liquid fertilizer is also nonlinear. Therefore, how to quickly and accurately adjust the pH value of fertilizer to an appropriate range in the process of fertilization is the key research field of water fertilizer integration technology [3]. E. Ali et al. [4] proposed an adaptive PI algorithm that uses a simple process model to predict the pH closed-loop response and its sensitivity to PI parameter settings, and finally, the obtained information was directly used to adjust the PI controller parameters on-line. G.Balasubramanian et al. [5] proposed an adaptive control scheme based on Recurrent Neural Network (RNN). The scheme included an on-line adaptive RNN estimator and RNN controller, and the estimator weights were updated recursively by back-propagation algorithm. The controller weights were corrected by the steepest descent method. The proposed scheme was compared with the model-based IMC controller, and the results showed that the RNN-based controller had better performance in the nonlinear pH neutralization process.

Homero J. Sena et al. [6] used a real-time adaptive algorithm based on the Extended Kalman Filter (EKF) to correct the artificial neural network predictions at process runtime, which reduced the sum of squared errors in pH by 64.3% compared to the MPC of the artificial neural network without model adaptation. Douglas Alves Goulart and Renato Dutra Pereira [7] developed a Continuous Stirred-Tank Reactor (CSTR) neutralization simulator and an adaptive Particle Swarm Optimization (PSO) algorithm for automatic selection of Reinforcement Learning hyperparameters. During regulation and servo operation, the controller stabilized the effluent pH in the neutral range better than the PID controller. Shahin Salehi et al. [8] proposed an adaptive control scheme for pH value based on a fuzzy logic system and verified the effectiveness of the controller through simulation and experimental research. The results showed that the controller performed well in setpoint tracking and was much better than the PI controller. Hui Wu et al. [9] proposed a predictive control method based on Decentralized Fuzzy Inference (DFIPC), which locally linearized the nonlinear object model and predicted the future output of the control object according to its step response model. The method was applied to the pH neutralization process. The results showed that the method had better robustness than traditional model predictive control.

In the industrial field, Smith prediction compensation is mainly used to solve the pure lag of the system [10]. Guangda Chen [11] proposed a Smith predictor combined with Linear Active Disturbance Rejection Control (LADRC) and analyzed the stability of the Smith + LADRC time-delay control system from a theoretical point of view. Simulation and experimental results showed that the algorithm was superior to the traditional method in terms of overshoot and response time. Mahmoud Gamal et al. [12] combined the classical Smith predictor and the adaptive Smith predictor in a networked control system and compared it with other delay compensation schemes. The results showed that the scheme significantly improved the performance of the networked control system and reduced the impact of delay on the system. Chenkun Qi et al. [13] proposed a hybrid Smith predictor and phase lead compensation method. This method can achieve higher simulation fidelity with less convergence. The effectiveness of the compensation method was verified by the simulation of the undamped elastic contact process. Yonghui Nie et al. [14] proposed an optimal wide-area damping controller considering delay, using the Smith predictor to provide delay compensation and using particle swarm optimization to further improve the controller. The simulation results showed that the method improved the delay tolerance of the closed-loop system and improved the dynamic stability of the power system.

In this paper, according to the characteristics of pH value regulation of liquid fertilizers, the prediction compensation controllers based on Smith, PID-Smith and BP-PID-Smith were designed respectively. They were simulated and analyzed under the conditions of model matching and mismatching, and step response curves were obtained respectively. The performance of the three controllers was evaluated from four aspects: rise time, peak time, maximum overshoot, and adjustment time [15]. The results showed that the control effect of the BP-PID-Smith controller was the best. On this basis, an experimental platform was built to verify the practicability of the algorithm. The results showed that the BP-PID-Smith predictor compensator can effectively solve the adverse effects of the time delay and nonlinearity of the system in the fertilization process, and meet the control requirements for precise regulation of the pH value of liquid fertilizers.

The purpose of this paper is to design a BP-PID-Smith predictive compensation control algorithm, which can quickly adjust the pH value of water and fertilizer to the set value, and effectively solve the problems caused by factors such as time lag and nonlinearity in the pH adjustment process.

The contents of this paper are as follows: The first part introduces the research status of precise pH control and Smith prediction compensation. The second part explains the working principle of the pH regulation system, establishes the dynamic and static model of the pH regulation process, and reveals the nonlinear and time-delay characteristics of the pH regulation process. In the third part, the formulas of the Smith predictor compensator algorithm and BP neural network algorithm are derived, and the simulation models based on Smith, PID-Smith, and BP-PID-Smith predictor compensation are established by using the Simulink module in MATLAB. In the fourth part, the above three models are simulated and analyzed, respectively, and the models are evaluated according to the results. In the fifth part, experiments are carried out to verify the practicability of the controller. The sixth part gives the conclusion.

2. Introduction of pH Value Regulation System and Analysis of Regulation Process

2.1. pH Control System Structure Composition

Figure 1 is the structural block diagram of the pH value regulation system of liquid fertilizer. The regulation system includes a water storage tank, solenoid valve, fertilizer tank, regulating liquid tank, flowmeter, pH value sensor, hose pump, mixing tank, and other main devices. The liquid in the water storage tank, fertilizer tank, and regulating liquid tank finally flows into the mixing tank and is stirred inside. A pH sensor is installed inside the mixing tank to monitor the pH value. The outlet of the mixing tank is connected with the field drip irrigation belt to transport the adjusted fertilizer to the field. The dilution ratio of liquid fertilizer is set to 1:8, the pH value of fertilizer before dilution is 7.5, and dilute hydrochloric acid with a concentration of 0.2 mol/L is used as the regulating liquid. Flowmeter and pressure gauge are installed at the inlet and outlet of the mixing tank. The hose pump is used as the conveying device of the regulating system. The three-phase asynchronous motor is connected with the pump body of the hose pump. The pressure formed by squeezing the hose by the roller is used to transport the materials. The system adjusts the outlet flow of the hose pump by changing the frequency of the frequency converter connected with the hose pump, to accurately adjust the pH value.

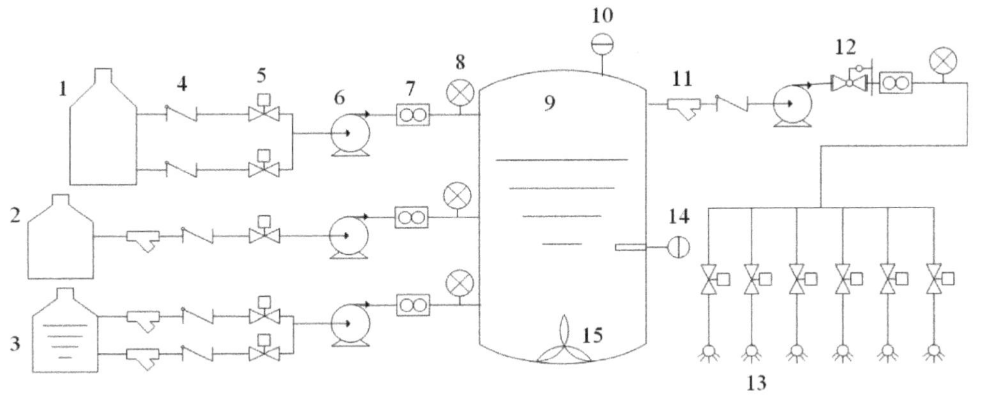

Figure 1. Structural block diagram of pH value regulation system of liquid fertilizer: 1, regulating liquid tank; 2, fertilizer tank; 3, water storage tank; 4, check valve; 5, solenoid valve; 6, hose pump; 7, flowmeter; 8, pressure gauge; 9, mixing tank; 10, liquid level gauge; 11, Y-type filter; 12, pressure holding valve; 13, drip irrigation belt; 14, pH sensor; 15, mixing pump.

In this paper, the STM32F103ZET6 microcontroller is used as the control core, and the BP-PID-Smith prediction compensation algorithm is written in it. The set pH value is used as the input value, and the pH value of the liquid fertilizer collected by the pH sensor is used as the actual feedback value for calculation. The required regulating liquid flow is calculated, and the flow is converted into the working frequency of the hose pump, to achieve the purpose of efficiently regulating the pH value of liquid fertilizer.

When the pH value regulation system of liquid fertilizer is working, the monitoring end will input the set pH value of liquid fertilizer and fertilization flow into the system, the hose pump at the fertilizer tank and the hose pump at the water storage tank will pump liquid fertilizer and water to the mixing tank in corresponding proportion for mixing. When the regulation system monitors the pH value in the mixing tank and the set value, the solenoid valve opens and the hose pump at the regulating liquid tank starts to run to extract the regulating liquid into the mixing tank, and the stirring pump stirs the liquid in the mixing tank, and when the regulation system monitors the pH value of liquid fertilizer in the mixing tank reaches the set value, the system maintains a stable state.

2.2. Analysis of Liquid Fertilizer pH Regulation Process

Usually, water and liquid fertilizer are weakly basic, while the regulating liquid generally uses dilute hydrochloric acid. Therefore, the mixing of fertilizer, regulating liquid, and water can be considered a neutralization process of a strong acid and a weak base [16], which can be represented by a static pH equation describing the neutralization titration curve and a dynamic equation describing the state variables.

2.2.1. Objective Function and Design Variables

The ionization process of m-membered acid is described below.

First level ionization

$$H_mA \rightleftharpoons H_{m-1}A^- + H^+$$
$$K_{a1} = \frac{[H^+][H_{m-1}A^-]}{[H_mA]} \tag{1}$$

Secondary ionization

$$H_{m-1}A^- \rightleftharpoons H_{m-2}A^{2-} + H^+$$
$$K_{a2} = \frac{[H^+][H_{m-2}A^{2-}]}{[H_{m-1}A^-]} \tag{2}$$

m-level ionization

$$HA^{(m-1)-} \rightleftharpoons A^{m-} + H^+$$
$$K_{am} = \frac{[H^+][A^{m-}]}{[HA^{(m-1)-}]} \tag{3}$$

The ionization process of n-membered bases in liquid fertilizer can be expressed as:

First level ionization

$$B(OH)_n \rightleftharpoons B(OH)_{n-1}{}^+ + OH^-$$
$$K_{b1} = \frac{[B(OH)_{n-1}{}^+][OH^-]}{[B(OH)_n]} \tag{4}$$

Secondary ionization

$$B(OH)_{n-1}{}^+ \rightleftharpoons B(OH)_{n-2}{}^{2+} + OH^-$$
$$K_{b2} = \frac{[B(OH)_{n-2}{}^{2+}][OH^-]}{[B(OH)_{n-1}{}^+]} \tag{5}$$

n-level ionization

$$B(OH)^{(n-1)+} \rightleftharpoons B^{n+} + OH^-$$
$$K_{bn} = \frac{[B^{n+}][OH^-]}{[B(OH)^{(n-1)+}]} \tag{6}$$

where $K_{b1}, K_{b2}, \ldots, K_{bn}$ are the ionization equilibrium constants.

The ionization equilibrium of water can be expressed as:

$$H_2O \rightleftharpoons H^+ + OH^-$$
$$K_w = [H^+][OH^-] \tag{7}$$

where $K_w = 10^{-14}$.

Let x_i be the total ionic concentration of the acid or the total ionic concentration of the base in the fertilizer mixture, then

When i is an acid:

$$x_i = [H_m A] + \left[H_{m-1}A^-\right] + \cdots + \left[A^{m-}\right] \tag{8}$$

When i is a base:

$$x_i = [B(OH)_n] + [B(OH)_{n-1}{}^+] + \cdots + \left[B^{n+}\right] \tag{9}$$

Since the solution must always remain electrically neutral, the charge balance equation yields:

$$\begin{aligned} &\sum_{i=acid} \{[H_{p_i-1}A^-] + 2[H_{p_i-2}A^{2-}] + \cdots + p_i[A^{p_i-}]\} + [OH^-] \\ &\quad = \sum_{i=base} \left\{\left[B(OH)^+_{n_i-1}\right] + 2\left[B(OH)^{2+}_{n_i-2}\right] + \cdots + n_i[B^{n_i+}]\right\} + [H^+] \end{aligned} \tag{10}$$

The pH equation can be derived from Equations (1)–(10) as:

$$\sum_{i=1}^{n} a_i([H^+])x_i + [H^+] - \frac{K_w}{[H^+]} = 0 \tag{11}$$

When i is an acid:

$$a_i([H^+]) = -\frac{m_i + (m_i-1)\frac{[H^+]}{K_{am_i}} + \cdots + \frac{[H^+]^{m_i-1}}{K_{a2_i}K_{a3_i}\cdots K_{am_i}}}{1 + \frac{[H^+]}{K_{am_i}} + \cdots + \frac{[H^+]^{m_i-1}}{K_{a2_i}K_{a3_i}\cdots K_{am_i}} + \frac{[H^+]^{m_i}}{K_{a1_i}K_{a2_i}\cdots K_{am_i}}} \tag{12}$$

When i is a base:

$$a_i([H^+]) = \frac{n_i[H^+]^{n_i} + (n_i-1)\frac{K_w}{K_{bn_i}}[H^+]^{n_i-1} + \cdots + \frac{K_w^{n_i-1}[H^+]}{K_{b2_i}K_{b3_i}\cdots K_{bn_i}}}{[H^+]^{n_i} + \frac{K_w}{K_{bn_i}}[H^+]^{n_i-1} + \cdots + \frac{K_w^{n_i-1}[H^+]}{K_{b2_i}K_{b3_i}\cdots K_{bn_i}} + \frac{K_w^{n_i}}{K_{b1_i}K_{b2_i}\cdots K_{bn_i}}} \tag{13}$$

Since the whole pH neutralization process can be considered as the neutralization of a strong acid and a weak base, thus

$$a_1([H^+]) = -1 \tag{14}$$

$$a_2([H^+]) = \frac{[H^+]}{[H^+] + \frac{K_w}{K_b}} = \frac{1}{1 + \frac{[OH^-]}{K_b}} \tag{15}$$

The above analysis leads to the fact that the pH equation can be rewritten as:

$$-x_1 + \frac{1}{1 + 10^{pK_b + pH - 14}}x_2 + 10^{-pH} - 10^{pH-14} = 0 \tag{16}$$

where $pK_b = -\log K_b$, x_1, x_2 are the total ionic concentrations of acid and base in the fertilizer mix, respectively.

This equation is the static model of pH neutralization.

As can be determined from Equation (16), the control process of pH described by this static equation is strongly nonlinear.

2.2.2. Dynamic Model

Assuming that the liquid volume in the mixing tank is constant and uniformly mixed, ignoring the influence of liquid temperature on pH, the dynamic process of pH neutralization reaction can be represented by a CSTR process [17–19]. The dynamic process of acid-base neutralization is shown in Figure 2.

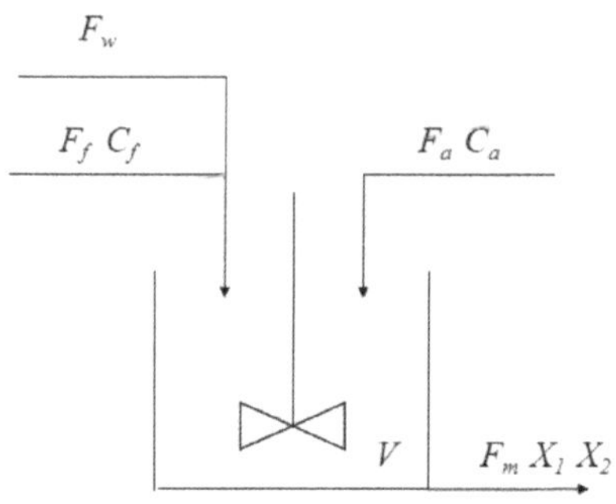

Figure 2. The dynamic process of acid-base neutralization.

In the figure, V is the mixing tank volume; F_w is the irrigation water flow rate; C_w is the irrigation water input concentration; F_a is the acid flow rate input into the mixing tank; C_a is the regulating liquid input concentration; F_f is the liquid fertilizer flow rate input into the mixing tank; C_f is the liquid fertilizer input concentration; F_m is the fertilizer mixture flow rate output from the mixing tank; x_1, x_2 are the acid and alkali concentrations in the output fertilizer mixture, respectively.

According to the principle of material conservation, the dynamic equation of each state variable is obtained when the fertilizer mixing process reaches equilibrium.

$$\begin{cases} V\frac{dx_1}{dt} = F_aC_a - F_mx_1 \\ V\frac{dx_2}{dt} = F_fC_f + F_wC_w - F_mx_2 \end{cases} \tag{17}$$

The input and output flows should be dynamically balanced, then:

$$F_m = F_a + F_w + F_f \tag{18}$$

Equations (16)–(18) together form a mathematical model of the pH neutralization process.

$$\begin{cases} -x_1 + \frac{1}{1+10^{pK_b+pH-14}}x_2 + 10^{-pH} - 10^{pH-14} = 0 \\ V\frac{dx_1}{dt} = F_aC_a - F_mx_1 \\ V\frac{dx_2}{dt} = F_fC_f + F_wC_w - F_mx_2 \\ F_m = F_a + F_w + F_f \end{cases} \tag{19}$$

From Equation (19), it can be seen that the mathematical model of the pH neutralization process consists of a static model and a dynamic model, and the dynamic process exhibits a slight nonlinearity, which is negligible when it is assumed that the input flow rate of the liquid fertilizer in the mixing tank is much larger than the input flow rate of the conditioning fluid.

However, Equation (16) expresses the inherent nonlinearity of the pH neutralization process. And in the actual fertilizer mixing process, besides the stirring and mixing process, the delay factors such as the slow flow of liquid in the pipeline and the time lag in the measurement link also have an impact on the pH regulation process.

Therefore, the pH regulation process is characterized by nonlinearity and time lag, which place higher demands on the performance of the controller.

2.2.3. Determination of the System Transfer Function

In this paper, the pH regulation system of liquid fertilizer is studied to analyze the process of pH regulation when mixing fertilizers.

The pH regulation characteristics and the complexity of the model are taken into account, and the mathematical model of the pH regulation system is described using a first-order system transfer function with a delay link [20].

$$G(s) = \frac{Ke^{-\tau s}}{Ts+1} \tag{20}$$

A step response with a pH value of 6.8 was used as the input of the open-loop system. The sampling time interval of the system was set to 1 s. The initial pH value of the liquid in the mixing tank was 7.5, and the data of pH value changing with time was obtained. The first-order approximation method was used to input data into the computer and fit the step response curve of the system. The gain coefficient K of the system is 1.02, the delay time is obtained τ is 7.5 and the time constant T is 1.78. Therefore, the pH control process of liquid fertilizer has a time lag.

3. Design and Simulation of BP-PID-Smith Based Controller for pH Regulation System

3.1. Design of Time Lag Compensation for Smith's Prognosticator Model

PID control can adjust the size of the control quantity in time according to the error between the actual value and the desired value so that the actual value gradually approaches the desired value, which is a kind of closed-loop control with high reliability and robustness and is widely used in industry [21]. The PID closed-loop control system consists of two parts, the PID controller and the controlled object, as shown in Figure 3.

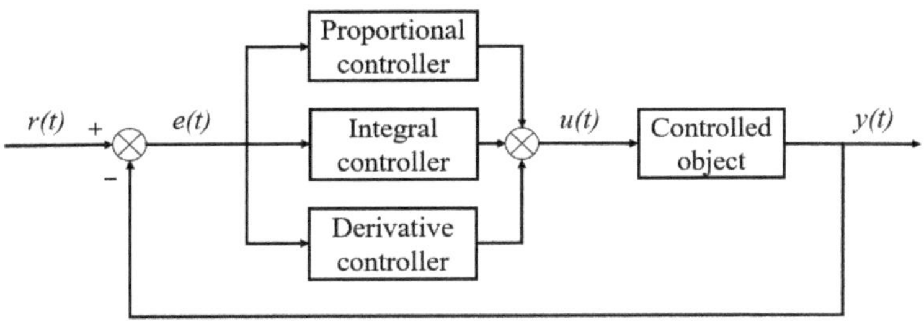

Figure 3. PID control structure diagram.

The error value $e(t)$ between the expected value $r(t)$ and the actual output value $y(t)$ is obtained; in the second step, the error values obtained are subjected to proportional, integral, and differential operations, and the closed-loop control quantity $u(t)$ is obtained after linear combination; in the third step, the controlled object receives the control quantity $u(t)$, and the output value $y(t)$ approaches the expected value $r(t)$ to complete the control of the controlled object within the error allowance. The mathematical expression of the control principle is:

$$u(t) = K_p\left[e(t) + \frac{1}{T_i}\int_0^t e(\tau)d\tau + T_d\frac{de(t)}{dt}\right] \tag{21}$$

where K_p is the proportionality constant, T_i is the integration time constant, and T_d is the differential time constant.

To discretize Equation (21), let T be the sampling period, perform k consecutive samples, and replace the continuous time t with the discrete sampling time point kT

$$\begin{cases} t \approx kT \\ \int_0^t e(t)dt \approx T\sum_{j=0}^{k} e(jT) = T\sum_{j=0}^{k} e(j) \\ \frac{de(t)}{dt} \approx \frac{e(kT)-e[(k-1)T]}{T} \end{cases} \tag{22}$$

Bringing Equation (22) into Equation (21) and assuming that T is sufficiently short, Equation (21) can be simplified to:

$$u(k) = K_p e(k) + K_i \sum_{j=0}^{k} e(j) + K_d[e(k) - e(k-1)] \quad (23)$$

where K_p, K_i, and K_d are proportional, integral, and differential coefficients, $K_i = K_p \frac{T}{T_i}$, $K_d = K_p \frac{T_d}{T}$.

In the actual design of the controller, there are inevitably delays, including control delays and sensor delays, which may cause controller instability when the delays are relatively large. Therefore, the effect of delay needs to be considered when designing a controller. Smith predictor, as a classical solution, can offset and compensate for the delay effect of the system, significantly improve the control performance of the time-lag system, and reduce the instability of the system [22].

Smith's predictive compensation control structure is shown in Figure 4.

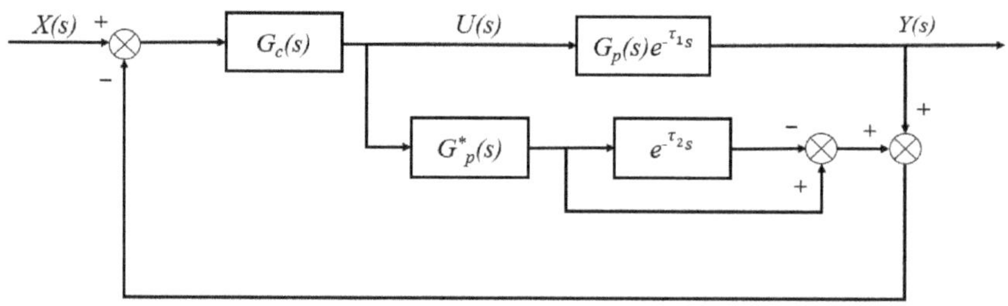

Figure 4. Smith prediction compensation control structure diagram.

Where $G_p^*(s)e^{-\tau_2 s}$ is the introduced Smith's prediction compensation transfer function, $G_p(s)e^{-\tau_1 s}$ is the system with delay model, and $G_c(s)$ is the main controller transfer function, when the model matches exactly, $G_p^*(s) = G_p(s)$ and $\tau_1 = \tau_2 = \tau$. The overall closed-loop transfer function of the system at this point is:

$$\frac{Y(s)}{X(s)} = \frac{G_c(s)G_p(s)e^{-\tau s}}{1 + G_c(s)G_p(s)} \quad (24)$$

The characteristic equation of the system is:

$$D(s) = 1 + G_c(s)G_p(s) = 0 \quad (25)$$

In this paper, the Ziegler–Nichols parameter rectification method is used for the initial rectification of the proportional, integral, and differential constants of the PID, as shown in Equation (26).

$$\begin{cases} Kp = 1.2\frac{T}{K\times\tau} \\ T_i = 2.2\tau \\ T_d = 0.5\tau \end{cases} \quad (26)$$

The controlled object model in this paper is shown in Equation (20), where $K = 1.02$, $T = 1.78$, $\tau = 7.5$, and the parameters are brought into Equation (26) to obtain the preliminary rectified values of the proportional, integral, and differential constants calculated by the Ziegler–Nichols method as $Kp= 0.28$, $K_i= 0.02$, $K_d = 1.05$, respectively.

The simulation model of the pH regulation system based on Smith's prediction compensation is shown in Figure 5.

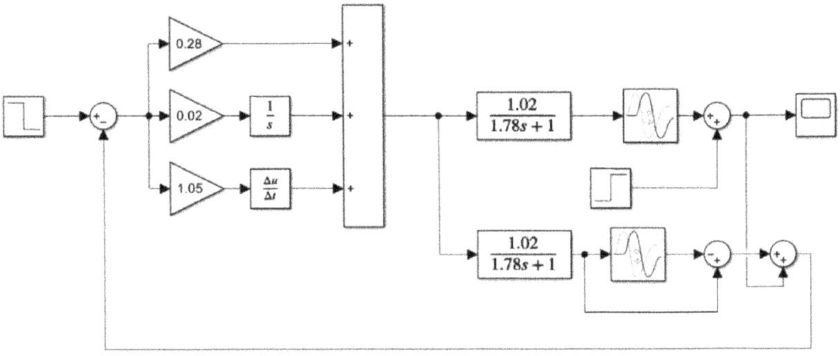

Figure 5. Simulation model of pH regulation system based on Smith's prediction compensation.

3.2. Design of PID-Smith Prediction Compensator

When the model cannot be matched exactly, at this time $G_p^*(s) \neq G_p(s)$, $\tau_1 \neq \tau_2$, the overall closed-loop transfer function of the system is:

$$\frac{Y(s)}{X(s)} = \frac{G_c(s)G_p(s)e^{-\tau_1 s}}{1 + G_c(s)\left[G_p^*(s) + G_p(s)e^{-\tau_1 s} - G_p^*(s)e^{-\tau_2 s}\right]} \tag{27}$$

From the above equation, it can be seen that when the deviation between the actual model and the theoretical model is large, there is a lag term in the characteristic equation of the system, which makes the Smith controller unable to correct the error between the actual model and the theoretical model even though, which may eventually cause the output signal of the system to oscillate and diverge. Therefore, Smith's prediction compensation is not suitable for the case where the theoretical model has a large deviation from the actual model.

The PID-Smith prediction control adds a suitable PID compensation controller to avoid the time lag term in the closed-loop characteristic equation and the model mismatch, which eventually leads to the oscillation and divergence of the output signal, thus reducing the effect of the time lag term on the system, and the structure of the PID-Smith prediction compensator is shown in Figure 6.

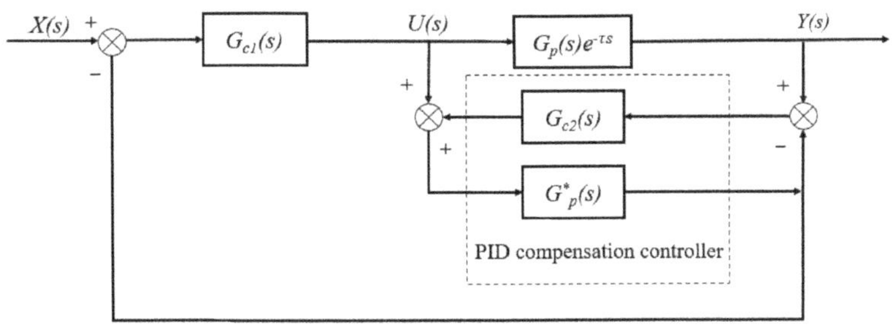

Figure 6. Structure of PID-Smith predictive compensator.

Where $G_{c2}(s)$ and $G_p^*(s)$ together form the compensation controller, $G_p(s)e^{-\tau s}$ is the system with delay model, and $G_{c1}(s)$ is the main controller model.

From Figure 6, the closed-loop transfer function of the system is:

$$\frac{Y(s)}{X(s)} = \frac{G_{c1}(s)G_p(s)e^{-\tau s}}{1 + G_{c1}(s)G_p^*(s)\frac{1+G_p(s)G_{c2}(s)e^{-\tau s}}{1+G_{c2}(s)e^{-\tau s}G_p^*(s)}} \tag{28}$$

The characteristic equation of the system is:

$$D(s) = 1 + G_{c1}(s)G_p^*(s)\frac{1 + G_p(s)G_{c2}(s)e^{-\tau s}}{1 + G_{c2}(s)e^{-\tau s}G_p^*(s)} = 0 \tag{29}$$

If the mode of $G_{c2}(s)$ is chosen to be small enough, then:

$$1 + G_p(s)G_{c2}(s)e^{-\tau s} \approx 1 \tag{30}$$

$$1 + G_{c2}(s)e^{-\tau s}G_p^*(s) \approx 1 \tag{31}$$

At this point the system characteristic equation simplifies to

$$D(s) = 1 + G_{c1}(s)G_p^*(s) = 0 \tag{32}$$

From the above equation, it can be seen that the system stability is not affected by the time lag of the compensating controller and the controlled object, and the system characteristic equation does not contain the time lag term.

The simulation model of the pH regulation system based on PID-Smith prediction compensation is shown in Figure 7.

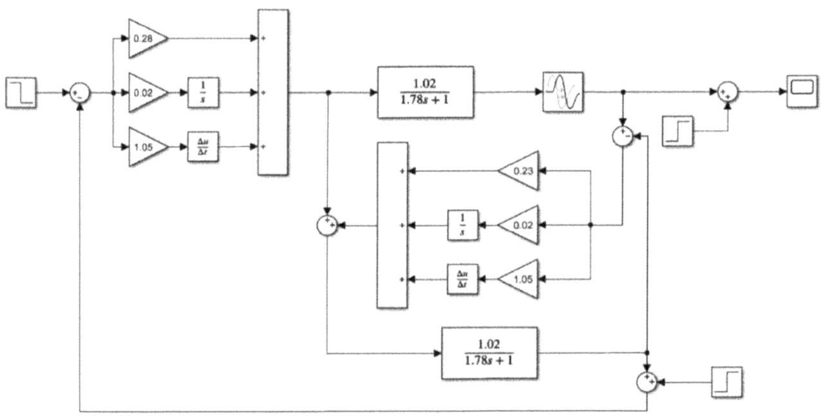

Figure 7. Simulation model of pH regulation system based on PID-Smith prediction compensation.

3.3. Design of BP-PID-Smith Prediction Compensator

Because the pH control system in this paper is a large time-delay system, and the traditional PID control is easily affected by time-delay, it cannot be optimized and adjusted according to the overall changes of the system. Therefore, this paper combines BP neural network with traditional PID control, designs a predictive compensator based on BP-PID-Smith, reduces the impact of time delay on the system, improves its learning efficiency by optimizing PID controller parameters, and realizes the parameter tuning of the control system.

The structure of the PID control system based on the BP neural network is shown in Figure 8.

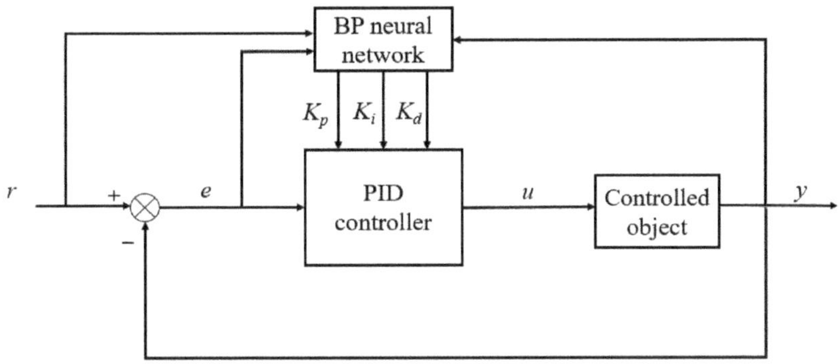

Figure 8. Structure of PID control system based on BP neural network.

The structure consists of a conventional PID and a BP neural network. The conventional PID realizes the closed-loop feedback control of the controlled object, and the BP neural network finally obtains the optimal PID control parameters of the system by continuously updating iterations according to the system state and learning algorithm. The neural network structure in Figure 8 is shown in Figure 9 below.

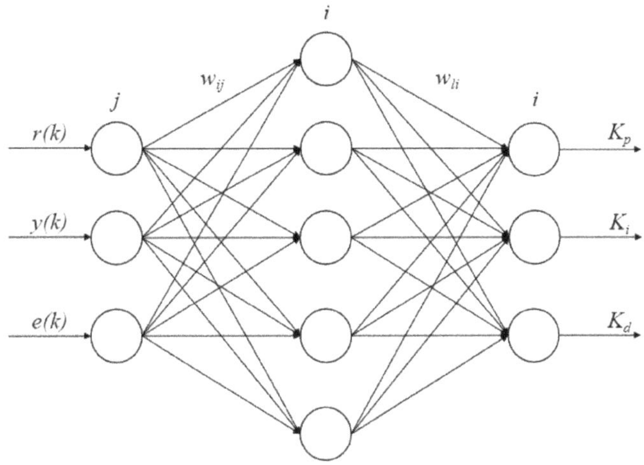

Figure 9. Optimization of PID parameters based on BP neural network.

The neural network consists of an input layer, an implicit layer, and an output layer, in which the input layer contains three neurons, and the inputs are $r(k)$, $y(k)$, and $e(k)$; to reduce the complexity of the system and improve the learning efficiency, the number of neurons in the implicit layer is set to five; the output layer contains three neurons, and the outputs correspond to the three parameters of the PID controller, K_p, K_i, and K_d, respectively.

A neural network structure based on 3-5-3, where the input layer inputs are:

$$O_j^{(1)} = x(j) \quad (j = 1, 2, \ldots, M) \tag{33}$$

M is the number of neurons in the input layer.

The implicit layer inputs and outputs are:

$$net_i^{(2)} = \sum_{j=0}^{M} w_{ij}^{(2)} O_j^{(1)} \tag{34}$$

$$O_i^{(2)}(k) = f\left(net_i^{(2)}(k)\right) \quad (i = 1, 2, \ldots, Q) \tag{35}$$

where Q is the number of neurons in the hidden layer and $w_{ij}^{(2)}$ is the hidden layer connection weight, $f(x) = \tanh(x) = \frac{e^x - e^{-x}}{e^x + e^{-x}}$.

The output layer inputs and outputs are:

$$net_l{}^{(3)}(k) = \sum_{i=0}^{Q} w_{li}{}^{(3)} O_i{}^{(2)}(k) \tag{36}$$

$$O_l^{(3)}(k) = g\left(net_l^{(3)}(k)\right) \quad (l = 1, 2, \ldots, N) \tag{37}$$

where N is the number of neurons in the output layer and $w_{li}^{(3)}$ is the output layer connection weight, $g(x) = \frac{1}{2}[1 + \tanh(x)] = \frac{e^x}{e^x + e^{-x}}$.

The control quantity $u(t)$ of the PID controller is calculated according to Equation (23), K_p, K_i, and K_d are $O_1^{(3)}(k)$, $O_2^{(3)}(k)$, and $O_3^{(3)}(k)$ as found in Equation (37). The selected performance metrics have the following functional form:

$$E(k) = \frac{1}{2}[r(k) - y(k)]^2 \tag{38}$$

The gradient descent method is used to continuously and iteratively update the connection weights between the neurons in the neural network so that the error signal decreases in the negative gradient direction. In addition, to speed up the convergence of the BP neural network algorithm and to obtain better dynamic properties, an inertia term is added to obtain a new update of the output layer connection weights when the learning rate is η.

$$\Delta w_{li}^{(3)}(k) = -\eta \frac{\partial E(k)}{\partial w_{li}^{(3)}} + \alpha \Delta w_{li}^{(3)}(k-1) \tag{39}$$

where α is the inertia factor.

$\frac{\partial E(k)}{\partial w_{li}^{(3)}}$ can be split into:

$$\frac{\partial E(k)}{\partial w_{li}^{(3)}} = \frac{\partial E(k)}{\partial y(k)} \times \frac{\partial y(k)}{\partial \Delta u(k)} \times \frac{\partial \Delta u(k)}{\partial O_l^{(3)}(k)} \times \frac{\partial O_l^{(3)}(k)}{\partial net_l^{(3)}(k)} \times \frac{\partial net_l^{(3)}(k)}{\partial w_{li}^{(3)}(k)} \tag{40}$$

After performing a series of simplifications, the updated output layer connection weights after learning by the neural network are obtained as:

$$\Delta w_{li}^{(3)}(k) = \alpha \Delta w_{li}^{(3)}(k-1) + \eta \delta_l^{(3)} O_i^{(2)}(k) \tag{41}$$

$$\delta_l^{(3)} = e(k) sgn\left(\frac{\partial y(k)}{\partial \Delta u(k)}\right) \frac{\partial \Delta u(k)}{\partial O_l^{(3)}(k)} g'\left(net_l^{(3)}(k)\right) \quad (l = 1, 2, \ldots, N) \tag{42}$$

Similarly, the update of the connection weights of the hidden layer after learning can be obtained as:

$$\Delta w_{ij}^{(2)}(k) = \alpha \Delta w_{ij}^{(2)}(k-1) + \eta \delta_i^{(2)} O_j^{(1)}(k) \tag{43}$$

$$\delta_i^{(2)} = f'\left(net_i^{(2)}(k)\right) \sum_{l=1}^{3} \delta_l^{(3)} w_{li}^{(3)}(k) \quad (i = 1, 2, \ldots Q) \tag{44}$$

The mathematical model for adding BP neural network is established above.

The PID main controller parameters are fine-tuned and the compensation controller is written using the BP-PID algorithm with the S-Function in the Simulink module, and the error between the actual model and the theoretical model is used as the input to the BP-PID algorithm.

The simulation model of the pH regulation system based on BP-PID-Smith prediction compensation is shown in Figure 10.

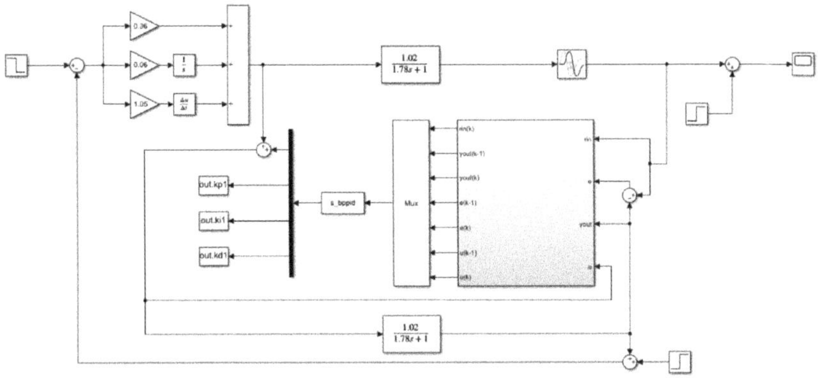

Figure 10. Simulation model of pH regulation system based on BP-PID-Smith prediction compensation.

4. Analysis of Simulation Results

4.1. Model Match

In this paper, based on BP neural network algorithm and Smith prediction compensation principle, the liquid fertilizer pH adjustment system based on the Smith prediction compensator, PID-Smith prediction compensator, and BP-PID-Smith prediction compensator were designed, respectively, and the final obtained simulation results were compared and analyzed. According to the actual situation, the initial pH value of liquid fertilizer was 7.5, the pH value of adjusted liquid fertilizer was set to 6.8, and the simulation time was set to 500 s. The focus of this paper is on the dynamic process of the system, and this process is described by the dynamic performance metrics, which are composed of rise time, peak time, regulation time, and overshoot of the controller. The rise time is the time required for the response to rise from zero to the final value for the first time; the peak time is the time required for the response to exceeding its final value to reach the first peak; the regulation time is the minimum time required for the response to reach and remain within ±5% (or ±2%) of the final value; and the overshoot is the percentage of the ratio of the maximum deviation of the response $c(t_p)$ to the final value $c(\infty)$, i.e., $\sigma\% = \frac{C(t_p)-C(\infty)}{C(\infty)} \times 100\%$. The rise time, peak time and regulation time are important indicators to evaluate the response speed of the system, and the overshoot reflects the stability of the system control process. The response curves of the three controllers when the models are matched are shown in Figure 11, the training error curves of the BP-PID-Smith controller are shown in Figure 12, and the dynamic performance comparisons are shown in Table 1.

Table 1. Comparison of the dynamic performance of the three controllers for model match.

Controller Type	Rise Time(s)	Peak Time(s)	Regulation Time (s)	Maximum Overshoot
Smith	397.541	397.541	178.197	0
PID-Smith	115.902	163.947	83.770	0.13%
BP-PID-Smith	56.712	66.891	36.230	0.03%

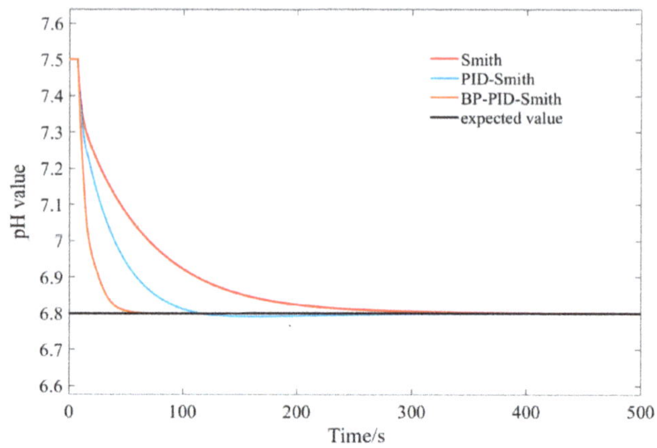

Figure 11. Step response curves of the three controllers under model matching.

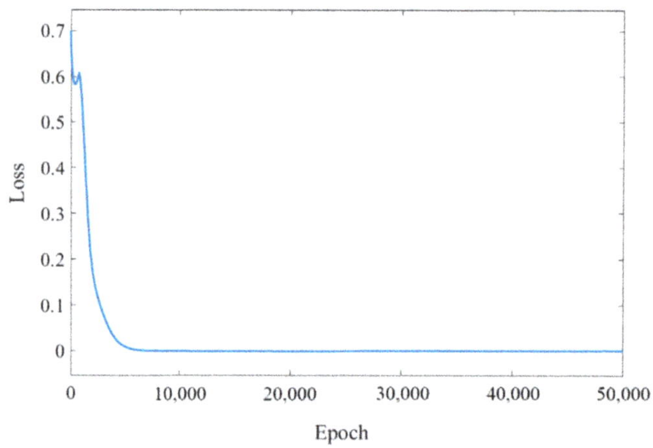

Figure 12. Training error curve of BP-PID-Smith controller under model matching.

According to the data in Table 1, the dynamic performance index of the BP-PID-Smith controller is significantly better than the other two controllers, so the BP-PID-Smith controller has a faster response and better steady-state performance compared with the PID-Smith controller and Smith controller. The Smith controller has a slow regulation time and rise time, reflecting its slow response time. The rise time and regulation time of the modified PID-Smith controller become shorter compared with the Smith controller, but at the same time, a certain amount of overshoot is generated, and the overall control effect is not as good as that of the BP-PID-Smith controller.

4.2. Model Mismatch

The above prediction compensator is based on the accurate mathematical model of the controlled object, however, in the actual process of pH regulation of liquid fertilizer, the accurate mathematical model of the system was difficult to obtain because of the complexity inside the system, which led to the actual model's inability to completely match with the theoretical controlled object model, thus seriously affecting the control effect of the controller and making the controller unable to operate stably. In this paper, three controllers were simulated under the conditions of model mismatch to examine the performance of the controllers.

In this paper, the actual model was set to $K = 1.5$, $T = 1.5$, $\tau = 12$, and a perturbation of amplitude 0.1 was added. The step response curves of the three controllers under the model mismatch condition are shown in Figure 13, the training error curves of the BP-PID-Smith controller are shown in Figure 14, and the dynamic performance comparison is shown in Table 2.

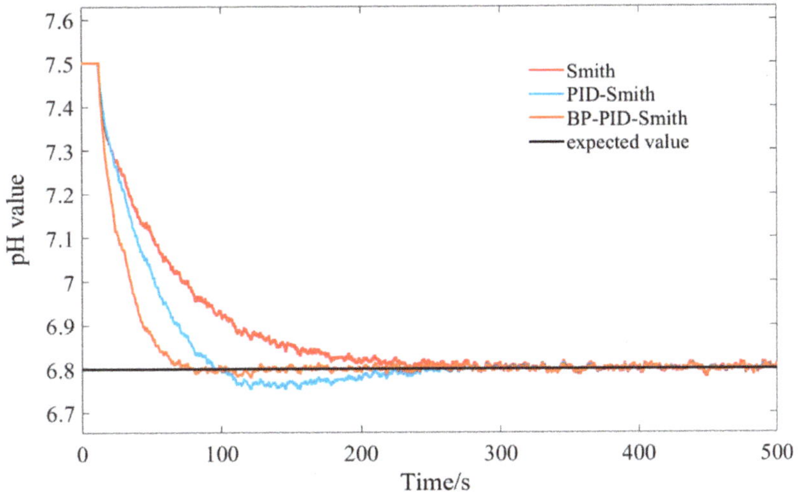

Figure 13. Step response curves of the three controllers under model mismatch.

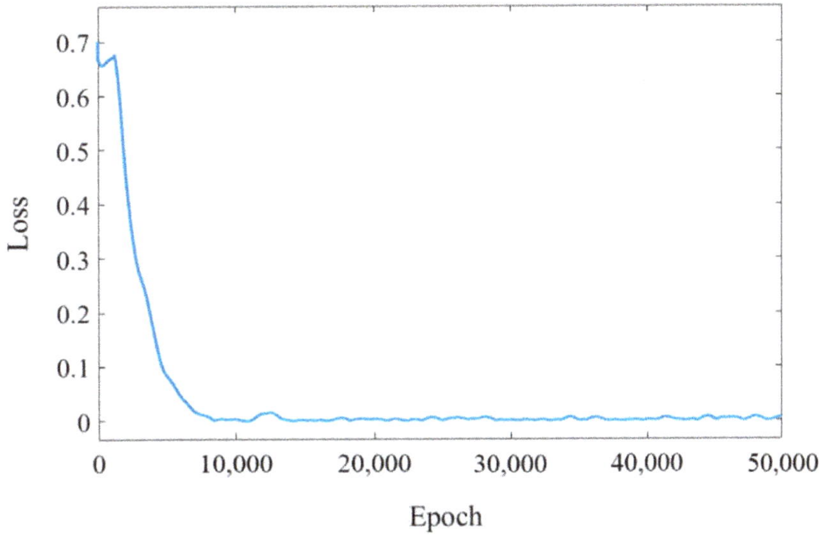

Figure 14. Training error curve of BP-PID-Smith controller under model mismatch.

Table 2. Comparison of the dynamic performance of the three controllers for the model mismatch.

Controller Type	Rise Time(s)	Peak Time(s)	Regulation Time (s)	Maximum Overshoot
Smith	232.228	449.377	170.216	0.28%
PID-Smith	96.494	141.383	84.043	0.53%
BP-PID-Smith	73.231	91.354	57.831	0.13%

According to the data in Table 2, even though the Smith controller had a smaller maximum overshoot of 0.28%, its response speed was the slowest, the PID-Smith controller had a great improvement in response speed and stability compared to the Smith controller but there was still a higher overshoot, and the BP-PID-Smith controller had a better performance than the other two controllers both in terms of response speed and stability. The performance of the BP-PID-Smith controller was better than the other two controllers in terms of response speed and stability, and it had a smaller overshoot. Therefore, the BP-PID-Smith controller can meet the control requirements in the actual regulation process.

5. pH Regulation Experiment

5.1. Construction of the Experimental Platform for pH Regulation

In this paper, a pH regulation platform was built according to the system structure to further test the practicality of the BP-PID-Smith control algorithm. The STM32F103ZET6 microcontroller was used as the control core to receive the signal from the pH sensor using the I/O port, and the microcontroller adjusted the output frequency of the inverter by changing the magnitude of the analog voltage, thus changing the flow rate at the hose pump of the regulating liquid tank. The delivery flow rate of the hose pump was 1 m^3/h, rated power was 1.5 kW, and rated voltage was 380 V. The inverter's rated power was 2.2 kW, the output frequency can be adjusted from 0 to 400 Hz, rated voltage was 380 V. A pH sensor of RMD-ISSF-5 was used, with an accuracy of 0.01. The volume of the fertilizer mixture was kept at 40 L during the experiment. The pH adjustment experiment platform is shown in Figure 15.

Figure 15. pH regulation experimental platform.

The regulation platform used USB-1252A to collect the data needed for the regulation process. The collector was equipped with an advanced measurement and control system with 16 analog input channels, 12-bit vertical resolution, and up to 500 kSa/s analog acquisition capability. The schematic diagram of the data acquisition and control system is shown in Figure 16.

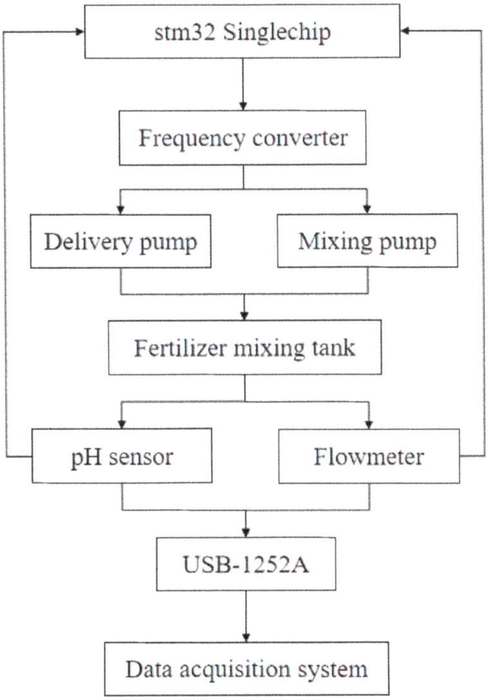

Figure 16. Schematic diagram of data acquisition and control system.

5.2. Analysis of Experimental Results

The fertilizer was diluted using dilute hydrochloric acid to maintain the pH of the fertilizer mixture at 7.5, and the flow rate of the hose pump at the fertilizer tank was adjusted to 0.35 m^3/h, 0.58 m^3/h, and 0.73 m^3/h. The performance of the three controllers was tested. The experimental results are shown in Figures 17–19, and the performance indexes of the three controllers are shown in Tables 3–5.

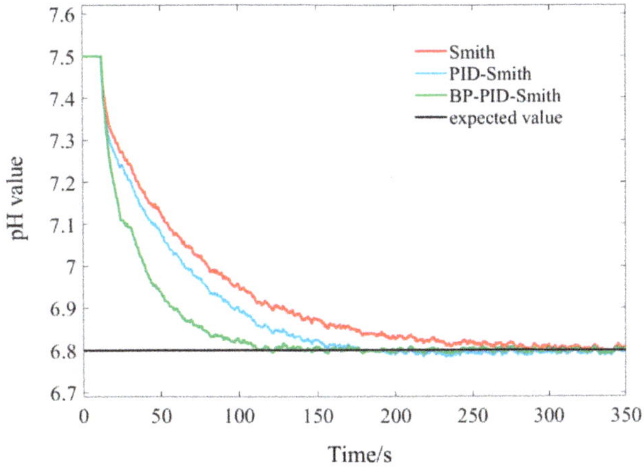

Figure 17. Regulation curves of three controllers for a fertilizer flow rate of 0.35 m^3/h.

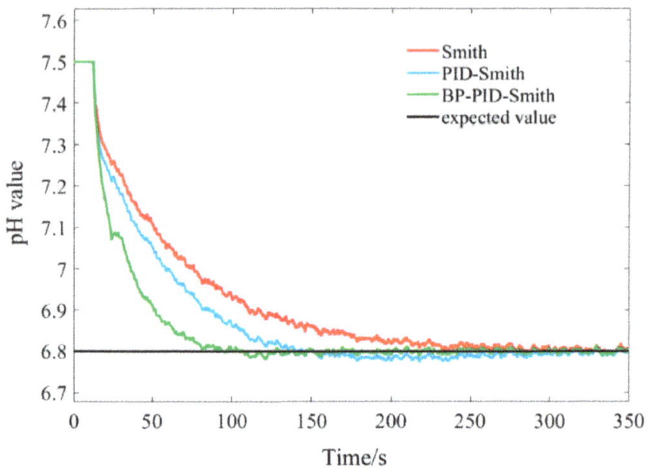

Figure 18. Regulation curves of three controllers for a fertilizer flow rate of 0.58 m^3/h.

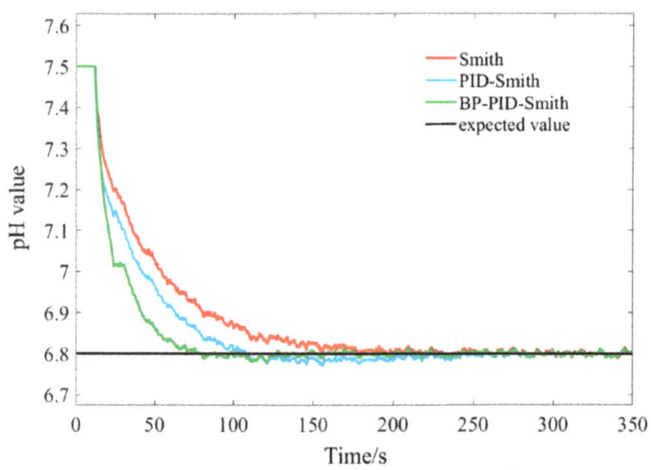

Figure 19. Regulation curves of the three controllers for a fertilizer flow rate of 0.73 m^3/h.

Table 3. Comparison of the dynamic performance of the three controllers at a fertilizer flow rate of 0.35 m^3/h.

Controller Type	Rise Time (s)	Peak Time (s)	Regulation Time (s)	Maximum Overshoot
Smith	285.395	333.145	178.644	0.06%
PID-Smith	176.957	218.060	131.623	0.25%
BP-PID-Smith	111.479	179.137	80.893	0.13%

Table 4. Comparison of the dynamic performance of the three controllers at a fertilizer flow rate of 0.58 m^3/h.

Controller Type	Rise Time (s)	Peak Time (s)	Regulation Time (s)	Maximum Overshoot
Smith	235.547	288.110	174.819	0.05%
PID-Smith	139.638	178.810	110.445	0.37%
BP-PID-Smith	94.403	121.180	70.695	0.31%

Table 5. Comparison of the dynamic performance of the three controllers at a fertilizer flow rate of 0.73 m^3/h.

Controller Type	Rise Time(s)	Peak Time(s)	Regulation Time (s)	Maximum Overshoot
Smith	178.436	235.560	111.320	0.02%
PID-Smith	107.892	156.370	80.897	0.46%
BP-PID-Smith	80.949	111.602	62.578	0.38%

From Tables 3–5, it can be found that the performance of all three controllers improved as the flow rate of fertilizer increased. Although the Smith controller had the smallest overshoot, its rise time was slow and it could not respond quickly to the input pH setting value; the PID-Smith controller was a significant improvement in terms of rise time, peak time, and adjustment time compared with the Smith controller, but with the increase of flow rate, the overshoot amount also increased gradually. The BP-PID-Smith controller had the advantages of the other two controllers: it can reach the set pH value in a short time and the overshoot was also reduced compared with the PID-Smith controller; therefore, it can meet the control demand well even in the case of high flow rate.

6. Conclusions

For the liquid fertilizer pH regulation system, this paper fitted its mathematical model, got the transfer function of the pH regulation system, combined BP neural network algorithm with the Smith prognosticator, designed a BP-PID-Smith prognostication compensation controller, and compared the performance of three controllers, BP-PID-Smith, PID-Smith, and Smith, in both simulation and practical application. The results showed that the BP-PID-Smith predictive compensation controller was able to bring the pH to the set value at a faster rate in both cases with a smaller overshoot compared to the other two controllers.

According to the experiments, the BP-PID-Smith controller showed excellent dynamic performance at different fertilizer application flow rates and was able to respond to the input signal at a faster rate and achieve the desired target pH value. The average maximum overshoot was 0.27% and the average regulation time was 71.39 s, which was significantly better than the PID-Smith and Smith controllers.

The BP-PID-Smith predictive compensator can reduce the adverse effects of the system in the fertilization process due to time lag and nonlinearity in practical applications while possessing excellent dynamic performance and robustness to meet the control requirements in practical applications.

Author Contributions: This study was conceptualized by Z.M. and L.Z. The software was designed by Z.M. and validated by H.L., R.Z. and R.M. Z.M. provided resources and Z.M. curated data. The original draft of the manuscript was prepared by Z.M., Y.S. and X.M. reviewed and edited the manuscript. H.B., R.Z. and R.M. assisted with project administration. X.M. and H.L. managed funding acquisition. All authors have read and agreed to the published version of the manuscript.

Funding: This research was funded by the National Natural Science Foundation of China, grant number 52065055.

Institutional Review Board Statement: Not applicable.

Informed Consent Statement: Not applicable.

Data Availability Statement: All relevant data presented in the article are stored according to institutional requirements and, as such, are not available on-line. However, all data used in this Manuscript can be made available upon request to the authors.

Conflicts of Interest: The authors declare no conflict of interest.

References

1. Kuzman, B.; Petković, B.; Denić, N.; Petković, D.; Ćirković, B.; Stojanović, J.; Milić, M. Estimation of optimal fertilizers for optimal crop yield by adaptive neuro fuzzy logic. *Rhizosphere* **2021**, *18*, 100358. [CrossRef]
2. Ajdary, K.; Singh, D.K.; Singh, A.K.; Khanna, M. Modelling of nitrogen leaching from experimental onion field under drip fertigation. *Agric. Water Manag.* **2007**, *89*, 15–28. [CrossRef]
3. Hermansson, A.W.; Syafiie, S. Model predictive control of pH neutralization processes: A review. *Control Eng. Pract.* **2015**, *45*, 98–109. [CrossRef]
4. Ali, E. pH Control Using PI Control Algorithms with Automatic Tuning Method. *Chem. Eng. Res. Des.* **2001**, *79*, 611–620. [CrossRef]
5. Balasubramanian, G.; Hariprasad, K.; Sivakumaran, N.; Radhakrishnan, T. Adaptive Control of Neutralization Process Using Recurrent Neural Networks. *Instrum. Sci. Technol.* **2009**, *37*, 383–396. [CrossRef]
6. Sena, H.J.; Silva, F.V.D.; Fileti, A.M.F. ANN model adaptation algorithm based on extended Kalman filter applied to pH control using MPC. *J. Process Control* **2021**, *102*, 15–23. [CrossRef]
7. Alves Goulart, D.; Dutra Pereira, R. Autonomous pH control by reinforcement learning for electroplating industry wastewater. *Comput. Chem. Eng.* **2020**, *140*, 106909. [CrossRef]
8. Salehi, S.; Shahrokhi, M.; Nejati, A. Adaptive nonlinear control of pH neutralization processes using fuzzy approximators. *Control Eng. Pract.* **2009**, *17*, 1329–1337. [CrossRef]
9. Wu, H.; Yan, F.; Wang, G.; Lv, C. A predictive control based on decentralized fuzzy inference for a pH neutralization process. *J. Process Control* **2022**, *110*, 76–83. [CrossRef]
10. Chen, H.; Zouaoui, Z.; Chen, Z. Neuro-Fuzzy Modified Smith predictor for IPDT and FOPDT Processes Control. *IFAC Proc. Vol.* **2013**, *46*, 839–844. [CrossRef]
11. Chen, G.D.; Liu, D.J.; Mu, Y.X.; Xu, J.; Cheng, Y. A Novel Smith Predictive Linear Active Disturbance Rejection Control Strategy for the First-Order Time-Delay Inertial System. *Math. Probl. Eng.* **2021**, *2021*, 5560123. [CrossRef]
12. Gamal, M.; Sadek, N.; Rizk, M.R.; Abou-Elsaoud, A.K. Delay compensation using Smith predictor for wireless network control system. *Alex. Eng. J.* **2016**, *55*, 1421–1428. [CrossRef]
13. Qi, C.; Gao, F.; Zhao, X.; Wang, Q.; Ren, A. Hybrid Smith predictor and phase lead based divergence compensation for hardware-in-the-loop contact simulation with measurement delay. *Acta Astronaut.* **2018**, *147*, 175–182. [CrossRef]
14. Nie, Y.; Zhang, P.; Cai, G.; Zhao, Y.; Xu, M. Unified Smith predictor compensation and optimal damping control for time-delay power system. *Int. J. Electr. Power Energy Syst.* **2020**, *117*, 105670. [CrossRef]
15. Jin, X.; Chen, K.; Zhao, Y.; Ji, J.; Jing, P. Simulation of hydraulic transplanting robot control system based on fuzzy PID controller. *Measurement* **2020**, *164*, 108023. [CrossRef]
16. Wright, R.A.; Kravaris, C. Nonlinear control of pH processes using the strong acid equivalent. *Ind. Eng. Chem. Res.* **1991**, *30*, 1561–1572. [CrossRef]
17. Heckenthaler, T.; Engell, S. Fuzzy Logic Controller Design for Ph-Control in a CSTR. *IFAC Proc. Vol.* **1995**, *28*, 27–32. [CrossRef]
18. Sira-Ramírez, H.; Verónica Pernía-Espinoza, A. On the pH Control of a CSTR System: An Invariant Stabilization Approach. *IFAC Proc. Vol.* **2001**, *34*, 1337–1342. [CrossRef]
19. Tyagi, R.; Davison, E.J. Control of pH in a Continuous Stirred Tank Reactor (CSTR). *IFAC Proc. Vol.* **1993**, *26*, 115–126. [CrossRef]
20. Estofanero, L.; Edwin, R.; Claudio, G. Predictive Controller Applied to a pH Neutralization Process. *IFAC-PapersOnLine* **2019**, *52*, 202–206. [CrossRef]
21. Bu, Q.; Cai, J.; Liu, Y.; Cao, M.; Dong, L.; Ruan, R.; Mao, H. The effect of fuzzy PID temperature control on thermal behavior analysis and kinetics study of biomass microwave pyrolysis. *J. Anal. Appl. Pyrolysis* **2021**, *158*, 105176. [CrossRef]
22. Mejía, C.; Salazar, E.; Camacho, O. A comparative experimental evaluation of various Smith predictor approaches for a thermal process with large dead time. *Alex. Eng. J.* **2022**, *61*, 9377–9394. [CrossRef]

applied sciences

MDPI

Article

Smart Irrigation System Considering Optimal Energy Management Based on Model Predictive Control (MPC)

Wilmer Quimbita, Edison Toapaxi and Jacqueline Llanos *

Electrical and Electronic Department, Universidad de las Fuerzas Armadas ESPE, Sangolquí 171103, Ecuador; wequimbita@espe.edu.ec (W.Q.); ertoapaxi@espe.edu.ec (E.T.)
* Correspondence: jdllanos1@espe.edu.ec; Tel.: +593-3-2810-206

Abstract: Traditional irrigation techniques usually cause the wasting of water resources. In addition, crops that are located in rural areas require water pumps that are powered by environmentally unfriendly fossil fuels. This research proposes a smart irrigation system energized by a microgrid. The proposal includes two stages: the first generates the daily irrigation profile based on an expert system for the adequate use of the water. Then, considering the irrigation profile, the power required for the water pump is measured—the optimal daily profile of electricity demand is determined in the second stage. The energy system is a microgrid composed of solar energy, a battery energy storage system (BESS) and a diesel generator. The microgrid is managed by an energy management system (EMS) that is based on model predictive control (MPC). The system selects the optimal start-up time of the water pump considering the technical aspects of irrigation and of the microgrid. The proposed methodology is validated by a simulation with real data from an alfalfa crop in an area of Ecuador. The results show that the smart irrigation proposed considers technical aspects that benefit the growth of the crops being studied and also avoids the waste of water.

Keywords: model predictive control; energy management system; renewable energy; smart irrigation; agriculture 4.0

Citation: Quimbita, W.; Toapaxi, E.; Llanos, J. Smart Irrigation System Considering Optimal Energy Management Based on Model Predictive Control (MPC). *Appl. Sci.* **2022**, *12*, 4235. https://doi.org/10.3390/app12094235

Academic Editors: Paweł Kiełbasa, Tadeusz Juliszewski and Sławomir Kurpaska

Received: 18 March 2022
Accepted: 12 April 2022
Published: 22 April 2022

1. Introduction

Agriculture in the world plays an important role in feeding the population. Agriculture has recently been threatened by the availability of water and has generated concern about the use of water in the world, leaving much to be desired in agriculture as it has generated an unsustainable situation [1] For this reason, smart irrigation systems were created to optimize water use in crops and improve productivity [2].

The development of the 21st century is still largely based on agriculture with activities in agribusiness, taking into consideration climate change, soil and irrigation factors in almost all regions, for this, a study [3] was conducted using artificial neural networks (ANN) that analyzes the smart agriculture dataset with parameters such as temperature, soil moisture, wind speed, solar radiation and soil water tension.

Agriculture has played an important role in the development of every country, therefore, automation has been increased the area of irrigation with smart technology in some cases through embedded systems that concentrates on controlling the irrigation process automatically using a Raspberry Pi device through Python programming with the help of moisture sensors that monitor the soil [4]. The level of irrigation depends on the soil moisture content and crop type, which reduces the overall energy consumption and optimizes the use of water reserves, a low-cost Arduino-based smart irrigation solution [5].

The advancement in automation techniques allows improving crop yields, making them more profitable and reducing irrigation wastage by having sensors deployed in an agricultural field, which sends data through a microprocessor via Internet of Things (IoT) devices with the cloud, through a decision tree algorithm, which sends an alert by mail to farmers and allows the decisions on water supply to be made in advance [6]. Automation through irrigation scheduling has been achieved by several methods, such as the measurement of environmental variables used for control, simulation of irrigation strategies that allows forecasting conditions for crops without damaging the crops and then being applied in the field [7].

Automated irrigation is based on parameters monitored from a meteorological station that indicates parameters such as crop water demand, soil, evapotranspiration, precipitation, climatic conditions, etc., which are used to analyze the data in various scenarios and to propose strategies and control methods where irrigation efficiency analysis can be applied [8]. Through simulation, the performance of proposals can be validated, so [9] shows a control algorithm with a database that forecasts crop yield and manages water resources, closed-loop control is performed and through feedback from weather conditions the model anticipates the crop water demand and to generate the irrigation scheduling of water consumption [10].

Although water resources are available in many countries, global warming, pollution and losses due to the misapplication of water are considerable and inappropriate types of risks have been generated, affecting the efficient use of water [11]. Agricultural irrigation is mostly done manually/mechanically without considering the amount of water needed for the crop. With the help of technology, it is possible to create smart irrigation systems that focus on saving water and energy and obtaining a higher quality in the final product [12].

Different projects have been carried out in different sectors of smart irrigation systems with fuzzy logic based on rules, by sensing soil moisture and meteorological variables and managing the flow of water through an agronomic design [12]. The smart irrigation design in vegetables implemented in a university is managed through a graphical interface, allowing the automation of irrigation using fuzzy logic compared to manual drip irrigation [13]. In rural areas, irrigation control is performed by comparing environmental parameters with sensors located along the crop, electrovalves as actuators that control the water flow and thus obtain greater efficiency and less water waste [14]. Although these studies have contributed to the improvement of irrigation, they do not consider the future state of irrigation variables, so their decisions are not the most optimal.

Optimal irrigation has not been fully addressed with smart tools, such as model-based predictive control (MPC), which allow the improvement of agricultural production at low costs by taking advantage of optimally managed natural resources and analyzing the technical feasibility of future events [15].

Agriculture is one of the main energy demanders in society through the extraction of water for irrigation [16], and it should be taken into account that crops interact with solar radiation, which means that the higher the radiation, the greater the water needs of the crop. A viable solution is solar pumping in off-grid systems to supply water to different communities [17,18].

On the other hand, irrigation systems require energy for their operation and the most common is that irrigations are located in areas far from the power grid and require diesel-based generation units, in few cases based on renewable resources [19]; however, microgrids can be an alternative in order to supply the energy to this system in an environmentally friendly way.

In developing countries, in rural areas that do not have access to electricity supply, agricultural production is affected. Therefore, the combination of a water management systems for irrigation water requirements with an energy management system considering climatic conditions optimizes the use of energy and water in agricultural production [20].

Microgrids are useful for supplying electricity demand in isolated locations, considering the characteristics of the loads and managing energy [21]. The surplus energy after supplying the local demand can be stored in batteries to reduce the consumption of energy based on fossil sources [22].

Work has been done on proposals for smart irrigation; however, this is not enough—techniques that also use water optimally are required, supporting the reduction of water resources and minimizing the effect of global warming. In addition, including energy sources to supply the demand of irrigation systems based on natural resources and managing them properly contributes to the reduction of carbon footprints.

Currently, there are problems in food production in the agricultural sector due to the increase in the world population growth [1,23] as well as pollution problems due to the use of pumps based on fossil resources in isolated areas, far the electricity grid. These aspects have increased the costs of agricultural production [24]. In order to reduce these problems, some studies have researched how to meet the demand for water by reducing energy consumption [25,26]. Thus, it is necessary to create strategies to save water resources and energy and minimize pollution. In this context, the main aim of this article is to propose a smart irrigation system considering optimal energy management based on model predictive control (MPC), which achieves efficient cultivation with minimum water consumption at the same time that the electrical demand is supplied with a microgrid based on renewable resources, which is optimally managed.

This research proposes a system that allows smart irrigation and, at the same time, achieves an optimal management of water resources. In addition, optimal energy management is achieved by improving agricultural production while respecting technical conditions and guaranteeing clean, environmentally friendly and low-cost energy. The proposal considers two stages, the first stage corresponds to the smart irrigation through the water balance, which provides the knowledge of the water needs in the root zone of the crop, thus generating the optimal irrigation profile, considering climatic variables. After understanding the irrigation needs, the necessary electrical demand is dimensioned, which is considered in the second stage. The second stage corresponds to the optimal energy management system of the microgrid that is responsible for managing the storage system, the diesel generator, to adequately turn on the pumps, with previous knowledge of the daily solar power forecast, giving priority to the generation of energy through natural resources in order to supply the necessary irrigation to the crop, modifying the water balance and closing the control loop.

2. Irrigation System Modeling

For the use of water resources in an irrigation area, the water consumption of the crop in its different phenological stages must be known, so that the water supply allows the plants to replenish the water lost by crop evapotranspiration. Soil moisture is important for crop yield, which depends largely on irrigation to maintain the extreme levels necessary to prevent soil saturation and water stress in the plant [27].

The units in the different variables of the irrigation system are represented in mm, which expresses the water lamina in the soil, the functionof the depth of the root of the plant, the most frequent units of the water lamina are: mm, $\mathrm{ltr/m^2}$, and $\mathrm{m^3/ha}$, and its equivalence is: $1\ \mathrm{mm} = 10\ \mathrm{m^3/ha} = 1\ \mathrm{ltr/m^2}$ $1\ \mathrm{mm} = 10\ \mathrm{m^3/ha} = 1\ \mathrm{ltr/m^2}$ [28].

2.1. Water Balance Modeling

The water balance estimates the water content in the root zone at a given time, water increases in the presence of rain, irrigation and capillary rise (see inputs in Figure 1), while the amount of water in the soil decreases due to evapotranspiration, runoff and deep percolation (see outputs in Figure 1). The AD represents the available water between the field capacity and the permanent wilting point. Within these limits there is a threshold which allows the plant to have readily available water (*RAW*) in the root zone [10].

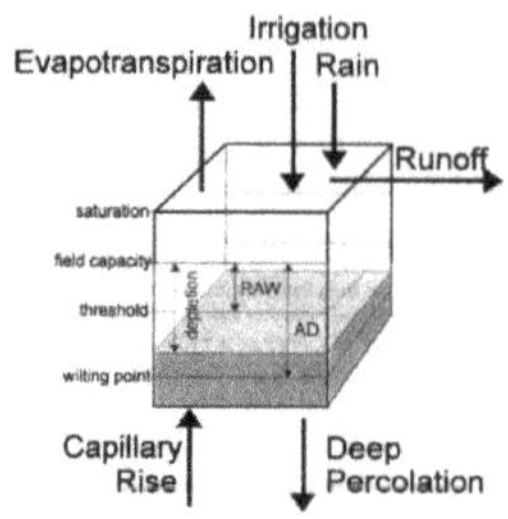

Figure 1. Water balance [29].

To predict the water content in the soil, the daily moisture balance for the root zone of the soil is needed. The dynamics of water content in the root zone is described by Equation (1), which represents the water balance [10,27].

$$H(t+1) = H(t) + P(t) + R(t) - ET_c(t) - PP(t) \quad (1)$$

where $H(t+1)$ is the water lamina in the root zone at the end of the day in mm, $H(t)$ is the water content in the root zone at the beginning of the day in mm, $P(t)$ is the precipitation occurring in the day in mm, $R(t)$ is the irrigation applied in mm, $ET_c(t)$ is the crop evapotranspiration in mm, and $PP(t)$ is deep percolation in mm.

In the root zone, root growth, root depth, crop evapotranspiration estimation, rainfall and irrigation must be taken into account to determine the moisture in the soil, which is limited by field capacity and permanent wilting point [27].

2.2. Field Capacity and Permanent Wilt Point

The amount of water retained in well-drained soil is the field capacity (Cc) and the permanent wilting point (Pmp). When the crop is unable to extract water, it will have reached the permanent wilting point [28]. To estimate soil moisture sheets for root depth, Equations (2) and (3) should be taken into account [27].

$$L_{Cc} = Cc \cdot D_a \cdot Z_r \quad (2)$$

$$L_{Pmp} = Pmp \cdot D_a \cdot Z_r \quad (3)$$

where L_{Cc} is the stored sheet at field capacity in mm, Cc is the moisture content at field capacity (%), L_{PMP} is the stored sheet at permanent wilting point in mm, Pmp is the moisture content at permanent wilting point (%) and Z_r is the depth of the root zone.

Available water (AD) or useful water is the amount of usable soil water for the plant and is between the field capacity limits (Cc) and the permanent wilting point (Pmp), located at different depths of the irrigated soil. To determine the available water sheet see Equation (4) [28].

$$L_{AD} = (Cc - Pmp) \cdot D_a \cdot Z_r \quad (4)$$

where L_{AD} is the available water lamina in mm and D_a is the bulk density of the soil.

The plant absorbs water from the soil without effort when the water is at field capacity and its yield is maximum, so it should not be expected that the available water is close to the permanent wilting point, for which the threshold p is taken into account, which depends on the type of crop and allows the water to be readily available water (RAW) for the crop, ensuring optimum plant performance [28]. If the water content in the soil is below the permanent wilting point, the crop will begin to suffer water stress [29]. The readily available water table (L_{AFD}) is obtained with Equation (5).

$$L_{AFD} = (Cc - Pmp) \cdot D_a \cdot Z_r \cdot p \quad (5)$$

where L_{AFD} is the readily available water sheet in mm and p is the irrigation threshold.

Regarding the precipitation (P) produced in a day, if the rainfall is less than 0.2 mm it is assumed that it will evaporate in its entirety. For an optimal irrigation scheduling, the precipitation that may occur should be considered, as well as the possible lamina contributed by such precipitation [28,29].

Deep percolation (PP) appears if in irrigation the water applied to crops is greater than required, according to the field capacity of the soil, and this water moves beyond the roots, causing water loss by deep percolation [27].

2.3. Crop Evapotranspiration

Crop evapotranspiration (ET_c) is the combination of two processes, when water is lost from the soil by evaporation and crop transpiration. Evaporation is when liquid water is converted to water vapor and removed from the surface by evaporating. Transpiration is the vaporization of liquid water contained in plant tissues into the atmosphere [29].

Factors affecting evapotranspiration are subject to climate conditions and crop characteristics. According to the crop coefficient approach the crop evapotranspiration is determined by the product of the reference crop evapotranspiration, ET_o and the crop coefficient K_c, as shown in Equation (6) [29].

$$ET_c = K_c \cdot ET_o \tag{6}$$

where ET_c is the crop evapotranspiration in mm per day, K_c is the dimensionless single crop coefficient and ET_o is the reference crop evapotranspiration in mm per day.

Figure 2 represents, in a general way, the curve of the crop coefficient K_c, according to the development stage of a crop, where the evolution of the plant from the initial growth stage, is the crop coefficient $K_{c\ ini}$, the $K_{c\ med}$ is the crop coefficient during the middle stage of crop growth and $K_{c\ fin}$ is the crop coefficient at the final stage [29].

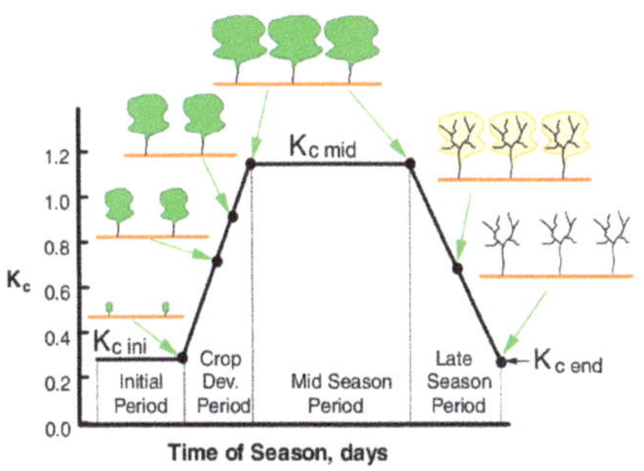

Figure 2. Generic curve of the crop coefficient (K_c).

There are different approaches given by the Food and Agriculture Organization of the United Nations (FAO) that allow the calculation of reference evapotranspiration by means of equations that use meteorological variables. As a standard method, the empirical method of Equation (7), corresponding to the FAO Penman–Monteith equation for the calculation of ET_o, is recommended [27,29].

$$ET_o(t) = \frac{0.408\Delta(t)\cdot(R_n(t) - G(t)) + \gamma\frac{900}{T(t)+273}u_2(t)\cdot(e_s(t) - e_a(t))}{\Delta(t) + \gamma(1 + 0.34u_2(t))} \tag{7}$$

where ET_o is the reference evapotranspiration of mm per day, R_n is the net radiation at the crop surface $[\mathrm{MJ/m^2/day}]$, G is the heat flux density $[\mathrm{MJ/m^2/day}]$, T is the daily average air temperature at 2 m height [°C], u_2 is the wind speed at 2 m height [m/s], e_s is the saturation vapor pressure [KPa], e_a is the actual real vapor pressure in [KPa], Δ is the temperature vapor saturation slope [KPa/°C] and γ psychrometric constant [KPa/°C].

3. Smart Irrigation System Considering Optimal Energy Management with Predictive Control

The proposed smart irrigation system considering optimal energy management with model predictive control MPC is shown in Figure 3. It contains two stages. The initial stage focuses on obtaining the daily irrigation demand profile in a smart way through an expert system. This system is programmed to automatically obtain the daily moisture by means of the water balance, i.e., the water gain and loss in the root zone of the crop by means of Equation (1), for which the forecasted data of one day of precipitation $(\hat{P})$ are needed as water input to the system. Similarly, it is necessary to know the water losses that occur in the crop by evapotranspiration. Using Equation (6), the evapotranspiration of the crop (ET_c) is determined and by means of meteorological data the referenced evapotranspiration is obtained with Equation (7). The energy demand is the result of the irrigation time for the daily turning on of the pumps throughout the day, this demand profile obtained is transmitted to the next stage which is called energy manager system (EMS) of the microgrid for the irrigation system; in this stage an optimization problem is solved and is responsible for the efficient management of the energy system to determine the energy needed to turn on the water pump, taking advantage of renewable resources for power generation. In this second stage (EMS), the availability of energy from the hybrid energy system formed of photovoltaic panels, storage systems and a diesel generator is guaranteed.

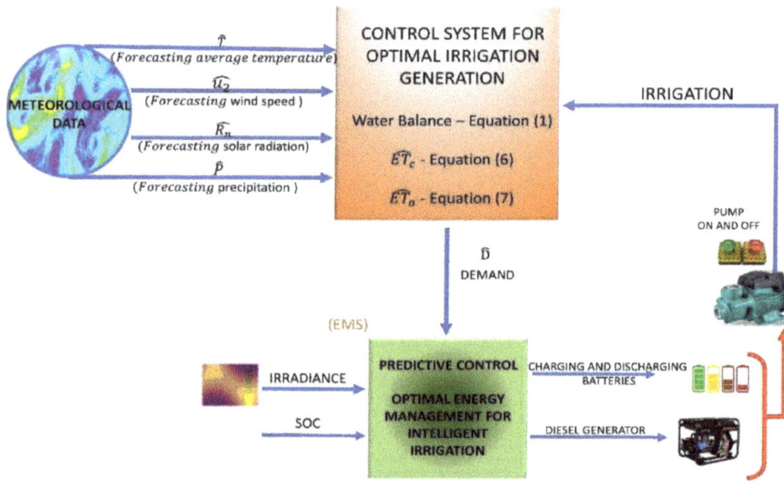

Figure 3. Proposal for a smart irrigation system, considering optimal EMS.

3.1. *Expert System for the Generation of Optimal Irrigation—(First Stage)*

In order to guarantee optimum irrigation, we start with the study of the type of soil and the knowledge of its characteristics that help to maintain sufficient moisture for the crop without exceeding the limits of the field capacity and permanent wilting point. The analysis of the water balance is analyzed daily throughout the crop, the limits depend on the depth of the roots and how much water was contributed to or removed from the crop.

The upper part of Figure 3 shows the forecasting data needed to determine the optimum irrigation (average temperature $\hat{T}$, wind speed $\hat{u}_2$, Solar radiation $\hat{R}_n$ and precipitation $\hat{P}$), irrigation is automatically determined by an automatic expert system, which executes the following steps: (i) meteorological data and prediction models with a control horizon of one day during the cultivation time are obtained to determine the reference evapotranspiration (ET_o) with Equation (7); (ii) the expert system determines the crop evapotranspiration (ET_c) in one day with Equation (6), which depends on the value of the crop coefficient (Kc) in its different growth stages, determining the amount of water consumed or removed in the root zone; (iii) by means of the water balance of Equation (1), the soil moisture at the end of the day is known, thus determining the need for irrigation by means of the input of forecasts involved in the water balance.

In the automatic expert system, the lamina at the field capacity (L_{Cc}) of the plants is determined by Equation (2), which depends on the depth of the crop root (Z_r) and the bulk density of the soil (D_a). The lamina at field capacity is the maximum limit of water required in the crop and at this level water is optimally absorbed. The lower limit is lamina at permanent wilting point (L_{Pmp}); at this point the amount of water adheres only to the soil particles, which makes it impossible for the plant to obtain water and would generate water stress; therefore, to prevent the crop from reaching this point, the water content is determined at a suitable threshold (Pw), which results from the sum of the value of L_{Pmp} with L_{AFD}, thus preventing the water content at the root from reaching L_{Pmp}.

The irrigation required each day occurs when $H \leq Pw$, i.e., when the initial soil moisture H is less than or equal to (Pw), it is necessary to provide water to the crop, and the system has the capacity to automatically determine the necessary amount of irrigation lamina (R) by means of Equation (8), where the irrigation lamina (R) must be less than the field capacity lamina $R \leq L_{Cc}$, and the forecasted precipitation $(\hat{P})$ is also taken into account to determine the amount of effective irrigation lamina (R), so that the crop does not suffer soil saturation.

$$R(t) = L_{Cc} - H(t) - P(t+1) \tag{8}$$

Once the water requirement of the crop is determined, the gross or total irrigation lamina is determined to ensure sufficient penetration of water in the root zone, by Equation (9), since not all the water will be consumed by the plant, it is, therefore, necessary to apply a greater amount of water in irrigation with respect to the efficiency of the irrigation system [28].

$$L_b = \frac{R}{Ef} \tag{9}$$

In Equation (9), L_b is the gross lamina in mm, R is the required irrigation lamina in mm and Ef is the efficiency of the irrigation system.

To determine the all-day demand, the switch-on time of the pump is obtained by Equation (10) [30].

$$t_r = \frac{L_b}{Pms} \tag{10}$$

In Equation (10), t_r is the irrigation time in hours and Pms is the average rainfall of the system.

According to the irrigation time, the necessary electrical demand for the day is obtained by multiplying the irrigation time t_r with the pump power, thus sending the forecasting or necessary demand for the whole day to the next stage, which is in charge of supplying electric power with the optimal and efficient management of the pump start-up throughout the whole day.

3.2. Microgrid Controlled by Energy Management System (EMS) for Irrigation System (Second Stage)

The energy consumption to achieve smart irrigation can be supplied by a microgrid, which requires optimal management. When different isolated energy sources can be managed to supply energy in a coordinated and reliable way, the microgrid is composed by a diesel generator, solar panels and a battery energy storage system (BESS).

The second stage consists of the integration of smart irrigation with optimal energy management for the proposed microgrid. The operation of pumps are driven by generation units such as batteries, photovoltaic power or diesel generator.

The control is constituted by an optimization problem that, by means of the objective function of Equation (11), minimizes the diesel generation cost ($Cost_{GD}$), cost for energy not supplied ($Cost_{NS}$) and cost for solar discharge ($Cost_{VS}$) in order to take advantage of the solar energy and always preserve the useful life of the batteries. The total costs are obtained by multiplying the cost of generation by technology by the power generated by each unit as the diesel generator power (P_{GD}), which is presented when the optimizer detects that the available energy from the batteries or solar generation is not able to supply energy to the system. The power not supplied (P_{NS}) occurs when the three energy sources present in the system are not able to meet the required demand. The solar curtailment power (P_{VS}) occurs when the solar power is greater than the electrical demand.

$$min \sum_{t=1}^{T} (Cost_{GD} \cdot P_{GD}(t) + Cost_{NS} \cdot P_{NS}(t) + Cost_{VS} \cdot P_{VS}(t)) \tag{11}$$

The optimization problem is subject to equality and inequality constraints, referring to the generation units involved in the energy system.

The energy system consists of a battery energy storage system (BESS), which operates in discharge mode (P_{BD}) when it provides energy to the system and in charge mode (P_{BC}) when the batteries are recharged, being part of the energy consumption.

The first equality constraint is the energy balance, which is represented in Equation (12). The energy system needs to satisfy the energy balance, where the sum of the energy generated by the solar panels (P_{GS}), the diesel generator (P_{GD}) and the battery power in discharge mode (P_{BD}), is equal to the sum of the electrical demand of electronics devices (D), electrical demand to water pump (D_B), the solar dump power (P_{VS}), the unsupplied power and the battery power in charge mode (P_{BC}).

$$P_{GD}(t) + P_{GS}(t) + P_{BD}(t) = D(t) + D_B(t) + P_{VS}(t) - P_{NS}(t) - P_{BC}(t) \tag{12}$$

The diesel generator has min power, $P_{GD_{min}}$, and max power, $P_{GD_{max}}$, operating limits, which are multiplied by a binary variable (X_{GD}). The binary variable decides whether to start the diesel generator or not and the diesel generator limit constraints are expressed in Equation (13).

$$P_{GD_{min}}(t) \cdot X_{GD}(t) \leq P_{GD}(t) \leq P_{GD_{max}}(t) \cdot X_{GD}(t) \tag{13}$$

The BESS model considers two operating modes: loading or unloading. For each mode, an equal loading and unloading efficiency coefficient ($n_{BD} = n_{Bc}$) is considered. Equation (14) gives the available energy of the BESS at the first instant $E(t = 1)$, which is equal to the initially available energy (E_0). The available energy of the BESS can start at a certain percentage, then the energy contributed by the BESS when in charging mode is subtracted and finally the energy of the BESS when in discharging mode is added, which is shown in Equation (14).

$$E(t = 1) = E_0 - n_{BC} \cdot P_{BC}(t = 1) - \frac{1}{n_{BD}} P_{BD}(t = 1) \tag{14}$$

Equation (15) represents the point when the energy of the BESS starts to respond dynamically after the initial condition.

$$E(t) = E(t-1) - n_{BC} \cdot P_{BC}(t) - \frac{1}{n_{BD}} P_{BD}(t) \tag{15}$$

The available battery energy is less than or equal to the installed battery power (P_B_inst) represented in Equation (16), the model requires the state of charge (SOC), which estimates the amount of battery energy represented in Equation (17), and through the BESS usage policies limits the charging and discharging of the BESS, thus lengthening the BESS lifespan expressed in Equations (18) and (19).

$$E(t) \leq P\_B\_inst(t) \tag{16}$$

$$SOC(t) = \frac{E(t)}{P\_B\_inst(t)} \tag{17}$$

$$SOC(t) \geq 0.1 \tag{18}$$

$$SOC(t) \leq 0.9 \tag{19}$$

The binary variables or optimization variables X_{BC} and X_{BD} take the value of zero or one to define the state of the BESS, so as to not turn on the BESS to charging or discharging mode at the same time, as defined by Equation (20).

$$X_{BC}(t) + X_{BD}(t) \leq 1 \tag{20}$$

The constraints represented by Equations (21) and (22) estimate the limit of the BESS power in charging and discharging mode.

$$0 \geq P_{BC}(t) \geq -P_{B_{inst}}(t) \cdot X_{BC}(t) \tag{21}$$

$$0 \leq P_{BD}(t) \leq P_{B_{inst}}(t) \cdot X_{BD}(t) \tag{22}$$

The model solar generator power (P_{GS}) is estimated by panel efficiency, inverter efficiency (n_{inst}) with panel area A_{panel}, number of panels $\left(num_{panel}\right)$ and irradiance (Ir), expressed by Equation (23).

$$P_{GS}(t) = n_{panel} \cdot n_{inst} \cdot A_{panel} \cdot num_{panel} * Ir(t) \tag{23}$$

It must be ensured that solar dumping power (P_{VS}) is less than solar generation power (P_{GS}), as represented by Equation (24). This ensures that maximum utilization of the solar resource is achieved.

$$P_{VS}(t) \leq P_{GS}(t) \tag{24}$$

4. Results

This section presents the results obtained by the application of the smart irrigation system, considering the proposed optimal energy management with predictive control. For the irrigation modeling, the meteorological data forecasts relevant weather for crops on daily basis, the acquired data being temperature, solar radiation, wind speed, relative soil moisture and precipitation. These records are obtained from the 'POWER Single Point Data Access' (https://power.larc.nasa.gov/data-access-viewer/) accessed on 11 August 2022, which provides solar and meteorological datasets from NASA research to support renewable energy, building energy efficiency and agricultural needs. The obtained climatological data were recorded in an hourly computational database from July 1 through August 29, totaling 60 days, which corresponds to a full growing period of the alfalfa crop.

To determine the reference evapotranspiration, a database was programmed and by means of the FAO Penman–Monteith equation (Equation (7)) the daily reference evapotranspiration (ET_o) was determined in the Excel add-in HF Irrigation.

The solar irradiance data were obtained from 'POWER Single Point Data Access' records, the irradiance data are used for the EMS with a daily prediction horizon totaling 24 samples per hour, which allows the optimization of energy management using the FICO® Xpress Workbench software, as it has the ability to solve optimization problems.

4.1. Case Study

The present research was carried out by the sizing of a plot of 1173 m^2 in Ecuador, located in the province of Cotopaxi, in the canton Latacunga, parish Ignacio Flores, with a latitude of −0.935061 and longitude of −78.603145, where the terrain is sandy loam and the reference crop is alfalfa [19].

The crop in this research is alfalfa and, according to FAO-56 [26], alfalfa development time is 60 days, distributed over 10 days in the initial stage, 20 days of development, 2 days in the middle stage and 10 days the final stage, with its corresponding crop coefficient of $K_{c\ ini} = 0.4$, $K_{c\ med} = 1.2$ and $K_{c\ fin} = 1.15$.

The soil properties are represented in Table 1, which indicate the values used in the development of the research.

Table 1. Soil properties.

Properties	Value
Field capacity (Cc)	0.09 (m^3/m^3)
Permanent wilting point (Pmp)	0.04 (m^3/m^3)
Soil bulk density (D_a)	1.3 (gr/m^3)

The required irrigation lamina or net irrigation lamina represents the amount that can be easily extracted by the plants, according to Equation (5). Where the average fraction of the total water available in the soil is the factor that indicates the sensitivity of the crop to a threshold, whether it is very delicate or of high economic value, such as vegetables or flowers, this factor adopts a value between 0.3 and 0.4 (30–40%) [30].

The irrigation system consists of 12 sprinklers with a spacing between sprinklers (E_{lat}, E_{asp}) of 10 m, a pressure of 20 PSI and a flow rate of 1.1 GPM. With Equation (25) the average rainfall of the system of 2.5 mm/h is obtained [19].

$$Pms = \frac{227.12 \cdot Q}{E_{lat} \cdot E_{asp}} \tag{25}$$

$$Pms = \frac{227.12 \cdot (1.1)}{(10) \cdot (10)}$$

$$Pms = 2.5 \text{ mm/h}$$

The energy consumption of each piece of equipment that makes up the smart irrigation system is detailed in Table 2:

Table 2. Energy consumption of the smart irrigation system.

Equipment	Power Consumption (W)
Pump	750
PLC S7-1200	11
I/O module	11
24 W power supply	48
Total	820

Table 3, below, details the technical data of the hybrid energy system of the sources available for energy management.

Table 3. Technical power generation data.

Equipment	Power W	Efficiency
Installed Solar Power	1095	18.82%
BESS	1920	90%
Diesel generator	2500	

4.2. *Irrigation Technique to Evaluate and Compare*

Three irrigation techniques were evaluated and compared: (i) traditional irrigation, (ii) technified irrigation, and (iii) proposed smart irrigation.

4.2.1. Traditional Irrigation

Traditional or empirical irrigation is an intuitive irrigation, which applies water resources with a pump and by means of a sprinkler system, without considering technical parameters. Using this irrigation method, the pump is turned on for one hour daily (at night), throughout the whole growing period of 60 days of the alfalfa crop.

4.2.2. Technified Irrigation

The type of technified irrigation is based on an agronomic design, the water requirements of the crop are determined by calculating the irrigation calendar, taking into account the duration of the 60-day crop, corresponding to the alfalfa growing period from July 1 to August 29. By means of the meteorological data 'POWER Single Point Data Access' of the NASA and the extension of Excel HF irrigation, the reference evapotranspiration in the month of July $ET_o = 2.336$ mm/day and August $ET_o = 2.5609$ mm/day is obtained.

The crop coefficient K_c of alfalfa, with the duration of each stage and root depth, is shown in Table 4.

Table 4. Crop Parameters.

Crop: Alfalfa			
	K_c	Days	Z (mm)
Initial	0.4	10	100
Development	1.05	20	275
Intermediate	1.2	20	800
Final	1.15	10	800

Equation (26) determines the frequency of technical irrigation that results from the division of the lamina of readily available water from the crop evapotranspiration ET_c.

$$Fr = \frac{L_{RAW}}{ET_c} \tag{26}$$

Table 5 determines the irrigation schedule according to the irrigation needs, taking into account the reference evapotranspiration ET_o; crop coefficient K_c; crop evapotranspiration ET_c, according to Equation (6); irrigation frequency from Equation (25); irrigation numbers; gross lamina from Equation (8) and irrigation time (tr).

Table 5. Technified Irrigation Schedule.

Date	Stage	ET_o (mm/day)	K_c	ET_c (mm/day)	L_{RAW} (mm)	Fr (days)	N° Irrigation	L_b (mm/ Irrigation)	tr (h)
1 July 10 July	Initial	2.336	0.4	0.9345	1.95	2.08	5	2.59	1
11 July 30 July	Development	2.336	1.05	2.45311	5.363	2.18	10	7.15	3
31 July 19 August	Intermediate	2.5609	1.2	3.07308	15.6	4.91	8	10.05	4
20–29 August	Final	2.5609	1.15	2.68895	15.6	4.91	4	10.05	4

4.2.3. Proposed Smart Irrigation

The proposal uses the daily water balance to determine the amount of water in the root zone and, through irrigation, to modify the amount of water, respecting the limits that must be kept lower than the field capacity (Cc) and higher than the permanent wilting point (Pmp). The balance is based on the input and output of water in the root zone, and the proposed system in Figure 3 of the first stage is applied to generate the optimal daily demand.

The initial moisture is taken into account from the date of planting or when the first irrigation performed; the precipitation is the rainfall that occurs per day and the amount of water it brings to the soil; the irrigation is the amount of water that the controller will bring, depending on the initial moisture and limits; the crop evapotranspiration depends on climatic conditions such as wind, solar radiation, average temperature and relative soil moisture to obtain the reference evapotranspiration and, finally, the crop coefficient depends on the stage of development of alfalfa.

The proposed smart irrigation determines the upper limit or maximum moisture limit, which is the lamina at field capacity (L_{Cc}) with Equation (2) and the minimum limit of water content at a suitable threshold (Pw). In this way, the plant does not suffer water stress and does not reach below the point of permanent wilting lamina (L_{Pmp}). These limits depend on the depth of the root until the first cut, then the root maintains its depth, keeping these limits constant. R is the irrigation required when the initial moisture is less than or equal to Pw.

Table 6 shows the irrigation schedule applied to the proposed smart irrigation during the first 10 days in the intermediate stage of the crop. From day 31 to day 38 there is no precipitation (P), therefore the system generates irrigation (R).

Table 6. Irrigation schedule for 10 days at the intermediate stage.

	Day 31	Day 32	Day 33	Day 34	Day 35	Day 36	Day 37	Day 38	Day 39	Day 40
$H(t)$ (mm)	63.91	60.82	57.89	54.92	59.19	55.96	60.26	56.81	61.25	64.00
L_{Cc} (mm)	93.6	93.6	93.6	93.6	93.6	93.6	93.6	93.6	93.6	93.6
Pw (mm)	57.20	57.20	57.20	57.20	57.20	57.20	57.20	57.20	57.20	57.20
ET_o (mm)	2.58	2.44	2.48	2.68	2.69	2.66	2.88	2.54	2.11	2.39
K_c (mm)	1.2	1.2	1.2	1.2	1.2	1.2	1.2	1.2	1.2	1.2
ET_c (mm)	3.10	2.93	2.97	3.22	3.23	3.20	3.46	3.05	2.53	2.87
P (mm)	0	0	0	0	0	0	0	0	5.27	0
R (mm)	0.00	0.00	0.00	7.49	0.00	7.49	0.00	7.49	0.00	0.00
H $(t+1)$	60.82	57.89	54.92	59.19	55.96	60.26	56.81	61.25	64.00	61.13

The maximum irrigation input is 7.49 mm since a maximum irrigation time of 4 h per day is considered.

4.3. Moisture Analysis When Evaluating the Different Irrigation Techniques

In this section the three irrigation techniques are compared and the moisture that the crop will have is analyzed to verify that it does not exceed the moisture limits. In Figures 4–6 the yellow line representing the moisture level at the end of each day is identified, the blue line represents the maximum water content level at field capacity, the orange line is the minimum level that the crop can obtain and, finally, the gray line is the lower limit Pw and the readily available water (RAW). The abscissa axis (y) represents the water sheet or moisture in the root zone and the coordinate axis (x) represents the development of alfalfa in a 60-day cycle.

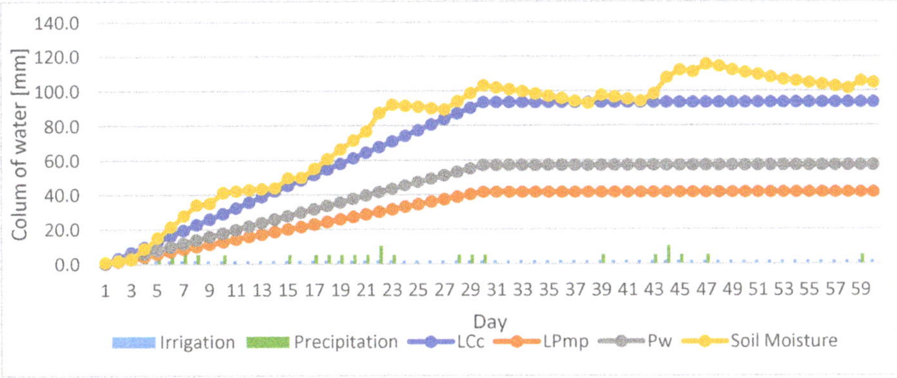

Figure 4. Moisture with traditional irrigation.

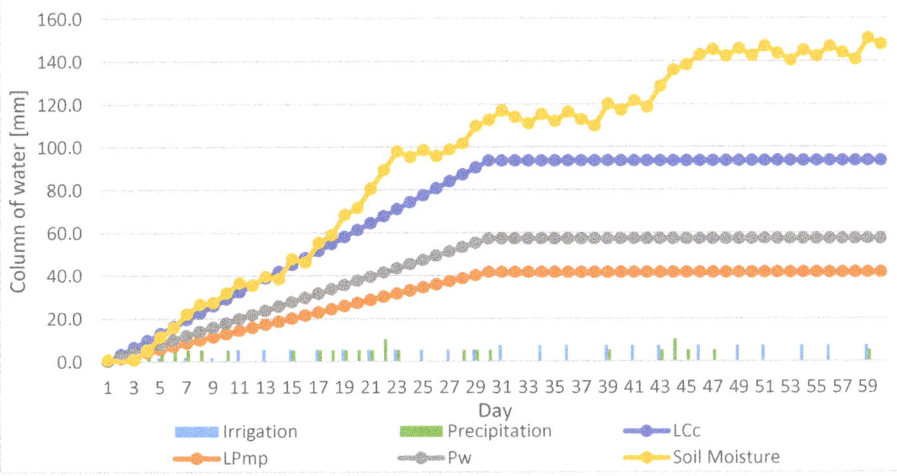

Figure 5. Moisture with technical irrigation.

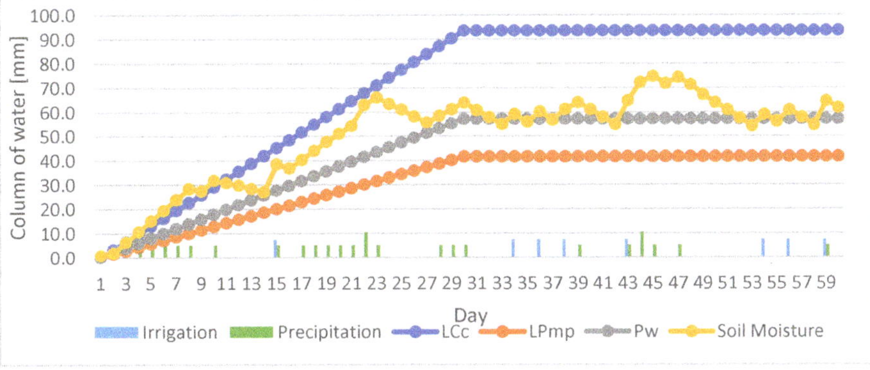

Figure 6. Proposed smart irrigation.

Figure 4 shows the moisture with traditional irrigation where the moisture curve, (H), represented in yellow color, exceeds the curve (L_{Cc}) indicated in blue color, where it can be observed that there is an excess of water in the different stages of the crop during the 60 days, and the excess water is lost by runoff or by deep percolation.

Figure 5 shows the soil moisture in the technified irrigation technique, carried out by means of an agronomic design, when observing the soil moisture, (H), represented by a yellow curve, it is evident that there is an excess of water resources because it exceeds the blue curve (L_{Cc}) throughout the whole period of the crop, but this excess water is lost and is not usable, so there is a waste of water resources.

The third irrigation technique is the proposed smart irrigation that uses the daily water balance with a maximum irrigation time of 4 h, shown in Figure 6. The soil moisture, (H), is represented by a yellow curve and is maintained within the range of the allowed limits without causing water stress or saturation of the soil due to excess water

When comparing and analyzing Figures 4–6 of the three irrigation techniques, it should be noted that the smart irrigation proposal maintains the necessary moisture without causing water stress or soil saturation due to excess water, while the traditional and technified irrigation techniques produce water losses, while the smart irrigation technique makes optimal use of the water resources.

4.4. Economic Analysis of Irrigation Techniques

Once the irrigation performance of the crop has been evaluated, the economic cost of irrigation is now analyzed, for which two scenarios are analyzed and applied to the three irrigation techniques.

4.4.1. Scenario 1

This scenario is applied to the three irrigation techniques where the only source of energy to supply the electrical demand for irrigation is a diesel generator, whose tank capacity is approximately one gallon with an estimated autonomy of two hours. This scenario is used since diesel generators is currently used in many rural areas with crops, because the electrical grid does not supply energy to different crop areas in the rural regions.

To determine the costs for diesel generation in the three irrigation techniques, the value of diesel fuel in Ecuador is used, with a value of USD 1.90 per gallon. The number of gallons to be used in each technique is determined by using Equation (27) and the cost generated by diesel consumption by Equation (28)

$$\text{number gallons} = \frac{\text{total irrigation time}}{\text{autonomy}} \tag{27}$$

$$\text{Costo} = \text{number gallons}\cdot\text{gallon price} \tag{28}$$

Table 7 summarizes the cost generated by each irrigation technique, considering the irrigation times during the 60 days of the crop, the total time in the traditional irrigation is 60 h, in the technified irrigation the total irrigation time is 83 h, while in the smart irrigation method proposed the irrigation time is 37 h, as determined by Equation (10), the irrigation time resulting from the quotient between the gross lamina L_b and the average rainfall of the sprinkler irrigation system Pms.

Table 7. Cost of generation Scenario 1.

Irrigation	Type Crop	Time Irrigation	Gallons	Cost
Traditional	60 days	60 h	30	$57 USD
Technified	60 days	83 h	41.5	$78.85 USD
Smart proposed	60 days	37 h	18.5	$35.15 USD

Analyzing the costs generated in Table 7 for each irrigation technique, it is shown that the traditional irrigation would generate a cost of USD 57, the technified irrigation generates a cost of USD 78.85, while the proposed smart irrigation cost is USD 35.15 during the cultivation time. Therefore, it can be concluded that the cost for technified irrigation and traditional irrigation is high compared to the proposed smart irrigation, which is the most economical and also allows the compliance with the best technical characteristics, regarding irrigation.

4.4.2. Scenario 2

The proposed smart irrigation, in conjunction with energy management, requires 37 h of irrigation, of which the days that the pump spends more time on is 4 h, performing energy optimization and thus not using the diesel generator to turn on the pump and obtaining zero cost for power generation. Our proposal, because it takes the future into account and is based on it, can make decisions based on optimal solutions and always takes the most optimal on and off point for the pump, whether there is sun or not, and thus it will always be the most efficient.

4.5. Analysis of Energy Management in the Three Irrigation Types

Figure 7 shows the energy management with the proposal of a predictive control, based on MPC models with a control horizon of one day for the proposed optimal irrigation, where the energy or energy demand of the pumps (Demand_b) sent by the expert system is the 4 h turn-on, represented by the purple line, the optimizer decides the pump start-up time automatically and thus optimizes the sources that provide energy to the system, identifying the best time for the batteries to operate in charge or in discharge mode. The available energy of the BESS system (Energy_B) is represented by the red curve. It can be observed that there is no use of the diesel generator (P_diesel), i.e., it does not generate cost in the system, keeping the brown curve at zero. The solar power (P_GS), represented by the lead curve, indicates that it is used to recharge the batteries when conditions are more opportune. The yellow line represents when the batteries are in charge mode (P_Battery_Charge), and the light blue curve indicates when the BESS system is in discharge mode (P_Battery_Discharge), while the blue curve represents the energy not supplied (P_Nosupplied), which is kept at zero, i.e., the system is maintained with the necessary energy all the time. It can also be observed that the green curve is the solar shedding (P_solar_shedding) the excess solar energy when more energy is produced than is demanded.

After reviewing the energy consumption with the irrigation techniques of the crop in Scenario 2, the curves of energy consumption without management can be analyzed, i.e., the pumps are turned on according to a fixed schedule and the system must supply energy whatever the source (traditional irrigation, technified irrigation), whereas in the proposed irrigation with optimal energy management, the optimizer has the flexibility to see the most appropriate time to turn on the pump, taking advantage of solar energy and batteries, to avoid the ignition of the diesel generator as much as possible.

There are no proposals describing optimal irrigation simultaneously with the energy management used to supply the electrical irrigation demand. However, a comparative analysis was performed concerning a smart irrigation system proposed in [31], in which the control system is based on fuzzy logic. The ignition pump was analyzed in order to establish the differences found in our proposal regarding energy consumption. In [31], the water pump powered up for 7 h of growing period (three days). Compared to our proposed smart irrigation, the pump powers up for 3 h during the growing period, achieving a good performance in the crop. The powering up of the pump is related to the cost of energy consumption. Therefore, [31] presents a higher cost. Moreover, unlike [31], our proposal includes the management of a microgrid based on renewable sources.

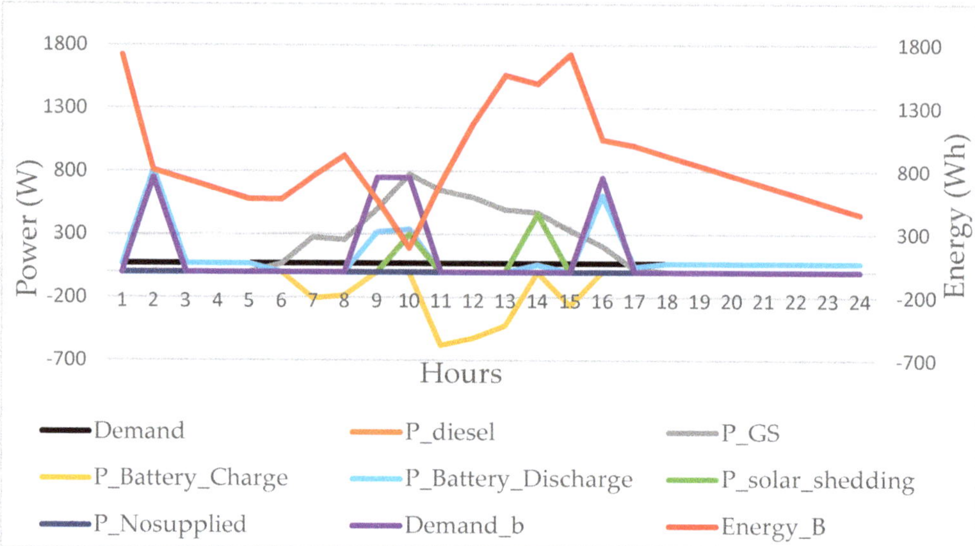

Figure 7. Energy management of a day with the proposed optimal irrigation.

5. Conclusions

In this research, three irrigation techniques were evaluated and compared, the first one known as traditional or empirical that is applied by many farmers. The second one was called technified irrigation, which applies agricultural knowledge to create the irrigation schedule, and, finally, the smart irrigation proposed in this work. The smart irrigation proposal plans irrigation automatically and intelligently, considering energy management. When comparing the three techniques, it can be observed that traditional irrigation and technical irrigation exceed the moisture limit range, which indicates a waste of the water resources that can generate soil saturation, damaging its maximum growth. Whereas with the intelligent irrigation proposal, the moisture is within the limit band, which benefits plant growth and avoids wasting water.

The smart irrigation system proposed, when compared to traditional irrigation, decreases diesel consumption by 38.4% and saves 55.42% compared to technified irrigation; by implementing the EMS of the microgrid, the cost for electricity generation creates a saving of 100% compared to technified irrigation. Being more economical, the proposed smart irrigation system optimizes the efficient use of water in the crop and maximizes the use of renewable resources in energy management.

Finally, the smart irrigation proposal generates the irrigation hours in a crop and supplies the required electrical demand through the use of a microgrid based on renewable resources, at the same time that the environment is respected. The pump on/off system is according to the technical and optimal irrigation conditions. Also, the proposed determines the battery operation set points by taking into account the limitations of the *SOC*. If necessary, the diesel generator power is defined. The decisions made by the EMS maximizes the use of natural resources. This proposal ensures that water resources is not wasted by keeping the moisture in the desired range, allowing for optimal crop growth and an uninterrupted power supply.

Author Contributions: Conceptualization, W.Q., E.T. and J.L.; methodology, W.Q., E.T. and J.L.; software, W.Q. and E.T.; validation, W.Q., E.T. and J.L.; formal analysis, W.Q. and E.T.; investigation, W.Q., E.T. and J.L.; resources, W.Q. and E.T.; writing—original draft preparation, W.Q., E.T. and J.L.; writing—review and editing, W.Q., E.T. and J.L.; visualization, W.Q., E.T. and J.L.; project administration, J.L.; funding acquisition, J.L. All authors have read and agreed to the published version of the manuscript.

Funding: This research received no external funding.

Institutional Review Board Statement: Not applicable.

Informed Consent Statement: Not applicable.

Data Availability Statement: Not applicable.

Acknowledgments: This work was supported in part by the Universidad de las Fuerzas Armadas ESPE and KU Leuven through the Project "MIRA-ESTE: Specific, Innovative Microgrids Solutions (Accounting for Environmental, Social, Technological and Economic Aspects) for isolated rural areas of Ecuador" under Project 2020-EXT-007, and in part by the Vlaamse Interuniversitaire Raad (VLIR)-Universitaire Ontwikkelingssamenwerking (UOS) and the Belgium Development Cooperation (DGD) under Project EC2020SIN322A101. In addition, the authors would like to thank the Subdirección de Redes, Estrategia y Conocimiento (REC), Agencia Nacional de Investigación y Desarrollo (ANID) of Chile, through the Project "Electric microgrids for an energetically sustainable agriculture", 2021–2022.

Conflicts of Interest: The authors declare no conflict of interest.

References

1. Maya, J.; Reinoso, C. Incidencia de la Agricultura en el Crecimiento y Desarrollo Económico del Ecuador del 2006 al 2012. 2013. Available online: http://walker.dgf.uchile.cl/Explorador/Eolico2/ (accessed on 25 February 2022).
2. Montalvo, A.; Bajana, L. Comercialización de Sistema de Riego Inteligente Basado en Sensores y Módulos de Radio Frecuencia para Transmisión y Sistema de Control. February 2018. Available online: http://www.dspace.espol.edu.ec/handle/123456789/42504 (accessed on 25 February 2022).
3. Sury, S.; Mustika, I.; Nugroho, L. Optimized Back-propagation Artificial Neural Network Algorithm for Smart Agriculture Applications. In Proceedings of the 2018 4th International Conference on Science and Technology (ICST), Yogyakarta, Indonesia, 7–8 August 2018; pp. 1–5. [CrossRef]
4. Namala, K.; Prabhu, K.; Math, A.; Kumari, A.; Kulkarni, S. Smart irrigation with embedded system. In Proceedings of the 2016 IEEE Bombay Section Symposium (IBSS), Baramati, India, 21–22 December 2016; pp. 1–5. [CrossRef]
5. Jain, P.; Kumar, P.; Palwalia, D. Irrigation management system with micro-controller application. In Proceedings of the 2017 1st International Conference on Electronics, Materials Engineering and Nano-Technology (IEMENTech), Kolkata, India, 28–29 April 2017; pp. 1–6. [CrossRef]
6. Pratyush, K.; Roopa, Y.; Rajeev, K.; Nandan, N. IoT based Smart Agriculture using Machine Learning. In Proceedings of the 2020 Second International Conference on Inventive Research in Computing Applications (ICIRCA), Coimbatore, India, 15–17 July 2020; pp. 130–134. [CrossRef]
7. Capraro, F.; Tosetti, S.; Vita Serman, F. Laboratorio Virtual y Remoto para Simular, Monitorizar y Controlar un Sistema de Riego por Goteo en Olivos. *Rev. Iberoam. Automática Inf. Ind. RIAI* **2010**, *7*, 73–84. [CrossRef]
8. Marinescu, T.; Arghira, N.; Hossu, D.; Fagarasan, L.; Stamate, L.; Calofir, G.; Iliescu, S. Advanced control strategies for irrigation systems. In Proceedings of the 2017 9th IEEE International Conference on Intelligent Data Acquisition and Advanced Computing Systems: Technology and Applications (IDAACS), Bucharest, Romania, 21–23 September 2017; pp. 843–848. [CrossRef]
9. Negi, D.; Kumar, A.; Kadam, P.; Savant, B.N. Smart Harvest Analysis Using Raspberry Pi Based on Internet of Things. In Proceedings of the 2018 Fourth International Conference on Computing Communication Control and Automation (ICCUBEA), Pune, India, 16–18 August 2018; pp. 1–5. [CrossRef]
10. Balbis, L.; Jassim, A. Dynamic Model of Soil Moisture for Smart Irrigation Systems. In Proceedings of the 2018 International Conference on Innovation and Intelligence for Informatics, Computing, and Technologies (3ICT), Sakhier, Bahrain, 18–20 November 2018; pp. 1–4. [CrossRef]
11. Garcés, L.M. Plan Nacional de Riego y Drenaje 2019–2027. p. 160. Available online: https://prefecturadeesmeraldas.gob.ec/docs/8_plan_nacional_de_riego_y_drenaje.pdf (accessed on 2 March 2022).
12. Fierro, A.; Torres, J. Fuzzy Logic That Determines Sky Conditions as a Key Component of a Smart Irrigation System. In Proceedings of the 2019 Sixth International Conference on eDemocracy eGovernment (ICEDEG), Quito, Ecuador, 24–26 April 2019; pp. 230–235. [CrossRef]

13. Salazar, L.J. Diseño de un Sistema de Riego Inteligente Para Cultivos de Hortalizas Basado en Fuzzy Logic en la Granja la Pradera de la Universidad Técnica del Norte 2019. Available online: http://repositorio.utn.edu.ec/handle/123456789/9137 (accessed on 2 March 2022).
14. Chacho, P.; Maza, J.; Icaza, D.; Arias, P.; Diaz, J.; Amón, X. Wireless System for the Control and Real Time Monitoring of Agricultural Areas of Arenillas Canton in Ecuador. In Proceedings of the 2019 IEEE CHILEAN Conference on Electrical, Electronics Engineering, Information and Communication Technologies (CHILECON), Valparaiso, Chile, 13–27 November 2019; pp. 1–8. [CrossRef]
15. Sarria, C.A. Simulación de Sistemas Híbridos Fotovoltaico/Biogás/Propano Para Generación de Energía Eléctrica de Una Vivienda Empleando la Herramienta Computacional Homer [Recurso Electrónico] Thesis. 2014. Available online: https://bibliotecadigital.univalle.edu.co/handle/10893/7565 (accessed on 2 March 2022).
16. Omaña, J.M.; Antonio, A.; Medrano, J. Bombeo solar de alta potencia con presión y caudal constante. Caso de éxito. In Proceedings of the XXXIII Congreso Nacional de Riegos, Valencia, Spain, 16–18 June 2015; pp. 432–439. [CrossRef]
17. Marisquirena, L. Bombeo solar, experiencias en Uruguay y la región. Perspectivas de futuro. *Rev. Ing. Agrícola* **2018**, *8*, 1.
18. Meunier, S.; Heinrich, S.; Quéval, L.; Cherni, J.; Vido, L. A validated model of a photovoltaic water pumping system for off-grid rural communities. *Appl. Energy* **2019**, *241*, 580–591. [CrossRef]
19. Montaluisa, T.; Vargas, F.G. Diseño e Implementación de un Sistema de Riego Inteligente Para Incrementar el Rendimiento en los Cultivos Basado en Energía Solar Fotovoltaica. Available online: http://repositorio.espe.edu.ec/jspui/handle/21000/24271 (accessed on 25 February 2022).
20. Roje, T.; Sáez, D.; Muñoz, C.; Daniele, L. Energy–Water Management System Based on Predictive Control Applied to the Water–Food–Energy Nexus in Rural Communities. *Appl. Sci.* **2020**, *10*, 7723. [CrossRef]
21. Liu, Z.; Yang, J.; Jiang, W.; Wei, C.; Zhang, P.; Xu, J. Research on Optimized Energy Scheduling of Rural Microgrid. *Appl. Sci.* **2019**, *9*, 4641. [CrossRef]
22. Zhang, X.; Sharma, R.; He, Y. Optimal energy management of a rural microgrid system using multi-objective optimization. In Proceedings of the 2012 IEEE PES Innovative Smart Grid Technologies (ISGT), Washington, DC, USA, 16–20 January 2012; pp. 1–8. [CrossRef]
23. Lopez, J.; Vande, A.; Quijano, N. Dynamic Modeling of Crop–Soil Systems to Design Monitoring and Automatic Irrigation Processes: A Review with Worked Examples. *Water* **2022**, *14*, 889. [CrossRef]
24. Raza, F.; Tamoor, M.; Miram, S.; Arif, W.; Kiren, T.; Amjad, W.; Hussain, M.; Lee, G. The Socio-Economic Impact of Using Photovoltaic (PV) Energy for High-Efficiency Irrigation Systems: A Case Study. *Energies* **2022**, *15*, 1198. [CrossRef]
25. Scarlatache, F.; Grigoras, G.; Scarlatache, V.-A.; Neagu, B.-C.; Ivanov, O.A. Hybrid Methodology Based on Smart Management Energy Consumption in Irrigation Systems. *Electronics* **2021**, *10*, 2864. [CrossRef]
26. Calero-Lara, M.; López-Luque, R.; Casares, F. Methodological Advances in the Design of Photovoltaic Irrigation. *Agronomy* **2021**, *11*, 2313. [CrossRef]
27. Ojeda Bustamante, W.; Sifuentes Ibarra, E.; González Camacho, J.; Guillén González, J.; Unland Weiss, H. Pronóstico del Riego en Tiempo Real. Instituto Mexicano de Tecnología del Agua. 2000. Available online: http://repositorio.imta.mx/handle/20.500.12013/1660 (accessed on 2 March 2022).
28. Consorcio de Gobiernos Autonomos Provinciales del Ecuador. *Hablemos de Riego*. 2017. Available online: http://www.congope.gob.ec/?publicacion=hablemos-de-riego (accessed on 23 February 2022).
29. Allen, R.G.; Pereira, L.; Smith, M. Evapotranspiración del Cultivo Rome 2006. Available online: https://www.fao.org/3/x0490s/x0490s00.htm (accessed on 23 February 2022).
30. Guerrero, M. Diseño de un Sistema de Riego por Aspersión con Automatización Para el Sector de Ugñag, Cantón Penipe. July 2013. Available online: http://dspace.espoch.edu.ec/handle/123456789/2477 (accessed on 2 March 2022).
31. Krishnan, S.; Julie, G.; Robinson, Y.; Raja, S.; Kumar, R.; Thong, P.; Son, L. Fuzzy Logic based Smart Irrigation System using Internet of Things. *J. Clean. Prod.* **2020**, *252*, 119902. [CrossRef]

applied sciences

MDPI

Article

A Low-Cost Global Navigation Satellite System Positioning Accuracy Assessment Method for Agricultural Machinery

Dorijan Radočaj *, Ivan Plaščak, Goran Heffer and Mladen Jurišić

Faculty of Agrobiotechnical Sciences Osijek, Josip Juraj Strossmayer University of Osijek, Vladimira Preloga 1, 31000 Osijek, Croatia; iplascak@fazos.hr (I.P.); hgoran@fazos.hr (G.H.); mjurisic@fazos.hr (M.J.)
* Correspondence: dradocaj@fazos.hr; Tel.: +385-31-554-879

Abstract: The high-precision positioning and navigation of agricultural machinery represent a backbone for precision agriculture, while its worldwide implementation is in rapid growth. Previous studies improved low-cost global navigation satellite system (GNSS) hardware solutions and fused GNSS data with complementary sources, but there is still no affordable and flexible framework for positioning accuracy assessment of agricultural machinery. Such a low-cost method was proposed in this study, simulating the actual movement of the agricultural machinery during agrotechnical operations. Four of the most commonly used GNSS corrections in Croatia were evaluated in two repetitions: Croatian Positioning System (CROPOS), individual base station, Satellite-based Augmentation Systems (SBASs), and an absolute positioning method using a smartphone. CROPOS and base station produced the highest mean GNSS positioning accuracy of 2.4 and 2.9 cm, respectively, but both of these corrections produced lower accuracy than declared. All evaluated corrections produced significantly different median values in two repetitions, representing inconsistency of the positioning accuracy regarding field conditions. While the proposed method allowed flexible and effective application in the field, future studies will be directed towards the reduction of the operator's subjective impact, mainly by implementing autosteering solutions in agricultural machinery.

Keywords: real-time kinematic (RTK); precision agriculture; ISO standard; global positioning system (GPS); GLONASS; agricultural tractor

Citation: Radočaj, D.; Plaščak, I.; Heffer, G.; Jurišić, M. A Low-Cost Global Navigation Satellite System Positioning Accuracy Assessment Method for Agricultural Machinery. *Appl. Sci.* **2022**, *12*, 693. https://doi.org/10.3390/app12020693

Academic Editors: Tadeusz Juliszewski, Sławomir Kurpaska and Paweł Kiełbasa

Received: 17 December 2021
Accepted: 9 January 2022
Published: 11 January 2022

1. Introduction

The advancement of positioning and navigation technology using global navigation satellite systems (GNSSs) ensures its growing use in both precision agriculture and conventional farming [1]. These systems improved multiple aspects of agricultural production since their inception, most notably regarding the accuracy of agrotechnical operations, enabled working hours under lower visibility conditions, as well as lower fatigue for the workers [2]. Moreover, it became a cornerstone of all agrotechnical operations in precision agriculture, allowing precise navigation and implementation of pre-made crop sowing, fertilization, and crop protection prescription maps [3]. With Global Positioning System (GPS) and GLONASS fully operational on a global scale, as well as Galileo and Beidou being at the high level of operability, the future of GNSS application in farming will gradually offer more capabilities in the future [4].

The relative observation techniques using GNSS, most notably real-time-kinematic (RTK), provide corrections up to 2 cm for horizontal and up to 4 cm for vertical positioning accuracy [5]. This positioning accuracy ensures high performance for any agrotechnical operation in precision agriculture [6]. In Croatia, this service is provided by the Croatian Positioning Service (CROPOS), distributed nationwide and based on the 33 reference GNSS base stations [7]. However, due to its operating issues near country border areas due to the occasional loss of GPRS connection, as well as the availability of similar commercial positioning solutions, farmers tend to have difficulty in selecting the optimal solution for their needs. Among the present solutions, RTK corrections distributed via radio signal using base stations by individual farmers or local agricultural machinery suppliers are commonly used, as well as the Satellite-based Augmentation Systems (SBASs) data [8]. Among the agrotechnical operations, precision sowing usually demands the highest positioning accuracy [9], while agrotechnical operations, such as fertilization, tend to result in high-performance even with slightly lower positioning accuracy [10]. High requirements for precise positioning were also noted in previous studies for soil tillage and crop harvesting [11]. Therefore, it is mandatory for farmers to know the necessary positioning accuracy per agrotechnical operation and the exact capabilities of available GNSS corrections in their local area for maximum cost efficiency [12].

The latter often highly varies and is subjected to various factors, such as the condition of the GNSS receiver and communication with the base station for transmission of RTK corrections [9], as well as the distance from the base station [13]. Some of these parameters vary within the country or relative location to the base station, so these should be evaluated within the local operating range of farmers. In order to ensure the widespread availability and affordability of this procedure, standardized and low-cost frameworks for GNSS positioning and navigation for agricultural machinery should be developed [14]. Previous studies successfully improved low-cost GNSS hardware solutions [15] and fusion with complementary data sources and methods [16]. However, there is no standardized approach for the accuracy assessment of GNSS positioning for agricultural machinery other than ISO standards, such as ISO 17123-8 for the GNSS RTK field measurement systems [17]. While these standards are useful in some disciplines, such as geodesy [18], they cannot be performed for GNSS receivers mounted on agricultural machinery without their detachment nor do they present actual field conditions during the agrotechnical operations. These standards also do not simulate the real trajectory during agrotechnical operations, containing both straight and curved sections. Therefore, to keep pace with the development of precise GNSS solutions and the advancement of precision agriculture, it is necessary to establish a widely available and low-cost accuracy assessment method for the positioning of agricultural machinery. To ensure repeatability of these methods, flexibility regarding the field capabilities for its implementation is also mandatory [19], as agricultural areas differ based on the agricultural land management systems worldwide. The low-cost property of such methods is important to provide a possibility of regular GNSS evaluation, even for small farmers [20], so they should not have additional costs besides GNSS receivers on agricultural machinery and RTK corrections.

The aim of this study was to propose such a low-cost, flexible, and straightforward method for the accuracy assessment of positioning using GNSS receivers mounted on agricultural machinery. Additionally, the objective was to develop a method that simulates the actual trajectory during agrotechnical operations, as well as it being easily performed regardless of the field conditions or location in the world.

2. Materials and Methods

2.1. Study Area

The study area is located in the municipality of Koška in eastern Croatia, which dominantly contains agricultural land traditionally used for intensive agricultural production. The trajectory for the assessment of GNSS positioning was projected on a paved track used for agricultural purposes, connected with an improvised roundabout (Figure 1). These properties simulated the actual movement of the agricultural machinery during the agrotechnical operations, evaluating the GNSS positioning accuracy on both straight and curved sections [21]. The total length of the projected trajectory was 506.97 m. All GNSS observations were georeferenced in the Croatian Terrestrial Reference System (HTRS96/TM).

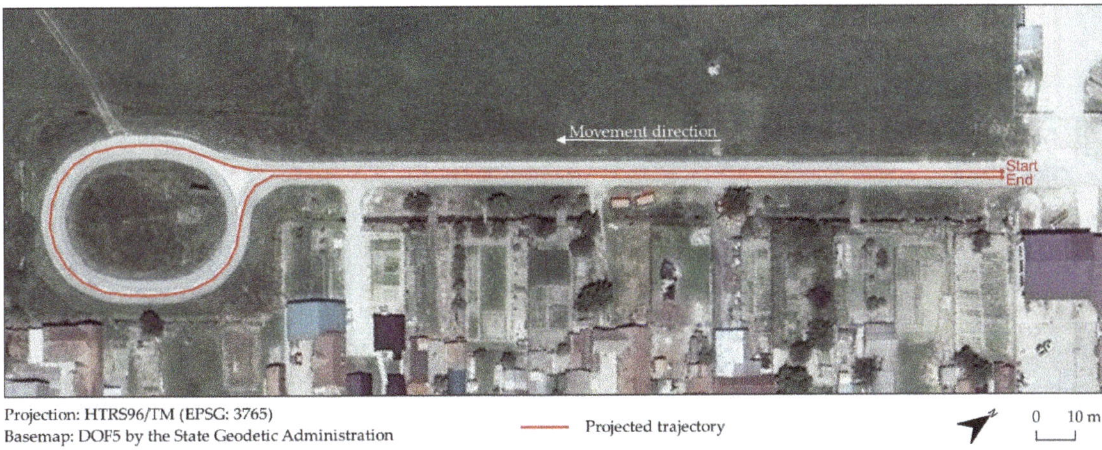

Figure 1. Study area and the trajectory for the accuracy assessment of GNSS RTK corrections.

2.2. GNSS RTK Corrections and Field Data Acquisition

A Trimble Ag25 dual-frequency GNSS receiver with a Trimble CFX-750 GNSS controller (Trimble, Sunnyvale, CA, USA) using GPS and GLONASS satellite data were used for GNSS observations (Figure 2). The GNSS positioning and navigation system was equipped on the agricultural tractor Case IH CS 105 PRO (Case IH, Racine, WI, USA), while the Trimble EZ-steer system (Trimble, Sunnyvale, CA, USA) ensured assisted guidance of the vehicle. The same assisted guidance solution minimized subjective steering errors by the operator, with the GNSS receiver dominantly producing positioning errors with a major variability in a previous study [11]. While inferior to the more expensive fully integrated autosteering solution, its practical effectiveness and consistency combined with RTK corrections in a similar application were noted [8]. EZ-steer was integrated and compatible with all evaluated GNSS corrections, applied for both straight and curved sections of the trajectory. The movement speed of the agricultural tractor was constantly 4 km h^{-1} for all repetitions, simulating the tractor movement speed during the agrotechnical operations, such as sowing and planting [22].

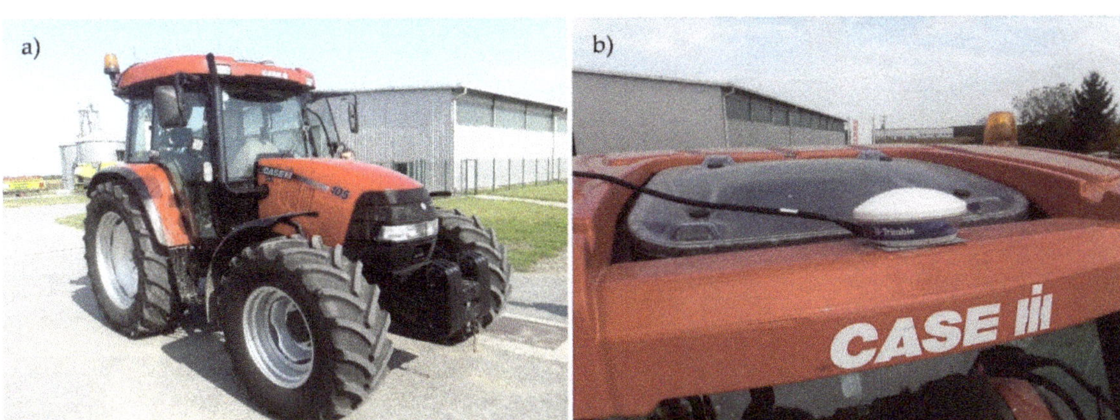

Figure 2. The equipment used for the research: (**a**) Case IH CS 105 PRO agricultural tractor, (**b**) Trimble Ag25 GNSS receiver with the CFX-750 controller.

Three GNSS corrections and an absolute positioning method with the smartphone receiving GPS, GLONASS, and Beidou signals were used in the research. All GNSS observations representing the actual trajectory of the agricultural tractor were performed in one epoch, simulating the actual use of GNSS positioning and navigation during agrotechnical operations. Each positioning variant was performed in two repetitions to ensure repeatability of the proposed method, the lower effect of the operator's subjective assessment, and lower effect of a particular constellation of GNSS satellites [23,24]. Previous studies successfully addressed these conditions either by performing fieldwork in a similar duration to this study under varying satellite constellations [4] or by applying few shorter consecutive repetitions [9].

The three GNSS corrections used included CROPOS, base station, and SBAS. CROPOS with High-Precision Positioning Service (VPPS) was used for RTK observation with the declared horizontal accuracy of 2 cm and vertical accuracy of 4 cm [7]. The GNSS positioning was performed using the 33 permanent base stations in Croatia, distributing mobile RTK corrections in RTCM 3.1 format. Base station variant represented GNSS positioning using the single base station, distributing radio RTK corrections [25]. A commercial solution by the major local agricultural company was used, being 18 km away from the study area. Similar solutions implemented in previous studies achieved a horizontal positioning accuracy up to 2 cm, while a relatively large distance from the base station to the rover is expected to produce a slightly lower accuracy [26]. SBAS is arguably the most widely implemented GNSS solution for agricultural machinery in Croatia due to its low cost, allowing horizontal precision accuracy in the range of 15–25 cm [27]. The European Geostationary Navigation Overlay Service (EGNOS) (European Space Agency, Paris, France) corrections were used in this study, while SBAS solutions are widely available worldwide depending on the location by similar services [28]. Mobile device GNSS positioning using the absolute positioning method was the last variant, as the most widely available method requiring only a smartphone and a free Android tracking app. It was used as a reference for the comparison to other GNSS corrections, with a declared positioning accuracy of more than 3 m [29].

The field observations were conducted on 12 September 2019, with the starting times of the repetitions shown in Table 1. The mission was planned with at least a two-hour difference between repetitions using the respective GNSS corrections to reduce the effect of a particular constellation on the positioning accuracy [30]. Geometric Dilution of Precision (GDOP) values were under a maximum tolerance of six during the entire study [31]. Mission planning was conducted using the Trimble GNSS Planning Online website (https://www.gnssplanning.com/, accessed on 12 October 2021). During the fieldwork, the number of visible GPS + GLONASS satellites ranged from 13 to 16, while four SBAS satellites were visible during the same time. An elevation mask of 10° was applied for GNSS observations. The total electron content (TEC) ranged from 10.19 to 11.21 during this study, with minimal variation during the individual repetitions.

Table 1. Start time periods and GDOP for the used RTK corrections in two repetitions according to Trimble GNSS Planning Online.

Repetition	GNSS Corrections	Starting Time (UTC + 01:00)	Number of Satellites			GDOP	TEC
			GPS	GLONASS	Total		
1st	CROPOS	10:00	8	5	13	2.14	10.19
	Base station	10:15	8	6	14	1.83	10.54
	SBAS	10:30	10	6	16	1.71	10.65
	Mobile device	10:45	9	6	15	1.61	10.75
2nd	CROPOS	13:00	12	4	16	2.30	10.81
	Base station	13:15	11	5	16	1.84	10.97
	SBAS	13:30	11	5	16	1.81	10.99
	Mobile device	13:45	11	5	16	1.68	11.04

2.3. Determination of the Projected Trajectory and Bias for the Actual Trajectory

The reference line was defined in the middle of the white road lane on the paved track, being measured every 5 m using 30-epoch GNSS RTK observations with CROPOS corrections (Figure 3). The average coordinates from the 30 epochs were calculated for each location to achieve the highest possible positioning accuracy of the projected trajectory while maintaining time and cost efficiency [32]. The projected trajectory was determined according to the properties of the chosen track and agricultural machinery, based on the offset from the reference line (1):

$$o_{total} = o_{track} + o_{machinery} \tag{1}$$

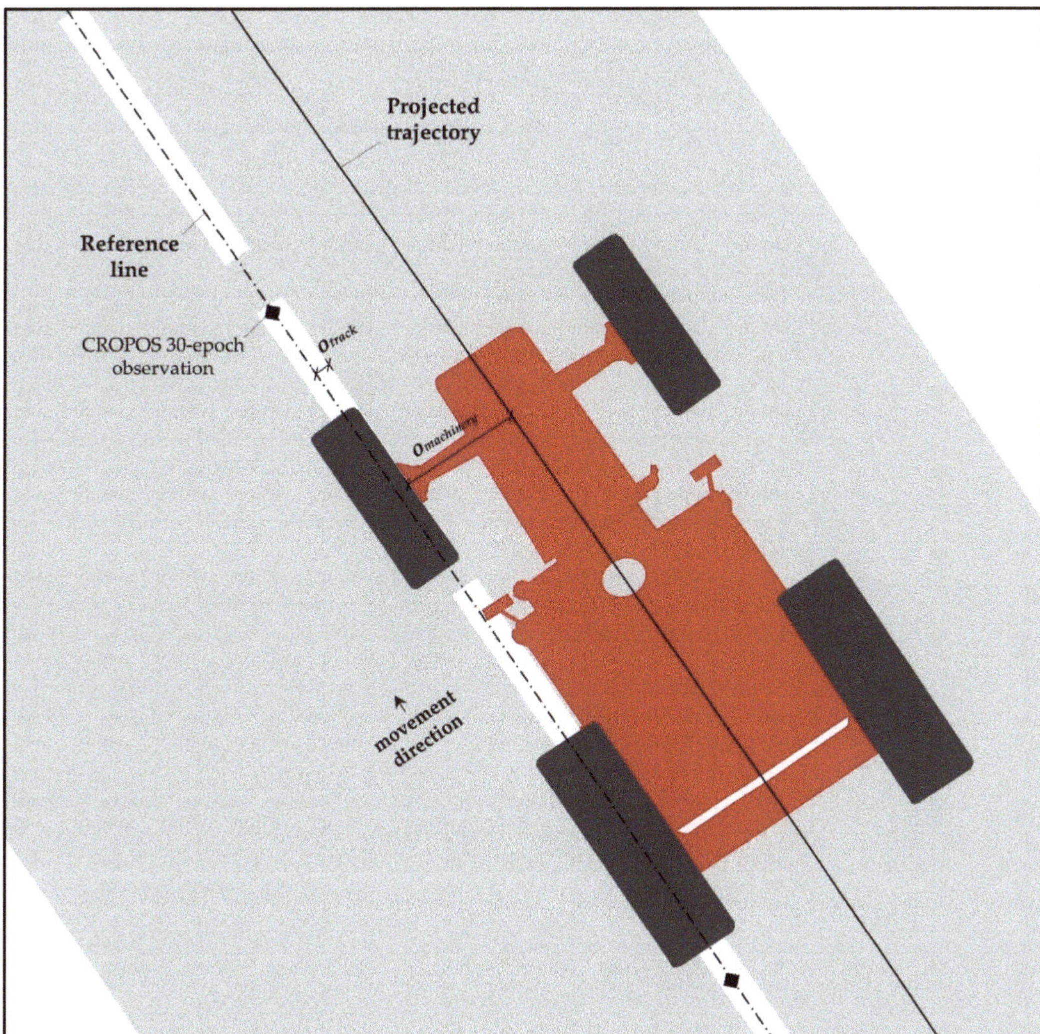

Figure 3. Definition of the projected trajectory according to the reference line.

The o_{track} is the sum of corrections specific for the offset of the projected trajectory according to the properties of the study area. In this case, it represented half of the width of the central dividing line for unclassified tracks in Croatia, amounting to 0.060 m. The $o_{machinery}$ is the sum of corrections specific for the offset caused by the dimensions of the agricultural machinery, its mechanical properties relative to the operator, and the placement of the GNSS antenna. A reference for the steering of the agricultural machinery in this study was determined using the width of the front left tire. Therefore, $o_{machinery}$ equaled the length of the front semi-axle corrected by the tire width, which amounted to 0.675 m. A projected trajectory was determined according to the total offset o_{total} from the reference line, which was calculated as the sum of previously determined corrections and amounted to 0.735 m.

Fifty auxiliary points with equal relative distance were generated 10 m apart from each other on the projected trajectory and were used as a reference for the accuracy assessment of GNSS positioning (Figure 4). Accordingly, the positioning error per observation ($bias_i$) was determined according to Equation (2):

$$bias_i = \sqrt{\Delta E_i{}^2 + \Delta N_i{}^2} = \sqrt{(E_{pi} - E_{qi})^2 + (N_{pi} - N_{qi})^2}, \tag{2}$$

where i represents one epoch of the GNSS observation, E and N represent the eastern and northern coordinates in the HTRS96/TM projection, p represents the observed location on the actual trajectory, and q represents the perpendicular location on the projected trajectory closest to the observed location p.

The statistical analysis of the observed GNSS field results was performed using R v4.0.3 in RStudio v1.3 (R Foundation, Vienna, Austria). The fundamental properties of two samples of bias determined for each of the repetitions were analyzed using descriptive statistics. Median values were used to evaluate the relationship of the actual GNSS accuracy in the field with the positioning accuracy declared by the provider of the corrections. Minimum and maximum values were used to establish a range of positioning accuracy, while the coefficient of variation (CV) quantified the variability of the positioning accuracy within the repetition. The lower CV values indicated a more uniform accuracy, which benefits the application of a particular GNSS correction in the field due to the increased repeatability [33].

The normal distribution of input samples was evaluated using the Shapiro-Wilk test to determine the suitability of a parametric (t-test) or non-parametric test (Wilcoxon test). The null hypothesis of the Shapiro-Wilk test that the tested GNSS positioning bias is normally distributed was rejected for every p value lower than 0.05 [34]. In these cases, the Wilcoxon non-parametric test was determined as optimal instead of the default selection of the parametric t-test. The difference from the median values of two samples created from the GNSS observations in two repetitions was tested to assess the repeatability of the positioning accuracy for the tested GNSS corrections [35]. A correlation matrix containing Pearson's correlation coefficients for the evaluated pairs of GNSS positioning bias was used to determine the relationship of the tested GNSS corrections, allowing integrated observation of the positioning accuracy on straight and curved sections of the projected trajectory.

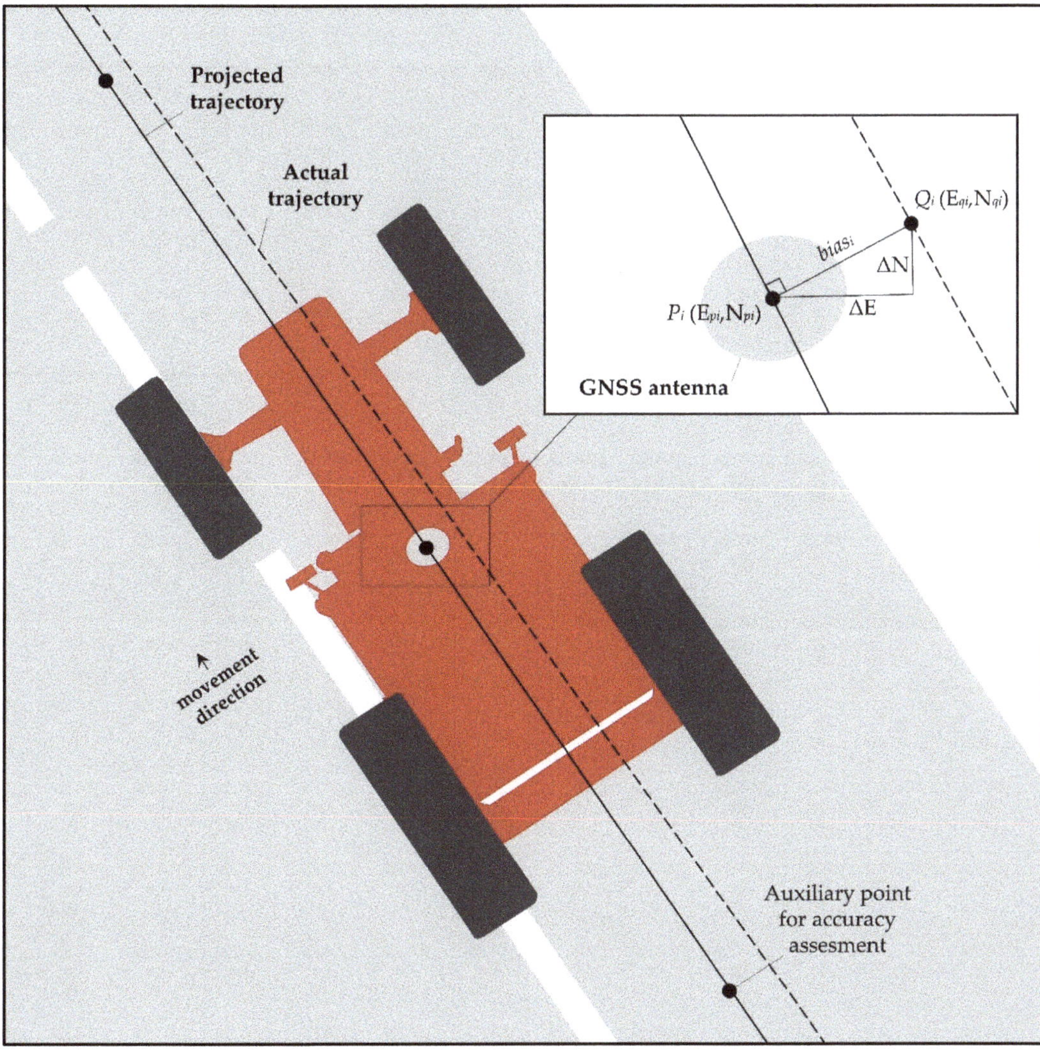

Figure 4. Determination of GNSS positioning accuracy according to the projected trajectory.

3. Results

The descriptive statistics of the GNSS observations for the total bias and its components per coordinates in HTRS96/TM are presented in Table 2. CROPOS corrections produced the most accurate GNSS positioning in both repetitions, followed by the base station. The smaller deviation from the projected trajectory for these corrections was observed for N coordinates, as the trajectory had a higher range of E coordinates. Tested GNSS correction variants produced higher positioning accuracy in the first repetition characterized with the lower GDOP, with the exception of SBAS. While moderate variability of the CROPOS and base station observations was noted, SBAS produced the lowest CV values of the positioning bias in both repetitions, indicating the most consistent GNSS observation. The mobile device produced the least consistent observations, having a positioning accuracy range from sub-centimeter accuracy up to more than 3 m. A comparative visual representation of the deviations of the tested GNSS positioning variants from the projected trajectory is presented in Figure 5.

Table 2. Descriptive statistics of the bias produced by the used GNSS RTK corrections from the projected trajectory.

	GNSS Corrections	First Repetition					Second Repetition				
		n	Median (m)	CV	Min (m)	Max (m)	*n*	Median (m)	CV	Min (m)	Max (m)
bias	CROPOS	50	0.019	0.604	0.003	0.043	50	0.029	0.511	0.002	0.052
	Base station	50	0.027	0.496	0.003	0.049	50	0.032	0.572	0.002	0.045
	SBAS	50	0.263	0.174	0.120	0.323	50	0.223	0.172	0.118	0.269
	Mobile device	50	0.842	0.979	0.007	3.611	50	1.340	0.762	0.028	3.739
ΔE	CROPOS	50	0.015	0.671	0.000	0.035	50	0.023	0.538	0.001	0.045
	Base station	50	0.022	0.542	0.001	0.041	50	0.016	0.647	0.000	0.036
	SBAS	50	0.212	0.199	0.066	0.264	50	0.179	0.190	0.096	0.222
	Mobile device	50	0.581	0.803	0.003	1.788	50	1.006	0.759	0.024	3.358
ΔN	CROPOS	50	0.011	0.642	0.000	0.026	50	0.016	0.594	0.000	0.042
	Base station	50	0.016	0.513	0.002	0.033	50	0.012	0.577	0.000	0.028
	SBAS	50	0.155	0.154	0.097	0.207	50	0.131	0.187	0.069	0.183
	Mobile device	50	0.548	1.334	0.007	3.536	50	0.815	0.933	0.014	3.170

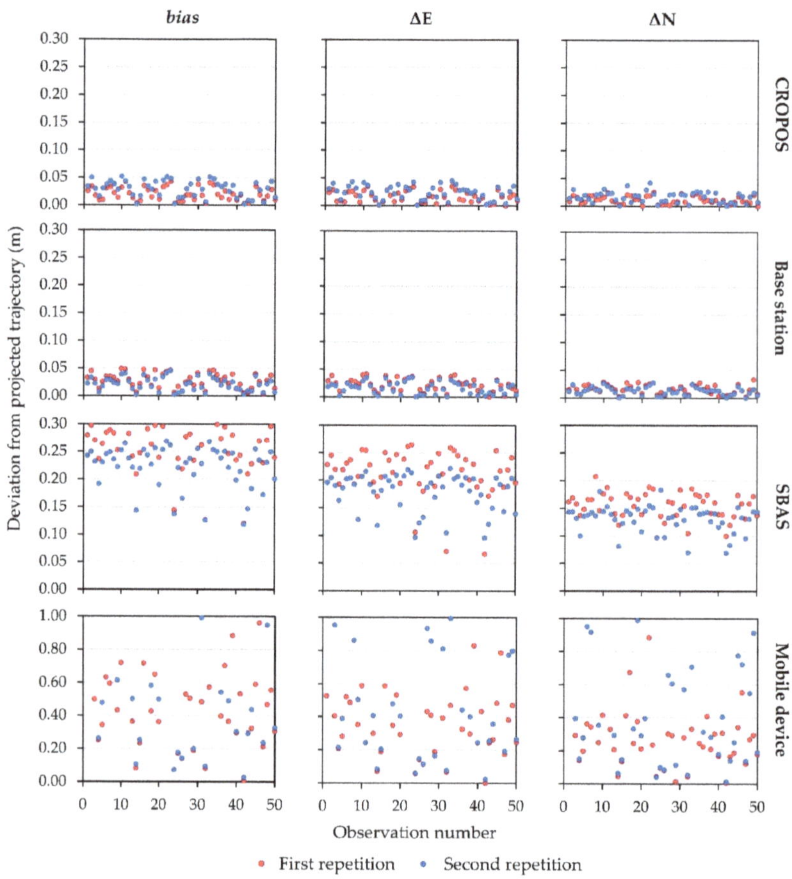

Figure 5. A comparative visual representation of deviations from the projected trajectory for four tested GNSS positioning variants.

The null hypothesis of the Shapiro-Wilk test was rejected for all tested samples, indicating that neither pair of samples observed with the same GNSS correction possessed a normal distribution of values (Table 3). Accordingly, a non-parametric Wilcoxon statistical test was selected for the evaluation of all GNSS observation samples.

Table 3. Shapiro-Wilk normality test results for GNSS positioning bias.

	GNSS Corrections	1st Repetition		2nd Repetition		Normality Observed
		W	*p*	*W*	*p*	
bias	CROPOS	0.9411	0.0149	0.9462	0.0239	no
	Base station	0.9531	0.0456	0.9633	0.1218	no
	SBAS	0.8815	0.0001	0.8610	>0.0001	no
	Mobile device	0.7884	>0.0001	0.9287	>0.0001	no
ΔE	CROPOS	0.9413	0.0151	0.9630	0.1186	no
	Base station	0.9621	0.1085	0.9428	0.0174	no
	SBAS	0.8310	>0.0001	0.8646	>0.0001	no
	Mobile device	0.8640	>0.0001	0.9296	>0.0001	no
ΔN	CROPOS	0.9442	0.0198	0.9703	0.2376	no
	Base station	0.9622	0.0497	0.9756	0.3852	no
	SBAS	0.9688	0.2058	0.9158	0.0017	no
	Mobile device	0.6329	>0.0001	0.8658	>0.0001	no

The Wilcoxon test indicated that neither of the repetitions from respective GNSS corrections belong in the same population regarding the positioning bias and its components on E and N coordinates (Table 4). A display of the actual trajectories per GNSS correction and their relative location according to the projected trajectory is presented in Figure 6. The actual trajectories of all four GNSS variants were generally closer to the projected trajectory on straight sections, with the least matching on the most curved sections of the projected trajectory.

Table 4. Wilcoxon test results for GNSS positioning bias.

	GNSS Corrections	*W*	*p*	Significantly Different Medians
bias	CROPOS	784	0.0013	yes
	Base station	1587	0.0204	yes
	SBAS	1962	>0.0001	yes
	Mobile device	3712	0.0017	yes
ΔE	CROPOS	790	0.0015	yes
	Base station	1578	0.0240	yes
	SBAS	1940	>0.0001	yes
	Mobile device	3559	0.0004	yes
ΔN	CROPOS	850	0.0059	yes
	Base station	1570	0.0276	yes
	SBAS	1894	>0.0001	yes
	Mobile device	3814	0.0038	yes

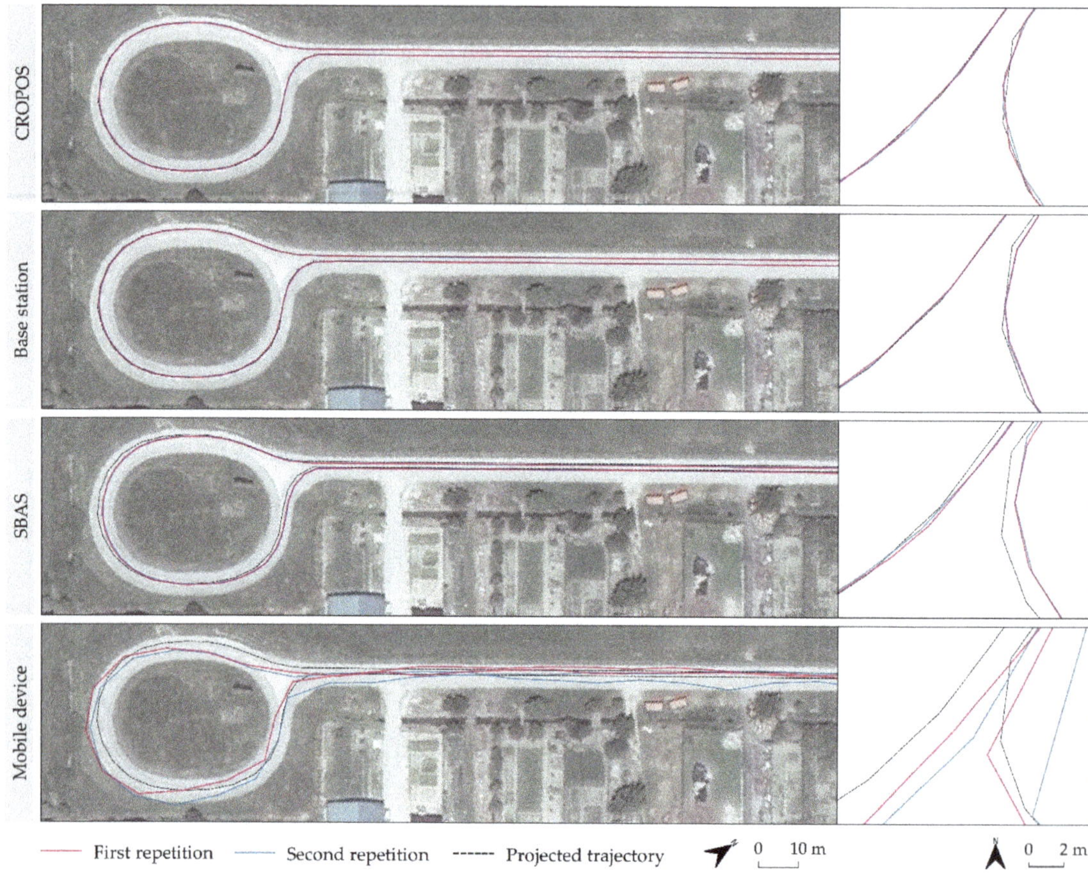

Figure 6. Actual trajectories per GNSS correction in two repetitions and their location according to the projected trajectory, with an enlarged display of the connection of straight and curved sections.

Positioning bias from the repetitions of particular GNSS corrections resulted in a high correlation for all four varieties, constantly producing values higher than 92% (Table 5). The strongest mutual relationship was observed between CROPOS and Base station observations, with a mean Pearson's correlation coefficient of 0.948. SBAS also produced a relatively high correlation with both high-precision GNSS RTK corrections, while the mobile device produced a wide range of lower correlation values, indicating low repeatability of the positioning accuracy results.

Table 5. A correlation matrix of positioning bias per GNSS correction and their repetitions.

	C1	C2	B1	B2	S1	S2	M1	M2
C1	1.000							
C2	0.921	1.000						
B1	0.923	0.965	1.000					
B2	0.952	0.952	0.958	1.000				
S1	0.851	0.908	0.884	0.881	1.000			
S2	0.833	0.915	0.905	0.867	0.958	1.000		
M1	0.660	0.623	0.626	0.617	0.571	0.551	1.000	
M2	0.815	0.785	0.798	0.783	0.722	0.709	0.938	1.000

C1 and C2: CROPOS repetitions, B1 and B2: Base station repetitions, S1 and S2: SBAS repetitions, D1 and D2: Mobile device repetitions.

4. Discussion

This study presented a low-cost method of GNSS positioning accuracy assessment for agricultural machinery, based on the commonly used GNSS corrections in Croatia. Its properties of straightforwardness and adaptability under various conditions in the study area meet the requirements of a potential basis for widespread global application [36]. Unlike available ISO standards for the evaluation of GNSS RTK receivers [17], it simulates the actual movement of the agricultural machinery during the agrotechnical operations, which has more practical importance for the farmers in their work. The majority of technological improvements and applications in precision agriculture are still largely focused on GNSS positioning and navigation, as the basis of all precise agrotechnical operations [37]. Its importance is described in many previous studies and its importance is expected to grow further [38]. Despite its availability and mentioned advantages, the proposed method at the current stage still has two main drawbacks:

- The retained minor subjective impact of the operator on the GNSS positioning accuracy;
- The lack of newly available GNSS corrections and multiple study areas.

The first disadvantage of the operator's subjective impact on the reliability of GNSS observations occurs due to the imperfections regarding steering aim and reaction time [39]. The selection of EZ-steer justifies the selection of two repetitions due to assisted guidance and removal of the vast majority of steering subjective errors caused by the operator. This solution is cost-efficient but not perfectly reliable, which could be resolved with the more expensive use of an autosteering solution in future studies. Additionally, the fewer repetitions are user-friendly to farmers, being time-efficient regarding fieldwork and data processing, allowing the widespread and efficient implementation of the proposed method. However, more repetitions would produce more reliable results and farmers who prioritize additional accuracy over time efficiency should opt for more than two repetitions. These errors have a random character and a previous study indicated that these can be minimized in the process of statistical analysis [40]. The minor subjective assessment of the operator remains a disadvantage of the proposed method but ensures a time-efficient and low-cost property, which can be a priority to farmers. The proposed method is applicable with the autosteering solution, as D'Antonio et al. [8] noted the maximum retention of positioning accuracy by applied GNSS correction but also its practical constraints, because of which it might not be available to a wide range of farmers.

As another more expensive option, the possible upgrade of this method is the implementation of an automated procedure, simultaneously using two GNSS receivers mounted on the agricultural machinery. One GNSS receiver with maximum available accuracy GNSS RTK corrections could be used for an actual reference positioning, which corresponds to CROPOS in this study. The second receiver with GNSS corrections, which are evaluated in the process, coupled with autosteering to remove the effect of human error, could round up an upgraded version of the proposed method. Since the positioning accuracy of the reference receiver equals 2 cm horizontally [7], this approach will be optimal for the assessment of the performance of corrections up to 10 cm, such as SBAS. This approach does not require setting up a projected trajectory, which makes it even more flexible for widespread application. However, this is not a low-cost method and is time-expensive due to the necessity of coupling two GNSS receivers in the same agricultural machinery [41].

While this study covered the most commonly used GNSS corrections for farming purposes in Croatia, the second disadvantage is based on the current lack of testing of the reliability of the proposed method using the additional corrections and study areas. The novel GNSS corrections offer upgrades to the ones tested in this study regarding the global availability and no need for a second GNSS receiver other than the rover [42]. The potential of Trimble RTX [6] and Precise Point Positioning (PPP) [4] was recognized in previous studies for GNSS positioning and navigation in farming. Therefore, their implementation is planned to be evaluated using the proposed method in future studies. The positioning accuracy for tested GNSS corrections declared by its suppliers was determined under conditions that do not correspond to the ones in the field during agricultural operations [43]. Among several factors, this refers to the observation of multiple epochs using the stationary GNSS receiver and observation under uniform conditions. Since the GNSS positioning accuracy under common field conditions in farming is subjected to other secondary effects, such as the availability of mobile Internet [44], or distance to the base station [45], every farmer should at times test the performance of GNSS receivers mounted on agricultural machinery. The distance from the rover to the base station exceeding the one recommended by producers is a particularly frequent case in eastern Croatia, as was the case in this study. While CROPOS solves this issue by calculating the virtual reference stations, the commercial solutions provided by the local agricultural suppliers often rely on a simpler approach, which does not negate the effects of relatively large base-rover distances [46]. The GNSS constellation also affected GNSS positioning accuracy in this study, as observations using CROPOS and base station in the first repetition produced higher positioning accuracy with lower GDOP and TEC compared to the second repetition. The GNSS positioning accuracy results from all four GNSS variations indicated that the farmers could not expect either a positioning accuracy as declared by its suppliers nor uniform and constant values. Therefore, to ensure reliability for agrotechnical operations requiring higher positioning accuracy, such as sowing and planting, high-precision corrections are recommended [4], such as using CROPOS and base station.

5. Conclusions

To match the technological improvement of the low-cost GNSS positioning and navigation solutions regarding hardware development and advanced processing methods, this study proposed the low-cost positioning accuracy assessment method. Besides its affordability for farmers worldwide, its flexibility and straightforwardness allow for the regular examination of GNSS receivers mounted on agricultural machinery without its detachment and in actual field conditions. Present ISO standards for GNSS RTK accuracy assessment cannot be fulfilled adequately and are not suitable for the receiver mounted on agricultural machinery nor do they represent actual field conditions during agrotechnical operations. While the proposed method still has a slight disadvantage of the operator's subjective impact, various approaches are possible to further reduce this issue. The most suitable ones are an increased number of repetitions and longer trajectory to retain the low-cost property of the method, while the implementation of autosteering would completely remove this impact but requires additional investments.

As the positioning accuracy assessment results based on the four of the most commonly used GNSS corrections in Croatia indicated, the GNSS observations under actual field conditions during agrotechnical operations can produce lower accuracy compared to their declared positioning accuracy. Additionally, some site-specific conditions, such as the availability of mobile Internet signal and the distance to the base station, could produce heterogenous GNSS positioning accuracy, even within a county or a municipality. Therefore, greater emphasis should be put on the local properties and multiple study areas for the assessment of GNSS positioning accuracy in both professional farming activities and future scientific work. Resolving these issues will be a subject of future studies to provide more reliable insight into the current solutions for GNSS positioning and navigation for agricultural machinery under various field conditions.

Author Contributions: Conceptualization, D.R. and I.P.; methodology, D.R.; software, D.R.; validation, D.R.; formal analysis, D.R., I.P., G.H. and M.J.; investigation, D.R. and I.P.; resources, D.R. and I.P.; data curation, D.R.; writing—original draft preparation, D.R.; writing—review and editing, D.R., I.P., G.H. and M.J.; visualization, D.R.; supervision, I.P., G.H. and M.J.; project administration, M.J.; funding acquisition, D.R. and M.J. All authors have read and agreed to the published version of the manuscript.

Funding: This research received no external funding.

Institutional Review Board Statement: Not applicable.

Informed Consent Statement: Not applicable.

Data Availability Statement: Not applicable.

Acknowledgments: The authors thank "Jerković d.o.o." company for the cooperation during the fieldwork and for the agricultural tractor equipped with the GNSS receiver and necessary corrections for the research. This work was supported by the Faculty of Agrobiotechnical Sciences Osijek as a part of the scientific project "AgroGIT—technical and technological crop production systems, GIS and environment protection".

Conflicts of Interest: The authors declare no conflict of interest.

References

1. Tayebi, A.; Gomez, J.; Fernandez, M.; Saez de Adana, F.; Gutierrez, O. Low-Cost Experimental Application of Real-Time Kinematic Positioning for Increasing the Benefits in Cereal Crops. *Int. J. Agric. Biol. Eng.* **2021**, *14*, 194–199. [CrossRef]
2. Ayerdi Gotor, A.; Marraccini, E.; Leclercq, C.; Scheurer, O. Precision Farming Uses Typology in Arable Crop-Oriented Farms in Northern France. *Precis. Agric.* **2020**, *21*, 131–146. [CrossRef]
3. Toriyama, K. Development of Precision Agriculture and ICT Application Thereof to Manage Spatial Variability of Crop Growth. *Soil Sci. Plant Nutr.* **2020**, *66*, 811–819. [CrossRef]
4. Guo, J.; Li, X.; Li, Z.; Hu, L.; Yang, G.; Zhao, C.; Fairbairn, D.; Watson, D.; Ge, M. Multi-GNSS Precise Point Positioning for Precision Agriculture. *Precis. Agric.* **2018**, *19*, 895–911. [CrossRef]
5. Zhao, B.; Li, J.; Wang, L.; Shi, Y. Positioning Accuracy Assessment of a Commercial RTK UAS. In *Proceedings of the Autonomous Air and Ground Sensing Systems for Agricultural Optimization and Phenotyping V*; Thomasson, J.A., TorresRua, A.F., Eds.; SPIE-International Society Optical Engineering: Bellingham, UK, 2020; Volume 11414, p. 1141409.
6. Carballido, J.; Perez-Ruiz, M.; Emmi, L.; Agüera, J. Comparison of Positional Accuracy between RTK and RTX GNSS Based on the Autonomous Agricultural Vehicles under Field Conditions. *Appl. Eng. Agric.* **2014**, *30*, 361–366. [CrossRef]
7. Republic of Croatia, State Geodetic Administration CROPOS Users' Manual. Available online: https://www.cropos.hr/files/docs/cropos_users-manual.pdf (accessed on 22 November 2021).
8. D'Antonio, P.; D'Antonio, C.; Evangelista, C.; Doddato, V. Satellite Guidance Systems in Agriculture: Experimental Comparison between EZ-Steer/RTK and AUTOPILOT/EGNOS. *J. Agric. Eng.* **2013**, *44*, 173–177. [CrossRef]
9. Scarfone, A.; Picchio, R.; del Giudice, A.; Latterini, F.; Mattei, P.; Santangelo, E.; Assirelli, A. Semi-Automatic Guidance vs. Manual Guidance in Agriculture: A Comparison of Work Performance in Wheat Sowing. *Electronics* **2021**, *10*, 825. [CrossRef]
10. Zhang, J.; Liu, G.; Huang, J.; Zhang, Y. A Study on the Time Lag and Compensation of a Variable-Rate Fertilizer Applicator. *Appl. Eng. Agric.* **2021**, *37*, 43–52. [CrossRef]
11. Perez-Ruiz, M.; Carballido, J.; Agueera, J.; Gil, J.A. Assessing GNSS Correction Signals for Assisted Guidance Systems in Agricultural Vehicles. *Precis. Agric.* **2011**, *12*, 639–652. [CrossRef]
12. Catania, P.; Comparetti, A.; Febo, P.; Morello, G.; Orlando, S.; Roma, E.; Vallone, M. Positioning Accuracy Comparison of GNSS Receivers Used for Mapping and Guidance of Agricultural Machines. *Agronomy* **2020**, *10*, 924. [CrossRef]
13. Alkan, R.M.; Erol, S.; İlçi, V.; Ozulu, İ.M. Comparative Analysis of Real-Time Kinematic and PPP Techniques in Dynamic Environment. *Measurement* **2020**, *163*, 107995. [CrossRef]
14. Gomez-Gil, J.; Alonso-Garcia, S.; Gómez-Gil, F.J.; Stombaugh, T. A Simple Method to Improve Autonomous GPS Positioning for Tractors. *Sensors* **2011**, *11*, 5630–5644. [CrossRef] [PubMed]
15. Valente, D.S.M.; Momin, A.; Grift, T.; Hansen, A. Accuracy and Precision Evaluation of Two Low-Cost RTK Global Navigation Satellite Systems. *Comput. Electron. Agric.* **2020**, *168*, 105142. [CrossRef]
16. Kaivosoja, J.; Linkolehto, R. GNSS Error Simulator for Farm Machinery Navigation Development. *Comput. Electron. Agric.* **2015**, *119*, 166–177. [CrossRef]
17. Garrido-Carretero, M.S.; de Lacy-Perez de los Cobos, M.C.; Borque-Arancon, M.J.; Ruiz-Armenteros, A.M.; Moreno-Guerrero, R.; Gil-Cruz, A.J. Low-Cost GNSS Receiver in RTK Positioning under the Standard ISO-17123-8: A Feasible Option in Geomatics. *Measurement* **2019**, *137*, 168–178. [CrossRef]
18. Preseren, P.P.; Mencin, A.; Stopar, B. Analysis of Gnss-Rtk Instruments Testing on the Iso 17123-8 Instructions. *Geod. Vestn.* **2010**, *54*, 607–626. [CrossRef]
19. Paziewski, J.; Wielgosz, P. Investigation of Some Selected Strategies for Multi-GNSS Instantaneous RTK Positioning. *Adv. Space Res.* **2017**, *59*, 12–23. [CrossRef]
20. Wang, Q.; Yang, C.; Wang, Y.; Wu, S. Application of Low Cost Integrated Navigation System in Precision Agriculture. *Intell. Autom. Soft Comput.* **2020**, *26*, 1419–1428. [CrossRef]
21. Passalaqua, B.P.; Molin, J.P. Path Errors in Sugarcane Transshipment Trailers. *Eng. Agric.* **2020**, *40*, 223–231. [CrossRef]
22. Shi, Y.; Xi, X.; Gan, H.; Shan, X.; Zhang, Y.; Shen, H.; Zhang, R. Design and Experiment of Row-Controlled Fertilizing–Weeding Machine for Rice Cultivation. *Agriculture* **2021**, *11*, 527. [CrossRef]
23. de Silva, T.M.A.; de Mayrink, G.O.; Valente, D.S.M.; Queiroz, D.M. Integration of a Low-Cost Global Navigation Satellite System to a Single-Board Computer Using Kalman Filtering. *Eng. Agric.* **2019**, *39*, 323–330. [CrossRef]

24. Akkamis, M.; Keskin, M.; Sekerli, Y.E. Comparative Appraisal of Three Low-Cost GPS Speed Sensors with Different Data Update Frequencies. *AgriEngineering* **2021**, *3*, 423–437. [CrossRef]
25. Armenteros, J.A.; Gil, A.J. A Methodology for Creating Rtk Positioning Coverage Maps Via a Radio Modem Link to Cors Stations. *Surv. Rev.* **2010**, *42*, 406–411. [CrossRef]
26. He, K.; Xu, T.; Forste, C.; Petrovic, S.; Barthelmes, F.; Jiang, N.; Flechtner, F. GNSS Precise Kinematic Positioning for Multiple Kinematic Stations Based on A Priori Distance Constraints. *Sensors* **2016**, *16*, 470. [CrossRef] [PubMed]
27. Li, L.; Jia, C.; Zhao, L.; Cheng, J.; Liu, J.; Ding, J. Real-Time Single Frequency Precise Point Positioning Using SBAS Corrections. *Sensors* **2016**, *16*, 1261. [CrossRef]
28. Nie, Z.; Zhou, P.; Liu, F.; Wang, Z.; Gao, Y. Evaluation of Orbit, Clock and Ionospheric Corrections from Five Currently Available SBAS L1 Services: Methodology and Analysis. *Remote Sens.* **2019**, *11*, 411. [CrossRef]
29. Liu, Q.; Gao, C.; Peng, Z.; Zhang, R.; Shang, R. Smartphone Positioning and Accuracy Analysis Based on Real-Time Regional Ionospheric Correction Model. *Sensors* **2021**, *21*, 3879. [CrossRef]
30. Meneghini, C.; Parente, C. Advantages of Multi GNSS Constellation: GDOP Analysis for GPS, GLONASS and Galileo Combinations. *Int. J. Eng. Technol. Innov.* **2017**, *7*, 1–10.
31. Pereira, F.; Selva, D. Exploring the Architecture Trade Space of NextGen Global Navigation Satellite Systems. In Proceedings of the 2019 IEEE Aerospace Conference, Big Sky, MT, USA, 2–9 March 2019; IEEE: New York, NY, USA, 2019.
32. Ding, W.; Sun, W.; Gao, Y.; Wu, J. Carrier Phase-Based Precise Heading and Pitch Estimation Using a Low-Cost GNSS Receiver. *Remote Sens.* **2021**, *13*, 3642. [CrossRef]
33. Scott, D.N.; Brogan, D.J.; Lininger, K.B.; Schook, D.M.; Daugherty, E.E.; Sparacino, M.S.; Patton, A.I. Evaluating Survey Instruments and Methods in a Steep Channel. *Geomorphology* **2016**, *273*, 236–243. [CrossRef]
34. Pepe, M.; Costantino, D.; Vozza, G.; Alfio, V.S. Comparison of Two Approaches to GNSS Positioning Using Code Pseudoranges Generated by Smartphone Device. *Appl. Sci.* **2021**, *11*, 4787. [CrossRef]
35. Kazmierski, K.; Hadas, T.; Sośnica, K. Weighting of Multi-GNSS Observations in Real-Time Precise Point Positioning. *Remote Sens.* **2018**, *10*, 84. [CrossRef]
36. Park, K.W.; Park, J.-I.; Park, C. Efficient Methods of Utilizing Multi-SBAS Corrections in Multi-GNSS Positioning. *Sensors* **2020**, *20*, 256. [CrossRef] [PubMed]
37. Mahato, S.; Rakshit, P.; Santra, A.; Dan, S.; Tiglao, N.C.; Bose, A. A GNSS-Enabled Multi-Sensor for Agricultural Applications. *J. Inform. Optim. Sci.* **2019**, *40*, 1763–1772. [CrossRef]
38. Marucci, A.; Colantoni, A.; Zambon, I.; Egidi, G. Precision Farming in Hilly Areas: The Use of Network RTK in GNSS Technology. *Agriculture* **2017**, *7*, 60. [CrossRef]
39. Wu, C.; Zhang, W.; You, X.; Du, N. Which Accuracy Levels of Positioning Technologies Do Drivers Really Need in Connected Vehicle Settings for Safety? *Accid. Anal. Prev.* **2021**, *157*, 106106. [CrossRef]
40. Dvulit, P.D.; Savchuk, S.; Sosonka, I. The Processing of GNSS Observation by Non-Classical Error Theory of Measurements. *Geodynamics* **2020**, *28*, 19–28. [CrossRef]
41. de Bakker, P.F.; Tiberius, C.C.J.M. Real-Time Multi-GNSS Single-Frequency Precise Point Positioning. *GPS Solut.* **2017**, *21*, 1791–1803. [CrossRef]
42. Atiz, O.F.; Shakor, A.Q.; Ogutcu, S.; Alcay, S. Performance Investigation of Trimble RTX Correction Service with Multi-GNSS Constellation. *Surv. Rev.* **2021**, 1–11. [CrossRef]
43. Dabrowski, P.S.; Specht, C.; Felski, A.; Koc, W.; Wilk, A.; Czaplewski, K.; Karwowski, K.; Jaskolski, K.; Specht, M.; Chrostowski, P.; et al. The Accuracy of a Marine Satellite Compass under Terrestrial Urban Conditions. *J. Mar. Sci. Eng.* **2020**, *8*, 18. [CrossRef]
44. Mayer, P.; Magno, M.; Berger, A.; Benini, L. RTK-LoRa: High-Precision, Long-Range, and Energy-Efficient Localization for Mobile IoT Devices. *IEEE Trans. Instrum. Meas.* **2021**, *70*, 3000611. [CrossRef]
45. Deng, J.; Wang, S.L. Divisional Ambiguity Resolution for Long Range Reference Stations in Network RTK. *Surv. Rev.* **2015**, *47*, 272–278. [CrossRef]
46. Berber, M.; Arslan, N. Network RTK: A Case Study in Florida. *Measurement* **2013**, *46*, 2798–2806. [CrossRef]

applied sciences

MDPI

Article

Mobile Internet Technology Adoption for Sustainable Agriculture: Evidence from Wheat Farmers

Nawab Khan [1], Ram L. Ray [2], Hazem S. Kassem [3] and Shemei Zhang [1,*]

1 College of Management, Sichuan Agricultural University, Chengdu 611100, China; nawabkhan@stu.sicau.edu.cn
2 College of Agriculture and Human Sciences, Prairie View A&M University, Prairie View, TX 77446, USA; raray@pvamu.edu
3 Department of Agricultural Extension and Rural Society, King Saud University, Riyadh 11451, Saudi Arabia; hskassem@ksu.edu.sa
* Correspondence: 14036@sicau.edu.cn

Abstract: Mobile internet technology (MIT) is considered a significant advancement in information and communication technology (ICT), due to its crucial impact on the financial system and social life. In addition, it is an essential technology to overcome the digital divide between urban and rural areas. In terms of agricultural advancement, MIT can play a key role in data collection and the implementation of smart agricultural technologies. The main objectives of this study were to (i) investigate MIT adoption and use in sustainable agriculture development among selected wheat farmers of Pakistan and (ii) examine the crucial factors influencing MIT adoption. This study selected 628 wheat farmers from four districts of Khyber Pakhtunkhwa Province (KPK), Pakistan, for sampling. This study used a bivariate probit method for sampling wheat farmers. The analysis of wheat farmer's data showed farmer's age, farm size, farm location, and knowledge about Internet technology (IT) are strongly correlated with MIT adoption in sustainable agriculture development. Results showed on average, 65% of wheat farmers have mobile devices supporting these Internet technologies, and 55% use MIT in agricultural environments. Since the extant research on MIT adoption for agriculture production in Pakistan is sparse, this study helps advance MIT adoption-based studies. These outcomes may draw the attention of decision-makers dealing with IT infrastructure and agricultural equipment who can support farmers adopting MIT.

Keywords: sustainable; agriculture; rural; mobile internet technology; bivariate probit model; agricultural modernization; Pakistan; wheat; smart

Citation: Khan, N.; Ray, R.L.; Kassem, H.S.; Zhang, S. Mobile Internet Technology Adoption for Sustainable Agriculture: Evidence from Wheat Farmers. *Appl. Sci.* **2022**, *12*, 4902. https://doi.org/10.3390/app12104902

Academic Editors: Paweł Kiełbasa, Tadeusz Juliszewski and Sławomir Kurpaska

Received: 21 March 2022
Accepted: 9 May 2022
Published: 12 May 2022

1. Introduction

Agriculture is not only a source of food but also a source of employment and opportunities, especially in rural areas. Therefore, the development and growth of agriculture, significantly advancing technology, must be highly prioritized, especially in developing countries [1,2]. This is because food security is one of the main goals of the governments of these developing countries. It is worth noting that the United Nations defined food security as all people, at all times, having physical, social, and economic access to sufficient, safe, and nutritious food, meeting their food preferences and dietary needs for an active and healthy life [3]. Agriculture plays a pivotal role in poverty alleviation and economic development. The link between agricultural productivity growth and poverty reduction is well documented by Raza et al. [4]. Advancing the agricultural system is critical to ensure food security and reduce poverty. However, in developing countries, small farmers frequently encounter several challenges that prevent them from advancing the agricultural system. These barriers include, for instance, asymmetric information regarding importing traders and exporting consumers, high transaction expenses, insufficient agricultural services, and insufficient access to credit resources [5,6]. Especially due to information asymmetry (when one party or group, or community has more information than others), small farmers, particularly rural farmers, are not equally informed and equipped with advanced technologies which help advance their skills and knowledge of farming [7]. Due to a lack of appropriate training and skills, they may not be able to adopt/use the latest equipment such as digital technologies (e.g., mobile phone, internet, and computer) and inputs (e.g., seeds, fertilizers, fungicides, and pesticides or effective use of existing agricultural inputs) to increase agricultural productivity [8]. Therefore, the crop yields and incomes of these farmers are low, which is not conducive to their livelihoods and rural development [9–12]. Hence, advanced methods used to reduce information asymmetry are worthwhile, especially when using information and communication technology (ICT) to improve farm performance and overall agricultural productivity.

Technological innovations are becoming increasingly crucial in agricultural development and productivity. The use of ICTs in agriculture provides a more efficient and cost-effective way of sharing and exchanging knowledge more widely. Farmers benefit from access to key information such as pest reports, weather conditions, and market prices [13]. The use of Internet technology (IT) can help reduce information asymmetry as it disseminates information fast and at a low cost. Mobile phones have a long history of only being used for voice communication and text message. Recent advancements in mobile phones, including smartphones, revolutionized using the internet on mobile phones. The essential primary features of mobile phones allow users to access the internet without using a computer. Internet access through a mobile phone influenced us, including farmers [14]. Previous evidence has proven that the adoption/use of mobile Internet technology (MIT) can improve the accessibility of financial and agricultural services for smallholder farmers [8,15], the availability of input and output markets [16], and the promotion of income activities (such as non-agricultural commodities). Due to the significant advantages of MIT, some developing countries have adopted many Internet-based agendas to implement farms better for rural development [17–19]. For example, the internet + agriculture + finance model, rural e-commerce, and farmer's field school are Internet-based agendas used in China [20].

The ICT has drastically modified communication, sales and information methods [21]. Several studies have investigated the effects of Internet use by ICTs, such as mobile phones and computers on-farm performance and rural household's productivity income [22–24]. These studies focused on the selection bias of IT used through the application of various methods such as instrumental variables (IV), endogenous therapy regression (ETR), and propensity score matching (PSM) models [20]. By estimating the PSM model, Issahaku et al. [25] found that mobile phone use significantly improves agricultural productivity in Ghana. Ma et al. [26] estimated an ETR model and showed that Internet use significantly increases household income and expenditure in rural China. Kelemu [27] found a significant impact of mobile phones and mobile communications on improving wheat productivity and efficiency in Ethiopia. In addition, Quandt et al. [28] examined perceptions about the effects of mobile phone use on agricultural productivity in rural Tanzania, East Africa. They reported that about 47% of respondents stated that mobile phone usage had reduced the amount of time they spent buying inputs or selling crops, and 50% reported that mobile phone usage had reduced the amount of money they spent on-farm activities. Further, 64% reported that the mobile phone usage has increased profits from farming compared to when the respondent did not have a mobile phone.

The adoption/use of MIT may affect the production of crops because it may influence farmer's production behaviors in combining and using different inputs (e.g., labor, capital assets, fertilizers, and pesticides) [29]. Technical efficiency refers to the ratio of farmer's observed output to the maximum realizable output given the existing inputs [30], indicating the use efficiency of different agricultural inputs. Several studies show that the adoption/use of MIT will significantly affect the behavior of farmers to use seeds and fertilizers [18] and land expansion [31]. To the best of our knowledge, apart from work completed for Zambia by Mwalupaso et al. [32], sparse findings have investigated the influence of MIT use on crop production. Mwalupaso et al. [32] studied the impact of mobile phone access to IT on corn production in Zambia. They observed that mobile phones have greatly improved the technical efficiency of farmers.

This article aims to examine MIT adoption by wheat farmers in Pakistan. A bivariate probit model is employed using a representative data set of 628 wheat farmers from four districts of Khyber Pakhtunkhwa (KPK) province in Pakistan for sample selection. We are concerned about wheat farmers for two purposes. First, Pakistan is the eighth leading producer of global wheat production, after China, India, Russia, the United States, Canada, Australia, and Ukraine (Figure 1). Despite the high total yield, Pakistan's wheat yield is only three tons per hectare, ranking 15th globally [33]. Since lower wheat yield is reducing the farmer's income and weakening the effectiveness of the Pakistani wheat industry in the global market, the development of wheat production must be promoted. Secondly, Pakistan's broadband Internet has developed rapidly in recent years, especially in remote areas. Currently, Internet users using any device have reached 76.38 million. This means that 35% of the population is using IT. The use of the Internet plays an important role in farmer's lives and agricultural production. In addition, technological innovation has greatly shaped agriculture. Humans have developed new ways to make farming more efficient and grow more food [29]. As mentioned above, due to the urgent need to increase wheat productivity, this research can reveal whether the adoption/use of the MIT can affect farmer's decision to utilize inputs wisely to increase wheat yield in Pakistan.

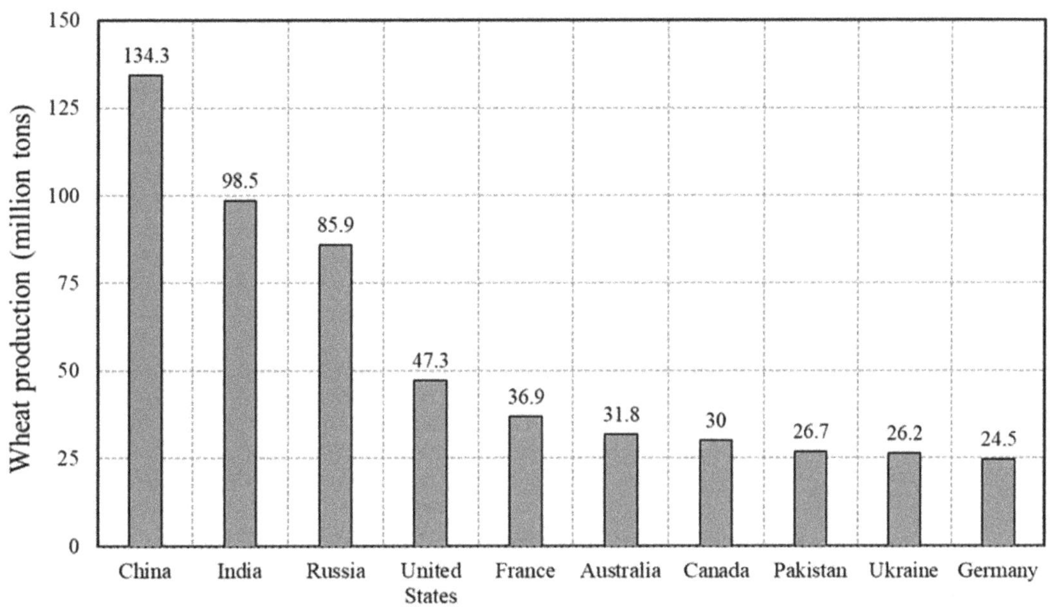

Figure 1. Total wheat production in the top ten countries in 2021. (Data source: FAOSTAT).

In order to determine the important factors that influence the MIT adoption in agricultural advancement, it is assumed that this set of factors can affect the mobile devices (MDs) adoption, including farmer's characteristics, MIT, and farm characteristics. Moreover, factors that influence the knowledge provided by MIT, which may be useful to decision makers and companies, are also identified. This information can be implemented in the marketing policies of agricultural systems that depend on MIT and equipment such as precision and intelligence. The main objectives of this study were to (i) investigate MIT adoption and use in sustainable agriculture development among selected wheat farmers of Pakistan and (ii) examine the crucial factors influencing MIT adoption.

The remainder of the article is organized as follows. Section 2 provides various study hypotheses based on a review of extant literature. Section 3 introduces the model specification, study area, sampling strategy, and data collection. Section 4 describes the results and discussion of the study hypothesis. Section 5 presents the research conclusions.

2. Literature Review

A growing number of researches focus on the impact of ICTs on farmers. The dissemination and adoption of MIT in developing countries have attracted the attention of economists. The broad growth in MIT advancements over the past decade offers innovative prospects to overcome these search and deal prices and the potential to enhance agricultural productivities and incomes [34,35]. In particular, the theory proposes that MIT affects many of the main channels through which smallholder's incomes are affected. First, MIT may reduce the search cost for farmers, obtaining price information in more markets and selling in the market with the highest price after deducting transportation costs [36,37]. Second, improved access to information may enhance the bargaining position of farmers with traders in the absence of sales in different markets [38]. Third, MIT may allow farmers to use mobile phones to complete sales, thereby reducing the uncertainty associated with sales in remote markets [39]. Fourth, if information technology increases farmer's commodity prices, and since agriculture is price elastic, this will increase the future production of such commodities [40,41].

From an arbitrage perspective, past studies have shown that if farmers could obtain selling price information in other markets through ICT, and the transportation cost is lower than the price difference between these two markets, farmers would go to other markets to sell their products for better profits. As a result, arbitrage that occurs between markets will decrease, and Pareto efficiency will be achieved [34,42]. Along this line, Jensen [43] observed the mobile phone's role in market arbitrage in local agricultural markets in Indian states. The key outcomes suggest that the introduction of mobile phones stimulated cross-market arbitrage, reduced price differences across markets, and eliminated the oversupply of sardines in individual local markets. As a result, both the producer's profit and the consumer's surplus increase. Meanwhile, Aker [39] examined mobile phones' impact on the grain market in Niger and initiated that price differences between markets have decreased. A significant additional study looked at the ICT impact on agricultural development and market choices, focusing on mobile internet and radio technology [34].

Especially in developing nations, farmers often can sell their products to merchants who shuttle between villages and markets or transport produce to the nearest market themselves. Uncertainty about market prices is usually high for farmers due to remote communities and poor market communication. Courtois and Subervie [44] demonstrated the conditions under which market information services are beneficial to farmers and examined efficiency issues related to information asymmetry. The causal impact of the mobile phone market information services program on farmer's marketing performance in Ghana showed that farmers who benefited from a market information service program received significantly higher prices for peanut and maize: about 7% higher for peanut and 10% higher for maize than what they could have received had they not participated in the market information services program. Additionally, Zanello [45] used a new dataset of 393 households in northern Ghana with comprehensive information on market transactions, agricultural development, and ICT use through a transaction cost context. The results showed that obtaining market info through mobile technology had a positive and substantial influence on market participation, with a better effect on households with surplus food crops and agricultural progress.

In China, the advancement of modern technology increased China's rural non-farm employment by 10–20% and ultimately upsurged the income of rural people [34]. In conclusion, the expansion of ICT in developing countries provides new technologies and opportunities for farmers to access information. Remarkably, the rapid development of computers and MIT have introduced a new search technology that offers numerous advantages over other alternatives in terms of cost, geographic coverage, etc. Compared to broadcasts and newspapers with one-way communication systems, information can be obtained more efficiently using modern information systems. Especially in developing countries, the rapid growth of MIT access and computers have ushered in an era in which everything is changed by information technology, which has an equal effect on rural farmers who have traditionally been far away from the development of information technology [46].

Specifically, timely access to market information through communication networks can help farmers decide what crops to grow, where to sell their products, and how to improve input efficiency. ICTs can also provide unprecedented access to rural finance, while financial and information service networks can provide microfinance opportunities for locals and small businesses [34,46–48]. As mentioned earlier, statistics and some empirical studies can provide general patterns for mobile telegraphy. However, research on mobile Internet technology development patterns and their consequences based on raw data is particularly significant.

3. Study Hypothesis for the MIT Adoption

According to Savari and Gharechaee [49], the widespread dissemination of innovation theory has been used in some scientific disciplines to describe advanced technology adoption through individuals, society, or organizations. This hypothesis reflects many variables expected to affect the adoption of modern technologies. The variable set contains adapters and innovations as well as firm characteristics. To the best of our knowledge, there is a lack of understanding of MIT used in agricultural development among small farmers in developing countries, including Pakistan. Therefore, these assumptions are also generally considered from the literature on adopting information technology and MIT. The study also reviewed the literature on farmers' adoption/use of computers and the Internet.

3.1. Farmer's Characteristics

Savari and Gharechaee [49] analyzed the MIT by American farms and determined that the adoption rate of the Internet and computers is declining with age. Indeed, as the descriptive statistics show, young people have a higher Internet access rate. Bort-Roig et al. [50] found that the use of MIT is higher among young people. Therefore, in terms of the adoption and usage of MIT, it is not surprising that some studies have shown that the adoption rate decreases with age [51]. Therefore, the following conditions are assumed:

Hypothesis 1a. *(Age): Older farmers are less likely to adopt MIT.*

Gender plays a crucial role in information technology adoption decisions [49,52]. Compared with male, female farmers are less likely to adopt modern technologies [49,53]. However, when it comes to adopting the internet for agriculture, the results are mixed. For example, Mendes et al. [54] found no association between gender and Internet adoption, while Khan et al. [29] found that gender was correlated with internet adoption. Specifically, male farmers are more likely to become MIT adopters. Haq et al. [55] pointed out that more and more men use MDs to access the internet. Regarding the adoption of MIT, Mayzelle et al. [56] found that the adopter is more likely to be male. Hence, the following assumption was made:

Hypothesis 1b. *(Gender): Male farmers are more likely to adopt MIT.*

Education develops the ability of individuals to understand, learn and adopt new technologies, such as the internet [29], and therefore is a significant socio-economic factor for information technology adoption [57–59]. Based on this insight, Fabregas et al. [60] pointed out that more formal schooling is optimistically associated with the use of IT in agricultural development. Generally speaking, well-educated adults are more likely to access the internet [20]. In addition, MIT adoption and education are also positively correlated [23]. Additionally, Nie et al. [61] believed that education is an important factor influencing IT adoption. Therefore, the following assumption was made:

Hypothesis 1c. *(Education): farmers with higher education are more likely to adopt MIT.*

Innovation is manifested in the willingness to check the latest technology [62] and is an important factor affecting the latest technology adoption. Haile et al. [53] revealed that innovation is positively associated with precision agriculture adoption in the agriculture field. Regarding MIT, Singh et al. [63] pointed out evidence that innovation is associated with MIT adoption. Therefore, the following assumption was made:

Hypothesis 1d. *(Innovativeness): More advanced farmers are more likely to adopt MIT.*

3.2. Mobile Internet Technology Characteristics

Sekabira and Qaim [64] indicated that some farmers did not utilize IT for safety reasons. In addition, Rehman et al. [65] pointed out that overall, security issues are the main obstacle to the adoption of IT. O'Leary et al. [66] found that security and privacy risks are the main issues when using mobile information services. For example, IT can spread diseases, where hackers can intercept signals, thereby imperiling the safety of individuals' data transmitted. Consequently, the following assumption was made:

Hypothesis 2. *(Awareness of IT risks): farmers aware of IT risks are unlikely to adopt MIT.*

3.3. Farm Characteristics

Because of several factors, such as the high demand for information on the farm and the complexity of the organization, large farm size is positively associated with MIT adoption [67]. Generally, the size of the server farm will affect the information technology application [68,69], and it is also believed that large-scale farms are more likely to adopt MIT. Consequently, the following research hypothesis was made:

Hypothesis 3a. *(Farm size): Large farm farmers are more likely to adopt the MIT.*

Internet-based regional obstacles are usually the outcome of the geographic environment of digital communication infrastructure and the absence of digital connection [26]. Regarding agricultural and IT adoption [60,63,67], evidence showed that the regional farm location is strongly related to higher IT adoption rates in the United States and Greece. About MIT, Matassa et al. [70] indicated that due to the decentralized digital infrastructure, the adoption of MIT could be influenced by the location of farm and agricultural activities. According to the information specified via the government of Pakistan, the MIT coverage in northwestern Pakistan seems to be relatively underdeveloped. Considering all those facts, the following hypothesis was assumed:

Hypothesis 3b. *(Region): The farm's location in the northwestern area is adversely associated with MIT adoption.*

The impact of farm diversification on agricultural technologies, including computers and the internet, has shown various interesting findings. For example, Alvarez et al. [71] and Roco [72] found no connection between farm diversification and equipment adoption, while Sekabira and Qaim [64] found a negative correlation between farm diversification and the adoption of IT. According to Roco [72], the diversification of farms is positively related to IT and MIT adoption. The owners of relatively diversified farms must gain more knowledge to make agricultural decisions, hence more likely to MIT adoption. According to the findings of Kaila and Tarp [73], diversified companies/farms have higher requirements for information technology. Although the outcomes obtained are different in the past study, we assumed the following hypothesis:

Hypothesis 3c. *(Farm diversification): Farms diversification is correlated to MIT adoption.*

4. Material and Methods

4.1. Model Specification

The dependent variables can be divided into two different sections. The first section is the MDs adoption (selection stage), which is the binary result of deciding whether the farmer adopts MDs that support the internet ($y_1 = 1$; otherwise, $y_1 = 0$); the second section (outcome stage) is the binary result, which determines whether the farmer adopt MIT ($y_2 = 1$; otherwise $y_2 = 0$). The probit method is a common econometric model which contains dependent variables and obtains binary results through maximum probability assessment [74]. By estimation, the two probit models for MDs adoption and MIT disregard the apparent association between the two. The bivariate probit method considers this association, which expands the probit technique [75]. However, the bivariate probit approach does not fully consider the predictable association between MDs adoption and MIT. More specifically, the MDs' adoption ultimately determines the probability of adoption of the MIT. Therefore, the selection stage result can only be observed if the farmers adopt the MDs. Hence, selection of sample bias may appear if the second result observation is not a random sample from the population [76].

The econometric model used to solve this issue is the bivariate probit method for sample selection. This model is based on the well-known Heckman selection model [77]. Alemi et al. [78,79] utilized the bivariate probit method with sample selection to examine smartphone use and mobile broadband adoption in Sweden. The following three types of outcomes were observed in this study [80]:

(i) Wheat farmers do not adopt MDs ($y_1 = 0$);
(ii) Wheat farmers adopts MDs but do not use MIT ($y_1 = 1$; $y_2 = 0$);
(iii) Wheat farmers adopts MDs and use MIT ($y_1 = 1$; $y_2 = 1$).

Therefore, the following probabilities apply to all three categories of potential findings in the sample:

$$Y_1 = 0,\ \Pr(Y_1 = 0) = \Phi(-X_1\beta_1) \tag{1}$$

$$Y_1 = 1,\ Y_2 = 0,\ \Pr(y_1 = 1, y_2 = 0) = \Phi(x_1\beta_1) - \Phi_2(x_1\beta_1, x_2\beta_2, \rho) \tag{2}$$

$$Y_1 = 1,\ Y_2 = 1,\ \Pr(y_1 = 1, y_2 = 1) = \Phi_2(x_1\beta_1, x_2\beta_2, \rho) \tag{3}$$

The following log-likelihood function can be created by taking these probabilities into account:

$$\mathrm{InL} = \sum_{i=1}^{N} Y_{i1}Y_{i2}\mathrm{In}\Phi_2(X_1\beta_1, X_2\beta_2, \rho) + Y_{i1}(1 - Y_{i2})\ \mathrm{In}[\Phi(X_1\beta_1) - \Phi_2(X_1\beta_1, X_2\beta_2, \rho)] + (1 - Y_{i1})\ \mathrm{In}\Phi(-X_1\beta_1) \tag{4}$$

where Pr represents the probability that the farmer makes a binary decision, Y denotes the dependent variable of the selection and outcome equation, X denotes the independent variable vector for both farts, β is the estimated coefficient of each independent variable, and Φ is the unit normal distribution function. ρ represents the association between the two equation errors. Figure 2 shows the sample collection method of the proposed model.

According to Sartori [81], the instrumental variable is usually used to estimate the Heckman model in the selection phase, and this variable is not used in the outcome phase. No additional variables are required to determine the coefficients in the model. However, since the identification is based only on the parameter assumptions of bivariate normality and there are no missing variable deviations, Sartori [81] emphasizes that estimation techniques with similar descriptive variables are not recommended in the two equations.

Therefore, researchers have no other options other than to unblock an additional explanatory variable for the select equation or just classify it from the distribution hypothesis regarding the residuals [81]. In order to solve this problem, the outcomes of the proposed method are compared with the findings of the probit model, the bivariate probit model of the sample selection without instrumental variables, and the Sartori selection model [80,81].

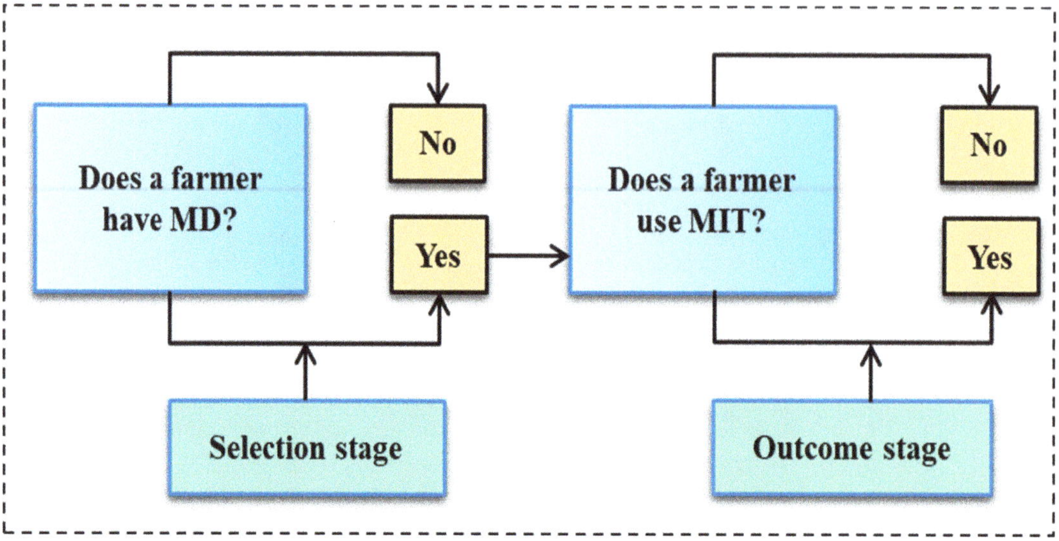

Figure 2. The bivariate probit model for sample selection for MIT adoption via wheat farmers. Note: MD = Mobile device; MIT = Mobile internet technology.

4.2. *Study Area, Sampling Strategy, and Data Collection*

4.2.1. Wheat Production in the Study Area

Wheat is mainly produced in Punjab, Sindh, Balochistan, and the Khyber Pakhtunkhwa provinces of Pakistan. In this research, KPK province was selected as a study area. The study area has favorable climatic conditions for wheat and can produce high-quality wheat. However, wheat production in this region mainly relies on rainwater, and about 40% of farmers use irrigation. Small farmers, as cash crops, dominate wheat production. The sector needs modern technologies (e.g., MIT, MDs, and the Internet of Things) to support wheat production. However, in this study area, the mechanization rate is still relatively low, challenging wheat production. Wheat production is affected by pests and diseases globally, and Pakistan is no exception. Therefore, farmers usually replace wheat plants with new plants after 3–4 years of high yield to eliminate specific types of bugs that only attack wheat, better soil health and crop yield, etc. In 2020, wheat was planted at 772.3 thousand hectares of cropland in KPK and produced about 1400.5 thousand tons of wheat.

4.2.2. Sampling Strategy and Data Collection

The study was conducted in KPK Province, Pakistan, from January to March 2021. In total, 650 questionnaires were distributed to the wheat farmers: a total of 628 questionnaires were considered fairly reliable to collect the data needed for this study, and 22 questionnaires were not considered in this study because they were not complete. This study used multistage random sampling techniques to collect the essential information from wheat farmers face to face (hard copy questionnaire distributed to the farmers in person). In order to understand the first stage of the MIT adoption by wheat farmers in KPK province, data were collected from four districts, namely, Dera Ismail Khan, Charsadda, Mansehra, and Swat, depending upon the share of agriculture production in these areas (Figures 3 and 4).

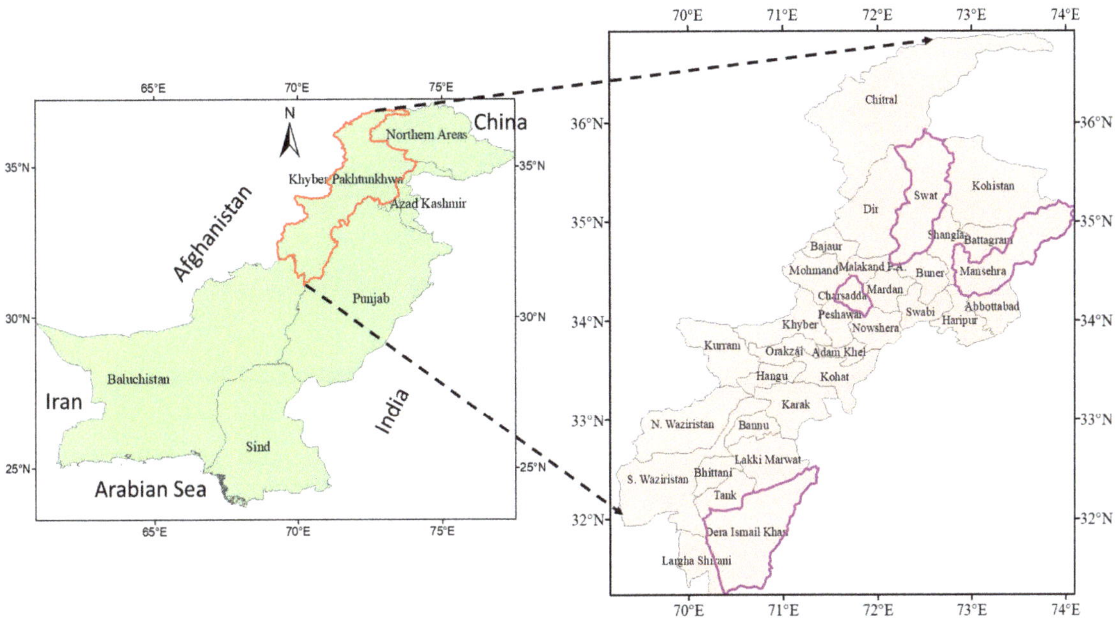

Figure 3. Map of the study area. The selected four district's boundaries are shown in pink color [37].

In the second stage, one tehsil was selected to fill out questionnaires, and in the third stage, one union council was targeted from each tehsil. In the fourth stage, four villages were focused randomly on each selected union council, and finally, the essential data were collected from wheat farmers from the studied villages. The questionnaires used in this study were divided into different parts. The first portion of the organized questionnaire contained the demographic and socio-economic characteristics of the respondents. The rest of the questionnaire aimed to obtain information about the MIT adoption by wheat farmers. The questionnaire was originally written in English and later translated into Urdu for the ease of the interviewees.

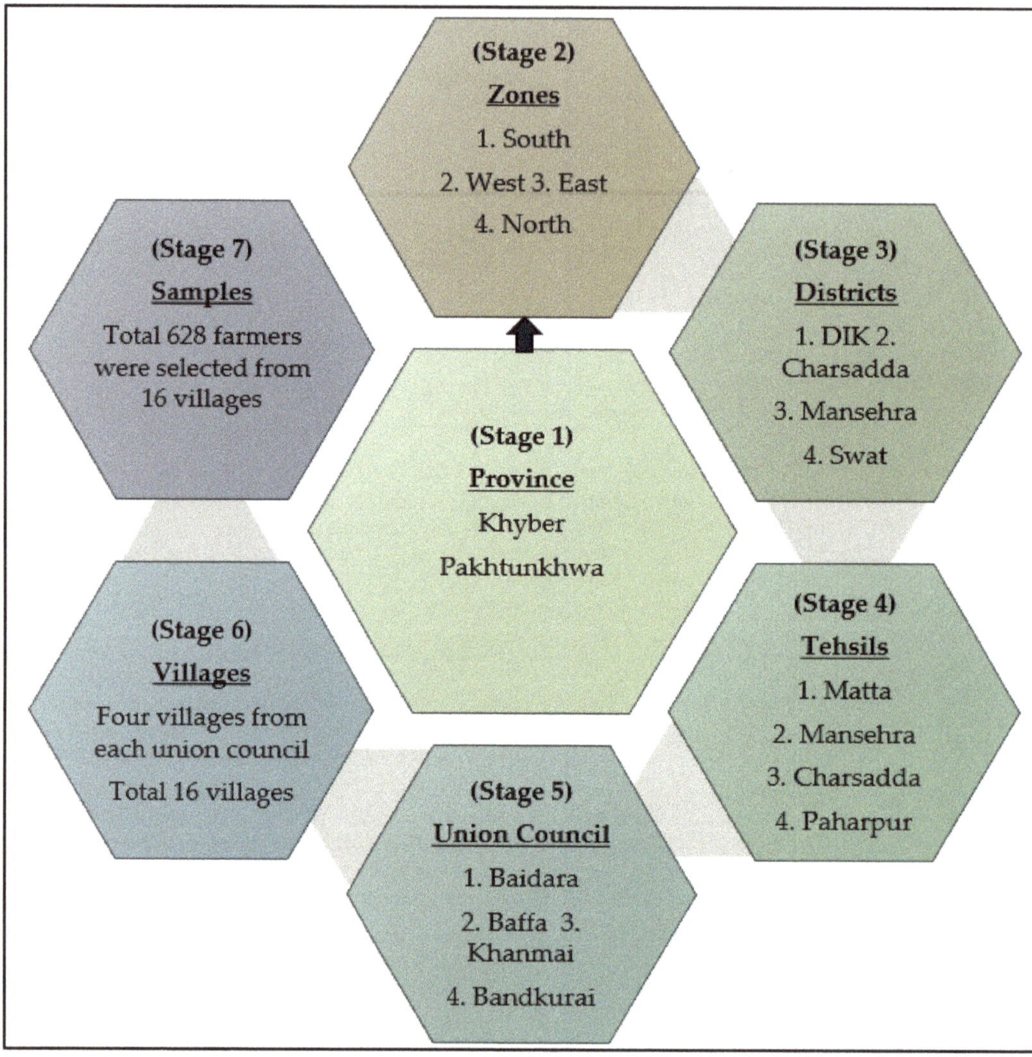

Figure 4. Sample distribution.

5. Results and Discussion

Table 1 indicates a statistics summary of the entire variables in the econometric assessment. On average, 65% of wheat farmers have MDs supporting these Internet technologies, and 55% use MIT in agricultural environments. In 2021, 61.64 million of the Pakistani population have accessed MIT via smartphones and/or tablets, which is less similar to the proportion of Pakistani farmers who use MIT through such MDs. Regarding sociodemographic variables, the average age of the respondents was 47.88 years old, 88% were males, and 17% of farmers had university degrees. The farm area varies from 5 to 20 ha, with an average of 7.17 ha. Table 1 also includes data regarding the area distribution of farms across Pakistan. For example, 24% of farms are located north of KPK. The farmers were also queried to utilize an isometric five-point Likert scale to express agreement or disagreement with two declarations. In order to measure innovation, farmers were questioned whether they were interested in using modern technologies or innovations, and on average, they slightly refused this (2.26%). Next, the interviewees were asked whether they had enough knowledge to avoid harm to IT; on average, they also slightly rejected it (2.70%). In accordance with the framework of Adapa [82], the frequency of fixed IT usage in the MD adoption step (not in the result step) was calculated to illustrate the sample selection technique introduced in the prior portion. Moreover, 74% of the farmers surveyed use the fixed internet every day.

Table 1. Variables descriptions and descriptive statistics.

Variables	Hypothesis	Explanation of Research Hypothesis	Mean (SD)
Dependent variables			
MIT		1 if farmers use MIT; 0 otherwise	0.55 (0.04)
MD		1 if the farmer has Internet-enabled MD; 0 otherwise	0.65 (0.02)
Independent variables			
Age	Hypothesis 1a	Age of the farmers (years)	47.88 (11.77)
Gender	Hypothesis 1b	1 if the farmer is male; 0 otherwise	0.88 (0.03)
Education	Hypothesis 1c	1 if the farmers hold a university degree; 0 otherwise	0.17 (0.03)
Innovativeness	Hypothesis 1d	Once a new technological innovation arrives on the market, I will be interested in assessing it	2.26 (1.07)
Awareness of IT risks	Hypothesis 2	Aware to avoid the IT risks	2.70 (1.12)
Farm size	Hypothesis 3a	Farm size (ha)	7.17 (6.01)
Region	Hypothesis 3b		
North		The farmhouse is situated in the north of KPK	0.24
West		The farmhouse is situated in the west of KPK	0.26
South		The farmhouse is situated in the south of KPK	0.42
East		The farmhouse is situated in the east of KPK	0.07
Farm diversification	Hypothesis 3c	Measurement of the farmhouse diversification	0.25 (0.22)
Control variable			
IT usage (Regular)		1 if farmer use IT regularly; 0 otherwise	0.74

Note. Standard errors (S.E.), Mean, and SD is shown as the ratio for MDs = 1.

Table 2 displays the evaluation results of the bivariate probit method for sample selection. Software version STATA 14 was used for analysis. The selected model Wald test was statistically significant ($p < 0.01$) and rejects the null hypothesis of simultaneous equality to zero of selected coefficients. The likelihood ratio test of $p = 0$ was rejected at the 1% significance level. The outcome shows that it was necessary to use the sample selection method.

The outcomes of the selection phase indicate that farmers who utilize the fixed IT daily are more likely to use MDs that support IT because the coefficient is statistically substantial and positive. This is consistent with the outcomes of Adapa [82]. In addition, young farmers with university degrees are more likely to use/adopt MDs. Additionally, similar outcomes were obtained by Hou et al. [34] for the adoption of tablets and smartphones. No association was found between the adoption rate of MDs and gender, farm size, and farm diversification.

Table 2. Outcomes of the bivariate probit model, including sample selection of MDs and MIT adoption for wheat farmers.

Variables	Hypothesis	MD Adoption (SS)	MIT Adoption (OS)
		Coefficient (S.E)	Coefficient (S.E)
IT use (Regular)	-	0.7802 *** (0.0986)	-
Age	Hypothesis 1a	−0.0218 *** (0.0044)	−0.0201 *** (0.0054)
Gender	Hypothesis 1b	0.0202 (0.1578)	0.0617 (0.1677)
Education	Hypothesis 1c	0.3177 ** (0.1335)	0.0299 (0.1347)
Innovativeness	Hypothesis 1d	-	0.192 8 *** (0.0485)
Awareness of IT risks	Hypothesis 2	-	0.0955 ** (0.0463)
Farm size	Hypothesis 3a	0.033 (0.015)	0.010 * (0.005)
Region		Hypothesis3b	
North		-	0.3344 *** (0.1277)
West		-	0.4798 *** (0.1263)
South		-	0.3135 (0.1134)
East		-	0.0740 (0.2090)
Farm diversification	Hypothesis 3c	−0.1066 (0.2106)	−0.1102 (0.1347)
Constant		0.8708 *** (0.2890)	0.4165 (0.3383)
Atanh (p)		-	−1.3407 *** (0.2809)
p		-	−0.8718 (0.0674)
Likelihood ratio test for $p = 0$		21.42 ***	-
Wald x^2		42.55 ***	-
Log-likelihood		−777.88	-

Note: SE, standard error, * $p < 0.10$, ** $p < 0.05$, and *** $p < 0.01$. (SS) denotes the selection stage, and (OS) outcome stage.

5.1. Study Hypothesis 1a–d

Figure 5 summarizes the hypothesis test outcomes. H1a indicates the age impact on the adoption of MIT, and the coefficient has a negative sign in statistics, indicating that, ceteris paribus, the older age is negatively correlated with the MIT adoption. Therefore, H1a could not be rejected. This outcome is consistent with the aforementioned research on MIT and the internet in sustainable agriculture development. Young farmers may be more interested in the use of innovative technologies, as [83,84] also pointed out. In addition, skills used in conjunction with information technology and MDs are generally well among young farmers [85]. It is consistent with the previous findings. For example, Woodburn et al. [86] found that older farmers have less computer and smartphones experience. However, young farmers have less agricultural experience [87]. Young farmers may use MIT as a source of additional knowledge for decision making. In short, young farmers are more likely to adopt MIT.

H1b assumes gender variances in the MIT adoption among wheat farmers. The measurement has anticipated indication but is not statistically positive at the 10% significance level. Therefore, the analysis cannot support H1b that gender is associated with MIT adoption. This is consistent with the regression outcomes of Yang et al. [88]. Although the past study shows that men are generally enthusiastic about modem technologies [53,89], especially mobile phones [90], their research shows that by comparing the statistics on the adoption of MIT over time, the gender difference is narrowing quickly [50]. By comparing the regression outcomes in 2017 and 2019, it was found that the statistical indication of gender is no longer significant for MIT adoption. This may describe the fact that there is also no association between the farmer's gender and the MIT adoption in the study. Therefore, men and women farmers have equal opportunities in the MIT adoption.

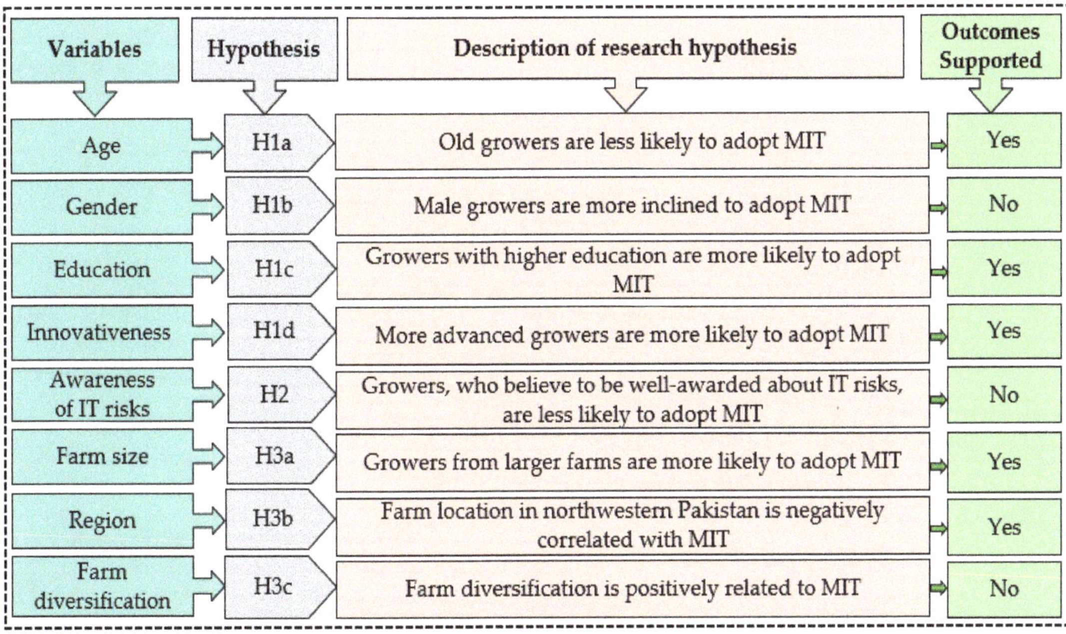

Figure 5. Explanation of the research hypothesis assessment outcomes.

According to H1c, the impact of farmer's education on MIT adoption was verified. The coefficient has an anticipated indication, but it is significant at the five percent level of significance. Therefore, the research cannot confirm H1c, which is that farmer's schooling is associated with the adoption of MIT (ceteris paribus). The outcomes are consistent with Islam et al. [91] and Khan et al. [92], but contrary to the previous research results, the latter shows an optimistic correlation between schooling and MIT adoption. However, as presented in the third column of Table 2, there is a significant correlation between education and MDs adoption. Therefore, if a farmer can generally utilize a smartphone or tablet due to his/her formal schooling, he/she already has the right to use the MIT. Furthermore, the educational impact on MIT adoption is possible because education makes it easier for farmers or anyone to process information [93]. In addition, well-educated farmers may also have more requirements for modern information [94]. Therefore, farmers with higher education can make more use of MIT to collect information. However, according to our results, education and MIT adoption are highly related.

H1d tested the impact of farmer's innovation on MIT adoption. Predictably, innovative farmers with ceteris paribus are more likely to adopt the MIT, because of the positive coefficient and has high statistical significance. Therefore, the analysis cannot reject H1d. These outcomes are consistent with the results of Hoang et al. [95]. Their outcomes indicate that the innovation of farmers' self-reports is significantly associated with their optimistic attitudes to mobile use [96]. This is also consistent with research on the use of technologies other than agriculture [97], because innovators adopt/use the latest/advanced skills and products faster than others. Therefore, innovative farmers adopt MIT faster than other non-innovative farmers.

5.2. Study Hypothesis 2

H2 shows the perception of farmers being informed regarding the dangers of IT. This coefficient has no estimated positive indicator and is statistically substantial. Therefore, the analysis could not deny H2 that farmer's awareness of IT danger (e.g., cyberbullying) is associated with MIT adoption (ceteris paribus). While it may not cause monetary loss, infringement of user privacy is still the main concern of numerous internet users [98]. In addition, most farmers use the Internet and its applications for private purposes and primarily for commercial reasons [20,67]. Hence, they can process not only personal data but also commercially associated data that is considered very sensitive on IT. Intuitively, farmers can foresee the dangers of IT and therefore do not want to adopt/utilize MIT because they want to preserve their privacy and maintain the security of business-related information. The outcomes of our research can be elucidated as follows: The well-informed farmers may also understand how to develop suitable methods to safeguard their privacy when utilizing MIT and are thus more likely to adopt MIT.

For example, educated or informed farmers screen apps or websites before using the certificates. It has been indicated that certificates can support individuals, thereby raising the likelihood of online acquisitions [99]. Fecke et al. [100] also indicated that agribusiness utilizing e-commerce would consider ascertaining seals or certificates to upsurge trust. In addition, the outcomes also indicate that digital training courses for farmers would elucidate risks of internet use and how to develop suitable security instruments to promote farmers to use mobile information services.

5.3. Study Hypothesis 3a–c

H3a–c addresses the impact of farm characteristics on MIT adoption, and the farm size is significantly associated with the adoption of MIT. However, the coefficient has an optimistic indication and is statistically substantial at a significant level of 10%. Therefore, the research could not refuse H3a, which is that the size of the farm is related to MIT adoption. Large farms may face more multi-faceted decision making and higher complexity of the organization [101]. Thus, the MIT could be utilized to establish robust strategies for the effective business of the farm, such as banking and the acquisition of operating resources for sustainable agriculture development. In addition, it is possible to contact employees and consultants through an MIT-based messenger facility [67,102], which also supports the fact that farmers having huge farms may have a critical need for the latest technologies. Therefore, farmers on large farms may use MIT to gather data quickly. Specifically, the MIT allows farmers to obtain the price and weather information that is varied both in location and time. In summary, farmers who manage large farms have a higher chance of using the MIT.

It is estimated that the location of the farm is associated with MIT adoption verified with H3b. After estimation, the joint significance test shows that the coefficient of farm location is equivalent to zero. The assessment is statistically substantial (x2 (3) = 16.17, $p < 0.01$) and rejects the assumption that zero-coefficient is statistically insignificant. Therefore, the study could not refuse H3b. The northwestern region was set as a basic category in the initial econometric examination. The model expresses no statistically substantial variance between farmers whose farmhouses are situated in the eastern and northwestern regions. However, compared to the north and west fin, southern Pakistan is less likely to adopt MIT. Roco [72] proposed that the difference in digital infrastructure is the reason for the behavioral change in MIT adoption.

In addition, Bellon-Maurel et al. [103] indicated that the location could be understood as a proxy for internet access. Regarding the coverage of mobile broadband, the information provided by the current study shows that the coverage and long-term evolution of general mobile telecommunications services in the east of KPK are much smaller than in other regions, which may explain the results. In addition, cultural variances (for example, most of the farmers in Pakistan are more conservative) may become an obstacle to adopting innovations such as the MIT. Though this study did not explicitly consider this dimension, it can be inferred that the farm location will affect the adoption of MIT.

Finally, H3c addresses the farm diversification influence on MIT adoption, and the selected model shows that farm diversification is not correlated to MIT adoption. Therefore, this research could not facilitate H3c that farm diversification is significantly related to the adoption of MIT. This outcome is consistent with Roco [72] view on the adoption of computers. However, diversified farms may have a higher need for skills and MIT usage for several production functions and information collection. It is conceivable that there is a positive correlation between farm diversification and the use of MIT. However, the outcomes show that MIT adoption has nothing to do with the diversification of farms.

6. Conclusions

The study analyzed a representative data set of 628 wheat farmers to understand the MIT adoption and use in agriculture development. This study used the bivariate probit method for sample selection to examine the crucial factors influencing MIT adoption. The outcomes indicate that farmer's age (H1d) and farm's size (H3a) correlate with MIT adoption. In addition, the study findings suggest that farm location (H3b) is related to farmer's MIT adoption rate. Additionally, educated (H1c) and innovative (H1d) farmers are most likely to adopt MIT. However, the outcomes showed gender (H1b), awareness of IT risks (H2), and farm diversification (H3c) are not correlated with the MIT adoption.

Farmers who fully understand the IT danger are more inclined to adopt MIT, which is counterintuitive. The fact that well-informed farmers may have put safety measures to deal with possible risks may explain this result. The implication of this result has two aspects: the apprenticeship should include the digital aspect so that farmers are aware of the potential risks of IT use. Correspondingly, information on measures to ensure online security will also be provided. In addition, agricultural equipment providers that depend on MIT (such as smart agricultural technology) should know farmer's safety issues and strive to clarify the dangers related to MIT to decrease unwillingness in the adoption process. This may also be achieved by providing a certificate or stamping.

The innovations of large farms and young farmers are the focus groups of marketing pursuits because they are expected to become MIT adopters. For example, suppliers and providers of cutting-edge and modern agricultural technology can focus on marketing mobile phones to achieve their focus group. These findings indicate that the young farmers of large farms are expected to utilize MIT. Therefore, providers can emphasize the possibility of integrating MDs and IT with these tools for this focus group. Moreover, this research benefits the farmers to increase their agricultural production by adopting MIT, which ultimately enhances their skills for modern agriculture. On the other hand, this study also benefits policymakers to understand the advantage of MIT for modern agriculture and their role in supporting farmers through IT infrastructure development. However, policymakers of this region/country should consider expanding the coverage of MIT in remote areas. The results also highlight the need for wheat farmers to have MIT services in the study area. To gain more insights, consider the farmer's location and satisfaction with mobile broadband coverage, not just the farm's location.

This study has some limitations. Firstly, the current research was conducted during the COVID-19 pandemic issues. Secondly, due to financial concerns, this study only focused on four districts in the KPK province of Pakistan. Hence, an inclusive study is needed in future research. The data essentially cover one province in Pakistan, so it is difficult to generalize the conclusions at the national level. Therefore, future studies should use more representative samples, which may have wider implications for rural Pakistan. Possible research must examine how and to what extent MIT use promotes farmer's income and market participation to facilitate a supply chain asset and financial efficiency perspective. However, this study provides different starting points for other research projects. For instance, this research can be applied to other developed and developing countries. An in-depth analysis of farmer's awareness and familiarity with specific MIT risks, such as phishing, should also be carried out. It is also worth examining how farmers specifically integrate MIT and related internet content or associated applications into their farm business responsibilities. Additionally, it may be interesting to see how mobile and fixed internet technologies differ in retrieving farm business-related information.

Author Contributions: N.K., R.L.R., H.S.K. and S.Z. developed and outlined this concept, including method and approach to be used; N.K. and R.L.R., developed and outlined the manuscript; N.K. and S.Z. contributed to the methodology and revision of this manuscript; N.K. and R.L.R. wrote the article. All authors have read and agreed to the published version of the manuscript.

Funding: The authors extend their appreciation to the researchers supporting the project of (No: RSP-2021/403) King Saud University, Riyadh, Saudi Arabia.

Institutional Review Board Statement: Not applicable.

Informed Consent Statement: Not applicable.

Data Availability Statement: The data that support our research findings are available from the corresponding author on request.

Conflicts of Interest: The authors declare no conflict of interest.

Abbreviations

MIT	Mobile Internet Technology
MPU	Mobile Phone Usage
BPM	Bivariate Probit Method
ICTs	Information Communication Technologies
IT	Internet Technology
AM	Agricultural Modernization
H	Hypothesis
PAK	Pakistan
RH	Research Hypothesis
MDs	Mobile Devices
SS	Selection Stage
OS	Outcome Stage
KPK	Khyber Pakhtunkhwa
MITA	Mobile Internet Technology Adoption
DIK	Dera Ismail Khan
UC	Union Council

References

1. Khan, N.; Ray, R.L.; Kassem, H.S.; Hussain, S.; Zhang, S.; Khayyam, M.; Ihtisham, M.; Asongu, S.A. Potential Role of Technology Innovation in Transformation of Sustainable Food Systems: A Review. *Agriculture* **2021**, *11*, 984. [CrossRef]
2. Khan, N.; Ray, R.L.; Sargani, G.R.; Ihtisham, M.; Khayyam, M.; Ismail, S. Current Progress and Future Prospects of Agriculture Technology: Gateway to Sustainable Agriculture. *Sustainability* **2021**, *13*, 4883. [CrossRef]
3. Pérez-Escamilla, R. Food security and the 2015–2030 sustainable development goals: From human to planetary health: Perspectives and opinions. *Curr. Dev. Nutr.* **2017**, *1*, e000513. [CrossRef] [PubMed]
4. Reza Anik, A.; Rahman, S.; Sarker, J.R. Five Decades of Productivity and Efficiency Changes in World Agriculture (1969–2013). *Agriculture* **2020**, *10*, 200. [CrossRef]
5. Reganold, J.P.; Wachter, J.M. Organic agriculture in the twenty-first century. *Nat. Plants* **2016**, *2*, 15221. [CrossRef] [PubMed]
6. Tang, K.; Hailu, A.; Kragt, M.E.; Ma, C. The response of broadacre mixed crop-livestock farmers to agricultural greenhouse gas abatement incentives. *Agric. Syst.* **2018**, *160*, 11–20. [CrossRef]
7. Gray, B.; Babcock, L.; Tobias, L.; McCord, M.; Herrera, A.; Cadavid, R. *Digital Farmer Profiles: Reimagining Smallholder Agriculture*; Grameen Foundation: Washington, DC, USA, 2018.
8. Trendov, M.; Varas, S.; Zeng, M. *Digital Technologies in Agriculture and Rural Areas: Status Report*; FAO: Rome, Italy, 2019.
9. Pender, J.; Gebremedhin, B. Determinants of agricultural and land management practices and impacts on crop production and household income in the highlands of Tigray, Ethiopia. *J. Afric. Econo.* **2008**, *17*, 395–450. [CrossRef]
10. Wu, J.; Ma, C.; Tang, K. The static and dynamic heterogeneity and determinants of marginal abatement cost of CO_2 emissions in Chinese cities. *Energy* **2019**, *178*, 685–694. [CrossRef]
11. Hayden, J.; Rocker, S.; Phillips, H.; Heins, B.; Smith, A.; Delate, K. The importance of social support and communities of practice: Farmer perceptions of the challenges and opportunities of integrated crop–livestock systems on organically managed farms in the northern US. *Sustainability* **2018**, *10*, 4606. [CrossRef]
12. Khan, N.; Siddiqui, B.N.; Khan, N.; Khan, F.; Ullah, N.; Ihtisham, M.; Ullah, R.; Ismail, S.; Muhammad, S. Analyzing mobile phone usage in agricultural modernization and rural development. *Int. J. Agric. Ext.* **2020**, *8*, 139–147. [CrossRef]

13. Baumüller, H. Towards smart farming? Mobile technology trends and their potential for developing country agriculture. In *Handbook for ICT in Developing Countries: 5G Perspectives*; Skouby, K.E., Williams, I., Gyamfi, A., Eds.; River Publishers: Gistrup, Denamrk, 2017; pp. 191–201.
14. Sołtysik-Piorunkiewicz, A. The development of mobile Internet technology and ubiquitous communication in a knowledge-based organization. *Online J. Appl. Knowl. Manag. (OJAKM)* **2013**, *1*, 29–41.
15. United Nations. *World Population Projected to Reach 9.8 Billion in 2050, and 11.2 Billion in 2100*; United Nations: Rome, Italy, 2017.
16. Xydis, G.A.; Liaros, S.; Avgoustaki, D.-D. Small scale Plant Factories with Artificial Lighting and wind energy microgeneration: A multiple revenue stream approach. *J. Clean. Prod.* **2020**, *255*, 120227. [CrossRef]
17. Tang, K.; He, C.; Ma, C.; Wang, D. Does carbon farming provide a cost-effective option to mitigate GHG emissions? Evidence from China. *Aust. J. Agric. Resour. Econ.* **2019**, *63*, 575–592. [CrossRef]
18. Xiang, T.; Malik, T.H.; Nielsen, K. The impact of population pressure on global fertiliser use intensity, 1970–2011: An analysis of policy-induced mediation. *Technol. Forecast. Soc. Chang.* **2020**, *152*, 119895. [CrossRef]
19. Notarnicola, B.; Sala, S.; Anton, A.; McLaren, S.J.; Saouter, E.; Sonesson, U. The role of life cycle assessment in supporting sustainable agri-food systems: A review of the challenges. *J. Clean. Prod.* **2017**, *140*, 399–409. [CrossRef]
20. Zheng, H.; Ma, W.; Wang, F.; Li, G. Does internet use improve technical efficiency of banana production in China? Evidence from a selectivity-corrected analysis. *Food Policy* **2021**, *102*, 102044. [CrossRef]
21. Mondal, A.; Chakrabarti, A.B. Information and Communication Technology Adoption Strategies of Emerging Multinationals From India. *J. Glob. Inf. Manag. (JGIM)* **2021**, *29*, 161–175. [CrossRef]
22. Twumasi, M.A.; Jiang, Y.; Zhou, X.; Addai, B.; Darfor, K.N.; Akaba, S.; Fosu, P. Increasing Ghanaian fish farms' productivity: Does the use of the internet matter? *Mar. Policy* **2021**, *125*, 104385. [CrossRef]
23. Zhu, X.; Hu, R.; Zhang, C.; Shi, G. Does Internet use improve technical efficiency? Evidence from apple production in China. *Technol. Forecast. Soc. Chang.* **2021**, *166*, 120662. [CrossRef]
24. Morrar, R.; Abdeljawad, I.; Jabr, S.; Kisa, A.; Younis, M.Z. The role of information and communications technology (ICT) in enhancing service sector productivity in Palestine: An international perspective. *J. Glob. Inf. Manag. (JGIM)* **2019**, *27*, 47–65. [CrossRef]
25. Issahaku, H.; Abu, B.M.; Nkegbe, P.K. Does the use of mobile phones by smallholder maize farmers affect productivity in Ghana? *J. Afr. Bus.* **2018**, *19*, 302–322. [CrossRef]
26. Ma, W.; Nie, P.; Zhang, P.; Renwick, A. Impact of Internet use on economic well-being of rural households: Evidence from China. *Rev. Dev. Econ.* **2020**, *24*, 503–523. [CrossRef]
27. Kelemu, K. Impact of Mobile Telephone on Technical Efficiency of Wheat Growing Farmers in Ethiopia. *Int. J. Res. Stud. Agric. Sci.* **2016**, *2*, 1–9.
28. Quandt, A.; Salerno, J.D.; Neff, J.C.; Baird, T.D.; Herrick, J.E.; McCabe, J.T.; Xu, E.; Hartter, J. Mobile phone use is associated with higher smallholder agricultural productivity in Tanzania, East Africa. *PLoS ONE* **2020**, *15*, e0237337.
29. Khan, N.; Ray, R.L.; Kassem, H.S.; Ihtisham, M.; Asongu, S.A.; Ansah, S.; Zhang, S. Toward Cleaner Production: Can Mobile Phone Technology Help Reduce Inorganic Fertilizer Application? Evidence Using a National Level Dataset. *Land* **2021**, *10*, 1023. [CrossRef]
30. Tang, K.; Kragt, M.E.; Hailu, A.; Ma, C. Carbon farming economics: What have we learned? *J. Environ. Manag.* **2016**, *172*, 49–57. [CrossRef]
31. Rousta, I.; Olafsson, H.; Moniruzzaman, M.; Zhang, H.; Liou, Y.A.; Mushore, T.D.; Gupta, A. Impacts of drought on vegetation assessed by vegetation indices and meteorological factors in Afghanistan. *Remote Sens.* **2020**, *12*, 2433. [CrossRef]
32. Mwalupaso, G.E.; Wang, S.; Rahman, S.; Alavo, E.J.-P.; Tian, X. Agricultural informatization and technical efficiency in maize production in Zambia. *Sustainability* **2019**, *11*, 2451. [CrossRef]
33. Price, C.J.A.C. Wheat: Outlook to 2020–21. *Agric. Commod.* **2016**, *6*, 29–45.
34. Hou, J.; Huo, X.; Yin, R. Does computer usage change farmers' production and consumption? Evidence from China. *China Agric. Econ. Rev.* **2018**, *11*, 387–410. [CrossRef]
35. Aker, J.C.; Ksoll, C. Can mobile phones improve agricultural outcomes? Evidence from a randomized experiment in Niger. *Food Policy* **2016**, *60*, 44–51. [CrossRef]
36. Tack, J.; Aker, J.C. Information, mobile telephony, and traders' search behavior in Niger. *Am. J. Agric. Econ.* **2014**, *96*, 1439–1454. [CrossRef]
37. Khan, N.; Ray, R.L.; Kassem, H.S.; Ihtisham, M.; Siddiqui, B.N.; Zhang, S. Can Cooperative Supports and Adoption of Improved Technologies Help Increase Agricultural Income? Evidence from a Recent Study. *Land* **2022**, *11*, 361. [CrossRef]
38. Zanello, G.; Srinivasan, C.S. Information sources, ICTs and price information in rural agricultural markets. *Eur. J. Dev. Res.* **2014**, *26*, 815–831. [CrossRef]
39. Aker, J.C.; Mbiti, I.M. Mobile phones and economic development in Africa. *J. Econ. Perspect.* **2010**, *24*, 207–232. [CrossRef]
40. Deichmann, U.; Goyal, A.; Mishra, D. Will digital technologies transform agriculture in developing countries? *Agric. Econ.* **2016**, *47*, 21–33. [CrossRef]
41. Aker, J.C.; Fafchamps, M. Mobile phone coverage and producer markets: Evidence from West Africa. *World Bank Econ. Rev.* **2015**, *29*, 262–292. [CrossRef]

42. Khan, N.; Ullah, R.; Khan, N.; Naseem, M.; Farhatullah, M.R.; Khan, N.U.; Muhammad, S.; Khayyam, M.; Ihtisham, M.; Siddiqui, B.N. Agricultural Extension Worker Role in the Transfer of Garlic Production Technology: A Case Study of Balochistan Province, Pakistan. *Int. J. Biosci. (IJB)* **2021**, *18*, 26–35.
43. Jensen, R.T. Information, efficiency, and welfare in agricultural markets. *Agric. Econ.* **2010**, *41*, 203–216. [CrossRef]
44. Courtois, P.; Subervie, J. Farmer bargaining power and market information services. *Am. J. Agric. Econ.* **2015**, *97*, 953–977. [CrossRef]
45. Zanello, G. Mobile phones and radios: Effects on transactions costs and market participation for households in Northern Ghana. *J. Agric. Econ.* **2012**, *63*, 694–714. [CrossRef]
46. Min, S.; Liu, M.; Huang, J. Does the application of ICTs facilitate rural economic transformation in China? Empirical evidence from the use of smartphones among farmers. *J. Asian Econ.* **2020**, *70*, 101219. [CrossRef]
47. Zaremohzzabieh, Z.; Samah, B.A.; Muhammad, M.; Omar, S.Z.; Bolong, J.; Hassan, S.B.H.; Mohamed Shaffril, H.A. Information and communications technology acceptance by youth entrepreneurs in rural Malaysian communities: The mediating effects of attitude and entrepreneurial intention. *Inf. Technol. Dev.* **2016**, *22*, 606–629. [CrossRef]
48. Büyükbay, E.O.; Gündüz, O. An investigation on computer and internet use for agricultural development in rural areas: A case study for Tokat Province in Turkey. *Afric. J. Biotechn.* **2011**, *10*, 11879–11886.
49. Savari, M.; Gharechaee, H. Application of the extended theory of planned behavior to predict Iranian farmers' intention for safe use of chemical fertilizers. *J. Clean. Prod.* **2020**, *263*, 121512.
50. Bort-Roig, J.; Gilson, N.D.; Puig-Ribera, A.; Contreras, R.S.; Trost, S.G. Measuring and influencing physical activity with smartphone technology: A systematic review. *Sports Med.* **2014**, *44*, 671–686. [CrossRef]
51. Bai, X.; Wang, Y.; Huo, X.; Salim, R.; Bloch, H.; Zhang, H. Assessing fertilizer use efficiency and its determinants for apple production in China. *Ecol. Indic.* **2019**, *104*, 268–278. [CrossRef]
52. Fan, L.; Yuan, Y.; Ying, Z.; Lam, S.K.; Liu, L.; Zhang, X.; Liu, H.; Gu, B. Decreasing farm number benefits the mitigation of agricultural non-point source pollution in China. *Environ. Sci. Pollut. Res.* **2019**, *26*, 464–472. [CrossRef]
53. Liu, Y.; Sun, D.; Wang, H.; Wang, X.; Yu, G.; Zhao, X. An evaluation of China's agricultural green production: 1978–2017. *J. Clean. Prod.* **2020**, *243*, 118483.
54. Mendes, J.; Pinho, T.M.; Neves dos Santos, F.; Sousa, J.J.; Peres, E.; Boaventura-Cunha, J.; Cunha, M.; Morais, R. Smartphone applications targeting precision agriculture practices—A systematic review. *Agronomy* **2020**, *10*, 855. [CrossRef]
55. Huq, M.E.; Fahad, S.; Shao, Z.; Sarven, M.S.; Al-Huqail, A.A.; Siddiqui, M.H.; ur Rahman, M.H.; Khan, I.A.; Alam, M.; Saeed, M.; et al. High arsenic contamination and presence of other trace metals in drinking water of Kushtia district, Bangladesh. *J. Environ. Manag.* **2019**, *242*, 199–209. [CrossRef]
56. Mayzelle, M.; Santibañez, M.-P.; Schweiger, J.; Jallo, C. *Assessment of Information and Communication Technologies in Afghan Agricultural Extension*; Government of the United States of America, University of California: Oakland, CA, USA, 2015.
57. Mittal, S.; Mehar, M. Socio-economic factors affecting adoption of modern information and communication technology by farmers in India: Analysis using multivariate probit model. *The J. Agric. Edu. Ext.* **2016**, *22*, 199–212. [CrossRef]
58. Razaque, A.; Sallah, M. The use of mobile phone among farmers for agriculture development. *Int. J. Sci. Res.* **2013**, *2*, 95–98.
59. Khan, N.; Ray, R.; Ihtisham, M.; Siddiqui, B.; Khayyam, M.; Anjum, R.; Asongu, S. Socioeconomic Determinants of the awareness and adoption of apple production practices: A case study of Balochistan, Pakistan. *Sarhad J. Agric.* **2022**, *38*, 322–330. [CrossRef]
60. Fabregas, R.; Kremer, M.; Schilbach, F. Realizing the potential of digital development: The case of agricultural advice. *Science* **2019**, *366*, eaay3038. [CrossRef]
61. Nie, P.; Ma, W.; Sousa-Poza, A. The relationship between smartphone use and subjective well-being in rural China. *Electron. Commer. Res.* **2020**, *21*, 983–1009. [CrossRef]
62. Zhao, Q.; Pan, Y.; Xia, X. Internet can do help in the reduction of pesticide use by farmers: Evidence from rural China. *Environ. Sci. Pollut. Res.* **2021**, *28*, 2063–2073. [CrossRef]
63. Singh, R.K.; Luthra, S.; Mangla, S.K.; Uniyal, S. Applications of information and communication technology for sustainable growth of SMEs in India food industry. *Resour. Conserv. Recycl.* **2019**, *147*, 10–18. [CrossRef]
64. Sekabira, H.; Qaim, M. Mobile money, agricultural marketing, and off-farm income in Uganda. *Agric. Econ.* **2017**, *48*, 597–611. [CrossRef]
65. Rehman, A.; Jingdong, L.; Khatoon, R.; Hussain, I.; Iqbal, M.S. Modern agricultural technology adoption its importance, role and usage for the improvement of agriculture. *Life Sci. J.* **2016**, *14*, 70–74.
66. O'Leary, D.; Zimmermann, R.; Grahn, A.; Poarch, D.; Cook, M.; Pirc, J. Mobile Device Security in the Workplace: 5 Key Risks and a Surprising Challenge. 2017. Available online: http://focus.forsythe.com/articles/55 (accessed on 20 March 2022).
67. Yuan, F.; Tang, K.; Shi, Q. Does Internet use reduce chemical fertilizer use? Evidence from rural households in China. *Environ. Sci. Pollut. Res.* **2021**, *28*, 6005–6017. [CrossRef] [PubMed]
68. Cunha, C.R.; Peres, E.; Morais, R.; Oliveira, A.A.; Matos, S.G.; Fernandes, M.A.; Reis, M.J.C.S. The use of mobile devices with multi-tag technologies for an overall contextualized vineyard management. *Comp. Electr. Agric.* **2010**, *73*, 154–164. [CrossRef]
69. Salahuddin, M.; Alam, K.; Ozturk, I. The effects of Internet usage and economic growth on CO2 emissions in OECD countries: A panel investigation. *Renew. Sustain. Energy Rev.* **2016**, *62*, 1226–1235. [CrossRef]
70. Matassa, S.; Batstone, D.J.; Hülsen, T.; Schnoor, J.; Verstraete, W. Can direct conversion of used nitrogen to new feed and protein help feed the world? *Environ. Sci. Technol.* **2015**, *49*, 5247–5254. [CrossRef] [PubMed]

71. Alvarez, J.; Nuthall, P. Adoption of computer based information systems: The case of dairy farmers in Canterbury, NZ, and Florida, Uruguay. *Comput. Electron. Agric.* **2006**, *50*, 48–60. [CrossRef]
72. Roco, L.; Engler, A.; Bravo-Ureta, B.E.; Jara-Rojas, R. Farmers' perception of climate change in mediterranean Chile. *Reg. Environ. Change* **2015**, *15*, 867–879. [CrossRef]
73. Kaila, H.; Tarp, F. Can the Internet improve agricultural production? Evidence from Viet Nam. *Agric. Econ.* **2019**, *50*, 675–691. [CrossRef]
74. Verbeek, M. *A Guide to Modern Econometrics*; John Wiley & Sons: Hoboken, NJ, USA, 2008.
75. Jarrett, A.; Gan, J.; Johnson, C.; Munn, I.A. Landowner awareness and adoption of wildfire programs in the southern United States. *J. For.* **2009**, *107*, 113–118.
76. Heckman, J.J. Sample selection bias as a specification error. *Econom. J. Econom. Soc.* **1979**, *47*, 153–161. [CrossRef]
77. Van de Ven, W.P.; Van Praag, B.M. The demand for deductibles in private health insurance: A probit model with sample selection. *J. Econom.* **1981**, *17*, 229–252. [CrossRef]
78. Kongaut, C.; Bohlin, E. Investigating mobile broadband adoption and usage: A case of smartphones in Sweden. *Telemat. Inform.* **2016**, *33*, 742–752. [CrossRef]
79. Alemi, F.; Circella, G.; Mokhtarian, P.; Handy, S. What drives the use of ridehailing in California? Ordered probit models of the usage frequency of Uber and Lyft. *Transp. Res. Part C Emerg. Technol.* **2019**, *102*, 233–248. [CrossRef]
80. Michels, M.; Fecke, W.; Feil, J.H.; Musshoff, O.; Pigisch, J.; Krone, S. Smartphone adoption and use in agriculture: Empirical evidence from Germany. *Precis. Agric.* **2020**, *21*, 403–425. [CrossRef]
81. Sartori, A.E. An estimator for some binary-outcome selection models without exclusion restrictions. *Political Anal.* **2003**, *11*, 111–138. [CrossRef]
82. Adapa, S. Indian smart cities and cleaner production initiatives–Integrated framework and recommendations. *J. Clean. Prod.* **2018**, *172*, 3351–3366. [CrossRef]
83. Abdullah, F.A.; Samah, B.A. Factors impinging farmers' use of agriculture technology. *Asian Soc. Sci.* **2013**, *9*, 120. [CrossRef]
84. Uduji, J.I.; Okolo-Obasi, E.N.; Asongu, S.A. Does growth enhancement support scheme (GESS) contribute to youth development in informal farm entrepreneurship? Evidence from rural communities in Nigeria. *J. Enterprising Communities People Places Glob. Econ.* **2021**, *15*, 451–476. [CrossRef]
85. Ogbeide, O.A.; Ele, I. Smallholder farmers and mobile phone technology in Sub-Sahara Agriculture. *Mayfair J. Inf. Technol. Manag. Agric.* **2015**, *1*, 1–19.
86. Woodburn, M.; Ortmann, G.; Levin, J.B. Computer use and factors influencing computer adoption among commercial farmers in Natal Province, South Africa. *Comput. Electron. Agric.* **1994**, *11*, 183–194. [CrossRef]
87. Adamowicz, M.; Szepeluk, A. Support to young farmers as part of agricultural policy of the European Union. *Probl. Agric. Econ.* **2016**, *3*, 106–127.
88. Yang, L.; Tang, K.; Wang, Z.; An, H.; Fang, W. Regional eco-efficiency and pollutants' marginal abatement costs in China: A parametric approach. *J. Clean. Prod.* **2017**, *167*, 619–629. [CrossRef]
89. Hong, I.B. Understanding and predicting behavioral intention to adopt mobile banking: The Korean experience. *J. Glob. Inf. Manag. (JGIM)* **2019**, *27*, 182–202. [CrossRef]
90. Khan, N.; Ray, R.L.; Zhang, S.; Osabuohien, E.; Ihtisham, M. Influence of mobile phone and internet technology on income of rural farmers: Evidence from Khyber Pakhtunkhwa Province, Pakistan. *Technol. Soc.* **2022**, *68*, 101866. [CrossRef]
91. Islam, M.S.; Grönlund, Å. Factors influencing the adoption of mobile phones among the farmers in Bangladesh: Theories and practices. *Inter. J. Advan. ICT Emer. Reg.* **2011**, *4*, 4–14. [CrossRef]
92. Khan, N.A.; Qijie, G.; Sertse, S.F.; Nabi, M.N.; Khan, P. Farmers' use of mobile phone-based farm advisory services in Punjab, Pakistan. *Inf. Dev.* **2020**, *36*, 390–402. [CrossRef]
93. Salamzada, K.; Shukur, Z.; Bakar, M.A. A framework for cybersecurity strategy for developing countries: Case study of Afghanistan. *Asia-Pac. J. Inf. Technol. Multimed.* **2015**, *4*, P1–P10. [CrossRef]
94. Krone, M.; Dannenberg, P.; Nduru, G. The use of modern information and communication technologies in smallholder agriculture: Examples from Kenya and Tanzania. *Inf. Dev.* **2015**, *32*, 1503–1512. [CrossRef]
95. Hoang, H.G. Determinants of the adoption of mobile phones for fruit marketing by Vietnamese farmers. *World Dev. Perspect.* **2020**, *17*, 100178. [CrossRef]
96. Matous, P.; Todo, Y.; Ishikawa, T. Emergence of multiplex mobile phone communication networks across rural areas: An Ethiopian experiment. *Netw. Sci.* **2014**, *2*, 162–188. [CrossRef]
97. Akudugu, M.A.; Guo, E.; Dadzie, S.K. Adoption of modern agricultural production technologies by farm households in Ghana: What factors influence their decisions? *J. Biol. Agric. Health* **2012**, *2*, 1–14.
98. Malhotra, N.K.; Kim, S.S.; Agarwal, J. Internet users' information privacy concerns (IUIPC): The construct, the scale, and a causal model. *Inf. Syst. Res.* **2004**, *15*, 336–355. [CrossRef]
99. Mohmand, A.M.; Marjan, A.; Sangin, A. Developing e-government in Afghanistan. In Proceedings of the 4th International Conference on Theory and Practice of Electronic Governance, Beijing, China, 25–28 October 2010; pp. 43–48.
100. Fecke, W.; Danne, M.; Mußhoff, O. Online-Einkauf von Pflanzenschutzmitteln: Ein Discrete Choice Experiment mit landwirtschaftlichen Unternehmern in Deutschland; Diskussionsbeitrag: 2018. Available online: http://hdl.handle.net/10419/190683 (accessed on 20 March 2022).

101. Ghimire, R.; Huang, W.; Poudel, M.P. Adoption intensity of agricultural technology: Empirical evidence from smallholder maize famers in Nepal. *Int. J. Agric. Innov. Res.* **2015**, *4*, 139–146.
102. Narula, S.A. Revolutionizing food supply chains of Asia through ICTs. In *Sustainability Challenges in the Agrofood Sector*; Bhat, R., Ed.; Wiley: Hoboken, NJ, USA, 2017; p. 212.
103. Bellon-Maurel, V.; Peters, G.M.; Clermidy, S.; Frizarin, G.; Sinfort, C.; Ojeda, H.; Roux, P.; Short, M.D. Streamlining life cycle inventory data generation in agriculture using traceability data and information and communication technologies—Part II: Application to viticulture. *J. Clean. Prod.* **2015**, *87*, 119–129. [CrossRef]

applied sciences

MDPI

Article

Theory of Movement of Machine-Tractor Unit with Trailer Haulm Harvester Machine

Volodymyr Bulgakov [1], Aivars Aboltins [2,*], Semjons Ivanovs [2], Hristo Beloev [3], Volodymyr Nadykto [4], Yevhen Ihnatiev [4] and Jüri Olt [5]

1 Department of Mechanics, Faculty of Construction and Design, National University of Life and Environmental Sciences of Ukraine, 15 Heroiv Oborony Str., 03041 Kyiv, Ukraine; vbulgakov@meta.ua
2 Faculty of Engineering, Institute of Agricultural Machinery, Latvia University of Life Sciences and Technologies, Cakstes Blvd. 5, 3001 Jelgava, Latvia; semjons.ivanovs@apollo.lv
3 Department of Agricultural Machinery, Angel Kanchev University of Ruse, 5 Studentska Str., 7017 Ruse, Bulgaria; hbeloev@uni-ruse.bg
4 Department of Machine-Using in Agriculture, Dmytro Motornyi Tavria State Agrotechnological University, 18B Khmelnytsky Ave., 72310 Melitopol, Ukraine; volodymyr.nadykto@tsatu.edu.ua (V.N.); yevhen.ihnatiev@tsatu.edu.ua (Y.I.)
5 Institute of Technology, Estonian University of Life Sciences, 56 Kreutzwaldi Str., 51006 Tartu, Estonia; jyri.olt@emu.ee
* Correspondence: aivars.aboltins@llu.lv; Tel.: +371-29-134-846

Abstract: Harvesting sugar and fodder beet tops is a complex technological process that requires the use of special harvesting machines. Trailed harvesters of different rows, which together with aggregate tractors form symmetric or asymmetric machine-tractor units, the movement of which in the horizontal plane is not always stable, are widely used. The purpose of this study is to determine the parameters of stable plane-parallel motion of asymmetric harvester machine-tractor unit based on numerical computer simulation of the obtained analytical dependencies. According to the results of the analytical study, the values of the amplitude and phase-frequency characteristics of the turning angle tractor's oscillations were obtained. They reflect the reproduction by the angle rotation fluctuations of the haulm harvester machine in the horizontal plane. Calculations have shown that reducing the value of the input resistance coefficient of pneumatic tires of the driving wheels of the aggregating tractor increases its sensitivity to the action of disturbing influences. The greater the sensitivity, the closer the wheels of the power tool are to the attachment point of the trailed haulm harvester. In qualitative terms, increasing the speed of the machine-tractor unit from 1.5 to 2.5 m·s^{-1} leads to an undesirable increase in the amplitude-frequency response and desired increase in the phase-frequency response when reproducing its external disturbing effects in the form of oscillations of the angle of rotation of the harvester.

Keywords: asymmetric machine-tractor unit; motion; stability; resistance coefficients; amplitude-frequency characteristic; phase-frequency characteristic

Citation: Bulgakov, V.; Aboltins, A.; Ivanovs, S.; Beloev, H.; Nadykto, V.; Ihnatiev, Y.; Olt, J. Theory of Movement of Machine-Tractor Unit with Trailer Haulm Harvester Machine. *Appl. Sci.* **2022**, *12*, 3901. https://doi.org/10.3390/app12083901

Academic Editors: Paweł Kiełbasa, Tadeusz Juliszewski, Sławomir Kurpaska and Francesca Scargiali

Received: 31 January 2022
Accepted: 8 April 2022
Published: 12 April 2022

1. Introduction

The most significant technological operations in the sugar beet harvesting process are digging the root crops out of the soil and removing the uncut leaves [1–3]. Sugar beet haulms have suitable nutritional properties and can be used as feed in animal husbandry in green form and silage [3,4]. Sugar beet haulms are now widely used as a raw material in the production of biogas and organic fertilizer, which can be applied to the soil immediately after its cutting and disintegrating.

The use of asymmetric machine-tractor units in harvesting sugar beet and fodder beet haulm is quite justified. In this case, the stable motion of such a harvesting unit should be understood as the one in which the deviation of its trajectory under the influence of random disturbing factors (perturbed trajectory) from the trajectory without the action of these factors (unperturbed trajectory) is within the limits of permissible values. These permissible values can be considered as such agrotechnical requirements, which ensure the quality of the technological process of harvesting the haulm while reducing energy costs, reliability of operation, and the same.

The theoretical study of motion stability of agricultural machines and machine units is based on the basic provisions of motion stability of complex tracking dynamic systems. It is based on the study of perturbed and unperturbed motions, described by systems of differential equations of their motion in the horizontal plane [5–8].

The papers [5,8–17] are devoted to studying the motion stability of agricultural machine-tractor units using computational mathematical models of their motion (functioning). The theoretical study of the plane-parallel motion of asymmetric machine-tractor unit and some questions concerning the stability of its motion in the horizontal plane are thoroughly considered in the article [18,19].

The effective use of trailed haulm harvesters in the harvesting of haulm is caused by the need to ensure their stable movement in the horizontal plane because, in this case, such harvesting unit is usually asymmetrical. The solution to this issue is an urgent scientific and technical problem, which still does not have a sufficiently complete solution. Part of the difficulty in solving it lies in how the trailed haulm harvester is connected to the energy vehicle (tractor) used. Of the two possible options for connecting a trailed harvester to the rear tractor unit, rigid and articulated, both lead to the creation of asymmetric machine-tractor units, each of which has its features and disadvantages. Therefore, improving the trajectory performance and efficient operation is relevant for all trailed haulm harvesters.

The study aims to determine the parameters of stable plane-parallel motion of asymmetric haulm machine and tractor unit based on numerical computer simulation of the obtained analytical dependencies.

2. Materials and Methods

2.1. Method for Compiling the Mathematical Model of the Harvesting Unit

This study uses higher mathematics methods to transform systems of differential equations, the theory of stability of motion, automatic regulation, behavioral studies of tracking systems, compiling computer programs, conducting numerical calculations, building graphical dependencies, and their analysis.

We have developed a new haulm harvester that performs solid cutting of the main mass of the green haulm from sugar beet root crops (or forage) with a rotary haulm harvester, which can be attached behind a wheeled tractor that is aggregating. Thus, the aggregating wheeled tractor moves on the harvested area of the field, and the rear trailed topper cuts the green mass of the tops from the rows of beet crops.

The cut mass is either distributed on the harvested area of the field or loaded into the body of a vehicle moving next to the working harvesting machine. A diagram of such a unit is shown in Figure 1.

The construction of this harvester, regardless of the number of rows of sugar or fodder beetroot crops, includes two supporting, copying wheels, which move in between the rows and have mechanisms to set the height of the green mass haulm to be cut. We have developed the basic provisions of the theory of plane-parallel motion of asymmetric haulm harvester machine-tractor unit consisting of a wheel tractor of nominal traction force 14 kN and a rear trailed haulm harvester of our construction.

First, an equivalent diagram of such a machine-tractor unit was built, shown in Figure 2. It shows a wheeled tractor in an arbitrary position, with a harvester machine attached to the rear implement in a certain row.

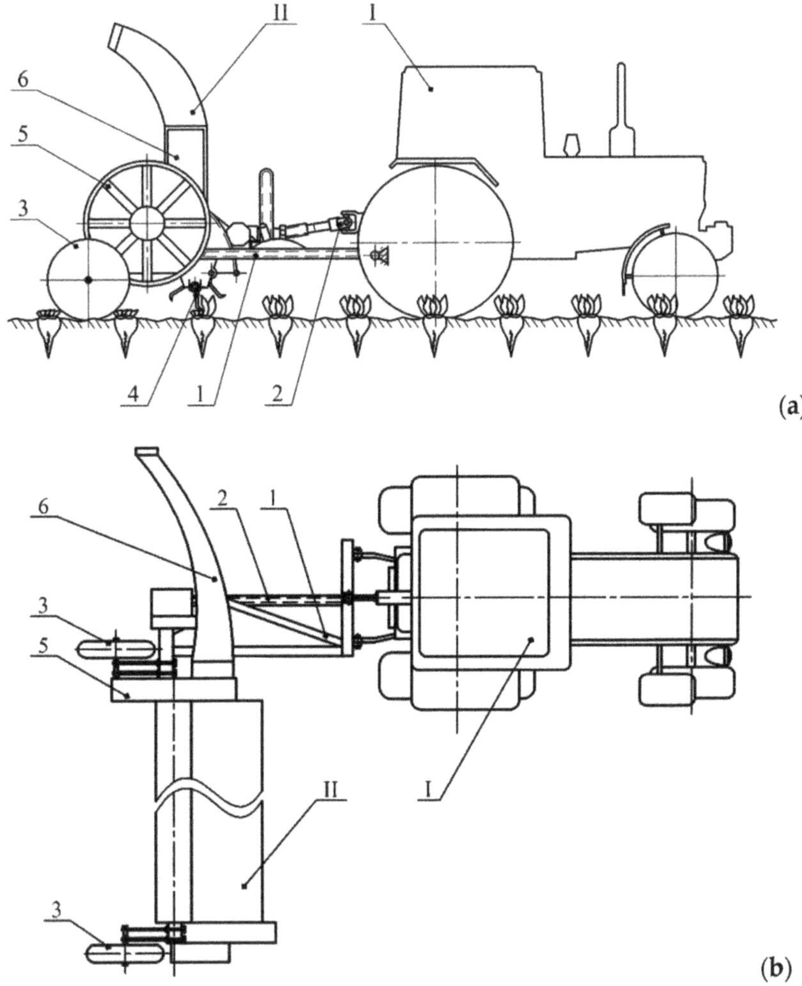

Figure 1. Construction and technological scheme of asymmetric of the haulm harvester (**a**)—side view; (**b**)—top view: I—wheeled tractor; II—trailed asymmetric haulm harvester: 1—frame; 2—drive; 3—support-copying pneumatic wheel; 4—rotary haulm cropper; 5—transporting working device; 6—loading device.

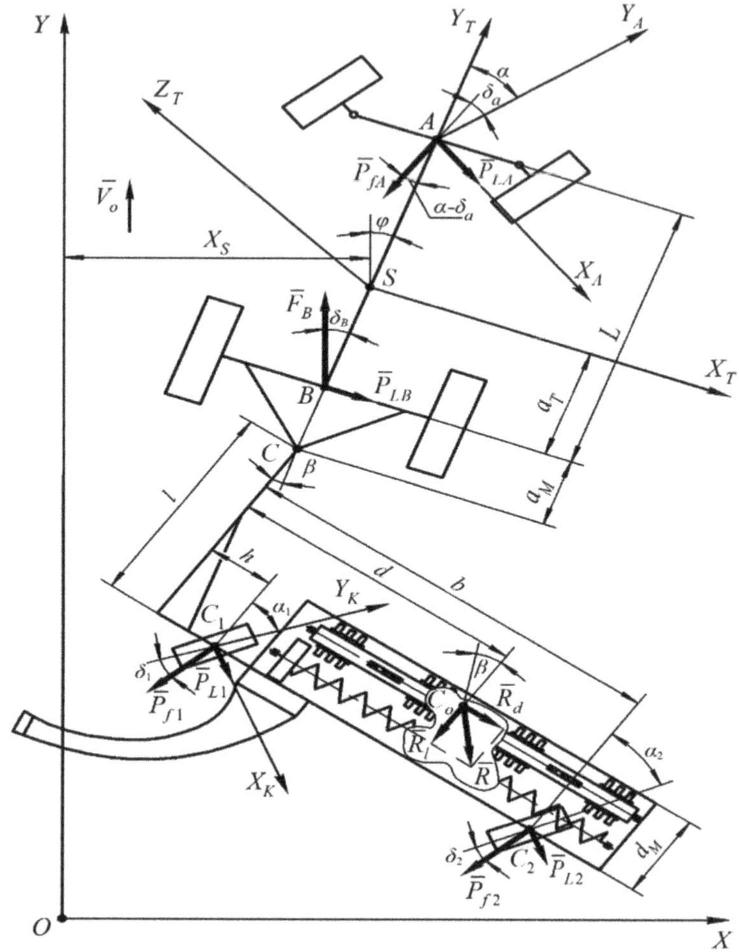

Figure 2. Equivalent scheme of asymmetric haulm harvester machine-tractor unit [7].

In addition, during the theoretical study of the mentioned unit motion, some of its properties were formalized, and appropriate assumptions were made. These assumptions in no way distorted the real process but greatly simplified the analytical research. Thus, the main assumptions were as follows:

1. The surface of the field over which this harvesting machine is moving is horizontal, and therefore the roll and trim of this machine have been rejected;
2. The wheeled aggregating tractor was represented as a solid body with a longitudinal plane of symmetry passing through its mass center;
3. Fluctuations of traction resistance of the trailed haulm harvester do not significantly affect the machine's forward speed, so in the first approximation, they had constant values;
4. Interaction of pneumatic tires of driving wheels of the aggregating tractor and supporting copying wheels of the trailed mower harvester with the soil surface, especially in the lateral directions, fully met the hypothesis of "lateral input" of pneumatic wheel tires;
5. Since the gyroscopic and stabilizing moments of the pneumatic tires of the aggregating tractor and the supporting copying wheels of the harvester are small, as well as the

moments of their twisting relative to the vertical axes, they were not taken into account during the movement of this unit;

6. The entry angles of the pneumatic tires of the running wheels of the aggregating tractor, which are located on the same geometrical axis and the lateral forces acting on them, were considered small enough;
7. The turning angles of the left and right steering wheels of the aggregating tractor were also considered small and equal to each other since the main movement of the machine when performing the technological process of harvesting sugar beet tops is rectilinear.

The equivalent diagram shows the characteristic points, the applied external forces on the links of the machine and tractor unit, the adopted systems of stationary and moving coordinates, and also shows the specified linear and angular dimensions.

Taking into account the accepted assumptions, as well as based on the analysis of the types of motions of the given harvester machine and tractor unit, the generalized coordinates (X_S, φ, β) were taken and based on the initial equations of motion in the Lagrange form of the second kind the system of the second-order linear differential equations describing the motion of the asymmetric machine and tractor unit in the horizontal plane was obtained as follows:

$$\left.\begin{array}{l} A_{11} \cdot \ddot{X}_S + A_{12} \cdot \dot{X}_S + A_{13} \cdot \dot{\varphi} + A_{14} \cdot \varphi + A_{15} \cdot \beta = f_{11} \cdot \alpha, \\ A_{21} \cdot \ddot{\varphi} + A_{22} \cdot \dot{\varphi} + A_{23} \cdot \varphi + A_{24} \cdot \dot{X}_S + A_{25} \cdot \beta = f_{21} \cdot \alpha, \\ A_{31} \cdot \ddot{\beta} + A_{32} \cdot \dot{\beta} + A_{33} \cdot \beta + A_{34} \cdot \dot{\varphi} + A_{35} \cdot \varphi + A_{36} \cdot \dot{X}_S = f_{31}, \end{array}\right\} \tag{1}$$

where $A_{11} = M_T$;

$$A_{12} = \frac{k_A + k_B + P_{fA} - F_B}{V_o};$$

$$A_{13} = \frac{\left(k_A + P_{fA}\right) \cdot (L - a_T) + (F_B - k_B) \cdot a_T}{V_o};$$

$$A_{14} = F_B - k_A - k_B - P_{fA};$$

$$A_{15} = P_{KR};$$

$$A_{22} = \frac{\left(k_A + P_{fA}\right) \cdot (L - a_T)^2 + (k_B - F_B) \cdot a_T^2}{V_o};$$

$$A_{21} = J_S;$$

$$A_{23} = -A_{13} \cdot V_o;$$

$$A_{24} = A_{13};$$

$$A_{25} = -P_{KR}(a_T + a_M);$$

$$A_{31} = J_C;$$

$$A_{32} = \frac{l \cdot \left[\left(k_1 + P_{f1}\right) \cdot (l^2 + h^2)^{\frac{1}{2}} + \left(k_2 + P_{f2}\right) \cdot (l^2 + b^2)^{\frac{1}{2}}\right]}{V_o};$$

$$A_{33} = R_l(l - d_M) + l\left(k_1 + k_2 + P_{f1} + P_{f2}\right);$$

$$A_{34} = \frac{l \cdot (a_T + a_M) \cdot \left(k_1 + k_2 + P_{f1} + P_{f2}\right)}{V_o};$$

$$A_{35} = l\left(k_1 + k_2 + P_{f1} + P_{f2}\right);$$

$$A_{36} = -\frac{A_{35}}{V_o};$$

$$f_{11} = k_A;$$
$$f_{21} = (L - a_T) \cdot k_A;$$
$$f_{31} = R_l \cdot d + P_{f1} \cdot h + P_{f2} \cdot b - \alpha_1 \cdot l \cdot k_1 - \alpha_2 \cdot l \cdot k_2.$$

In this system of differential Equation (1), the following designations were used: M_T and J_S—the mass of the wheeled aggregate tractor and its moment of inertia relative to the center of mass (point S); V_o—velocity of the harvester machine and tractor unit movement; k_A, k_B—resistance coefficients of pneumatic tires of the tractor front and rear axles, respectively; J_C —the moment of inertia of the haulm harvester along the vertical axis passing through the point (i.e., the pivot point of the trailed haulm harvester and the coupling tractor); k_1, k_2—resistance coefficients of pneumatic tire input in accordance with the left and two right copying wheels of the trailed haulm harvester; α_1, α_2—angles of installation (discrepancies) of the left and right wheels of the haulm harvester; L—tractor base; a_T—longitudinal coordinate of the tractor's center mass; a_M—distance from the tractor's rear axle to the point of the harvester attachment; l—length of the harvester hook-on hitch; h, d—placement coordinates of the left and right harvester wheels; b—placement coordinates of the harvester center resistance; and d_M—haulm harvester machine width.

2.2. Method for Checking the Mathematical Model for the Adequacy

The verification of the mathematical model (1) for adequacy was carried out by comparing the theoretical (A_t) and experimental (A_e) amplitude-frequency characteristics of the dynamic system when it was working out the disturbing effect in the form of the harvester machine rotation angle β.

The experimental characteristic A_e was found from the equation:

$$A_e = \frac{\sigma_\varphi}{\sigma_\beta}\sqrt{\frac{S_\varphi}{S_\beta}}, \tag{2}$$

where $\sigma_\varphi, \sigma_\beta$ are the standard deviations of the tractor's yaw angle and the harvester's rotation angle, respectively; S_φ, S_β—normalized spectral densities of the parameters' φ and β oscillations, respectively.

The experimental field was divided into sites, each 250 m long. On the site, the harvesting unit moved in the direction "there" and "back" at a given speed. During the unit's movement for at least 60 s, two parameters were recorded: the yaw angle φ of the tractor and the angle β of the harvester rotation. For this, gyroscopes and accelerometers GY-521 based on MPU-6050 (InvenSence Inc., Sunnyvale, CA, USA) were used. The electrical signals received from them were recorded on a Laptop using an Arduino UNO (Roma, Italy). The calculation of Equation (2) parameters was carried out using the MathCad 15.

The theoretical amplitude-frequency characteristic of the harvesting unit was calculated using a mathematical model (1) and a transfer function (8).

For the numerical solution of the system of differential Equation (1), we will further use methods of operational calculus based on the Laplace transform [7,20,21]. For this purpose, we will carry out the transition from the original functions to their reflections by introducing an operator of this form:

$$s = \frac{d}{dt}.$$

Using this operator allows us to go from a system of differential Equation (1) to a system of algebraic equations, which will look as follows:

$$\left.\begin{array}{l} K_{11} \cdot X_S(s) + K_{12} \cdot \varphi(s) + K_{13} \cdot \beta(s) = F_{11} \cdot \alpha(s) + F_{12} \cdot 1(s), \\ K_{21} \cdot X_S(s) + K_{22} \cdot \varphi(s) + K_{23} \cdot \beta(s) = F_{21} \cdot \alpha(s) + F_{22} \cdot 1(s), \\ K_{31} \cdot X_S(s) + K_{32} \cdot \varphi(s) + K_{33} \cdot \beta(s) = F_{31} \cdot \alpha(s) + F_{32} \cdot 1(s), \end{array}\right\} \tag{3}$$

where

$K_{11} = A_{11} \cdot s^2 + A_{12} \cdot s;$

$K_{12} = A_{13} \cdot s + A_{14};$

$K_{22} = A_{21} \cdot s^2 + A_{22} \cdot s + A_{23};$

$K_{23} = A_{25};$

$K_{31} = A_{36} \cdot s;$

$K_{32} = A_{34} \cdot s + A_{35};$

$F_{12} = F_{22} = F_{31} = 0;$

$K_{13} = A_{15};$

$K_{21} = A_{24} \cdot s;$

$K_{33} = A_{31} \cdot s^2 + A_{32} \cdot s + A_{33};$

$F_{11} = f_{11};$

$F_{21} = f_{21};$

$F_{32} = f_{31};$

$I(s)$—single step change of input parameter.

Thus, the system of Equation (3) is a computational mathematical model of the asymmetric machine-tractor harvester in operator's form, which we will use in the following to study the stability of its motion. The input variables in the system of Equation (3) are the steering action in the form of the steering angle α of the tractor wheels and the unit step effect of each disturbing moment, which are components of the total coefficient f_{31}. Since a change in each of the components of this coefficient leads to a change in the value of the disturbing moments, they can all act as a single-step impact. The output parameters of this dynamic system are the transverse displacement X_S of the center of mass and the course angle φ of the tractor and the rotation angle β of the trailed asymmetric harvester in the horizontal plane of projections.

From the system of Equation (3), it is easy to obtain a mathematical model of the movement of a different kind of unit, provided that the rotation angle β of the trailed mulcher has a perturbing effect. Then we get:

$$\left.\begin{aligned} K_{11} \cdot X_S(s) + K_{12} \cdot \varphi(s) &= -K_{13} \cdot \beta(s) + F_{11} \cdot \alpha(s), \\ K_{21} \cdot X_S(s) + K_{22} \cdot \varphi(s) &= -K_{23} \cdot \beta(s) + F_{21} \cdot \alpha(s). \end{aligned}\right\} \tag{4}$$

With the help of model (4), you can trace the effect of changes in the rotation angle β of the trailed harvester on the fluctuations of the course angle φ of the wheeled tractor.

To calculate the corresponding amplitude-frequency and phase-frequency characteristics characterizing the effect of angle β changes on the oscillations of the tractor course angle φ a transfer function should be composed of the following form:

$$W_0 = \frac{D_3}{D_0}. \tag{5}$$

where

$$D_0 = \begin{vmatrix} K_{11}K_{12} \\ K_{21}K_{22} \end{vmatrix}. \tag{6}$$

$$D_3 = \begin{vmatrix} K_{11} - K_{13} \\ K_{21} - K_{23} \end{vmatrix}. \tag{7}$$

Substituting in the determinants D_0 and D_3 in expressions (6) and (7), values of the coefficients K_{ij} from the system of Equation (3), after appropriate transformations, we obtain an expression for calculating the necessary transfer function in the following form:

$$W_0\,(s) = \frac{F_1 \cdot s + F_0}{C_2 \cdot s^2 + C_1 \cdot s + C_0}, \tag{8}$$

where $C_2 = A_{11} \cdot A_{21};$

$$C_1 = A_{12} \cdot A_{21} + A_{11} \cdot A_{22};$$

$$C_0 = A_{12} \cdot A_{22} + A_{11} \cdot A_{23} - A_{13} \cdot A_{24};$$

$$F_1 = -A_{11} \cdot A_{25};$$

$$F_0 = A_{15} \cdot A_{24} - A_{12} \cdot A_{25}.$$

Based on the transfer function $W_0(s)$ obtained by the method outlined in the work of [21], it is possible to calculate the amplitude-frequency and phase-frequency characteristics of the asymmetric tractor unit during the development of its disturbing influences. We have carefully considered the influence of the rotation angle β of the trailed haulm harvester on the oscillation of the course angle φ of the tractor. It is known [15] that the amplitude-frequency response should be as small as possible when the dynamic system works out any perturbation. Ideally, it should be equal to zero. At the same time, the phase-frequency shift (i.e., the delay in the response of the dynamic system to the disturbing influence) should be as large as possible. As a result, when the dynamic system is working out the perturbation, the desired amplitude-frequency characteristics should be equal to 0. The phase-frequency characteristics, on the contrary, should tend to infinity.

Those parameters and operating modes of the investigated asymmetric harvester machine-tractor unit, which, in the operating frequency range of oscillations of controlling and disturbing input influence maximally approximate the actual amplitude-frequency and phase-frequency characteristics to the desired ones, will thus be considered optimal. The construction parameters of this asymmetric toothed machine-tractor unit required calculating the theoretical amplitude and phase-frequency characteristics as follows:

M_T = 4250 kg, J_S = 4.6 kN·m·s^2, P_{KR} = 7.3 kN, P_{fA} = 1.7 kN, L = 2.45 m, a_T = 0.98 m, a_M = 1.20 m, k_A = 80 kN·rad^{-1}, k_B = 120 kN·rad^{-1}, F_B = 9.55 kN.

3. Results and Discussion

When comparing the theoretical and experimental amplitude-frequency characteristics (AFC), it was found that in the entire frequency range of the angle β, the maximum difference between the calculated and experimental data does not exceed 4% (Figure 3).

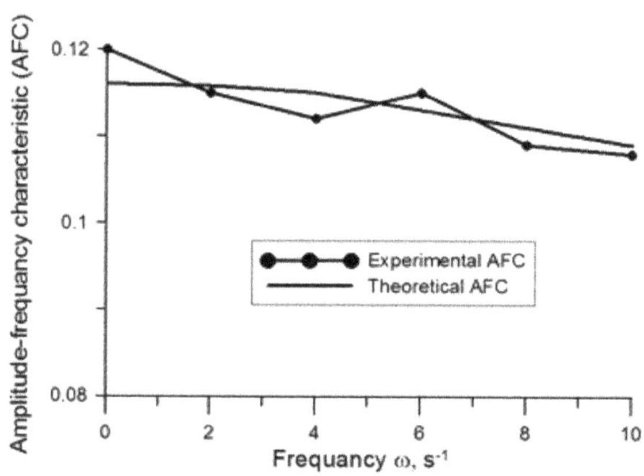

Figure 3. Experimental and theoretical amplitude-frequency characteristics of the dynamic system.

This result indicates that the developed mathematical model (1) of the harvesting unit movement in the horizontal plane is adequate, and the results of theoretical calculations using it can be considered quite reliable.

After compiling the program of numerical calculations and solving the obtained mathematical model on the PC, we plotted the amplitude-frequency and phase-frequency characteristics that allow for evaluating the stability of the motion of this asymmetric machine-tractor harvester (Figures 3–7). Based on the obtained graphical dependencies, let us analyze how some construction and technological factors of the given haulm harvester affect the amplitude and phase-frequency characteristics of oscillations of the course angle φ of the wheeled aggregate tractor when changing the oscillation frequency of the angle β of the trailed haulm harvester in the horizontal plane. First, let us consider the effect of the forward speed V_o of a given harvesting machine. The analysis of the obtained amplitude-frequency characteristics testifies to the following. First, as the frequency of the disturbing oscillations (i.e., angle β) increases, the amplification factor of the considered dynamic system of this input impact at each speed mode of the harvesting machine motion gradually decreases (Figure 4).

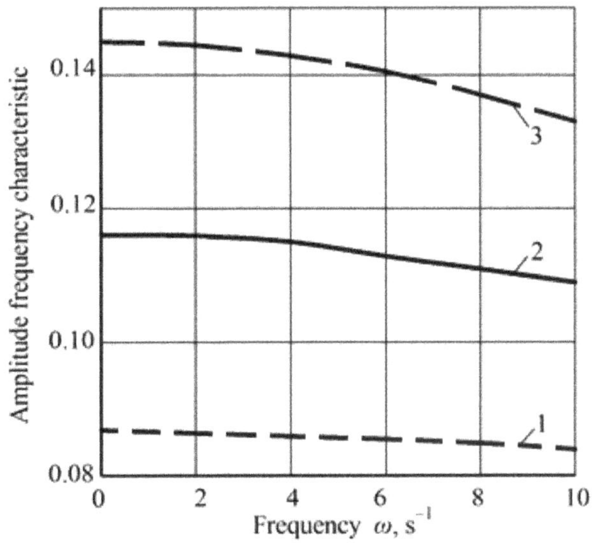

Figure 4. Amplitude-frequency characteristic of the tractor course angle φ when working out perturbations by it in the form of oscillations of the haulm harvester's steering angle β at various speeds of the machine: 1—V_o = 1.5 m·s−1; 2—V_o = 2.0 m·s−1; 3—V_o = 2.5 m·s−1.

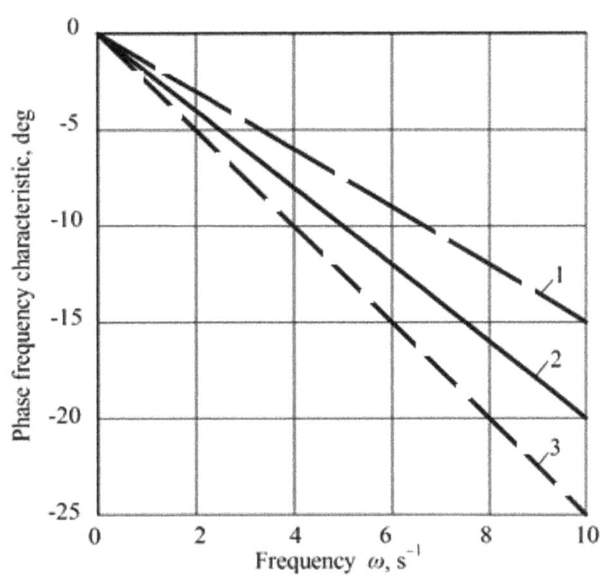

Figure 5. Phase-frequency characteristic of the tractor course angle φ when working out perturbations by it in the form of oscillations of the haulm harvester's steering angle β at different unit movement speeds: 1—V_o = 1.5 m·s−1; 2—V_o = 2.0 m·s−1; 3—V_o = 2.5 m·s−1.

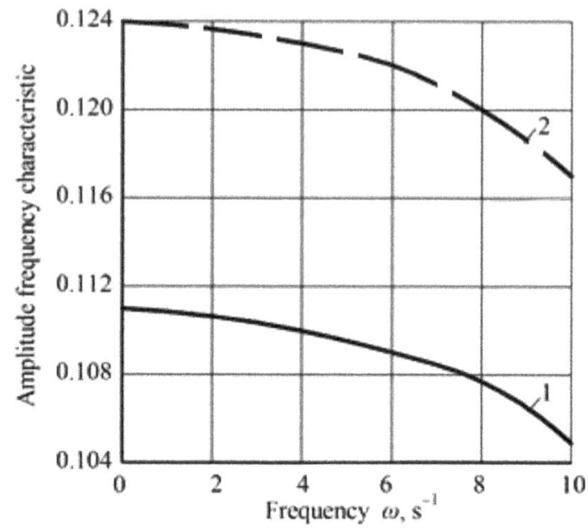

Figure 6. Amplitude-frequency characteristic of the course angle φ of the tractor when it works out disturbances in the form of oscillations of the angle β of rotation of the harvester at different values of the input resistance coefficient k_A of pneumatic tires of its front steerable wheels: 1–120 kN·rad^{-1}; 2–80 kN·rad^{-1}.

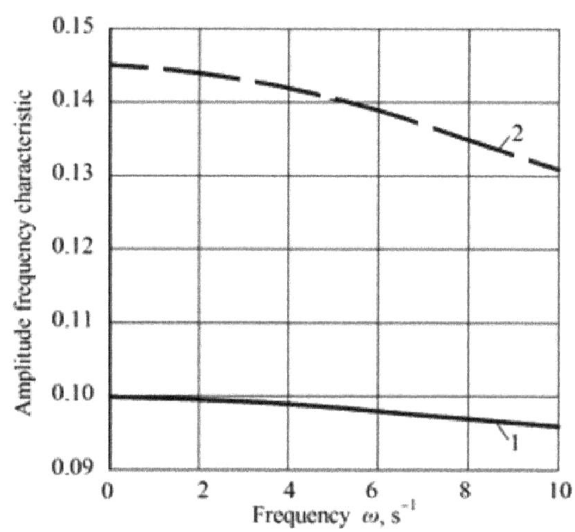

Figure 7. Amplitude-frequency characteristic of the tractor course angle φ when working out disturbances in the form of oscillations of the harvester machine rotation angle β at different values of the input resistance coefficient k_B of pneumatic tires of rear traction wheels: 1–150 kN·rad^{-1}; 2–90 kN·rad^{-1}.

There is every reason to believe that this result is logical since the greater the frequency of oscillations of perturbation ω, the greater the stabilizing role played by the inertial properties of the dynamic system, which is considered. Secondly, increasing the speed V_o of the machine-tractor unit movement becomes more sensitive to disturbing influences. For example, at the angle β oscillation frequency ω = 4 s^{-1} and the speed of the machine 1.5 m·s^{-1}, its amplitude-frequency characteristic is 0.09 (curve 1, Figure 4). In simplified form for understanding, this means that for the amplitude of oscillations of the disturbance (angle β) at 5° (and this is appreciable), the amplitude of oscillations of the course angle φ of the aggregating wheeled tractor is only 0.45°, which is almost imperceptible.

When the speed V_o of motion of the haulm, the harvester is 2.5 m·s^{-1}, and with the same perturbation frequency ω = 4 s^{-1}, the amplitude-frequency response of the dynamic system increases to 0.14 (curve 3, Figure 4). Compared with the previous velocity mode (when the amplitude-frequency response is equal to 0.09), this is almost 1.6 times greater. However, according to the same amplitude of fluctuations of the angle β = 5°, the amplitude of fluctuations of the course angle φ of the aggregate tractor does not exceed 0.7°. As we can see, in qualitative terms, the increase in the velocity V_o of the haulm harvester, considered, leads to an undesirable increase in the amplitude-frequency response when reproducing the external disturbing influence in the form of oscillations of the angle β of rotation of the haulm harvester. In quantitative terms, this influence is such that it cannot worsen the practical stability of the movement of a given asymmetric machine-tractor harvester.

Now, what concerns the delayed response of the considered dynamic system to perturbing influences. Under the condition of increasing its frequency, the phase-frequency characteristic at each speed mode of motion of the given harvester machine increases (Figure 5).

The greater the value of the oscillation frequency ω, the more influential the inertial properties of this machine, and the greater is its delay (in this case, the phase shift) to the action of disturbing influences. The same inertial properties of the machine-tractor unit cause the fact that the time of its response to perturbations increases with the increase in the velocity V_o. So, if at a speed of forwarding movement equal to V_o = 1.5 m·s^{-1} and ω = 10 s^{-1}, the phase shift of the given dynamic system (i.e., phase-frequency characteristic) makes −15° (curve 1, Figure 5), then already at an increase in the speed up to V_o = 2.5 m·s^{-1} and at the same frequency ω this parameter increases to a mark −25° (curve 3, Figure 5). The phase shift difference, in this case, is 10° or 0.17 rad.

Next, let us estimate more carefully the value of the specified indicator. At speed V_o of forwarding motion of the given harvester machine-tractor unit at a level of 2.5 m·s^{-1} and frequency of oscillations of disturbance ω = 10 s^{-1}, it makes 0.017 s. The significance of such a result is noteworthy if it were not for the essential invariance of the aggregate to the amplitude of oscillations of external disturbing influences. As a result, as we can see, when changing the speed mode of this machine-tractor unit in the range of 1.5–2.5 m·s^{-1}, it reacts sensitively to fluctuations in the angle β of rotation of the harvester and satisfactorily reproduces changes in its amplitude.

It should be separately emphasized that the character of stability of motion of this asymmetric machine-tractor unit when it reproduces external disturbances as a dynamic system significantly depends on the values of the input resistance coefficients of pneumatic tires of the wheeled aggregate tractor. In our case, these are the coefficients k_A and k_B. As the results of mathematical modeling show, the increase in the input resistance coefficient k_A of the front pneumatic tires of the aggregating tractor wheels lead to a decrease in the value of the corresponding amplitude-frequency response (Figure 6).

In a range of frequency changes ω = 0–10 s^{-1}, the specified increase on the average makes 11%. This result can be explained as follows. The greater the value of the coefficient k_A (which can be achieved by a corresponding increase in the air pressure in the pneumatic tires of the wheels of the aggregating tractor), the greater the resistance of the wheel to disturbing influences. In this case, it is the turning moment due to the turning of the topper by an angle β. On the contrary, the lower the air pressure in pneumatic tires of the driving wheels, the lower the value of their input resistance coefficient k_A, and the greater is the sensitivity of the wheel aggregate tractor to the action of disturbing influences in the form of turning moment. The result obtained should be taken into account when choosing the air pressure in the tractor tires since the value of this pressure determines the values of the coefficients k_A and k_B.

The qualitatively similar situation is associated with the change in the coefficient k_B of the resistance of the pneumatic tires input of the rear running wheels of the aggregating tractor (Figure 7). The difference is that the rear wheels of the aggregating tractor are closer to the attachment point of the trailed harvester to the tractor than the front ones.

As a result, they are quicker and more sensitive to disturbing influences due to fluctuations in the angle β we consider. Based on the numerical calculations obtained on the PC, we will make a more thorough analysis of the obtained result and compare at the same value of the frequency of vibration of the disturbance $\omega = 2\ \text{s}^{-1}$ two amplitude-frequency characteristics of this harvester machine obtained for sufficiently close values of the input resistance coefficients $k_A = 80$ kN·rad^{-1} (Figure 6) and $k_B = 90$ kN·rad^{-1} (Figure 7).

So, in the first case (Figure 6), the value of this characteristic almost reaches the mark 0.124. In the second case (Figure 7), the value of the amplitude-frequency response is about 0.145, which is almost 17% more. Thus, it is evident that the smaller the value of the amplitude-frequency response of the machine-tractor unit when working out the disturbance, the higher the stability of its movement in the horizontal plane.

The above analytical studies were carried out on the assumption that the energy base of the machine-tractor unit under consideration is universal tillage wheeled tractor of traction force of 14 kN. As it is known, this tractor has the classical layout, according to which about 40% of its weight is on the front and the rest is on the back of the tractor.

The results of the conducted mathematical modeling show that in qualitative terms, an increase in the speed of a given asymmetric machine-tractor unit from 1.5 to 2.5 m·s^{-1} leads to an undesirable increase in the amplitude-frequency response and desired increase in the phase-frequency response when they play an external disturbance in the form of oscillations of the rotation angle β of the rear-mounted harvester.

4. Conclusions

1. Based on the system of differential equations, the stability of motion of asymmetric machine-tractor unit when it performs the technological process of cutting the beet tops has been studied. Quantitative indices of this stability are the values of amplitude and phase-frequency characteristics of oscillations of the course angle of the tractor when this machine works out an external disturbance in the form of oscillations of the turn angle of the haulm harvester;
2. As evidenced by the results of mathematical modeling, in qualitative terms, an increase in the speed of the topper-tractor unit from 1.5 to 2.5 m·s^{-1} led to an undesirable increase in the amplitude-frequency response and desired increase in the phase-frequency response when they reproduce the disturbing external influence in the form of oscillations of the rotation angle of the haulm harvester unit. In quantitative terms, the change in the indicated amplitude-frequency and phase-frequency responses practically does not impair the stability of the dynamic system under consideration. It, therefore, cannot act as a limiting factor when choosing the high-speed movement mode of this machine-tractor harvester unit;
3. Found that reducing the value of the input resistance coefficient of the front tractor wheels from 120 to 80 kN·rad^{-1} and rear tractor wheels from 150 to 90 kN·rad^{-1} increases their sensitivity to the action of disturbing influences. This sensitivity is greater in rear tractor wheels because they are closer to the point of attachment of the trailed mower harvester.

Author Contributions: Conceptualization, V.B.; methodology, S.I. and V.N.; software, Y.I.; validation, A.A. and V.B.; formal analysis, V.B. and J.O.; investigation, V.B., S.I., V.N. and J.O.; data curation, A.A., V.B. and J.O; writing—original draft preparation, V.B.; writing—review and editing, A.A. and V.B.; visualization, Y.I. and V.N.; project administration, V.B.; funding acquisition, H.B. All authors have read and agreed to the published version of the manuscript.

Funding: This research received no external funding.

Institutional Review Board Statement: Not applicable.

Informed Consent Statement: Not applicable.

Data Availability Statement: Not applicable.

Conflicts of Interest: The authors declare no conflict of interest.

References

1. Ivančan, S.; Sito, S.; Fabijanić, G. Factors of the quality of performance of sugar beet combine harvesters. *Bodenkultur* **2002**, *53*, 161–166.
2. Vilde, A. Development of technologies and machinery for production of sugar beet in Latvia. In Proceedings of the Conference Safe and Economical Agricultural Technologies, Priekuļi, Latvia, 25–26 July 2002; pp. 62–66.
3. Nadykto, V.T.; Krizhachkovsky, N.L.; Kyurchev, V.N.; Abdula, S.L. *New Mobile Energy Facilities of Ukraine*; Theoretical Foundations of Use in Agriculture: Melitopol, Russia, 2006; 337p.
4. Maxakov, V. *The Haulm, Pulp and Melissa When Feeding Animals*; Urozhai: Kyiv, Ukraine, 1993; 145p. (In Russian)
5. Bhatia, N.P.; Szegő, G. *Stability Theory of Dynamical Systems*; Springer: Heidelberg, Germany, 2002.
6. Bulgakov, V.; Ivanovs, S.; Ruzhylo, Z.; Golovach, I. Theoretical investigations in cleaning sugar beet heads from remnants of leaves by cleaning blade. In Proceedings of the 15th International Scientific Conference Engineering for Rural Development, Jelgava, Latvia, 23–25 May 2016; pp. 1090–1097.
7. Bulgakov, V.; Pascuzzi, S.; Nadykto, V.; Ivanovs, S. A mathematical model of the plane-parallel movement of an asymmetric machine-and-tractor aggregate. *Agriculture* **2018**, *8*, 151. [CrossRef]
8. Merkin, D.R. *Introduction to the Theory of Stability of Motion*; Nauka: Moscow, Russia, 1971; 312p. (In Russian)
9. Anche, G.M.; Velmurugan, M.A.; Kumar, A.; Subramanian, S.C. Model Based Compensator Design for Pitch Plane Stability of a Farm Tractor with Implement. *IFAC-Paper* **2018**, *51*, 208–213. [CrossRef]
10. Bietresato, M.; Mazzetto, F. Increasing the safety of agricultural machinery operating on sloping graunds by perfoming static and dynamic tests of stability on a new-concept fasility. *Int. J. Saf. Secur. Eng.* **2018**, *8*, 77–89.
11. Bulgakov, V.; Ivanovs, S.; Adamchuk, V.; Boris, A. Mathematical model for determination of losses of sugar bearing-mass when sugar beet tops are removed. In Proceedings of the 14th International Scientific Conference, Engineering for Rural Development, Proceedings, Jelgava, Latvia, 23–25 May 2015; Volume 14, pp. 441–451.
12. Bulgakov, V.M.; Kaletnik, G.N.; Grinik, I.V.; Lezhenkin, A.N. *Dynamics of Grain Harvesting Units. Monograph*; Agrarian Science: Kyiv, Ukraine, 2010; 276p.
13. Demšar, I.; Bernik, R.; Duhovnik, J. A mathematical model and numerical simulation of the static stability of a tractor. *Agric. Conspec. Sci.* **2012**, *77*, 143–150.
14. Hac, A.; Fulk, D.; Chen, H. Stability and control considerations of vehicle-trailer combination. *SAE Int. J. Passeng. Cars Electron. Electr. Syst.* **2009**, *1*, 925–937. [CrossRef]
15. Szakács, T. Developing stability control theories for agricultural transport systems. *Acta Polytech. Hung.* **2010**, *7*, 25–37.
16. Szczepaniak, J. Simulation calculation of tractor-potato planner combination model. In Proceedings of the 7th International Workshop on Modeling and Applied Simulation, Held at the International Mediterranean Modeling Multiconference, Genova, Italy, 17–19 September 2008; pp. 240–245.
17. Sun, C.; Nakashima, H.; Shimizu, H.; Miyasaka, J.; Ohdo, K. Physics engine application to overtuning dynamics analysis on banks and uniform slopes for an agricultural tractor with a rollover protective structure. *Biosyst. Eng.* **2019**, *185*, 150–160. [CrossRef]
18. Gyachev, L.V. *Stability of Movement of Agricultural Machines and Aggregates*; Mashinostroenie: Moscow, Russia, 1981; 206p. (In Russian)
19. Vasilenko, P. *Introduction to Agricultural Mechanics*; Agricultural Education: Kyiv, Ukraine, 1996; 252p.
20. Dreizler, R.M.; Ludde, C.S. *Theoretical Mechanics*; Springer: Heidelberg, Germany, 2010; 402p.
21. Lipschutz, S.; Spiegel, M.R.; Liu, J. *Mathematical Handbook of Formulas and Tables, Schaum's Outline Series*, 3rd ed.; McGraw-Hill: New York, NY, USA, 2009; p. 183, ISBN 978-0-07-154855-7.

applied sciences

MDPI

Article

Design of an Adaptive Algorithm for Feeding Volume–Traveling Speed Coupling Systems of Rice Harvesters in Southern China

Lexing Deng, Tianyu Liu *, Ping Jiang, Fangping Xie, Junchi Zhou, Wenhan Yang and Aolin Qi

College of Mechanical and Electrical Engineering, Hunan Agricultural University, Changsha 410128, China
* Correspondence: liutianyu@hunau.edu.cn

Abstract: We developed an adaptive algorithm to reduce rice loss in harvesting, promote threshing and improve the quality and efficiency of small- and medium-sized rice harvesters operating in southern China's hilly and mountainous areas. Using a fuzzy PID control algorithm, the harvester adapts to the rice harvesting conditions in southern China, and monitors rice feed volume changes and instantly adjust the traveling speed to optimize feed volume levels and threshing quality. We compared and analyzed the algorithm and the traditional PID control regulation effect in the simulation experiment. The algorithm had a quicker response speed and stable accuracy. In the field trial, the average error rate was 3.4%, and the maximum error rate was 5.1%, with most data points centered around the ideal feeding rate of 3.2 kg/s. Our results showed that the algorithm's stability, accuracy, and real-time performance met the threshing loss reduction requirements of southern China's rice harvesting operations.

Keywords: southern rice; threshing loss; fuzzy PID; adaptive algorithm; hilly mountainous areas

Citation: Deng, L.; Liu, T.; Jiang, P.; Xie, F.; Zhou, J.; Yang, W.; Qi, A. Design of an Adaptive Algorithm for Feeding Volume–Traveling Speed Coupling Systems of Rice Harvesters in Southern China. *Appl. Sci.* **2023**, *13*, 4876. https://doi.org/10.3390/app13084876

Academic Editor: César M. A. Vasques

Received: 1 March 2023
Revised: 30 March 2023
Accepted: 11 April 2023
Published: 13 April 2023

1. Introduction

As an important grain cash crop in southern China, research on threshing loss reduction, referring to effectively reducing grain loss and improving agricultural production efficiency by adopting appropriate harvesting equipment and control methods, is of great significance to improving farmers' profitability and guaranteeing national food security [1,2].

Threshing losses are usually analyzed from two aspects. On the one hand, during the operation of the harvester, a considerable portion of the rice grain will fall into the field or be discharged with the stem, resulting in direct losses; on the other hand, the harvested rice can be partially broken, leading to losses such as mold and spoilage during storage and transportation. The proper feeding volume is the key to improving the threshing quality, reducing losses and achieving the optimal performance of the harvester. Under current traveling speed adaptive system regulations, the traveling speed of the harvester and its feeding volume should have a good coupling relationship [3]. When the traveling speed changes, the feeding volume changes accordingly. This direct positive correlation reflects the close coupling between the traveling speed and the feeding volume. Therefore, the feed amount can be controlled by adjusting the traveling speed.

Many developed countries have researched smart traveling speed controls based on traditional manual operations, making extensive use of various types of electronic instruments and monitoring, alarms and other devices [4,5]. Sood et al. investigated a machine vision system for automatic plant division adapted to different light conditions, which can be used to monitor crop density in real time during harvester operations [6]. Federico et al. presented a control system for the modeling, control design and experimental automatic calibration of a combine harvester leveling system [7]. Nevavuori et al. used deep convolutional neural network techniques to predict crop yields and thus guide harvesting [8]. Contemporary research results are more reliable and widely referenced, and mainstream products are suitable for large-scale farming operations. However, the corresponding purchase cost is relatively high and does not align with China's agricultural situation of diversified farming types and family farming [9]. Therefore, it is difficult to implement in China.

Numerous Chinese scholars have researched combine harvester traveling speed controls. Li et al. established an online monitoring system for combine harvesters' hydraulic actuators using the AMESim simulation platform, and experimentally verified its high accuracy, providing harvester drivers with a reference for controlling traveling speed [10]. Yang et al. electrically modified a harvester and tested functions such as sequential start/stop, reel speed tracking and blockage detection, with good results [11]. Wang et al. designed a new adaptive control system for combine harvesters that considered cutting and tool advancement speeds [12]. Their experiments showed that the system controlled the tool's cutting and advancement speeds more effectively. Chen et al. proposed a method for the real-time monitoring of grain breakage rates of rice combine harvesters based on machine vision, which can be used for the optimization and adjustment of traveling speeds of an intelligent harvester [13]. Some deep-learning related algorithms have inspired the development of intelligent agricultural machinery [14–17].

However, the current research results have certain limitations: (1) most of the studies are based on ideal conditions; that is, the field operation terrain is flat, or the undulating degree is minor, the operation area is large enough that the harvester does not need to change direction frequently, or the harvesting needs of scattered rice fields in the southern area's hilly, mountainous terrain are not being met [18,19]; (2) the design does not consider the impact caused to the system's detection accuracy and control effects by the high moisture characteristics of rice harvesting materials from southern China; (3) in addition to the high cost of promoted automatic control technology, some complex integrated control systems struggle to meet the current market demand for affordable, stable, or modifiable small- and medium-sized harvester-traveling speed control systems [20].

To solve the above limitations, we designed an adaptive algorithm for the feeding volume–traveling speed coupling system for a rice harvester in southern China. We calculated the feeding volume by detecting the torque and speed of the power shaft of the harvester's tilting conveyor. We also developed a traveling speed optimization strategy to calculate the optimal traveling speed, which was used as an input quantity to automatically adjust the stepper electric actuator action online based on fuzzy PID. We discussed and optimized the signal analysis method and controller algorithm design for rice harvesting conditions in southern China. The stability, accuracy, and real-time performance of the algorithm were verified through simulation experiments and field trials. Our results showed that this can provide a reference for rice harvesters' feeding volume–traveling speed coupling systems in southern China.

In summary, the main contributions of this paper are as follows: (1) We proposed an optimal speed optimization strategy based on torque and detection of the tilting conveyor power shaft to improve the stability and accuracy of the harvester's traveling speed control; (2) We developed an adaptive algorithm for feeding volume–traveling speed coupling systems that adapts to rice harvesting conditions in southern China; (3) A control system with low hardware requirements was designed. The system has a high practical value. It can be used to develop new small- and medium-sized rice harvesters in hilly and mountainous areas. It can also be used for the modification of traditional harvesters. We propose that the variation of traveling speed during harvesting causes threshing loss. Our experimental results can provide rice harvesters with a reference for designing threshing and loss reduction systems in southern China.

The rest of the paper is organized as follows: Section 2 discusses and determines the feed rate detection scheme. Section 3 outlines the overall framework of the control system. Section 4 presents the design of the algorithm. Section 5 presents experiment-related content and discusses the simulation and field trial results. Section 6 summarizes our proposed approach and the direction of future work.

2. Methods

2.1. Detection Scheme Selection

Currently, considering the detection location, sensors are mainly placed in four parts of the tilt conveyor, screw conveyor, threshing drum, and active shaft of the cutting platform [21–23]. Figure 1 shows the installation position of the sensor test. In terms of detection stability under the harsh harvesting conditions in southern China, which are mountainous with high temperatures and humidity, it is difficult to maintain the working stability of the cutting platform's active shaft sensor. Regarding real-time detection, the threshing drum is used after the threshing process, and the detection results of the sensor at this location lag significantly. Regarding detection practicality, there is less space for the screw conveyors of small- and medium-sized harvesters, which is not conducive to sensor installation. Installing sensors at the inclined conveyor position ensures operational stability and better real-time performance, and it is simple and easy to retrofit.

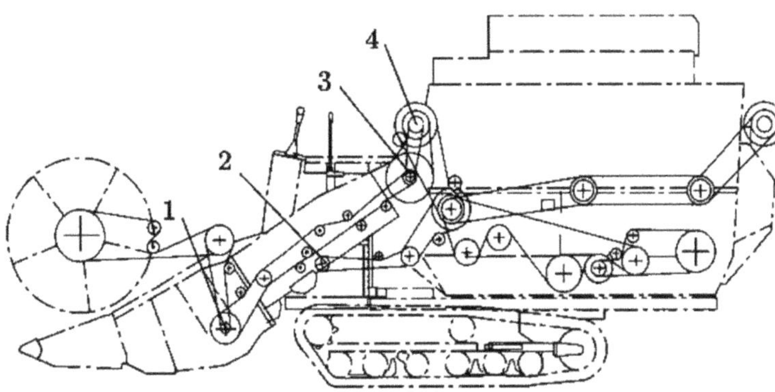

Figure 1. Sensor installation location diagram: 1: active shaft sensor for cutting platform, 2: screw conveyor shaft sensors, 3: tilt conveying active shaft sensor, 4: threshing drum active shaft sensor.

There are two main categories of feeding detection parameters: torque detection and power detection. Although the torque can theoretically reflect the feed variation more effectively, measurement accuracy and stability are greatly influenced by the angle of the tilted conveyor, the width of the cutting platform, etc. The actual detection accuracy is not ideal when applied to small- and medium-sized harvesters [24]. As a linear quantity, shaft power has fewer variables affecting detection accuracy and better stability, making it more suitable for feeding volume detection in southern China.

In summary, we selected a feeding volume detection scheme based on the power consumed by the tilting conveyor's power shaft. When the harvester was in operation, the feed crop was transported by the tilting conveyor to the threshing drum for threshing. The shaft power consumption of the tilting conveyor was then measured. A mathematical model of the feeding volume and total power consumption detected the feeding volume in real time.

2.2. Detection Principle and Calculation Method

The two main feed detection devices were torque sensors and Hall-type speed sensors. The former was used to measure the power shaft torque of the tilting conveyor, T_s. The latter was used to measure the speed of the main shaft of the cutting platform at n_1.

Torque strain gauges were pasted in the 45 and $-45°$ directions along the tilted conveyor's power axis. Adjacent strain gauges were arranged at $90°$ equal intervals. Figure 2 shows the principle of torque sensor detection. When the harvester is in operation, the power shaft of the tilting conveyor will deform to different degrees with the change in feeding volume. The resistance of the strain gauge will then change, causing the output voltage to change. The output voltage signal of the bridge will convert to digital signals through the signal amplification and acquisition circuit, and will then be output to the industrial control machine. The power shaft torque is obtained by measuring the voltage signal.

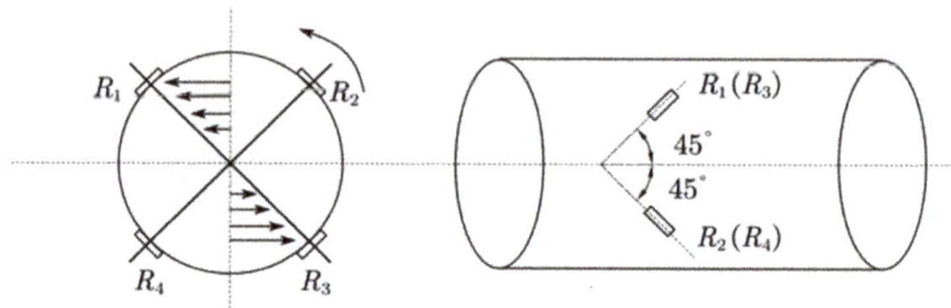

Figure 2. Schematic diagram of the torque sensor detection principle. R_{1-4} represent torque strain gauges.

Several magnets were attached to the side of the main shaft of the cutting platform. When the main shaft rotated, the Hall effect occurred when the magnets alternately passed through the Hall element, forming a pulse signal. The number of pulses generated per unit of time and the number of magnets on the surface of the main shaft can be used to calculate the corresponding speed of the main shaft of the cutting platform n_1. The calculation formula for the speed of the power shaft of the tilting conveyor n_s is as follows:

$$n_s = kn_1 \tag{1}$$

where n_s is the speed of the harvester tilting conveyor power shaft, r/min; k is the transmission ratio between the tilt conveyor power shaft and the main shaft of the cutting platform; and n_1 is the speed of the cutting platform active shaft.

The formula for calculating the power shaft of a harvester tilt conveyor is as follows:

$$P_s = \frac{T_s n_s}{9550} \tag{2}$$

where P_s is the power of the power shaft of the harvester tilt conveyor, kW; T_s is the torque of the harvester tilting conveyor power shaft, N-m.

Figure 3 shows a schematic diagram of the operation of the combine harvester tilt conveyor.

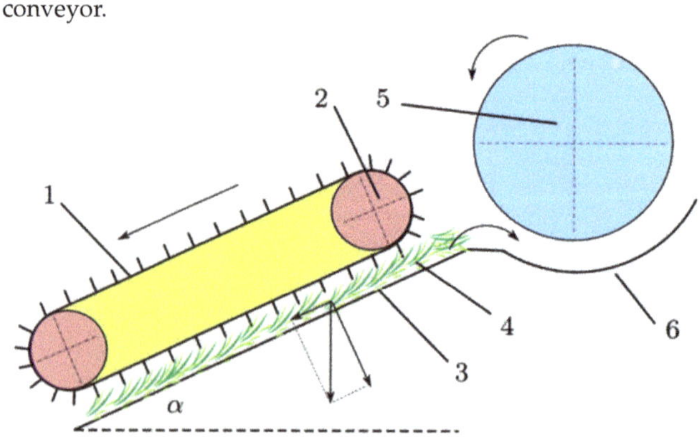

Figure 3. Working diagram of combine harvester tilt conveyor: 1: crossbar strip chain rake, 2: tilt conveyor active shaft, 3: tilting conveyor bottom plate, 4: straw, 5: threshing drum, 6: concave plate.

Assume that the tilt conveyor feeding volume at the moment t is q (kg/s). Through the chain rake, the active shaft of the inclined conveyor exerts a traction force on the grain that is F (N). The pressure on the grain is M (N). For the analysis of the tilting conveyor in the grain force, it can be seen that, in order to make the grain along the conveyor bottom plate display an upward movement, the following conditions must be met:

$$F \geq mgsin\alpha + \mu(M + mg)cos\alpha \quad (3)$$

where α is the inclination angle of the inclined conveyor, rad; μ is the friction factor between the grain and the bottom plate of the conveyor; and m is the mass of the grain conveyed by the conveyor, kg.

m is calculated as follows:

$$m = a_p l h_p \gamma \quad (4)$$

where a_p is the width of the conveyor, m; l is the length of the conveyor, m; h_p is the average thickness of the material in the down-strike chain rake conveyor, m; and γ is the density of the transported grain layer, kg/m^3.

$$h_p = \frac{q}{v a_p \gamma \epsilon} \quad (5)$$

where v is the speed of the conveyor chain, m/s; ϵ is the filling factor of the material in the space formed between the lower working part of the conveyor and the inclined surface (ϵ = 0.6 to 0.8).

Substituting Equation (3) into Equation (2), we can obtain:

$$m = \gamma a_p l \frac{q}{v a_p \gamma \epsilon} = \frac{ql}{v\epsilon} \quad (6)$$

The pressure M of the chain harrow on the grain is approximately proportional to the diameter h_p, so the following relationship exists:

$$M = kh_p = k\frac{q}{v a_p \gamma \epsilon} \quad (7)$$

where k is the proportionality constant.

Substituting Equations (6) and (7) into Equation (3), we can obtain the following equation when the inclined conveyor is moving at a constant speed:

$$F = q\left[\frac{l}{v\epsilon}gsin\alpha + \mu\left(\frac{k}{v a_p \gamma \epsilon} + \frac{l}{v\epsilon}g\right)cos\alpha\right] \quad (8)$$

When feeding uniformly and at a constant speed, neglecting the influence of certain secondary factors, the power balance equation of the inclined conveyor according to the principle of transmission dynamics is as follows:

$$P_s - P_c = Fv \quad (9)$$

where P_c is the unwanted power when the inclined conveyor is operating.

Substituting Equation (8) into Equation (9), we can finally obtain the calculation formula for the inclined conveyor:

$$q = \frac{a_p \gamma \epsilon}{a_p \gamma \lg(sin\alpha + \mu cos\alpha) + \mu k cos\alpha}(P_s - P_c) \quad (10)$$

3. System Design

We selected the original 4LZ-3.2Z full-feed grain combine harvester from Nongyou Agricultural Equipment Co., Ltd. in Hunan Province, China, as the test prototype for this paper. The shape of the harvester is shown in Figure 4. The harvester has a rated feeding capacity of 3.2 kg/s, a theoretical operating speed of 2.2 m/s, a working width of 1700 mm on the cutting platform, and a rated engine power of 55.2 kW. The driving method is fully hydraulic, with highly integrated electronic controls. The traveling speed is regulated by an electronic control handle, which is conducive to the design and modification of the hardware system. Therefore, we based the design of the adaptive coupling system on this harvester. We also based the relevant parameters of the adaptive coupling system on this harvesting platform's performance index.

Figure 4. The 4LZ-3.2Z Whole Grain Combine.

3.1. Control System Hardware Framework

The harvester's feeding volume–traveling speed coupling system consists of a sensing module, a control unit, an electric actuation module, and a power supply module. The control principle is that the sensor module detects the torque and speed of the tilting conveyor's power shaft to calculate the real-time feeding volume. It obtains the deviation value $e(t)$ by comparing the calculated value $c(t)$ to the ideal feeding volume $r(t)$ provided by the system. The fuzzy PID controller receives and processes the error signal, and the corresponding control signal is output. The signal is multi-stage amplified by the drive module to control the stepping electric actuator action, adjusting the harvester's stopper position, i.e., controlling the traveling speed toward deviation elimination until the deviation tends to zero, thus realizing the feeding volume–traveling speed self-adaptation. The hardware framework of the control system was designed as shown in Figure 5.

The control unit is responsible for communication management, analyzing all information and implementing logical operations. The sensing module detects parameters such as the torque and rotation speed of the tilt conveyor's power shaft during harvesting operations. The power supply module provides power to all modules.

3.2. Electric Actuator Module

The stepper electric actuator was mounted on the self-propelled chassis frame, and the output shaft was flexibly connected to the hydraulic transmission joystick. The actuator was lightweight, small in size, simple in principle, and met the modification requirements of small- and medium-sized harvesters. Figure 6 shows a schematic diagram of the electric actuator module.

Under the control of the control system, the size of the stepping electric actuator extension or contraction Δl (mm) will drive the angle of the joystick swing θ (rad). The speed of the harvester is controlled by adjusting the hydraulic transmission to realize the control of the traveling speed of the harvester. Assuming that the illustrated mechanism occurs while operating the actuator, if the dotted line n indicates the starting position of the joystick, the formula for calculating the angle of the joystick position would be:

$$\theta = \arcsin\left\{ \frac{h\Delta l}{\left[\left(\Delta l + l - \sqrt{d^2 - h^2}\right)^2 + h^2\right]\left[h^2 + \left(\sqrt{d^2 - h^2} - l\right)^2\right]} \right\} \tag{11}$$

where d and h are the linear and vertical distances between the center of the stepper motor actuator base and the stopper pivot point, respectively ($d > h$); l is the original length of the electric actuator.

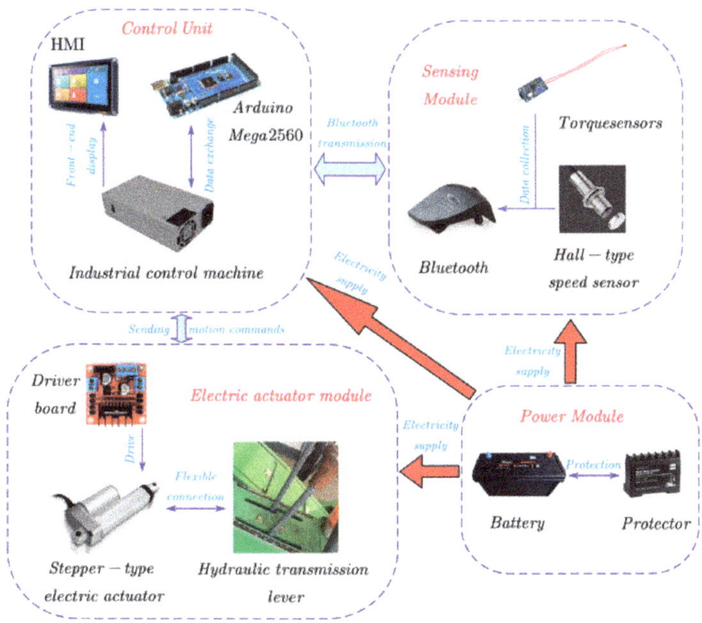

Figure 5. Hardware block diagram of the control system.

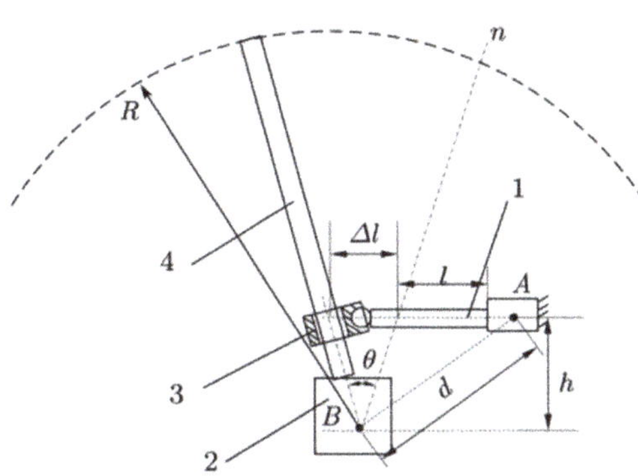

Figure 6. Schematic diagram of the electric actuator module: 1: stepping electric actuator, 2: hydraulic transmission, 3: kinematic pair, 4: gear lever.

4. Algorithm Design

4.1. Optimal Traveling Speed Calculation Strategy

When the feeding volume q exceeds the rated feeding volume Q of the harvester, the traveling speed should be reduced appropriately. The traveling speed should be increased when the feeding volume q is too low. The ideal target speed can be calculated from real-time feed rates while ensuring rice harvester efficiency and quality and reducing equipment failure.

$$v_n = v_0 + \frac{Q - q}{3\sigma} \tag{12}$$

where v_n and v_0 are the ideal target velocity and real-time velocity (in m/s), respectively.

Considering the many contingent factors when rice harvesters operate in southern China's complex field environments, the target speed calculated from the input quantity alone was not well adapted to the system. Therefore, we propose the traveling speed optimization formula, as shown in Equation (13). In the following exposition, the traveling speed of the harvester is optimized based on the tilt conveyor's power shaft torque T_s and rotational speed n_s.

- When T_s and n_s are both in a reasonable state, the absolute value of the relative error between its value and the rated value belongs to [0, 0.1]; then, the traveling speed at this time is v_n;
- When there is an abnormality between T_s and n_s, i.e., the absolute value of the relative error between at least one set of values and the rated value belonging to [0.1, 0.2], adjust the speed to $0.8v_n$ and observe the subsequent values. If all values return to a reasonable state within 3 s, adjust the speed to v_n; otherwise, decelerate to $0.4v_n$;
- When the harvester fails, i.e., the absolute value of the relative error between any value and the rated value is constantly >20%, the speed is immediately reduced to 0 m/s and the machine is stopped for inspection.

$$V = \begin{cases} v_n, 0 \leq |n_i| \leq 0.1N_i \\ 0.8v_n, 0.1N_i < |n_i| \leq 0.2N_i; T \leq 5 \\ 0.4v_n, 0.1N_i \langle |n_i| \leq 0.2N_i; T \rangle 5 \\ 0, |n_i| > 0.2N_i \end{cases} \tag{13}$$

where V is the optimal traveling speed of the harvester, m/s; i = 1,2, n_i are the error values of T_s and n_s, respectively; N_i are the rated values of T_s and n_s, respectively; and T is the data update time, s.

4.2. Adaptive Fuzzy PID for Coupling System

In this paper, we used a fuzzy PID control strategy combining fuzzy control and classical PID to achieve an adaptive coupling system. It uses the PID algorithm with fuzzy inference based on fuzzy rules to adjust PID parameters and achieve optimal control [25]. It has the advantages of simplicity and high accuracy in traditional PID regulation while avoiding the disadvantages, such as poor adaptability and flexibility. This paper uses a PID type controller, and Figure 7 shows the schematic. Figure 8 shows the schematic block diagram of the fuzzy PID controller.

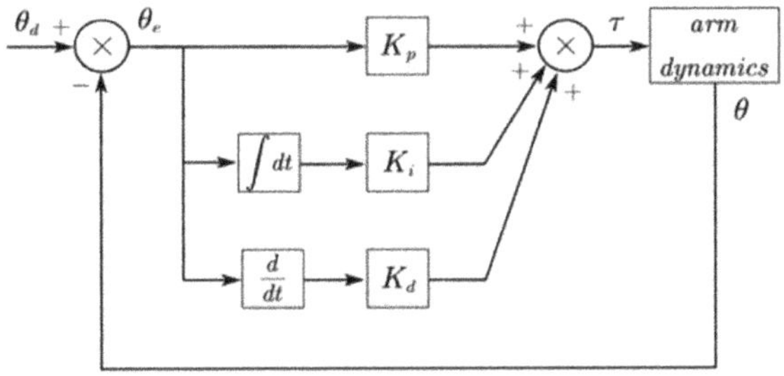

Figure 7. Schematic diagram of PID type controller.

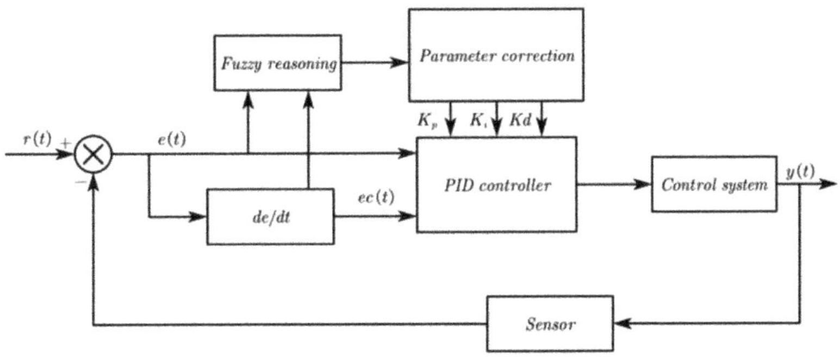

Figure 8. Principle block diagram of the fuzzy PID controller.

4.2.1. Domain of Discourse and Fuzzy Subset Determination

In this paper, we defined the traveling speed deviation e and the rate of change in deviation e_c as the input variables. We also defined the corrections of the three PID control parameters ΔK_p, ΔK_i and ΔK_d as the output variables. The constant parameters of the conventional PID controller were as follows: K_p = 3, K_i = 0.055 and K_d = 0.01. This was designed according to the transfer function and dynamic characteristics of the system. Additionally, MATLAB was used for trial control and optimization.

We modified the experimental vehicle with a feed detection function. Simulation tests were performed with the manual feeding of rice. At the same time, we performed a conventional PID simulation of the feeding volume–traveling speed coupling system. By referring to the variation characteristics of the input variables in the simulation tests, the basic domain of each variable was finally determined (Table 1).

After determining the fuzzy domain, the variables were divided into several fuzzy sets using fuzzy linguistic values. Each variable value corresponded to a set of fuzzy language values, and the number of fuzzy sets was selected according to the control accuracy. The more accurate the number, the more complex the corresponding control rules. In this paper, we selected seven levels of fuzzy language values {NB, NM, NS, ZO, PS, PM, and PB} based on the actual harvesting conditions.

4.2.2. Affiliation Function Selection

The continuous domain fuzzy sets were expressed in the functional form [26]. To filter the random perturbations that often exist in the input variables e and e_c, we selected the isosceles triangular shape as its affiliation function, which we set to a Gaussian affiliation curve to improve the response sensitivity, considering that the basic domain edges of the input variables belonged to the large deviation category. To enhance the smoothness of the controller output, the output variables ΔK_p, ΔK_i, and ΔK_d were chosen to be of Gaussian type as the affiliation function. The affiliation function and distribution of each variable are shown in Figure 9, where the horizontal coordinate is the fuzzy domain and the vertical coordinate is the affiliation degree.

Table 1. The basic domain.

Variables	The Basic Domain	Based on
e	[−6, 6]	The traveling speed variation fluctuated greatly during the test, and the limit deviation was 6 m/s.
e_c	[−6, 6]	The test's rate of change in deviation was significant; the range was determined by simulation tests.
ΔK_p	[−6, 6]	Affects the speed at which the system adjusts. In mountainous hilly areas, the speed and direction of the machine needed to be adjusted frequently, so the value should be extended. Determine the range by simulation test.
ΔK_i	[−0.3, 0.3]	Affects the system's steady-state error; too large will lead to overshoot. Determine the range by simulation test.
ΔK_d	[−0.3, 0.3]	Affects the system response speed; too large will lead to a long response time. Determine the range by simulation test.

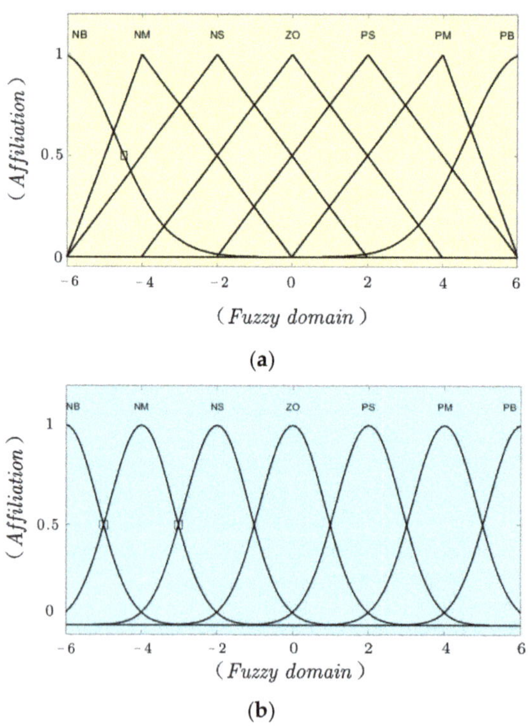

Figure 9. *Cont.*

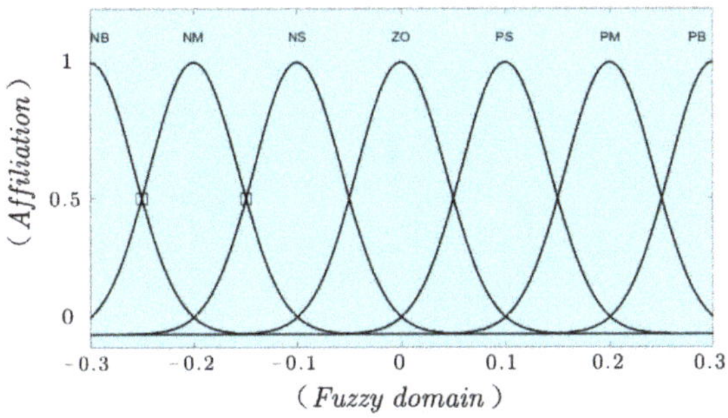

(c)

Figure 9. (**a**) The affiliation function and distribution of e and e_c; (**b**) the affiliation function and distribution of ΔK_p; (**c**) the affiliation function and distribution of ΔK_i and ΔK_d.

4.2.3. Fuzzy Rules Design

The fuzzy rules for the design output quantities ΔK_p, ΔK_i and ΔK_d are summarized based on expert experience [27,28], as shown in Table 2.

Table 2. Fuzzy rules of ΔK_p, ΔK_i and ΔK_d.

e_c \ e	NB	NM	NS	ZO	PS	PM	PB
NB	PB/NB/PS	PB/NB/NS	PM/NM/NB	PM/NM/NB	PS/NS/NB	ZO/ZO/NM	ZO/ZO/PS
NM	PB/NB/PS	PB/NB/NS	PM/NM/NB	PS/NS/NM	PS/NS/NM	ZO/ZO/NS	NS/ZO/ZO
NS	PM/NB/ZO	PM/NM/NS	PM/NS/NM	PS/NS/NM	ZO/ZO/NS	NS/PS/NS	NS/PS/ZO
ZO	PM/NM/ZO	PM/NM/NS	PS/NS/NS	ZO/ZO/NS	NS/PS/NS	NM/PM/NS	NM/PM/ZO
PS	PS/NM/ZO	PS/NS/ZO	ZO/ZO/ZO	NS/PS/ZO	NS/PS/ZO	NM/PM/ZO	NM/PB/ZO
PM	PS/ZO/PS	ZO/ZO/PS	NS/PS/PS	NM/PS/PS	NM/PM/PS	NM/PB/PS	NB/PB/PB
PB	ZO/ZO/PB	ZO/ZO/PM	NM/PS/PM	NM/PM/PM	NM/PM/PS	NB/PB/PS	NB/PB/PB

The fuzzy inference surface of each output quantity is shown in Figure 10. It can be concluded that, when |e| is large, ΔK_p should be considered larger, ΔK_d as smaller, and ΔK_i needs no adjustment. When |e| is medium, ΔK_p should be considered smaller, and ΔK_i, and ΔK_d should be considered medium. When |e| is small, ΔK_p and ΔK_i should be increased appropriately and ΔK_d should be kept as a medium.

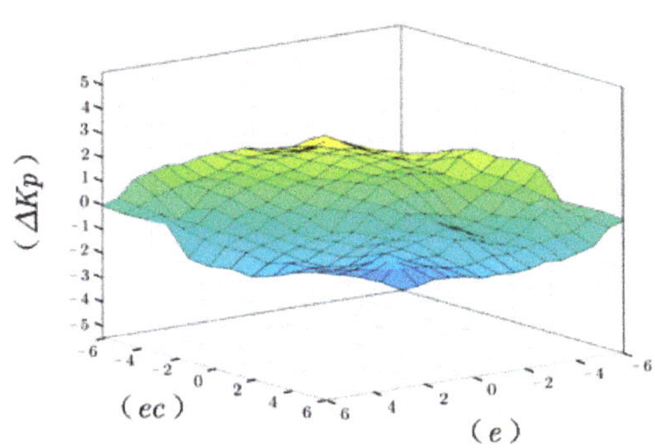

(a)

Figure 10. *Cont.*

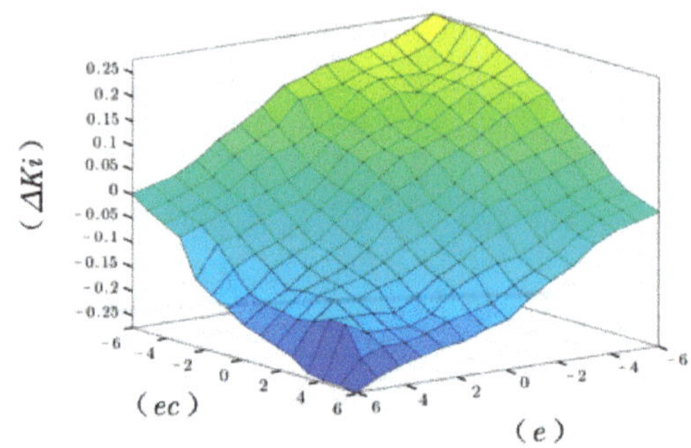

(b)

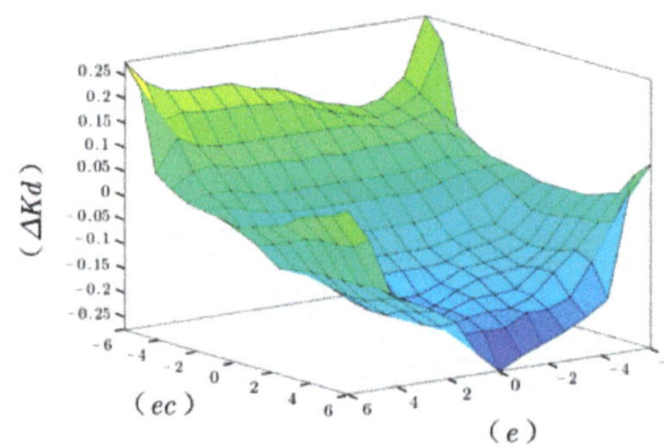

(c)

Figure 10. (**a**) The fuzzy inference surface of ΔK_p; (**b**) the fuzzy inference surface of ΔK_i; (**c**) the fuzzy inference surface of ΔK_d.

According to the affiliation function of each fuzzy variable and fuzzy rule table, we designed the PID parameter adjustment algorithm using the fuzzy inference theory to obtain more accurate control parameters and improve the performance of the fuzzy control system. If ΔK_p, ΔK_i and ΔK_d are the pre-tuned values of the PID controller's three parameters, then the formulae for calculating the three are:

$$K_p = K_p' + \Delta K_p \tag{14}$$

$$K_i = K_i' + \Delta K_i \tag{15}$$

$$K_d = K_d' + \Delta K_d \tag{16}$$

4.2.4. Inference and Defuzzification

We used the Zadeh approximation inference method based on fuzzy rules to complete the inference and obtain the fuzzy control quantity [29]. Due to its aggregate form, the system must be defuzzified to receive a clear control signal that drives the actuator. For a smooth control effect and convenient operation, we adopted the weighted average method and the calculation formula is as follows:

$$Z_0 = \frac{\sum_{i=1}^{n} \mu_C(Z_i) * (Z_i)}{\sum_{i=1}^{n} \mu_C(Z_i)} \tag{17}$$

where Z_0 is the exact output value; Z_i is the value corresponding to the fuzzy control quantity in the theoretical domain; and $\mu_C(Z_i)$ is the value of the affiliation degree corresponding to Z_i.

5. Experiment and Analysis

5.1. Simulation Experiments

5.1.1. Pusher Voltage–Motor Speed Coupling Model

Electric actuators are a class of devices that convert the rotational motion of a motor into linear motion and comprise a motor, gear structure, screw, and other components. Hu et al. studied electric actuators' synchronous control systems and derived the electric actuator simulation structure that forms the basis for the coupled actuator voltage–motor speed model in this paper (Figure 11) [30].

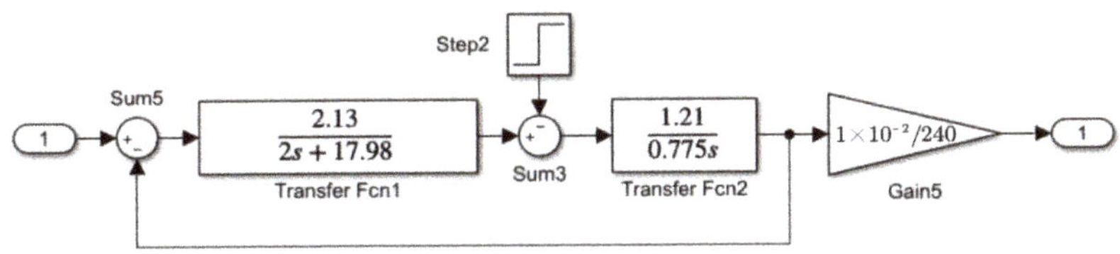

Figure 11. Electric actuator simulation model.

After the system outputs the motor speed, it converts it into the push rod retraction, joystick tilt angle and, finally, the harvester traveling speed. A second-order system to describe the controlled object can be approximated by neglecting secondary factors under certain accuracy conditions [31]. After data processing, the system transfer function is expressed as follows:

$$G(s) = \frac{4531}{(s^2 + 0.335s + 0.0108)} \tag{18}$$

The Bode plot in Figure 12 was plotted using MATLAB according to the transfer function. The ideal suppression of harmonics was in the high-frequency region, and resonant spikes were successfully suppressed at resonant frequencies. The gain margin of the system was approximately infinite and the phase margin was close to satisfying $\varnothing m \approx \pi - \beta\pi/2$, β was the system damping ratio, and the above transfer function corresponds to a damping ratio of 1.675. Therefore, we could conclude that this closed-loop system's accuracy and stability are high.

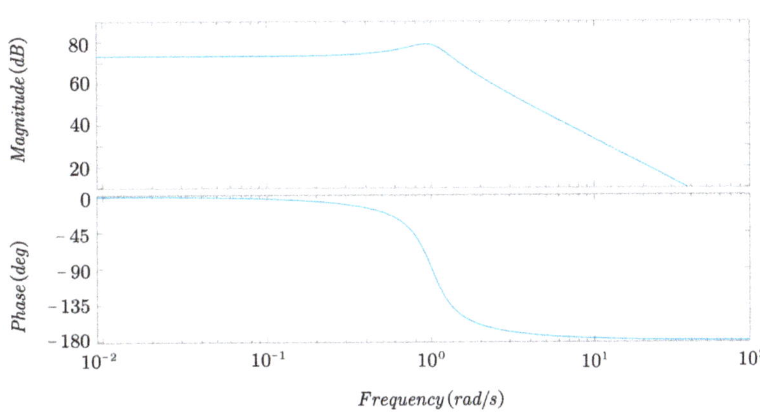

Figure 12. Bode Plot.

5.1.2. Simulation Comparison Test between Fuzzy PID and a Traditional PID System

The model for building a conventional PID and fuzzy PID controller is shown in Figure 13.

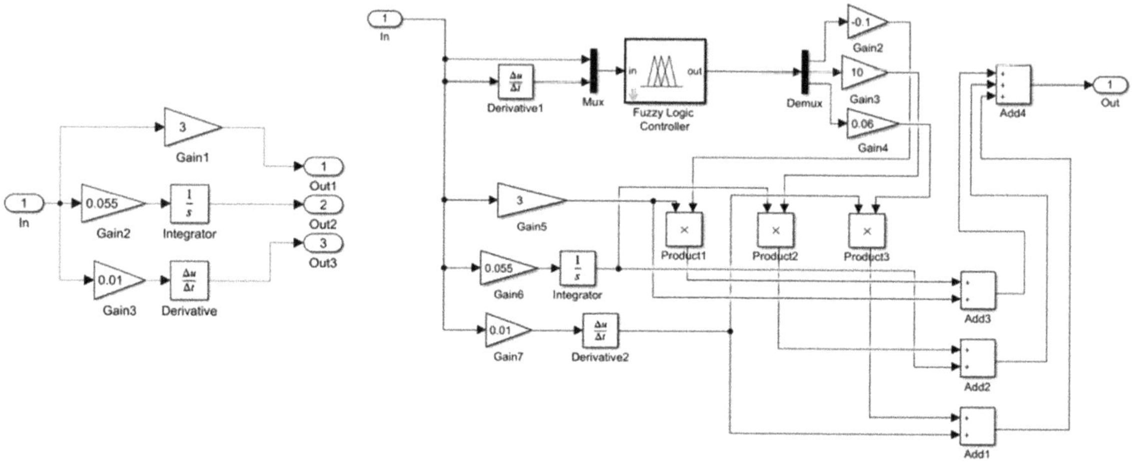

(a) (b)

Figure 13. (**a**) Traditional PID simulation model; (**b**) fuzzy PID simulation model.

Based on the traditional PID and fuzzy PID controllers, we built the simulation model of the harvester's adaptive traveling speed under the action of the two controllers, as shown in Figure 14.

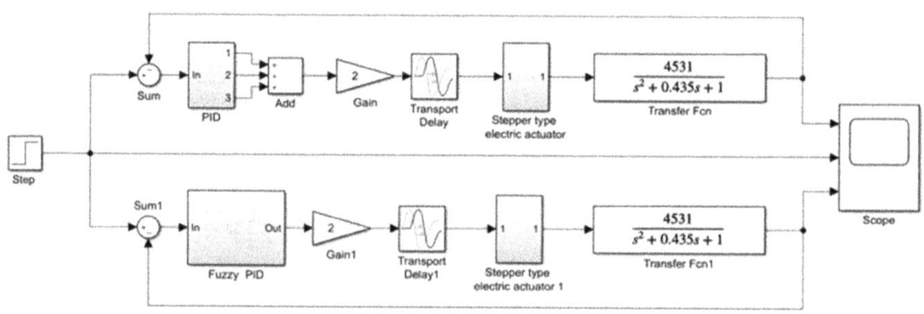

Figure 14. Simulation model of the harvester traveling speed adaptive regulation comparison test system.

The simulation test received a traveling speed step signal of 1 m/s as input, and we set the simulation time to 30 s. Figure 15 shows the adaptive simulation results of the traveling speed under two control algorithms. The results show the harvester's traveling speed variation rise time under the conventional PID algorithm t_r = 3.2 s, peak time t_m = 5.2 s. The time to reach the steady state was slower at about 20 s, with a significant overshoot. The rise time corresponding to the fuzzy PID t_r = 1.8 s, peak time t_m = 3.2 s, had a shorter time to steady state, at about 5.5 s, with almost no oscillation.

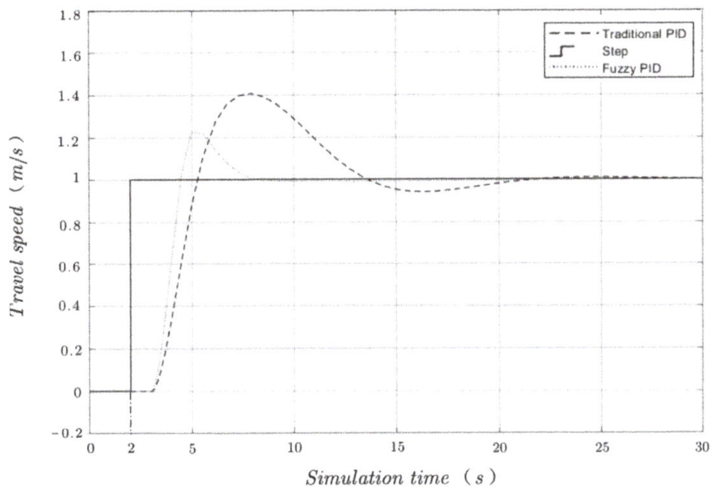

Figure 15. Simulation results of adaptive adjustment to the harvester's traveling speed.

5.2. Field Trials

5.2.1. Trial Conditions

To verify the regulation effect of the Coupling System's adaptive algorithm, a field trial was conducted in July 2022 in Yintang Township, Shuangfeng County, Loudi City, Hunan Province, China. The test area was 400 m^2, and a buffer zone of 5 m was set up. We investigated the field conditions in the area before the test. The rice variety was Jingliangyou 1125, with plant rows spaced about 30 cm apart and distributed evenly. It had an average height of about 970 mm, no overturning, and approximately the same water content. The terrain was flat, with no obvious potholes. The harvesting trial site is shown in Figure 16.

Figure 16. Harvesting trial site.

5.2.2. Experiment and Analysis

We used the following experimental method, since we only completed the electrification of the electric actuator and related hardware arrangement and did not make functional modifications to other modules. A person with rice harvesting experience drove the modified car to the test area and performed operations such as changing direction and stabilizing the height of the cutting platform while ensuring that changes in the harvester's feeding volume were mainly affected by traveling speed. The 4LZ-3.2 Z full-feed grain combine harvester had a screen display function that displayed the traveling speed in real-time during operation. The industrial control computer read the harvester bridge's output voltage signal and the Hall-type speed sensor's pulse signal and substituted them into the power shaft of the tilting conveyor model to obtain the current feeding volume.

The test was timed with a stopwatch. Rice was harvested in 130.43 s from the measurement area with no blockage in the machinery's operation process. Data starting at 0.43 s were excluded and rounded to 130 s to facilitate the subsequent data processing and analysis. The time domain diagram of the traveling speed is shown in Figure 17. The time domain diagram of the feeding volume is shown in Figure 18.

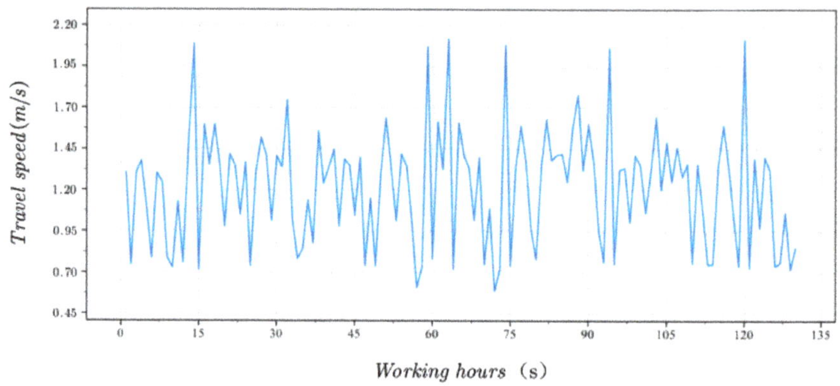

Figure 17. Time domain diagram of traveling speed.

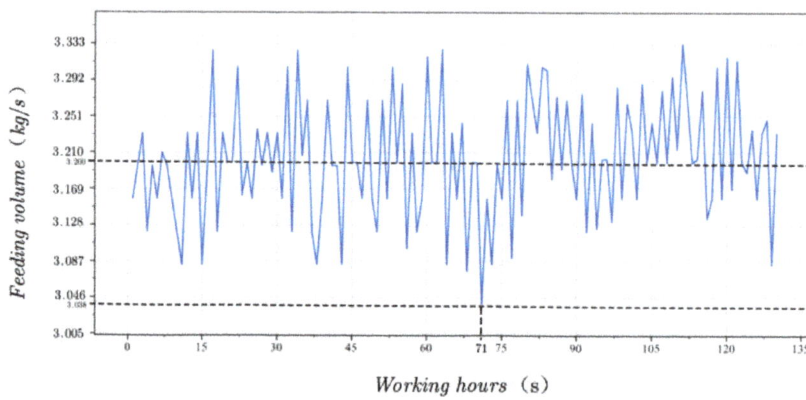

Figure 18. Time domain diagram of feeding volume.

The response accuracy, speed and stability of the control system were evaluated by analyzing the time-domain curve characteristics of the feeding volume. In Figure 18, the maximum deviation of the feed curve from the rated feed (3.2 kg/s) occurred at 71 s with a value of 3.038 kg/s, corresponding to a maximum error of 0.162, an error rate of 5.2%, and an average error rate of 3.4% for the whole process. These results indicate the system's high response accuracy. As seen in the curve change trend, most data points were between 3.17 and 3.23, indicating the system's good stability. Compared to Figure 17, the feeding volume and traveling speed changes during the operation were nearly matched, indicating the system's quick and effective response.

6. Conclusions and Outlook

In this paper, we proposed an adaptive algorithm based on fuzzy PID controls for coupled feeding volume–traveling speed systems in rice harvesters in southern China. It is suitable for rice harvesting needs in southern China, where harvesting areas are scattered across hilly and mountainous areas. The harvester's speed could be adjusted instantly according to changes in rice feeding volume during the operation, which it subsequently optimized. Our experimental analysis showed that the harvester's adaptive traveling speed algorithm based on fuzzy PID controls was better than the traditional PID control method in many control indexes. Our field test analysis showed that a harvester regulated by the algorithm could adapt to changing field operation conditions more quickly and stabilized the feeding volume in a reasonable range to meet the needs of practical applications. Our study can provide rice harvesters with a reference for designing threshing and loss reduction equipment in southern China.

The harsh operating environment in the field challenged the sensor's stability, and the feeding volume detection error was significant. In the future, we must continue researching methods to improve the stability of sensor detection, adjust the acquisition circuit's filtering effect, optimize data analysis and processing methods, and improve detection accuracy.

The system is suitable for modifying small- and medium-sized combine harvesters in hilly mountainous areas. Considering the lighter mass of the small- and medium-sized combine harvesters, as well as the limited optimization of their structures, the noise and vibration were more severe during the driving process, causing deviances in the system detection and expanding control. In the future, system improvement should focus on reducing the influence of vibration and noise on the system.

In order to improve the performance of rice harvesters and develop more effective and efficient control methods, a long-term task of this research project is to compare the fuzzy PID control strategies proposed in this paper with each other and with other control methods. We plan to conduct both quantitative and qualitative studies to achieve this goal. For the quantitative study, we will design experiments to compare the performance of the fuzzy PID control system with other control methods, such as sliding mode control and model predictive control. We will measure various performance metrics, including fuel consumption, harvest efficiency and crop damage, to assess the effectiveness of each control method. For the qualitative study, we will conduct surveys and interviews with farmers and operators to gather feedback on ease of use, reliability and their overall satisfaction with each control method.

Author Contributions: Conceptualization, L.D. and T.L.; methodology, T.L.; software, L.D.; validation, L.D. and J.Z.; formal analysis, L.D. and W.Y.; investigation, L.D. and A.Q.; resources, T.L., P.J. and F.X.; data curation, L.D. and J.Z.; writing—original draft preparation, L.D.; writing—review and editing, L.D. and T.L.; visualization, L.D.; supervision, T.L. and F.X.; project administration, T.L. and P.J.; funding acquisition, T.L., P.J. and F.X. All authors have read and agreed to the published version of the manuscript.

Funding: This research was funded by the Hunan Education Department Scientific Research Project (No. 21C0126), the Hunan Agricultural Machinery Equipment and Technological Innovation R&D Project (Xiang Cai Nong Zhi [2020] No.107) and the Hunan Agricultural Machinery Equipment and Technological Innovation R&D Project (Xiang Cai Nong Zhi [2021] No.47).

Institutional Review Board Statement: Not applicable.

Informed Consent Statement: Not applicable.

Data Availability Statement: Not applicable.

Conflicts of Interest: The authors declare no conflict of interest.

References

1. Xu, F.; Zhang, L.; Zhou, X.; Guo, X.; Zhu, Y.; Liu, M.; Xiong, H.; Jiang, P. The ratoon rice system with high yield and high efficiency in China: Progress, trend of theory and technology. *Field Crops Res.* **2021**, *272*, 108282. [CrossRef]
2. Li, C.; Bremer, P.; Harder, M.K.; Lee, M.S.; Parker, K.; Gaugler, E.C.; Mirosa, M. A systematic review of food loss and waste in China: Quantity, impacts and mediators. *J. Environ. Manag.* **2021**, *303*, 114092. [CrossRef] [PubMed]
3. Li, Z.; Wang, X.; Li, J.; Zhang, Q. Deep attributed network representation learning of complex coupling and interaction. *Knowl.-Based Syst.* **2021**, *212*, 106618. [CrossRef]
4. Da Silveira, F.; Lermen, F.H.; Amaral, F.G. An overview of agriculture 4.0 development: Systematic review of descriptions, technologies, barriers, advantages, and disadvantages. *Comput. Electron. Agric.* **2021**, *189*, 106405. [CrossRef]
5. Sood, S.; Singh, H. Computer vision and machine learning based approaches for food security: A review. *Multimed. Tools Appl.* **2021**, *80*, 27973–27999. [CrossRef]
6. Sabzi, S.; Abbaspour-Gilandeh, Y.; Javadikia, H. Machine vision system for the automatic segmentation of plants under different lighting conditions. *Biosyst. Eng.* **2017**, *161*, 157–173. [CrossRef]
7. Federico, D.; Matteo, C.; Daniele, D.A.; Andrea, A.; Francesco, T.; Matteo, S.S. Modeling, control design and experimental automatic calibration of a leveling system for combine harvesters. *Control Eng. Pract.* **2023**, *132*, 105411.
8. Nevavuori, P.; Narra, N.; Lipping, T. Crop yield prediction with deep convolutional neural networks. *Comput. Electron. Agric.* **2019**, *163*, 104859. [CrossRef]
9. Ma, W.; Zhu, Z.; Zhou, X. Agricultural mechanization and cropland abandonment in rural China. *Appl. Econ. Lett.* **2022**, *29*, 526–533. [CrossRef]
10. Li, R.; Cheng, Y.; Xu, J.; Li, Y.; Ding, X.; Zhao, S. Research on on-line monitoring system of hydraulic actuator of combine harvester. *Processes* **2022**, *10*, 35. [CrossRef]
11. Yang, Z.; Wang, H.; Sun, H.; Wang, P.; Cao, Q. Experimental Study on Electric Harvesting of Combine Harvester. *J. Phys. Conf. Ser.* **2022**, *2218*, 012064. [CrossRef]
12. Wang, G.; Wang, L.; Yang, Y.; You, X. Optimal Design of Harvesting Speed and Forward Speed of Harvester Based on Adaptive Control System. *J. Sens.* **2022**, *2022*, 4176942. [CrossRef]
13. Chen, J.; Lian, Y.; Zou, R.; Zhang, S.; Ning, X.; Han, M. Real-time grain breakage sensing for rice combine harvesters using machine vision technology. *Int. J. Agric. Biol. Eng.* **2020**, *13*, 194–199. [CrossRef]
14. Kounalakis, T.; Triantafyllidis, G.A.; Nalpantidis, L. Deep learning-based visual recognition of rumex for robotic precision farming. *Comput. Electron. Agric.* **2019**, *165*, 104973. [CrossRef]
15. Saleem, M.H.; Potgieter, J.; Arif, K.M. Automation in agriculture by machine and deep learning techniques: A review of recent developments. *Precis. Agric.* **2021**, *22*, 2053–2091. [CrossRef]
16. Darwin, B.; Dharmaraj, P.; Prince, S.; Popescu, D.E.; Hemanth, D.J. Recognition of bloom/yield in crop images using deep learning models for smart agriculture: A review. *Agronomy* **2021**, *11*, 646. [CrossRef]
17. Liu, T.; Ma, Y.; Yang, W.; Ji, W.; Wang, R.; Jiang, P. Spatial-temporal interaction learning based two-stream network for action recognition. *Inf. Sci.* **2022**, *606*, 864–876. [CrossRef]
18. Looh, G.A.; Xie, F.; Zhang, Z.; Mangeh Iii, F.C. Development of Small Agricultural Machines in China: 4LZ–0.8 Mini Combine Harvester. *Development* **2019**, *6*, 9849–9854.
19. Li, H. Research and Prospect of crawler chassis of agricultural machinery in mountainous and hilly areas. In Proceedings of the 2021 International Conference on Control and Intelligent Robotics, Guangzhou, China, 18–20 June 2021.
20. Chapagain, T.; Raizada, M.N. Agronomic challenges and opportunities for smallholder terrace agriculture in developing countries. *Front. Plant Sci.* **2017**, *8*, 331. [CrossRef]
21. Wang, H.; Li, Y.; Qing, Y. Current status and prospect of research on combine harvester header for rape. *IOP Conf. Ser. Earth Environ. Sci.* **2021**, *742*, 012001. [CrossRef]
22. Li, W.; Shuo, L.; Dianji, L.; Ghay, E. Multi-Sensor Signal Acquisition and Data Processing Analysis of Combine Harvester. *INMATEH-Agric. Eng.* **2021**, *63*, 335–344.
23. Jiang, T.; Guan, Z.; Li, H.; Mu, S.; Wu, C.; Zhang, M.; Wang, G.; Chen, X. A feeding quantity monitoring system for a combine harvester: Design and experiment. *Agriculture* **2022**, *12*, 153. [CrossRef]
24. Sun, Y.; Liu, R.; Ou, H.; Zhang, Z.; Zhang, M.; Li, H. Analysis and comparison of feed rate detection methods of combine harvester based on power detection. *Trans. Chin. Soc. Agric. Mach.* **2020**, *5*, 5.
25. Zhao, R.; Qin, D.; Chen, B.; Wang, T.; Wu, H. Thermal Management of Fuel Cells Based on Diploid Genetic Algorithm and Fuzzy PID. *Appl. Sci.* **2023**, *13*, 520. [CrossRef]
26. Borase, R.P.; Maghade, D.K.; Sondkar, S.Y.; Pawar, S.N. A review of PID control, tuning methods and applications. *Int. J. Dyn. Control* **2021**, *9*, 818–827. [CrossRef]
27. Maghfiroh, H.; Ramelan, A.; Adriyanto, F. Fuzzy-PID in BLDC Motor Speed Control Using MATLAB/Simulink. *J. Robot. Control JRC* **2022**, *3*, 8–13. [CrossRef]
28. Chi, K.; Hsiao, Y.; Chen, C. Robust Feedback Linearization Control Design for Five-Link Human Biped Robot with Multi-Performances. *Appl. Sci.* **2023**, *13*, 76. [CrossRef]
29. Li, F.; Shang, C.; Li, Y.; Yang, J.; Shen, Q. Approximate reasoning with fuzzy rule interpolation: Background and recent advances. *Artif. Intell. Rev.* **2021**, *54*, 4543–4590. [CrossRef]

30. Hu, H.; Zha, M.; Chang, L.; Kong, W. Design of Double Redundancy Controller for Electric Actuator Based on Dual three-phase PMSM. In Proceedings of the 2021 IEEE 2nd China International Youth Conference on Electrical Engineering (CIYCEE), Chengdu, China, 15–17 December 2021.
31. Sain, D.; Mohan, B.M. Modeling, simulation and experimental realization of a new nonlinear fuzzy PID controller using Center of Gravity defuzzification. *ISA Trans.* **2021**, *110*, 319–327. [CrossRef]

applied sciences

MDPI

Article

Analytical Method for Assessing Stability of a Counterbalanced Forklift Truck Assembled with Interchangeable Equipment

Leonardo Vita * and Davide Gattamelata

Department of Innovation Technologies, National Institute for Insurance against Accidents at Work (INAIL), Monte Porzio Catone, 00078 Rome, Italy
* Correspondence: l.vita@inail.it; Tel.: +39-0694181566

Featured Application: The proposed analytical method is suitable to check the stability (lateral and longitudinal) of the combination of a counterbalanced forklift truck with an interchangeable equipment and to assess if the interchangeable equipment could be safely assembled to the forklift truck.

Abstract: Counterbalanced forklift trucks (FLT) are frequently used in combination with interchangeable equipment in order to handle loads in different manners. The main risks which may arise after assembling interchangeable equipment to a FLT are related to the loss of stability of the assembly. Actually, the presence of interchangeable equipment and the associated payload may change in a significant way the overall centre of gravity with respect to the FLT in its basic configuration with forks. Therefore, the stability limits of the assembly, based on the same footprints on the ground of the FLT alone, are affected by the position of the overall centre of gravity. Thus, the presence of interchangeable equipment could reduce the functionality (e.g., lifting capability, lifting height, etc.) of the FLT in order to continue its stability during use. Often, interchangeable equipment is placed on the market by manufacturers other than the FLT manufacturer. In these cases, the correct and safe coupling of the interchangeable equipment with the FLT is the responsibility of the manufacturers of interchangeable equipment, including the stability risk assessment. Thus, the interchangeable equipment manufacturer should have access to the relevant information of the FLT concerning operative and structural features and its configuration as a procedure for assessing the correct and safe coupling. Otherwise, he should perform experimental stability tests for each model of FLT so that its interchangeable equipment can be fitted. Specific research activity is developed in order to define an analytical procedure to assess the stability of FLT when assembled with interchangeable equipment. Specific typologies of FLTs and interchangeable equipment have been selected in order to better characterise the case study. The analytical equations mimic the static stability tests. The results achieved have been compared to experimental data in order to optimise the procedure. The results attained by the application of the analytical procedure to all the combinations of main typologies of FLTs and the interchangeable equipment selected showed good agreement with experimental tests.

Keywords: analytical procedure; forklift truck; interchangeable equipment; static stability assessment

Citation: Vita, L.; Gattamelata, D. Analytical Method for Assessing Stability of a Counterbalanced Forklift Truck Assembled with Interchangeable Equipment. *Appl. Sci.* **2023**, *13*, 1206. https://doi.org/10.3390/app13021206

Academic Editors: Paweł Kiełbasa, Tadeusz Juliszewski and Sławomir Kurpaska

Received: 8 November 2022
Revised: 7 January 2023
Accepted: 13 January 2023
Published: 16 January 2023

1. Introduction

The forklift truck (FLT) is one of the most popular in-plant transport machinery. It also represents a substantial part of the severe occupational injury problem in the manufacturing, agricultural, and transport industries. Many studies focus their attention on FLT in order to identify the main safety issues related to its use. In particular, a study conducted in the US in 1998 reported that 95,000 workers were injured, and 100 were killed each year as a result of FLT incidents [1]. The primary cause of fatalities was tip-over. At the beginning of 2000 in Sweden [2] and in Australia [3], similar investigations revealed that the tip-over of FLTs still remains one of the most important causes of accidents among workers. The causes for a loss of stability in the FLT could be attributed to travel speed and steering angle [4], with respect to lateral stability, overloading [5,6], and human error [7] for longitudinal stability. With respect to this, the presence of the interchangeable equipment and the associated load could change in a significant way the overall centre of gravity position. Therefore, the stability limits of the assembly, based on the same footprints on the ground of the FLT alone, are affected by the presence of the interchangeable equipment and its load. This may lead to the necessity of reducing functionality (e.g., lifting capability, lifting height, etc.) of the FLT in order to continue to grant its stability during use. According to this, it is important to investigate the stability behavior of FLT. It is possible to use analytical [8] or virtual prototyping [4,9–13] approaches to predict the loss of stability of the FLT with additional experimental validation. Moreover, the last decade has registered an increment in the number of research activities and patents on active systems for increasing the stability of the FLT [14]. However, these systems are not yet in production, and the mentioned approaches are not used to determine the effect of interchangeable equipment and its eventual load on the stability of the FLT. In addition, an analytical approach applied to the FLT's stability is not able to take into account the deflection of the FLT's tires, frame, mast, carriages, and forks. For this reason, the stability of FLT is generally calculated by means of experimental tests. The main target of the proposed methodology is to define an analytical method to predict the stability of the FLT in combination with interchangeable equipment and its intended load. In particular, starting from the technical data of FLT, which is easily accessible from its instruction for use, and taking into account the results of the stability tests developed for FLT with a fork by its manufacturer, it is possible to check if the interchangeable equipment may lead to instability during its use while assembled with FLT. In this way, the effects of the aforementioned deflections of the structural elements of FLT and tires are already included in the results of stability tests, and it is not necessary to mimic them. This simplified approach may be used by the manufacturer of the interchangeable equipment as an alternative to experimental tests for its risk assessment with respect to loss of stability in the complex. The results achieved seem to be in good agreement with the experimental tests, even if, in some cases, the analytical procedure appears more restrictive than the experimental one. This encourages researchers to continue this research activity in order to optimise the procedure and make it much more reliable.

2. Materials and Methods

In order to assess the stability of FLT, the standard ISO 22915-2 [15] can be used. The standard specifies the tests for verifying the stability of counterbalanced trucks with masts when equipped with fork arms or with load handling attachments. The standard requires that the FLT is positioned on a tilt table in specific configurations (e.g., orientation with respect to the tilt axis, the direction of the test, with or without a load at the load centre of gravity, lift height, etc.). Then, the minimum value of stability to be achieved during the test with the load at the load centre of gravity is specified as follows:

- 4% longitudinal direction during stacking/retrieving with load at maximum lift height and actual capacity less than 5000 kg;
- 3.5% longitudinal direction during stacking/retrieving with load at maximum lift height and actual capacity, not less than 5000 kg;
- 18% longitudinal direction during travelling with load at lift height for travelling (maximum 500 mm);
- 6% lateral direction during stacking/retrieving with load at maximum lift height.

The test load should have a mass that is equivalent to the actual capacity so that the truck can elevate to the corresponding height when acting through the centre of gravity and nominally positioned at the standard load centre distance (see Figure 1) both horizontally from the front face of the fork arm shank, and vertically from the upper face of the fork arm blade.

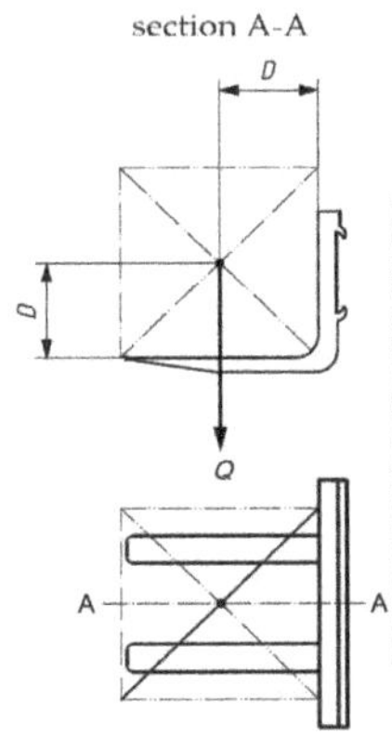

Rated load, Q kg		Standard load centre distance, D mm				
		400	500	600	900	1 200
0	< 1000	X		X[a]		
⩾ 1000	< 5000		X	X[b]		
⩾ 5000	⩽ 10,000			X	X	
> 10,000	< 20,000			X	X	X
⩾ 20,000	< 25,000				X	X
⩾ 25,000						X

[a] 600 mm is used in the USA.
[b] 600 mm is used in the USA, Asia and Australia.

Figure 1. Standard load centre distance.

Considering that the additional weight due to the interchangeable equipment itself and the different position of the load centre of gravity might affect the original stability of the truck, the proposed analytical procedure for assessing the stability of FLT in combination with interchangeable equipment is intended to replicate the minimum stability conditions of the FLT as per stability tests in ISO 22915-2. In particular, it is necessary that the longitudinal/transversal moment introduced by the interchangeable equipment does not exceed the longitudinal/transversal moment due to the test load on the fork of the FLT.

2.1. Longitudinal Stability

With reference to the longitudinal stability of the FLT in combination with interchangeable equipment, it is necessary to acquire the following data from the instruction handbook and/or the marking plate on the FLT:

- The rated capacity (Q_{ij}) with forks according to EN ISO 3691-1 [16]. It can also be defined by means of a load chart depending on the specified height at a specified load centre distance;
- The type of tyres the load chart refers to (if any);
- The fork arm blade thickness(s).
- Similarly, the following data concerning the interchangeable equipment should be acquired in the instruction handbook:
- The total mass of the interchangeable equipment (PCA);
- The distance (L) measured horizontally from the fork carrier to the centre of gravity for the unladen interchangeable equipment;
- The distance (D_{ai}) measured horizontally from the fork carrier to the centre of gravity of the load;
- The maximum load permitted by the manufacturer (Q_i).

Once known, the aforementioned parameters are necessary to calculate the maximum operating longitudinal moment (OLM) according to each value of the actual capacity (Q_{ij}) of the FLT and with respect to the lift height (H_j), as follows (see Figure 2):

$$OLM = Q_{ij} \times D_i, \quad (1)$$

where D_i is the sum of the fork arm blade thickness(s) and the specified load centre distance.

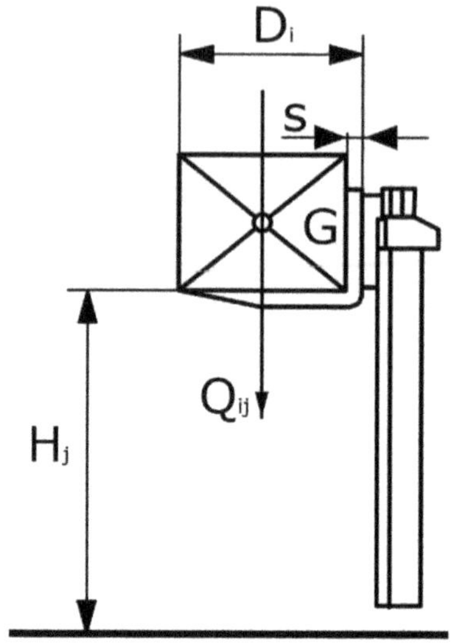

Figure 2. Evaluation of OLM: nomenclature. Key: D_i—load centre distance from the fork carrier; H_j—lift height; Q_{ij}—rated capacity according to D_i and H_j; s—fork arm blade thickness.

On the other hand, for interchangeable equipment, it is necessary to calculate:

a. The maximum load at the fork carrier (CI) as the sum of the maximum load permitted by the manufacturer (Q_i) and the total mass of the interchangeable equipment (PCA);
b. The overall longitudinal moment at the fork carrier (MLA) according to Equation (2) and nomenclature in Figure 3.

$$MLA = PCA \times L + Q_i \times D_{ai}. \quad (2)$$

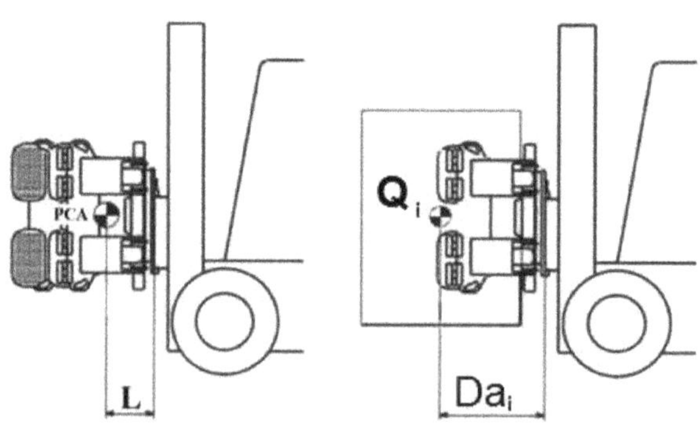

Figure 3. Evaluation of MLA: nomenclature. Key: PCA—total mass of the interchangeable equipment; L—distance measured horizontally from the fork carrier to the centre of gravity of the unladen interchangeable equipment; Da_i—distance measured horizontally from the fork carrier to the centre of gravity of the load; Q_i—maximum load permitted by the manufacturer at Da_i.

The last step is to verify for each condition the lift height and load distance which is foreseen in the FLT load chart as:

$$CI \le Q_{ij}, \quad (3)$$

and

$$MLA \le OLM. \quad (4)$$

Thus, the actual capacity of the FLT, when combined with interchangeable equipment, is the maximum load that simultaneously satisfies Equations (3) and (4). For this purpose, it could be necessary to limit the value of the maximum load which can be handled with interchangeable equipment. If there are no values of the loads which simultaneously satisfy Equations (3) and (4) for any combination of the lift height and load distance, then the interchangeable equipment is not compatible with the FLT.

2.2. Lateral Stability

With reference to the lateral stability of the FLT, and in combination with interchangeable equipment, it is necessary to acquire the following additional data from the instruction handbook and/or the marking plate on the interchangeable equipment:

- The maximum lateral displacement S [mm];
- The total mass (PCAS) of the elements FOR the interchangeable equipment which takes part in the lateral displacement [kg].

On the basis of the acquired data, it is possible to:

- Evaluate the maximum operating transversal moment (OTM) of the truck as per Equation (5) according to the lift height (H_j in Figure 2). With reference to Figure 4, the load is supposed to be on the fork arm of the FLT and is positioned according to the test condition for lateral stability (Test 3) required by ISO 22915-2 at a slope of 6% ($\alpha = 3.4°$).

$$OTM = Q_{ij} \times (S + H_j \sin\alpha). \quad (5)$$

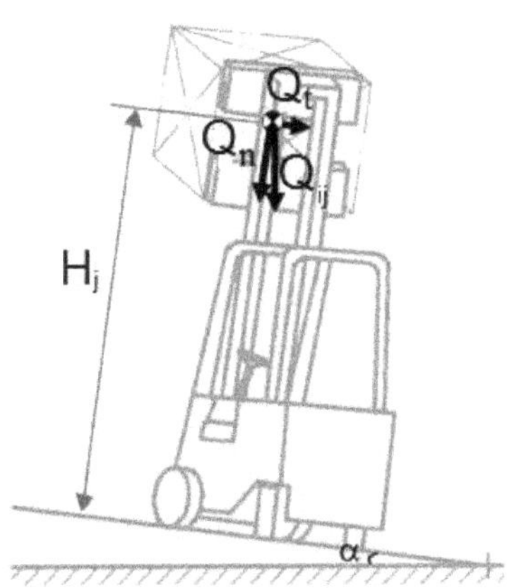

Figure 4. Evaluation of OTM: nomenclature. Key: H_j—lift height; Q_{ij}—rated capacity according to D_i and H_j; Q_n—rated capacity component normal to tilting platform; Q_t—rated capacity component tangent to tilting platform; α—slope.

- It is important to evaluate the total transversal moment of the interchangeable equipment (MTA) according to the lift height (H_j) as per Equation (6). The FLT is supposed to be positioned according to the test condition for lateral stability required by ISO 22915-2 (Test 3) at a slope of 6% ($\alpha = 3.4°$). The value of lateral displacement should be the maximum allowed by the interchangeable equipment by the nature of the load being handled or by the load handle device employed. The total load at the interface (CI_L) should be evaluated as the sum of the maximum load permitted by the manufacturer (Q_i in Figure 3) and the total mass of the elements of the interchangeable equipment which take part in the lateral displacement (PCA_S).

$$MTA = CI_L \times S + CI_L \sin\alpha \times H_j. \quad (6)$$

The last step is to verify for each condition the lift height and load distance as foreseen in the FLT load chart:

$$\text{MTA} \leq \text{OTM}. \tag{7}$$

Thus, the actual capacity of the truck coupled with the interchangeable equipment should be evaluated by considering the loads which allow Equation (7) to be satisfied. For this purpose, it could be necessary to reduce the value of the maximum load permitted for the interchangeable equipment (Q_i).

Interchangeable equipment which does not satisfy Equation (7) for any combination of loads is not compatible with FLT.

2.3. Lateral Displacement

The interchangeable equipment could be provided with a powered load-handling device, such as a side-shift, which can displace the centre of gravity to a substantial, predetermined extent from the longitudinal centre plane of the FLT. It is also possible that the centre of gravity of the load could be substantially offset from the longitudinal centre plane of the FLT because of the nature of the load being handled or by the load handle device employed. In both cases, a displacement is considered to be substantial when it is more than (Figure 5):

- At 100 mm for FLT with a rated capacity of <5000 kg;
- At 150 mm for FLT with a rated capacity of ≥5000 kg and ≤10,000 kg;
- At 250 mm for FLT with a rated capacity of >10,000 kg and <20,000 kg;
- At 350 mm for FLT with a rated capacity of ≥20,000 kg.

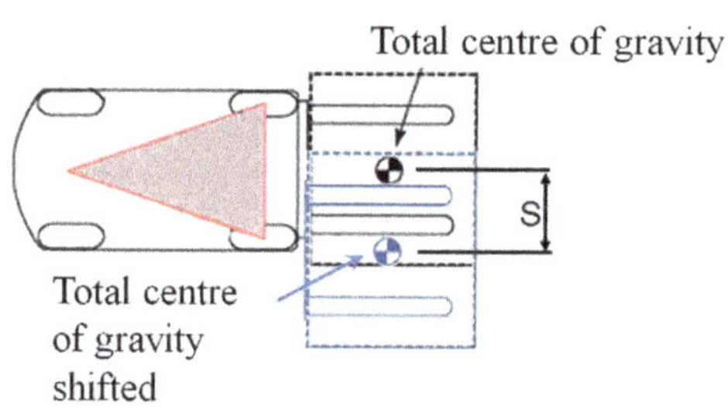

Figure 5. Lateral displacement.

Thus, for Equation (5), the value of lateral displacement S to be used should be equal to the value in the above list according to the FLT-rated capacity. For Equation (6), the value of lateral displacement S to be used should be equal to the value in the above list according to the FLT-rated capacity or greater in the event that the interchangeable equipment undergoes substantial lateral displacement. In this case, the value has to be the one defined by the manufacturer of interchangeable equipment or the maximum lateral displacement achieved by the interchangeable equipment. In any case, it should be more than the value listed above according to the FLT-rated capacity.

2.4. Rated Capacity of the Assembly

The analytical procedure for evaluating the rated capacity of the assembly (i.e., FLT + interchangeable equipment) is based on Equations (3), (4), and (7). With reference to the flowchart in Figure 6, as a first attempt, the maximum load permitted for the interchangeable equipment (Q_i) should be used to check if the longitudinal stability is satisfied according to Equations (3) and (4). If not, the value of the maximum load permitted for the interchangeable equipment should be reduced (Q_i^I) until the aforementioned equations are both fulfilled. Then, it is possible to check the lateral stability by means of Equation (7). If the updated maximum load (Q_i^I) also fulfills this last equation, it represents the rated capacity of the assembly. Otherwise, it should be reduced until Equation (7) is also fulfilled.

The final value (Q_i^{II}) is the rated capacity of the FLT combined with the interchangeable equipment.

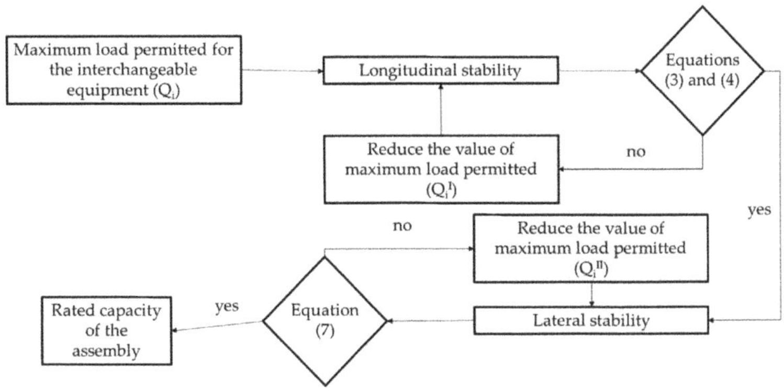

Figure 6. Analytical assessment of assembly-rated capacity.

3. Results

In order to validate the proposed analytical method, experimental tests were conducted. In particular, a total of six models of FLT and three interchangeable pieces of equipment were selected. The FLTs had a rated capacity of 2 ÷ 2.5 tons, a maximum lift height of 3 ÷ 5 m, and a load centre of gravity at a horizontal distance of 500 mm from the front face of the fork arm shank. With reference to Figure 7, the selected interchangeable equipment included:

- The lateral displacement—is used for the lateral displacement of the fork arms of a predetermined length. It has a rated capacity of 2500 kg, a horizontal distance of the load centre of gravity from the fork carrier of 600 mm, a weight of 55 kg, a horizontal distance of the interchangeable equipment centre of gravity from the fork carrier of 37 mm, and a lateral displacement of 100 mm.
- The clamp with parallel arms—it is used for handling cardboard bales, wool, garbage, reels of cellulose, foam blocks, concrete blocks, kegs, reels, etc. It allows the clamp to load by means of the parallel motion of its forks/arms. It has a rated capacity of 1900 kg, a horizontal distance of the load centre of gravity from the fork carrier of 637 mm, a weight of 448 kg, a horizontal distance of the interchangeable equipment centre of gravity from the fork carrier of 249 mm, and a lateral displacement of 313 mm.
- The clamp for reels—it is used for handling paper reels of a different nature. This attachment can also continuously rotate the load. It has a rated capacity of 2000 kg, a horizontal distance of the load centre of gravity from the fork carrier of 852 mm, a weight of 620 kg, a horizontal distance of the interchangeable equipment centre of gravity from the fork carrier of 235 mm, and a lateral displacement of 425 mm.

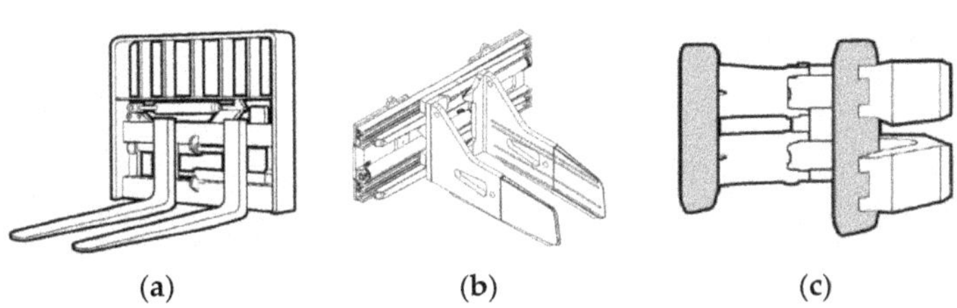

Figure 7. Interchangeable equipment: (**a**) Lateral displacement, (**b**) Clamp with parallel arms, and (**c**) Clamp for reels.

For each combination of the aforementioned FLTs and interchangeable equipment, the rated capacity of the assembly (FLT + interchangeable equipment) was evaluated by means of the analytical procedure that has been compared to the one obtained by the manufacturer of the FLT by means of experimental stability tests. The results are summarized in Table 1 for lateral displacement, in Table 2 for the clamp with parallel arms, and in Table 3 for the clamp for reels.

Table 1. Comparison between experimental tests and analytical procedure for lateral displacement.

FLT	Experimental Test [kg]	Analytical Method [kg]	Delta [%]
1	2300	2195	−4.5
2	2400	2195	−8.5
3	1740	1785	+2.5
4	1790	1785	−0.3
5	1840	1785	−3.0
6	2310	2235	−3.2

Table 2. Comparison between experimental tests and analytical procedure for clamp with parallel arms.

FLT	Experimental Test [kg]	Analytical Method [kg]	Delta [%]
1	1840	1802	−2.1
2	1900	1802	−5.4
3	1510	1502	−0.5
4	1580	1502	−4.9
5	1610	1517	−5.8
6	1900	1942	+2.2

Table 3. Comparison between experimental tests and analytical procedure for clamp for reels.

FLT	Experimental Test [kg]	Analytical Method [kg]	Delta [%]
1	1440	1515	+5.2
2	1630	1515	−7.0
3	1170	1080	−7.7
4	1170	1080	−7.7
5	1240	1095	−11.7
6	1630	1410	−13.5

4. Discussion

Basically, when an analytical model is developed, it is used to predict the stability factor of a basic machine. Thus, static and, in some cases, dynamic forces acting on the basic machine while facing an overturn are considered [17–20]. For instance, in [21], the zero moment point theory to calculate the maximum lifting capacity of the truck crane when considering the dynamic load is described in order to predict the overturning state of the crane and reduce its risk in operation. When referring to FLT, it has to be considered that the dynamic effects due to the load are minimised. In fact, the load is stabilised (e.g., by means of a pallet or similar devices) on the forks. However, the maximum lift capacity of a FLT is always based on its stability. Thus, knowing the correct location of its centre of gravity is crucial. In [22], a method for determining the center of gravity of a forklift truck is illustrated. A similar approach is described in [23], which points out in the discussions that the deflection of tires, frames, masts, carriages, forks, carriage looseness, and mast looseness have an effect on the stability of a fork truck and are virtually impossible to solve mathematically. Therefore, using a tilting platform for conducting stability tests would be realistic because, in this way, it is possible to simulate the effect of centrifugal forces that take place when maneuvering, and it automatically takes into consideration the aforementioned factors that affect the stability of a FLT. Standard ISO 22915-2 defines the stability tests that need to be performed. With reference to the aforementioned standard, in [8], the stability equations were derived with detailed derivations for the four kinds of stability tests. The effects of acceleration and centrifugal forces were also additionally included in the equations. However, according to the authors' knowledge, at present, the analytical procedures that are at one's disposal refer to FLTs with forks. No analytical procedures are available to evaluate the stability of FLT when combined with interchangeable equipment. A useful guide for identifying interchangeable equipment can be found in [24], where the European materials handling federation (FEM) proposes a classification for the attachments of the FLT with reference to the applicable legislation and on the basis of the documentation available to date and the technical knowledge of its members (industrial trucks manufacturers and attachment manufacturers). Hence, the analytical method herein proposed is focused on filling this gap. It is mainly based on the static equilibrium (force and moment) of a rigid body. Even if this simplified approach is not able to mimic the flexibility of the FLT structure and the dynamic effects of the lifting operations, it is able to capture the changes introduced by the interchangeable equipment in the rated capacity of the FLT with forks. The proposed analytical approach takes into account both the fact that FLT has to fulfill stability requirements defined in ISO 22915-2 and that the actual capacity at a maximum lift height with the load centre distance should accordingly be provided by the FLT's manufacturer. Thus, the main target is to verify if the centre of gravity of the interchangeable equipment plus the centre of gravity of the maximum intended load remains within the limits of the actual capacity of the FLT. If so, no changes are needed, and the FLT could be used with the interchangeable equipment and its maximum intended load according to the rated capacity of FLT with forks. If not, it is necessary to define which is the maximum load that the FLT may lift with the interchangeable equipment (i.e., the rated capacity of the assembly) that is equivalent to the rated capacity of FLTs with forks. As a consequence, it is not necessary to model the FLT (basic machine) in order to define its static stability limit. In fact, the main assumption is that the FLT is stable, at least at the slope required by the ISO 22915-2 standard, which should be experimentally verified by the FLT manufacturer. Thus, it is necessary to model only the deviation of FLTs to the centre of gravity when introduced by interchangeable equipment and its intended load.

According to the proposed analytical procedure, the rated capacity of the assembly is determined as the maximum value, which allows for the satisfaction of the contemporary Equations (3), (4), and (7). The analytical results summarised in Tables 1–3 are compared to the experimental tests developed by the manufacturer of the FLT assembled with the same interchangeable equipment. With the exception of FLT numbers five and six combined with the clamp for reels, where the most severe condition is the lateral stability, for all the other combinations, the longitudinal stability represents the worst-case condition. The rated capacity, evaluated by means of the analytical method, is, in most of the test cases, lower than the one obtained by the means of experimental tests. This is mainly due to the assumption of rigid bodies in the analytical model. However, a more restrictive value in terms of the rated capacity is in favor of safety. The average percentage of rated capacity reduction is 5.7, which is not so relevant in terms of FLT functionality. However, it is necessary to observe that three analytical results on eighteen are higher than the experimental ones. Considering that each of them refers to different FLTs in combination with different interchangeable equipment, it is not possible to clearly identify the parameter which may affect the results. In addition, it should be considered that even experimental tests are affected by boundary conditions, which may affect the experimental result. At this stage of the study, it is not possible to replicate them in a simple analytical model. In addition, the maximum delta registered on rated capacity is about +5%, which is in line with the tolerance of the instruments for detecting the static stability angle on the experimental test rig. The comparison with experimental tests encourages us to use this simplified analytical procedure as an alternative to extensive experimental tests on the FLT assembled with interchangeable equipment. In fact, it is able to capture the main changes in the centre of gravity location, which mainly affect the rated capacity of the assembly. Nevertheless, if there are no values of load able to fulfill Equations (3), (4), and (7), it is necessary to pay particular attention. In this case, the combination of the interchangeable equipment and the FLT may not be advisable due to the potential loss of stability during lifting operations.

5. Conclusions

The analytical procedure herein presented represents for manufacturers an alternative method to experimental tests in order to define the rated capacity of a FLT assembled with interchangeable equipment. This procedure is easier to perform, less costly, and less time-consuming with respect to experimental tests. The innovative approach relays on the assumption that the FLT with a fork should be stable at specific angles according to which its rated capacity is tested out. Thus, it is not necessary to model all the assemblies but only the variations introduced by the interchangeable equipment and its intended load. Moreover, this procedure is particularly useful for the stability risk assessment of FLTs and interchangeable equipment that are already in use and that are assembled by operators and not by engineers or trained technicians. The results achieved seem to be in good agreement with the experimental tests, even if, in some cases, the analytical procedure seems to reduce the rated capacity of the assembly to a greater extent than the one evaluated by means of experimental tests. However, it should be noted that this reduced rated capacity is in favor of safety. On the basis of the results of this study, the procedure needs to be refined. Increasing the number of numerical applications will allow for a larger set of data to be generated in order to compare the results with the manufacturer's database of the stability test measurements. This will allow us to optimise the procedure and make it much more reliable without increasing the level of information required. Moreover, once the procedure is refined and optimised, it can also be applied in other sectors, such as earth-moving machinery and agricultural machinery.

Author Contributions: Conceptualization, L.V. and D.G.; Data curation, L.V. and D.G.; Formal analysis, L.V. and D.G.; Investigation, L.V. and D.G.; Methodology, L.V. and D.G.; Validation, L.V. and D.G.; Writing—original draft, L.V. and D.G.; Writing—review and editing, L.V. All authors have read and agreed to the published version of the manuscript.

Funding: This research received no external funding.

Institutional Review Board Statement: Not applicable.

Informed Consent Statement: Not applicable.

Data Availability Statement: Data concerning analytical results and experimental tests will be be provided upon request to the corresponding author.

Conflicts of Interest: The authors declare no conflict of interest.

References

1. Swartz, G. Forklift tipover: A detailed analysis. *Prof. Saf.* **1998**, *43*, 20.
2. Larsson, T.J.; Oldertz, C. Hazardous Exposures and injury types associated with the use of industrial lift trucks in Sweden 2005–2007. *Saf. Sci. Monit.* **2011**, *15*, 237–242.
3. Saric, S.; Bab-Hadiashar, A.; Hoseinnezhad, R.; Hocking, I. Analysis of forklift accident trends within Victorian industry (Australia). *Saf. Sci.* **2013**, *60*, 176–184. [CrossRef]
4. Rebelle, J.; Mistrot, P.; Poirot, R. Development and validation of a numerical model for predicting forklift truck tip-over. *Veh. Syst. Dyn.* **2009**, *47*, 771–804. [CrossRef]
5. Górny, A. Identification of occupational accident causes by use the Ishikawa diagram and Pareto principles. *Econ. Manag.* **2017**, *1*, 384–388.
6. Kozińska, M.; Radek, N.; Rozlach, A.; Mazur, M. Risk estimation with the application of the matrix method during the operation of the forklift. *Syst. Saf. Hum.-Tech. Facil.-Environ.* **2021**, *3*, 337–346.
7. Duarte, J.; Marques, A.T.; Santos Baptista, J. Occupational accidents related to heavy machinery: A systematic review. *Safety* **2021**, *7*, 21. [CrossRef]
8. Kim, J.B.; Shin, W.; Park, J.H. Stability analysis of counterbalanced forklift trucks. *J. Korean Soc. Saf.* **2015**, *30*, 1–8.

9. Gardella, M.; Martini, A. Multibody models and simulations to assess the stability of counterbalance forklift trucks. In *European Congress on Computational Methods in Applied Sciences and Engineering*; Springer: Cham, Switzerland, 2019; pp. 526–533.
10. Xia, G.; Xia, Y.; Tang, X.; Zhao, L.; Hu, J. Anti-rollover control based on stable zone partition of counterbalanced forklift trucks. *Int. J. Automot. Technol.* **2021**, *22*, 1529–1543. [CrossRef]
11. Lee, S.-H.; Bae, Y.-C. Design of simulator for rollover prevention of forklift truck. *J. Korea Inst. Electron. Commun. Sci.* **2021**, *16*, 571–576.
12. Sunday, B.A.K.O. Stability analysis of a semi-trailer articulated vehicle: A review. *Int. J. Automot. Sci. Technol.* **2021**, *5*, 131–140.
13. Lemerle, P.; Höppner, O.; Rebelle, J. Dynamic stability of forklift trucks in cornering situations: Parametrical analysis using a driving simulator. *Veh. Syst. Dyn.* **2011**, *49*, 1673–1693. [CrossRef]
14. Berdiev, A.; Bahadirov, G.; Zhang, D.; Wang, X.; Li, Q. Analysis of the design of lifting and transporting vehicles with a variable center of gravity—A literature and patent overview. *Int. J. Eng. Trends Technol.* **2021**, *69*, 56–65. [CrossRef]
15. *ISO 22915-2*; Industrial Trucks—Verification of Stability—Part 2: Counterbalanced Trucks with Mast. International Organization for Standardization: Geneva, Switzerland, 2018.
16. *EN ISO 3691-1*; Industrial Trucks—Safety Requirements and Verification—Part 1: Self-Propelled Industrial Trucks, Other than Driverless Trucks, Variable-Reach Trucks and Burden-Carrier Trucks. European Committee for Standardization: Brussels, Belgium, 2020.
17. Troyanovskaya, I.; Grebenshchikova, O.; Erofeev, V. Static stability of articulated front loader. *AIP Conf. Proc.* **2022**, *2503*, 050033.
18. Troyanovskaya, I.; Shepelev, S.D. Design of Agricultural Loader with Articulated Frame. In Proceedings of the 6th International Conference on Industrial Engineering (ICIE 2020), Sochi, Russia, 18–22 May 2020.
19. Robertson, L.S.; Maloney, A. Motor vehicle rollover and static stability: An exposure study. *Am. J. Public Health* **1997**, *87*, 839–841. [CrossRef] [PubMed]
20. Huston, R.L.; Kelly, F.A. Another look at the static stability factor (SSF) in predicting vehicle rollover. *Int. J. Crashworthiness* **2014**, *19*, 567–575. [CrossRef]
21. Zhou, Q.; Liu, Y.; Xie, Y.; Zhang, X. Analysis on anti-overturning stability of a truck crane based on zero moment point theory. *DYNA* **2022**, *97*, 281–287. [CrossRef] [PubMed]
22. Kasahara, M.; Mori, Y. Relation between overturn and the center of gravity of a forklift truck. In Proceedings of the 56th Annual Conference of the Society of Instrument and Control Engineers of Japan (SICE), Kanazawa, Japan, 19–22 September 2017; pp. 135–140.
23. Puder, R.W.; Bogar, W.H.; Jay, R.D. Fork-truck stability—A study in moments. *Trans. ASME* **1957**, *79*, 1591–1596. [CrossRef]
24. FEM European Materials Handling Federation. *Guidance on the Classification of Attachments for Industrial Trucks According to Machinery Directive 2006/42/EC*, 1st ed.; FEM 4.107; FEM: Brussels, Belgium, 2021.

applied sciences

MDPI

Article

Measuring System Design and Experiment for Ground Pressure on Seeding Skateboard of Rice Direct Seeding Machine

Yuan Gao, Guozhong Zhang *, Hongchang Wang, Abouelnadar Salem, Jianwei Fu and Yong Zhou

School of Engineering, Huazhong Agricultural University, Wuhan 430072, China;
gaoyuan@mail.hzau.edu.cn (Y.G.); wanghc84@mail.hzau.edu.cn (H.W.);
abouelnadar@webmail.hzau.edu.cn (A.S.); fjwtap@mail.hzau.edu.cn (J.F.); zhyong@mail.hzau.edu.cn (Y.Z.)
* Correspondence: zhanggz@mail.hzau.edu.cn; Tel.: +86-186-7278-3365

Featured Application: Intelligent agricultural machinery equipment, wireless measurement technology of the Internet of Things.

Abstract: Acquiring real-time ground pressure measurements from the surface of the soil in working parts of paddy fields is a challenging task. The real-time data can be used to monitor the changing state of the ground pressure of the working parts in a paddy field. To effectively reduce the accumulation of choked mud at the front end of the seeding skateboard and the contact adhesion between the skateboard and the paddy soil, a ground pressure measuring device suitable for paddy fields was designed. The device uses an Arduino controller, combined with Internet of things technology and wireless measurement technology. It can measure the pressure from 16 measuring points at the same time and transmit the measurement data to the computer remotely through the Internet of things technology, which greatly reduces the labor intensity of measuring personnel in the muddy paddy field. Analysis of the data showed that the forward tilt angle, ground pressure, and forward resistance of the seeding skateboard also increased with the increase of forward speed and vertical load. In addition, the distribution law of the ground pressure between the skateboard and the paddy soil was obtained. The conclusions show that the ground pressure measurement system can work stably in the paddy field and the measured data can be wirelessly transmitted to the computer and mobile phone.

Keywords: ground pressure; paddy soil; seeding skateboard; internet of things; wireless measurement system

Citation: Gao, Y.; Zhang, G.; Wang, H.; Salem, A.; Fu, J.; Zhou, Y. Measuring System Design and Experiment for Ground Pressure on Seeding Skateboard of Rice Direct Seeding Machine. *Appl. Sci.* **2021**, *11*, 10024. https://doi.org/10.3390/app112110024

Academic Editors: Paweł Kiełbasa, Tadeusz Juliszewski and Sławomir Kurpaska

Received: 26 September 2021
Accepted: 22 October 2021
Published: 26 October 2021

1. Introduction

The total land area of paddy fields in China is 30 million hm^2 [1], accounting for about 1/4 of the cultivated land in China [2]. Using a direct seeding machine for mechanized direct seeding of rice can effectively improve the degree of mechanization of rice production [3]. When the rice direct seeding machine is working, the seeding skateboard with larger grounding area is often pulled by the power head to flatten the surface in order to make the seeding mud surface smooth [4]. There is often a large amount of soil accumulation in front of the seeding skateboard due to the squeezing of the skateboard on the paddy field soil [5], which directly affects the operation efficiency, traction and adhesion performance, fuel consumption, and power consumption of the rice direct seeding machine, and also reduces the passability of the machine [6]. The serious choking of the mud also causes the seeds on the adjacent sides to be covered and affects the flatness of the paddy field [7]. The occurrence of the above problems is closely related to the contact between the planter skateboard and the paddy soil. Studying the ground pressure at the bottom of the skateboard is helpful to reduce the mud adhered to the surface of the skateboard and reduce the resistance of the skateboard during sliding [8].

The paddy soil used for transplanting and direct seeding has high water content, a fluid state, and remarkable rheological properties [9]. In order to accurately analyze the interaction process between seeding skateboard and paddy soil, it is necessary to study the rheological mechanical properties of paddy soil [10]. In order to understand the rheological properties of lacustrine soft soil at different stress levels, Deng, Z. carried out a one-dimensional indoor consolidation creep experiment on soft soil in the Dongting Lake area. A rheological constitutive model under lateral confinement conditions was then constructed, and the analytical solution was obtained under one-dimensional consolidation conditions according to the fitted rheological constitutive model. The results indicated that soft soil in the Dongting Lake area has obvious nonlinear creep characteristics [11]. Ram, R. B. performed uniaxial compression experiments on soil samples and plotted rheological curves, thus establishing rheological models and rheological equations [12]. Zhuoliang He established an air–water–soil coupling model, used ANSYS/LS-DYNA to conduct a fluid–solid coupling simulation, and applied the rheological parameter research to the design and parameter optimization of a paddy soil boat tractor [13].

Domestic and foreign experts have carried out a lot of theoretical and experimental research regarding the ground pressure measurement technology. Guozhen Wu and others installed a self-made diaphragm ground pressure sensor at the bottom of a machine ploughing ship and measured the ground pressure at the bottom of the ship [14]. Bijuan Yan invented a device that can dynamically measure the distributed ground pressure from a crawler device of heavy equipment [15]. Moreover, Eisenkolb designed a kind of ground pressure measuring device and a set of evaluation methods. The measuring device is embedded in the ground, then, when the car passes from above, the sensor and measuring device can detect the dynamic change of the ground pressure in the contact area between the tire and the ground, and analyze and evaluate the record [16]. In order to reduce the mud of the skateboard of the rice direct seeding machine, Ya Zou used a thin film sensor to measure and analyze the ground pressure between the skateboard and the soil in order to explore the force relationship between the skateboard and the soil [17]. This characteristic of paddy field soil strongly affects the operation performance and work efficiency of paddy field operation machinery. Therefore, studying the rheological characteristics of paddy field soil is of great significance in exploring the relationship between paddy field soil and agricultural machinery and farming machinery, revealing the law of its interaction, and providing a theoretical basis for the design and manufacture of farmland operation machinery.

This study mainly focused on effectively reducing the accumulation of choked mud at the front end of the seeding skateboard and the contact adhesion between the skateboard and the paddy soil. Therefore a specific pressure measurement system for the seeding skateboard of a rice direct seeding machine was designed. Taking the skateboard of Datong 2BD-8A (DXD830) rice direct seeding machine (Figure 1) as the object, the ground pressure and forward resistance between the skateboard and the soil were measured and analyzed. The distribution law of ground pressure between the skateboard and paddy soil was preliminarily obtained in order to provide reference for the optimal design for adhesion and resistance reduction of soil contact.

Figure 1. Datong 2BD-8A (DXD830) rice direct seeding machine.

2. Materials and Methods

2.1. Working Principle

The ground pressure measurement system consists of two parts: hardware and software. Among them, the hardware system includes a pressure sensor, 9-axis attitude sensor, GPS module, Arduino controller, signal amplifier, wireless serial communication module, Internet of things wireless module, and computer equipped with Arduino software. In addition, the software system mainly includes technology for data acquisition, data filtering, wireless data transmission, data display, data storage, and other functions. The pressure sensor converts the force of the soil on the skateboard of the rice direct seeding machine into a resistance signal, and the signal amplifier converts the resistance signal into an analog voltage signal [18]. The Arduino controller uploads the data to the mobile phone and computer through the wireless transmission module and the Internet of things module after filtering and analyzing the voltage signal. Figure 2 shows the structure of the ground pressure measurement system. Figure 3 shows the flow chart of the ground pressure measurement system. The detailed information of the components used and measurement instrumentations has been provided in Table 1.

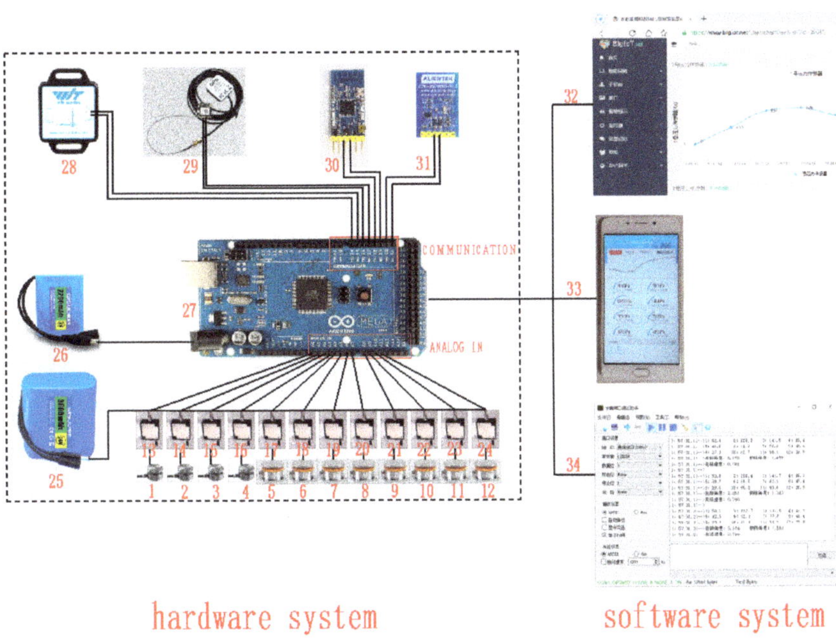

Figure 2. Structure of the ground pressure measurement system. 1~4: Forward resistance sensor, 5~12: ground pressure sensor, 13~24: signal amplifier, 25: 24 V power supply, 26: 12 V power supply, 27: Arduino controller, 28: 9-axis attitude sensor, 29: GPS module, 30: wireless serial port transceiver module, 31: Internet of things wireless module, 32: Internet of things measurement system—computer terminal, 33: Internet of things measurement system—mobile terminal, 34: wireless serial communication measurement system.

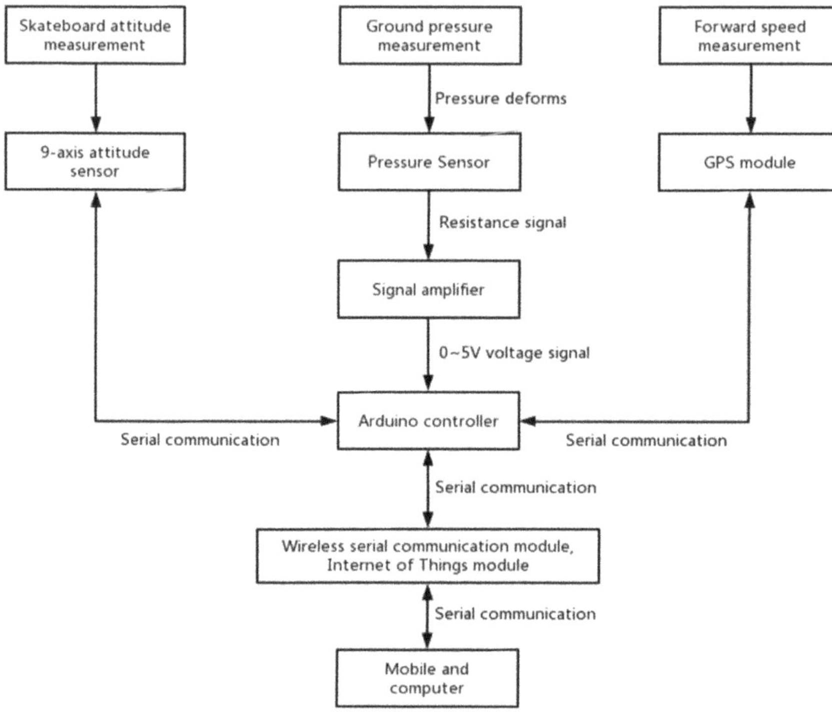

Figure 3. Flow chart of ground pressure measurement system.

Table 1. Brief information about components used and measurement instrumentations.

Sensor	Model	Operating Voltage	Output	Measuring Range	Measurement Accuracy
Forward resistance sensor	JHBM-H3	5–10 V	350 ± 3Ω	0–5 kg	±0.1%
Ground pressure sensor	JHBS-1–5KG	5–10 V	350 ± 3Ω	0–5 kg	±0.1%
Signal amplifier	BSQ-3	24 V	0 ± 5 V	0~700 Ω	±0.05%
9-axis attitude sensor	WTGAHRS1	3.0–5.5 V	Serial communication	±180°	±0.1°
GPS module	ATGM332D	3.0–5.5 V	Serial communication	0~515 m/s	<0.1 m/s
Wireless serial communication module	CC2530	3.0–5.5 V	Serial communication	250 m	±0.1%

According to the experiment requirements, the measurement system can remotely set the sensor parameters, control the data acquisition frequency, initialize the sensor, etc. After the data acquisition starts, the Arduino controller will remotely upload the data to the mobile phone and computer, display the current measurement data in real time, and allow the user to view historical data and save measurement data. Figure 4 shows the software flow of the ground pressure measurement system.

The ground pressure measurement system has 12 independent pressure data acquisition channels. The data from each acquisition channel do not interfere with each other. Nos. 1–4 sensors are forward resistance sensors with a height of 10 cm from the bottom plane of the skateboard, which can effectively measure the resistance caused by the mud at the front end of the skateboard. Nos. 5–8 are the front pressure sensors and Nos. 9–12 are the rear pressure sensors. These are mainly used to measure the front and rear ground pressure and analyze the combination with forward inclination angle and forward resistance. Figures 5 and 6 show the pressure sensor installation location.

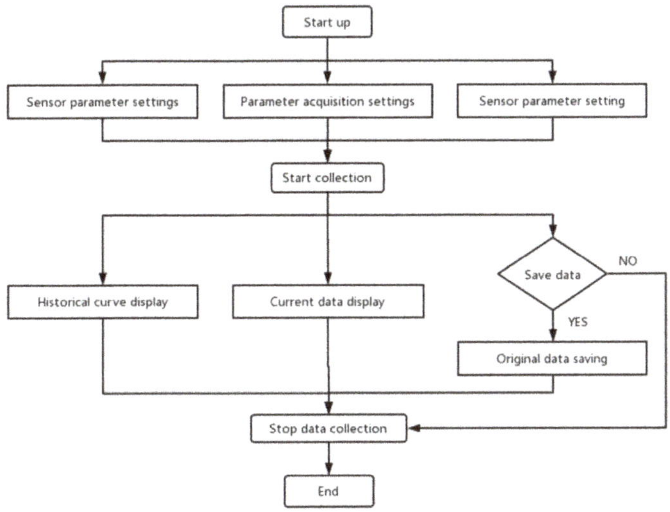

Figure 4. Flow chart of measurement system software.

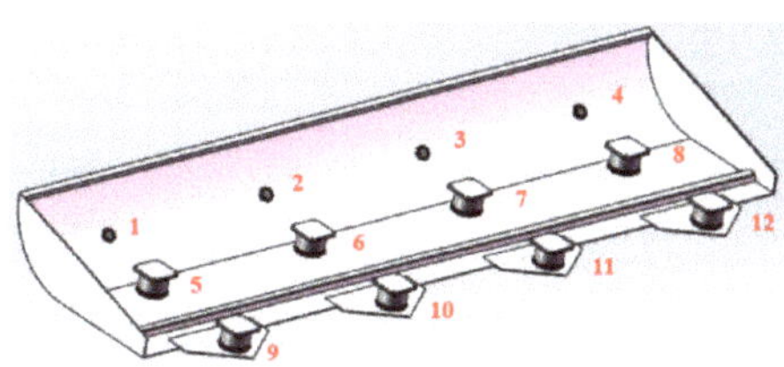

Figure 5. Pressure sensor installation location.

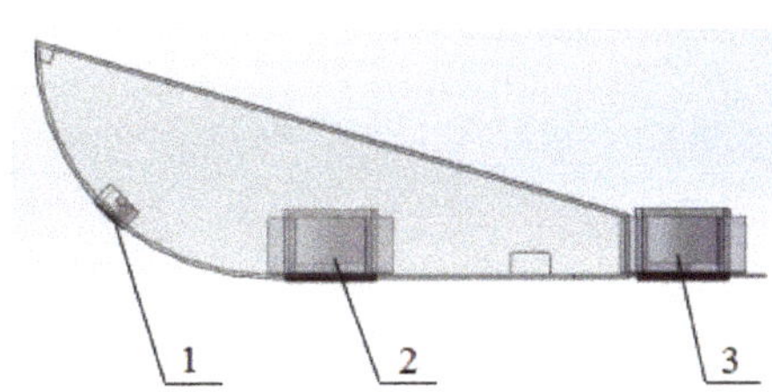

Figure 6. Pressure sensor installation location (Side view). 1. Forward resistance sensor. 2. Front pressure sensor. 3. Rear pressure sensor.

2.2. Ground Pressure Calibration Device

Due to the fact that there is no mature product for measuring the ground pressure of paddy soil in the market, a calibration device combing ground pressure measurement systems was designed and manufactured in order to detect the error between the ground pressure measured by the measuring device and the actual ground pressure and to calibrate and correct the measurement accuracy of the whole ground pressure measurement system.

The pressure sensor is connected with the signal amplifier. The signal amplifier converts the resistance signal into a voltage signal of 0–5 V and inputs it to the Arduino controller. In addition, the Arduino controller filters the data and sends it to the computer through the wireless serial port transceiver module. The number of counterweight plates is gradually increased, and the ground pressure is measured in turn. Figure 7 shows the ground pressure combined calibration system.

Figure 7. Ground pressure calibration device. 1. 12 V power supply. 2. Arduino controller. 3. Wireless serial port transceiver module. 4. Signal amplifier. 5. Pressure sensor. 6. Mounting base. 7. Counterweight plate. 8. Bottom plate.

The experimental results show that when the number of counterweight plates is between 0 and 10, the actual ground pressure (q_p) value is linearly related to the total mass (m).

$$q_p = \frac{m \cdot g}{S_p} \tag{1}$$

where S_p represents the contact area between pressure sensor and soil.

When the ground pressure is in the range of 0~2000 Pa, the actual ground pressure value measured by the measuring system is compared with the theoretical value. The error between the ground pressure values is within 2%, which meets the actual demand for ground pressure measurement accuracy. The ground pressure calibration data are shown in Table 2.

Table 2. Ground pressure calibration data.

Number of Counterweights	Total Mass (kg)	Theoretical Ground Pressure (Pa)	Actual Ground Pressure (Pa)	Error (%)
0	3.854	306.69	310.32	1.18%
1	6.258	498.00	505.46	1.50%
2	8.662	689.30	694.15	0.70%
3	11.066	880.60	887.13	0.74%
4	13.470	1071.91	1079.45	0.70%
5	15.874	1263.21	1270.33	0.56%
6	18.278	1454.52	1462.81	0.57%
7	20.682	1645.82	1648.17	0.14%
8	23.086	1837.13	1831.71	−0.30%
9	25.490	2028.43	2019.19	−0.46%
10	27.894	2219.73	2203.88	−0.71%

2.3. *Design of Key Components*

2.3.1. Wireless Serial Communication Measurement System

The Arduino controller is used for wireless serial port transmission of the measurement data. In addition, the serial port debugging software is used on the computer to receive the measurement data and display the measurement data in real time. In the meanwhile, the values of 12 pressure sensors are recorded, as well as forward tilt angle, roll angle, forward speed, time, and so on. Figure 8 shows the measurement software. The measurement data are saved by the measurement software and it can be processed and analyzed immediately.

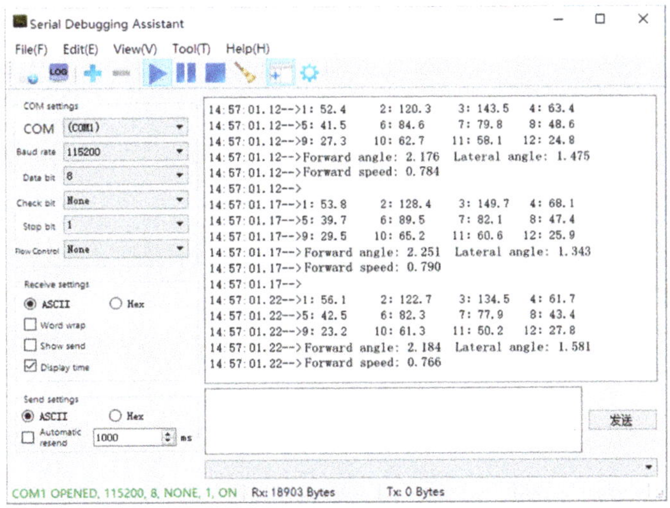

Figure 8. Wireless serial communication measurement system.

2.3.2. Internet of Things Measurement System

The ground pressure measurement system uses the shell Internet of things cloud platform, which can communicate with the Arduino controller, chat with the Arduino controller in the form of dialogue, remotely control the system, etc., through the Internet, using mobile phone apps or web pages to send instructions, view real-time data and historical data, control remotely, and so on. The pressure measurement data can be uploaded remotely to the Internet of things platform and displayed after the Arduino controller is connected to the Internet of things. When entering the web page of the Internet of things platform, you can connect an interface for each pressure sensor. In addition, you can view the measurement data of the interface in real time, and you can also look back at the historical data through this interface. When the measured data reach the threshold, the automatic alarm can be set to the mobile phone reminder. Figure 9 shows the computer-end ground pressure measurement system.

In the meanwhile, the mobile phone app and WeChat mini programs can be used to monitor the measured data in real time and the ground pressure measurement system can be remotely controlled to gradually approach the intelligence of agricultural machinery. Figure 10 shows the mobile phone terminal of the ground pressure measurement system.

2.4. Field Experiment

2.4.1. Experimental Conditions

The experimental field of Huazhong Agricultural University was selected. The field surface was irrigated to the height of 3–6 cm above the soil surface, soaked, rotary ploughed, scraped, and settled for 48 h before the experiment. The red wireframe in Figure 11 shows the location of the field, longitude: 114.3631, latitude: 30.4643. The experiment was conducted on the afternoon of 23 May 2020; the weather was cloudy to sunny and the temperature was 15 °C~23 °C.

Figure 9. Internet of things measurement system—computer terminal.

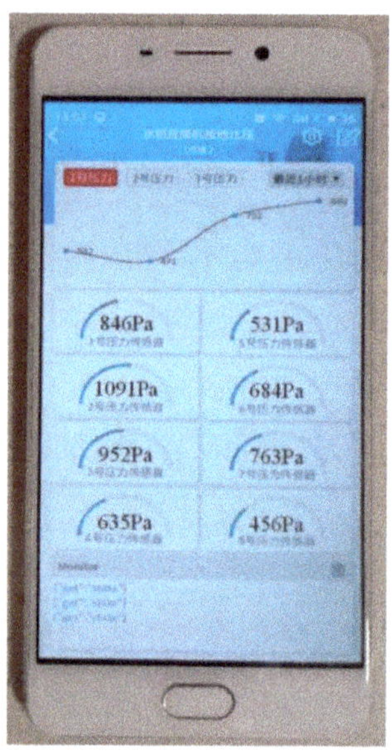

Figure 10. Internet of things measurement system—mobile terminal.

Figure 11. Experimental field.

There was no obvious stagnant water on the surface of the experimental paddy field. In addition, there was a small amount of stagnant water in a few areas, which is in line with the seeding environment of the rice direct seeding machine. Figure 12 shows the actual experimental environment.

Figure 12. Experimental environment.

The moisture content of soil samples was measured by a MB45 halogen moisture meter, which uses the principle of pyrolysis weight to measure the quality of soil samples at the beginning of drying, followed by heating and air-drying the soil samples through the heating unit and rapid water evaporator in the instrument. The average moisture content of the experimental field was 52.37%.

The soil firmness measurements of the surface layer of the water area and the non-water area were quite different, and the depth of the seeding skateboard was less than 200 mm because the surface of the paddy soil was uneven. As a consequence, 200 mm below the soil surface was selected as the measuring depth. Moreover, the average soil firmness of the experimental field was 278 kPa under the condition of 200 mm depth.

2.4.2. Experimental Design

The main factors affecting the ground pressure of the seeding skateboard model are forward speed and vertical load. As a consequence, the forward speed and vertical load were selected as the experimental factors, and the horizontal parameter range of each factor was as follows:

Generally speaking, the higher the forward speed, the higher the working efficiency, and the running speed of the rice transplanter is 0–1.5 m/s. Considering the operational control of the measuring system, according to the scale of the maximum cruising speed of the rice transplanter, the scale was divided into 4 equal points and marked. Cruise driving at a constant speed was carried out according to the scale mark. The average speeds measured by GPS module under different experiment conditions were 0.175, 0.544, 0.919 and 1.283 m/s.

The vertical load is an important factor affecting the ground pressure of the seeding skateboard, as well as the sliding resistance and sinking depth of the skateboard in the process of sliding. The weight of the experiment skateboard was 80 kg. The average ground pressure was about 1000 Pa. The counterweight is the standard counterweight of a Dongfanghong 954 tractor, and its weight was 40 kg according to the weight of the experiment skateboard divided by the grounding area, 0.8 m^2, of the experiment skateboard. Considering these factors comprehensively, the vertical load was set to 0, 40, 80, 120, and 160 kg.

2.4.3. Experimental Process

Power was used for the ground pressure measurement system at the start of the experiment and to make the Internet of things and wireless serial port transceiver module run. In addition, the seeding skateboard of the rice transplanter is pulled by the transplanter to start the experiment. In order to ensure the wireless and stable transmission of data, and to prevent the excessive amount of repeated data from affecting the calculation and filtering effect of the Arduino controller, the experimental data acquisition frequency was set at 20 Hz. The experimental results show that this frequency can ensure the accuracy and reliability of the data. Cruising was carried out with different speeds and different loads, and the data were measured by the ground pressure measurement system and wirelessly uploaded to the computer end. The system was used to change the number of counterweights or forward speed, to regularly check whether the working condition of each sensor was stable, and to try to ensure that the experimental conditions of each group were consistent after each group of experiments. The experiment scenario is shown in Figure 13.

Figure 13. Wireless measurement system.

3. Results

3.1. The Relationship between Forward Resistance, Ground Pressure, and Forward Speed

Multi-level experiments were carried out under variable forward speeds (v = 0.175, 0.544, 0.919, and 1.283 m/s). In addition, the forward resistance and ground pressure values at different forward speeds were measured. Nos. 1–4 are the forward resistance sensors, Nos. 5–8 are the front-end pressure sensors, and Nos. 9–12 are the back-end pressure sensors. The variation curves are shown in Figures 14 and 15.

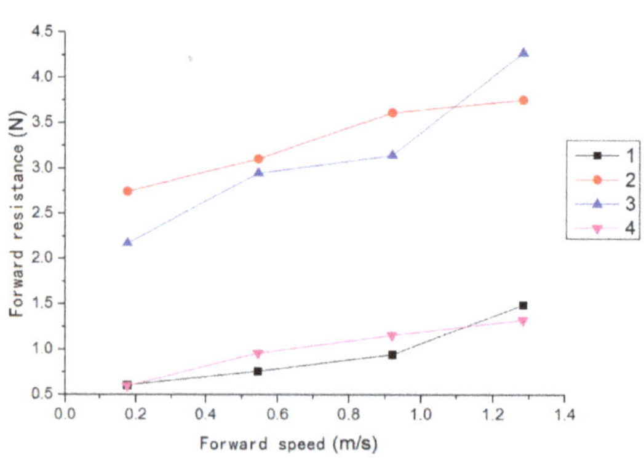

Figure 14. Relationship between forward resistance and forward speed.

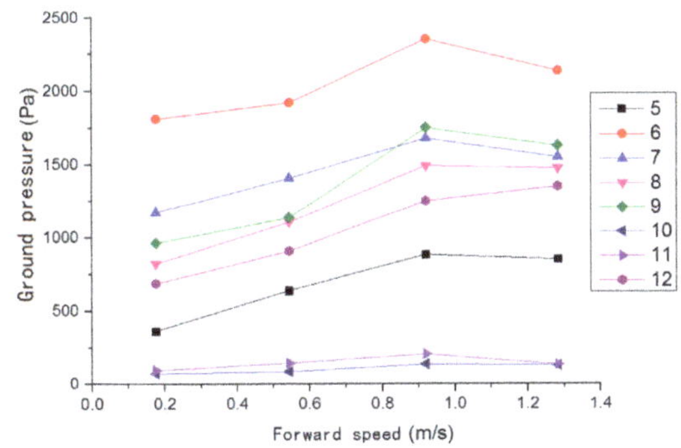

Figure 15. Relationship between ground pressure and forward speed.

The following conclusions can be drawn based on the analysis of the above figure:

The ground pressure and forward resistance of the skateboard show an upward trend when other conditions remain unchanged and the forward speed of the skateboard increases. When the forward speed v increases, the traction force F and the reaction force Nx of the skateboard increase, forming a torque that makes the skateboard tilt forward, so that the forward tilting angle of the skateboard increases. When the skateboard tilts forward, the ground-specific pressure at the front and back ends of the skateboard will decrease slightly. In addition, Figure 16 shows the main force on the skateboard.

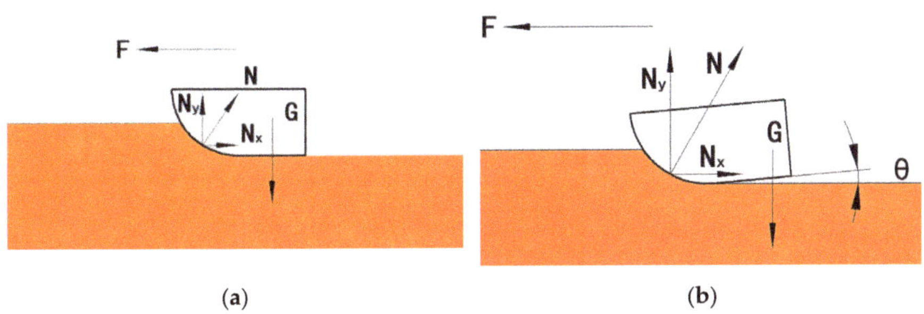

Figure 16. Force analysis of skateboard at different speeds. (**a**) At slow speed; (**b**) at fast speed.

3.2. Relationship between Forward Resistance, Ground Pressure, and Vertical Load

Multi-horizontal experiments were carried out under the following vertical load conditions: m = 0, 40, 80, 120, and 160 kg. Moreover, the values of forward resistance and ground pressure under different vertical loads were measured. Figures 17 and 18 show the variation curves.

The ground pressure and forward resistance of the skateboard show an upward trend, and the increasing range is similar when other conditions remain unchanged and the vertical load of the skateboard increases. With the increase of the counterweight, the accumulation of mud in front of the skateboard becomes more serious. Figure 19 shows the schematic diagram of skateboard force under obstructive mud accumulation. With the increase of mud, the interaction area between skateboard and paddy soil increases, and the total soil force N of skateboard increases. It leads to the increase of ground pressure and forward resistance of the skateboard.

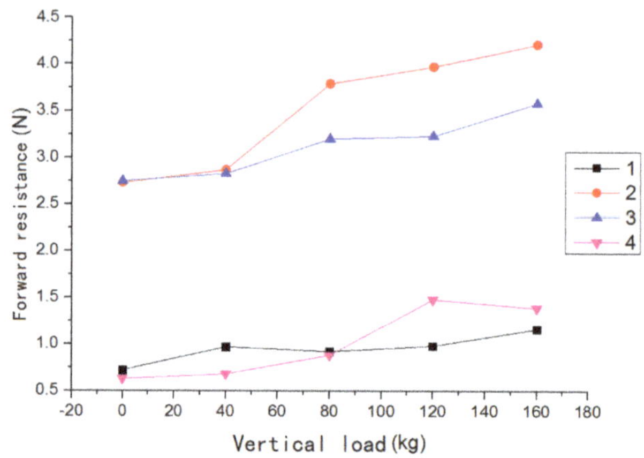

Figure 17. Relationship between forward resistance and vertical load.

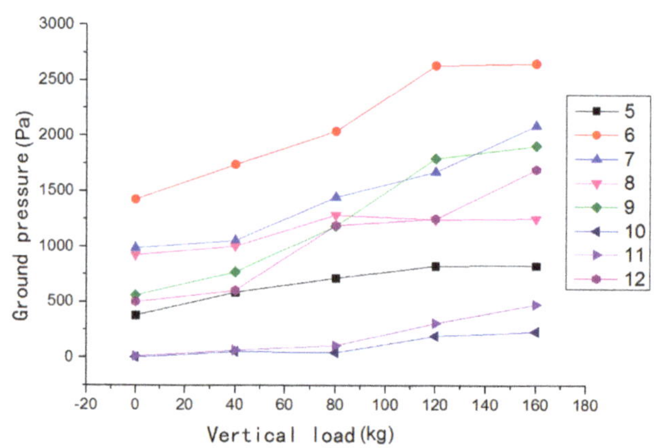

Figure 18. Relationship between ground pressure and vertical load.

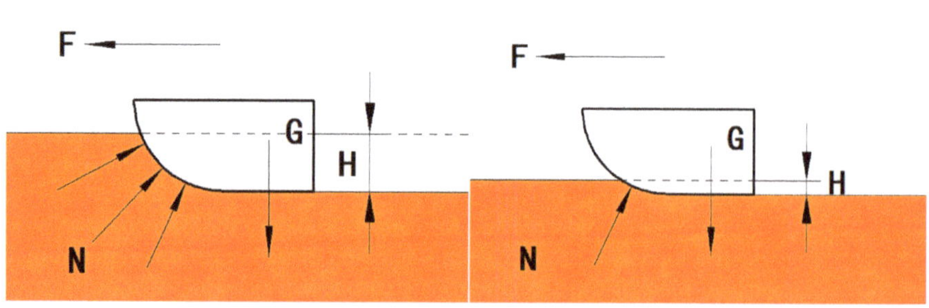

Figure 19. Force analysis of the skateboard under different accumulation thicknesses (H) of mud.

3.3. The Relationship between Forward Inclination Angle, Forward Speed and Vertical Load

The seeding skateboard falls flat on the paddy soil and reaches the equilibrium state when the rice transplanter is in the static state and the vertical load of the seeding skateboard is 0 kg. At this point, the forward tilting angle is recorded as zero. Two-factor and multi-level experiments were carried out under varying conditions of forward speed (v = 0.175, 0.544, 0.919, and 1.283 m/s) and vertical loads (m = 0, 40, 80, 120, and 160 kg).

The values of forward tilt angles under different forward speeds and vertical loads were measured. The variation curves are shown in Figures 20 and 21.

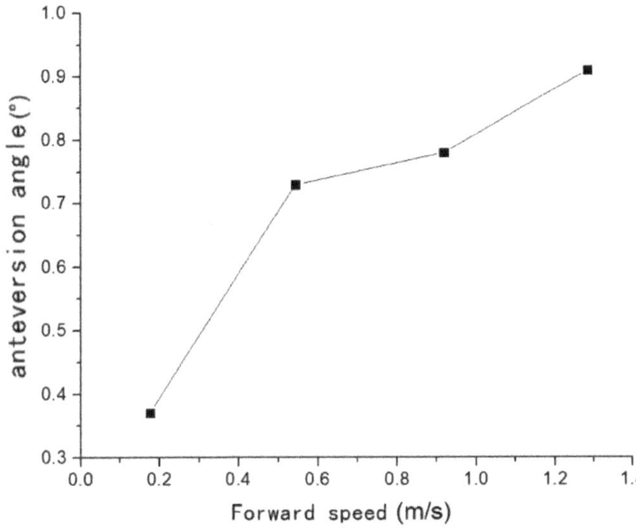

Figure 20. Relationship between forward tilt angle and forward speed.

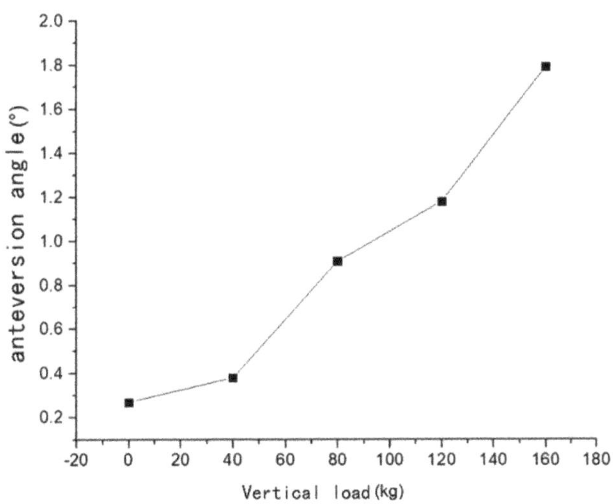

Figure 21. Relationship between forward tilt angle and vertical load.

The forward tilting angle of the skateboard increases with the increase of forward speed when other conditions are constant. The forward tilting angle of the skateboard also shows an upward trend with the increase of vertical load. The main reason is that the traction force of the skateboard increases with the increase of speed and vertical load, and the contact state between the skateboard and soil changes and the increase of tractive force increases the acting moment, which makes the skateboard tilt forward. The experimental results are consistent with the actual field operation, suggesting that attention should be paid to the change of the forward tilting state of the seeding skateboard when the direct seeder works at a high speed.

3.4. Distribution Law of Ground Pressure at the Bottom of Skateboard

Two-factor and multi-level experiments were carried out under varying conditions of forward speed (v = 0.175, 0.544, 0.919, and 1.283 km/h) and vertical loads (m = 0, 40, 80, 120, and 160 kg). Moreover, the front-end ground pressure and rear-end ground pressure under different forward speed and load conditions were measured and the variation curves are shown in Figures 22 and 23.

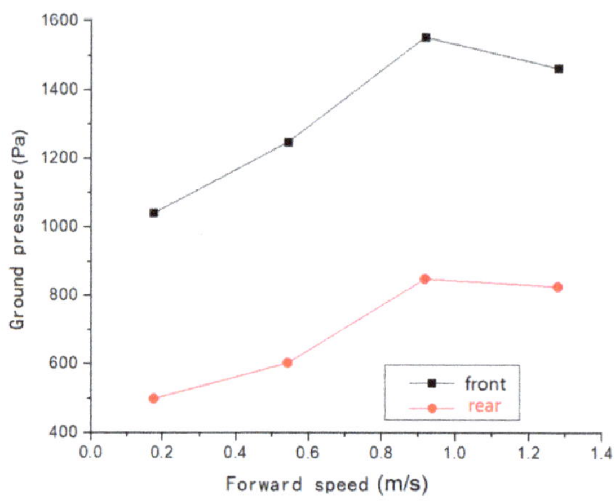

Figure 22. Relationship between the front and rear ground pressure and forward speed.

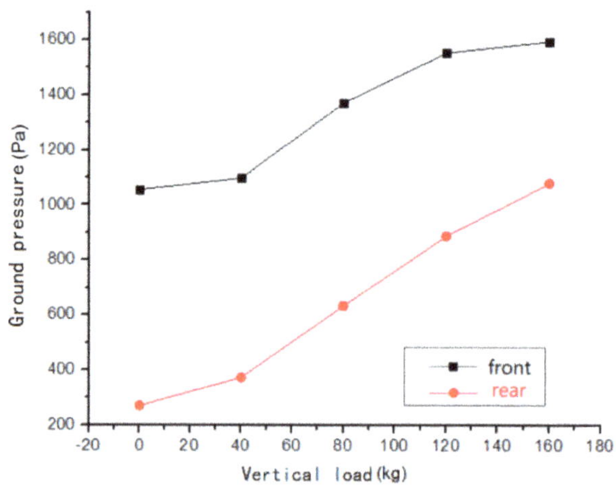

Figure 23. Relationship between the front and rear ground pressure and vertical load.

The value of the ground pressure sensor at the bottom of the skateboard is larger than that at the back end due to the forward tilt of the skateboard. The forward tilting angle increases gradually, and the ground pressure at the front and back ends increases gradually with the increase of vertical load. When the forward speed increases to a certain value, the resistance and traction force on the skateboard increase, forming a torque that makes the skateboard tilt forward, which increases the forward tilting angle of the skateboard. The ground pressure at the front and back ends will decrease slightly when the skateboard tilts forward.

There are 12 pressure sensors at the bottom of the skateboard. The pressure distribution law of each sensor was obtained, as shown in Figure 24, by analyzing and calculating the values of each pressure sensor. Red represents larger pressure and green represents smaller pressure.

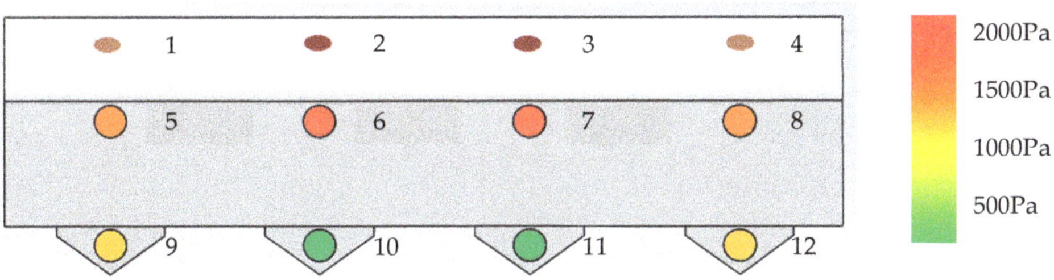

Figure 24. Pressure distribution of sensors at the bottom of skateboard.

The contact force between the bottom of the skateboard and the paddy field soil is mainly concentrated in the area between the forward resistance sensor and the front pressure sensor. It is large in the middle and small at both ends of the skateboard. The main reason is that the choking mud accumulated at both ends of the skateboard is more easily removed from the outside of the skateboard than that in the middle, which leads to the accumulation of more mud in the middle and less on both sides, as shown in Figure 25. The actual situation is consistent with the skateboard force diagram in the simulation, as shown in Figure 26.

Figure 25. Mud accumulation in front of skateboard.

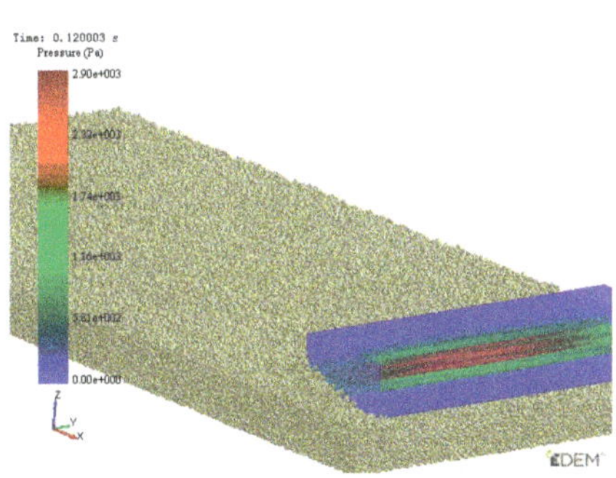

Figure 26. Force diagram when the skateboard is sliding.

Taking the seeding skateboard of rice direct seeding machine as the object, an experimental study was carried out on the variation law of ground pressure and forward resistance under different speeds and loads by selecting the outdoor paddy field environment and using the self-developed ground pressure measurement system. According to the experimental purpose and experimental index, a multi-factor and multi-level experiment scheme was designed and the experiment results were analyzed. The results show that the ground pressure measurement system can work stably in the paddy field, and the measured data can be wirelessly transmitted to the computer and mobile phone. The forward tilt angle, ground pressure, and forward resistance of the seeding skateboard also increase with the increase of forward speed and vertical load.

4. Discussion

Combined with networking technology and wireless serial port transceiver technology, a ground pressure measurement system was designed. It includes a ground pressure measurement module, an Internet of things wireless data transmission module, a data display and storage module, and so on. Its hardware mainly includes a Mega2560 Arduino controller, JHBS-1–5KG foil pressure sensor, and JHBM- H3 column pressure sensor. It is calibrated by combination, and the experimental results show that the ground pressure measurement system can meet the requirements set by the measurement experiment for forward resistance and ground pressure of the sliding seeder of the rice direct seeding machine. It has the characteristics of Internet of things remote wireless measurement, high precision, sensitive response, stable performance, simple operation, wide practicability, and so on.

An experimental field study on the ground pressure measuring system was carried out. The single-factor multi-horizontal experiment of ground pressure was carried out by taking the seeding skateboard of the rice direct seeding machine as the research object and taking the traveling speed, v, and vertical load, G, as the influencing factors. The experimental results show that with the increase of vertical load, G, and traveling speed, v, the forward tilting angle of the skateboard, the ground pressure of the skateboard, and the forward resistance all showed an upward trend. The ground pressure measurement system can work continuously and stably in the paddy field. In addition, the measured data can be wirelessly transmitted to the computer and mobile phone stably at a high speed.

By observing the distribution law of the stagnant mud generated by the skateboard, it was found that a large amount of stagnant mud is generated in the middle of the skateboard, which increases the forward resistance and ground pressure of the skateboard, and reduces the working efficiency of the rice direct seeding machine. In order to reduce the mud generated by the seeding slide of the rice direct seeding machine and improve the work efficiency of the rice transplanter, the surface of the slide can be treated to reduce the sliding resistance, for example, by adding a water retention support plate to the bottom of the slide to form a muddy channel under the skateboard. It is also possible to use a slide material with a small friction coefficient and to design a separate skateboard so that the mud in the middle of the skateboard can be quickly discharged.

Author Contributions: Conceptualization, Y.G. and G.Z.; methodology, G.Z.; software, Y.G.; validation, Y.G., A.S. and J.F.; formal analysis, Y.G.; investigation, Y.Z.; resources, Y.Z.; data curation, Y.G.; writing—original draft preparation, Y.G.; writing—review and editing, H.W.; visualization, G.Z.; supervision, G.Z.; project administration, Y.Z.; funding acquisition, G.Z. All authors have read and agreed to the published version of the manuscript.

Funding: The APC was funded by Innovative Research Group Project of the National Natural Science Foundation of China (No. 51775220).

Institutional Review Board Statement: Not applicable.

Informed Consent Statement: Not applicable.

Data Availability Statement: No new data were created or analyzed in this study. Data sharing is not applicable to this article.

Conflicts of Interest: The authors declare no conflict of interest.

References

1. Xin, F.; Xiao, X.; Dong, J.; Zhang, G.; Li, B. Large increases of paddy rice area, gross primary production, and grain production in Northeast China during 2000–2017. *Sci. Total. Environ.* **2019**, *711*, 135–183. [CrossRef] [PubMed]
2. Zhou, Y.; Li, X.; Liu, Y. Cultivated land protection and rational use in China. *Land Use Policy.* **2021**, *106*, 1–3. [CrossRef]
3. Yang, X.; Zhang, W.; Zhang, X.; Ma, Z. The Development of Budded Rice Direct Seeding Machine. *China. Agric. Mech.* **2005**, *6*, 7–12.
4. Lv, X.R.; Lv, X.L.; Ren, W. Experimental study on working performance of rice rope direct seeding machine. *Agric. Sci. China.* **2010**, *9*, 275–279.
5. Ye, H.; Chen, H.; Zhu, D.; Chen, Y.; Xu, Y. Yield and Application of direct seeding mechanization technology of hybrid rice. *China. Rice.* **2009**, *5*, 38–40.
6. Luo, X.; Li, X.; Liu, T.; Tan, M. Present situation and development prospect of rice mechanical direct seeding. In Proceedings of the Annual Meeting of Chinese Society of Agricultural Engineering, Wuhan, China, 24–27 October 2016.
7. Liu, Y. Soil Trafficability Study Based on Mechanical Properties of Paddy Soil. Master's Thesis, Nanjing Agricultural University, Nanjing, China, June 2014.
8. Liu, K. Paddy field operation skills of rotary tiller. *Hunan Agric. Mach.* **2011**, *12*, 28.
9. Lei, W.; Yi, L.; Cong, W.; Ling, W. Driving performance of bearing pressure-sinkage plates based on SPH. *J. Agric. Mech. Res.* **2021**, *43*, 5–7.
10. Zheng, P.; Fang, C.; Juan, H. The modified rheological model for paddy soils in South China after remoulding. *J. Terramechanics* **1990**, *27*, 4–7.
11. Deng, Z.; Zhu, Z.; Zeng, X.; Kong, G. Study of rheological properties of the lacustrine soft soil under different stress level. *Ind. Constr.* **2014**, *44*, 67–72.
12. Ram, R.B.; Gupta, C.P. Relationship Between rheological coefficients and soil parameters in compression test. *Trans. ASAE* **1972**, *15*, 1054–1058. [CrossRef]
13. He, Z. Experiment of Structures and Working Parameters Optimization for Paddy Field Boat-Type Tractor. Master's Thesis, Zhejiang University, Hangzhou, China, June 2018.
14. Wu, G.A. Preliminary study on the measurement of hull grounding pressure of tractor. *J. Zhejiang Univ. Agric. Life Sci.* **1984**, 351.
15. Yan, B.; Liu, Z.; Ma, L.; Yang, Z.; Li, W.; Fan, X.A. Dynamic Measuring Device for Specific Pressure of Crawler Grounding. China Patent No.202010951243.9.
16. Eisenkolb, U.; Eckhoff, D.; Kaspari, A. Ground Pressure Measuring Device and Method for Evaluation. European Patent No.1722185, 31 December 2008.
17. Zou, Y. Design and Experiment of the Test System for Ground Pressure of Seeding Pallet of Rice Direct Seeder. Master's Thesis, Huazhong Agricultural University, Wuhan, China, June 2017.
18. Soy, H.; Toy, İ. Design and implementation of smart pressure sensor for automotive applications. *Measurement* **2021**, *176*, 109–184. [CrossRef]

applied sciences

MDPI

Article

Comparing Performances of CNN, BP, and SVM Algorithms for Differentiating Sweet Pepper Parts for Harvest Automation

Bongki Lee [1], Donghwan Kam [1], Yongjin Cho [2,3,*], Dae-Cheol Kim [2,3] and Dong-Hoon Lee [4,*]

1 Institute of Biotechnology and Bioengineering, Sungkyunkwan University, Suwon-si 16419, Korea; dkways@skku.edu (B.L.); kamdh@skku.edu (D.K.)
2 Department of Bio-Industrial Machinery Engineering, Jeonbuk National University, Jeonju 54896, Korea; dckim12@jbnu.ac.kr
3 Institute for Agricultural Machinery & ICT Convergence, Jeonbuk National University, Jeonju 54896, Korea
4 Department of Biosystems Engineering, Chungbuk National University, Cheongju 28644, Korea
* Correspondence: choyj@jbnu.ac.kr (Y.C.); leedh@cbnu.ac.kr (D.-H.L.); Tel.: +82-63-270-2615 (Y.C.)

Abstract: For harvest automation of sweet pepper, image recognition algorithms for differentiating each part of a sweet pepper plant were developed and performances of these algorithms were compared. An imaging system consisting of two cameras and six halogen lamps was built for sweet pepper image acquisition. For image analysis using the normalized difference vegetation index (NDVI), a band-pass filter in the range of 435 to 950 nm with a broad spectrum from visible light to infrared was used. K-means clustering and morphological skeletonization were used to classify sweet pepper parts to which the NDVI was applied. Scale-invariant feature transform (SIFT) and speeded-up robust features (SURFs) were used to figure out local features. Classification performances of a support vector machine (SVM) using the radial basis function kernel and backpropagation (BP) algorithm were compared to classify local SURFs of fruits, nodes, leaves, and suckers. Accuracies of the BP algorithm and the SVM for classifying local features were 95.96 and 63.75%, respectively. When the BP algorithm was used for classification of plant parts, the recognition success rate was 94.44% for fruits, 84.73% for nodes, 69.97% for leaves, and 84.34% for suckers. When CNN was used for classifying plant parts, the recognition success rate was 99.50% for fruits, 87.75% for nodes, 90.50% for leaves, and 87.25% for suckers.

Keywords: NDVI; image processing; SURF; SIFT; SVM; BP algorithm; performance; sweet pepper; deep neural network

Citation: Lee, B.; Kam, D.; Cho, Y.; Kim, D.-C.; Lee, D.-H. Comparing Performances of CNN, BP, and SVM Algorithms for Differentiating Sweet Pepper Parts for Harvest Automation. *Appl. Sci.* **2021**, *11*, 9583. https://doi.org/10.3390/app11209583

Academic Editor: Francesco Bianconi

Received: 23 September 2021
Accepted: 12 October 2021
Published: 14 October 2021

1. Introduction

In South Korea, the need for automation of physically demanding agricultural labor is increasing because of the aging agricultural workforce and the growing proportion of women among the agricultural workforce [1]. In agricultural work such as weeding and pruning, the technology to distinguish and classify plant parts is necessary for automation of various agricultural tasks. Classification of plant parts can be performed using an image recognition technology. Once an image of a recognition target is obtained, pre-processing to extract the interested region is performed. The process of recognizing an object and transforming it into a suitable format to be processed with a computer is then performed by dividing the features of the recognition target within the area of interest [2].

Machine vision is a useful tool for plant recognition and identification [3]. To extract leaves and canopies, various algorithms such as Active Shape Model [4], Color Segmentation [5] using an RGB-D (Red, Green, Blue Depth) Camera, Support Vector Machines [6], Clustering Algorithm [7], Watershed Algorithm [8–10], and genetic algorithm [11] have been proposed and successfully used to recognize plant parts in the image. Machine recognition has been effectively used to analyze the shape and growth status of crops in fields such as plant phenomics. PhenoAIxpert (LemnaTec, Aachen, Germany) is using machine recognition technology to acquire plant phenotypic information in an image-spectral acquisition chamber. In addition, PlantScreen (Photon Systems Instruments, Drásov, Czech Republic) acquires an image and spectral information of plants and uses machine recognition technology to select excellent plants based on plant shape and spectral response information. However, plant parts recognized via images have been acquired in a controlled experimental environment rather than in the field where plants are actually grown. In the actual farm field, the canopy, including leaves, does not exist alone. A group of plants can form a colony to affect image recognition, making it difficult to clearly differentiate plant parts [12].

Various methods such as vegetation index [13] have been used to measure crop productivity on a farm and to diagnose crop growth. Silva et al. (2016) conducted an experiment to confirm the water stress of soybeans through NDVI (Normalized Difference Vegetation Index) images and confirmed a significant correlation between water stress and the NDVI of soybeans from each experiment [14]. To introduce plant shape recognition technology through images, it is necessary to distinguish targeted plant parts from the surrounding background. In a real farm field, it is easy to misread the contrast of an object because of lighting, sunlight, and shadow, which significantly lower the recognition rate of an object [15]. To improve the recognition rate of an object in an image, various methods such as Bayesian-classifier [16] and Clustering [17] using machine learning have been proposed. Methods for improving the recognition rate of objects in images have been evaluated for their accuracies in a competition called ILSVRC (ImageNet Large Scale Visual Recognition Challenge) [18]. Prior to 2012, when machine learning was predominant, the average recognition rate was below 75%. In 2012, Krizhevsky et al. [19] proposed a convolutional neural network (CNN), an artificial neural network model, and showed a recognition rate of 85%, exceeding the existing recognition rate. Since 2012, various artificial neural network models have been proposed, showing recognition accuracy exceeding 95%, which is comparable to human recognition accuracy [20]. An artificial neural network is a pattern recognition method inspired by the interconnection of neurons in the human nervous system. A general artificial neural network consists of an input layer, a hidden layer, and an output layer. Each layer includes neurons [21]. Neurons between neighboring layers are connected by weights. Input values are learned by changing weights through repeated learning. The larger the number of hidden layers, the more complex the data that can be modeled. When the number of hidden layers is two or more, it is called a deep neural network [22]. It is necessary to apply an artificial neural network model to recognize the shape of a plant through an image. Pound et al. [23] proposed a method of classifying wheat grains, nodes, and leaves using a CNN (Convolution Neural Network) but analyzed wheat as a learning factor in a controlled indoor environment. To recognize the shape of plants in an environment where actual plants are grown, color-based 3-dimensional fruit recognition [24], fruit recognition using an LED reflected light and color model [25], color information obtained from an RGB-D camera, and recognition of the fruit stalk by classifying the surface normal vector and curvature into Support Vector Machines [26] have been reported. However, the shape of the plant based on color information could not be recognized because nodes and leaves had the same color information.

Therefore, the purpose of this study was to develop an image processing algorithm that classifies plant parts, and the target of classification was paprika, which has various colors of fruits such as red, yellow, orange, and green. Sweet pepper is one of the most cultivated greenhouse crops in South Korea [27]. This study aimed to develop image processing algorithms to classify plant parts for automation of sweet pepper farming tasks such as weeding, pruning, and fruit thinning. Another aim of this study was to evaluate performances of these algorithms. Detailed study objectives are as follows:

1. To recognize the shape of sweet pepper in a farm environment, classify each part, and develop an image processing algorithm to classify sweet pepper parts for Plenty

(red sweet pepper), President (yellow sweet pepper), and Derby (orange sweet pepper) varieties.

2. To classify sweet pepper parts with the NDVI (Normalized Difference Vegetation Index) by dividing the targeted area using k-means clustering and morphological skeletonization followed by extraction of local features using SIFT (Scale-Invariant Feature Transform) and SURF (Speeded-Up Robust Features).
3. To evaluate performances of developed algorithms such as the BP (Backpropagation) algorithm and the SVM (Support Vector Machine) algorithm for classifying each part (leaf, node, stem, and fruit) compared to a deep neural network algorithm.

2. Materials and Methods

2.1. Hardware Composition

The sweet pepper greenhouse maintained a temperature between 20 and 30 °C and a relative humidity between 60 and 80% [28]. A Blackfly BFLY-U3-13S2C-C color camera (Teledyne FLIR LLC, Wilsonville, OR, USA) was selected to obtain images in the sweet pepper greenhouse (Table 1). The operating temperature range of the selected camera was 0–45 °C. The humidity was 20–80%. Most images were in the 400–950 nm range to analyze the NDVI. The image sensor used in this study acquired images from 400 to 950 nm (Figure 1).

Table 1. Specifications of the Blackfly BFLY-U3-13S2C-CS color camera.

Model	Imaging Sensor	Resolution	Operating Temperature	Operating Humidity
Blackfly BFLY-U3-13S2C-CS	Sony ICX445, 1/3″, 3.75 µm	1288 × 964 (30 FPS)	0–45 °C	20–80%

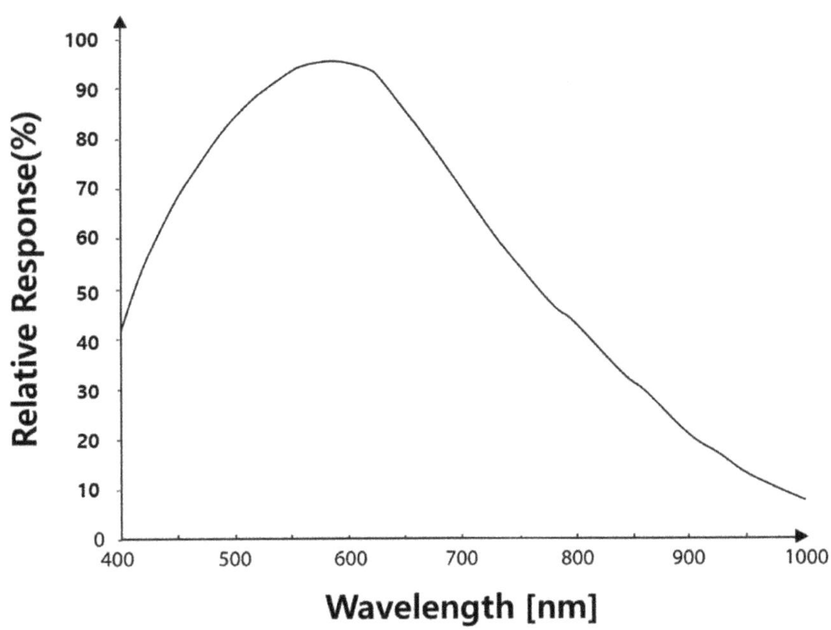

Figure 1. Spectral sensitivity of the imaging sensor, Sony ICX445.

The standard size of the passage width of the sweet pepper greenhouse was 1.6 m. The width between crops not interrupted by sweet pepper crops when moving from one passage to another was 1.0 m [29]. The planting density ranged from 3.3 to 3.5 stems/m^2 [30]. The spacing of sweet pepper nutrient solution beds based on the planting density was within 0.3 m. An 8 mm lens was selected to acquire images of the sweet peppers planted at intervals of 0.3 (the minimum) to 1.0 m (the maximum). The size of the Sony ICX445 image sensor on the selected camera was 1/3″. The field of view (FOV) according to the working distance using an 8 mm lens is shown in Figure 2. The distance between the pupils of human eyes ranges from about 60 to 75 mm [31]. Thus, the distance between cameras was set to 70 mm to acquire sweet pepper images (Figure 3). According to Figure 2, the area of the imaging system in this study overlapped from a maximum of 530 to a minimum of 230 mm. The average length of sweet pepper leaves was 223 mm. The average leaf width was 121 mm [32] and the average diameter of sweet pepper fruit was 80.92 mm, suitable for obtaining overlapping images of each sweet pepper part at the corresponding distance.

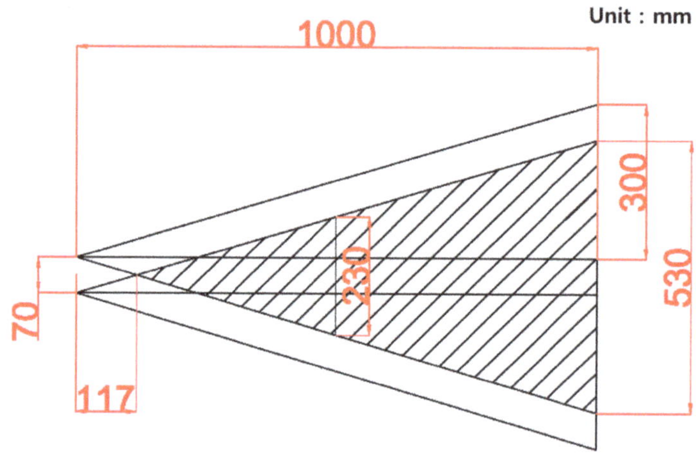

Figure 2. FOV of 8 mm lens and overlapping area of vision system.

(a) (b)

Figure 3. A structure for image acquisition (**a**) and image acquisition in the greenhouse (**b**).

The device for obtaining images of the sweet peppers was designed as shown in Figure 3a. Figure 3b shows image acquisition. The sweet pepper was planted in nutrient solution beds. The minimum height of the nutrient solution beds was 0.1 m and the camera's FOV at 1 m was 0.6 m. Thus, the height of the camera from the ground was set at 0.7 m. In addition, halogen lamps were selected as image light sources. Three lamps each were installed on the upper side and the lower side for lighting. A halogen lamp 44870 WFL (Osram, Munich, Bavaria, Germany) was selected as an image light source to assist image acquisition in order to obtain an image from the 435–950 nm region. The selected lamp had a brightness of 680 lm of 50 W and a color temperature of 3000 K. Halogen lamps have a broad spectrum from visible light to infrared regions [33]. For image analysis using the NDVI, a band-pass filter from the 435 to 950 nm region of BP-Series (Midwest Optical Systems, Palatine, Illinois, USA) was used. Specifications of the band-pass filter used are shown in Table 2.

Table 2. Specifications of the band-pass filter used by the system.

Model	Useful Range	FWHM [1]	Tolerance	Peak Transmission
BP470	435−495 nm	85 nm	+/−10 nm	>90%
BP500	440−555 nm	248 nm	+/−10 nm	>85%
BP505	485−550 nm	90 nm	+/−10 nm	>90%
BP525	500−555 nm	80 nm	+/−10 nm	>90%
BP635	610−650 nm	65 nm	+/−10 nm	>90%
BP660	640−680 nm	65 nm	+/−10 nm	>90%
BP695	680−720 nm	65 nm	+/−10 nm	>90%
BP735	715−780 nm	90 nm	+/−10 nm	>90%
BP800	745−950 nm	315 nm	+/−10 nm	>90%

[1] FWHM: Full-Width Half-Maximum.

2.2. Sweet Pepper Part Classification Process

When growing sweet pepper, to promote its growth, reduce damage caused by pests and diseases, and increase fruit yield, it is necessary to remove old leaves at the lower part and suckers that occur between central stems and nodes [28]. In order to automate agricultural tasks such as pruning and harvesting, it is important to detect the branching point of the central stem and nodes, the sucker that is a shoot of new growth from the nook where a branch splits in two and a piece of the plant that gardeners remove in the pruning process. In this study, local features of sweet pepper were used for the classification of plant parts such as fruits, stems, leaves, and nodes from obtained images to differentiate them by each part. First, the harvesting image was pre-processed through the NDVI to remove noise such as the background image. Fruit, stem, leaf, and node were segmented through k-means clustering. Skeletonization of each segmented part was performed through a morphological method. Local features from the skeletonized image were then figured out using SIFT and SURF algorithms. Numbers of local features of fruits, stems, leaves, and nodes were then compared to confirm local features. The extracted local features were classified through the SVM and BP classifier algorithms. Their classification performances were compared. In addition, a CNN (Convolutional Neural Network), a type of deep neural network, was implemented to differentiate fruits, nodes, leaves, and suckers of sweet pepper. Its classification performance was compared to that of the SVM or BP classifier algorithm. Figure 4 shows the whole classification process for plant parts.

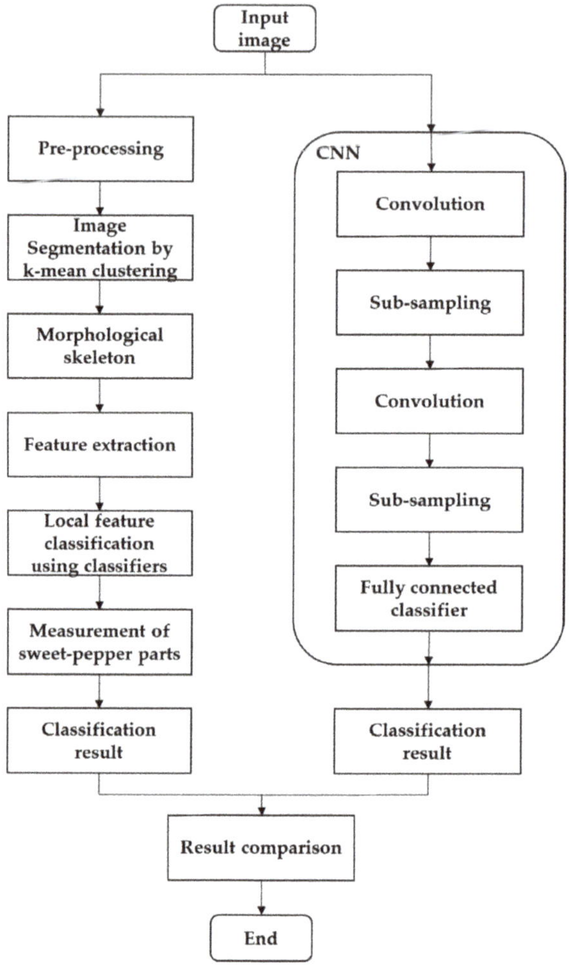

Figure 4. The classification process of sweet pepper plant parts.

2.2.1. Pre-Processing for Differentiation of Stems and Leaves from Backgrounds Using the NDVI

Silva et al. [14], Story et al. [34], and Yu [35] analyzed plants using their differences in light transmittance and reflectance according to plant wavelength. Figure 5 shows an image in the visible spectrum band and images using band-pass filters of 695, 735, and 800 nm.

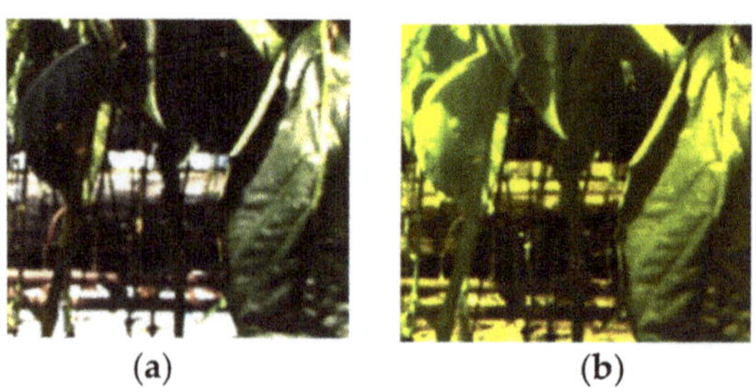

Figure 5. *Cont.*

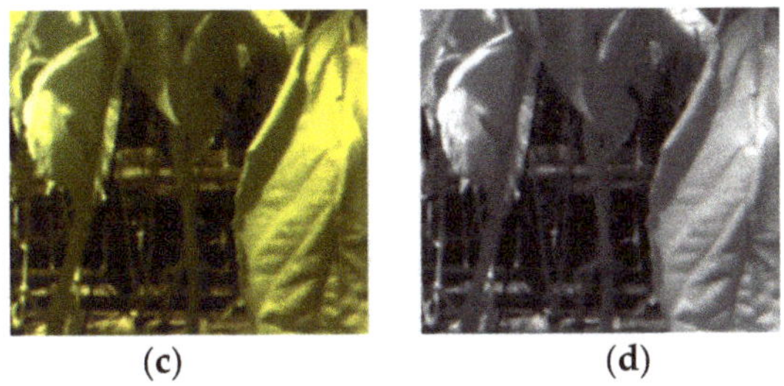

Figure 5. Spectrum images using band-pass filters of (a) visible spectrum, (b) 695, (c) 735, and (d) 800 nm.

Plants have a different reflectance for each wavelength band. Sweet pepper stems and leaves were separated from the background by analyzing images from 470 to 800 nm. The NDVI used the difference in reflectance for each wavelength band to distinguish plants from non-plants. As shown in Figure 6, 695, 735, and 800 nm sectors (hereinafter, group A) and 470, 500, 505, 525, 635, and 660 nm sectors (hereinafter, group B) were analyzed to select a wavelength band suitable for the imaging system of this study. Using Formula (1) based on the NDVI, differences between A and B sectors were investigated.

$$f = \frac{A - B}{A + B} \times 255 \quad (1)$$

where A = image at 695, 735, and 800 nm; B = image at 470, 500, 505, 525, 635, and 660 nm, and f = NDVI image (Rouse et al. [36]).

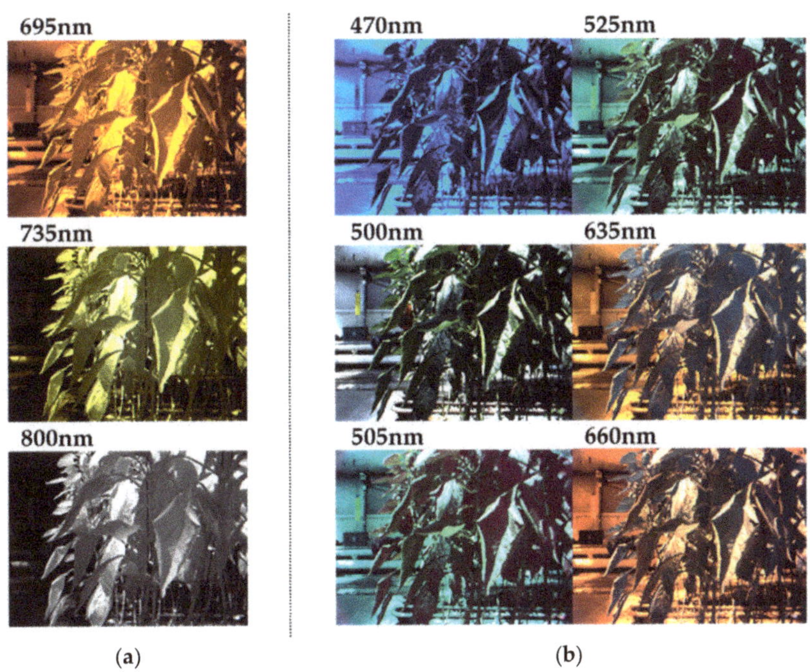

Figure 6. Image comparison between (a) images of 695, 735, and 800 nm wavelength and (b) images at 470, 500, 505, 525, 635, and 660 nm wavelength.

After obtaining the ratio of the difference and the sum of images of groups A and B, it was multiplied by 255, normalized to a size between 0 and 255, and then visualized. The quality of the normalized image was evaluated using Peak Signal-to-Noise Ratio (PSNR), the maximum signal-to-noise ratio. Figure 7a is an image in the visible ray region, which is a reference for comparing image quality and differences of images by wavelength bands. Figure 7b shows the loss degree by PSNR.

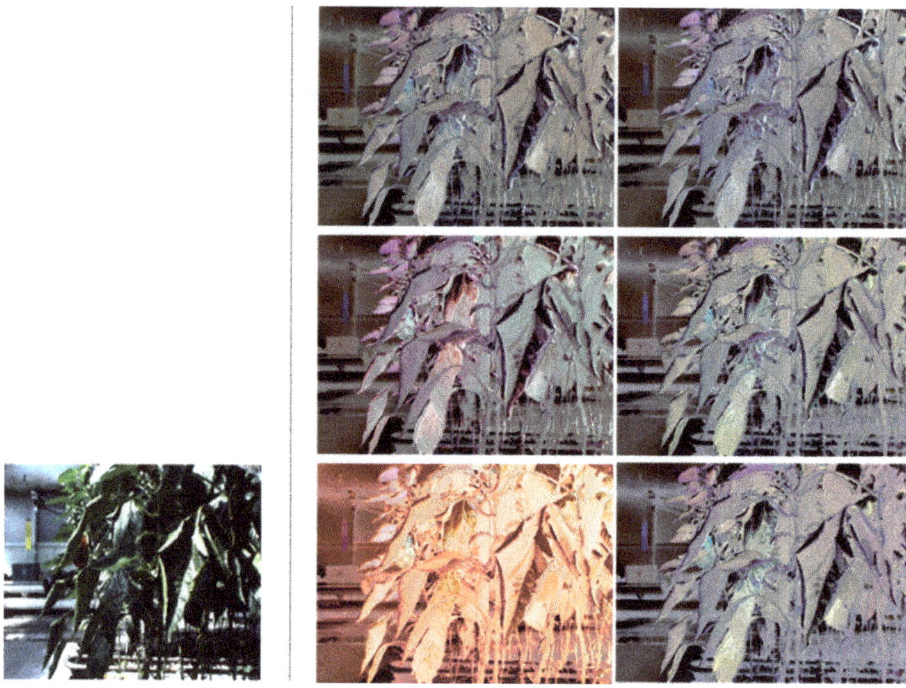

(a) (b)

Figure 7. Reference image for PSNR comparison (**a**) and PSNR images by difference of wavelength (**b**).

PSNR was measured using a total of 180 NDVI images (10 images from each band). A value with a high average of PSNR from each band difference was selected as an image for the NDVI. An image with a higher PSNR value meant less distortion during image conversion to image of higher quality. Leaves and stems were separated in the sweet pepper harvesting image using differences between groups A and B with high PSNR values. Local features of leaves, stems, and nodes were then extracted from separated regions to detect sweet pepper leaves, stems, and nodes within the image.

2.2.2. Image Segmentation by K-Means Clustering

In this paper, the image that underwent the pre-processing of separating targeted objects from the background was segmented through k-means clustering, which was first proposed by Lloyd [37]. In the farm image, colors of the same objects can have various color distributions depending on nutrition and growth conditions, lighting, sunlight, and shadows. Therefore, if the image was segmented using a simple binarization method, it was difficult to accurately classify targeted objects. When segmentation by k-means clustering was performed, color segmentation of targeted objects was possible rather than performing segmentation using a single threshold value.

$$min_{m_k} \sum_{k=1}^{k} \sum_{x \in c_k} (x - m_k)^2 \tag{2}$$

where x = image pixels to be clustered; m_k = the center of clusters; k = the number of randomly assigned clusters at the beginning.

In the clustering algorithm (Equation (2)), k-means clustering made a starting point at an arbitrary position as many as the initial number of k. Based on this, the distance between each data point was calculated using the Euclidean distance and clustered at the nearest initial point. A position at which the average point of the image pixels divided into clusters was obtained. It became a new center point. This process was repeated until the average position of all clusters did not change by performing clustering. In this paper, the number of clusters, k, was designated as 3 using HSV color information known to be stable upon lighting changes rather than RGB (Red, Green, Blue) color information. Three clusters were then obtained for fruits, stems, and leaves images with the background separated. For sweet pepper under cultivation, 400 sweet pepper cultivation images of 1288 × 964 pixels were obtained at 680–800 mm. The ratio of width and length was equally divided into 4 parts as shown in Figure 8. Numbers of fruits, stems, leaves, and suckers lost during image segmentation in each area were then measured.

Figure 8. Pictures of each part before image segmentation (**a**) and after image segmentation (**b**).

2.2.3. Extracting Local Features

Chatbri et al. [38] compared local features of objects generated by contour detection and skeletonization of the image to extract the morphological features of objects in the binary image segmented by color. In this paper, SIFT [39] proposed by David G. Lowe and SURF [40] proposed by H. Bay were used to extract local features from 40 skeletonized images through a morphological method. By extracting and comparing the number of local features of each method, local features suitable for differentiating sweet pepper parts were selected. In this study, the curvature threshold of SIFT was set to be 10 and the control threshold was set to be 0.04. SURF constructed a window using Haar wavelets in the surrounding area based on the direction and intensity of key points found in the scale space. This window had local features that were strong against rotation and scale changes. Once SIFT local features were used, a 128-dimensional vector was obtained as a descriptor representing the point using an image gradient of 4 × 4 size in 8 directions. Using local features of the SURF, the gradients in the x-direction and y-direction were classified into 8 categories and accumulated to obtain a 128-dimensional vector as a descriptor representing key points. To use as learning factors for SVM and BP algorithms, coordinates, size, angle, response, octave, and class identification number found using SIFT and SURF were stored as key point information. The descriptor was stored as 128 real numbers. Among them, 128 descriptors of each key point were used as learning factors for SVM and BP algorithms. Figure 9 shows a sweet pepper cultivation image segmented by k-means clustering with local features extracted using skeletonization, SIFT, and SURF.

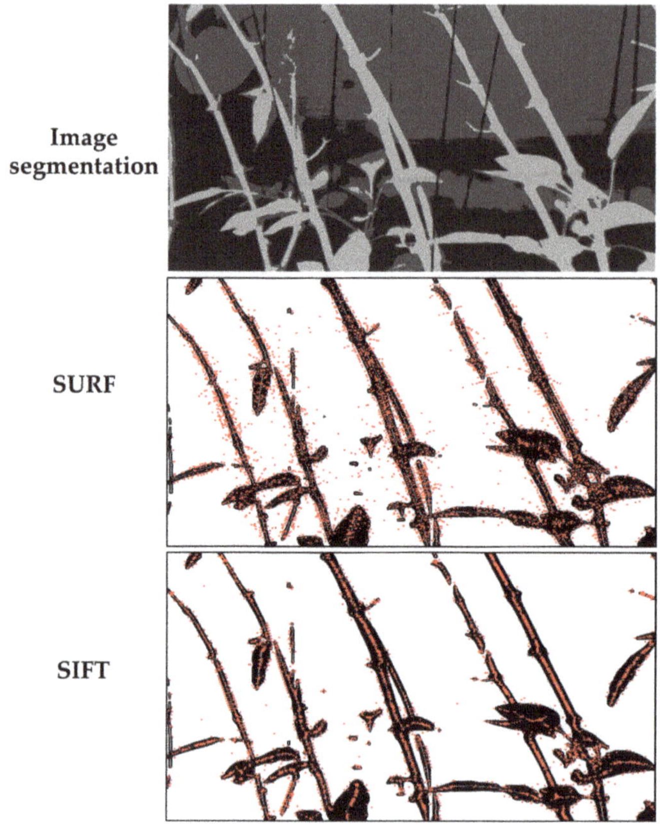

Figure 9. The image of local feature extraction processes using skeletonization, SIFT, and SURF.

2.2.4. Local Features Classification Using the SVM Algorithm and BP Algorithm

The method of classifying local features was determined by training and recognizing local features obtained in Section 2.2.3 using the SVM (Support Vector Machine) algorithm proposed by Vapnik [41] and the BP (Back Propagation) algorithm proposed by P. Werbos [42].

In this paper, the RBF (Radial Basis Function) kernel was used as the kernel function of SVM. The RBF (Radial Basis Function) kernel is expressed as Equation (3):

$$K(x,y) = exp\left(-\frac{\|x-y^2\|}{2\sigma^2}\right) \tag{3}$$

where σ = Gaussian window width; x = input vector; y = input pattern, $K(x,y)$ = RBF kernel.

SVM is basically a binary classifier that solves two kinds of problems. Methods for expanding the M-class SVM include a one-to-many classification method and a one-to-one method. In this paper, M binary classifiers were performed using the one-to-many classification method. Here, the i-th binary classifier classified the class and the remaining M-1. In this way, classification for M times was performed and the highest value was classified.

$$k = \underset{i}{argmax}\; d_i(x) \tag{4}$$

where $d_i(x)$ = the decision hyperplane of the i-th binary classifier.

The k in Equation (4) indicates that the class with the highest value is classified from the M-class SVM classifiers. To classify local features of sweet pepper using the SVM algorithm, a total of 400 images were used (100 images each were used to determine local features of fruits, stems, leaves, and suckers). For each class, 60 were used for training and 40 were used for validation.

Figure 10 shows the structure of the BP neural network used in this paper. The connection strength w was initially set to a value ranging from −0.5 to maximum 0.5 and v was initially set to a random value ranging from minimum −0.5 to +0.5. v was initialized by Equation (5) at the start of learning:

$$v^{new} = \frac{\beta v^{old}}{\left\| v^{old} \right\|}, \quad \beta = 0.7^{n}\sqrt{p} \tag{5}$$

where n = the number of nodes at input layer; p = the number of nodes at hidden layer.

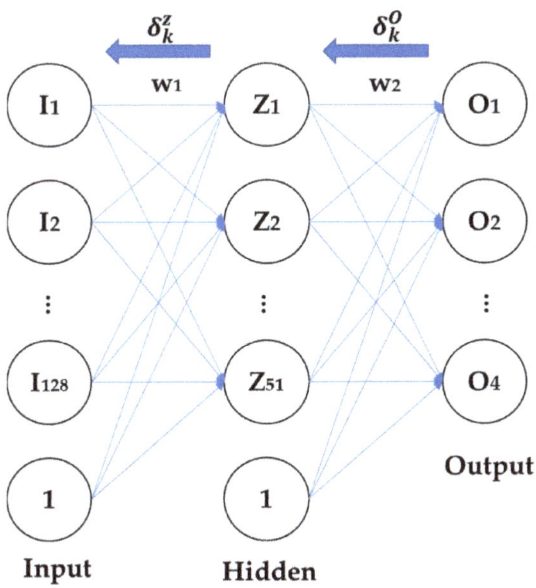

Figure 10. The structure of the backpropagation algorithm.

The number of nodes of the input layer was 129, including bias nodes and 128 descriptors of local features calculated by the gradient histogram of local features. If the input pattern space is n-dimensional and the number of nodes in the hidden layer is p, the maximum number M of linearly separable regions can be obtained through Equation (6) when k is smaller than p. If n is greater than or equal to p, it can be obtained through Equation (7). Therefore, the number p of nodes in the hidden layer for solving the problem of M linearly separable regions can be obtained using Equation (8):

$$(case\ 1)\ if\ k < p,\ \ M = \sum_{k=0}^{n} {}_{p}C_{k}, \tag{6}$$

$$(case\ 2)\ if\ n \geq p,\ \ M = 2^{p}, \tag{7}$$

$$p = \log_2 M \tag{8}$$

where n = the number of nodes at input layer; p = the number of nodes at hidden layer; M = the maximum number of regions capable of linear separation.

The number of nodes (p) in the hidden layer was set at 51, including bias nodes. The number of nodes in the output layer was set at 4 because there should be one node for each class of fruits, stems, leaves, and suckers. The activation function used a bipolar sigmoid function with a value between −1 and 1. The learning rate was set to be 0.001. The maximum error was set to be 0.1. Learning was terminated when an error of less than 0.1 was obtained. In case the error did not become smaller than 0.1, learning was terminated when the number of learning generations (epoch) exceeded 500. A total of 400 images were used to classify local features of sweet peppers using the BP algorithm. Images were classified using the descriptor of key points for images of fruits, nodes, leaves, and suckers as learning factors of the SVM and the BP algorithm. Classification results were evaluated using precision, recall, F-measure, and accuracy.

2.2.5. Partial Classification Performance Experiment Using Deep Neural Network

In this paper, a convolutional neural network (CNN), a kind of deep neural network, was implemented to differentiate fruits, stems, leaves, and suckers of plants and classify each part of the sweet pepper. The CNN in this paper had 16 hidden layers using the focal loss method of Lin et al. [43]. The bias was 0 and the weight was a random Gaussian weight except for the final layer. The bias of the last layer was calculated using Equation (9):

$$bias = -\log\left(\frac{1-\pi}{\pi}\right) \tag{9}$$

where π = the confidence value for the foreground of labels of all anchors in the initial learning (0.01).

In Equation (10), the focal loss function was calculated with γ = 2 as the rectified linear unit (ReLU) activation function, and the learning rate was set to be 0.01.

$$FL(p_t) = -(1-p_t)^{\gamma}\log(p_t)\,, \quad \gamma \geq 0$$
$$p_t = \begin{cases} p & if\, y = 1 \\ 1-p & otherwise \end{cases} \tag{10}$$

where y = the ground-truth class (1 or -1); p = the class probability predicted by the model (if $y = 1$); γ = the focal loss function.

In order to learn each part of the sweet pepper, as shown in Figure 11, a total of four types of sweet pepper were classified into fruits, stems, leaves, and suckers. A total of 300 brightness images were used (75 for each class mentioned above). A total of 100 images of sweet pepper cultivation were used to verify the performance of the deep neural network that was learned. Classification results were evaluated using precision, recall, F-measure, and accuracy. In order to confirm the classification performance for suckers, stems, leaf, and fruits of sweet peppers, a good method for differentiating plant parts between the SVM and BP algorithms was selected. The performance of the better one of the two ways (SVM and BP algorithms) was then compared to that of the deep neural network.

Figure 11. Example of learning data for leaf, sucker, fruit, and node.

3. Results and Discussion

3.1. Part Classification Results

3.1.1. Results of Pre-Process for Differentiation of Stems, Leaves, and Backgrounds Using NDVI

Figure 12 shows the transformation of the difference between images in 695, 735, and 800 nm bands and the images in the 470, 500, 505, 525, 635, and 660 nm bands through Equation (1). The quality of the transformed image was evaluated through the PSNR (Peak Signal-to-Noise Ratio) value. The higher the PSNR value, the lower the loss. Table 3 shows the average of PSNR values of images in the 695, 735, and 800 nm bands and the difference images of 10 images in the 470, 500, 505, 525, 635, and 660 nm bands. Among the average PSNR values measured in the 10 images with a total of 18 band differences, the image obtained by transforming the difference between 735 and 660 nm showed the highest value of 871,625. The higher the PSNR value, the lower the loss compared to the original image. Therefore, the difference between 735 and 660 nm confirmed image with the least image loss compared to the visible light region. Stems, leaves, and the background were distinguished using the transformed image difference between 735 and 660 nm.

Figure 12. Differences of visible spectrum image at 695 (**a**), 735 (**b**), and 800 nm (**c**) wavelength by 470, 500, 525, 635, 660 nm.

Table 3. PSNR values at different wavelengths.

Wavelength	695 nm		735 nm		800 nm	
	Mean	SD [1]	Mean	SD	Mean	SD
470 nm	8.33309	0.0229	5.46441	0.0949	8.62191	0.0339
500 nm	8.20943	0.0294	6.51014	0.0837	7.81063	0.0306
505 nm	8.30828	0.0118	7.75226	0.0999	8.68881	0.0673
525 nm	8.2368	0.0971	7.57739	0.0283	7.53663	0.0667
635 nm	7.98845	0.0248	7.62452	0.0393	7.66518	0.0538
660 nm	7.84794	0.0298	8.71625	0.0315	7.70154	0.0178

[1] SD: standard deviations.

3.1.2. Image Segmentation Results by K-Means Clustering

Plenty (a red sweet pepper variety), President (a yellow sweet pepper variety), and Derby (an orange sweet pepper variety) varieties were used in this study. In the average image of the fruit area, the fruit and the background were differentiated by back-projecting the histogram of color and saturation. For stems and leaves, the canopy of sweet pepper and the background were differentiated using the NDVI for the images of 735 and 660 nm bands.

The result of image segmentation using k-means clustering with the number 3 clusters separated from the background is shown in Figure 13. To determine losses of fruits, stems, leaves, and suckers in the image through image segmentation, an image of a sweet pepper at 1288 × 964 pixels was divided into four parts horizontally and vertically. Numbers of fruits, stems, leaves, and suckers damaged in the area were measured.

Figure 13. Image segmentation of steam, leaves, suckers (**a**), and fruits (**b**) by k-means clustering.

Table 4 shows actual numbers of fruits, stems, leaves, and suckers in the cultivation image and numbers of fruits, stems, leaves, and suckers obtained through the image segmentation process. As a result, in the case of fruits, through the image segmentation process, the images of 6 fruits, 1421 stems, 2188 leaves, and 37 suckers were removed from the background.

Table 4. Numbers of parts retained after image segmentation.

Parts	Actual Number	Segmented Number
Fruit	682	676
Sucker	345	308
Leaf	8169	5981
Node	11,856	10,435

3.1.3. Results of Local Feature Extraction

Using k-means clustering, the targeted region and the background were image segmented. The image was modified so that local features could be obtained through morphological skeletonization. Numbers of local features of SIFT and SURF were compared in images of 80 × 80 pixels in width and length of 30 fruits, nodes, leaves, and suction regions, respectively. Figure 14a shows numbers of local features extracted using SIFT. For fruits, the median number and average number of local features were 121 and 123, respectively, with a maximum value of 129 and a minimum value of 111. In the case of suckers, the median number and average number of local features were 105 and 101, respectively, with a maximum value of 123 and a minimum value of 81. In the case of leaves, the median number and average number of local features were 103.5 and 104, respectively, with a maximum value of 115 and a minimum value of 92. For stems, the median, the average, the maximum, and the minimum were 111.5, 113, 133, and 99, respectively.

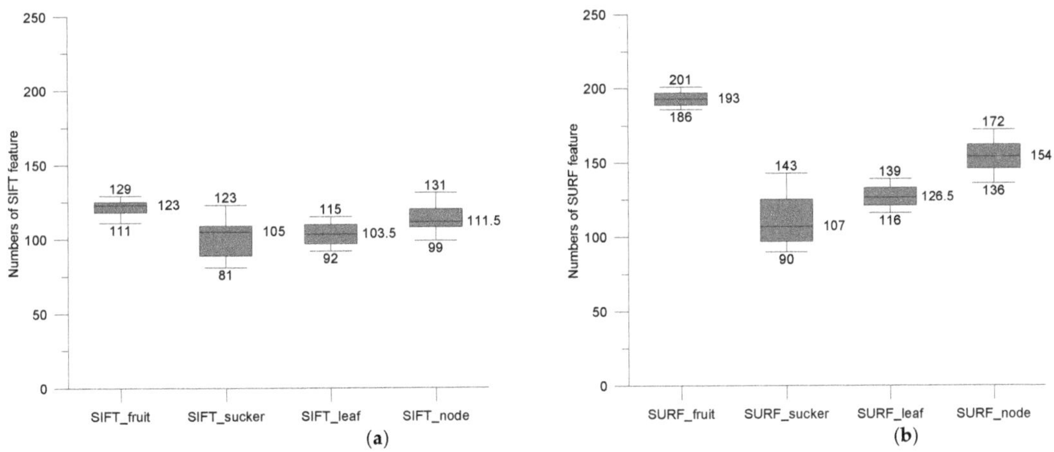

Figure 14. Numbers of SIFT features (**a**) and SURF features (**b**).

Figure 14b shows the number of local features extracted using SURF. For fruits, the median number and the average number of local features were the same at 193, with a maximum value of 201 and a minimum value of 186. In the case of suckers, the median number and average number of local features were 107 and 112, respectively, with a maximum value of 143 and a minimum value of 90. In the case of leaves, the median number and average number of local features were 126.5 and 127, respectively, with a maximum value of 139 and a minimum value of 116. For stems, the median, the average, the maximum, and the minimum were 154, 153, 172, and 136, respectively. In the case of suckers, numbers of local features of SIFT and SURF were extracted from overlapping ranges. The average, maximum, and minimum values showed high SURF. In the case of fruits, leaves, and stems, the number of local features of SURF was higher than that of SIFT without overlapping ranges. Therefore, the SURF algorithm was selected as the classification method for extracting local features of sweet pepper fruits, stems, leaves, and suckers.

3.1.4. Classification Results of Local Features Using SVM and BP Algorithms

Fruit and background were differentiated by back-projection of the histogram of color and saturation of the average image of the fruit area. In the case of stems and leaves, the canopy and the background of sweet pepper were differentiated using the NDVI operation of images at 735 and 660 nm band wavelengths. After segmenting the image, local features of fruits, stems, leaves, and suckers were extracted using SURF from skeletonized images and differentiated using the SVM algorithm and the BP algorithm. Performances of these two algorithms were then compared.

A total of 100 images of each region were used for features of the SURF region of fruits, stems, leaves, and suckers. As a result of classifying a total of 400 images through the SVM algorithm using the RBF kernel, the accuracy was 63.75%. Table 5 shows results of classification using SVM for fruits, stems, leaves, and suckers. The precision for fruit was 75% and the recall rate was 73.5%. The precision for stems was 68% and the recall rate was 61.26%. The precision for leaves was 60% and the recall rate was 57.69%. The precision for suckers was 52% and the recall rate was 62.65%. F-measure results for fruits, stems, leaves, and suckers were 74.26, 64.46, 58.82, and 56.83%, respectively.

Table 5. Results of SVM classifier using RBF kernel.

Parts		True Condition			
		Fruit	Node	Leaf	Sucker
Predicted condition	Fruit	75	0	25	0
	Node	0	68	11	21
	Leaf	27	3	60	10
	Sucker	0	40	8	52

Table 6 shows results of 178,750 features collected from 400 images of each class of fruit, stem, leaf, and sucker into a training set and a validation set used to train the BP algorithm. SURF region features were extracted from 400 images including fruits, stems, leaves, and suckers. Of a total of 178,750 features, 10% were used for training. As a result of repeating the learning for 500 generations (epoch) with the learning rate set to be 0.001, the final learning result was 98.46%.

Table 6. Number of features used for learning.

Parts	Training Feature	Test Feature
Fruit	694	6250
Node	10,720	96,478
Leaf	6144	55,298
Sucker	317	2849
Total feature	17,875	160,875

The accuracy was 95.95% as a result of classifying through the BP algorithm using a total of 400 images with 100 images for each class targeting SURF local features of fruits, stems, leaves, and suckers. Table 7 shows results of classification using the BP algorithm for local features of fruits, stems, leaves, and suckers. The accuracy for local features of fruits was 95.27% and the recall rate was 96.18%. The accuracy for local features of stems was 96.26% and the recall rate was 97.60%. The accuracy for local features of leaves was 95.57% and the recall rate was 93.42%. The accuracy for local features of suckers was 94.47% and the recall rate was 90.94%. F-measure results for local features of fruits, stems, leaves, and suckers were 95.72, 96.93, 94.48, and 92.67%, respectively.

Table 7. Results of the BP algorithm.

Parts		True Condition			
		Fruit	Node	Leaf	Sucker
Predicted condition	Fruit	6616	1	325	2
	Node	102	103,196	3758	142
	Leaf	154	2416	58,718	154
	Sucker	7	115	53	2991

The accuracy of the BP algorithm was 95.95%, which was higher than that of the SVM at 63.75%. Accuracy, recall, and F-measure were all higher with the BP algorithm than those with the SVM. When the BP algorithm was used to classify SURF local features of fruits, stems, leaves, and suckers, it showed a higher performance than the SVM.

3.1.5. Results of Local Feature Performance Experiment Using the Deep Neural Network

When the deep neural network was trained repeatedly for five generations using the focal loss method, the accuracy reached 99.9%. As a result of learning through the learning system described in this study, it took an average of 2000 s to repeat the first generation. Of a total of 400 images, 75% were used for training.

As a result of classifying through a deep neural network using 400 images (100 images for each class of fruit, stem, leaf, and sucker), the accuracy was 92.93%. Table 8 shows results of classification using a deep neural network for fruits, stems, leaves, and suckers.

The accuracy for fruits was 99.5% and the recall rate was 97.79%. The accuracy for stems was 87.75% and the recall rate was 87.31%. The accuracy for leaves was 90.50% and the recall rate was 99.45%. The accuracy for suckers was 87.25% and the recall rate was 87.69%. F-measure results for fruits, stems, leaves, and suckers were 98.63, 87.53, 94.76, and 87.47%, respectively. In the process of acquiring the 3D shape as a two-dimensional image, the shape of the sucker and the node was not clearly distinguished, so 49 out of 400 nodes were classified as suckers, and 51 out of 400 suckers were classified as nodes. In the case of leaves, 29 leaves in images were not recognized as leaves. The more samples used for training with the CNN algorithm, the better the performance. Therefore, in order to solve the problem of accuracy deterioration due to confusion between nodes and sucker and the problem of not recognizing leaves that occurred in this paper, it is necessary to acquire images of nodes, suckers, and leaves from various directions and angles and increase the number of samples used for training.

Table 8. Result of the CNN algorithm.

Parts		True Condition			
		Fruit	Node	Leaf	Sucker
Predicted condition	Fruit	398	0	2	0
	Node	0	351	0	49
	Leaf	9	0	362	0
	Sucker	0	51	0	349

3.2. Results of Comparing Classification Performances between BP and CNN for Sweet Pepper Parts

Table 9 shows results of comparing classification performances for sweet pepper parts. The classification performance of the BP algorithm was the result of summing proportions of each region lost in the image segmentation process. In the case of plant part classification using the BP algorithm, the recognition success rate was 94.44% for fruits, 84.73% for stems, 69.97% for leaves, and 84.34% for suckers. In the case of plant part classification using the deep neural network, the recognition success rate was 99.50% for fruits, 87.75% for stems, 90.50% for leaves, and 87.25% for suckers. In the case of plant part classification using the BP algorithm, the separation of fruit and background and separation of canopy and background were performed before fruit classification. In this process, 99.12% of fruits, 88.01% of stems, 73.22% of leaves, and 89.28% of suckers were preserved during image segmentation. In the image segmentation process, some regions were removed together with the background. For this reason, the performance of plant part classification using the BP algorithm might be lower than the performance of classifying local features of each part using the BP algorithm. Therefore, when the classification performance of the BP algorithm was compared to that of the deep neural network, the performance of the deep neural network was judged to be superior to that of the BP algorithm.

Table 9. Classification performance comparison between the BP algorithm and CNN.

Performance	Fruit (%)	Node (%)	Leaf (%)	Sucker (%)
BP Algorithm	94.44	84.73	69.97	84.34
CNN	99.5	87.75	90.50	87.25

4. Conclusions

As a pre-processing step for plant part classification, a method of separating stem and leaf regions from the background using the NDVI based on the difference between a 735 nm band image and a 660 nm band image was applied. In addition, the color of the region of interest separated from the background was classified using k-means clustering. Through this, it was possible to segment the image of the region of interest. To find local features in an image in which the interested region was segmented using k-means clustering, morphological skeletonization was performed for the image to find the skeleton of the region of interest. As a result of extracting local features using SIFT and SURF, it was confirmed that local feature extraction using SURF could lead to more local features. Therefore, the SURF method was selected to detect local features of fruits, stems, leaves, and suckers in this study.

To classify SURF local features of fruits, stems, leaves, and suckers, local features were classified using SVM and BP algorithms with the radial basis function kernel. Their performances were then compared. Results confirmed that the accuracy of the BP algorithm was 95.96%, which was higher than that of the SVM at 63.75%.

Fruits, stems, leaves, and suckers were classified using the CNN (Convolutional Neural Network), a kind of deep neural network. Its performance was then compared to that of the BP algorithm. As a result of comparing the performances for classifying each part of sweet pepper, it was confirmed that the performance of the CNN was superior to that of the BP algorithm. If further research, such as stereo vision, to calculate distance and the mechanism of a robot-arm to perform agricultural works is carried out, the proposed method in this study could be applied to an image processing algorithm for the multipurpose cultivation robot system.

Author Contributions: Conceptualization, B.L., D.K., Y.C., D.-C.K. and D.-H.L.; Methodology, B.L. and D.-H.L.; Software, B.L.; Validation, B.L., D.K. and Y.C.; Formal analysis, B.L.; Investigation, B.L., D.K. and Y.C.; Resources, B.L., D.K. and Y.C.; Data curation, Y.C.; Writing—original draft preparation, B.L., D.K. and Y.C.; Writing—review and editing, B.L., Y.C. and D.-H.L.; Visualization, Y.C.; Supervision, Y.C.; Project administration, Y.C.; Funding acquisition, Y.C. All authors contributed almost equally to all aspects of this research. All authors have read and agreed to the published version of the manuscript.

Funding: This work was supported by Korea Institute of Planning and Evaluation for Technology in Food, Agriculture and Forestry (IPET) through the Agriculture, Food and Rural Affairs Convergence Technologies Program for Educating Creative Global Leader Program funded by the Ministry of Agriculture, Food and Rural Affairs (MAFRA) (320018-2). It was also supported by a grant (NRF-2021R1G1A1012778) of the National Research Foundation of Korea (NRF) funded by the Korean government (MSIT, Ministry of Science and ICT). This research was supported by the "Research Base Construction Fund Support Program" funded by Jeonbuk National University in 2020.

Institutional Review Board Statement: Not applicable.

Informed Consent Statement: Not applicable.

Data Availability Statement: Not applicable.

Conflicts of Interest: The authors have no conflict of interest to disclose relevant to this study.

References

1. Han, S. A study on development of the Korea agricultural population forecasting model and long-term prediction. *J. Korea Acad. -Ind. Coop. Soc.* **2015**, *16*, 3797–3806. [CrossRef]
2. Gonzalez, R.; Woods, R. *Digital Image Processing*, 3rd ed.; Prentice Hall: Hoboken, NJ, USA, 2006.
3. Meyer, G.E. Machine Vision Identification of Plants. In *Recent Trends for Enhancing the Diversity and Quality of Soybean Products*; Krezhova, D., Ed.; InTech: Rijeka, Croatia, 2011; pp. 401–420. Available online: https://www.intechopen.com/chapters/22613 (accessed on 14 September 2021). [CrossRef]
4. Xia, C.; Lee, J.; Li, Y.; Song, Y.; Chung, B. Plant leaf detection using modified active shape models. *Biosyst. Eng.* **2013**, *116*, 23–35. [CrossRef]
5. Uchiyama, H.; Sakurai, S.; Mishima, M.; Arita, D.; Okayasu, T.; Shimada, A.; Taniguchi, R. An Easy-to-Setup 3D Phenotyping Platform for KOMATSUNA Dataset. In Proceedings of the IEEE International Conference on Computer Vision, Venice, Italy, 22–29 October 2017; pp. 2038–2045. [CrossRef]
6. Peter, W.; Zhang, S.; Chikkerur, S.; Little, S.; Wing, S.; Serre, T. Computer vision cracks the leaf code. *Proc. Natl. Acad. Sci. USA* **2016**, *113*, 3305–3310. [CrossRef]
7. Sruti, D.; Goswami, S.; Bashyam, S.; Awada, T.; Samal, A. Automated Stem Angle Determination for Temporal Plant Phenotyping Analysis. In Proceedings of the IEEE International Conference on Computer Vision, Venice, Italy, 22–29 October 2017; pp. 2022–2029. [CrossRef]

8. Lee, W.; Slaughter, D. Recognition of partially occluded plant leaves using a modified watershed algorithm. *ASAE* **2004**, *47*, 1269–1280. [CrossRef]
9. Wang, X.; Huang, D.; Du, J.; Xu, H.; Heutte, L. Classification of plant leaf images with complicated background. *Appl. Math. Comput.* **2008**, *205*, 916–926. [CrossRef]
10. Tang, X.; Liu, M.; Zhao, H.; Tao, W. Leaf Extraction from Complicated Background. In Proceedings of the 2nd International Congress on Image and Signal Processing, Tianjin, China, 17–19 October 2009; pp. 1–5. [CrossRef]
11. João, C.; Meyer, G.; Jones, D. Individual leaf extractions from young canopy images using Gustafson-Kessel clustering and a genetic algorithm. *Comput. Electron. Agric.* **2006**, *51*, 66–85. [CrossRef]
12. Dey, D.; Mummert, L.; Sukthankar, R. Classification of Plant Structures from Uncalibrated Image Sequences. In Proceedings of the IEEE Workshop on the Applications of Computer Vision, Breckenridge, CO, USA, 9–11 January 2012; pp. 329–336. [CrossRef]
13. Dworak, V.; Selbeck, J.; Dammer, K.; Hoffmann, M.; Zarezadeh, A.; Bobda, C. Strategy for the development of a smart NDVI camera system for outdoor plant detection and agricultural embedded systems. *Sensors* **2013**, *13*, 1523–1538. [CrossRef]
14. Silva, G.; Goncalves, A.; Silva, C.; Nanni, M.; Facco, C.; Cezar, E.; Silva, A. NDVI response to water stress in different phenological stages in culture bean. *J. Agron.* **2016**, *15*, 1–10. [CrossRef]
15. Robby, T. Visibility in Bad Weather from a Single Image. In Proceedings of the IEEE Conference on Computer Vision and Pattern Recognition, Anchorage, AK, USA, 23–28 June 2008; pp. 1–8. [CrossRef]
16. Friedman, N.; Geiger, D.; Goldszmidt, M. Bayesian network classifiers. *Mach. Learn.* **1997**, *29*, 131–163. [CrossRef]
17. Steinbach, M.; George, K.; Vipin, K. A Comparison of Document Clustering Techniques. Available online: https://conservancy.umn.edu/handle/11299/215421 (accessed on 14 September 2021).
18. Deng, J.; Dong, W.; Socher, R.; Li, L.; Li, K.; Li, F. ImageNet: A Large-Scale Hierarchical Image Database. In Proceedings of the IEEE Conference on Computer Vision and Pattern Recognition, Miami, FL, USA, 20–25 June 2009; pp. 248–255. [CrossRef]
19. Krizhevsky, A.; Sutskever, I.; Hinton, G. ImageNet classification with deep convolutional neural networks. *Adv. Neural Inf. Process. Syst.* **2012**, *25*, 1097–1105. [CrossRef]
20. Hu, J.; Shen, L.; Sun, G. Squeeze-and-Excitation Networks. In Proceedings of the IEEE Conference on Computer Vision and Pattern Recognition, Salt Lake City, UT, USA, 18–23 June 2018; pp. 7132–7141.
21. Hopfield, J. Neural networks and physical systems with emergent collective computational abilities. *Proc. Natl. Acad. Sci. USA* **1982**, *79*, 2554–2558. [CrossRef]
22. Collobert, R.; Weston, J. A Unified Architecture for Natural Language Processing: Deep Neural Networks with Multitask Learning. In Proceedings of the 25th International Conference on Machine learning, Helsinki, Finland, 5–9 July 2008; pp. 160–167. [CrossRef]
23. Pound, M.; Atkinson, J.; Wells, D.; Pridmore, T.; French, A. Deep Learning for Multi-Task Plant Phenotyping. In Proceedings of the IEEE International Conference on Computer Vision, Venice, Italy, 22–29 October 2017; pp. 2055–2063. [CrossRef]
24. Barnea, E.; Mairon, R.; Ben-Shahar, O. Colour-agnostic shape-based 3D fruit detection for crop harvesting robots. *Biosyst. Eng.* **2016**, *146*, 57–70. [CrossRef]
25. Kitamura, S.; Oka, K. A recognition method for sweet pepper fruits using led light reflections. *SICE J. Control. Meas. Syst. Integr.* **2009**, *2*, 255–260. [CrossRef]
26. Sa, I.; Lehnert, C.; English, A.; Mccool, C.; Dayoub, F.; Upcroft, B.; Perez, T. Peduncle detection of sweet pepper for autonomous crop harvesting-combined color and 3-d information. *IEEE Robot. Autom. Lett.* **2017**, *2*, 765–772. [CrossRef]
27. Ha, J. Demand Survey of Agricultural Products Export Information and Prioritization Study on Smart Farms. Ph.D. Thesis, Chonnam National University, Gwangju, Korea, 2017.
28. RDA (Rural Development Administration). Press Releases. Available online: https://www.nihhs.go.kr/pdf/farmerUseProgram/smart_greenhouse_guideline.PDF (accessed on 14 September 2021).
29. Ryu, H.; Yu, I.; Cho, M.; Um, Y. Structural reinforcement methods and structural safety analysis for the elevated eaves height 1-2w type plastic green house. *J. Bio-Environ. Control.* **2009**, *18*, 192–199.
30. Lee, J.; Kim, H.; Jeong, P.; Ku, Y.; Bae, J. Effects of supplemental lighting of high pressure sodium and lighting emitting plasma on growth and productivity of paprika during low radiation period of winter season. *Hortic. Sci. Technol.* **2014**, *32*, 346–352. [CrossRef]
31. Kang, S.; Hong, H. A study on comparision of the quantity of phoria as way to separation of binocular fusion. *J. Korean Ophthalmic Opt. Soc.* **2014**, *19*, 331–337. [CrossRef]
32. Um, Y.; Choi, C.; Seo, T.; Lee, J.; Jang, Y.; Lee, S.; Oh, S.; Lee, H. Comparison of growth characteristics and yield by sweet pepper varieties at glass greenhouse in reclaimed land and farms. *J. Agric. Life Sci.* **2013**, *47*, 33–41. [CrossRef]
33. Wang, Q.; Yan, Y.; Zeng, Y.; Jiang, Y. Experimental and numerical study on growth of high-quality ZnO single-crystal microtubes by optical vapor supersaturated precipitation method. *J. Cryst. Growth* **2016**, *468*, 638–644. [CrossRef]
34. Story, D. Design and implementation of a computer vision-guided greenhouse crop diagnostics system. *Mach. Vis. Appl.* **2015**, *26*, 495–506. [CrossRef]
35. Yu, Y. A Study on Index of Greenness Measurement Using Near-Infrared Photograph and Vegetation Index. Ph.D. Thesis, Seoul National University, Seoul, Korea, 2017.
36. Rouse, J.W.; Haas, R.H.; Schell, J.A.; Deering, D.W. Monitoring vegetation systems in the great plains with ERTS. *NASA Spec. Publ.* **1974**, *351*, 309–317.

37. Lloyd, S. Least squares quantization in PCM. *IEEE Trans. Inf. Theory* **1982**, *28*, 129–137. [CrossRef]
38. Chatbri, H.; Kaymeyama, K.; Kwan, P. A comparative study using contours and skeletons as shape representations for binary image matching. *Pattern Recognit. Lett.* **2015**, *76*, 59–66. [CrossRef]
39. Lowe, D.G. Distinctive image features from scale-invariant keypoints. *Int. J. Comput. Vis.* **2004**, *60*, 91–110. [CrossRef]
40. Bay, H.; Ess, A.; Tuytelaars, T.; Gool, L. Speeded-up robust features. *Comput. Vis. Image Underst.* **2008**, *110*, 346–359. [CrossRef]
41. Vapnik, V. *The Nature of Statistical Learning Theory*, 2nd ed.; Springer-Verlag New York: New York, NY, USA, 2000. [CrossRef]
42. Werbos, P. Beyond Regression: New Tools for Prediction and Analysis in the Behavioral Sciences. Ph.D. Thesis, Harvard University, Cambridge, MA, USA, 1974.
43. Lin, T.; Goyal, P.; Girshick, R.; He, K.; Dollár, P. Focal Loss for Dense Object Detection. In Proceedings of the IEEE International Conference on Computer Vision, Venice, Italy, 22–29 October 2017; pp. 2980–2988. [CrossRef]

applied sciences

MDPI

Article

CFD Analysis and Validation of a Foreign Material Winnowing Machine for Pepper Harvester

Seo-Yong Shin [1], Myoung-Ho Kim [1,2], Yongjin Cho [1,2,*] and Dae-Cheol Kim [1,2,*]

[1] Department of Bioindustrial Machinery Engineering, Jeonbuk National University, Jeonju 54896, Korea; ssy9970@naver.com (S.-Y.S.); myoung59@jbnu.ac.kr (M.-H.K.)
[2] Institute for Agricultural Machinery & ICT Convergence, Jeonbuk National University, Jeonju 54896, Korea
* Correspondence: choyj@jbnu.ac.kr (Y.C.); dckim12@jbnu.ac.kr (D.-C.K.)

Abstract: The winnowing machine of chili pepper harvesters was developed to reduce the potential problem of low pepper stem and fruit separation. The developed winnowing machine was combined with two impellers and a center bearing to prevent a strain on the drive shaft and to ensure durability. The terminal velocity of chili pepper was measured, and an aerodynamic analysis was conducted based on this winnowing machine. A CFD (Computational Fluid Dynamics, Ansys Fluent 2020 R1) analysis was conducted for three levels of discharge port guide form (0, 3, and 5 guides) and three levels of rotating speed (1600, 1800, and 2000 RPM) of a winnowing machine designed utilizing aerodynamic analysis results. A validation test was conducted by fabricating a winnower test device. As for aerodynamic analysis conducted using measured values of terminal velocity, chili pepper fruits were collected at an outlet wind speed lower than 17.5 m/s and chili pepper branches were separated at a speed higher than 12.5 m/s. As a result of CFD analysis, the wind speed deviation at outlets of the 0-, 3-, and 5-guide depending on the rotating speed appeared to be 15.8, 1.4, and 1.0 m/s on average, respectively. The result of the CFD analysis showed values higher than wind speeds of the actual winnower test device by a minimum of 0 and a maximum of 2.4 m/s. Through the CFD analysis and the wind speed validation test of the winnower test device, optimal conditions to separate foreign materials were found to be a winnowing machine at a rotating speed of 1800 RPM with a discharge port having three guides or a winnowing machine at a rotating speed of 2000 RPM with a discharge port having five guides.

Keywords: CFD (Computational Fluid Dynamics); terminal velocity; winnowing machine separation; chili pepper harvester

Citation: Shin, S.-Y.; Kim, M.-H.; Cho, Y.; Kim, D.-C. CFD Analysis and Validation of a Foreign Material Winnowing Machine for Pepper Harvester. *Appl. Sci.* **2022**, *12*, 6134. https://doi.org/10.3390/app12126134

Academic Editors: Paweł Kiełbasa, Tadeusz Juliszewski and Sławomir Kurpaska

Received: 19 May 2022
Accepted: 14 June 2022
Published: 16 June 2022

1. Introduction

As a vegetable crop, chili pepper has an output ranked the seventh largest among vegetables globally. It contains capsaicin with a spicy taste. It is widely used as an industrial material and a seasoning ingredient for various foods. Major chili pepper-producing countries are China, India, Thailand, Bangladesh, Myanmar, Vietnam, USA, Peru, and Korea, most of which are concentrated in Southeast Asia [1]. Chili pepper is the second most important crop for farm household income after rice in Korea. It is an important seasoning vegetable that accounted for 33% of condiment vegetables in 2020. However, its cultivation area and output are decreasing due to the aging of chili pepper growing farms, low mechanization rate, and labor cost burden with an increase in the production cost. It has been reported that chili pepper cultivation area, output, and production cost per 10a of Korea have been changed from 74,471 ha, 193,786 tons, and 1,126,000 KRW in 2000 to 31,146 ha, 60,076 tons, and 3,707,000 KRW in 2020, respectively. Additionally, the report showed decreases of 41.8%p and 31.0%p in cultivation area and output, and an increase of 329.2%p in production cost [2,3].

The chili pepper harvesters have been researched and developed in many countries to increase chili pepper cultivation area and output [4–6]. As additional human resources are put in to separate foreign materials because existing chili pepper harvesters have a high rate of foreign materials in the collection tank after harvest, the foreign material separation system needs to be improved [7–10]. The separation of small chili pepper branches, leaves, soil, and so on that are moved to the collection section during mechanical harvesting of chili pepper occupies an important part of low transportation cost, processing cost, and simplification of the post-treatment process [11–13]. As a result of developing a small size chili pepper harvester and conducting a performance comparison test, Kim et al. [14] have reported foreign material adulteration rates of 17.6 to 31.0% on average and presented the necessity for a foreign material separation device to reduce additional separation workforce after harvest. Studies on the separation of foreign materials use the method where foreign materials are separated by square plates installed at a constant interval on a rotating shaft and the winnower method where foreign materials are separated using the air based on the difference in terminal velocity crop. In a study on rotating square plates, Lenker et al. [15] have removed 65% of foreign materials by putting harvested chili pepper on a wire belt with a number of 3.1-cm high fingers installed at an interval of 2.5 cm and making small chili pepper branches and leaves fall through the space between the fingers while the belt vibrates and rotates. Wolf et al. [16] have removed 42% of foreign materials due to fabricating three star-shaped star wheels that rotate in a single direction and one square rubber plate and using them to separate seeds, small branches, and leaves. Eaton et al. [17] have achieved chili pepper separation rates of 90% to 95% and a foreign material separation rate of 93.5% under conditions of a rotating card cleaner size of 6 inches, an interval of 22.2 mm, an inclination angle of 20°, and rotating speeds of 50 to 100 RPM after mechanical harvesting of a chili pepper species cultivated in New Mexico State. Kong et al. [18] have conducted a test after designing and fabricating a separation system using star wheels and air current based on morphological features and aerodynamic characteristics of a chili pepper species cultivated in Xinjiang, China. When the interval of star wheel plates was 108 to 200 mm and the distance between shafts was 180 to 200 mm, it was found that 29.5% of branches and 75.0% of foreign materials were separated. Jo et al. [19] designed and fabricated a card cleaner device and conducted a performance test. As a result of the test, the foreign material separation rate was 53.5% under a straight-line card arrangement, an inclination angle of 15°, and a rotation speed of 50 RPM. Statistical analysis showed that the foreign material separation rate was significantly related to the inclination angle and the rotating speed. Byun et al. [20] have established optional conditions of a card cleaner device by grasping its mechanical and physical properties using the discrete element method (EDEM) and found that conditions are similar to those of the factorial performance experiment of Jo et al. [19].

In a study on the aerodynamic winnowing method, Bilanski et al. [21] have experimentally obtained and induced terminal velocity of freely falling grains using a time-displacement technique to measure terminal velocities of seven kinds of grains and seeds such as wheat about that of a plastic sphere. In an aerodynamic study on the air stream separation of grains, Lee et al. [22,23] analyzed the effect of rice and barley sizes and moisture content on the measurement of terminal velocity. Their measurement results showed that the terminal velocity of rice was 4.6 to 4.8 m/s and that of barley was 4.9 to 5.5 m/s. Chung et al. [24] have experimentally investigated aerodynamic characteristics of the separation section in a combine. As a result of making measurements at a winnower rotating speed of 1150 RPM, it was found that air separation of wheat, branch straw, and leave straw was possible at 4.1 m/s, 2.4 m/s, and 1.3 m/s, respectively. To confirm the physical, mechanical, and aerodynamic properties required for separation after the mechanical harvesting of peanuts, Kim et al. [25] have compared the results of a modeling calculation and an actual test. Their test results showed that the breaking power of peanut hulls was irrelevant to their shape, size, or moisture content, with a drag coefficient range of 0.67 to 0.93. Choi [26] has conducted a test after fabricating a device to separate chili pepper using a winnowing fan and a vibrating separation sieve. As a result of the separation, the chili pepper separation rate was 68.2% when the wind speed of the winnowing machine was 6.2 m/s, the frequency of the separating sieve was 6.9 Hz, the amplitude was 40 mm, and the thickness and interval of the separation sieve rods were 6 mm and 29 mm, respectively. As a result of conducting a test of removing foreign materials from chili peppers by organizing a fan absorption winnowing device, a discharge conveyor, and a foreign material collecting bin, Hong et al. [27] have found that chili pepper leaves are absorbed in a winnowing machine wind speed of 12 to 15 m/s and the foreign material removal rate is 95%. Noh et al. [28] have evaluated the separation rate and loss rate at the foreign material separation winnowing machine's rotation speed of 180 to 300 RPM to develop a sesame harvester. As a result of the test, separation rates of sesame were 99.7% and 99.8% at 220 and 240 RPM, respectively. These were reported as the optimal RPMs of the winnowing machine with the least loss.

Lee et al. [29] have analyzed the fluid flow of a peanut harvester's winnowing machine by conducting a CFD analysis. They found that the fluid flow was adequate when the direction was counterclockwise and the rotating speed and the flow field velocity were 1500 RPM and 8 to 17 m/s, respectively. The separation system of the peanut harvester was investigated in a factor experiment using the analyzed result [30]. Yuan et al. [31] have analyzed the separation performance using the CFD-DEM (Computational Fluid Dynamics—Discrete Element Method model) to improve the separation performance of a combine and save the cost. Separation and performance efficiencies were 98.9% and 87.7%, respectively, at a wind speed of 11.0 m/s. The error in the separation rate was 5% or less. Those in the loss rate and separation efficiency were 9.8% and 2.6%, respectively, through the analysis result and validation test. Shin et al. [32] have designed and fabricated a winnower test device for a chili pepper harvester and conducted a performance test. The conditions that achieved the least loss of chili pepper and the highest separation rate of foreign materials were a winnower inclination angle of 34° and a rotating speed of 2000 RPM. Winnowing machines presented in the existing literature showed the following trend: the higher the rotating speed of the winnowing machine, the larger the loss of chili pepper fruit and the higher the foreign material separation rate. It is required to find a way to minimize the loss of chili pepper fruits and secure adequate foreign material separation at the same time. The optimization and application of CFD solution were studied to evaluate fluid flow around a target object at a low Reynolds number and numerical study on the drag coefficient. Additionally, the CFD simulation was used to evaluate the flow patterns and indicate the important numerical solutions. To optimize parameters of winnowing equipment for machine-harvested *Lycium barbarum* L., a discrete element method (DEM) simulation of ripe fruit in the airflow was performed using EDEM software [33]. Aerodynamic analysis and CFD simulation were conducted to evaluate the model on temperature and relative humidity or predict the terminal velocity of agricultural granular materials or perform the wind tunnel in the agriculture field [34–36].

Thus, the objective of the present study was to design a foreign material separation winnowing machine for specializing work in chili pepper harvesters through an aerodynamic analysis and validate the same using CFD (Computational Fluid Dynamics, Ansys Fluent 2020 R1) software. Detailed objectives are below.

1. The first objective is to measure the terminal velocities of chili pepper species and conduct an aerodynamic analysis. The terminal velocities of chili pepper fruit, chili pepper stem, and chili pepper leaf move distance by wind are each derived through aerodynamic analysis.
2. The second objective is to validate the winnowing machine through a CFD analysis and test device. The wind speed of CFD flow analysis is verified as a flow test of separation equipment. In this result, the optimal rotation speed and discharge out guides for foreign material separation are decided, which will be applied to the chili pepper harvester.

2. Materials and Methods

2.1. Aerodynamic Properties of Mechanically Harvested Chili Pepper

In the present study, species 'Jeokyoung' and 'AR Legend' were selected. Terminal velocities of their chili pepper fruits, chili pepper branches, and chili pepper leaves were measured. 'Jeokyoung' and 'AR Legend' are species with large fruits developed by the National Institute of Horticultural and Herbal Science (NIHHS). They have high chromaticity, lots of chili pepper fruits, and strong resistance against diseases and pests such as anthrax bacterium. These chili peppers were transplanted from an NIHHS experimental field on 8 May 2019. Measurements were made on 18 August 2019 (after 103 days) during the harvesting period.

2.2. Terminal Velocity Measurement

The terminal velocity of the chili peppers was measured by controlling the wind speed. Chili pepper fruits and chili pepper branches were put in a floating state at the height of 10 to 20 cm using an upward air current with a uniform velocity [37,38]. Figure 1a shows the look of the terminal velocity device used for the test. Figure 1b shows the wind speed meter used. As for the measurement method, velocity values were measured by inserting the hot wire probe of a wind speed meter into the hole of the transparent tube with the chili pepper fruits and chili pepper branches in a floating state [19,20]. Each chili pepper fruit and chili pepper branch were collected 10 times, and the average value and standard deviation were calculated by three repetitions of the test. Detailed specifications of the wind speed meter used for the test are shown in Table 1. Table 2 shows the results of measuring terminal velocities of chili pepper fruits, chili pepper branches, and chili pepper leaves.

(a) (b)

Figure 1. Terminal velocity experiment bench (**a**) and wind velocity measuring instrument (**b**).

Table 1. Specifications of wind velocity measuring equipment.

Item	Specification
Model number	TESTO 440
Company/Nation	TESTO/GERMANY
Measuring range	0–30 m/s
Resolution	0.1 m/s
Accuracy	0.04%

Table 2. Experiment results of pepper terminal velocity.

Pepper Part	Variety	V_t [1] (m/s) AVE [2]	SD [3]
Fruit	Jeokyoung	13.7	1.1
	AR legend	13.5	0.4
Branch	Jeokyoung	11.2	0.5
	AR legend	11.5	1.6
Leaf	Jeokyoung	4.3	0.5
	AR legend	4.2	0.2

[1] V_t: terminal velocity; [2] AVE: average; [3] SD: standard deviation.

2.3. Aerodynamic Analysis of Winnowing

The chili pepper harvester developed by Shin et al. [10,32] was arranged to separate foreign materials such as chili pepper branches and chili pepper leaves and collect only the harvested chili pepper fruits in a collection tank. Figure 2 shows the construction of the foreign material winnowing system developed by Shin et al., in addition to the chili pepper harvester [10,32] they developed. It is organized to transfer harvested chili pepper fruits and foreign materials such as chili pepper branches and leaves to a conveyor belt and separate foreign materials outside of the collection tank by the wind of the winnowing machine when they fall into the collection tank. Terminal velocities of chili pepper fruit, chili pepper branch, and chili pepper leaf were set to be 13.5, 11.5, and 4.2 m/s, respectively, using the terminal velocity measurement result. Chili pepper leaf was excluded from the aerodynamic analysis considering it could be separated at a wind speed lower than that of a chili pepper branch. In the aerodynamic analysis of winnowing, travel distances of chili pepper fruits and chili pepper branches with time were derived from to the wind speed discharged at an inclination angle of 34°. At this time, it was assumed that the wind speed was constant only in one direction. If a wind speed is applied to chili pepper fruits and chili pepper branches when they fall freely into the collection tank from a height of 400 mm, forces of drag and gravity are generated in the direction of F_{Dx}, F_{Dy}, mg (Figure 3). F_{Dx} is the drag force moving in the x-axis direction and F_{Dy} is the drag force moving in the y-axis direction. Conditions used to calculate of measured material properties and travel distances are shown in Table 3.

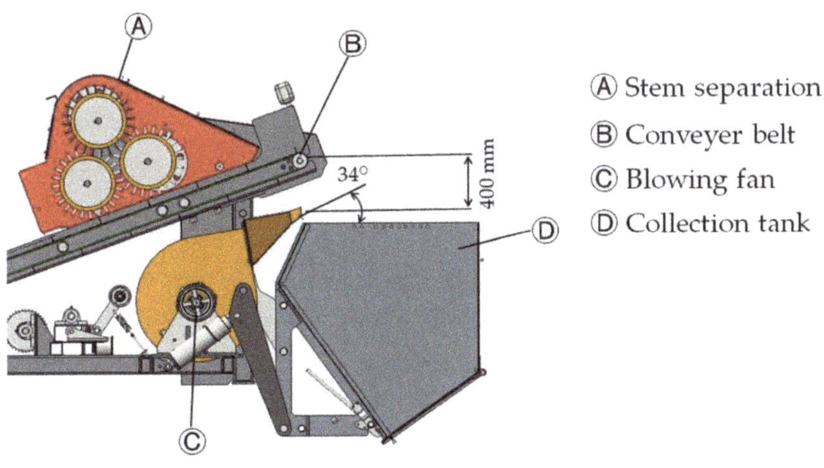

Figure 2. Components of a separation system for foreign materials.

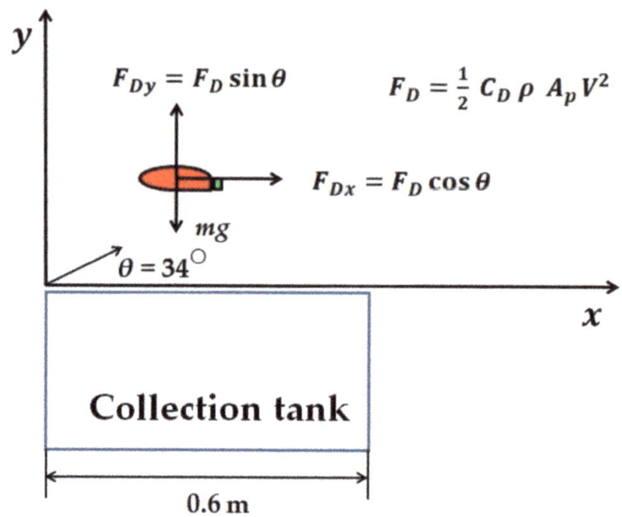

Figure 3. Force acting on the pepper fruit by winnowing machine velocity.

Table 3. Calculation formula parameter for pepper fruit and branch in the blowing fan.

Parameter	Pepper Fruit	Branch	Unit
Length (L_1)	0.11	0.2	m
Diameter (L_2)	0.017	0.004	m
Weight (m)	0.012	0.004	kg
Terminal velocity (V_t)	13.5	11.5	m/s
Air density (ρ)	1.27	1.27	kg/m^3
Acceleration of gravity (g)	9.81	9.81	m/s^2
Velocity (V_{air})	10, 12.5, 15, 17.5	10, 12.5, 15, 17.5	m/s
Winnowing machine angle	34	34	°
x-axis start direction (S_{x0})	0	0	m
y-axis start direction (S_{y0})	0.4	0.4	m
Analysis time interval (t)	0.1	0.1	s

Chili pepper fruit and chili pepper branch are assumed to be aspherical ellipsoids. Therefore, the projected area is shown in Equation (1) [39]:

$$A_p = \frac{\pi\, L_1\, L_2}{4} \tag{1}$$

where A_p = projected area (m^2); L_1 = length of object (m); and L_2 = diameter (thickness) of object (m).

When the projected area and terminal velocity value are used, the drag coefficient of chili pepper fruit and chili pepper branch is shown in Equation (2) [39]:

$$C_D = \frac{2\, m\, g}{\rho A_p\, V_t^2} \tag{2}$$

where C_D = drag coefficient; ρ = air density (kg/m^3); V_t = terminal velocity of object (m/s); m = weight of object (kg); and g = acceleration of gravity (m/s^2).

The drag applied to chili pepper fruit and chili pepper branch in the x-axis direction, which is expressed using the projected area value as shown in Equation (3) [40]. The acceleration (a_x) applied in the x-axis direction is shown in Equation (4) [40]. The wind speed (V_{air}) is the relative velocity of the chili fruit and chili branch's speed (V_{xp}). When the initial time (t) is 0, (V_{xp}) becomes 0. The acceleration is assumed to be constant at the analysis time interval of 0.1 s.

$$F_{Dx} = m\, a_x \tag{3}$$

where F_{Dx} = drag force of x-axis direction (N); and a_x = acceleration of x-axis direction (m/s^2).

$$a_x = \frac{F_{Dx}}{m} = \frac{1}{2}\frac{C_D\, \rho\, A_p\, \left(V_{air}\, cos\, \theta - V_{xp}\right)^2}{m} \tag{4}$$

where V_{air} = wind velocity (m/s); and V_{xp} = x-axis direction velocity of object (m/s).

The drag applied to chili pepper fruit and chili pepper branch in the y-axis direction is affected by the gravitational acceleration, which is shown in Equation (5) [40]. The acceleration applied in the y-axis direction (a_y) is shown in Equation (6) [40].

$$F_{Dy} = -m\, a_y + m\, g \tag{5}$$

where F_{Dy} = drag force of y-axis direction (N); and a_y = acceleration of y-axis direction (m/s^2).

$$a_y = g - \frac{F_{Dy}}{m} = g - \frac{1}{2}\frac{C_D\, \rho\, A_p\, \left(V_{air}\, sin\, \theta - V_{yp}\right)^2}{m} \tag{6}$$

where V_{yp} = y-axis direction velocity of object (m/s).

Velocities of chili pepper fruit and chili pepper branch in the x-axis direction, V_x, and in the y-axis direction, V_y, are expressed using acceleration values (a_x) and (a_y) and calculated at the analysis time (t) of 0.1 s as shown in Equation (7) [40]. Based on the analysis time (t), the initial velocity (V_{n-1}) of V_x or V_y is the velocity value previously calculated. It becomes the initial value.

$$V_n = V_{n-1} + a_n\, t \tag{7}$$

where V_n = velocity of an object (V_x or V_y) (m/s); V_{n-1} = previous value for velocity of object (m/s); a_n = acceleration (a_x or a_y) (m/s^2); and t = time interval of analysis (s).

The travel distance in the x-axis direction, S_x, expressed using the acceleration in the x-axis direction and the y-axis direction (a_x, a_y) and velocity (V_x, V_y) values calculated at each time (t) are shown in Equation (8) [40]. The travel distance in the y-axis direction, S_y, is shown in Equation (9) [40]. The initial travel distance S_{x0} is 0 m. S_{y0} is fixed to the fall height of 0.4 m. As to the analysis time (t), the calculation is made at an interval of 0.1 s for up to 5 s.

$$S_x = S_{x0} + V_x\, t + \frac{1}{2}\, a_x\, t^2 \tag{8}$$

where S_x = travel distance of x-axis direction (m) and S_{x0} = initial travel distance of x-axis direction (0 m).

$$S_y = S_{y0} + V_y\, t + \frac{1}{2}\, a_y\, t^2 \tag{9}$$

where S_y = travel distance of y-axis direction (m) and S_{y0} = initial travel distance of y-axis direction (0.4 m).

2.4. Flow Analysis of Winnowing Machine

In this chapter, a flow analysis is conducted based on the rotating speed of the winnower and the shape of the discharge port guide designed using the Ansys FLUENT 2020R1 program, a commercial CFD (Computational Fluid Dynamics) code. The winnower comprised of two impellers was designed in Figure 4 to select an optimal flow field based on the discharge port guide level 3. Detailed data are shown in Table 4.

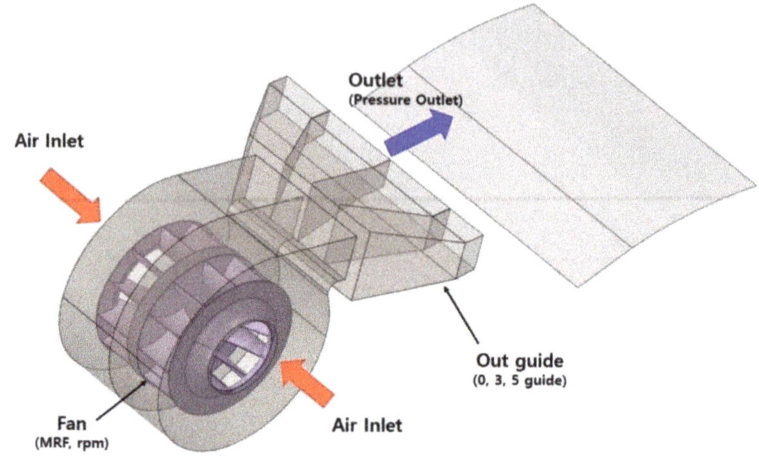

Figure 4. Schematic diagram showing boundary conditions and internal flow field of blowing fan for analysis model.

Table 4. Specifications of the blowing fan model.

Design Specifications	Value
Model Size: length, width, high (mm)	631 × 510 × 602
Out guide type	0, 3, 5
Impeller	Centrifugal type (2ea)
Impeller: diameter × width × wing number (mm, ea)	330 × 137 × 12

2.5. CFD Modeling and Simulation Setup

In the flow analysis, a coupled-type algorithm with excellent convergence was used. The MRF (Multi Reference Frame) method was used to simulate the rotation of a fan. The CFD numerical analysis procedure was performed in the order of geometry modeling, mesh, physics, solver settings, computing solution, and post-processing. Modification of the model was considered when reviewing the analysis results [41]. Detailed boundary conditions for numerical analysis are shown in Table 5. Depending on the winnower fan rotating speed (1600, 1800, 2000 rpm) and the outlet guide shape (0-, 3-, 5-guide), analysis conditions for the winnower rotating speed and the discharge port guide were set, referring to the reported factorial experiment of the foreign material separation [32,42]. Internal parallel processing with six cores is used to solve the simulation. Regarding the grid system used for the flow analysis, a tetrahedron grid system was applied. It was comprised of a total of 1,259,970 nodes and 5,004,653 grids. The maximum element size for the rotating area is set to 0.002 m. The length of the element at the interface between the fixed and rotating area is set to 0.005–0.01 m. For grid generation, the surface mesh used a triangle, and the volumetric mesh used a tetrahedron and prisms. The satisfy condition with orthogonal quality was 0.1 to 0.2, and the validation with a skewness value was between 0.8 and 0.9. The solver setting was confirmed as the stable state in the conditions of time step 0.5 s and residual 10^{-3} less [43]. The housing of the winnower to which the grid system was applied is shown in Figure 5a. The impeller to which a dense grid system was applied is shown in Figure 5b.

Table 5. Numerical analysis method for boundary conditions.

Parameter	Value
CFD method	Moving reference frame
Fluid	Air
Inlet, outlet	Static pressure
Turbulence model	Standard k-epsilon/Realizable/Standard Wall Functions
Wall	No Slip condition
Mesh Nodes/Elements	1,259,970/5,004,653
Mesh minimum orthogonal quality	0.13
Mesh maximum quality	0.87817363
Time step	0.5 s
Iterations	Hybrid, 500

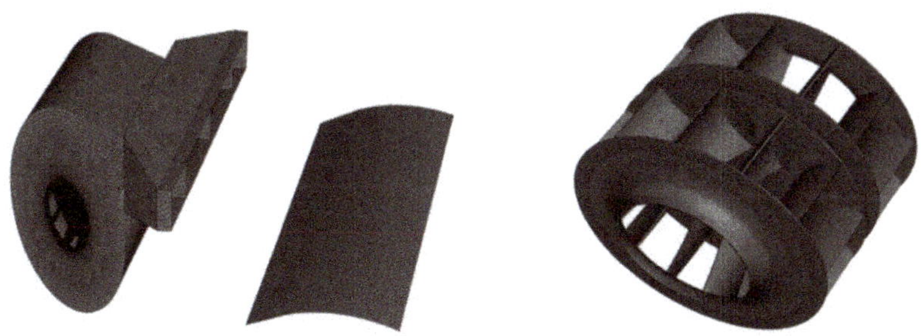

(**a**) (**b**)

Figure 5. Mesh composition of (**a**) a blowing fan and (**b**) the main impeller.

2.6. Fabrication and Evaluation of Winnowing Machine

The test device by discharge port guide that affects the performance of the winnowing system is shown in Figure 6a,b. Depending on the rotating speed of the winnower and the discharge port guide shape, measurements were made at each position using a wind speed meter (TESTO-440, Lenzkirch, Germany).

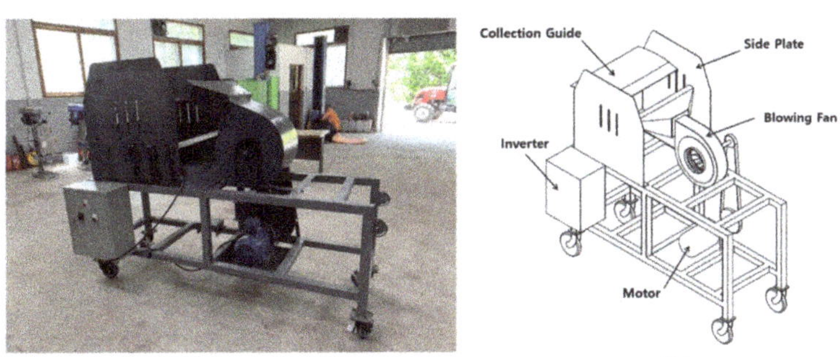

(**a**) (**b**)

Figure 6. Picture (**a**) and schematic view (**b**) of the experiment bench for a separation equipment.

3. Results and Discussion

3.1. Result of Aerodynamic Analysis

Results of calculating the travel distances of chili pepper fruit and chili pepper branch depending on the wind speed at the winnower discharge port are shown in Figures 7 and 8, respectively. Chili pepper fruits are shown in Figure 7a; those which fell freely were found to have been collected in the collection tank at wind speeds of 10.0, 12.5, and 15.0 m/s. At a wind speed of 17.5 m/s, chili pepper fruits were found to have flown outside the collection tank, causing a loss to occur. As for the loss of the chili pepper fruit, it was found that the moving distance of the pepper fruit was over 0.6 m at a wind speed of 17.5 m/s. Chili pepper branches, as shown in Figure 7b, were found to have been separated, flying out of the collection tank at wind speeds of 15.0 and 17.5 m/s. For the separated chili pepper branches, it was found that the moving distance of pepper branch was over 0.6 m at a wind speed of 15 and 7.5 m/s. As a result of conducting an aerodynamic analysis, it was determined that chili pepper fruits were collected in the collection tank at a wind speed lower than 17.5 m/s. Furthermore, the chili pepper branches were separated to be outside of the collection tank at a wind speed higher than 12.5 m/s.

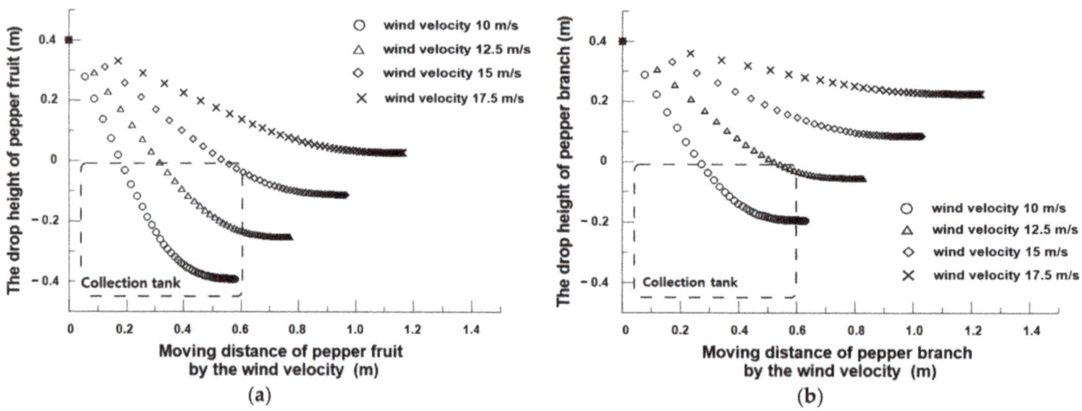

Figure 7. Graph of calculated moving distance for the pepper fruit (**a**) and the pepper branch (**b**) by velocity equation.

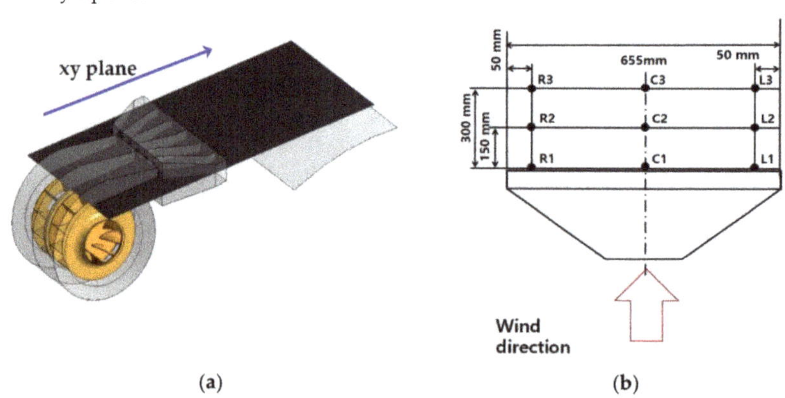

Figure 8. The xy plane on 3D model (**a**) and measuring position of blower on top view (**b**).

3.2. Flow Analysis Results

Wind speed measurement positions depending on the shape of the discharge port guide are shown in Figure 8a,b. Drift velocity distributions of the upper xy plane for the discharge port with 0-, 3-, and 5-guide at a winnower rotating speed of 1600 RPM are shown in Figure 9a–c, respectively. As a result of flow field velocity distribution, the 0-guide showed an average of 19.9 m/s at C1 and 4.3 m/s at L1 and R1. The 3-guide showed an average of 17.7 m/s at C1 and 15.9 m/s at L1 and R1. The 5-guide showed an average of 15.5 m/s at C1 and 14.5 m/s at L1 and R1. Flow velocity deviations at the left and right points, L1 and R1, from the center, C1, of the discharge port were shown to be 15.6, 1.8, and 1.0 m/s in the case of the 0-, 3-, and 5-guide, respectively. The velocity deviation of the 5-guide was found to be the least.

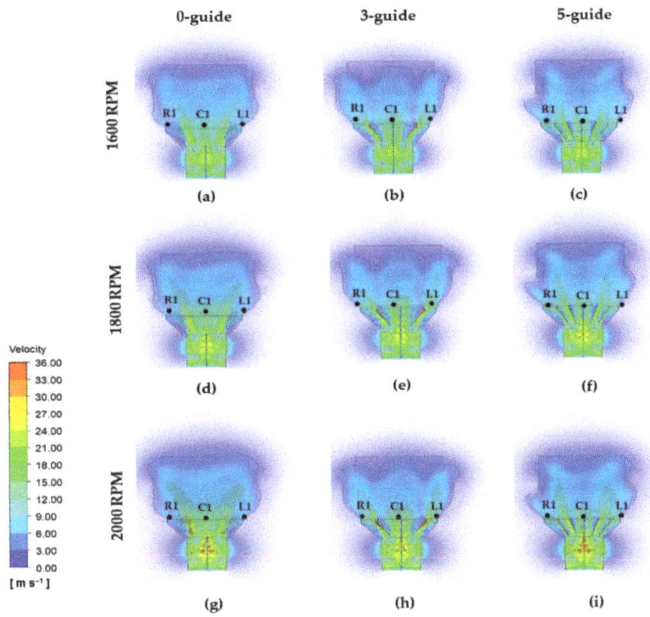

Figure 9. Shape analysis of flow field for foreign materials separating part of xy plane on conditions of 0-guide (**a**), 3-guide (**b**), 5-guide (**c**) at 1600 RPM, 0-guide (**d**), 3-guide (**e**), 5-guide (**f**) at 1800 RPM, and 0-guide (**g**), 3-guide (**h**), 5-guide (**i**) at 2000 RPM.

Figure 9d–f show the flow velocity distributions for the discharge port with 0-, 3-, and 5-guide at a winnower rotating speed of 1800 RPM, respectively. As a result of the flow field velocity distribution, the 0-guide showed an average of 22.6 m/s at C1 and an average of 5.0 m/s at L1 and R1. The 3-guide showed an average of 20.6 m/s at C1 and 23.2 m/s at L1 and R1. The 5-guide showed an average of 19.4 m/s at C1 and 16.7 m/s at L1 and R1. The flow velocity deviations at the left and right points, L1 and R1, from the center, C1, of the discharge port were shown to be 17.7, 2.6, and 2.7 m/s in the case of 0-, 3-, and 5-guide, respectively. The velocity deviation of the 3-guide was found to be the least.

Figure 9g–i show velocity distributions for discharge port with 0-, 3-, and 5-guide at a winnower rotating speed of 2000 RPM, respectively. As a result of the flow field velocity distribution, the 0-guide showed an average of 24.1 m/s at C1 and 5.8 m/s at L1 and R1. The 3-guide showed an average of 22.9 m/s at C1 and 25.6 m/s at L1 and R1. The 5-guide showed an average of 18.6 m/s at C1 and 19.0 m/s at L1 and R1. The flow velocity deviation at the left and right points, L1 and R1, from the center, C1, of the discharge port was shown to be 18.3, 2.7, 0.4 m/s in the case of 0-, 3-, and 5-guide, respectively. The velocity deviation of the 5-guide was found to be the least.

3.3. Results of Comparing Winnower Speed Measurement and Flow Analysis

Results of measuring wind speeds in the winnower test device and wind speeds determined by CFD analysis are shown in Table 6. Wind speeds measured in the winnower test device were lower than those determined by CFD analysis at 9 points by 0 to 2.4 m/s. Figures 10 and 11 show wind speed for actual measurement results of the winnower test device and the CFD analysis to match the top view of Figure 8b. The Kriging method was used. It could predict the attribute value at each point through the linear summation of each measured wind speed based on Table 6 [44]. The minimum unit of the grid was set to be 5.55×5.55 mm and the total number of nodes was set to be 5500. The 0-guide in Figures 10a and 11a showed that wind speed results of the winnower test device and the CFD analysis were high at points C1, C2, and C3 of the discharge port. Wind speeds of the winnower test device were lower than those of the CFD analysis by 0.1 to 0.7 m/s. As for the 3-guide shown in Figures 10b and 11b, wind speed results of the winnower test device and the CFD analysis showed constant values at distances from 0 to 300 mm from the discharge port. Wind speeds of the winnower test device were lower than those of the CFD analysis at 9 points by 0.1 to 1.6 m/s. Although the 5-guide shown in Figures 10c and 11c had uniform wind speeds at 9 points about the discharge port, and values at L1, C1, and R1 were lower than those of the 3-guide by 1.2 m/s on average. Wind speeds of the winnower test device were lower than those of the CFD analysis at 9 points by 0.4 to 1.8 m/s. The 0-guide shown in Figures 10d and 11d revealed that wind speed results of the winnower test device and the CFD analysis at 1800 RPM were high at points C1, C2, and C3 of the discharge port. Wind speeds of the winnower test device were lower than those of the CFD analysis by 0.8 to 1.2 m/s. As for the 3-guide shown in Figures 10e and 11e, wind speed results of the winnower test device and the CFD analysis had uniform values. Wind speeds of the winnower test device were lower than those of the CFD analysis at 9 points by 0.1 to 2.1 m/s. Although the 5-guide shown in Figures 10f and 11f had uniform wind speeds at 9 points about the discharge port, and values at L1, C1, and R1 were lower than those of the 3-guide by 4.8 m/s on average. Wind speeds of the winnower test device were lower than those of the CFD analysis at 9 points by 0.5 to 2.4 m/s. The 0-guide in Figures 10g and 11g showed that wind speed results of the winnower test device and the CFD analysis at 2000 RPM were high at points C1, C2, and C3 of the discharge port. Wind speeds of the winnower test device were lower than those of the CFD analysis by 0.9 to 1.4 m/s. As for the 3-guide shown in Figures 10h and 11h, wind speed results of the winnower test device and the CFD analysis had uniform values. Wind speeds of the winnower test device were lower than those of the CFD analysis at 9 points by 0.2 to 2.1 m/s. However, the 5-guide shown in Figures 10i and 11i had uniform wind speeds at 9 points about the discharge port, and values at L1, C1, and R1 were lower than those of the 3-guide by 6.3 m/s on average. Wind speeds of the winnower test device were lower than those of the CFD analysis at 9 points by 0.2 to 2.2 m/s, although the 5-guide showed that mean wind speeds at discharge port points L1, C1, and R1 were lower than those of the 3-guide by 6.3 m/s, respectively; the wind speed deviation was reduced by 1.4 m/s. However, as the internal resistance of the winnowing machine generates noise and vibration, additional studies on the way to reduce noise and vibration are required to be conducted.

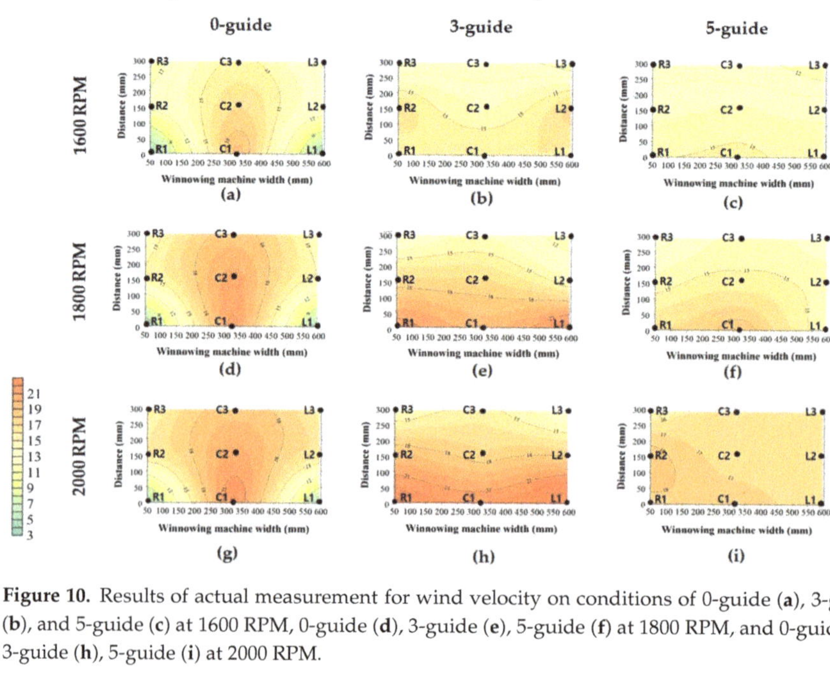

Figure 10. Results of actual measurement for wind velocity on conditions of 0-guide (**a**), 3-guide (**b**), and 5-guide (**c**) at 1600 RPM, 0-guide (**d**), 3-guide (**e**), 5-guide (**f**) at 1800 RPM, and 0-guide (**g**), 3-guide (**h**), 5-guide (**i**) at 2000 RPM.

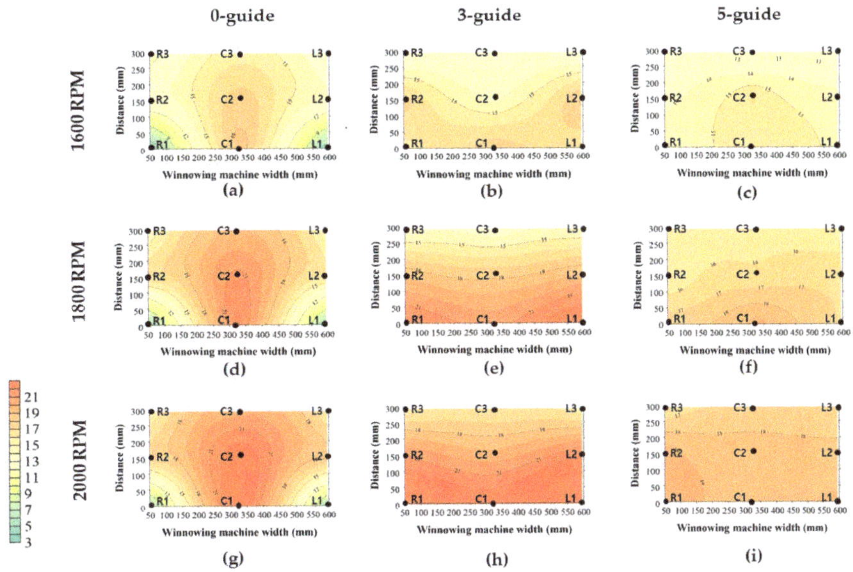

Figure 11. Results of CFD analysis for wind velocity on conditions of 0-guide (**a**), 3-guide (**b**), 5-guide (**c**) at 1600 RPM, and 0-guide (**d**), 3-guide (**e**), 5-guide (**f**), at 800 RPM, and 0-guide (**g**), 3-guide (**h**), 5-guide (**i**) at 2000 RPM.

Table 6. Experimental results of wind speed by the shape of the winnowing machine and measurement position.

Shape of Winnowing Machine	Distance (mm)	Rotating Speed (RPM)	Measured Value L1 (m/s)	Measured Value C1 (m/s)	Measured Value R1 (m/s)	Analysis Value L1 (m/s)	Analysis Value C1 (m/s)	Analysis Value R1 (m/s)
Existing winnowing machine (0-guide)	0	1600	3.3	19.8	3.6	4.1	19.9	4.4
		1800	3.7	21.6	4.2	4.9	22.6	5.0
		2000	4.9	22.7	4.7	5.8	24.1	5.8
	150	1600	12.6	17.0	13.4	13.8	17.6	14.3
		1800	15.3	20.6	14.8	16.9	21.4	16.2
		2000	16.3	20.5	16.6	17.5	23.2	18.4
	300	1600	10.6	15.2	11.4	11.1	15.5	12.2
		1800	13.4	18.4	13.7	13.8	19.1	14.6
		2000	14.6	18.5	15.4	15.0	19.6	15.8
Winnowing machine with three wind guides (3-guide)	0	1600	14.9	16.1	15.6	15.6	17.7	16.2
		1800	21.9	19.5	23.9	22.3	20.6	24.0
		2000	23.4	22.1	25.8	24.4	22.9	26.8
	150	1600	17.3	13.6	17.8	17.5	13.9	18.3
		1800	17.3	17.2	17.6	18.9	17.8	19.2
		2000	19.5	17.2	18.0	21.1	19.3	20.0
	300	1600	11.2	12.1	10.6	12.1	12.6	12.4
		1800	12.1	13.8	11.0	13.3	14.3	13.1
		2000	13.3	15.6	12.1	15.4	15.8	13.9
Winnowing machine with five wind guides (5-guide)	0	1600	15.0	15.3	14.4	14.2	15.5	14.8
		1800	17.0	18.9	13.9	17.5	19.4	15.9
		2000	17.7	17.9	16.4	19.0	18.6	18.2
	150	1600	13.6	14.2	13.9	13.8	16.1	14.4
		1800	14.4	15.4	14.7	15.0	16.2	16.8
		2000	18.9	16.1	16.1	19.5	18.0	18.3
	300	1600	12.5	12.2	10.9	13.2	12.6	12.3
		1800	12.4	14.3	13.1	14.8	15.6	15.1
		2000	15.1	16.1	16.7	16.1	17.0	17.3

4. Conclusions

The present study measured terminal velocities of chili pepper fruits, branches, and leaves of two species. Aerodynamic analysis of winnower was conducted to develop a foreign material winnowing machine to be used after harvesting chili peppers. The performance of the winnower designed through the analysis was validated by conducting a CFD analysis and using a winnower test device. For terminal velocity measurement, species Jeokyoung and AR Legend suitable for mechanical harvesting were used as test samples. An aerodynamic analysis was conducted for chili pepper fruits and chili pepper branches freely falling into a collection tank at different wind speeds. The analysis result showed that chili pepper fruits were collected into the collection tank at a wind speed lower than 17.5 m/s and that chili pepper branches and chili pepper leaves were separated at speeds higher than 12.5 m/s. A winnower was designed using the analysis result. Results of a CFD analysis and a winnower test device were compared to evaluate the rotating speed and the performance of 0-, 3-, and 5-guide at the discharge port. Although the CFD analysis showed that discharged wind speeds were higher than those of the winnower test device by 0 to 2.4 m/s, it was thought to be an effective validation due to the simplification of the design model. As a result of comparing the CFD analysis of the winnower developed in the present study and the performance of the test device, the 3-guide and 5-guide conditions were found to be suitable with a rotating speed of 1800 RPM and a rotating speed of 2000 RPM. However, with the 5-guide, since the winnowing machine generates internal noise and vibration, the guide type must be selected, taking into account the 3-guide or the 5-guide when it is applied to a chili pepper harvester. The winnowing machine developed in this study can be verified through terminal velocity, aerodynamic analysis, CFD flow analysis, testing equipment, and then applied to the chili pepper harvester to reduce the additional workforce in the separation of foreign substances after mechanical harvesting. Therefore, the conditions analyzed in the present study must be verified through additional field tests after applying them to foreign material separation units of chili pepper harvesters.

Author Contributions: Conceptualization, S.-Y.S. and Y.C.; Experiments and data analyses, S.-Y.S. and D.-C.K.; Methodology, M.-H.K.; Writing—original draft preparation, S.-Y.S. and Y.C.; Project administration, Y.C.; Writing—review and editing, Y.C., D.-C.K. and M.-H.K. All authors have read and agreed to the published version of the manuscript.

Funding: This work was supported by a grant (716001-7) of the Korea Institute of Planning and Evaluation for Technology in Food, Agriculture, and Forestry (IPET) through the Agriculture, Food, and Rural Affairs Convergence Technologies Program for Educating Creative Global Leader Program funded by the Ministry of Agriculture, Food, and Rural Affairs (MAFRA), Republic of Korea. It was also supported by a grant (NRF-2021R1G1A1012778) of the National Research Foundation of Korea (NRF) funded by the Korean government (MSIT, Ministry of Science and ICT). This research was supported by "Research Base Construction Fund Support Program" funded by Jeonbuk National University in 2022.

Institutional Review Board Statement: Not applicable.

Informed Consent Statement: Not applicable.

Data Availability Statement: Not applicable.

Conflicts of Interest: The authors declare no conflict of interest.

References

1. Food and Agriculture Organization-World Health Organization. Production Share of Chilies and Peppers, Dry by Region. Available online: https://www.fao.org/faostat/en/#data/QCL/visualize (accessed on 12 January 2022).
2. Statistics Korea. Production of Chili Pepper. Available online: http://kostat.go.kr/portal/eng/pressReleases/1/index.board?bmode=read&aSeq=415243 (accessed on 12 January 2022).
3. Statistics Korea. Current State of Farm Mechanization. Available online: http://www.index.go.kr/potal/main/EachDtlPageDetail.do?idx_cd=1288 (accessed on 7 February 2022).

4. Funk, P.A.; Walker, S.J. Evaluation of five green chili cultivars utilizing five different harvest mechanisms. *Appl. Eng. Agric.* **2010**, *26*, 955–964. [CrossRef]
5. Chen, Y.C.; Kong, L.J.; Wang, H.X. Research of mechanism and experiment of 4LZ-3.0 self-propelled pepper harvester. *Appl. Mech. Mater.* **2013**, *419*, 217–222. [CrossRef]
6. Nishanth, M.S.; Jayan, P.R.; Pankaj, M.; Ratnakiran, D.W. Design, development and evaluation of pepper harvester. *J. AgriSearch* **2020**, *7*, 82–85. [CrossRef]
7. Walker, S.J.; Funk, P.A. Mechanical harvest trials of New Mexican-type green chile (*Capsicum annuum L.*). *Hort. Sci.* **2010**, *45*, 145–146.
8. Calsoyas, I.; Walker, S.; Funk, P. Comparison of Six Green Chile (*Capsicum annuum* L.) Cultivars on Harvest Efficiency with Etgar®Picker. In Proceedings of the Annual Meeting of the American Society for Horticultural Science, Philadelphia, PA, USA, 3–7 January 2016; Volume 51, pp. 267–268.
9. Nam, J.S.; Kang, Y.S.; Kim, S.B.; Kim, D.C. Factorial experiment for drum-type secondary separating part of self-propelled pepper harvester. *J. Biosyst. Eng.* **2017**, *42*, 242–250. [CrossRef]
10. Shin, S.-Y.; Kim, M.-H.; Cho, Y.; Kim, D.-C. Performance testing and evaluation of drum-type stem separation device for pepper harvester. *Appl. Sci.* **2021**, *11*, 9225. [CrossRef]
11. Marshall, D.E. Mechanized pepper harvesting and trash removal. In Proceedings of the First International Conference on Fruit, Nut and Vegetable Harvesting Mechanization, Bet Dagan, Israel, 5–12 October 1983; American Society of Agricultural and Biological Engineers: St. Joesph, MI, USA, 1984; Volume 5, pp. 276–279.
12. Esch, T.A.; Marshall, D.E. Trash removal from mechanically harvested peppers. *Trans. ASAE* **1987**, *30*, 893–898. [CrossRef]
13. Funk, P.A.; Marshall, D.E. Pepper Harvest Technology. In *Peppers: Botany, Production and Uses*; Russo, V.M., Ed.; CABI: Wallingford, UK, 2012; pp. 227–240. [CrossRef]
14. Kim, T.-H.; Kim, D.-C.; Cho, Y. Performance comparison and evaluation of two small chili pepper harvester prototypes that attach to walking cultivators. *Appl. Sci.* **2020**, *10*, 2570. [CrossRef]
15. Lenker, D.H.; Nascimento, D.F. Mechanical harvesting and cleaning of chili peppers. *Trans. ASAE* **1982**, *25*, 42–46. [CrossRef]
16. Wolf, I.; Alper, Y. Mechanization of Paprika Harvest. In Proceedings of the First International Conference on Fruit, Nut and Vegetable Harvesting Mechanization, Bet Dagan, Israel, 5–12 October 1983; American Society of Agricultural and Biological Engineers: St. Joesph, MI, USA, 1984; Volume 5, pp. 265–275.
17. Eaton, F.E.; Wilson, C. *Refinement and Testing of Mechanical Cleaners for Red Chile. New Mexico State University Chile Task Force Report 22*; New Mexico State University Library: Las Cruces, NM, USA, 2005.
18. Kong, L.; Chen, Y.; Ma, L.; Duan, Y. Research and design of line pepper' cleaning and separating device, based on star wheel and airflow. *J. Chin. Agri. Mech.* **2013**, *34*, 102–105.
19. Jo, Y.J.; Kang, Y.S.; Nam, J.S.; Choe, J.S.; Inoue, E.; Okayasu, T.; Kim, D.C. Performance analysis for a card cleaner type separating system of pepper harvester. *J. Fac. Agric. Kyushu-Univ.* **2018**, *63*, 103–111. [CrossRef]
20. Byum, J.H.; Nam, J.S.; Choe, J.S.; Inoue, E.; Okayasu, T.; Kim, D.C. Analysis of the separating performance of a card cleaner for pepper harvester using EDEM software. *J. Fac. Agric. Kyushu-Univ.* **2018**, *63*, 347–354. [CrossRef]
21. Bilanski, W.K.; Collins, S.H.; Chu, P. Aerodynamic properties of seed grains. *Agric. Eng.* **1962**, *43*, 216–219.
22. Lee, C.H.; Cho, Y.J.; Kim, M.S. Aerodynamic Study on Pneumatic Separation of Grains(I) -An Experimental Study on The Vertical Wind Tunnel-. *J. Biosyst. Eng.* **1989**, *14*, 272–281.
23. Lee, C.H.; Cho, Y.J.; Kim, M.S. Aerodynamic Study on Pneumatic Separation of Grains(II) -The Measurement of the Terminal Velocities of Grains-. *J. Biosyst. Eng.* **1990**, *15*, 1–13.
24. Chung, C.J.; Nam, S.I.; Joo, B.C. Pneumatic Separation on Separating Unit of a Combine Harvester. *J. Biosyst. Eng.* **1988**, *13*, 32–43.
25. Kim, M.H.; Park, S.J.; Noh, S.H. Study on the Physical, Mechanical and Aerodynamic Properties of Peanut Pods. *J. Biosyst. Eng.* **1995**, *20*, 141–150.
26. Choi, Y. Development of Machine Harvester for Pepper. Ph.D. Thesis, Chonnam National University, Gwangju, Korea, 2006.
27. Hong, J.T.; Cho, K.H.; Cho, N.H.; Park, H.M.; Hong, S.K.; Choi, Y.; Shin, S.Y.; Cho, C.K. Study on Integrated Mechanization System for Harvest and Postharvest Operation of Once-over-harvest Pepper. In Proceedings of the Korean Society for Agricultural Machinery Conference, Suwon-si, Korea, 17–18 November 2006.
28. Noh, H.K.; Han, D.W.; Lee, E.S.; Kang, T.H. Effect of air blast velocity for separating efficiency of sesame thresher. In Proceedings of the Korean Society for Agricultural Machinery, Yesan-gun, Korea, 2–3 May 2013.
29. Lee, J.S.; Kim, B.J.; Kang, Y.S.; Kim, D.C. Analysis of flow for peanut harvest fan system using CFD. In Proceedings of the Korean Society for Agricultural Machinery, Jeju-si, Korea, 1–2 November 2013.
30. Kim, B.J.; Kang, Y.S.; Kim, H.G.; Inoue, E.; Okayasu, T.; Kim, D.C. Analysis of the separating performance of peanut harvester sorting system. *J. Fac. Agric. Kyushu-Univ.* **2015**, *60*, 209–214. [CrossRef]
31. Yuan, J.; Li, H.; Qi, X.; Hu, T.; Bai, M.; Wang, Y. Optimization of airflow cylinder sieve for threshed rice separation using CFD-DEM. *Eng. Appl. Comp. Fluid. Mech.* **2020**, *14*, 871–881. [CrossRef]
32. Shin, S.Y.; Kim, D.C.; Kang, Y.S.; Cho, Y. Factorial experiment for air blower of the pepper harvester. *J. Biosyst. Eng.* **2020**, *45*, 239–248. [CrossRef]
33. Zhao, J.; Sugirbay, A.; Liu, F.; Chen, Y.; Hu, G.; Zhang, E.; Chen, J. Parameter optimization of winnowing equipment for machine-harvested *Lycium barbarum* L. *Span. J. Agric. Res.* **2019**, *17*, e0203. [CrossRef]

34. Franco, A.; Valera, D.L.; Pena, A.; Pérez, A.M. Aerodynamic analysis and CFD simulation of several cellulose evaporative cooling pads used in Mediterranean greenhouses. *Comput. Electron. Agric.* **2011**, *76*, 218–230. [CrossRef]
35. Dousthaghi, M.H.; Minaei, S.; Khoshtaghaza, M.H. Computational fluid dynamics and analytical modeling for predicting effects of airflow temperature and relative humidity on the terminal velocity of agricultural granular materials. *J. Food Process Eng.* **2022**, *45*, e13941. [CrossRef]
36. Di Perta, E.S.; Agizza, M.A.; Sorrentino, G.; Boccia, L.; Pindozzi, S. Study of aerodynamic performances of different wind tunnel configurations and air inlet velocities, using computational fluid dynamics (CFD). *Comput. Electron. Agric.* **2016**, *125*, 137–148. [CrossRef]
37. Garrett, R.E.; Brooker, D.B. Aerodynamic drag of farm grains. *Appl. Eng. Agric.* **1965**, *26*, 45–52. [CrossRef]
38. Song, H.; Litchfield, J.B. Predicting method of terminal velocity for grains. *Appl. Eng. Agric.* **1991**, *34*, 225–231. [CrossRef]
39. Chakraverty, A.; Singh, R. *Postharvest Technology and Food Process Engineering*; CRC Press: Boca Raton, FL, USA, 2014.
40. Meriam, J.L.; Kraige, L.G. *Engineering Mechanics—Dynamics*, 9th ed.; John Wiley & Sons, Inc.: Hoboken, NJ, USA, 2020.
41. ANSYS Inc. *FLUENT Manual—ANSYS 2020 Version R1, Theory Guide*; ANSYS Inc.: Canonsburg, PA, USA, 2020.
42. Lee, M.-H.; Kang, Y.-S.; Kim, D.-C.; Choi, Y. A Study on the Performance Improvement of Foreign Materials Separating System for a Pepper Harvester. *J. Agric. Life Sci.* **2020**, *54*, 111–120. [CrossRef]
43. ANSYS Inc. *FLUENT Training Material—ANSYS 2020 Version R1, Lecture Transient Flow Modeling, Introduction to ANSYS Fluent*; ANSYS Inc.: Canonsburg, PA, USA, 2020.
44. Oliver, M.A.; Webster, R. Kriging: A method of interpolation for geographical information systems. *Int. J. Geogr. Inf. Syst.* **1990**, *4*, 313–332. [CrossRef]

applied sciences

MDPI

Article

A Digital Twin Architecture to Optimize Productivity within Controlled Environment Agriculture

Jesus David Chaux, David Sanchez-Londono and Giacomo Barbieri *

Department of Mechanical Engineering, Universidad de los Andes, Bogotá 111711, Colombia; jd.chaux10@uniandes.edu.co (J.D.C.); d.sanchezl@uniandes.edu.co (D.S.-L.)
* Correspondence: g.barbieri@uniandes.edu.co

Abstract: To ensure food security, agricultural production systems should innovate in the direction of increasing production while reducing utilized resources. Due to the higher level of automation with respect to traditional agricultural systems, Controlled Environment Agriculture (CEA) applications generally achieve better yields and quality crops at the expenses of higher energy consumption. In this context, Digital Twin (DT) may constitute a fundamental tool to reach the optimization of the productivity, intended as the ratio between production and resource consumption. For this reason, a DT Architecture for CEA systems is introduced within this work and applied to a case study for its validation. The proposed architecture is potentially able to optimize productivity since it utilizes simulation software that enables the optimization of: (i) Climate control strategies related to the control of the crop microclimate; (ii) treatments related to crop management. Due to the importance of food security in the worldwide landscape, the authors hope that this work may impulse the investigation of strategies for improving the productivity of CEA systems.

Keywords: controlled environment agriculture; digital twin; productivity; architecture; optimization

Citation: Chaux, J.D.; Sanchez-Londono, D.; Barbieri, G. A Digital Twin Architecture to Optimize Productivity within Controlled Environment Agriculture. *Appl. Sci.* **2021**, *11*, 8875. https://doi.org/10.3390/app11198875

Academic Editors: Tadeusz Juliszewski, Sławomir Kurpaska and Paweł Kiełbasa

Received: 10 June 2021
Accepted: 5 September 2021
Published: 24 September 2021

1. Introduction

The World Bank estimates there will be a global population of over 9.6 billion people by 2050 [1]. To feed this population, agricultural production should increase by about 50% with respect to today's levels [2]. However, several aspects complicate this context. These are next illustrated using predictions to 2050. Temperature rise due to global warming will result in reduced yields. For instance, rice, maize and soybean are estimated to reduce their yields by between 3.1% and 7.4% per each degree Celsius of increased temperature [3]. The growth of the world's population will mainly occur in urban areas, while the countryside population will remain stable, i.e., from 3.4 billion in 2018 to 3.1 billion in 2050 [1,4]. Considering that food is mainly produced in the countryside, the same amount of people available today will be required to produce the necessary food. Furthermore, the lands available for agriculture will decrease by between 8% and 20% due to land degradation, urbanization, and the use of crops for biofuel production [5]. Finally, it is likely that the current threats to freshwater will determine a sub-optimal supply of water to crops [6]. The combined effect of these factors might cause a deficit of between 5% and 25% in the availability of agricultural products, threatening the food security of the planet [5]. As such, agricultural production systems need to innovate in the direction of increasing production while reducing the utilized resources.

Controlled Environment Agriculture (CEA) refers to the control of plant growth and the surrounding environment with the objective of enhancing production efficiency, optimizing plant yields, and improving product quality [7]. CEA applications—such as plant factories and greenhouses [7,8]—may become an important tool to face the aforementioned challenges. Due to their ability to control the microclimate, CEA systems are robust to climate variability and temperature rise, enabling year-round production [9]. Furthermore, they are generally associated with soilless cultures that enable the recirculation of water and nutrients, and the utilization of inert soils [10]. In this way, the integration of CEA practices and soilless cultures provides intensive food production in limited areas with more efficient utilization of resources, including non-arable areas such as deserts and cities [11,12].

Different works have demonstrated that the increment in the level of automation determines better yield and quality of the harvested crops. Asseng et al. [13] estimated that the wheat grain cultivated through CEA vertical farmings may generate yields between 700 ± 40 and 1940 ± 230 ton/ha/yr with respect to the 3.2 ton/ha/yr obtained with open-field agriculture. Furthermore, Nicole et al. [14] detected an improvement in the lettuce food quality cultivated in plant factories with respect to open-field agriculture—in terms of color, nutrients and shelf life, amongst other things. However, the increment in the level of automation comes at the expense of higher energy consumption. Graamans et al. [15] compared greenhouses and plant factories showing that the production of 1 kg dry weight of lettuce requires an input of 247 kWhe in a plant factory, compared to 70, 111, 182 and 211 kWhe in greenhouses in, respectively, the Netherlands, the United Arab Emirates and Sweden ($\times 2$).Two Sweden greenhouses were utilized in the study: One with additional artificial illumination and the other without.

From the data reported above, it can be noticed that production (both in terms of yield and quality) and energy consumption are two conflicting goals for CEA systems. To ensure food security, there is the need to reach an optimization in between these two goals by maximizing productivity: The ratio between production and resource consumption.

Digital Twin may constitute a fundamental tool to reach the optimization of productivity. Digital Twin (DT) represents the next wave in modelling, simulation, and optimization technology [16]. According to Kritzinger et al. [17] and Negri et al. [18], DT "exploits sensed data, mathematical models and real-time data elaboration to forecast and optimise the behaviour of the physical asset at each life cycle phase, in real-time". DTs have been adopted in different domains such as manufacturing, aviation, hospital management and precision medicine and safety amongst others; see [19–21].

DTs are digital models enhanced with bilateral communication between the physical and the cyber space [17]. In traditional simulation, the digital representation of an existing physical asset does not use any form of automated data exchange between the physical asset and the digital one. In a DT, the data flow between an existing physical asset and a digital one is fully integrated in both directions. In this way, the digital model is synchronized with the status of the physical asset and the results of the simulation can be directly implemented to optimize the physical asset. In the context of CEA, the DT ability to integrate the real-time status of the physical asset into simulation may be adopted to guide the decision-making in crop management and microclimate control for the optimization of productivity.

To reach the aforementioned capabilities, the physical asset must be enhanced with a DT architecture consisting of (Figure 1):

- Physical Asset: Target system to optimize through the DT architecture.
- Digital Twin: Virtual test bed synchronized with the status of the physical asset that is responsible to evaluate the different 'what-if' scenarios that may optimize the system.
- Intelligence Layer: Hosts the rules and the knowledge to choose among the alternatives tested in the DT.

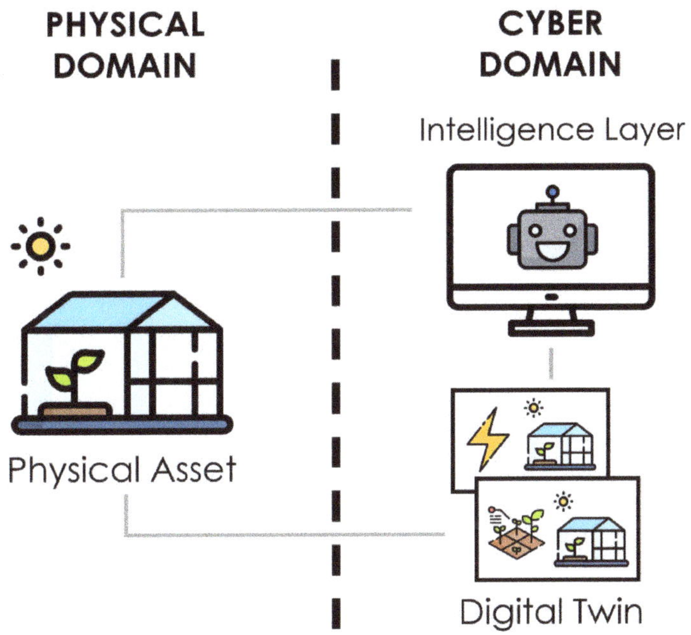

Figure 1. Actors involved in a DT architecture. The representation has been adopted from [22].

In the actual agricultural innovation systems, information-based management characterizes a technological phase called Farming 4.0 [23]. Few recent literature reviews can be found concerning the application of DT in agriculture. Pylianidis et al. [24] investigate the added-value of DTs for agriculture through the identification of 28 use cases, and their comparison with use cases from other disciplines. Based on their analysis, they examine the extent of the DT adoption in agriculture, shed light on the concept and the benefits it brings, and provide an application-based roadmap for a more extended adoption. Verdouw et al. [25] first review the concept of DT in agriculture by showing how DTs can advance farming practices and by developing a typology of different types of DTs. Then, they propose a conceptual framework for designing and implementing DTs. Sreedevi and Santosh [26] analyze ways in which DTs can contribute to most of the life cycle phases of hydroponic systems such as designing, operation, monitoring, optimization, and maintenance, amongst others.

In the aforementioned literature reviews, one of the potentialities of the DT for CEA applications is defined as the optimization of the productivity through the simulation and prediction of crop microclimate and growth. In their work, Rezvani et al. [27] review the basics of microclimate models for greenhouses and the results obtained from their application. Furthermore, a review of crop growth models and functional–structural plant models is also provided. Howard et al. [28] present the first advancements in the development of a DT that is intended to optimize the energy efficiency of greenhouses. Monteiro et al. [29] develop an IoT-enabled structure for vertical farming that has the objective of enabling sustainable CEA systems. Burchi et al. [30] introduce a multifunctional environment, equipped with sensors and monitoring systems that allows the acquisition of data and their processing using mathematical yield models to optimize crop management. To the best of the authors' knowledge, the presented works constitute the most relevant ones in the productivity optimization through DTs. However, it can be noticed that a DT architecture for CEA systems potentially able to optimize productivity due to the utilized simulation software is not available yet. This architecture—referred to as DT Architecture for CEA systems—constitutes the novelty presented in this work.

Given the above, the article is structured as follows: Section 2 resumes a previously introduced methodology for the design and verification of DT applications. Section 3 applies the methodology for the development of the DT architecture for CEA systems. Obtained results are discussed in Section 4, and the conclusion and future work are presented in Section 5.

2. Research Methodology

Barbieri et al. [31] proposed a methodology based on virtual commissioning—VC-based methodology—to retrofit manufacturing systems with DTs. The methodology consists of a stepwise approach in which the DT architecture is designed, integrated (to the retrofitted manufacturing system), and verified using a virtual environment before its implementation. In this work, part of the methodology is applied to develop the proposed DT architecture for CEA systems. Next, the research methodology utilized within this work is illustrated (Figure 2):

- Framework: The layered structure and functionalities of the DT architecture are identified without considering their implementation technologies.
- Technology: The technologies for instantiating the framework into an architecture are selected, and the actors that are interfaced within the architecture are specified.
- Digital Twin: The DT models are developed using the software and types of simulation identified in the previous phase.
- Intelligence Layer: The intelligence layer is designed starting from the defined functionalities and the selected implementation technologies. Within this phase, the interaction between the DTs and the intelligence layer is exploited with the aim to compare different optimization algorithms and to tune their parameters.
- Physical--Cyber Interface: The signals to be exchanged among the different actors within the DT architecture are identified. As depicted in Figure 1, signals are exchanged between: (i) Physical asset–intelligence layer; (ii) physical asset–digital twin; (iii) intelligence layer–digital twin. This phase also establishes the order in which signals are exchanged and which sequence of operations are implemented.
- Implementation: The architecture is implemented in the physical asset and verified.

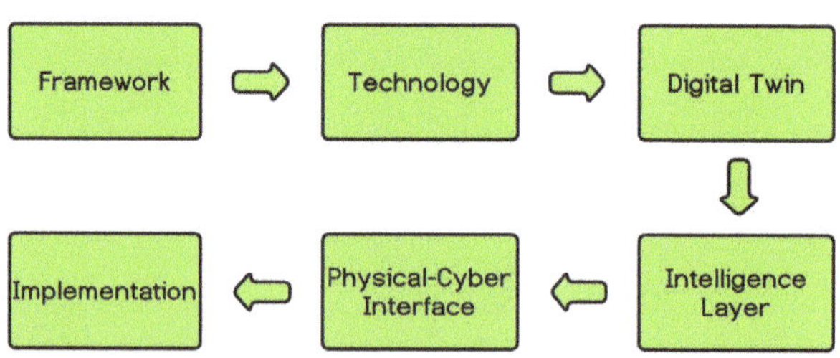

Figure 2. Methodology utilized for the development of the DT architecture for CEA systems. The representation has been adopted from [31].

3. Digital Twin Architecture

Next, the methodology illustrated in Section 2 is applied for the development of a DT architecture aimed to optimize the productivity of CEA systems. A prototype greenhouse (Figure 3) is utilized as case study for the design and verification of the DT architecture. The objective of this work is to build an architecture able to perform the bilateral communication typical of DTs using simulation models that can optimize the productivity. The utilization of the models for the identification of optimal crop treatments and climate control strategies is left as future work.

Finally, the following use case is defined for the architecture: 'The data necessary for the optimization must be available in the cloud and the user must download them in his/her local device. The optimization occurs in the local device and the optimal crop treatment and climate control strategy are communicated to the controller for its implementation through a gateway'.

Figure 3. Prototype automated greenhouse utilized as case study. In the figure, the different components are indicated with numbers.

3.1. Framework

In this phase, the layered structure and functionalities of the DT architecture are identified. The layered structure is depicted in Figure 4 and is illustrated next:

- Greenhouse: Physical asset to optimize; see Figure 3.
- Controller: Respectively, monitors and controls the greenhouse sensors and actuators. It also transmits the acquired sensor data to the gateway and receives the crop treatments and climate control strategies to implement.
- Gateway: Interface between the cyber and the physical domain; see Figure 1. It is responsible for transmitting sensor data to the storage layer and communicating the optimized crop treatments and climate control strategies to the controller for its implementation.
- Storage: Stores current and historical data that are utilized from the DTs for productivity optimization.
- Intelligence layer: Hosts the rules and the knowledge to choose among the different crop treatments and climate control strategies that may optimize the productivity of the greenhouse. It uses the DTs as virtual test beds to assess the evaluated alternatives.
- Digital twins: Utilizes current and historical data to assess the different crop treatments and climate control strategies received from the intelligence layer.

3.2. Technologies

The technologies for instantiating the framework into an architecture are selected. These are depicted in Figure 4 and illustrated next:

- Greenhouse $\longrightarrow$ two DHT11 sensors to, respectively, measure indoor and outdoor temperature and relative humidity, 12 V fan and exhaust fan, and a 12 V mini submersible pump.
- Controller $\longrightarrow$ Arduino Uno.
- Gateway $\longrightarrow$ Raspberry Pi 4: Communicates with the storage layer through wireless communication and with Arduino through serial communication.
- Storage $\longrightarrow$ phpMyAdmin: Administrator tool that manages a MySQL server for the data stored in the cloud.
- Intelligence Layer $\longrightarrow$ Visual Studio: Programmed in Python, it enables the communication with the cloud through MySQL and with the gateway through the MQTT communication protocol.

- Digital Twin 1⟶ EnergyPlus (energyplus.net): Builds energy software able to predict the microclimate within the greenhouse due to the ability to simulate the behaviour of heating, cooling, ventilation, and lighting systems, amongst others. It can communicate with Python-based IDEs through the EnergyPlus API (nrel.github.io/EnergyPlus/api/python).
- Digital Twin 2⟶ DSSAT (dssat.net): Agricultural decision support system that allows the simulation of growth, development, and yield as a function of "soil–plant–atmosphere dynamics" [32]. It can communicate with Python-based IDEs through TraDSSAT (github.com/julienmalard/traDSSAT).

EnergyPlus and DSSAT were selected as simulation software for the DTs since their integration has the premises to achieve the optimization of the productivity. EnergyPlus enables the optimization of control strategies related to the control of the crop microclimate [33], whereas DSSAT enables the optimization of treatments related to the crop management [34].

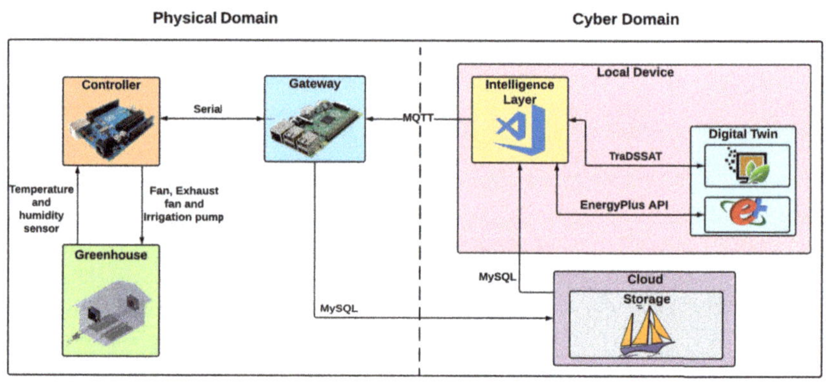

Figure 4. Framework and technologies utilized within the DT architecture for CEA systems.

3.3. *Digital Twin and Intelligence Layer*

After the definition of the framework and its implementation technologies, the DT models and the intelligence layer are developed. The tuning of these elements is outside the scope of the article since the objective of this work is to build a DT architecture and not to optimize the productivity of the prototype greenhouse. The demonstration of the effectiveness of the built DT architecture in optimizing the productivity is left as future work. Therefore, this section illustrates the communication between the intelligent layer and the DTs—referred to as optimization workflow—for the optimization of the productivity.

The optimization workflow is depicted in Figure 5 and consists in the following phases:

1. Generation of climate control strategies: The intelligence layer receives from the cloud: (i) Microclimate historical data (internal temperature and relative humidity); (ii) environmental historical data (external temperature and relative humidity); (iii) previous climate control strategies and crop treatments. Starting from these data, different alternatives of climate control strategies (CCSs) are generated. In this work, a CCS is defined as a control sequence of the greenhouse actuators to achieve a desired crop microclimate. Then, a prediction of the future environmental conditions is performed since EnergyPlus needs this information to assess the different CCSs.
2. Assessment of climate control strategies: The historical microclimate data, and the historical and predicted environmental data are transmitted to EnergyPlus. Then, all the generated CCSs are input to the software, and the predicted energy consumption and microclimate are computed for each CCS using the historical and predicted climate data.

3. Generation of crop treatments: The intelligence layer receives the predicted energy consumption and microclimate relative to each CCS. Then, different alternatives of crop treatments (TRTs) are generated, e.g., event to trigger the irrigation, volume delivered for irrigation, etc.
4. Assessment of crop treatments: The historical and predicted microclimate data are transmitted to DSSAT. Then, all the generated TRTs are input to the software, and the predicted production and resource consumption (e.g., water, nutrients, etc.) are computed for each TRT using the historical and predicted microclimate data.
5. Overall optimization: The intelligence layer receives the predicted production and resource consumption relative to each TRT. The productivity is computed for each pair of CCS and TRT—where the productivity is defined as the ratio between the production, and the sum of the energy and resource consumption. The best pair of climate control strategy (CCS^*) and crop treatment (TRT^*) is computed and transmitted to the gateway.

Given the interdependence between EnergyPlus and DSSAT, an overall optimization must be implemented. Starting from the domain knowledge, optimization algorithms and/or heuristics should be studied as future work to identify optimal solutions in acceptable computation time.

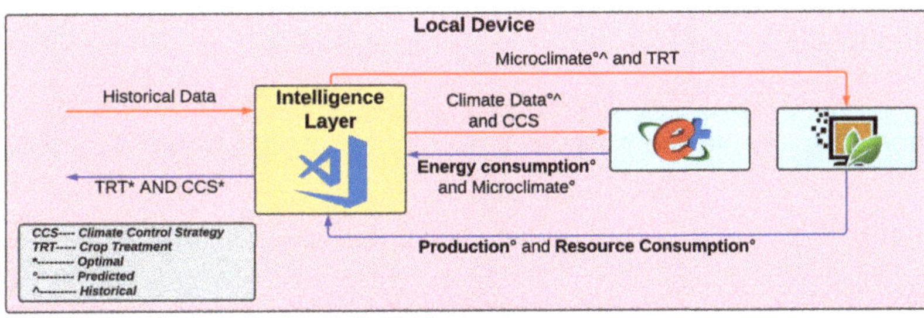

Figure 5. Optimization workflow: Communication between the intelligent layer and the DTs for the optimization of the productivity. Production and energy and resource consumption are indicated in bold since they are utilized for the calculation of the productivity. In the figure, climate data indicate the set of microclimate and environmental data.

3.4. Physical–Cyber Interface

After the design of the optimization workflow, the signals to be exchanged among the different actors within the DT architecture are identified. In this phase, the order in which signals are exchanged is also established. This information is depicted in Figure 6 through a sequence diagram. Sensor data are assessed from the controller and continuously uploaded to the cloud through the gateway. When an optimization is performed, historical data are sent to the local device and the optimization workflow illustrated in Section 3.3 is implemented. Once the optimal CCS^* and TRT^* have been identified, these are transmitted to the gateway. Finally, the gateway sends them to: (i) Cloud: To trace the implemented climate control strategies and crop treatments; (ii) Controller: To implement the optimal climate control strategy and crop treatment.

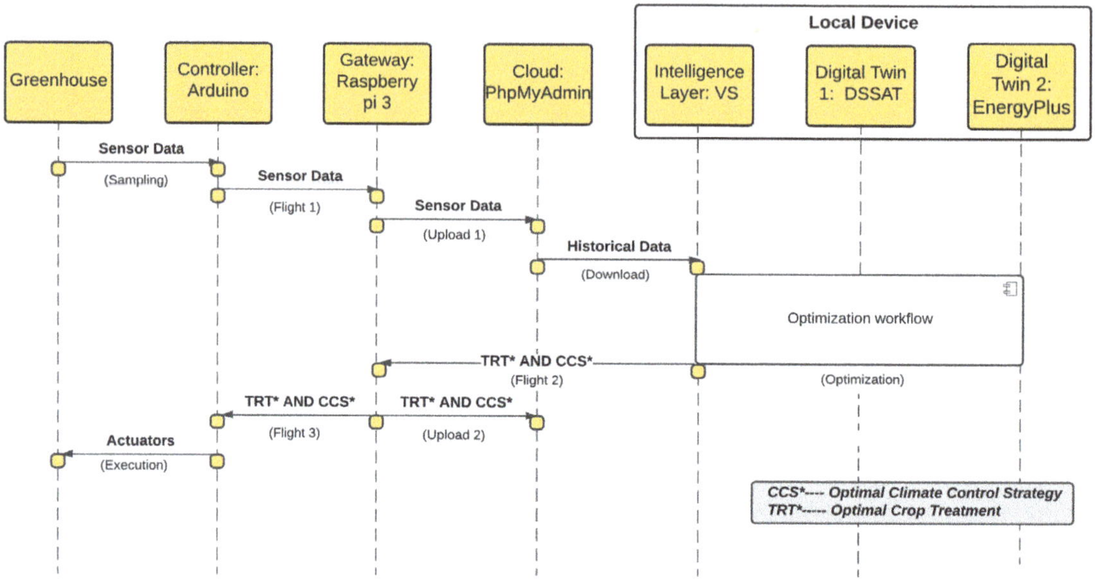

Figure 6. Sequence diagram showing the communications between the actors of the DT architecture. The name of each communication is indicated in parentheses.

3.5. Implementation

The presented DT architecture is applied to the prototype greenhouse illustrated in Figure 3. Then, latency tests are performed to verify the implementation of the communications depicted in Figure 6.

4. Results and Discussion

Latency tests were performed to verify the implementation of the DT architecture. In this article, a latency test refers to the assessment of the time taken for a message to travel in between two actors. It is worth noting that the minimization of the latency was not within the scope of this work and does not constitute a priority for CEA production systems since these are not hard real-time ones. Latency was assessed as a mean to certify the achievement of the communications indicated in Figure 6.

Figure 7 illustrates the latency taken within each communication. To emulate a feasible optimization scenario, a single data point was transmitted within each communication. The only exception was constituted by the 'download' communication in which one million data points were downloaded from the cloud. The communication sequence illustrated in Figure 6 was repeated 300 times to evaluate the latency variability through the random error formulation [35].

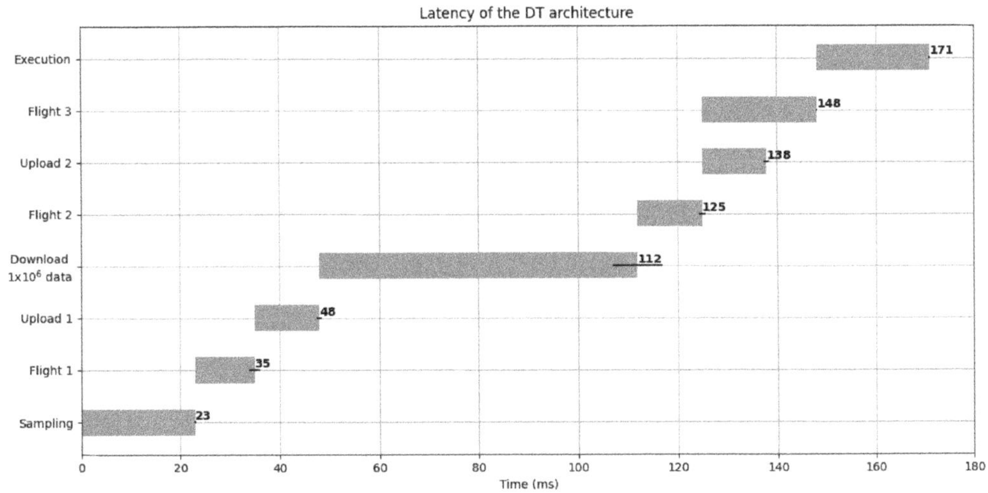

Figure 7. Timing diagram of the latency taken within each communication. Error bars quantify the uncertainty of each measurement. The nomenclature shown in Figure 6 is utilized for defining the different communications.

The latency data demonstrate that it was possible to achieve the bidirectional communication typical of DT architectures [17], i.e., from the physical domain to the cyber one and vice versa (Figure 1). Considering the selected simulators, a DT architecture potentially able to optimize the productivity was built.

From an industrial perspective, the proposed DT architecture constitutes a scalable architecture that may be applied to industrial systems by replacing the Arduino controller with the utilized PLCs (Programmable Logic Controllers). From an educational perspective, the illustrated low-cost prototype may be replicated from educational institutions for the generation of automation and agricultural hands-on laboratories, and for the investigation of approaches for the optimization of the productivity in CEA production systems.

Even if a functional DT architecture was built, several works are still missing before certifying its ability to optimize the productivity. In particular, a case study should be implemented and productivity should be optimized through the developed architecture. Different challenges will occur as the aforementioned definition of heuristics for the identification of close-to-optimal climate control strategies and crop treatments in acceptable computation time.

5. Conclusions and Future Work

To ensure food security, agricultural production systems should innovate in the direction of increasing production while reducing the utilized resources. Controlled environment agriculture and digital twins may represent fundamental tools to reach the optimization of productivity, thus contributing to the planet's food security.

With this in mind, the objective of this research work was the development of a DT architecture potentially able to optimize productivity in the context of CEA applications. The objective was reached by designing an architecture that utilizes (as DTs) simulation software that enables the optimization of: (i) Climate control strategies related to the control of the crop microclimate; (ii) treatments related to the crop management. The architecture was applied to a prototype greenhouse for its validation. Finally, communication latency was assessed as a means to test the achievement of the communications defined within the DT architecture.

This work contributes to the research on DT in CEA systems by proposing an architecture potentially able to optimize productivity. The methodological approach and the identified tools can be utilized by companies for retrofitting their CEA systems with the

DT functionality, and from universities for the generation of automation and agricultural hands-on laboratories, and for the investigation of approaches for the optimization of the productivity in CEA systems.

However, several works are still missing before the ability of the proposed architecture to optimize the productivity can be certified. In line with this goal, some future works are identified:

- Optimization workflow: The optimization workflow identified within this work involves a sequence of two simulation software. Starting from the domain knowledge, optimization algorithms and/or heuristics must be studied to identify optimal solutions in acceptable computation time.
- Case study: After the definition of heuristics, a case study must be implemented and productivity must be optimized to certify the ability of the proposed DT architecture to optimize productivity.
- Architecture improvement: Some improvements should be investigated as the movement of the intelligence layer and the DTs to the cloud, and the simplification of the physical domain with the implementation of smart sensors and actuators that would make the controller and gateway unnecessary.

Author Contributions: Conceptualization, J.D.C. and D.S.-L.; methodology, J.D.C. and G.B.; software, J.D.C.; validation, J.D.C. and D.S.-L.; formal analysis, J.D.C. and D.S.-L.; investigation, J.D.C., D.S.-L. and G.B.; resources, G.B.; data curation, J.D.C. and D.S.-L.; writing—original draft preparation, J.D.C., D.S.-L. and G.B.; writing—review and editing, G.B..; visualization, J.D.C. and D.S.-L.; supervision, G.B.; project administration, G.B.; funding acquisition, G.B. All authors have read and agreed to the published version of the manuscript.

Funding: This research received no external funding.

Institutional Review Board Statement: Not applicable.

Informed Consent Statement: Not applicable.

Data Availability Statement: Not applicable.

Conflicts of Interest: The authors declare no conflict of interest.

References

1. World Bank Group. Population Estimates And Projections. 2020. Available online: https://databank.worldbank.org/source/population-estimates-and-projections (accessed on 6 September 2021).
2. FAO. *The State of Food and Agriculture*; Food and Agriculture Organization of the United Nations: Rome, Italy, 2017.
3. Zhao, C.; Liu, B.; Piao, S.; Wang, X.; Lobell, D.B.; Huang, Y.; Huang, M.; Yao, Y.; Bassu, S.; Ciais, P.; et al. Temperature increase reduces global yields of major crops in four independent estimates. *Proc. Natl. Acad. Sci. USA* **2017**, *114*, 9326–9331. [CrossRef]
4. Department of Economic and Social Affairs. World Urbanization Prospects 2018 Highlights. 2019. Available online: https://population.un.org/wup/ (accessed on 6 September 2021).
5. Nellemann, C. *The Environmental Food Crisis: The Environment's Role in Averting Future Food Crises*; UNEP: Arendal, Norway, 2009.
6. Vörösmarty, C.J.; McIntyre, P.B.; Gessner, M.O.; Dudgeon, D.; Prusevich, A.; Green, P.; Glidden, S.; Bunn, S.E.; Sullivan, C.A.; Liermann, C.R.; et al. Global threats to human water security and river biodiversity. *Nature* **2010**, *467*, 555–561. [CrossRef]
7. Gómez, C.; Currey, C.J.; Dickson, R.W.; Kim, H.J.; Hernández, R.; Sabeh, N.C.; Raudales, R.E.; Brumfield, R.G.; Laury-Shaw, A.; Wilke, A.K.; et al. Controlled environment food production for urban agriculture. *HortScience* **2019**, *54*, 1448–1458. [CrossRef]
8. Engler, N.; Krarti, M. Review of energy efficiency in controlled environment agriculture. *Renew. Sustain. Energy Rev.* **2021**, *141*, 110786. [CrossRef]
9. R Shamshiri, R.; Kalantari, F.; Ting, K.; Thorp, K.R.; Hameed, I.A.; Weltzien, C.; Ahmad, D.; Shad, Z.M. Advances in greenhouse automation and controlled environment agriculture: A transition to plant factories and urban agriculture. *Int. J. Agric. Biol. Eng.* **2018**, *11*, 1–12. [CrossRef]
10. Resh, H.M. *Hydroponic Food Production: A Definitive Guidebook for the Advanced Home Gardener and the Commercial Hydroponic Grower*; CRC Press: Boca Raton, FL, USA, 2012.
11. Benke, K.; Tomkins, B. Future food-production systems: Vertical farming and controlled-environment agriculture. *Sustain. Sci. Pract. Policy* **2017**, *13*, 13–26. [CrossRef]
12. Raviv, M.; Lieth, J.H.; Bar-Tal, A. *Soilless Culture: Theory and Practice: Theory and Practice*; Elsevier: Amsterdam, The Netherlands, 2019.

13. Asseng, S.; Guarin, J.R.; Raman, M.; Monje, O.; Kiss, G.; Despommier, D.D.; Meggers, F.M.; Gauthier, P.P. Wheat yield potential in controlled-environment vertical farms. *Proc. Natl. Acad. Sci. USA* **2020**, *117*, 19131–19135. [CrossRef] [PubMed]
14. Nicole, C.; Charalambous, F.; Martinakos, S.; Van De Voort, S.; Li, Z.; Verhoog, M.; Krijn, M. Lettuce growth and quality optimization in a plant factory. In Proceedings of the VIII International Symposium on Light in Horticulture 1134, East Lansing, MI, USA, 22–26 May 2016; pp. 231–238.
15. Graamans, L.; Baeza, E.; Van Den Dobbelsteen, A.; Tsafaras, I.; Stanghellini, C. Plant factories versus greenhouses: Comparison of resource use efficiency. *Agric. Syst.* **2018**, *160*, 31–43. [CrossRef]
16. Rosen, R.; Von Wichert, G.; Lo, G.; Bettenhausen, K.D. About the importance of autonomy and digital twins for the future of manufacturing. *IFAC-PapersOnLine* **2015**, *48*, 567–572. [CrossRef]
17. Kritzinger, W.; Karner, M.; Traar, G.; Henjes, J.; Sihn, W. Digital Twin in manufacturing: A categorical literature review and classification. *IFAC-PapersOnLine* **2018**, *51*, 1016–1022. [CrossRef]
18. Negri, E.; Fumagalli, L.; Macchi, M. A review of the roles of digital twin in CPS-based production systems. *Procedia Manuf.* **2017**, *11*, 939–948. [CrossRef]
19. Barricelli, B.R.; Casiraghi, E.; Fogli, D. A Survey on Digital Twin: Definitions, Characteristics, Applications, and Design Implications. *IEEE Access* **2019**, *7*, 167653–167671. [CrossRef]
20. Agnusdei, G.P.; Elia, V.; Gnoni, M.G. Is Digital Twin Technology Supporting Safety Management? A Bibliometric and Systematic Review. *Appl. Sci.* **2021**, *11*, 2767. [CrossRef]
21. Liu, M.; Fang, S.; Dong, H.; Xu, C. Review of digital twin about concepts, technologies, and industrial applications. *J. Manuf. Syst.* **2021**, *58*, 346–361. [CrossRef]
22. Negri, E.; Berardi, S.; Fumagalli, L.; Macchi, M. MES-integrated digital twin frameworks. *J. Manuf. Syst.* **2020**, *56*, 58–71. [CrossRef]
23. Vecchio, Y.; Agnusdei, G.P.; Miglietta, P.P.; Capitanio, F. Adoption of precision farming tools: The case of italian farmers. *Int. J. Environ. Res. Public Health* **2020**, *17*, 869. [CrossRef] [PubMed]
24. Pylianidis, C.; Osinga, S.; Athanasiadis, I.N. Introducing digital twins to agriculture. *Comput. Electron. Agric.* **2021**, *184*, 105942. [CrossRef]
25. Verdouw, C.; Tekinerdogan, B.; Beulens, A.; Wolfert, S. Digital twins in smart farming. *Agric. Syst.* **2021**, *189*, 103046. [CrossRef]
26. Sreedevi, T.; Kumar, M.S. Digital Twin in Smart Farming: A Categorical Literature Review and Exploring Possibilities in Hydroponics. In Proceedings of the 2020 Advanced Computing and Communication Technologies for High Performance Applications (ACCTHPA), Kerala, India, 2–4 July 2020; IEEE: New York , NY, USA, 2020; pp. 120–124.
27. Rezvani, S.M.e.; Shamshiri, R.R.; Hameed, I.A.; Abyane, H.Z.; Godarzi, M.; Momeni, D.; Balasundram, S.K. Greenhouse Crop Simulation Models and Microclimate Control Systems, A Review. 2021. Available online: https://www.intechopen.com/chapters/76412 (accessed on 6 September 2021).
28. Howard, D.A.; Ma, Z.; Aaslyng, J.M.; Jørgensen, B.N. Data Architecture for Digital Twin of Commercial Greenhouse Production. In Proceedings of the 2020 RIVF International Conference on Computing and Communication Technologies (RIVF), Ho Chi Minh, Vietnam, 6–7 April 2020; IEEE: New York , NY, USA, 2020; pp. 1–7.
29. Monteiro, J.; Barata, J.; Veloso, M.; Veloso, L.; Nunes, J. Towards sustainable digital twins for vertical farming. In Proceedings of the 2018 Thirteenth International Conference on Digital Information Management (ICDIM), Berlin, Germany, 24–26 September 2018; IEEE: New York , NY, USA, 2018; pp. 234–239.
30. Burchi, G.; Chessa, S.; Gambineri, F.; Kocian, A.; Massa, D.; Milazzo, P.; Rimediotti, L.; Ruggeri, A. Information technology controlled greenhouse: A system architecture. In Proceedings of the 2018 IoT Vertical and Topical Summit on Agriculture-Tuscany (IOT Tuscany), Monteriggioni, Italy, 8–9 May 2018; IEEE: New York , NY, USA, 2018; pp. 1–6.
31. Barbieri, G.; Bertuzzi, A.; Capriotti, A.; Ragazzini, L.; Gutierrez, D.; Negri, E.; Fumagalli, L. A virtual commissioning based methodology to integrate digital twins into manufacturing systems. *Prod. Eng.* **2021**, *15*, 397–412. [CrossRef]
32. Hoogenboom, G.; Porter, C.; Boote, K.; Shelia, V.; Wilkens, P.; Singh, U.; White, J.; Asseng, S.; Lizaso, J.; Moreno, L.; et al. The DSSAT Crop Modeling Ecosystem. In *Advances in Crop Modeling for a Sustainable Agriculture*; Burleigh Dodds Science Publishing: London, UK, 2019.
33. Zhang, Y.; Kacira, M. Enhancing resource use efficiency in plant factory. In Proceedings of the XXX International Horticultural Congress IHC2018: III International Symposium on Innovation and New Technologies in Protected 1271, Istanbul, Turkey, 12–16 August 2018; pp. 307–314.
34. Bai, Y.; Gao, J. Optimization of the nitrogen fertilizer schedule of maize under drip irrigation in Jilin, China, based on DSSAT and GA. *Agric. Water Manag.* **2021**, *244*, 106555. [CrossRef]
35. Beckwith, T.G.; Marangoni, R.D.; Lienhard, J.H. *Mechanical Measurements*; Pearson: London, UK, 2009.

applied sciences

MDPI

Article

Design of an Intelligent Variable-Flow Recirculating Aquaculture System Based on Machine Learning Methods

Fudi Chen [1,2,†], Yishuai Du [1,2,†], Tianlong Qiu [1,2], Zhe Xu [3], Li Zhou [1,2], Jianping Xu [1,2], Ming Sun [1,2,4], Ye Li [1,2] and Jianming Sun [1,2,3,4,*]

1 CAS and Shandong Province Key Laboratory of Experimental Marine Biology, Center for Ocean Mega-Science, Institute of Oceanology, Chinese Academy of Sciences, Qingdao 266071, China; chenfudi@qdio.ac.cn (F.C.); duyishuai@qdio.ac.cn (Y.D.); oceanman@qdio.ac.cn (T.Q.); zhouli@qdio.ac.cn (L.Z.); xujianping@qdio.ac.cn (J.X.); sunming1@qdio.ac.cn (M.S.); liye@qdio.ac.cn (Y.L.)
2 Laboratory for Marine Biology and Biotechnology, Qingdao National Laboratory for Marine Science and Technology, Qingdao 266071, China
3 Dalian Huixin Titanium Equipment Development Co., Ltd., Dalian 116039, China; Xuzhe_DLHX@hotmail.com
4 Dalian Key Laboratory of Conservation of Fishery Resources, Liaoning Province Key Laboratory of Marine Biological Resources and Ecology, Liaoning Ocean and Fisheries Science Research Institute, Dalian 116023, China
* Correspondence: jianming_sun@outlook.com; Tel.: +86-532-82898031
† These authors contributed equally to this work.

Featured Application: The proposed classification models could be adapted to develop a recirculating aquaculture system with continuous variable-flow control technology.

Abstract: A recirculating aquaculture system (RAS) can reduce water and land requirements for intensive aquaculture production. However, a traditional RAS uses a fixed circulation flow rate for water treatment. In general, the water in an RAS is highly turbid only when the animals are fed and when they excrete. Therefore, RAS water quality regulation technology based on process control is proposed in this paper. The intelligent variable-flow RAS was designed based on the circulating pump-drum filter linkage working model. Machine learning methods were introduced to develop the intelligent regulation model to maintain a clean and stable water environment. Results showed that the long short-term memory network performed with the highest accuracy (training set 100%, test set 96.84%) and F1-score (training 100%, test 93.83%) among artificial neural networks. Optimization methods including grid search, cuckoo search, linear squares, and gene algorithm were proposed to improve the classification ability of support vector machine models. Results showed that all support vector machine models passed cross-validation and could meet accuracy standards. In summary, the gene algorithm support vector machine model (accuracy: training 100%, test 98.95%; F1-score: training 100%, test 99.17%) is suitable as an optimal variable-flow regulation model for an intelligent variable-flow RAS.

Keywords: recirculating aquaculture system; variable-flow regulation model; circulating pump-drum filter linkage working technique; machine learning methods; gene algorithm support vector machine

Citation: Chen, F.; Du, Y.; Qiu, T.; Xu, Z.; Zhou, L.; Xu, J.; Sun, M.; Li, Y.; Sun, J. Design of an Intelligent Variable-Flow Recirculating Aquaculture System Based on Machine Learning Methods. *Appl. Sci.* **2021**, *11*, 6546. https://doi.org/10.3390/app11146546

Academic Editor: Manuel Armada

Received: 12 June 2021
Accepted: 14 July 2021
Published: 16 July 2021

1. Introduction

With global economic growth, consumer demand for seafood products is also increasing. However, fishery productivity is facing a massive challenge of declining resources due to environmental pollution and overfishing [1]. The recirculating aquaculture mode is an effective solution to maintain the supply of seafood products and support the modern and sustainable development of the aquaculture industry while decreasing ecological impact [2]. A recirculating aquaculture system (RAS) can offer a high degree of environmental control and uses various technologies to carry out physical filtration, biofiltration, and disinfection for water recycling [3].

The core of an RAS is the water treatment system, which mainly includes micro-screen drum filters, biofilters, oxidation devices, and disinfection devices [4]. Suspended solids removal is a critical part of water treatment in the RAS. Suspended solid particles are composed mainly of feces, residual feed, and bacterial flocs [5–7]. Feed is the main source of suspended solids in the system, and studies have shown that 25% of feed is converted into suspended solids in an RAS [8]. Suspended solid particles have been proven to be the leading cause of high turbidity in aquaculture water, which can cause stress reactions and endanger the health of aquatic animals [9]. As residence time increases, the suspended solids block the breeding facilities and increase chemical oxygen demand. Organic solid waste can be mineralized and decomposed to increase ammonia and nitrite concentrations and increase the load on the nitrification function of the biofilter [10]. Suspended solids removal devices in an RAS can be roughly classified according to the particle size of the suspended matter: sedimentation separation devices, micro-mesh filtration devices, foam fractionators, and ozone generators. The micro-screen drum filter, which is a physical filter device widely used in RASs, has the characteristics of strong adaptability, minimal floor space, and a high level of automation [11]. In a drum filter, the screen is fixed on a rotating drum frame on the horizontal axis and partially submerged in water; water flows into the drum and radially through the straining cloth, which captures fine particles with a suitable mesh size [12]. The micro-screen is the central working part of the drum filter, and the mesh number can directly affect filtration performance. Gravdal Arve et al. [13] reported that the removal rate of particles larger than 60 μm by the drum filter could reach more than 68%. Su et al. [14] found that the removal rate rapidly increased when the mesh number was increased from 150 to 200. The effect was apparent when the screen mesh was 200; the TSS removal rate reached 54.90%. Generally, 200 mesh is the principal mesh size used, as it is the one with the most outstanding technical and economic advantages [11].

A high-power centrifugal pump and an oversized drum filter are generally used to ensure sufficient circulation flow and filtration ability in an RAS [15]. The water in a traditional fixed-flow RAS is highly turbid when the breeding animals are fed and when they defecate. However, at other times, the water is relatively clean and does not require high-power pumps to recirculate it, resulting in wasting resources. Compared with the traditional fixed-flow RAS, the variable-flow RAS can increase the total water circulation to accelerate the water treatment process when organic particles increase, and the ammonia and nitrite then can be eliminated from the source [16]. In addition, the variable-flow RAS consumes a low amount of electricity when the water is relatively clean. However, manual operation is often used to adjust the circulation pump frequency to determine the appropriate total water circulation in the variable-flow RAS. The manual operation experience may cause the water treatment efficiency to not match the actual situation, resulting in insufficient water processing efficiency or waste of electricity. Hence, an intelligent variable-flow RAS for culturing *Litopenaeus vannamei* was developed in the present study. Machine learning, which has emerged with Big Data technologies and created new opportunities in multidisciplinary aquaculture, was used to develop the intelligent variable-flow model. Currently, machine learning is applied in related fields, including environmental assessment, water management, animal welfare, disease detection, feeding control, and species recognition [17–23]. More data-intensive machine learning approaches have been reported, but model- and technology-intensive approaches have been infrequent [24,25]. For industrial control in recirculating aquaculture, in particular, there is an urgent need to apply machine learning models to improve instrument efficiency and promote the development of intelligent equipment applications.

The primary purpose of the present study was to develop the circulating pump-drum filter linkage working technique using machine learning methods. Water quality indicators and the backwash frequency of the drum filter were used as primary indicators in developing a variable-flow model. An intelligent variable-flow RAS can rapidly remove suspended solids and reduce ammonia and nitrite generation from the source.

2. Materials and Methods

2.1. Experimental RAS

The experimental RAS used the recirculating aquaculture system of Dalian Huixin Titanium Equipment Development Co., Ltd. (Dalian city, China) for breeding *L. vannamei*. Figure 1a shows the schematic of the experimental RAS control system. The control system collected the water quality indicators by connecting them with the sensors. Water quality changes can be monitored in real time, and the centrifugal pump was controlled by variable-frequency operation using a flow regulation model based on machine learning. The variable-flow circulation caused different trends in the drum filter backwash frequency during the unit period (0.5 h). The water quality indicators were used to train the regulation strategy model for variable-flow circulation. The types of water treatment equipment included biofilters, a micro-screen drum filter, an ultraviolet generator, ozone generators, foam fractionators, and oxygenation cones. Figure 1b shows the actual indoor workshop. The RAS contained 10 circular FRP tanks with a diameter of 1.8 m and a depth of 1.4 m, with a total water volume of 35 m^3. Shrimp were fed five times a day during the culture period with a 36% protein commercial feed (Dale 2# shrimp commercial feeds, Dale, Inc., Yantai, China). During the early stage of shrimp culture, the amount of feed accounted for 5–8% of the total biomass of shrimp. The amount of feed was reduced over time and accounted for 3.7–5% of the total biomass by the end of the culture process. The whole culture process lasted for 90 days, with a culture density of 800 individuals/m^3 and a final yield of 525 kg of shrimp.

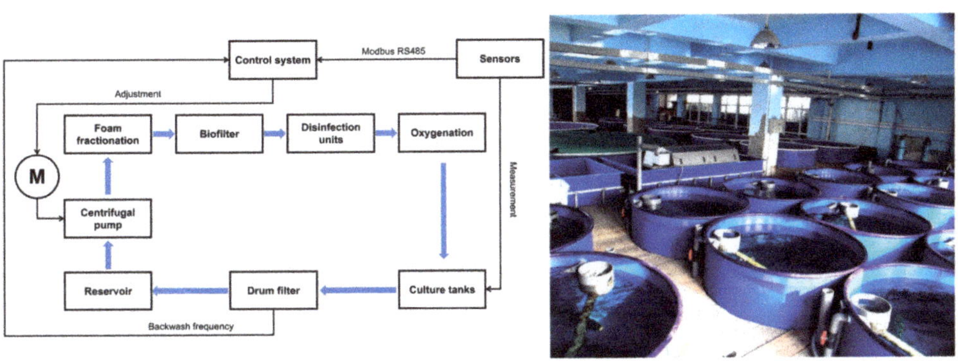

(a) (b)

Figure 1. The experimental recirculating aquaculture system: (**a**) the schematic of the experimental RAS control process; (**b**) the RAS in Dalian Huixin Titanium Equipment Development Co., Ltd.

2.2. Variable-Flow Experiment Design

Turbidity (NTU) is mainly influenced by water flow fluctuations and can only reflect the instantaneous transparency of the water body. This study proposes a technique for detecting turbidity in an RAS based on a micro-screen drum filter. The backwash frequency of the drum filter within a unit period (0.5 h) was used to represent overall RAS turbidity, and the variable-flow regulation model was constructed using the backwash frequency and various water quality data. The variable-flow regulation model can determine the operating frequency of the centrifugal pump for the next period using real-time data from the current period. The intelligent variable-flow RAS technology is implemented by controlling the RAS circulation rate by changing the circulating pump flow rate. The primary purpose of the variable-flow RAS is to implement a linkage control technology to model the relationship between the micro-screen drum filter backwash frequency and the circulation flow rate.

The total flow rate of the circulating pump was set to three levels: 55, 65, and 75 m^3/h. The circulation rate was operated with a cycle of 24 h. A cycle started with a circulation rate of 55 m^3/h and was adjusted to 65 m^3/h after an interval of 24 h and then to 75 m^3/h after the same interval (24 h). The drum filter controller collected backwash data every 0.5 h. Turbidity sensors were placed at the main return pipeline to monitor and record overall RAS water turbidity. Water quality indicators, including water temperature (T), dissolved oxygen (DO), pH, and salinity, were measured by sensors in real time using YSI ProPlus portable sensors. Total suspended solids (TSS), total ammonia nitrogen (TAN), and nitrite nitrogen (NO2-N) were measured daily with a Palintest 7500 water quality analyzer.

The circulating pump was set to three circulating levels: slow (55 m^3/h), medium (65 m^3/h), and fast (75 m^3/h). In the variable-flow RAS, the circulation rate was maintained at a medium level, and the control system read water quality indicators and backwash times from sensors at every unit period. The circulation rate for the next period could be adjusted to slow or fast levels. The circulation adjustment process could be operated in two ways: upshift and downshift. In the drum filter controller program, the backwash frequency was recorded for 48 periods in a day, using 0.5 h as a period. The circulating pump was utilized to determine the upshift/downshift for the next period by reading the current water quality sensors, current backwash frequency, and current circulating level. A water gauge controlled the drum filter backwash frequency; the backwash frequency reflects water turbidity in the RAS. Downshifts (-1) and upshifts (+1) of circulating pump frequency were used as indicators of circulation levels. The water quality indicators, current circulating pump frequency, and the drum filter backwash frequency were chosen as independent variables, and the downshifts (-1)/upshifts (+1) data were considered as the dependent variable. As the whole culture process lasted for 90 days in the RAS, the total circulation rate was set to 55 m^3/h for the first 30 days, 65 m^3/h for the middle 30 days, and 75 m^3/h for the last 30 days. Establishing a variable-flow circulation strategy was the core task of the experiment, and therefore the circulation rate regulation model was constructed using the optimal classification model based on machine learning to control the variable-flow circulation rate in the RAS.

As shown in Figure 2, the drum filter controller was used to collect the backwash frequency, circulation flow rate, and water quality data that were then uploaded to the industrial PC through the RS485 protocol. The embedded system was connected to the industrial computer. The dataset was processed with the optimal machine learning model in the industrial computer to regulate pump frequency for the next period and feed it back to the embedded system, so that the RAS circulation flow rate could be regulated intelligently.

2.3. Machine Learning Methods

2.3.1. Artificial Neural Networks (ANNs)

ANNs are statistical learning algorithms that possess prediction and approximation abilities given sufficient and considerable inputs [26]. ANNs are derived from the biological neural networks in the human brain. Interconnected artificial neural networks are usually composed of neurons that can deal with the inputs and follow various situations. ANNs are suitable not only for machine learning but also pattern recognition. Therefore, ANNs have become a popular way of indicating a function by observation in the case of complex data. Figure 3a shows a typical ANN structure, including input, hidden, and output layers.

In this study, several ANN methods, including the backpropagation neural network (BPNN), extreme learning machine (ELM), probabilistic neural network (PNN), and long short-term memory (LSTM) neural network, were used to develop variable-flow models. The BPNN and ELM are feedforward neural networks with no cycles or loops. Information propagates in one direction, forward from the input layer, through the hidden layer, and then to the output layer, in a feedforward neural network.

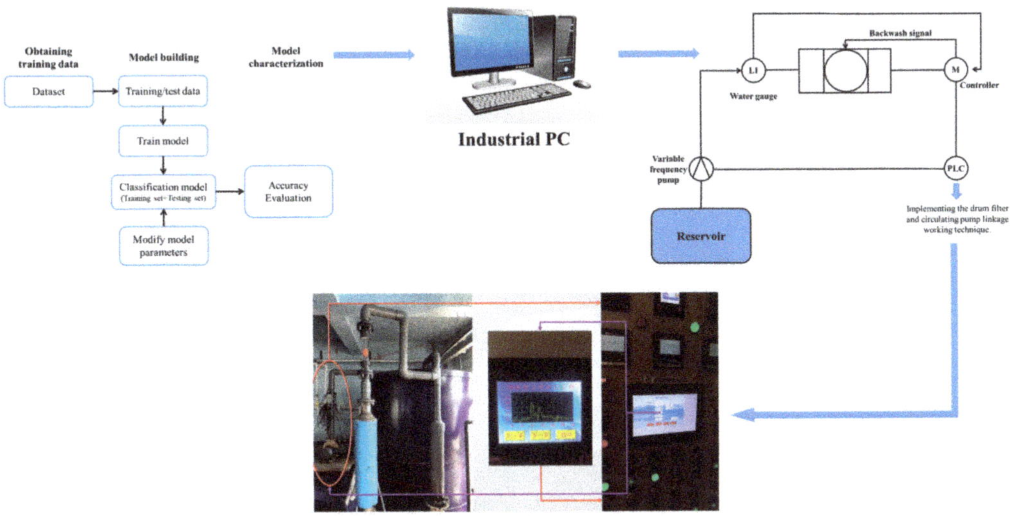

Figure 2. Design of control system in the variable-flow RAS.

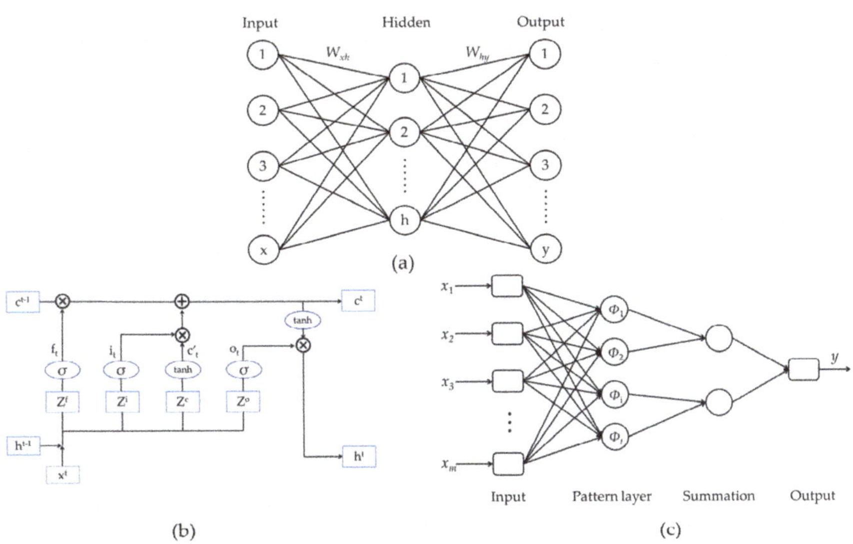

Figure 3. Artificial neural network architectures: (**a**) the typical structure of an ANN; (**b**) the LSTM architecture; (**c**) the PNN architecture.

The activation function can introduce a nonlinear factor to the neuron so that the ANN can approximate any nonlinear function. In the present study, a sigmoid function was adopted in the BPNN model and ELM model. For the sigmoid activation function, it holds that

$$f(z) = \frac{1}{1 + exp(-z)}, \tag{1}$$

where the output of the sigmoid function is between 0 and 1. For the binary classification task, the output of the sigmoid is divided into a positive class/negative class when the output satisfies a certain probability condition.

Figure 3b shows the schematic of the LSTM network. The LSTM network is a special RNN focusing on long sequences of data [27]. A standard LSTM unit comprises a cell, an input gate, an output gate, and a forget gate to solve the long-term dependency problem. Long-term memory information is stored during three steps (forgetting, remembering, and outputting) in an LSTM. In the present study, a rectified linear unit (ReLU) function was applied in the LSTM model. The ReLU function is described as

$$f(x) = \max(0, x), \tag{2}$$

which means that

$$\text{ReLU}(x) = \begin{cases} x, & x > 0 \\ 0, & x \leq 0 \end{cases}. \tag{3}$$

The convergence rate of the stochastic gradient descent obtained by the ReLU function is much faster than the tanh/sigmoid function. However, the learning rate should be set appropriately to prevent neurons in the network from losing their activation ability. In this study, the parameters of the LSTM training process were set as follows: sequence input layer = 9, initial learning rate = 0.01, learning rate drop factor = 0.1, batch size = 128, number of training epochs = 200, hidden layer = 1 (with 32 hidden units). Adaptive moment estimation (Adam) was chosen as the optimization method. The fully connected layer was set as 2 for the binary classification task.

Figure 3c shows the architecture of a typical PNN, which was first proposed by Dr. D.F. Specht [28]. As a branch of a radial basis network, PNN has the advantages of a simple learning process and fast training time. Therefore, PNN models can be well implemented in hardware since the neuron number in each layer is fixed. Generally, a PNN network contains four layers: input layer, pattern layer, summation layer, and output layer. The input layer simply distributes the input to the neurons in the pattern layer. The pattern layer neuron may compute its output by Gaussian function when receiving x from the input layer. It holds that

$$y_g(x;\sigma) = \frac{1}{l_g(2\pi)^{n/2}\sigma^n} \sum_{i=1}^{l_g} exp(-\sum_{j=1}^{n} \frac{(x_{ij}{}^{(g)} - x_j)^2}{2\sigma^2}), \tag{4}$$

where l_g denotes the total number of samples, n is the input feature, sigma represents the smoothing parameter, and x_{ij} represents the j-th data of the i-th neuron of the class g. The summation layer connects the pattern layer units of each class, and then the output layer is responsible for outputting the category with the highest score in the summation layer. K-fold cross-validation is useful for preventing models with small datasets from overfitting but is not used too frequently in deep learning. The dataset is equally divided into k parts. Every time a unique fold is used as a validation subset, the remaining pattern examples train the ANN. In this study, we introduced 4-fold cross-validation to evaluate the machine learning models. The evaluation indicators were all calculated by averaging the 4-fold cross-validation results.

2.3.2. Support Vector Machine (SVM)

An SVM has excellent generalization ability between model complexity and learning ability when dealing with limited sample information [29]. In SVM applications, choosing the appropriate kernel function and suitable parameters is crucial for prediction accuracy. As for the linear separable binary classification, finding the optimal hyperplane that divides all samples with maximum margin is the principal function of an SVM. For linear problems, the optimal classification hyperplane in separating two classes of training vector sets D is

$$D = \left\{ \left(x^1, y^1\right), \ldots, \left(x^l, y^l\right) \right\}, \; x \in R^n, y \in (-1, 1). \tag{5}$$

The plane was assumed as

$$< w, x > + b = 0, \tag{6}$$

When the optimal classification surface is generated, the vectors are classified without error, and when redundancy occurs, a typical hyperplane is assumed where w and b are constrained:

$$min_i \left| < w, x^i > + b \right| = 1. \tag{7}$$

The classification hyperplane in the regular form must satisfy the following constraints:

$$y^i \left[< w, x^i > + b \right] \geq 1, \; i = 1, \ldots, l. \tag{8}$$

The coordinate of the point x in the hyperplane at a distance $d(w, b; x)$ is

$$d(w, b; x) = \frac{\left| < w, x^i > + b \right|}{||w||}, \tag{9}$$

The final hyperplane that can satisfy the separated samples is the hyperplane that minimizes the data:

$$\Phi(w) = \frac{1}{2}||w||^2. \tag{10}$$

For nonlinear classification, the idea of SVM is to map the samples to a high-dimensional space, where the nonlinear problem is transformed into a linear solution using a kernel function, at which point the weight w is expressed as

$$w = \sum_{i=1}^{l} \alpha_i y_i \Phi(x_i). \tag{11}$$

Introducing the relaxation variable $\xi(\xi \geq 0)$ describing the function interval, the optimization equation under the kernel approach is expressed as

$$\min \alpha + C \sum_{i=1}^{l} \xi. \tag{12}$$

The model is described as

$$\begin{cases} y_l \left(\sum\limits_{j=1}^{l} \alpha_j y_j K(x_j, y_i) + b \right) \geq 1 - \xi_i, \; i = 1, \ldots, l \\ \alpha \geq \alpha_j, \; j = 1, \ldots, l \\ \alpha \geq -\alpha_j, \; j = 1, \ldots, l \\ \alpha, b \in R, \xi_i \geq 0, i = 1, \ldots, l \end{cases} . \tag{13}$$

In the present study, the SVM model was adopted to control the inverter frequency to improve circulating pump operating efficiency under different water quality conditions. The SVM is a kind of machine learning algorithm with a high generalization ability to classify and predict small samples. As upshifting and downshifting of the circulating pump is a binary problem, water quality indicators as variables can provide good generalization ability for the model. Support vector classification (SVC) can be used as the core algorithm for developing drum filter-circulating pump linkage technology. However, there is no international standard for selecting optimal parameters, and the parameter selection principles are based on dataset performance and the construction of a more reliable solution through cross-validation methods [30,31]. Here, we used the Gaussian kernel function in resolving the nonlinear support vector classification task:

$$K(x, z) = exp\left(-\frac{||x - z||^2}{2\sigma^2}\right), \tag{14}$$

For the SVM model, the penalty parameter C and RBF kernel parameter g need to be decided to improve the classification accuracy. In the present study, several optimizing algorithms, including grid search (GS), least squares method (LS), genetic algorithm (GA), and cuckoo search (CS) algorithm, were applied to improve the classification performance of the SVM model. The parameters of GA were set as follows: max generation = 300, population size = 50, generation gap = 0.9, range of parameter c = (0, 100), range of parameter g = (0, 1000). For the CS algorithm, the parameters were set as follows: iteration = 300, number of nests = 20, probability = 0.25. The best parameters of GS and LS methods were obtained through the traversal method; the ranges of c and g were set as (0, 100) and (0, 1000), respectively. K-fold cross-validation was utilized in the SVM models to prevent overfitting, and the evaluation indicators were calculated using averaging. The optimal SVM model can be determined by comparing the evaluation indicators of classification results from different algorithms.

3. Results

3.1. Data Processing for Variable-Flow Regulation

Ranges of the water quality data and backwash frequency from the measurements at three total circulation rates in RAS are shown in Table 1. The variable-flow regulation was decided by the frequency of the circulating pump. The upshifting and downshifting of the circulating pump inverter as two indicators of the classifier were labeled as 1 (upshift) and −1 (downshift) in the dataset. In order to develop the variable-flow regulation models based on the machine learning methods, water quality indicators, current circulation flow rate, and current backwash frequency were used as input variables, and regulating data (upshift/downshift) for the next period (0.5 h) were used as output variables. Upshift/downshift data were labeled by manual marking. The marking principal was decided from the variable-flow experiments under three circulation rates in RAS. The binary classification models can be applied for variable-flow regulation strategy, and the current data for water quality indicators and backwash frequency can be used to determine the total circulation rates for the next period through the classification models.

Table 1. Ranges of water quality indicators at three total circulation rate levels.

Indicator	55 m^3/h	65 m^3/h	75 m^3/h
Temperature (°C)	27.30~27.70	27.40~27.70	27.40~27.80
pH	6.90~7.50	6.91~7.50	6.91~7.49
DO (mg/L)	7.00~8.95	7.00~8.92	7.01~8.97
Salinity (‰)	30.50~30.90	30.50~30.90	30.50~30.90
TAN (mg/L)	0.18~2.68	0.20~1.36	0.19~1.13
Nitrite nitrogen (mg/L)	0.06~1.91	0.07~0.81	0.05~0.72
TSS (mg/L)	11.60~33.54	12.06~42.40	8.85~26.25
Turbidity (NTU)	3.52~8.98	3.63~13.68	2.83~8.87
Backwash frequency (Times per 0.5 h)	1~9	4~18	3~22

A total of 375 datasets were collected in the experiment, of which 280 were used as the training set and 95 as the test set. The training data were normalized after data pre-processing. The first step in developing the machine learning models was to simplify the explanatory variables by principal component analysis (PCA). PCA can reduce the complexity of the dataset and reveal hidden structures. The simplified principal components can be used as valid indicators to develop models. Figure 4 illustrates that the simplified variables reduced the original dataset from nine dimensions (water quality indicators) to three dimensions and could reflect 99% of the information in the original independent variables. However, the key components extracted from the original data were compressed and mapped to another space, and the simplified variables were not directly related to the original data [32]. Hence, in the present study, PCA successfully provided the optimal reduced representation for the data. The new dataset could then be used to develop machine learning models to reduce the complexity of the computation process.

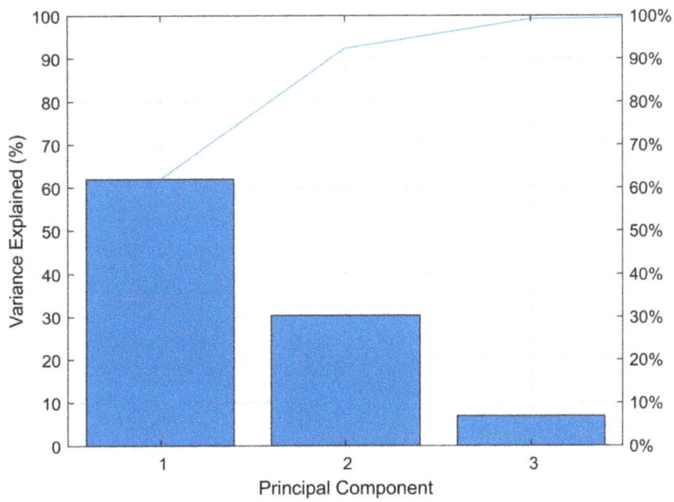

Figure 4. PCA process for reducing the complexity of water quality indicators.

3.2. *Intelligent Variable-Flow Models*

3.2.1. Results of the ANN Models

ANN classification models, including GA-BP, ELM, PNN, and LSTM, were used to adjust the circulating pump's frequency. The upshifting operation of circulating pump frequency was labeled as 1, and downshifting operation was labeled as −1. The classification process was regarded as a binary classification problem. The classification accuracy of both training set and test set data was calculated. For the BPNN model, the GA algorithm was applied to optimize the model performance. Models were tested by cross-validation to prevent the overfitting problem. ANN models were implemented by programming in Python 3.8.5 [33]. For the BPNN model, the maximum epoch was set to 1000 iterations, and the learning rate was set to 0.01 during the training process. The GA method optimized the BPNN model with the lowest error rate (2.59%) at 25 generations. The GA-BP model had the best validation performance (0.12) at epoch 142. For the LSTM model training process, loss and accuracy gradually converged after 350 iterations. The accuracy of the training set reached 100% when the loss was below 0.05.

Table 2 presents the evaluations of the ANN classification models. Results showed that the training accuracy of all the ANN models was beyond 90%. PNN and LSTM achieved the most accurate classification (100%). For the test set, the LSTM model had a 96.84% accuracy rate; however, the accuracy rates of other models were less than 90%. Thus, the optimal model was identified as the LSTM model, with the highest accuracy for both the training set (100%) and test set (96.84%) among the ANN models.

3.2.2. Results of the SVM Models

The SVM models were developed in Python 3.8.5. As classification accuracy is directly related to the optimal parameters of the SVM model, we used several optimizing methods to determine penalty parameter c and the kernel parameter g in the present study. Table 3 shows the optimizing methods for SVM models. The optimized parameters were determined by the grid search, least squares, cuckoo search, and gene algorithm.

Table 2. Classification accuracy of training sets and test sets from ANN models.

ANNs	Training Accuracy	Test Accuracy
GA-BP	92.14%	86.32%
ELM	95.00%	89.47%
PNN	100.00%	71.58%
LSTM	100.00%	96.84%

Table 3. Classification accuracy of training sets and test sets from SVM models.

Optimizations	Training Accuracy	Test Accuracy	Best Gamma	Best c
Grid Search	100.00%	97.89%	0.0039	48.50
Least Square	94.29%	96.84%	250.00	20.00
Cuckoo Search	100.00%	97.89%	0.56	25.51
Genetic Algorithm	100.00%	98.95%	0.33	66.43

As Table 3 shows, the accuracy rates of classification results of the SVM models were maintained at relatively high levels. The least squares method had 94.29% accuracy, and other methods all had 100% accuracy rates for the training set. The test set from the gene algorithm optimized support vector machine (GA-SVM) model had one set of data classified with the wrong label among 95 groups (accuracy 98.95%). The grid search optimized support vector machine (GS-SVM) and the cuckoo search optimized support vector machine (CS-SVM) both had two error sets (97.89%). For the least squares support vector machine (LS-SVM), the test set results exhibited lower accuracy (96.84%) than other methods. Thus, the GA-SVM was identified as the optimal SVM classification model through comprehensive comparison. Table 3 shows that the four searching algorithms optimized the parameters (penalty c and kernel radius g). Although the accuracy could be maintained at a high level, the ranges of the optimized parameters of the SVM models were quite different. Therefore, it was necessary to further select the SVM model through evaluation indicators.

3.3. Model Evaluation

The confusion matrix, which comprehensively reflects the performance of the classifiers, can derive many evaluation indicators. Here, the calculated evaluation indicators, including accuracy, precision, recall, and F1-score, were used to evaluate classification performance for the binary classifier. The SVM model was estimated by 4-fold cross-validation, and the indicators were computed by averaging the folds. Accuracy represents the ratio of correct samples to the total samples without considering the positive and negative. Recall refers to the ratio of the correctly classified positive samples to the total true positive samples, and precision refers to the ratio of correctly classified positive samples to all classified positive samples. The F1-score indicator is proposed based on precision and recall to evaluate the indicators as a whole. The F1-score can be used to comprehensively consider the pros and cons of the classification models.

Table 4 shows the results of model evaluation indicators for machine learning classifiers. Figure 5 shows the histograms of the evaluation indicators (accuracy and F1-score) of the training set and test set from machine learning classification models. According to the summaries of the model evaluation indicators, GA-SVM shows both higher accuracy and F1-score than other machine learning methods. Accuracy can reflect the classification correctness of the global results of the model. The F1-score can reflect the weighted average between precision and recall, and the results show that the GA-SVM classifier can be considered as an optimal model for drum filter-circulating pump linkage technology in a variable-flow RAS because the model indicators satisfied the criteria.

Table 4. Results of model evaluation indicators for machine learning classifiers.

Methods	Accuracy	Precision	Recall	F1-Score
GA-BP	92.14% (Train) 86.32% (Test)	95.21% (Train) 90.74% (Test)	92.98% (Train) 85.96% (Test)	94.08% (Train) 88.29% (Test)
ELM	95.00% (Train) 89.47% (Test)	96.41% (Train) 91.23% (Test)	95.27% (Train) 91.23% (Test)	95.84% (Train) 91.23% (Test)
PNN	100.00% (Train) 71.58% (Test)	100.00% (Train) 81.48% (Test)	100.00% (Train) 72.13% (Test)	100.00% (Train) 76.52% (Test)
LSTM	100.00% (Train) 96.84% (Test)	100.00% (Train) 94.34% (Test)	100.00% (Train) 100% (Test)	100.00% (Train) 93.83% (Test)
GS-SVM	100.00% (Train) 97.89% (Test)	100.00% (Train) 98.11% (Test)	100.00% (Train) 98.11% (Test)	100.00% (Train) 98.11% (Test)
LS-SVM	94.29% (Train) 96.84% (Test)	94.22% (Train) 96.55% (Test)	96.45% (Train) 98.25% (Test)	95.32% (Train) 97.39% (Test)
CS-SVM	100.00% (Train) 97.89% (Test)	100.00% (Train) 98.11% (Test)	100.00% (Train) 98.11% (Test)	100.00% (Train) 98.11% (Test)
GA-SVM	100.00% (Train) 98.95% (Test)	100.00% (Train) 98.36% (Test)	100.00% (Train) 100% (Test)	100.00% (Train) 99.17% (Test)

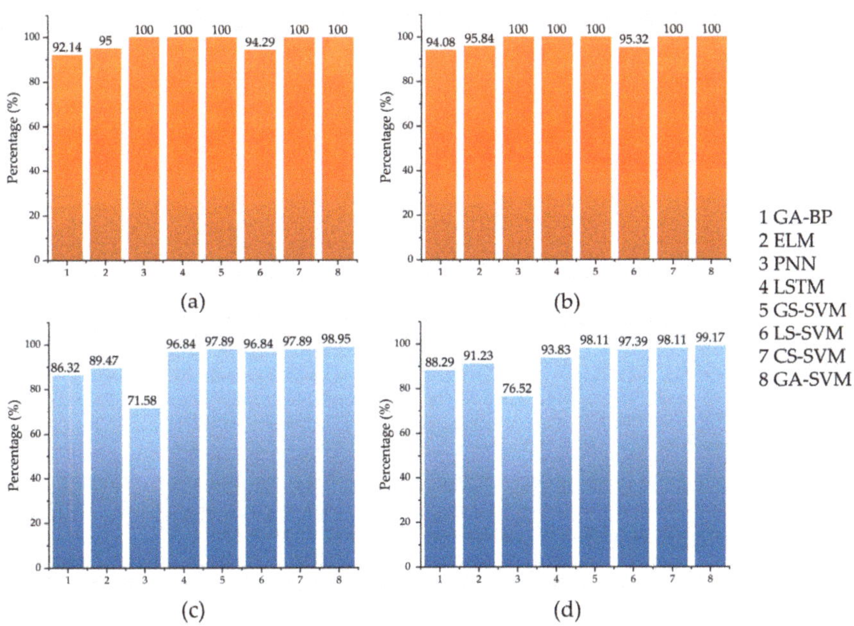

Figure 5. Evaluation indicators of classification models based on machine learning: (**a**) accuracy of the training sets; (**b**) F1-score of the training sets; (**c**) accuracy of the test sets; (**d**) F1-score of the test sets.

4. Discussion

Feces and residual feed may decompose to organic suspended solids, which further generate TAN and nitrite, harming breeding animals' health. Suspended solids in the RAS also provide surface area that can be colonized by bacteria. As circulation intensity increases, more particles accumulate, which may increase the bacterial carrying capacity of the system. Hence, rapid removal of solid waste is the most critical unit process in an RAS [34]. The traditional method of water quality regulation in an RAS is to act when water quality deteriorates. This approach leads to large fluctuations in the water environment, and the cost of water quality regulation becomes very high, often requiring many water exchanges to control water quality. This study proposes regulation of RAS circulation based on process control technology, relying on the microfilter backwash times in a unit period (0.5 h) as the main parameter to reflect the overall turbidity of the water body. The variable-flow RAS circulation strategy was designed to form microfilter-circulating pump linkage technology based on water quality parameters and backwash times at different flow rates. An intelligent variable-flow regulation model was developed to keep the water clean and quickly and dynamically remove suspended solids.

Related research has proven the significant differences in water quality between the high and low makeup water exchange treatment groups [35]. One study has shown that increasing RAS water circulation can effectively reduce ammonia and nitrite [36]; the higher the circulation level, the lower the ammonia and nitrite mass concentrations became. Moreover, the conversion of nitrite revealed a certain hysteresis, and the ammonia peak appeared earlier than the nitrite peak after feeding was stopped.

RAS solids come mainly from uneaten feed and fecal solids, and the decomposition and mineralization of these solids lead to elevated ammonia and nitrite levels in the RAS [10]. Data such as TAN, NO2-N, and TSS must be obtained by manual measurement and are challenging to obtain by sensors. According to Vinatea et al. [37], TSS tended to accumulate in the intensive *L. vannamei* culture and was eventually reflected in an increase in NTU. As both turbidity and TSS can reflect the clarity of a liquid, the turbidity parameter was used for modeling in this study. The principal component analysis (PCA) results for dimensionality reduction showed that turbidity, dissolved oxygen, pH, and temperature could be used as the leading indicators for modeling. The variable-flow regulation model obtains the current water quality indicators in real time and then applies these indicators to predict and classify the circulation rate for the next period. The turbidity sensor in turbulent flow had a measured data fluctuation that was too large, and the sensor arrangement position also caused measurement errors. An innovative point of this study is that the drum filter backwash frequency over a certain period was used as one of the critical factors for modeling instead of the momentary RAS water turbidity. Backwash times can effectively replace turbidity reading to reflect overall RAS water turbidity, avoiding the instability of the data collected by the turbidity sensor.

The application of machine learning methods in aquaculture-related research is focused mainly on the prediction, classification, and evaluation of water quality indicators such as dissolved oxygen, salinity, pH, ammonia, and nitrite [25]. In the present study, machine learning was used to model the variable-flow regulation strategy. Sensors collected water indicators, including DO, pH, temperature, and turbidity. In order to implement the variable-flow principle, the machine learning methods were introduced in the present study to develop the optimal variable-flow regulation model for RAS. The water quality indicators, the backwash frequency, and the circulating pump frequency were obtained through continuous monitoring. For the ANN methods, the LSTM model was identified as the optimal regulation model, since the accuracy and F1-score indicators reflected the strong ability of the LSTM classifier. The modeling data based on time series were collected from the continuously running RAS in the present study. The water quality indicators, backwash frequency, and total circulation rates were recorded through the fixed time interval during the whole rearing period. Research has shown that LSTM can indeed perform well in processing long time series sequences of data [38]. The optimal classification model needs to be relatively simple in order to be applied in the embedded devices. The variable-flow adjustment strategy in RAS also needs to respond quickly and satisfy the high standard of classification accuracy. All the evaluated indicators of the SVM models demonstrated better results compared with the LSTM model. The gene algorithm contributed the highest accuracy and F1-score among the four optimization algorithms in the classification task. As a supervised algorithm, GA-SVM can be applied to effectively adjust water refreshment in RAS.

In future work on variable-flow RAS regulation, the data-driven model needs to be improved to establish continuous variable-flow control technology by adjusting circulating pump frequency. A larger quantity of data from the running RAS can ensure higher availability and robustness for optimizing the intelligent variable-flow strategy. The continuous variable-flow control technology prerequisite is required for the indicators (water quality, backwash frequency, and rearing cycle) to correspond to the ideal circulation volume. Furthermore, the interaction effects between various indicators need to be revealed through experiments and analysis. The ultimate goal of the study is to achieve a precise circulation control strategy in the RAS and execute rapid water treatment without affecting the health of the reared animals.

5. Conclusions

A variable-flow regulation model was established in the present study to implement the circulating pump-drum filter linkage working technique. Classification models based on machine learning methods between the explanatory variables and the regulation strategy were developed based on experimental data. ANN models including GA-BP, LSTM, PNN, and ELM were established. The LSTM model had the highest accuracy (training set 100%, test set 96.84%) and F1-score (training 100%, test 93.83%) and was regarded as the best classification model among ANN methods. SVM models were developed and optimized using linear squares, grid search, cuckoo search, and gene algorithm. Results showed that SVM models required less training time and exhibited higher accuracy compared with ANN models. Finally, the optimal model was GA-SVM, with the highest classification accuracy (training 100%, test 98.95%) and F1-score (training 100%, test 99.17%). The model was tested under cross-validation with precise classification performance and used for the circulating pump-drum filter intelligent linkage working technique.

Author Contributions: F.C. conducted the experiments, analyzed the data, developed the models, and wrote the paper. Y.D. and T.Q. reviewed the manuscript. Z.X. designed the RAS. L.Z., J.X., M.S. and Y.L. conducted the shrimp rearing experiments. J.S. conceived and designed the experiments. All authors have read and agreed to the published version of the manuscript.

Funding: This research was funded by the Key Program for International Cooperation on Scientific and Technological Innovation, Ministry of Science and Technology of the People's Republic of China, grant number 2017YFE0118300, and also funded by the National Key R&D Programs of China, grant numbers 2019YFD0900800 and 2019YFD0900502.

Institutional Review Board Statement: Not applicable.

Informed Consent Statement: Not applicable.

Data Availability Statement: The data presented in this study are available on request from the corresponding author. The data are not publicly available due to privacy or ethical restrictions.

Acknowledgments: We are grateful to the reviewers and the editors for their valuable and insightful comments.

Conflicts of Interest: The authors declare no conflict of interest.

References

1. FAO. *The State of World Fisheries and Aquaculture 2016*; Publications of Food and Agriculture Organization of the United Nations: Rome, Italy, 2016; p. 200.
2. Zhang, S.Y.; Li, G.; Wu, H.B.; Liu, X.G.; Yao, Y.H.; Tao, L.; Liu, H. An integrated recirculating aquaculture system (RAS) for land-based fish farming: The effects on water quality and fish production. *Aquac. Eng.* **2011**, *45*, 93–102. [CrossRef]
3. Ebeling, J.M.; Timmons, M.B. *Recirculating Aquaculture*, 3rd ed.; Ithaca Publishing Company: New York, NY, USA, 2010; pp. 171–474.

4. Badiola, M.; Mendiola, D.; Bostock, J. Recirculating Aquaculture Systems (RAS) analysis: Main issues on management and future challenges. *Aquac. Eng.* **2012**, *51*, 26–35. [CrossRef]
5. Chen, S.; Stechey, D.; Malone, R.J.D. Suspended solids control in recirculating aquaculture systems. *Dev. Aquac. Fish. Sci.* **1994**, *27*, 61.
6. Noble, A.C.; Summerfelt, S.T. Diseases encountered in rainbow trout cultured in recirculating systems. *Annu. Rev. Fish. Dis.* **1996**, *6*, 65–92. [CrossRef]
7. Wedemeyer, G. *Physiology of Fish in Intensive Culture Systems*, 1st ed.; Springer Science & Business Media: Boston, MA, USA, 1996; pp. 36–80.
8. Cripps, S.J.; Bergheim, A.J.A. Solids management and removal for intensive land-based aquaculture production systems. *Aquac. Eng.* **2000**, *22*, 33–56. [CrossRef]
9. Alabaster, J.S.; Lloyd, R.S. *Water Quality Criteria for Freshwater Fish*, 2nd ed.; Elsevier: Amsterdam, The Netherland, 2013; pp. 4–6.
10. Chiam, C.k.; Sarbatly, R.J.S.; Reviews, P. Purification of aquacultural water: Conventional and new membrane-based techniques. *Sep. Purif. Rev.* **2011**, *40*, 126–160. [CrossRef]
11. Xiao, R.; Wei, Y.; An, D.; Li, D.; Ta, X.; Wu, Y.; Ren, Q. Review on the research status and development trend of equipment in water treatment processes of recirculating aquaculture systems. *Rev. Aquac.* **2019**, *11*, 863–895. [CrossRef]
12. Vilbergsson, B.; Oddsson, G.V.; Unnthorsson, R. Taxonomy of means and ends in aquaculture production—Part 2: The technical solutions of controlling solids, dissolved gasses and pH. *Water* **2016**, *8*, 387. [CrossRef]
13. Gravdal, A. Process and Means for the Treatment of Water in an Aquaculture System. U.S. Patent 7,052,601, 30 May 2006.
14. Su, M.; Liu, H.; Song, H.; Hu, B. Study on the TSS removal efficiency and energy consumption of micro-screen drum filter. *Fish. Mod.* **2008**, *35*, 9–12.
15. Malone, R.J.R. *Recirculating Aquaculture Tank Production Systems*; Southern Regional Aquaculture Center: Stoneville, MS, USA, 2013; p. 12.
16. Prabhu, P.A.J.; Kaushik, S.; Geurden, I.; Stouten, T.; Fontagne-Dicharry, S.; Veron, V.; Mariojouls, C.; Verreth, J.; Eding, E.; Schrama, J. Water exchange rate in RAS and dietary inclusion of micro-minerals influence growth, body composition and mineral metabolism in common carp. *Aquaculture* **2017**, *471*, 8–18. [CrossRef]
17. Di Nunno, F.; Granata, F.; Gargano, R.; de Marinis, G. Forecasting of extreme storm tide events using NARX neural network-based models. *Atmosphere* **2021**, *12*, 512. [CrossRef]
18. Di Nunno, F.; Granata, F.; Gargano, R.; de Marinis, G. Assessment, Prediction of spring flows using nonlinear autoregressive exogenous (NARX) neural network models. *Environ. Monit. Assess.* **2021**, *193*, 1–17. [CrossRef] [PubMed]
19. López-Cortés, X.A.; Nachtigall, F.M.; Olate, V.R.; Araya, M.; Oyanedel, S.; Diaz, V.; Jakob, E.; Ríos-Momberg, M.; Santos, L.S. Fast detection of pathogens in salmon farming industry. *Aquaculture* **2017**, *470*, 17–24. [CrossRef]
20. Zhou, C.; Lin, K.; Xu, D.; Chen, L.; Guo, Q.; Sun, C.; Yang, X. Near infrared computer vision and neuro-fuzzy model-based feeding decision system for fish in aquaculture. *Comput. Electron. Agric.* **2018**, *146*, 114–124. [CrossRef]
21. Chen, Y.; Fang, X.; Yang, L.; Liu, Y.; Gong, C.; Di, Y. Artificial Neural Networks in the Prediction and Assessment for Water Quality: A Review. *J. Phys. Conf. Ser.* **2019**, *1237*, 042051. [CrossRef]
22. Zhou, C.; Xu, D.; Chen, L.; Zhang, S.; Sun, C.; Yang, X.; Wang, Y. Evaluation of fish feeding intensity in aquaculture using a convolutional neural network and machine vision. *Aquaculture* **2019**, *507*, 457–465. [CrossRef]
23. Li, D.; Wang, Z.; Wu, S.; Miao, Z.; Du, L.; Duan, Y. Automatic recognition methods of fish feeding behavior in aquaculture: A review. *Aquaculture* **2020**, *528*, 735508. [CrossRef]
24. Li, H.; Zhang, Z.; Zhao, Z.-Z. Data-mining for processes in chemistry, materials, and engineering. *Processes* **2019**, *7*, 151. [CrossRef]
25. Chen, Y.; Song, L.; Liu, Y.; Yang, L.; Li, D. A review of the artificial neural network models for water quality prediction. *Appl. Sci.* **2020**, *10*, 5776. [CrossRef]
26. Li, H.; Liu, Z.; Liu, K.; Zhang, Z.J. Predictive power of machine learning for optimizing solar water heater performance: The potential application of high-throughput screening. *Int. J. Photoenergy* **2017**, *2017*. [CrossRef]
27. Granata, F.; Di Nunno, F. Forecasting evapotranspiration in different climates using ensembles of recurrent neural networks. *Agric. Water Manag.* **2021**, *255*, 107040. [CrossRef]
28. Specht, D.F. Probabilistic neural networks. *Neural Netw.* **1990**, *3*, 109–118. [CrossRef]
29. Vapnik, V.; Guyon, I.; Hastie, T. Support vector machines. *Mach. Learn.* **1995**, *20*, 273–297.
30. Chen, F.; Li, H.; Xu, Z.; Hou, S.; Yang, D.J. User-friendly optimization approach of fed-batch fermentation conditions for the production of iturin A using artificial neural networks and support vector machine. *Electron. J. Biotechnol.* **2015**, *18*, 273–280. [CrossRef]
31. Li, H.; Chen, F.; Cheng, K.; Zhao, Z.; Yang, D. Prediction of zeta potential of decomposed peat via machine learning: Comparative study of support vector machine and artificial neural networks. *Int. J. Electrochem. Sci.* **2015**, *10*, 6044–6056.
32. Ringnér, M. What is principal component analysis? *Nat. Biotechnol.* **2008**, *26*, 303–304. [CrossRef] [PubMed]
33. Bowles, M. *Machine Learning in Python: Essential Techniques for Predictive Analysis*, 1st ed.; John Wiley & Sons: Indianapolis, IN, USA, 2015; pp. 255–319.
34. Summerfelt, R.C.; Penne, C.R. Solids removal in a recirculating aquaculture system where the majority of flow bypasses the microscreen filter. *Aquac. Eng.* **2005**, *33*, 214–224. [CrossRef]

35. Davidson, J.; Good, C.; Welsh, C.; Summerfelt, S. The effects of ozone and water exchange rates on water quality and rainbow trout Oncorhynchus mykiss performance in replicated water recirculating systems. *Aquac. Eng.* **2011**, *44*, 80–96. [CrossRef]
36. Fivelstad, S.; Binde, M. Effects of reduced waterflow (increased loading) in soft water on Atlantic salmon smolts (*Salmo salar* L.) while maintaining oxygen at constant level by oxygenation of the inlet water. *Aquac. Eng.* **1994**, *13*, 211–238. [CrossRef]
37. Vinatea, L.; Gálvez, A.O.; Browdy, C.L.; Stokes, A.; Venero, J.; Haveman, J.; Lewis, B.L.; Lawson, A.; Shuler, A.; Leffler, J.W. Photosynthesis, water respiration and growth performance of Litopenaeus vannamei in a super-intensive raceway culture with zero water exchange: Interaction of water quality variables. *Aquac. Eng.* **2010**, *42*, 17–24. [CrossRef]
38. Siami-Namini, S.; Tavakoli, N.; Namin, A.S. A comparison of ARIMA and LSTM in forecasting time series. In Proceedings of the 2018 17th IEEE International Conference on Machine Learning and Applications (ICMLA), Orlando, FL, USA, 17–20 December 2018; IEEE: Piscataway, NJ, USA, 2018; pp. 1394–1401.

applied sciences

MDPI

Article

Generation of Typical Meteorological Sequences to Simulate Growth and Production of Biological Systems

Ousmane Wane [1,2,3], Luis F. Zarzalejo [1,*], Francisco Ferrera-Cobos [1], Ana A. Navarro [1], Alberto Rodríguez-López [1] and Rita X. Valenzuela [1]

1 CIEMAT Energy Department, Renewable Energy Division, Avda. Complutense 40, 28040 Madrid, Spain
2 E.T.S.I. Agronómica, Alimentaria y Biosistemas, Universidad Politécnica de Madrid, Avda. Puerta Hierro 2-4, 28040 Madrid, Spain
3 Department of Microbial and Plant Biotechnology, Center for Biological Research Margarita Salas–CSIC, Ramiro de Maeztu 9, 28040 Madrid, Spain
* Correspondence: lf.zarzalejo@ciemat.es

Abstract: Numerical simulation applied to agriculture or wastewater treatment (WWT) is a complementary tool to understand, a priori, the impact of meteorological parameters on productivity under limiting environmental conditions or even to guide investments towards other more relevant circular economic objectives. This work proposes a new methodology to calculate Typical Meteorological Sequences (TMS) that could be used as input data to simulate the growth and productivity of photosynthetic organisms in different biological systems, such as a High-Rate Algae Pond (HRAP) for WWT or in agriculture for crops. The TMS was established by applying Finkelstein-Schafer statistics and represents the most likely meteorological sequence in the long term for each meteorological season. In our case study, 18 locations in the Madrid (Spain) region are estimated depending on climate conditions represented by solar irradiance and temperature. The parameters selected for generating TMS were photosynthetically active radiation, solar day length, maximum, minimum, mean, and temperature range. The selection of potential sequences according to the growth period of the organism is performed by resampling the available meteorological data, which, in this case study, increases the number of candidate sequences by 700%.

Keywords: typical meteorological sequence; typical meteorological week; wastewater treatment; high-rate algae pond; solar irradiance; Finkelstein-Schafer statistics

Citation: Wane, O.; Zarzalejo, L.F.; Ferrera-Cobos, F.; Navarro, A.A.; Rodríguez-López, A.; Valenzuela, R.X. Generation of Typical Meteorological Sequences to Simulate Growth and Production of Biological Systems. *Appl. Sci.* **2023**, *13*, 4826. https://doi.org/10.3390/app13084826

Academic Editors: Paweł Kiełbasa, Tadeusz Juliszewski and Sławomir Kurpaska

Received: 16 February 2023
Revised: 31 March 2023
Accepted: 7 April 2023
Published: 12 April 2023

1. Introduction

The increase and change in the consumption pattern in the population is generating serious energy problems, which affect, among other things, food production and wastewater treatment (WWT). Simulations play an important role in the previous implementation of systems that contribute to controlling these issues, since they represent a long-term approximation of the technical economic viability, contributing to deciding the appropriate configuration for its implementation in reality. Furthermore, efficiency in the use of water and energy in agriculture is an increasingly important issue due to the growing scarcity of the former and the increasing costs of the latter [1]. Both constrain crop irrigation in many areas of the world, conditioning productivity. However, the need for WWT constitutes a challenge in any population and economic activity, especially in rural areas and developing countries, where the use of activated sludge treatment systems can produce high capital and operating costs. For that reason, nature-based technologies have been proposed in small populations [2]. One of these technologies is a High-Rate Algae Pond (HRAP), which consists of the use of microbial populations present in wastewater and inoculated microalgae in the medium to obtain a metabolic coupling that produces WWT [3]. Microalgae-based processes are much simpler, impose low CAPEX (capital expenditure), and maintenance costs are also easier than in conventional systems due to the less machinery required and less energy consumption [4]. Furthermore, the HRAP performance can be accurately described using only two variables: pH and dissolved oxygen [5]. However, microalgae are very sensitive to variations in climatic parameters, such as temperature and irradiance [6,7], particularly photosynthetically active radiation (PAR).

Therefore, to manage these biological systems, it is important to use climatic data and their long-term estimates to adopt adequate irrigation strategies in areas with water scarcity to optimize crop productivity or for WWT with HRAP systems. In this case, it is possible to estimate, with appropriate simulation programs, not only the microalgal productivity in the process and estimate the energy generation by means of biogas and its use as an energy source in the plant itself but also the efficiency of the WWT system.

On the other hand, a Typical Meteorological Year (TMY) [8] developed by Sandia National Laboratories is a time sequence widely used to describe the most likely meteorological conditions (including solar radiation, temperature, humidity, and others) in an arbitrary location. It is made up of 12 months statistically selected and concatenated from a series of years to generate a complete year [9], which offers a representative climatology at a location in the long term [10]. The variability of the meteorological series generated by this methodology is greater than that of a series of variables consisting of climatic averages. However, a TMY is not necessarily a good indicator of the climatic conditions of a specific year in the future or extreme meteorological events. TMYs have been used in the simulation and estimation of energy produced for different renewable energy technologies and energy efficiencies, such as SAM (https://sam.nrel.gov/) (accessed on 6 September 2022), PVSyst (https://www.pvsyst.com/) (accessed on 6 September 2022), ESP-r (https://www.strath.ac.uk/research/energysystemsresearchunit/applications/esp-r/) (accessed on 6 September 2022), DOE-2 (https://www.doe2.com/) (accessed on 6 September 2022), TRNSYS (http://www.trnsys.com/) (accessed on 6 September 2022), EnergyPlus (https://energyplus.net/) (accessed on 6 September 2022), and others.

As has been said before, variability due to changes in climatology has a great impact on crop productivity. In this way, agricultural production models would require a long-term description of the climatology of the location to obtain estimates of the crop yield. In this context, TMY has been used in studies on greenhouse designs [11–13], as well as for the development of an optimal irrigation scheme for different crops under external conditions [1,9,14]. Furthermore, it could be used in the hypothetical simulation of hybrid energy systems based on the use of semitransparent photovoltaic energy in microalgae production greenhouses [15] or to study the appropriate material for the greenhouse to reduce its energy consumption as much as possible [16].

As far as has been possible to review, the Typical Meteorological Sequence (TMS) concept has not been applied to study the behavior of a WWT in a HRAP, considering a representative typical meteorological series corresponding to the hydraulic retention time necessary for WWT. A TMS made up of seven consecutive days, then called a Typical Meteorological Week (TMW), could be analyzed as a period comparable to the water retention time to be treated in the HRAP (our case study), although the typical period can vary, depending on its application. Therefore, it may be useful in the long term to know in which seasons of the year the system can be in operation and when it can be stopped, depending on the climate of the location. It may also be useful in the identification of optimal sites for the implementation of this system, since it is expected that the process should not be stopped due to excess or a lack of solar irradiance or temperature.

The main objective of this paper is the development of a methodology to generate a Typical Meteorological Sequence, a week for our case study, representative of each meteorological season. For this, the work is divided into the following sections: (1) description of the Sandia methodology, (2) case study, with an indication of the locations where the TMW are generated in the Madrid region, (3) application of the TMW methodology for each season of the year, and finally, the (4) results and discussion.

2. Methodology

Since the 1980s, a considerable number of studies have been presented for the generation of TMY using equations from Finkelstein-Schafer (FS) statistics according to the methodology proposed by Sandia National Laboratories. These studies are mostly established with different climatic indices, weighting coefficients, and persistence criteria in the final process of selecting the appropriate sequences. In this work, to determine the importance of meteorological parameters in the growing period of microalgae in HRAP for WWT or plants in agriculture for crops, the Sandia methodology was used considering different scenarios of weighting coefficients and dividing the dataset into intervals to define the FS statistic.

2.1. Sandia National Laboratories Methodology

The Sandia methodology is widely present in the literature and turns out to be one of the most common methods for calculating a TMY [8,17–20]. The TMY is obtained from multiannual historical series, for instance, 30 years (climate cycle), of different meteorological parameters: among others, temperature (mean, maximum, minimum, and range) and solar irradiance (global horizontal irradiance). At first, these parameters were data measured at the study site (26 SOLMET stations) for 23 years beginning in 1953 and extending through 1975 [8]. From the available daily time series, the Sandia methodology selected 12 Typical Meteorological Months (TMM) to establish information on the annual variability of the parameters studied. Using the FS statistic, a TMM is chosen for each of the 12 calendar months of all the years available in the time series. This was done by assigning a weighting factor (wf) to the meteorological parameters considered, resulting in a reduction in the amount of data, losing the least amount of information as possible [21,22]. The wf can vary, depending on the importance of the variable [23]. The dataset achieved represents a typical year of reliable data in the simulation of energy of renewable energy technologies [20].

In addition to using FS statistics for generating a TMY, some studies introduced other approaches, such as the principal component analysis or genetic algorithms [24,25]. There are other methodologies, different from those listed above, based on the availability of meteorological data and the application of the generated sequence. Among them are the Test Reference Year (TRY) [26,27], the Design Reference Year (DRY) [28], and the Short Reference Years (SRY) [29]. To date, these methodologies have had remarkable results compared to average long-term weather data from meteorological stations [19,21,30,31].

2.2. Case of Study

Crop simulation is important to know the morphological characteristics of the crop according to the meteorological parameters and to anticipate in decision-making on agriculture, food security, climate change, energy saving, etc. [32,33]. To address the importance of meteorological influence, in this work, the application of a modified methodology to generate a typical weather sequence is applied; in this case, a TMW is applied in order to be used in the growth simulation of microalgae in a HRAP in the Madrid region. On the other hand, studies have been done on microalgae productivity as a raw material in the generation of high value-added products or as a source of energy. Therefore, some authors have used estimates of climate variables (Cligen) to incorporate them into microbial growth models to estimate microalgae production [34,35].

Microalgae are phototrophic microorganisms that grow rapidly and reproduce in hours. Therefore, microalgae generate a large amount of biomass in a relatively short time compared to other living species.

Biomass production and WWT are affected by uncontrollable meteorological parameters that vary throughout the cultivation period. Among these parameters, the temperature and solar irradiance between 400 and 700 nm (PAR) [36] are indispensable for microalgae growth [37–41]. The work carried out in [37] shows that the observed reduction in the mean daily PAR radiation entering the greenhouse affects the plant metabolism. The same effect is observed when the temperature stress is applied to the crop [42].

Therefore, due to the short hydraulic retention time for microalgae development, a TMS per meteorological season is studied using the data for the Madrid region. The four TMWs to be generated, one for each meteorological season, are based on the PAR and temperature in 18 wastewater treatment plants (WWTP) that already exist in the Madrid region, as shown in Figure 1.

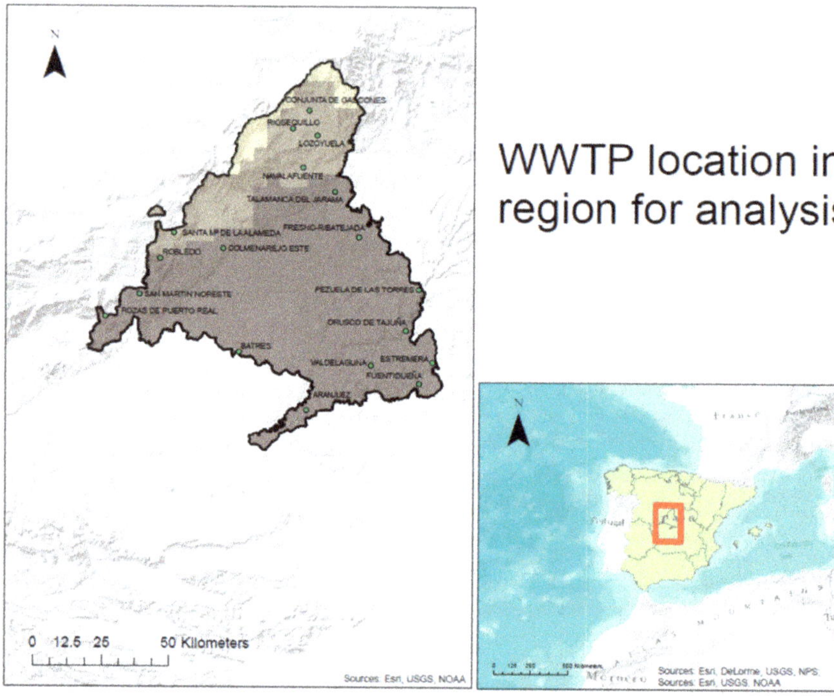

Figure 1. Location of WWTPs studied in the Madrid region.

Due to the availability of simultaneous PAR and temperature data in these locations of WWTPs, a 15-year set of PAR and daily mean, maximum, and minimum air temperature values was used. PAR has been obtained from Kato bands, provided by the spectral resolved irradiance (SRI) of the Satellite Application Facility on Climate Monitoring (CM-SAF), which belongs to the European Organization for the Exploitation of Meteorological Satellites (EUMETSAT) [43]. The daily mean temperature was obtained from the European Center for Medium-Range Weather Forecasts (ECMWF) [44]. The period used in the present study was 1991 to 2005, with a spatial resolution of 0.125° × 0.125°.

These four parameters are then grouped into a matrix in which two additional columns, the temperature range and the solar day length, were added by calculations. The latter is used to take into account the photoperiod; that is, the number of solar hours during which microalgae are exposed to PAR and the maximum possible duration of the solar day [45].

Madrid is almost located in the center of the Iberian Peninsula (between 41.15° N and 39.88° N latitude and between 3.05° W and 4.57° W longitude) on the Central Plateau, and the altitude ranges from 476 to 2428 m and the average is 678 m above sea level, with a surface area of approximately 8000 km^2. The orography of the Madrid region is characterized by the presence of the Central Mountain Range in the north and west of the territory, while the remaining areas are the plains and the Tajo River Valley. The climate of the region is strongly influenced by its orography. Therefore, in the range and its surroundings, there is a mountain climate (Dsc according to the Köppen-Geiger classification) [46–48] and an oceanic-Mediterranean climate (Csb). On the other hand, there is a typical Mediterranean climate (Csa) in the plains and a semi-arid climate (BSk) in the southern areas and around the Tajo River Valley. This variation in climate could be estimated by generating TMS that reflect this variability in the growth of microalgae.

2.3. Applying the TMS Methodology

The growth and productivity of microalgae is challenged by multiple cultivation parameters, such as pH, nutrients, light, temperature, agitation, cultivation medium, etc. For our case, only physical parameters, solar irradiance (PAR and solar day length), and temperatures are considered in this scientific approach to study the effects of both parameters on microalgae activity.

Therefore, the existence of values of these parameters at which the culture is at its optimum level or not leads us to consider different weight factor cases for temperature and irradiance. For this purpose, an approach based on TMY methodologies is used to determine the importance of the meteorological parameters in the growth period of microalgae. Moreover, as these two cultivation parameters change significantly between two meteorological seasons, a seasonal approach is adopted. Each season is examined separately, following a multistage process.

Firstly, the whole set of data available (15 years in our case study) has been distributed in the four seasons ($E_{w=1,2,3,4}$): 1 = spring (March–May), 2 = summer (June–August), 3 = autumn (September–November), and 4 = winter (December–February) in that order. These seasons are based on the annual temperature cycle and not on the astronomical seasons, so there is a clear transition between them. For each location, there is a time series corresponding to 14 seasons ($A_{y=1,2,\ldots,14}$) for each of these four weather seasons. Thereafter, a period of time is identified as follows E_wA_y. For example, E_1A_4 represents the spring season (E_1) of 1994 (A_4), which is the first spring of the 14 spring seasons that we have between March 1991 and May 2004. Additionally, E_4A_{14} represents the winter season that starts in December 2004 and ends in February 2005.

Since it is intended to characterize one week (S_p) over a season, and a week is a set of seven consecutive days (not necessarily beginning on Monday and ending on Sunday), the proposal is based on an increase in the available data so that the number of candidate weeks over the study period increases. In addition, for a given season, the weeks are constituted in such a way that there is a discontinuity when passing from one year to the next. In other words, in a sequence (week in our case), we cannot have days that come from two different years. This procedure will generate p weeks from the q available days per meteorological season (d_q) in the following way:

$$E_wA_y = \{d_1, d_2, \ldots, d_q\}$$
$$S_1 = \{d_1, d_2, \ldots, d_7\}, \quad S_2 = \{d_2, d_3, \ldots, d_8\}, \quad \ldots \quad S_p = \{d_{q-6}, d_{q-5}, \ldots, d_q\}$$

where $p = q - 6$. This represents an increase of nearly 700% in the number of candidate weeks for a season in each year (season) of the time series. The data obtained (all 7-day packages) represent, for example, all candidate weeks for all spring seasons between March 1991 and May 2004. The same has been done for the other meteorological seasons. Therefore, this procedure generated a good number (q) of candidate weeks for each of these four weather seasons: 1204 for spring, 1204 for summer, 1190 for autumn, and finally, 1184 for winter.

Thereafter, for the entire dataset corresponding to each season and for each week (of each season), a Cumulative Distribution Function (CDF), Equation (1), is determined for each one of the six selected meteorological parameters: PAR, solar day length, mean, maximum, minimum, and temperature range.

The CDF of each meteorological parameter (x) was calculated by classifying the dataset into equally sized intervals, often called lags, because the size of the long-term data is different from that of short-term data. This is why it is interesting to use lags to perform Equation (2). Thereafter, the number n of observations is equal to the number of lags (n).

Finally, the observations are arranged in ascending order $x_1, x_2, \ldots, x_n$. The CDF of each observation is given by a monotonically increasing step function defined by:

$$CDF(x) = \begin{cases} 0 & \text{for} \quad x < x_1 \\ \frac{(k-0.5)}{n} & \text{for} \quad x_k \leq x \leq x_{k+1} \\ 1 & \text{for} \quad x > x_n \end{cases} \quad (1)$$

where k is the order number from 1 to n − 1.

Then, the FS statistics of each sequence (in our case one week) for each given parameter (x) are obtained from the following Equation (2). In other words, the FS statistics for the candidate week are obtained by calculating the differences between the CDF (defined in Equation (1)) for this week (short-term) with the CDF for all the weeks contained in the corresponding weather season (long term) for each parameter and location.

$$\begin{gathered} FS = \frac{1}{n}\sum_{i=1}^{n} \delta_i \\ \delta_i = |CDF_{lt}(x_i) - CDF_{st}(x_i)| \end{gathered} \quad (2)$$

with CDF_{lt} and CDF_{st} as the long-term and short-term CDF of parameter x.

A weighted sum (WS) of the FS statistics corresponding to each parameter (FS_j) of each week is calculated by applying a weight factor (wf_j), where m is the number of meteorological parameters:

$$WS = \sum_{j=1}^{m} wf_j \cdot FS_j \quad (3)$$

The weighting factor chosen will depend on the importance that each parameter has on the growth of microalgae and must comply with:

$$\sum_{j=1}^{m} wf_j = 1 \quad (4)$$

Following the process, the 'best' candidate weeks (applying different options of wfs) are chosen according to a proportion determining the impact on the growth of microalgae. Thus, the proposal is to analyze the influence that these wfs have on the ranking of candidate weeks for a TMW.

Indeed, the generation of TMYs was done using different climate parameters and different weighting factors [10,17,25,45,49,50]. All these proposals are essentially similar; the main differences are the climate parameters to be included (type and quantity) and their corresponding weighting factors.

The studied parameters: temperature (mean, maximum, minimum, and range); PAR; and solar day length play an important role in the development of a TMS. However, in our case study, they do not have the same impact on microalgae productivity. Therefore, some meteorological parameters may be more important than others. The most influential parameters receive the highest weighting factor (wf_j), which is considered representative of their impact on microalgae growth.

In Table 1, nine scenarios with different wfs are proposed to test different options of wfs. This will allow us to check the robustness of the methodology. The idea is to give equal importance to the temperature parameters—maximum (Tmx), minimum (Tmn), mean (Tme), range (Trg), solar irradiance, PAR, and solar day length (Nsol).

Finally, the most representative sequence—week, in our case—among the five best-candidate weeks is obtained by determining the frequency of repetition of the candidate weeks, taking into account their persistence according to the different lags and wfs. Furthermore, the final decision on the choice of the TMW is also affected by its position in the particular season period. This position is validated by calculating the difference in Nsol between the day in the middle of the weather season and the fourth day of the candidate week, Equation (5). This was done to avoid extreme values for a season that could compromise expected results. The difference in Nsol is defined as follows:

$$\Delta \mathrm{Nsol}_{\mathrm{q}}^{\mathrm{w}} = \left| \mathrm{Nsol}_{\mathrm{q}/2}^{\mathrm{w}} - \mathrm{Nsol}_{4}^{\mathrm{w}} \right| \quad (5)$$

where $\mathrm{Nsol}_{\mathrm{q}/2}^{\mathrm{w}}$ represents the Nsol of the fourth day of the week in the middle of the considered weather season (w = 1, 2, 3, and 4 for spring, summer, autumn, and winter). $\mathrm{Nsol}_{4}^{\mathrm{w}}$ is Nsol of the fourth day of the given candidate week.

Table 1. Weighting factor (wf) options to obtain the weighted sum.

Parameters	wf_1	wf_2	wf_3	wf_4	wf_5	wf_6	wf_7	wf_8	wf_9
Tmx	0.10	0.05	0.05	0.10	0.05	0.05	0.05	0.05	0.10
Tmn	0.10	0.05	0.05	0.10	0.05	0.05	0.05	0.05	0.10
Tme	0.30	0.40	0.30	0.25	0.25	0.45	0.40	0.40	0.20
Trg	-	-	-	0.05	0.05	0.05	0.05	0.05	0.10
PAR	0.30	0.20	0.30	0.25	0.30	0.20	0.30	0.15	0.20
Nsol	0.20	0.30	0.30	0.25	0.30	0.20	0.15	0.30	0.30

3. Results and Discussion

Although microalgae growth is affected by several physicochemical parameters, such as temperature, light, pH, salinity, nutrients, and others, the results and discussion presented in this paper are limited to the input of solar energy and temperature.

The five weeks with the lowest WS values were selected for each WWTP. The selected TMW has a notable persistence and a low difference in $\Delta \mathrm{Nsol}_{\mathrm{q}}^{\mathrm{w}}$. The persistence of a week corresponds to the number of times it appears in the selection for different weighting factors and lag. By broadening the final choice criteria with more weighting factors and lags, a sequence of new candidate weeks emerges that might be different from their predecessors. In some cases, the same weeks are repeated but in a different order. Finally, the selected TMW is supposed to present the long-term characteristic properties of the meteorological data. The same process is adopted for each of the four meteorological seasons for each of the 18 WWPTs.

To facilitate the presentation of the results, only the details of the long-term and weekly statistics are presented in the following tables. The rest of the results are given in Appendix A. Table 2 shows the seasonal quarterly statistics for the Colmenarejo Este location during the period 1991 to 2005. It includes the mean and median of each of the six parameters for the different seasons, as well as the standard deviation, which provides information about the average dispersion of each of these. A standard deviation between 2.48 and 6.10 is observed for the temperature parameters. The PAR has a high standard deviation with values distributed over a range between 16.35 and 28.99, while the Nsol is about one for spring and autumn and less than one for summer and winter. The daily or seasonal variability of these growth parameters significantly affects the production of microalgae. Taking into account seasonal variations, it can be shown that winter has unfavorable meteorological conditions with less light, very low temperatures, and an inadequate photoperiod for microalgae growth. Summer offers the appropriate ranges of values to maximize productivity, while spring and autumn are unfavorable for microalgae metabolism. Extreme variations in these parameters can be observed throughout the year and can have inhibiting effects on microalgae.

Due to the large number of weeks to evaluate nine wf options (Table 1), the WS of the FS statistics for the different parameters are not presented in tabular form in this article.

Table 2. Long-term daily statistics in Colmenarejo Este.

	Spring					
		Temperature (°C)			Solar Irradiance (W/m^2)	
	Tmx	Tmn	Tme	Trg	PAR	Nsol
Mean	16.25	5.40	10.84	10.85	91.26	13.05
Median	16.05	5.11	10.45	10.98	92.99	13.13
Av. Std. Des	4.81	3.41	3.97	3.43	28.99	1.04
	Summer					
		Temperature (°C)			Solar irradiance(W/m^2)	
	Tmx	Tmn	Tme	Trg	PAR	Nsol
Mean	28.46	15.54	22.29	12.91	126.57	14.35
Median	29.21	16.00	22.88	13.30	130.40	14.58
Av. Std. Des	4.41	3.26	3.80	2.48	20.52	0.57
	Autumn					
		Temperature (°C)			Solar irradiance(W/m^2)	
	Tmx	Tmn	Tme	Trg	PAR	Nsol
Mean	17.27	8.00	12.39	9.27	61.25	10.91
Median	16.71	8.06	12.06	9.42	58.03	10.85
Av. Std. Des	6.10	4.54	5.18	3.35	26.51	1.02
	Winter					
		Temperature (°C)			Solar irradiance(W/m^2)	
	Tmx	Tmn	Tme	Trg	PAR	Nsol
Mean	8.61	0.50	4.10	8.11	39.68	9.63
Median	8.54	0.12	4.12	8.00	41.24	9.39
Av. Std. Des	2.91	2.98	2.65	3.07	16.35	0.55

Table 3 shows the results obtained for the spring season in the Colmenarejo Este location: each cell shows the number of the candidate week corresponding to a set of weight factors (column) and a number of lags (rows). For each pair of 'weighting factors set and lag size', we present the five best candidates ordered from top to bottom according to the minimum WS value. In the case of Colmenarejo Este, the total number of weeks analyzed is 1204, so that—for example—week number 235 represents the sequence of seven days that begins on 2 May 1993.

The TMW for each season and each of the 18 locations are given in Table A2. In this table, only the first day of the selected week is given. Although the years of the selected weeks are not identical, it can be observed that there is a slight difference in the periods (months and days) of the year for the sites studied. This difference could be due to the variations of the average temperatures that decrease with the latitude. However, the variations of the PAR between locations are very small and are due to the small differences in latitude from one point to another. The latter may have little influence on the expected results, especially since the RAP difference is very little between locations. Other factors that also play a very important role in the generation of weeks are the wfs assigned to the variables and the lag number (Table 3). If we focus on one column (wf value) from Table 3, we can see that almost the same weeks come back with different positions when we change the lag number. Likewise, when we look at a lag number, the trend of results also changes each time we change the distribution of wfs and in the same sense as previously mentioned, hence the interest in choosing several wf options that can compensate for the lack of information of the meteorological parameter with the most important impact on the growth of microalgae.

Table 3. The candidate weeks of the spring season in Colmenarejo Este presented for different sets of weighting factors and lag sizes.

lags \ wf	wf_1	wf_2	wf_3	wf_4	wf_5	wf_6	wf_7	wf_8	wf_9
lags = 10	1082	1082	642	641	642	235	642	235	235
	641	235	235	235	235	641	235	641	641
	642	641	641	642	641	642	641	642	642
	235	642	1082	1082	1081	1081	1081	1082	378
	507	1081	1081	378	507	1082	1082	1081	1082
lags = 20	642	642	642	642	642	642	642	642	642
	1082	1081	1081	641	641	235	641	235	235
	641	1082	641	235	235	1081	235	641	641
	496	641	235	1082	1081	641	1081	1081	1082
	235	235	507	496	507	1082	1082	1082	864
lags = 30	642	642	642	642	642	642	642	642	642
	1082	1082	507	235	235	1081	235	235	235
	496	1081	1081	641	641	235	1081	1081	641
	641	641	641	1082	507	641	641	939	378
	507	496	1082	496	1081	1082	507	641	1082

Table 4 shows the information for all generated candidate weeks for the spring season in Colmenarejo Este that are represented in Table 3. In Table 4, the number of repetitions of these generated candidate weeks is also shown, which are represented by their order number in the sequence of 1204 spring weeks. Once a candidate week is selected, its sequence number identifies the start of the week by giving the corresponding year, month, and day.

Table 4. Candidate weeks in the Colmenarejo Este location in the spring season.

Number of Week	Frequency	Year	Month	Day	$\mathbf{Nsol_4^w}$	$\mathbf{\Delta Nsol_q^w}$
235	24	1993	5	2	13.91	0.78
378	3	1995	4	3	12.64	0.49
496	5	1996	5	5	14.05	0.92
507	8	1996	5	16	14.41	1.28
641	27	1998	4	8	12.91	0.22
642	27	1998	4	9	12.96	0.18
864	1	2001	3	4	11.31	1.82
939	1	2001	5	18	14.44	1.31
1081	18	2003	4	18	13.35	0.21
1082	21	2003	4	19	13.39	0.26

The Nsol of the fourth day ($Nsol_4^w$) of each of these candidate weeks is also given in this table, as well as the absolute value of the difference $\left(\Delta Nsol_q^w\right)$ between the Nsol of the fourth day of the week in the middle of the season $\left(Nsol_{q/2}^w\right)$ and the Nsol of the candidate week ($Nsol_4^w$). This difference allows us to appreciate the position of the week in relation to the extremities of that season. As mentioned above, this avoids having a typical week with weather conditions closer to the earlier or later season. Therefore, the selected TMW has the highest frequency of occurrence. In the case event that this frequency of occurrence is equal, then the typical weather week would be the one with the lowest value of $\left(\Delta Nsol_q^w\right)$. For example, the Nsol on the fourth day of the week in the middle of the spring season in the Colmenarejo Este location is 12.96 h.

According to the different weeks presented in Table 4, the frequency of occurrence in week 641 is the same as in week 642. This coincidence in the number of occurrences can be explained by the fact that these two weeks differ by one day: one starts on 8 April 1998 and the other on 9 April 1998. Therefore, the final choice of the representative week of spring meteorological conditions in Colmenarejo Este is given by the week with the lowest $\left(\Delta Nsol_q^w\right)$ value. In the case of Colmenarejo Este, week 642 is the typical week that represents the spring weather conditions for the period 1991 to 2005.

The Madrid region is not very large, and consequently, a small difference of the order of magnitude for both PAR and temperature is observed in a given season when moving from one locality to another (Figures 2 and A1–A3). This can also be seen in Tables 2 and A1 (Appendix A), in which the statistics of long-term meteorological data are given for four locations.

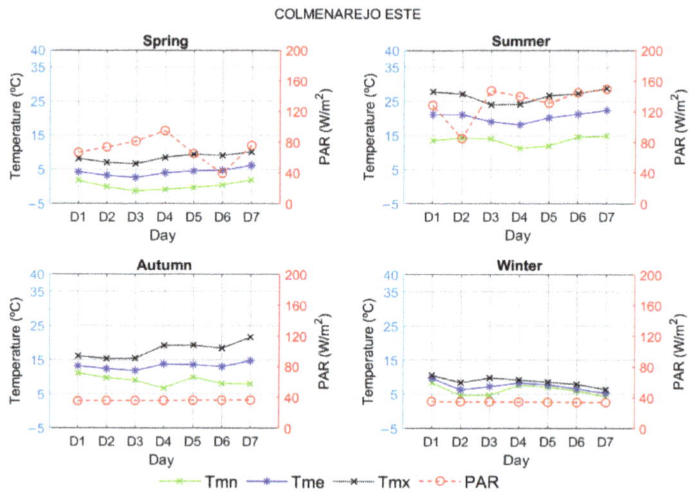

Figure 2. Daily variations of the daily mean values of temperature (mean, maximum, and minimum) and PAR for the selected TMWs for each season in Colmenarejo Este from 1991 to 2005.

The daily mean values of the TMW parameters for each season were obtained, and the variability of some parameters was plotted, which is illustrated in Figure 2, showing the daily variation of the mean values of PAR and temperature indices for the different seasons. Figures 2 and A1–A3 indicate that, for both meteorological parameters, there is inter-seasonal variability. The highest PAR and temperature values are observed during summer, and the lowest values are observed during winter. On the one hand, thermal oscillation is greater between summer and winter. Since Madrid is located in the central peninsular area, a possible explanation may come from the disappearance of the moderating effect of the sea, which decreases as one moves away from the coast. Furthermore, Figures 2 and A1–A3 show that the temperature range is narrower in winter. The spring and autumn seasons have approximately similar average daily temperatures. However, the daily average PAR is lower in autumn compared to spring, and its variability is sometimes similar to that of winter in certain localities. This can be explained by the predominance of cloudy and probably rainy skies at this time of year. In fact, cloudiness reduces insolation by obstructing solar radiation.

4. Conclusions

In this work, a methodology for the generation of TMS for the simulation of photosynthetic organism growth and productivity for WWT or agriculture is proposed. The selection of potential sequences according to the growth period of the organism is performed by resampling the available meteorological data, which, in our case study, increases the number of candidate sequences by 700%.

Prior knowledge of the impact of meteorological factors would allow the optimization of crop productivity, rational use of water, and evaluate the appropriate period during the year for WWT in a HRAP systems with microalgae. It is relevant to take into account the long-term variability of physical parameters among the seasons to develop sustainable systems. The advantage of TMS data is that they are suitable to overcome computational power limitations when multiple simulations are needed to have an overview of the biological system behavior as a function of local climatic conditions.

The TMS approach has allowed to generate a typical sequence called TMW intended to simulate the growth of microalgal biomass for biofuel production or sustainable wastewater treatment in a HRAP. For the generation of the TMW in our case study, once the most relevant climatic parameters were identified, a detailed exam of the different weight factors for each of the variables considered was performed to ensure the robustness of the methodology.

Author Contributions: Conceptualization, R.X.V. and L.F.Z.; methodology, L.F.Z., O.W. and A.A.N.; software, O.W.; validation, O.W. and L.F.Z.; formal analysis, L.F.Z., O.W. and F.F.-C.; investigation, O.W., A.R.-L., R.X.V., A.A.N., L.F.Z. and F.F.-C.; resources, R.X.V. and L.F.Z.; data curation, O.W., R.X.V., A.A.N., L.F.Z. and F.F.-C.; writing—original draft preparation, O.W., A.R.-L., R.X.V., A.A.N., L.F.Z. and F.F.-C.; writing—review and editing, O.W., R.X.V., A.A.N., A.R.-L., L.F.Z. and F.F.-C.; visualization, O.W., A.A.N. and A.R.-L.; supervision, R.X.V. and L.F.Z.; project administration, R.X.V. and L.F.Z.; and funding acquisition, R.X.V. All authors have read and agreed to the published version of the manuscript.

Funding: This research was funded by the Autonomous Community of Madrid, Spain, and financed by FEDER 'A way of making Europe' ALGATEC-CM (S2018/BAA-4532) and by the Spanish Ministry of Science and Innovation (MCIN/AEI/10.13039/501100011033) and the European Union "Next Generation EU"/PRTR, TEDDY (TED2021-130366B-I00).

Institutional Review Board Statement: Not applicable.

Informed Consent Statement: Not applicable.

Data Availability Statement: Data provided by ECMWF used in this study are openly available at https://www.ecmwf.int/en/forecasts/access-forecasts/access-archive-datasets, reference number [44] (accessed on 17 April 2021). CMSAF data used in this study are openly available at https://wui.cmsaf.eu/safira/action/viewProduktList?dId=2&d-1342877-p=6, reference numbers [43] (accessed on 14 April 2021).

Acknowledgments: The authors acknowledge the CYTED-Ibero American Program on Science and Technology for Development (RED RENUWAL P320RT0005 CYTED).

Conflicts of Interest: The authors declare that they have no conflict of interest.

Appendix A

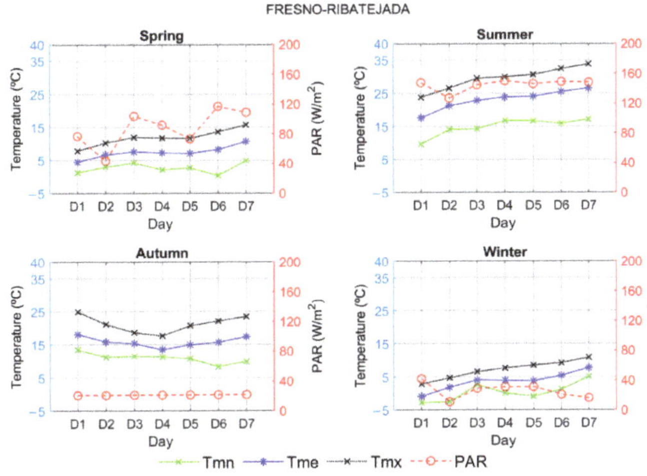

Figure A1. Daily variations of the daily mean values of temperature (mean, max, and min) and PAR for the selected TMWs for each season in Fresno-Ribatejada from 1991 to 2005.

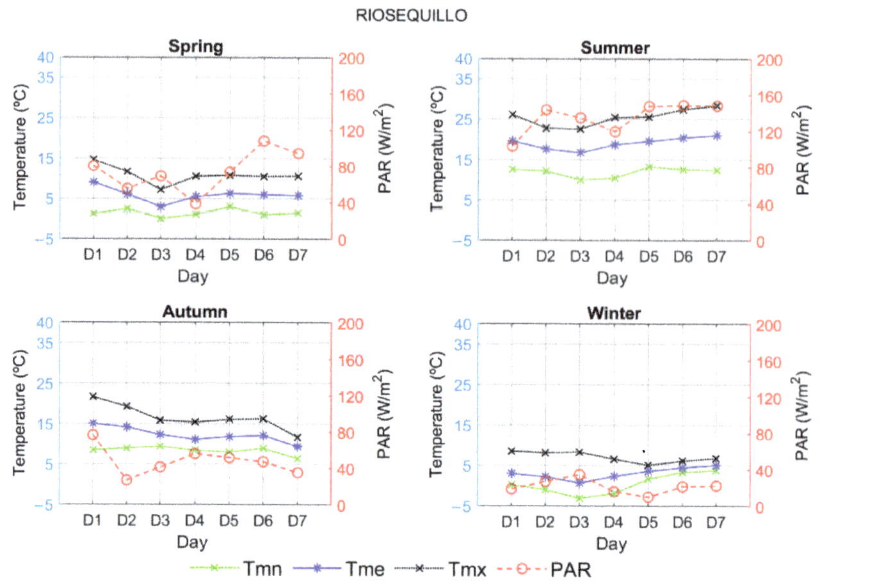

Figure A2. Daily variations of the daily mean values of temperature (mean, max, and min) and PAR for the selected TMWs for each season in Riosequillo from 1991 to 2005.

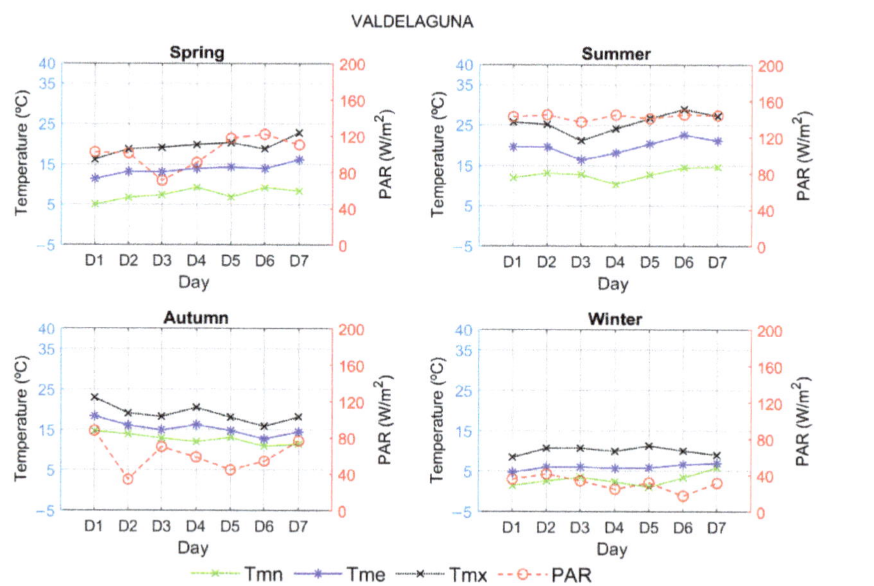

Figure A3. Daily variations of the daily mean values of temperature (mean, max, and min) and PAR for the selected TMWs for each season in Valdelaguna from 1991 to 2005.

Table A1. Long-term data statistics for four sites.

	Spring Mean	Spring Median	Spring Std. Dev.	Summer Mean	Summer Median	Summer Std. Dev.	Autumn Mean	Autumn Median	Autumn Std. Dev.	Winter Mean	Winter Median	Winter Std. Dev.
FRESNO-RIBATEJADA												
Tmx	16.36	16.20	4.83	28.55	29.41	4.38	17.34	16.75	6.09	8.73	8.62	2.91
Tmn	5.55	5.25	3.43	15.67	16.13	3.29	8.11	8.22	4.54	0.68	0.27	2.97
Tme	10.95	10.52	4.00	22.41	23.01	3.78	12.46	12.17	5.16	4.25	4.29	2.63
Trg	10.81	10.94	3.45	12.88	13.30	2.46	9.24	9.45	3.37	8.05	7.96	3.10
PAR	93.53	94.58	28.07	126.61	130.02	19.44	62.46	59.39	26.16	40.31	41.47	16.22
NSol	13.05	13.13	1.04	14.35	14.59	0.58	10.91	10.85	1.02	9.62	9.39	0.55
RIOSEQUILLO												
Tmx	15.21	14.99	4.76	27.06	27.75	4.44	16.23	15.69	6.02	7.80	7.73	2.95
Tmn	4.22	4.05	3.39	13.84	14.18	3.22	6.83	6.99	4.40	−0.24	−0.57	2.99
Tme	9.69	9.37	3.92	20.72	21.27	3.76	11.24	10.95	5.00	3.31	3.31	2.65
Trg	10.99	11.00	3.59	13.22	13.61	2.71	9.39	9.44	3.56	8.03	7.93	3.12
PAR	85.56	87.74	30.37	121.60	127.06	23.67	56.75	53.00	26.28	36.83	38.15	16.25
NSol	13.07	13.15	1.05	14.38	14.62	0.58	10.89	10.83	1.04	9.59	9.36	0.56
VALDELAGUNA												
Tmx	17.36	17.14	4.93	29.82	30.67	4.35	18.24	17.70	6.16	9.49	9.36	2.90
Tmn	6.49	6.14	3.51	17.10	17.63	3.35	9.00	9.01	4.73	1.25	0.92	3.06
Tme	11.92	11.43	4.09	23.76	24.35	3.80	13.37	12.99	5.32	4.92	5.00	2.66
Trg	10.87	11.18	3.44	12.71	13.07	2.31	9.24	9.51	3.28	8.24	8.27	3.18
PAR	94.36	95.49	27.66	127.23	130.60	18.73	63.75	60.96	25.94	41.55	42.45	15.88
NSol	13.04	13.12	1.02	14.32	14.55	0.57	10.92	10.86	1.01	9.66	9.43	0.55

Table A2. First day of the Typical Meteorological Week of each season for each of the EDAR stations.

Site	Season	Year	Month	Day	Site	Season	Year	Month	Day
ARANJUEZ	Spring	1994	03	20	ORUSCO DE TAJUÑA	Spring	2003	04	20
	Summer	1997	06	20		Summer	1995	06	19
	Autumn	1993	11	06		Autumn	1996	11	20
	Winter	1996	12	18		Winter	1992	12	18
BATRES	Spring	1993	05	02	PEZUELA DE LAS TORRES	Spring	1993	05	02
	Summer	1991	06	19		Summer	1997	06	15
	Autumn	1996	10	19		Autumn	1996	09	08
	Winter	1996	12	14		Winter	1994	12	16
COLMENAREJO ESTE	Spring	1998	04	09	RIOSEQUILLO	Spring	2002	04	09
	Summer	1994	06	16		Summer	1994	06	17
	Autumn	1996	09	20		Autumn	1994	10	13
	Winter	1996	12	18		Winter	2000	12	18
CONJUNTA DE GASCONES	Spring	2003	04	19	ROBLEDO	Spring	1995	04	03
	Summer	1991	06	19		Summer	1991	06	19
	Autumn	1996	11	21		Autumn	1996	09	20
	Winter	1995	12	14		Winter	1996	12	15
ESTREMERA	Spring	1993	03	17	ROZAS DE PUERTO REAL	Spring	1996	03	27
	Summer	1994	06	19		Summer	1998	06	19
	Autumn	1996	11	20		Autumn	2001	09	27
	Winter	1996	12	16		Winter	1996	12	15
FRESNO-RIBATEJADA	Spring	2002	04	11	SAN MARTIN NORESTE	Spring	2001	05	18
	Summer	1991	06	19		Summer	2000	06	15
	Autumn	1996	09	08		Autumn	1999	10	11
	Winter	1995	12	14		Winter	1996	12	15
FUENTIDUEÑA	Spring	2003	04	20	SANTA Mª DE LA ALAMEDA	Spring	1998	05	20
	Summer	1995	06	19		Summer	1994	06	16
	Autumn	1998	09	21		Autumn	1996	09	20
	Winter	1996	12	16		Winter	1996	12	15

Table A2. *Cont.*

Site	Season	Year	Month	Day	Site	Season	Year	Month	Day
LOZOYUELA	Spring	2002	04	09	TALAMANCA DEL JARAMA	Spring	2003	04	19
	Summer	1997	06	20		Summer	1997	06	20
	Autumn	1996	09	20		Autumn	1996	09	20
	Winter	2000	12	18		Winter	1994	12	16
NAVALAFUENTE	Spring	2003	04	19	VALDELAGUNA	Spring	1993	05	02
	Summer	1997	06	20		Summer	1997	06	20
	Autumn	1994	10	13		Autumn	1998	09	21
	Winter	1996	12	16		Winter	1992	12	18

References

1. Domínguez, A.; Martínez-Romero, A.; Leite, K.N.; Tarjuelo, J.M.; de Juan, J.A.; López-Urrea, R. Combination of Typical Meteorological Year with Regulated Deficit Irrigation to Improve the Profitability of Garlic Growing in Central Spain. *Agric. Water Manag.* **2013**, *130*, 154–167. [CrossRef]
2. Garfí, M.; Flores, L.; Ferrer, I. Life Cycle Assessment of Wastewater Treatment Systems for Small Communities: Activated Sludge, Constructed Wetlands and High Rate Algal Ponds. *J. Clean. Prod.* **2017**, *161*, 211–219. [CrossRef]
3. López-Sánchez, A.; Silva-Gálvez, A.L.; Aguilar-Juárez, Ó.; Senés-Guerrero, C.; Orozco-Nunnelly, D.A.; Carrillo-Nieves, D.; Gradilla-Hernández, M.S. Microalgae-Based Livestock Wastewater Treatment (MbWT) as a Circular Bioeconomy Approach: Enhancement of Biomass Productivity, Pollutant Removal and High-Value Compound Production. *J. Environ. Manag.* **2022**, *308*, 114612. [CrossRef] [PubMed]
4. Acién, F.G.; Gómez-Serrano, C.; Morales-Amaral, M.M.; Fernández-Sevilla, J.M.; Molina-Grima, E. Water Treatment Using Microalgae: How Realistic a Contribution Might It Be to Significant Urban Wastewater Treatment. *Appl. Microbiol. Biotechnol.* **2016**, *100*, 9013–9022. [CrossRef] [PubMed]
5. Robles, Á.; Capson-Tojo, G.; Galès, A.; Ruano, M.V.; Sialve, B.; Ferrer, J.; Steyer, J.P. Microalgae-Bacteria Consortia in High-Rate Ponds for Treating Urban Wastewater: Elucidating the Key State Indicators under Dynamic Conditions. *J. Environ. Manag.* **2020**, *261*, 110244. [CrossRef]
6. Arcila, J.S.; Buitrón, G. Influence of Solar Irradiance Levels on the Formation of Microalgae-Bacteria Aggregates for Municipal Wastewater Treatment. *Algal Res.* **2017**, *27*, 190–197. [CrossRef]
7. Iasimone, F.; Panico, A.; De Felice, V.; Fantasma, F.; Iorizzi, M.; Pirozzi, F. Effect of Light Intensity and Nutrients Supply on Microalgae Cultivated in Urban Wastewater: Biomass Production, Lipids Accumulation and Settleability Characteristics. *J. Environ. Manag.* **2018**, *223*, 1078–1085. [CrossRef]
8. Hall, I.J.; Prairie, R.R.; Anderson, H.E.; Boes, E.C. *Generation of Typical Meteorological Years for 26 SOLMET Stations*; Sandia Laboratories: Albuquerque, NM, USA, 1978.
9. Leite, K.N.; Martínez-Romero, A.; Tarjuelo, J.M.; Domínguez, A. Distribution of Limited Irrigation Water Based on Optimized Regulated Deficit Irrigation and Typical Metheorological Year Concepts. *Agric. Water Manag.* **2015**, *148*, 164–176. [CrossRef]
10. Marion, W.; Urban, K. *User's Manual for TMY2s Radiation Data Base*; National Renewable Energy Laboratory: Golden, CO, USA, 1995.
11. Fernández, M.D.; López, J.C.; Baeza, E.; Céspedes, A.; Meca, D.E.; Bailey, B. Generation and Evaluation of Typical Meteorological Year Datasets for Greenhouse and External Conditions on the Mediterranean Coast. *Int. J. Biometeorol.* **2015**, *59*, 1067–1081. [CrossRef]
12. Heinemann, P.H.; Walker, P.N. Effects of Greenhouse Surface Heating Water on Light Transmission. *Trans. ASAE* **1987**, *30*, 0215–0220. [CrossRef]
13. Marbis, J.M. *CO_2 Enrichment and Hot Water Heat in a Greenhouse as a Mean of Recovering Bioresources from Swine Waste*; North Carolina State University: Raleigh, NC, USA, 2001.
14. Martínez-Romero, A.; Domínguez, A.; Landeras, G. Regulated Deficit Irrigation Strategies for Different Potato Cultivars under Continental Mediterranean-Atlantic Conditions. *Agric. Water Manag.* **2019**, *216*, 164–176. [CrossRef]
15. Shen, Q.H.; Jiang, J.W.; Chen, L.P.; Cheng, L.H.; Xu, X.H.; Chen, H.L. Effect of Carbon Source on Biomass Growth and Nutrients Removal of Scenedesmus Obliquus for Wastewater Advanced Treatment and Lipid Production. *Bioresour. Technol.* **2015**, *190*, 257–263. [CrossRef] [PubMed]
16. Baneshi, M.; Gonome, H.; Maruyama, S. Wide-Range Spectral Measurement of Radiative Properties of Commercial Greenhouse Covering Plastics and Their Impacts into the Energy Management in a Greenhouse. *Energy* **2020**, *210*, 118535. [CrossRef]
17. Kalogirou, S.A. Generation of Typical Meteorological Year (TMY-2) for Nicosia, Cyprus. *Renew. Energy* **2003**, *28*, 2317–2334. [CrossRef]
18. Pissimanis, D.; Karras, G.; Notaridou, V.; Gavra, K. The Generation of a "Typical Meteorological Year" for the City of Athens. *Sol. Energy* **1988**, *40*, 405–411. [CrossRef]
19. Skeiker, K. Generation of a Typical Meteorological Year for Damascus Zone Using the Filkenstein-Schafer Statistical Method. *Energy Convers. Manag.* **2004**, *45*, 99–112. [CrossRef]

20. Huld, T.; Paietta, E.; Zangheri, P.; Pascua, I.P. Assembling Typical Meteorological Year Data Sets for Building Energy Performance Using Reanalysis and Satellite-Based Data. *Atmosphere* **2018**, *9*, 53. [CrossRef]
21. Cebecauer, T.; Suri, M. Typical Meteorological Year Data: SolarGIS Approach. *Energy Procedia* **2015**, *69*, 1958–1969. [CrossRef]
22. Sun, J.; Li, Z.; Xiao, F. Analysis of Typical Meteorological Year Selection for Energy Simulation of Building with Daylight Utilization. *Procedia Eng.* **2017**, *205*, 3080–3087. [CrossRef]
23. Georgiou, G.; Eftekhari, M.; Eames, P.; Mourshed, M. A study of the effect of weighting indices for the development of TMY used for building simulation. In Proceedings of the BS2013: 13th Conference of International Building Performance Simulation Association, Chambéry, France, 26–28 August 2013; pp. 922–929.
24. Yang, L.; Wan, K.K.W.; Li, D.H.W.; Lam, J.C. A New Method to Develop Typical Weather Years in Different Climates for Building Energy Use Studies. *Energy* **2011**, *36*, 6121–6129. [CrossRef]
25. Chan, A.L.S.; Chow, T.T.; Fong, S.K.F.; Lin, J.Z. Generation of a Typical Meteorological Year for Hong Kong. *Energy Convers. Manag.* **2006**, *47*, 87–96. [CrossRef]
26. De Miguel, A.; Bilbao, J. Test Reference Year Generation from Meteorological and Simulated Solar Radiation Data. *Sol. Energy* **2005**, *78*, 695–703. [CrossRef]
27. Lund, H. *Short Reference Years and Test Reference Years for EEC Countries*; Technical Report. (Final Report, Contract ESF-029-DK, Report EUR 10208 EN.); Thermal Insulation Laboratory, Technical University of Denmark: Lyngby, Denmark, 1985.
28. Lund, H. The Design Reference Year. In Proceedings of the 91th Conference, International Building Performance Simulation Association, Nice, France, 20–22 August 1991; pp. 600–606.
29. Liem, S.H.; Van Paassen, A.H. *Establishment of Short Reference Years for Calculation of Annual Solar Heat Gain or Energy Consumption in Residential and Commercial Buildings Part 1 and 2*; Technical Report. (Report EUR 8912 EN); Commission of the European Communities: Luxembourg, 1984; pp. 1–38.
30. Zang, H.; Xu, Q.; Bian, H. Generation of Typical Solar Radiation Data for Different Climates of China. *Energy* **2012**, *38*, 236–248. [CrossRef]
31. Rodríguez, F.; Castro, A.; Marín, F.; Roldán, G.; Viteri Moya, F. Typical Meteorological Year Based on the Precipitation of Nanegalito and Pacto-Ecuador. *Enfoque UTE* **2019**, *10*, 197–204. [CrossRef]
32. Holzworth, D.P.; Huth, N.I.; de Voil, P.G.; Zurcher, E.J.; Herrmann, N.I.; McLean, G.; Chenu, K.; van Oosterom, E.J.; Snow, V.; Murphy, C.; et al. APSIM—Evolution towards a New Generation of Agricultural Systems Simulation. *Environ. Model. Softw.* **2014**, *62*, 327–350. [CrossRef]
33. Kephe, P.N.; Ayisi, K.K.; Petja, B.M. Challenges and Opportunities in Crop Simulation Modelling under Seasonal and Projected Climate Change Scenarios for Crop Production in South Africa. *Agric. Food Secur.* **2021**, *10*, 1–24. [CrossRef]
34. Huesemann, M.; Crowe, B.; Waller, P.; Chavis, A.; Hobbs, S.; Edmundson, S.; Wigmosta, M. A Validated Model to Predict Microalgae Growth in Outdoor Pond Cultures Subjected to Fluctuating Light Intensities and Water Temperatures. *Algal Res.* **2016**, *13*, 195–206. [CrossRef]
35. Wigmosta, M.S.; Coleman, A.M.; Skaggs, R.J.; Huesemann, M.H.; Lane, L.J. National Microalgae Biofuel Production Potential and Resource Demand. *Water Res.* **2011**, *47*, 1–13. [CrossRef]
36. Darvehei, P.; Bahri, P.A.; Moheimani, N.R. *Modeling the Effect of Temperature on Microalgal Growth under Outdoor Conditions*; Elsevier Masson SAS: Îledefrance, France, 2018; Volume 43, ISBN 9780444642356.
37. Baxevanou, C.; Fidaros, D.; Katsoulas, N.; Mekeridis, E.; Varlamis, C.; Zachariadis, A.; Logothetidis, S. Simulation of Radiation and Crop Activity in a Greenhouse Covered with Semitransparent Organic Photovoltaics. *Appl. Sci.* **2020**, *10*, 2550. [CrossRef]
38. García-Rodríguez, A.; García-Rodríguez, S.; Granados-López, D.; Díez-Mediavilla, M.; Alonso-Tristán, C. Extension of PAR Models under Local All-Sky Conditions to Different Climatic Zones. *Appl. Sci.* **2022**, *5*, 2372. [CrossRef]
39. Viruela, A.; Murgui, M.; Gómez-Gil, T.; Durán, F.; Robles, Á.; Ruano, M.V.; Ferrer, J.; Seco, A. Water Resource Recovery by Means of Microalgae Cultivation in Outdoor Photobioreactors Using the Effluent from an Anaerobic Membrane Bioreactor Fed with Pre-Treated Sewage. *Bioresour. Technol.* **2016**, *218*, 447–454. [CrossRef] [PubMed]
40. Breuer, G.; Lamers, P.P.; Martens, D.E.; Draaisma, R.B.; Wijffels, R.H. Effect of Light Intensity, PH, and Temperature on Triacylglycerol (TAG) Accumulation Induced by Nitrogen Starvation in Scenedesmus Obliquus. *Bioresour. Technol.* **2013**, *143*, 1–9. [CrossRef] [PubMed]
41. Cabello, J.; Toledo-Cervantes, A.; Sánchez, L.; Revah, S.; Morales, M. Effect of the Temperature, PH and Irradiance on the Photosynthetic Activity by Scenedesmus Obtusiusculus under Nitrogen Replete and Deplete Conditions. *Bioresour. Technol.* **2015**, *181*, 128–135. [CrossRef]
42. Jabri, H.A.; Taleb, A.; Touchard, R.; Saadaoui, I.; Goetz, V.; Pruvost, J. Cultivating Microalgae in Desert Conditions: Evaluation of the Effect of Light-temperature Summer Conditions on the Growth and Metabolism of Nannochloropsis Qu130. *Appl. Sci.* **2021**, *11*, 3799. [CrossRef]
43. Müller, R.; Pfeifroth, U.; Träger-Chatterjee, C.; Trentmann, J.; Cremer, R. Digging the METEOSAT Treasure-3 Decades of Solar Surface Radiation. *Remote Sens.* **2015**, *7*, 8067–8101. [CrossRef]
44. Hersbach, H.; Bell, B.; Berrisford, P.; Hirahara, S.; Horányi, A.; Muñoz-Sabater, J.; Nicolas, J.; Peubey, C.; Radu, R.; Schepers, D.; et al. The ERA5 Global Reanalysis. *Q. J. R. Meteorol. Soc.* **2020**, *146*, 1999–2049. [CrossRef]
45. Zhang, J.; Zhao, L.; Deng, S.; Xu, W.; Zhang, Y. A Critical Review of the Models Used to Estimate Solar Radiation. *Renew. Sustain. Energy Rev.* **2017**, *70*, 314–329. [CrossRef]

46. Beck, H.E.; Zimmermann, N.E.; McVicar, T.R.; Vergopolan, N.; Berg, A.; Wood, E.F. Present and Future Köppen-Geiger Climate Classification Maps at 1-Km Resolution. *Sci. Data* **2018**, *5*, 180214. [CrossRef]
47. Köppen, W.; Geiger, R. *Das Geographische System Der Klimate*; Gebrüder, B., Ed.; Mit 14 Textflguren: Berlin, Germany, 1936.
48. Peel, M.C.; Finlayson, B.L.; McMahon, T.A. Updated World Map of the Köppen-Geiger Climate Classification. *Hydrol. Earth Syst. Sci.* **2007**, *11*, 1633–1644. [CrossRef]
49. Skeiker, K. Comparison of Methodologies for TMY Generation Using 10 Years Data for Damascus, Syria. *Energy Convers. Manag.* **2007**, *48*, 2090–2102. [CrossRef]
50. Zarzalejo, L.F.; Téllez, F.M.; Heras, M. Creation of TMY for Southern Spanish Cities. In Proceedings of the International Symposium Passive Cooling of Buildings, Athens, Greece, 19–20 June 1995; pp. 61–73.

applied sciences

MDPI

Article

Influence of Self-Compaction on the Airflow Resistance of Aerated Wheat Bulks (*Triticum aestivum* L., cv. 'Pionier')

Iris Ramaj *, Steffen Schock, Shkelqim Karaj and Joachim Müller

Tropics and Subtropics Group (440e), Institute of Agricultural Engineering, University of Hohenheim, 70599 Stuttgart, Germany
* Correspondence: i.ramaj@uni-hohenheim.de or info440e@uni-hohenheim.de; Tel.: +49-(0)-711-459-23119

Featured Application: Grain storage, design and analysis of cooling, aeration and low-temperature drying of in-store grain bulks, practical application.

Abstract: Aeration is a key post-harvest grain processing operation that forces air through the pore volume of the grain bulk to establish favorable conditions to maintain grain quality and improve its storability. However, during storage, grain bulk experiences self-compaction due to its dead weight, which alters the bulk properties and impedes the uniform flow of air during aeration. Thus, this study focused on investigating the effect of self-compaction on the pressure drop ΔP of wheat bulk (*Triticum aestivum* L., cv. 'Pionier', $X = 0.123$ kg·kg^{-1} d.b.) accommodated in a laboratory-scale bin ($V_b = 0.62$ m^3) at a coherent set of airflow velocities v_a. Pressure drop ΔP was measured at bulk depths H_b of 1.0, 2.0, 3.0 and 3.4 m and storage times t of 1, 65, 164 and 236 h. For the semi-empirical characterization of the relationship between ΔP and v_a, the model of Matthies and Petersen was used, which was proficient in describing the experimental data with decent accuracy ($R^2 = 0.990$, $RMSE = 68.67$ Pa, $MAPE = 12.50\%$). A tailored product factor k was employed for the specific grain bulk conditions. Results revealed a reduction of in-situ pore volume ε from 0.413 to 0.391 at bulk depths H_b of 1.0 to 3.4 m after 1 h storage time t and from 0.391 to 0.370 after 236 h storage time t, respectively. A disproportional increase of the pressure drop ΔP with bulk depth H_b and storage time t was observed, which was ascribed to the irreversible spatio-temporal behavior of self-compaction. The variation of pore volume ε was modeled and facilitated the development of a generalized model for predicting the relationship between ΔP and v_a. The relative importance of modeling parameters was evaluated by a sensitivity analysis. In conclusion, self-compaction has proven to have a significant effect on airflow resistance, therefore it should be considered in the analysis and modeling of cooling, aeration and low-temperature drying of in-store grain bulks.

Keywords: aeration; airflow resistance; pore volume; semi-empirical modelling; self-compaction; spatial and temporal; wheat

Citation: Ramaj, I.; Schock, S.; Karaj, S.; Müller, J. Influence of Self-Compaction on the Airflow Resistance of Aerated Wheat Bulks (*Triticum aestivum* L., cv. 'Pionier'). *Appl. Sci.* **2022**, *12*, 8909. https://doi.org/10.3390/app12178909

Academic Editor: José Miguel Molina Martínez

Received: 9 August 2022
Accepted: 2 September 2022
Published: 5 September 2022

1. Introduction

Cereal grains are among the most important and indispensable food sources for humans, with an annual global production of 3.0 billion tons in 2020 [1]. They account for 60 to 80% of the dietary calorie intake, which makes up a significant portion of human energy and nutrient requirements [2]. Storage technologies play a critical role in maintaining the nutritional quality and prolonging the shelf-life of cereal grains during the off-season. Grain temperature and moisture content are the two most important parameters impacting storage, with high values affecting the intrinsic quality of grains and promoting decay [3–5]. The interaction between these parameters during storage has resulted in losses of about 13.4% in the global production in 2018 [6]. Therefore, aeration is utilized to force air through the pore volume of stored grain to modify the bulk microclimate and create favorable conditions for quality preservation and improvement of storability. Aeration reduces the bulk temperature to a safe storage level to prevent insect and mite infestation, spontaneous heating and off-odors [7]. In addition, it inhibits the development of microflora by reducing the excess moisture and intergranular air humidity in isolated grain dump nests. Intergranular air humidity refers to the relative amount of water available in the air at a particular temperature described by sorption isotherms [8]. At sufficiently high levels (above 60%), the development of bacteria, fungi and yeasts is promoted, leading to the formation of toxins that are detrimental to humans [9,10]. Hence, ensuring adequate aeration throughout the grain bulk can be an important preservative measure to effectively control harmful substances.

During aeration, as the air flows through the pore volume of the grain bulk, it loses its kinetic energy due to intergranular friction and turbulence, resulting in airflow resistance known as pressure drop [11]. The grain species and cultivars, as well as their properties such as moisture content, physical and mechanical properties, surface roughness, bulk depth, pore volume configuration and extraneous impurity quantity, have a significant impact on the aeration process and uniformity of the airflow throughout the grain bulk [12–14]. They also influence the intergranular air pathways and associated inter-speed currents as well as the exchange of temperature and moisture in bulk [15]. Therefore, assessing the prevailing airflow resistance in grain bulks is essential for the energy-efficient design of ventilation systems, aeration management and grain quality retention [16].

Physical experiments are commonly used to assess the airflow resistance of grain bulks and serve as important means for the development of mathematical models. In this regard, Shedd [17] established an empirical model by fitting experimental data of pressure drop ΔP across the grain bulk and airflow velocity v_a for several grain types using a logarithmic scale. v_a referred to the hypothetical airflow velocity calculated from the volume flow rate in the free bulk cross-section area, also known as superficial velocity. However, Shedd's model was limited to a narrow range of airflow velocities, which was further enhanced by the model of Hukill and Ives [18]. Due to their ease of handling and simplicity, these models have been used in several studies [19–21]. In addition, Hunter [22] developed a lumped polynomial-based model capable of accurately anticipating the relationship of ΔP vs. v_a, but lacked insight into parameters affecting the airflow resistance. A modified version was proposed by Haque et al. [23] that included the moisture content as an input parameter. As these models were empirical in nature, they were tied to the same grain–air conditions and bin configurations for which they were created and therefore can outperform when those conditions vary widely.

To overcome the shortcomings of empirical models, semi-empirical models which use the grain's physical characteristics and air properties have been developed. Ergun [24] conducted a thorough data analysis to describe the relationship between ΔP and v_a of uniform spherical particles and developed a semi-empirical model based on the Kozeny-Carman [25] and Burke-Plummer relationships [26], making this model one of the most commonly analyzed and used in the literature. However, the Ergun model lacked adaptability to non-spherical shapes of particles, thus Patterson [27] and Li and Sokhansanj [28] suggested quantitative improvements to account for irregular and random-sized shapes of grains. A simplification of these models was proposed by Bern and Charity [29]. In addition, Leva [30] developed a semi-empirical model based on the Hagen-Poisseuille law for isothermal flow that contained a modified friction factor for the state-of-flow and a shape factor for non-spherical particles, while Matthies and Petersen [31] established another model for high grain bulks. Due to their theoretical underpinning, the semi-empirical models were able to determine the effect of different grains, moisture contents, filling methods, impurity concentration and airflow directions on the airflow resistance [12,13,23,32]. They were also viable in isolating and quantifying the grain bulk pore volume. A summary of the above-mentioned models for describing the relationship of ΔP vs. v_a in grain bulks is presented in Table 1. So far, the known models are limited to depicting the complexity and diversity of bulk pore structures [29].

Table 1. Models for describing the pressure drop ΔP as a function of airflow velocity v_a in grain bulks.

Source	Year	Model Type	Applicability		Comments
			Velocity v_a, $m \cdot s^{-1}$	Type of Grains	
Ergun [24]	1952	Semi-empirical	≥0.01	Maize, rice, sorghum, wheat	Covers both laminar and turbulent flow; suitable for spherical particles
Shedd [17]	1953	Empirical	0.005–0.30	Barley, maize, oat, rice, sorghum, soybean, wheat	Appropriate for low airflow velocities and uncompacted grain bulk; outperformance at high velocities
Hukill and Ives [18]	1955	Empirical	0.0003–1.0	Barley, maize, oat, rice, sorghum, soybean, wheat	Encompasses a wide range of velocities; limited to specific grain and air conditions
Leva [30]	1959	Semi-empirical	≥0.0001	Barley, maize, oat, wheat	Tedious to solve; includes a friction-factor for the state-of-flow and a shape factor for non-spherical shape of grains
Patterson [27]	1969	Semi-empirical	0.05–0.61	Beans, maize	Adjusted model for grains with different size distributions and shape irregularities
Matthies and Petersen [31]	1974	Semi-empirical	0.02–0.61	Barley, maize, rice, rye, wheat	Established for high bulks; considers several influencing parameters
Bern and Charity [29]	1975	Semi-empirical	0.015–0.60	Maize	Easy to solve; considers solely pore volume and airflow velocity; limited to maize
Haque, Ahmed and Deyoe [23]	1982	Empirical	0.01–0.22	Maize, sorghum, wheat	Includes the effect of moisture content on the calculation basis
Hunter [22]	1983	Empirical	0.006–0.21	Barley, maize, oat, rice, sorghum, soybean, wheat	Better fit compared to Shedd; considers the non-uniform nature of grain bulks; lacks insight into parameters affecting airflow resistance
Li and Sokhansanj [28]	1992	Semi-empirical	0.0001–0.90	Barley, maize, oat, wheat	Similar to Ergun; suitable for grains; established for a wide range of airflow velocities

During storage, grain bulk undergoes burden pressures imposed by its dead weight, contributing to self-compaction [33–35]. Hence, the bulk characteristics may change depending on the degree of compaction. According to Rocha et al. [36], the airflow resistance in aeration systems is significantly increased with the increase of compaction when higher pressures are applied. Therefore, the misestimation of airflow resistance due to compaction can lead to ineffective aeration strategies and grain quality problems [35]. To date, the literature offers limited coverage on the effect of compaction on the airflow resistance of stored grain bulks where controlled compaction systems or filling methods were used [36–38]. However, the influence of spontaneous temporal and spatial self-compaction on the airflow resistance of practical storage systems has not been considered so far. Therefore, the objectives of this study were: (i) to investigate the effect of self-compaction on the pressure drop during aeration at various sets of airflow velocities, bulk depths and storage times, (ii) to mathematically describe the relationship of ΔP vs. v_a using a semi-empirical modeling approach, (iii) to develop a generalized model with itemized product factor and variable bulk pore volume and (iv) to evaluate the influence of parameters in modeling of pressure drop through a sensitivity analysis.

2. Materials and Methods

2.1. Raw Material and Sample Preparation

A total quantity of 1000 kg wheat (*Triticum aestivum* L.), cultivar 'Pionier' (I.G. Pflanzenzucht GmbH, Ismaning, Germany) was obtained from the Heidfeldhof research farm of the University of Hohenheim, Stuttgart, Germany (48°42′56.54″ N, 9°11′23.07″ E). The non-cereal harvest impurities such as straw, chaff, dust and stones (8.86 ± 1.37%) of aggregate mass were removed using an automated cleaning machine (D-4950, Samatec Saatguttechnik & Maschinenbau GmbH, Bad Oeynhausen, Germany). The cleaned bulk was stored for 24 h at hygienically safe conditions (temperature T of 14.90 ± 1.50 °C and relative humidity φ of 52.09 ± 7.07%) before being used for measurement of physical properties and airflow resistance experiments.

2.2. Characterization of Grain Physical Properties

The moisture content X (kg·kg^{-1} d.b.) of wheat kernels was determined by the standard thermogravimetric analysis in a convective oven (UM 700, Memmert GmbH & CO. KG, Schwabach, Germany) at 105 ± 1 °C for 24 h and natural air circulation according to AOAC [39], where moisture of 0.123 ± 0.001 kg·kg^{-1} d.b. was observed (dry matter of 89.01 ± 0.01%).

The principal geometrical characteristics of kernels, length L (mm), width W (mm) and thickness T (mm) were measured via a digital Vernier caliper (Digi-Met IP 67, Helios-Preisser GmbH, Gammertingen, Germany) with a measuring resolution of ±0.01 mm. Measurements were carried out for a total of 100 randomly selected kernels. Shape-dependent geometric properties such as arithmetic diameter d_a (mm), geometric diameter d_g (mm), sphericity ϑ (%), aspect ratio R_a (-) and unit volume V (mm^3) were estimated from the basic geometrical characteristics, as described by Karaj and Müller [40]. The equivalent diameter d_e (mm) of kernels was calculated as:

$$d_e = \sqrt[3]{\frac{6\,V}{\pi}} \tag{1}$$

From Sirisomboon et al. [41], the surface area A (mm^2) of kernels was estimated as:

$$A = \frac{\pi\,W}{2}\left(W + \frac{L}{c}\text{arcsinc}\right) \text{ with } c = \sqrt{1 - \left(\frac{W}{L}\right)^2} \tag{2}$$

In addition, the gravimetric properties were assessed. Unit mass m (g) of kernels was measured by means of an analytical high-precision balance with an accuracy of ±0.10 mg (Sartorius BP221S, Sartorius AG, Göttingen, Germany). Solid density ρ_s (kg·m^{-3}) was defined based on the toluene displacement method in a 25 mL pycnometer (Blaubrand, Wertheim, Germany) as described by Mohsenin [42]. Toluene was utilized as a water-insoluble liquid. The solid density ρ_s was determined as:

$$\rho_s = \frac{m_{gr}\,\rho_{tol}}{m_{gr} + m_{fl,\,tol} - m_{gr,\,fl,\,tol}} \tag{3}$$

where m_{gr} (g) is the mass of kernel, $m_{fl,tol}$ (g) is the mass of pycnometer flask filled with toluene, $m_{gr,fl,tol}$ (g) is the mass of kernels soaked in toluene solution together with the flask and ρ_{tol} (kg·m^{-3}) is the toluene density. The default bulk density ρ_{b0} (kg·m^{-3}) was measured by freely pouring kernels into a cylindrical container (150 mm diameter, 100 mm height) by maintaining a natural flow rate until overflowing. Afterwards, the surplus mass was gently swiped off using a wooden striker from the brim of the container and weighted. Hence, the bulk density ρ_{b0} was calculated as:

$$\rho_{b0} = \frac{m_c}{V_c} \tag{4}$$

where m_c (kg) is amassed mass in the container and V_c (m^3) is the occupied volume. The default (uncompacted) pore volume ε_0 (-) was defined as the fraction of the volume of intergranular voids in the bulk and was calculated as a function of the solid density and bulk density:

$$\varepsilon_0 = 1 - \frac{\rho_{b0}}{\rho_s} \tag{5}$$

2.3. Experimental Test Bench

The test bench used to perform the airflow resistance experiments is shown schematically in Figure 1.

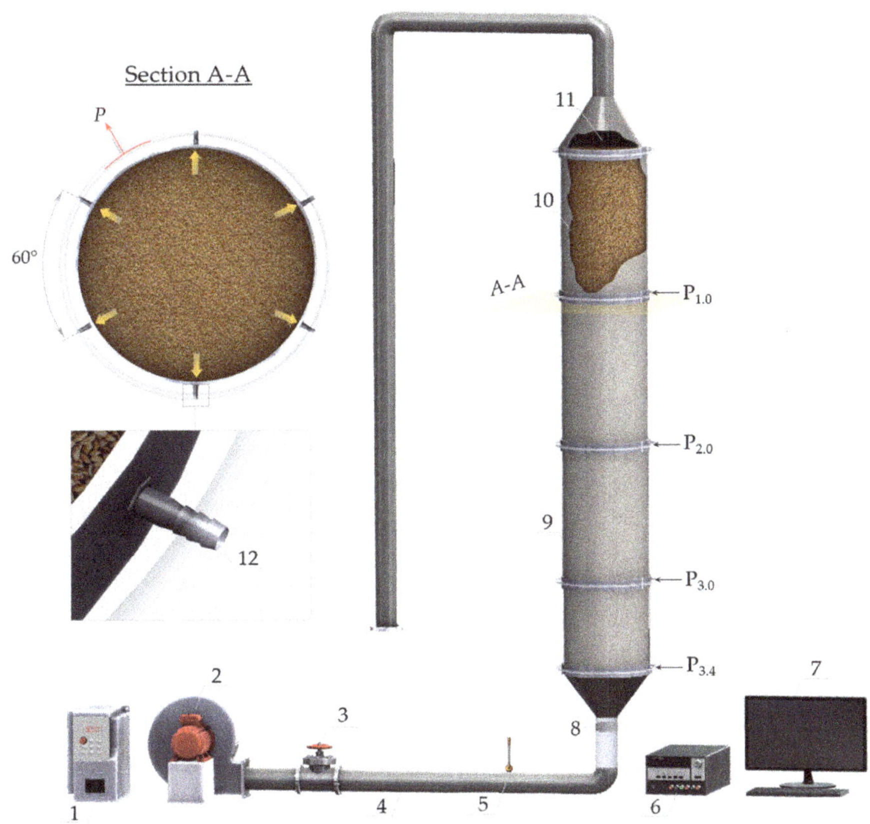

Figure 1. Test bench comprising (1) frequency inverter, (2) centrifugal fan, (3) safety valve, (4) air duct, (5) flow sensor, (6) data logger, (7) laboratory computer, (8) airflow straightener, (9) test bin, (10) wheat bulk, (11) sensor grid and (12) pressure taps. *P* denotes the equalized pressure of six pressure taps, whereas $P_{1.0}$, $P_{2.0}$, $P_{3.0}$ and $P_{3.4}$ represent the pressure data at bulk depths of 1.0, 2.0, 3.0 and 3.4 m, respectively.

A cylindrical acrylic-glass test bin (480 mm diameter, 3400 mm height and 5 mm wall thickness) with a wall friction coefficient of 0.32 ± 0.02 was used to accommodate approx. 500 kg of wheat kernels (V_b = 0.62 m^3). A perforated floor (3.80 mm apertures,

18.60% opening area) was installed at the bottom of the test bin to allow undisturbed upward movement of air within the pore volume of the grain bulk and to assert minimal resistance to airflow according to ASAE [43]. Air was supplied by a centrifugal fan (RD6-NRD80S/2, Elektror GmbH, Esslingen, Germany) with a maximal volumetric air capacity of 1230 $m^3 \cdot h^{-1}$, pressure of 2500 Pa at the nominal fan speed of 2890 min^{-1} and power consumption of 0.75 kW at 380 V/50 Hz. The fan speed was adjusted to the experimental requirements by a frequency inverter (ST 8100, Sourcetronic GmbH, Bremen, Germany).

An air duct (150 mm diameter, 2000 mm length) was employed to connect the fan with the test bin. A thermal flow sensor with an integrated transducer (TA10, Höntzsch Instrument, Waiblingen, Germany) with a measurement accuracy of ±2.0% was used to measure the airflow velocity at a distance of 10-fold diameter of the duct [44]. The flow sensor was calibrated using a bench wind tunnel (8390, TSI Incorporated, Shoreview, MN, USA). Based on the airflow velocity in the duct, the volume flow was calculated and subsequently the superficial velocity v_a in the test bin. At the duct end, a 90° bow rubber pipe and a honeycomb-shaped polycarbonate straightener (100 mm diameter, 50 mm thickness) were installed to prevent the fan propagating vibrations and ensure uniform flow conditions [12]. All joints of the test bench were examined for air leakage and were tightly sealed.

Pressure taps (2 mm diameter, 30 mm length) were attached in the wall of the test bin at depths of 1.0, 2.0, 3.0 and 3.4 m ($P_{1.0}$–$P_{3.4}$). The taps protruded 2 mm inside the bulk to reduce the possible wall effects during the pressure measurements. At each depth, six taps were evenly distributed around the circumference at a segment angle of 60° and connected by a loop of 4 mm diameter transparent polyethylene hose, which was also used to connect the loops with the pressure sensors (GMSD 25MR & GHM 3151-Ex, GHM Messtechnik GmbH, Remscheid, Germany) with an accuracy of ±0.50 Pa. Combined temperature/humidity sensors (SHT25, Sensirion AG, Zurich, Switzerland) were placed in the centerline of the test bin at the same depths as pressure sensors to measure the temperature T_a (±0.20 °C) and relative humidity φ_a (±1.80%) of intergranular air in the grain bulk. A data logger (Agilent 34901A, Agilent Technologies Inc., Santa Clara, CA, USA) was used to acquire data from all sensors and record them on a laboratory computer. Manual sampling and offline thermogravimetric analysis were also conducted to determine the grain moisture content in the bulk [39].

2.4. Experimental Procedure

A motor-driven screw conveyor (T206/4, Wolf Landtechnik GmbH, Petersberg, Germany) was used to fill the test bin and to ensure a practical filling procedure of grain bulk at a standard flow rate. The resulting bulk cone of approx. 30° was manually drawn off flush at the top edge [14]. After 1 h, during which the grain bulk rested, the fan was started with a frequency f of 10.0 Hz and gradually increased by 5.0 Hz intervals until 50.0 Hz, resulting in nine steps of airflow velocities. The fan speed was changed only when the fluctuations of the pressure readings were calming down to less than 2.0%. The pressure drop ΔP was estimated as the difference of pressure at H_b of 1.0 ($P_{1.0}$), 2.0 ($P_{2.0}$), 3.0 ($P_{3.0}$) and 3.4 m ($P_{3.4}$) to the pressure at the top of the test bin at H_b of 0.0 m bulk depth. In order to investigate the effect of self-compaction over time t, the same procedure was repeated after 65, 164 and 236 h. Table 2 shows the average fan speed ω, airflow velocity v_a, mass flow rate $\dot{m}$ and volume flow rate Q used for the experiments, which were chosen based on the practical recommendations for aeration and drying systems [45]. For the analysis of the relationship between pressure drop ΔP and airflow velocity v_a, a total of 15,760 data were gathered at different bulk depths H_b and storage times t. The intermittent forced aeration was applied only for the airflow resistance experiments, while the traditional storage without aeration was used for the rest of the time.

Table 2. Operating settings utilized for the airflow resistance experiments.

Frequency	Rotational Speed	Airflow Velocity	Mass Flow Rate	Volume Flow Rate
f, Hz	ω, min^{-1}	v_a, m·s^{-1}	$\dot{m}$, kg·h^{-1}	Q, m^3·h^{-1}
10.0	578	0.011 ± 0.000	8.57 ± 0.36	7.00 ± 0.29
15.0	867	0.021 ± 0.001	20.15 ± 0.96	16.46 ± 0.79
20.0	1156	0.037 ± 0.004	32.93 ± 3.40	26.90 ± 2.78
25.0	1445	0.056 ± 0.005	47.48 ± 4.25	38.79 ± 3.47
30.0	1734	0.074 ± 0.006	61.37 ± 4.66	50.14 ± 3.81
35.0	2023	0.092 ± 0.006	74.99 ± 5.04	61.27 ± 4.12
40.0	2312	0.109 ± 0.008	88.14 ± 6.28	72.01 ± 5.14
45.0	2601	0.125 ± 0.008	100.78 ± 6.61	82.34 ± 5.40
50.0	2890	0.141 ± 0.010	113.16 ± 8.37	92.45 ± 6.84

2.5. Semi-Empirical Modelling of Airflow Resistance

Out of the available semi-empirical models in literature, the Matthies and Petersen [31] was chosen as the most appropriate for modeling ΔP vs. v_a of grain bulks $H_b \geq 2.50$ m, which covers the irregular and random-sized shapes of wheat kernels and a wide range of airflow velocities. This model is expressed as:

$$\frac{\Delta P}{H_b} = k\,\zeta\,\frac{\rho_a\,{v_a}^2}{2\,d_e\,\varepsilon^4} \tag{6}$$

where ΔP (Pa) is the pressure drop in bulk, v_a (m·s^{-1}) is the airflow velocity, H_b (m) is the bulk depth, k (-) is the product factor related to the shape configuration, size and surface characteristic of wheat kernels, ζ (-) is the coefficient of air resistance, ρ_a (kg·m^{-3}) is the density of intergranular air and ε (-) is the pore volume of grain bulk. The product factor k was estimated by fitting the model to the experimental data, while the coefficient of air resistance ζ was determined as:

$$\zeta = \frac{47.92}{Re} + \left(\frac{1.18}{Re}\right)^{0.1} \tag{7}$$

where Re (-) is the Reynolds number, which was expressed as a function of the equivalent diameter of kernels d_e (mm):

$$Re = \frac{v_a\,\rho_a\,d_e}{\mu_a} \tag{8}$$

where μ_a (kg·m^{-1}·s^{-1}) is the dynamic viscosity of air in the pore volume of grain bulk. The thermodynamic characteristics of air μ_a, ρ_a and Reynolds number Re were calculated based on the temperature T_a and relative humidity φ_a of intergranular air of grain bulk during aeration. Therefore, the Matthies and Petersen [31] model (Equation (6)), by embedding the Re and ζ, can be written as:

$$\frac{\Delta P}{H_b} = k\left(\frac{23.96\;\mu_a v_a}{\varepsilon^4 d_e^2} + \frac{0.51\;\mu_a^{0.1}\rho_a^{0.9}{v_a}^{1.9}}{\varepsilon^4 d_e^{1.1}}\right) \tag{9}$$

The above-mentioned equation was used to fit the experimental data of ΔP vs. v_a at various bulk depths and storage times.

2.6. Statistical Analysis and Graphical Presentation

The graphical representation of data and the nonlinear least-squares fitting procedure at 95.0% significance level ($p \leq 0.05$) were carried out in MATLAB 2019a (MathWorks Inc., Natick, MA, USA). The Levenberg-Marquardt algorithm was used for the fitting of experimental data in a series of iterative steps with a convergence criterion of 1.0×10^{-6}. The coefficient of determination R^2, root mean square error $RMSE$ and mean absolute percentage error $MAPE$ were employed as statistical criteria to assess the goodness of fit, which were defined as follows:

$$R^2 = 1 - \frac{\sum_{i=1}^{n}\left(\Delta P_{obs} - \Delta P_{pred}\right)^2}{\sum_{i=1}^{n}\left(\Delta P_{obs} - \overline{\Delta P_{obs}}\right)^2} \quad (10)$$

$$RMSE = \sqrt{\frac{\sum_{i=1}^{n}\left(\Delta P_{obs} - \Delta P_{pred}\right)^2}{n}} \quad (11)$$

$$MAPE = \frac{100}{n}\sum_{i=1}^{n}\left|\frac{\Delta P_{obs} - \Delta P_{pred}}{\Delta P_{obs}}\right| \quad (12)$$

where ΔP_{pred} (Pa) is the predicted pressure drop, ΔP_{obs} (Pa) is the observed pressure drop ascertained from experiments and n (-) is the number of observations. The sensitivity analysis using Monte Carlo simulations in MATLAB/Simulink 2019a (MathWorks Inc., Natick, MA, USA) performed to evaluate the influence of modeling parameters on pressure drop. Furthermore, the CAD design of the experimental test bench was carried out in SOLIDWORKS 2019 (Dassault Systèmes, Waltham, MA, USA).

3. Results and Discussion

3.1. Characterization of Grain Physical Properties

Table 3 presents the summary of the geometric and gravimetric properties of wheat kernels (*Triticum aestivum* L.) cv. 'Pionier' at moisture content of 0.123 ± 0.001 kg·kg^{-1} d.b.

Table 3. Geometric and gravimetric properties of wheat (*Triticum aestivum* L.) cv. 'Pionier'.

Properties	Unit	Value
Length L	mm	6.87 ± 0.25
Width W	mm	3.75 ± 0.22
Thickness T	mm	3.11 ± 0.17
Arithmetic diameter d_a	mm	4.57 ± 0.15
Geometric diameter d_g	mm	4.30 ± 0.15
Equivalent diameter d_e	mm	4.49 ± 0.14
Aspect ratio R_a	-	0.55 ± 0.03
Sphericity ϑ	%	62.72 ± 1.94
Surface area A	mm^2	64.06 ± 5.42
Volume V	mm^3	41.95 ± 4.34
Unit mass m	g	0.06 ± 0.01
Bulk density ρ_{b0}	kg·m^{-3}	782.46 ± 6.68
Solid density ρ_s	kg·m^{-3}	1351.40 ± 4.62
Pore volume ε_0	-	0.421 ± 0.07

It can be discerned from Table 3 that a low standard deviation was exhibited from geometric properties, indicating that data were tightly clustered around the mean value. The length L, width W and thickness T were found to be consistent with the literature, with values falling within 5.78–7.45 mm, 2.36–3.93 mm, 2.56–3.27 mm reported by Tabatabaeefar [46], Karimi et al. [47], Molenda and Horabik [48] and Wang et al. [49] for other wheat varieties but similar moisture contents. Therefore, the shape-dependent properties such as arithmetic diameter d_a, geometric diameter d_g, equivalent diameter d_e, aspect ratio R_a, sphericity ϑ, surface area A and unit volume V were also in conformity with the same literature. However, kernel dimensions were slightly larger than those reported by Giner and Denisienia [12], Nelson [50], Petingco et al. [51] and Markowski et al. [52], which can be attributed to differences in sample origin, cultivar, specific growth conditions and moisture contents. Gürsoy and Güzel [53], on the other hand, reported lower values for width and thickness for similar kernel lengths, resulting in a lower aspect ratio R_a and sphericity ϑ.

In addition, the gravimetric properties are presented in Table 3. Due to the larger kernel dimensions, a larger unit mass m was observed compared to values reported by Gürsoy and Güzel [53] and Markowski, Żuk-Gołaszewska and Kwiatkowski [52]. The values of bulk density ρ_{b0}, solid density ρ_s and pore volume ε_0 were found to be in decent agreement with values reported by Molenda and Horabik [48], Haque et al. [23], Jayas and Cenkowski [54], Muir and Sinha [55] and Kraszewski [56]. However, higher values of bulk density ρ_{b0} and pore volume ε_0 were observed compared to Markowski, Żuk-Gołaszewska and Kwiatkowski [52], which can be ascribed to the cultivar and/or kernel moisture content, as well as container volume, size, quantity of impurities, filling procedure and filling height and rate, which in turn affected the bulk packing in the container [57,58]. In contrast to Giner and Denisienia [12], lower values of bulk density ρ_{b0} were obtained, resulting in a higher pore volume ε_0 for similar solid density ρ_s.

3.2. Bulk Conditions during Experimentation

During pressure drop experiments, variations in temperature T_a from 12.29 to 17.18 °C and relative humidity φ_a from 34.04 to 40.87% were observed for the intergranular air of the grain bulk at 1, 65, 164 and 236 h storage time t. The associated thermodynamic properties of air, in terms of viscosity μ_a and density ρ_a, were assessed according to White and Majdalani [59] and tabulated in Appendix A (Table A1). The observed data were utilized for the semi-empirical modeling of pressure drop. Despite fluctuations in temperature T_a and relative humidity φ_a, no significant differences were observed in moisture content X of the wheat bulk (0.123 ± 0.001 to 0.122 ± 0.001 kg·kg^{-1} d.b.) at $p \leq 0.05$ during the experiments, which means that possible effects of drying on the self-compaction were excluded.

3.3. Determination of the Product Factor k

Matthies and Petersen [31] found a product factor k ranging from 2.00 to 2.20 in their study for calculating the airflow resistance of stored wheat bulks. However, a different wheat variety with different moisture content, physical properties, kernel size distribution and filling method has been employed in this study, therefore the value of k was tailored to the specific grain bulk and experimental conditions. Herewith for the determination of k, the experimental data of ΔP vs. v_a for bulk depths of H_b 1.0, 2.0, 3.0 and 3.4 m and storage time t of 1 h were fitted by the Matthies and Petersen model (Equation (9)) and the pore volume ε at different depths H_b was determined, accordingly. The observed ε were afterwards fitted by a linear model to describe the relationship between ε and H_b and extrapolated to H_b = 0.1 m, which is the criterion for comparison with the default ε_0 (uncompacted) ascertained in the laboratory. First, the reported k were used and afterwards k was iteratively adjusted by intervals of 0.01 until the predicted pore volume matched the default ε_0 of 0.421 (Figure 2). By using this criterion, a value of k of 2.73 was found, which fell between values of 2.00 and 3.90 used for various grains by Matthies and Petersen [31]. This finding was consistent with findings of Bakker-Arkema et al. [60] and Patterson [27], who adjusted the factor k for the specific settings of their experiment and found out higher values of k than those reported by Matthies and Petersen [31].

The linear models and the goodness of fit acquired from individual fittings at k of 2.00, 2,10, 2.20 and 2.73 are presented in Table 4. An accuracy of $R^2 \geq 0.985$ was observed from fitting with the linear models, which indicated a high capability of the employed models to depict the relationship between ε and H_b at different k. Figure 2 displays graphically the variation of the pore volume ε influenced by factor k with respect to the default ε_0. It can be seen that the values of ε increased proportionally with the increase of k. The values proposed by Matthies and Petersen [31] resulted in underestimation of 7.31, 6.17 and 5.09% of default ε at H_b = 0.1 m for k of 2.00, 2.10 and 2.20, respectively. Moreover, these factor k yielded a *MAPE* of 26.82, 23.17 and 19.51% for pressure drop in fitting the experimental data. Therefore, a product factor k of 2.73 was used for prediction of pressure drop of wheat bulk cv. 'Pionier'.

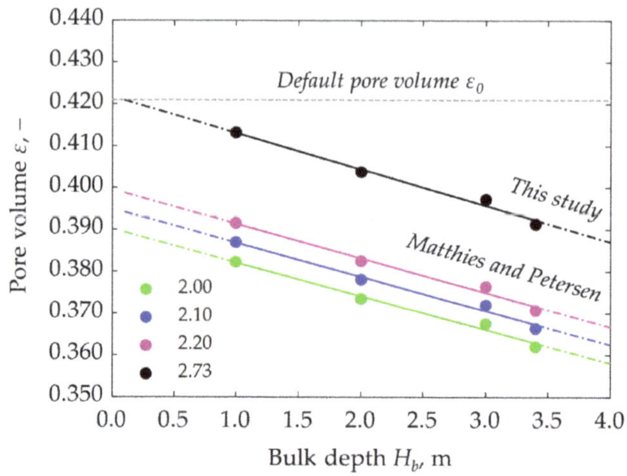

Figure 2. Pore volume ε vs. bulk depth H_b for product factor k of 2.00, 2.10, 2.20 and 2.73. Horizontal dashed line refers to the default pore volume ε_0 at bulk depth H_b of 0.1 m. Solid lines represent linear model fitting; dashed lines represent extrapolations beyond the dataset utilized for fitting.

Table 4. Linear models for describing the relationship between the pore volume ε and bulk depth H_b at different product factor k and coefficient of determination R^2.

Product Factor k, -	Mathematical Model	R^2, -
2.00	$y = -8.027 \times 10^{-3}\,x + 0.390$	0.986
2.10	$y = -8.141 \times 10^{-3}\,x + 0.395$	0.986
2.20	$y = -8.215 \times 10^{-3}\,x + 0.400$	0.985
2.73	$y = -8.693 \times 10^{-3}\,x + 0.422$	0.986

3.4. *Influence of Self-Compaction on the Airflow Resistance*

Figure 3a shows the experimental data of the pressure drop ΔP vs. airflow velocity v_a at bulk depths H_b of 1.0, 2.0, 3.0 and 3.4 m at 1 h storage time t. The overall variation of ΔP during the measurement cycle was relatively small, with the standard deviation ranging from 0.51 to 24.86 Pa, which indicated that the data were highly reproducible and tightly clustered around the mean values. The experimental data exhibited a progressive increase in pressure drop ΔP with increasing air velocity v_a and bed depth H_b, which were comparable to those of Giner and Denisienia [12] for similar moisture content and velocities smaller than 0.15 m·s^{-1}. This can be explained by the application of the same filling procedure that produced a dense bulk configuration as is typically used in practice. A similar trend of pressure drop for wheat was reported by Molenda et al. [37] for sprinkle filling. However, the results of ΔP vs. v_a were higher than those of Shedd [17] and Haque et al. [23], which can be attributed to the differences in wheat varieties and filling methods, resulting in higher resistance to airflow and higher pressure drops.

When fitting the Matthies and Petersen [31] to the experimental data and using the default pore volume ε_0 of 0.421 as obtained from laboratory measurements, where the curves were found to increasingly deviate with increasing airflow velocity v_a and bulk depth H_b. For H_b = 1.0 m, a decent fit is observed for $v_a \leq 0.10$ ms^{-1}, after which the model tends to underestimate the experimental data values by up to 11.78%. Notably, this tendency becomes more prominent with the increase of H_b to 2.0, 3.0 and 3.4 m, where differences increase with a deviation up to 19.62, 24.61 and 28.92%, respectively. Hence, the observed results were found to be irreconcilable with the homogeneous and isotropic consideration of grain bulk reported in literature, for which the pressure drop curves between depths H_b are linearly equidistant at a given velocity v_a, as the curves from prediction underestimated the behavior of the experimental data. Figure 3b shows the distribution of predicted ΔP_{pred} vs. observed ΔP_{obs} for the default pore volume ε_0.

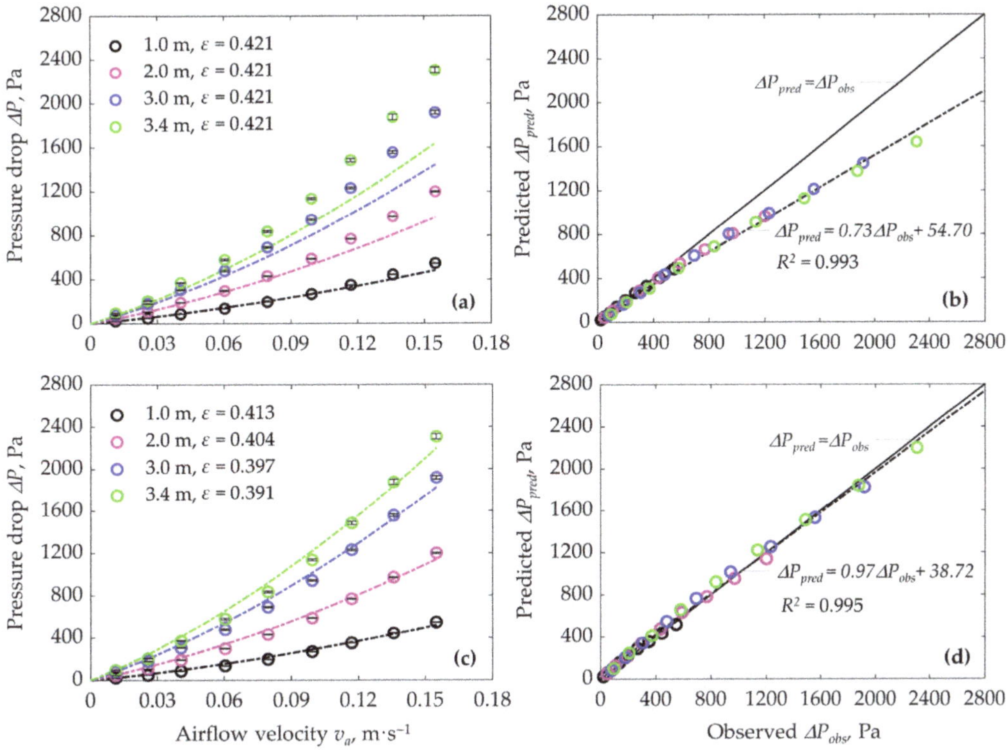

Figure 3. Experimental and predicted pressure drop ΔP vs. airflow velocity v_a at bulk depth H_b of 1.0, 2.0, 3.0 and 3.4 m and storage time t of 1 h storage time fitted with (**a**) default pore volume ε_0 of 0.421 and (**c**) adjusted pore volume ε between 0.391 and 0.413. Markers represent the experimental data points (±SD), dashed-dotted lines indicate fitting with the Matthies and Petersen model; Predicted pressure drop ΔP_{pred} vs. observed pressure drop ΔP_{obs} for (**b**) default and (**d**) variable pore volume.

It can be seen that the model exhibit an inferior performance with the data deviating from $\Delta P_{pred} = \Delta P_{obs}$ line and clustering towards the line $\Delta P_{pred} = 0.73\Delta P_{obs} + 54.70$ with $R^2 = 0.993$ and thus revealing an average underestimation of 22.0%. This disparity, however, is likely to increase as velocity v_a exceeds the limit used in this study.

To account for the spatial change of the pore volume in the grain bulk caused by self-compaction, the experimental data were refitted for each bulk depth H_b by adjusting the ε values. A reduction in pore volume ε of 0.413, 0.404, 0.397 and 0.391 was found from fitting analysis for H_b of 1.0, 2.0, 3.0 and 3.4 m. The variation of the pressure drop ΔP using the variable pore volumes ε is displayed in Figure 3c, where the fitted curves have accurately described the course of the experimental data. When using adjusted pore volume ε, the distribution of data was closely dispersed around the line $\Delta P_{pred} = 0.97\Delta P_{obs} + 38.72$ in close proximity with line $\Delta P_{pred} = \Delta P_{obs}$, hence revealing a high accuracy prediction with $R^2 = 0.995$ by the employed model (Figure 3d). These results were accredited to the vertical decrease of pore volume ε by 5.30%, which was in line with the findings of Cheng et al. [61] for compressive pressure levels ranging from 0 to 50 kPa. This behavior can be explained by the pressure of the overlying grain mass, which increases the in-situ intergranular stresses between kernels due to the dead weight of the overlying bulk [33]. Consequently, the pressure drop ΔP increased non-uniformly with the increase of the bulk depth H_b.

Since self-compaction is a dynamic process, the ΔP vs. v_a were fitted for different storage times t and the pore volume ε was also adjusted. Results of ε for bulk depths H_b of 1.0, 2.0, 3.0 and 3.4 m and storage time t of 1, 65, 164 and 236 h are shown in Figure 4. The default pore volume ε_0 was indicated by the grey plane in the chart.

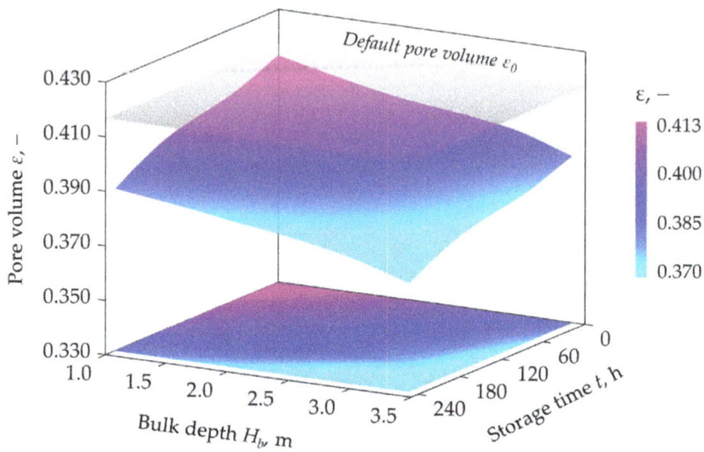

Figure 4. Variation of pore volume ε with bulk depth H_b and storage time t. The grey plane indicates the default pore volume ε_0.

For the different bulk depths H_b of 1.0 to 3.4 m, a temporal decrease of ε from 0.413 to 0.391, from 0.406 to 0.385, from 0.400 to 0.379 and from 0.390 to 0.370 was observed when storage time t increased from 1 to 65, 164 and 236 h. The resultant stresses are believed to initially influence the reorientation of the kernels and then cause the irreversible plastic deformation once the rupture force is attained, which eventually decreases the pore volume ε [33]. The compaction of the grain mass can be attributed to the visco-elastoplastic properties of kernels. However, Figure 4 shows that the grain mass did not settle completely and a longer time can be required to achieve the permanent equilibrium.

Figure 5 presents the pressure drop ΔP vs. airflow velocity v_a for storage time t predicted with the Matthies and Petersen model. It can be seen that the pressure drop manifested a temporal increase throughout 236 h of storage time t, which can be ascribed to the gradual and irreversible dynamic compaction of the bulk [62,63]. Hence, a variation of pressure drop ΔP from 1231.92 to 1536.97 Pa was estimated for 3.40 m bulk depth H_b at v_a = 0.10 m·s^{-1} once the storage time t increased from 1 to 236 h, which accounted for an increase of 24.76%. The higher pressure drops are mainly attributed to the dense fill created by the kernel packing due to the reduction of pore volume ε, which leads to increased kinetic energy dissipation due to friction and turbulence and higher intergranular resistances of the airflow. Similar outcomes were reported by Kumar and Muir [32], Molenda et al. [37] and Łukaszuk et al. [38], who found a considerable increase in pressure drop ΔP due to increase of ρ_b and reduction of ε obtained from the application of different filling methods and filling height. The same tendency has been also noted by Jayas et al. [64] for rapeseed, Kay et al. [65] for maize and Siebenmorgen and Jindal [66] for rice. However, they used the mean pore volume ε for modeling the relationship between ΔP and v_a and did not encounter the lateral variation of ε caused by self-compaction. Moreover, the results of this study are consistent with the outcomes of Haque [67], who confirmed the effect of non-homogeneous bulk of wheat on the pressure drop per unit of bulk depth due to self-compaction. Khatchatourian and Savicki [68] reported similar findings for soybeans. Despite consistency with published research, the effect of storage time t on ΔP has not been reported so far.

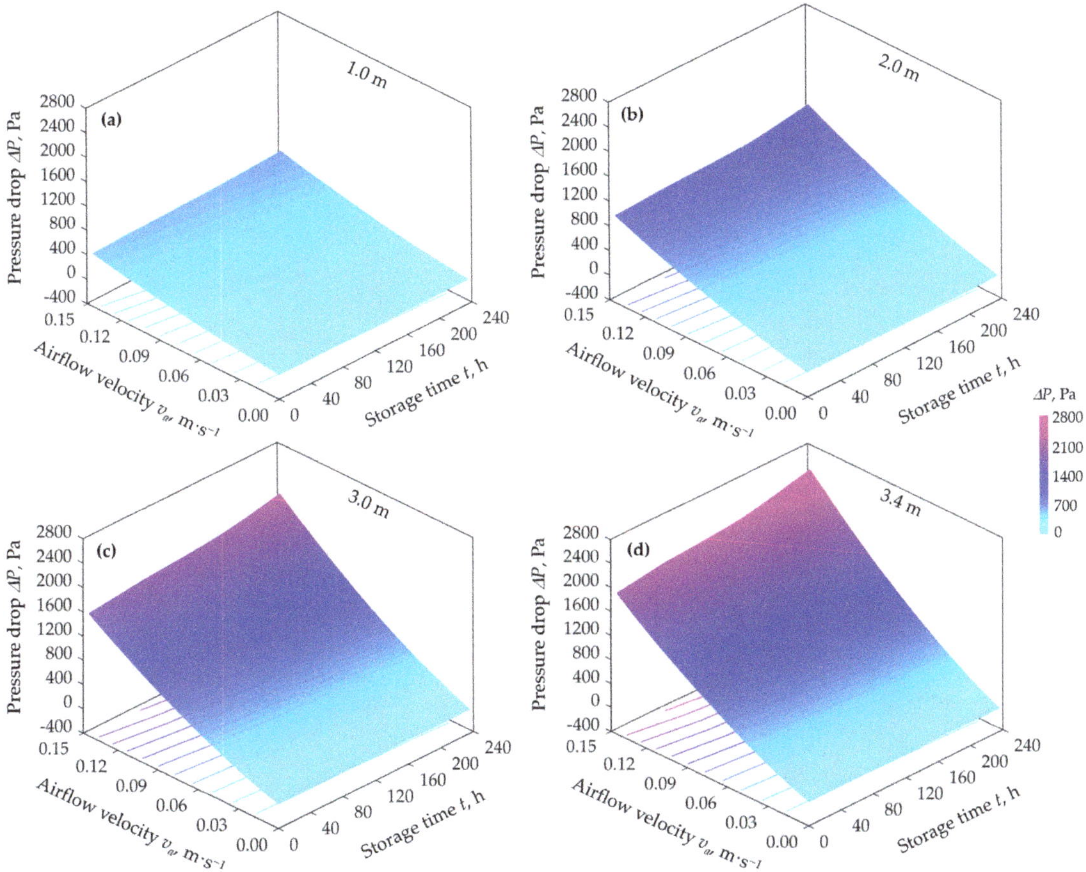

Figure 5. Pressure drop ΔP vs. airflow velocity v_a and storage time t predicted with the Matthies and Petersen model at bulk depths H_b of 1.0, 2.0, 3.0, 3.4 m (**a–d**).

A summary of adjusted pore volumes ε and statistical indicators attained from the fitting analysis is given in Table 5. The inspection of the statistical indicators revealed that the Matthies and Petersen [31] model was capable of depicting the course of ΔP vs. v_a at a decent accuracy. Particularly, R^2 between 0.983 and 0.996, *RMSE* between 15.18 and 123.77 Pa and *MAPE* between 8.21 and 16.16% were observed, respectively. However, a slight overestimation was observed at all predicted curves from 0.06 to 0.09 m·s^{-1}. Furthermore, from Table 5 can be discerned an increase in bulk density ρ_b which goes along with the decrease of pore volume ε. Particularly, a variation of 793.00–822.60 kg·m^{-3}, 802.33–830.98 kg·m^{-3}, 810.98–938.95 kg·m^{-3} and 825.03–851.65 kg·m^{-3} was observed for the bulk density ρ_b at the bulk depths H_b of 1.0 to 3.4 m and storage times t of 1, 65, 164, 236 h, respectively.

Figure 6a presents the data from pooling all predicted ΔP_{pred} and experimental ΔP_{obs} from the fitting analysis. It can be seen that the experimental data were satisfactorily anticipated by the model since they fell around the line of $\Delta P_{pred} = \Delta P_{obs}$, hence showing an appropriate accuracy of prediction of the Matthies and Pettersen model for the employed range of v_a, H_b and t with an $R^2 = 0.990$, $RMSE = 68.67$ Pa, $MAPE = 12.50\%$.

Table 5. Pore volume ε and bulk density ρ_b for different storage time t and bulk depth H_b as well as statistical indicators (R^2, $RMSE$, $MAPE$) observed from fitting the experimental data with Matthies and Peterson model.

Storage Time t, h	Bulk Depth H_b, m	Pore Volume ε, -	Bulk Density ρ_b, kg·m^{-3}	Statistical Indicators		
				R^2, -	$RMSE$, Pa	$MAPE$, %
1.0	1.0	0.413	793.00	0.995	15.18	8.21
	2.0	0.404	805.70	0.995	33.28	8.32
	3.0	0.397	814.62	0.995	52.39	7.19
	3.4	0.391	822.60	0.996	61.93	7.62
65.0	1.0	0.406	802.33	0.983	29.19	14.75
	2.0	0.400	811.38	0.983	62.74	15.12
	3.0	0.391	822.46	0.983	100.65	14.40
	3.4	0.385	830.98	0.982	123.77	13.97
164.0	1.0	0.400	810.98	0.987	26.58	16.16
	2.0	0.395	817.73	0.987	54.16	14.44
	3.0	0.385	831.11	0.987	89.30	13.28
	3.4	0.379	838.95	0.987	106.91	13.66
236.0	1.0	0.390	825.03	0.993	21.04	14.61
	2.0	0.383	833.27	0.993	45.24	15.11
	3.0	0.375	844.08	0.993	68.35	10.66
	3.4	0.370	851.65	0.993	84.17	12.75

The frequency distribution of residuals is shown in Figure 6b. The results indicate that the residuals follow a random distribution. This distribution was found to be unbiased and homoscedastic with non-constant variance, hence a reasonably symmetric and unimodal distribution of residuals around 0 was observed, which supported the validity of the engaged model. The values of residuals ranged from −169.24 to 233.55. However, 55.0% of data fell between −34.97 and 9.78. According to the Shapiro-Wilk test, the residuals indicated a significant likelihood of non-normal distribution at $p \leq 0.05$. Hence, a logistic model was employed to describe the distribution behavior of residuals.

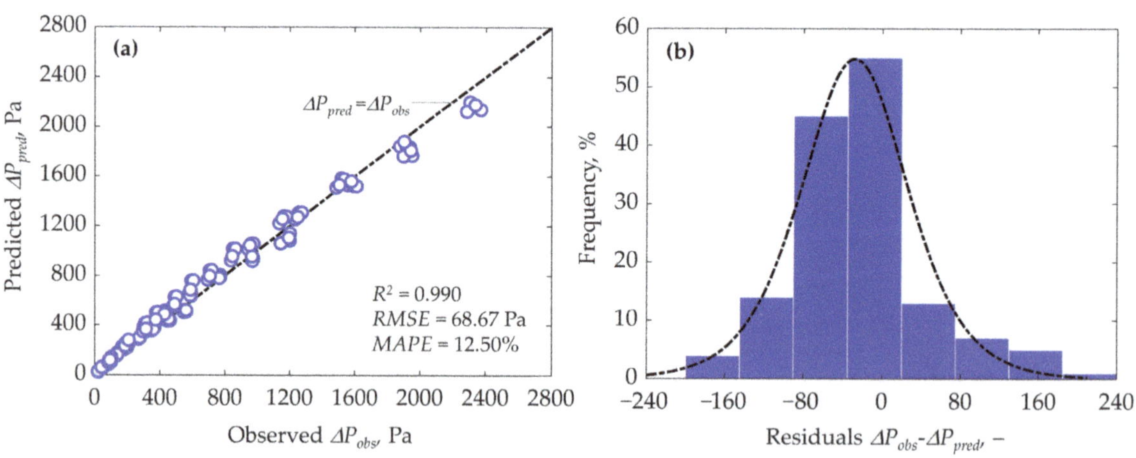

Figure 6. (**a**) Predicted pressure drop ΔP_{pred} vs. observed pressure drop ΔP_{obs} from pooled data of bulk depth H_b of 1.0, 2.0, 3.0, 3.4 m and storage times t of 1, 65, 164, 236 h; (**b**) Frequency distribution of residuals. Dashed-dotted line indicates the logistic probability distribution of residuals.

3.5. Modelling of Pore Volume Variation in Bulk

To establish a generalized semi-empirical model, pore volume ε was expressed as a function of bulk depth H_b analogous to the model proposed by Gao et al. [69] and Cheng et al. [35] for bulk density, which is given as:

$$1 - \frac{\varepsilon_0 - \varepsilon}{\varepsilon_0 - \varepsilon_{min}} = aH_b + c \quad (13)$$

where ε_0 (-) refers to the default pore volume and ε_{min} (-) refers to the minimal pore volume observed at the highest bulk depth H_b of 3.4 m, while a and c are the empirical constants observed from fitting analysis. To determine the pore volume ε, Equation (13) was rewritten as:

$$\varepsilon = \varepsilon_0 - (\varepsilon_0 - \varepsilon_{min}) \cdot (1 - aH_b - c) \quad (14)$$

Table 6 presents the equations and goodness of fit derived from the regression analysis for the different storage times t. The constants a and c were embodied in the equations. A variation of constant a from −0.292 to −0.156 and c from −1.026 to −0.557 was observed accordingly, hence revealing a decreasing trend of constants a and c with the increase of storage time t. The statistical indicators confirmed the capability of the employed model to predict closely the data with a high accuracy of $R^2 \geq 0.963$.

Table 6. Mathematical models for describing pore volume ε as function of bulk height H_b at different storage times t.

Storage Time t, h	Mathematical Model	R^2, -
1	$\varepsilon = \varepsilon_0 - (\varepsilon_0 - \varepsilon_{min}) \cdot (1 + 0.292H_b - 1.026)$	0.987
65	$\varepsilon = \varepsilon_0 - (\varepsilon_0 - \varepsilon_{min}) \cdot (1 + 0.238H_b - 0.851)$	0.977
164	$\varepsilon = \varepsilon_0 - (\varepsilon_0 - \varepsilon_{min}) \cdot (1 + 0.205H_b - 0.733)$	0.963
236	$\varepsilon = \varepsilon_0 - (\varepsilon_0 - \varepsilon_{min}) \cdot (1 + 0.156H_b - 0.557)$	0.980

In analogy with Equation (14), a model for describing pore volume ε as function of bulk height H_b and storage time t with an R^2 of 0.972 could be established:

$$\varepsilon = \varepsilon_0 - (\varepsilon_0 - \varepsilon_{min}) \cdot (1 + 0.016H_b + 0.002t - 1.013) \quad (15)$$

This allowed the inclusion of the influence of bulk depth H_b and storage time t in the Matthies and Petersen model (Equation (9)), hence yielding a generalized model:

$$\Delta P = 2.73 \cdot \left(\frac{23.96\ \mu_a v_a}{d_e^2} + \frac{0.51\ \mu_a^{0.1} \rho_a^{0.9} {v_a}^{1.9}}{d_e^{1.1}} \right) \cdot \left(\frac{H_b}{(\varepsilon_0 - (\varepsilon_0 - \varepsilon_{min}) \cdot (1 + 0.016H_b + 0.002t - 1.013))^4} \right) \quad (16)$$

The generalized model was able to depict the airflow resistance of wheat with an R^2 of 0.989, *RMSE* of 75.91 Pa and *MAPE* of 16.29%.

3.6. Sensitivity Analysis

The relative importance of parameters in modeling of pressure drop ΔP was determined through a sensitivity analysis, which was performed by generating a randomized combination of input parameters (v_a, H_b, ε, d_e, ρ_a, μ_a) of Equation (9) within their range of operating conditions and evaluating their impact on pressure drop ΔP. Figure 7 presents the standardized regression coefficients of sensitivity analysis, with parameters ranked by influence. Results of the analysis indicated that air velocity v_a is the most influential parameter, which significantly influences the pressure drop ΔP due to its contribution to energy dissipation of air pathways due to friction and turbulence. A value of 0.85 was obtained, indicating how decisive v_a is for the airflow resistance and aeration process of wheat bulk. Therefore, bulk depth H_b makes a considerable contribution in ΔP, where a value of 0.47 was observed, demonstrating a relatively weaker influence (−44.71%) compared to airflow velocity v_a. Noticeably, v_a and H_b are positively correlated with ΔP, therefore higher values of velocity or bulk depth result in the increase of pressure drop with the magnitude determined by the analysis. Pore volume ε and particle diameter d_e were identified as less decisive, which negatively affect ΔP with values of -0.18 and -0.15, respectively. They are responsible for the bulk configuration, therefore, their reduction increases the resistance to airflow and consequently increases the pressure drop ΔP. The parameters that had the least influence were air density ρ_a (0.05) and dynamic viscosity of air μ_a (0.01) which were affected by the minor variations of temperature T_a and relative humidity φ_a of the air passing through the grain bulk.

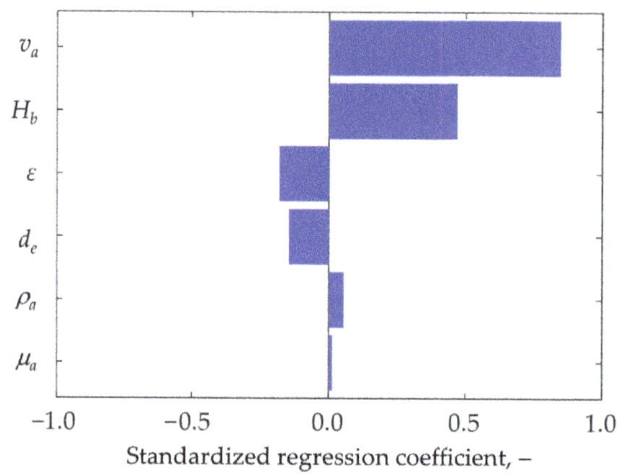

Figure 7. Standardized regression coefficients from sensitivity analysis including airflow velocity v_a, bulk depth H_b, pore volume ε, particle diameter d_e, air density ρ_a and dynamic air viscosity μ_a.

4. Conclusions

In this study, the resistance to airflow of a wheat grain bulk (*Triticum aestivum* L., cv. 'Pionier') under a set of air velocities, bulk depths and storage times was investigated. The physical characteristics of wheat kernels were experimentally assessed as a prerequisite for modeling the airflow resistance. For the characterization of ΔP vs. v_a relationship, the Matthies and Petersen model was employed, for which the product factor k was tailored for the specific wheat variety and experimental settings used in this study. From the fitting analysis, a goodness of fit with R^2 of 0.990, *RMSE* of 68.67 Pa and *MAPE* of 12.50% was observed for bulk depths ranging between 1.0 and 3.4 m and storage times between 1 and 236 h, which demonstrated a great potential of the employed model to describe the course of the experimental data with decent accuracy. Due to self-compaction, a spatial reduction of pore volume from 0.413 to 0.391 at bulk depths of 1.0 to 3.4 m after 1 h storage time and temporal reduction from 0.391 to 0.370 after 236 h storage time was observed, accordingly. Therefore, a disproportional increase of the pressure drop ΔP with bulk depth and storage time was observed, which was in contrast with the assumption of homogeneous and isotropic aerodynamic conditions in grain bulks often made in the literature. Thus, for practical application, higher power is required by the fan to maintain the required airflow velocity in bulk than when estimated from the default pore volume measured by the standard laboratory methods. The variation of pore volume ε was modeled and supported the development of a generalized model that could satisfactorily predict the airflow resistance of wheat bulk under self-compaction.

It could be shown that self-compaction plays a critical role in airflow resistance and therefore should be included in the design and analysis of cooling, aeration and low-temperature drying of in-store grain bulks. Further research should focus on the assessment of airflow resistances under self-compaction for other grains, moisture contents, bulk configurations and airflow velocity range. Moreover, the dynamics of grain compaction until permanent equilibrium should be further investigated. In addition, advanced numerical methods should be employed for an in-depth analysis of kernel and bulk behavior subjected to a wide range of loads encountered in practice during the storage of grain bulks.

Author Contributions: Conceptualization, I.R., S.S., S.K. and J.M.; methodology, I.R., S.S., S.K. and J.M.; software, I.R.; validation, I.R.; formal analysis, I.R.; investigation, I.R.; resources, I.R. and J.M.; data curation, I.R.; writing—original draft preparation, I.R.; writing—review and editing, I.R., S.S., S.K. and J.M; visualization, I.R.; supervision, J.M.; project administration, I.R., S.S., S.K. and J.M.; funding acquisition, J.M. All authors have read and agreed to the published version of the manuscript.

Funding: This research was financially supported by the German Federal Ministry for Economic Affairs and Energy (BMWi)—Project number KF2607404LT4.

Institutional Review Board Statement: Not applicable.

Informed Consent Statement: Not applicable.

Data Availability Statement: Not applicable.

Acknowledgments: The authors are grateful to the technical team of the Institute of Agriculture Engineering for their assistance in the construction of components required for the experimental investigation. Special gratitude goes to Simon Munder and Sabine Nugent for their contributions to sensor development and language editing, respectively. Furthermore, authors are thankful to the editors and reviewers for their constructive comments, which helped to improve the quality and scientific substance of manuscript.

Conflicts of Interest: The authors declare no conflict of interest.

Nomenclature

a, c	Empirical constants, -
A	Kernel surface area, mm^2
d_a	Kernel arithmetic diameter, mm
d_e	Kernel equivalent diameter, mm
d_g	Kernel geometric diameter, mm
$d.b.$	Dry basis, -
f	Fan frequency, Hz
H_b	Grain bulk depth, mm
L	Kernel length, mm
k	Product factor, -
$\dot{m}$	Air mass flow rate, kg·h^{-1}
m	Kernel unit mass, g
m_c	Aggregate mass of kernels in the container, g
m_{gr}	Mass of kernels soaked in toluene, g
$m_{fl,tol}$	Mass of pycnometer flask filled with toluene, g
$m_{gr,fl,tol}$	Mass of kernels, toluene solution and pycnometer flask, g
n	Number of observations, -
$MAPE$	Mean absolute percentage error, %
p	Probability level, -
Q	Air volume flow rate, m^3·h^{-1}
R^2	Coefficient of determination, -
R_a	Kernel aspect ratio, -
Re	Reynolds number, -
$RMSE$	Root mean square error, Pa
t	Elapsed storage time, h
T_a	Air temperature, °C
T	Kernel thickness, mm
v_a	Airflow velocity, m·s^{-1}
V	Kernel unit volume, mm^3
V_b	Test bin volume, m^3
V_c	Container volume, m^3

V_p	Pycnometer volume, mm^3
W	Kernel width, mm
X	Moisture content, -
x,y	Independent and dependent variables in linear models, -
x_a	Air absolute humidity, g·kg^{-1}
Δx_a	Air saturation deficit, g·kg^{-1}
P	Pressure, Pa
ΔP	Pressure drop, Pa
ΔP_{obs}	Observed pressure drop, Pa
ΔP_{pred}	Predicted pressure drop, Pa
ε	Bulk pore volume, -
ε_0	Default pore volume, -
ε_{min}	Pore volume at the highest bulk depth, -
ζ	Friction factor, -
ϑ	Kernel sphericity, %
φ_a	Air relative humidity, %
ω	Fan rotational speed, min^{-1}
μ_a	Air dynamic viscosity, kg·m^{-1}·s^{-1}
ρ_a	Air density, kg·m^{-3}
ρ_b	Bulk density, kg·m^{-3}
ρ_{b0}	Default bulk density, kg·m^{-3}
ρ_s	Solid density, kg·m^{-3}
ρ_{tol}	Toluene density, kg·m^{-3}

Appendix A

Table A1. Variations of air conditions in the pore volume of the grain bulk during pressure drop experiments.

Storage Time t, h	Moisture Content X, kg·kg^{-1} d.b.	Temperature T, °C	Relative Humidity φ_a, %	Absolute Humidity x_a, g·kg^{-1}	Saturation Deficit Δx_a, g·kg^{-1}	Viscosity $\mu_a \times 10^{-5}$, kg·m^{-1}·s^{-1}	Density ρ_a, kg·m^{-3}
1	0.123 ± 0.001	17.18 ± 0.09	39.83 ± 0.69	4.84 ± 0.07	2.85 ± 0.04	1.80 ± 0.00	1.22 ± 0.00
65	-	15.02 ± 0.10	39.54 ± 0.19	4.23 ± 0.02	2.65 ± 0.04	1.79 ± 0.00	1.23 ± 0.00
164	-	16.30 ± 0.02	40.87 ± 0.02	4.70 ± 0.01	2.72 ± 0.00	1.80 ± 0.00	1.22 ± 0.00
236	0.122 ± 0.001	12.29 ± 0.02	34.04 ± 0.02	3.01 ± 0.01	2.66 ± 0.00	1.78 ± 0.00	1.24 ± 0.00

References

1. FAO. Food and Agriculture Data. Available online: http://www.fao.org/faostat/en/#data (accessed on 28 June 2022).
2. Awika, J.M. Major cereal grains production and use around the world. In *Advances in Cereal Science: Implications to Food Processing and Health Promotion*; ACS Publications: Washington, DC, USA, 2011; pp. 1–13.
3. Ziegler, V.; Paraginski, R.T.; Ferreira, C.D. Grain storage systems and effects of moisture, temperature and time on grain quality—A review. *J. Stored Prod. Res.* **2021**, *91*, 101770. [CrossRef]
4. Anukiruthika, T.; Jian, F.; Jayas, D.S. Movement and behavioral response of stored product insects under stored grain environments—A review. *J. Stored Prod. Res.* **2021**, *90*, 101752. [CrossRef]
5. Collins, D.A. A review on the factors affecting mite growth in stored grain commodities. *Exp. Appl. Acarol.* **2012**, *56*, 191–208. [CrossRef] [PubMed]
6. Mesterházy, Á.; Oláh, J.; Popp, J. Losses in the Grain Supply Chain: Causes and Solutions. *Sustainability* **2020**, *12*, 2342. [CrossRef]
7. Navarro, S.; Noyes, R.T. *The Mechanics and Physics of Modern Grain Aeration Management*; CRC Press: Boca Raton, FL, USA, 2001.
8. Labuza, T.P. Sorption phenomena in foods: Theoretical and practical aspects. In *Theory, Determination and Control of Physical Properties of Food Materials*; Springer: Berlin, Germany, 1975; pp. 197–219.
9. Lenovich, L.M. Survival and death of microorganisms as influenced by water activity. In *Water Activity: Theory and Applications to Food*, 1st ed.; Routledge: Oxfordshire, UK, 2017; pp. 119–136.
10. Chakraverty, A.; Singh, R.P. Grain storage. In *Postharvest Technology and Food Process Engineering*; CRC Press: Boca Raton, FL, USA, 2014; pp. 131–157.
11. Brooker, D.B.; Bakker-Arkema, F.W.; Hall, C.W. *Drying and Storage of Grains and Oilseeds*, 1st ed.; Springer Science & Business Media: New York, NY, USA, 1992.

12. Giner, S.A.; Denisienia, E. Pressure drop through wheat as affected by air velocity, moisture content and fines. *J. Agric. Eng. Res.* **1996**, *63*, 73–85. [CrossRef]
13. Kashaninejad, M.; Maghsoudlou, Y.; Khomeiri, M.; Tabil, L. Resistance to airflow through bulk pistachio nuts (Kalleghochi variety) as affected by moisture content, airflow rate, bed depth and fill method. *Powder Technol.* **2010**, *203*, 359–364. [CrossRef]
14. Olatunde, G.; Atungulu, G.G.; Sadaka, S. CFD modeling of air flow distribution in rice bin storage system with different grain mass configurations. *Biosyst. Eng.* **2016**, *151*, 286–297. [CrossRef]
15. Neethirajan, S.; Karunakaran, C.; Jayas, D.S.; White, N.D.G. X-ray Computed Tomography Image Analysis to explain the Airflow Resistance Differences in Grain Bulks. *Biosyst. Eng.* **2006**, *94*, 545–555. [CrossRef]
16. Alagusundaram, K.; Jayas, D. Airflow resistance of grains and oilseeds. *Postharvest News Inf.* **1990**, *1*, 279–283. [CrossRef]
17. Shedd, C.K. Resistance of grains and seeds to air flow. *Agric. Eng.* **1953**, *34*, 616–619.
18. Hukill, W.; Ives, N. Radial airflow resistance of grain. *Agric. Eng.* **1955**, *36*, 332–335.
19. Gunasekaran, S.; Jackson, C.Y. Resistance to airflow of grain sorghum. *Trans. ASAE* **1988**, *31*, 1237–1240. [CrossRef]
20. Nimkar, P.M.; Chattopadhyay, P.K. PH—Postharvest Technology: Airflow Resistance of Green Gram. *Biosyst. Eng.* **2002**, *82*, 407–414. [CrossRef]
21. Agullo, J.O.; Marenya, M.O. Airflow Resistance of Parchment Arabica Coffee. *Biosyst. Eng.* **2005**, *91*, 149–156. [CrossRef]
22. Hunter, A.J. Pressure difference across an aerated seed bulk for some common duct and store cross-sections. *J. Agric. Eng. Res.* **1983**, *28*, 437–450. [CrossRef]
23. Haque, E.; Ahmed, Y.N.; Deyoe, C.W. Static pressure drop in a fixed bed of grain as affected by grain moisture content. *Trans. ASAE* **1982**, *25*, 1095–1098. [CrossRef]
24. Ergun, S. Fluid Flow Through Packed Columns. *Chem. Eng. Prog.* **1952**, *48*, 89–94.
25. Carman, P.C. Fluid flow through granular beds. *Chem. Eng. Res. Des.* **1997**, *75*, S32–S48. [CrossRef]
26. Burke, S.P.; Plummer, W.B. Gas flow through packed columns. *Ind. Eng. Chem.* **1928**, *20*, 1196–1200. [CrossRef]
27. Patterson, R.J. *Airflow-Pressure Drop Characteristics of Packed Beds of Biological Particles*; Michigan State University: East Lansing, MI, USA, 1969.
28. Li, W.; Sokhansanj, S. Generalized equation for airflow resistance of bulk grains with variable density, moisture content and fines. *Dry. Technol.* **1994**, *12*, 649–667. [CrossRef]
29. Bern, C.J.; Charity, L.F. Airflow resistance characteristics of corn as influenced by bulk density. *ASAE Pap.* **1975**, *48*, 1137–1145.
30. Leva, M. *Fluidization*; McGraw-Hill: New York, NY, USA, 1959.
31. Matthies, H.J.; Petersen, H. New data for calculating the resistance to air flow of stored granular materials. *Trans. ASAE* **1974**, *17*, 1144–1149. [CrossRef]
32. Kumar, A.; Muir, W.E. Airflow resistance of wheat and barley affected by airflow direction, filling method and dockage. *Trans. ASAE* **1986**, *29*, 1423–1426. [CrossRef]
33. Turner, A.P.; Montross, M.D.; McNeill, S.G.; Sama, M.P.; Casada, M.C.; Boac, J.M.; Bhadra, R.E.; Maghirang, R.G.; Thompson, S.A. Modeling the compressibility behavior of hard red wheat varieties. *Trans. ASABE* **2016**, *59*, 1029–1038. [CrossRef]
34. Thompson, S.; McNeill, S.; Ross, I.; Bridges, T. Packing factors of whole grains in storage structures. *Appl. Eng. Agric.* **1987**, *3*, 215–221. [CrossRef]
35. Cheng, X.; Zhang, Q.; Yan, X.; Shi, C. Compressibility and equivalent bulk modulus of shelled corn. *Biosyst. Eng.* **2015**, *140*, 91–97. [CrossRef]
36. Rocha, J.C.d.; Pohndorf, R.S.; Meneghetti, V.L.; Oliveira, M.d.; Elias, M.C. Effects of mass compaction on airflow resistance through paddy rice grains. *Biosyst. Eng.* **2020**, *194*, 28–39. [CrossRef]
37. Molenda, M.; Montross, M.D.; McNeill, S.G.; Horabik, J. Airflow resistance of seeds at different bulk densities using Ergun's equation. *Trans. ASAE* **2005**, *48*, 1137–1145. [CrossRef]
38. Łukaszuk, J.; Molenda, M.; Horabik, J.; Szot, B.; Montross, M.D. Airflow resistance of wheat bedding as influenced by the filling method. *Res. Agric. Eng.* **2008**, *54*, 50. [CrossRef]
39. AOAC. *Official Methods of Analysis*, 16th ed.; Association of Analytical Chemist: Arlington, VA, USA, 1998.
40. Karaj, S.; Müller, J. Determination of physical, mechanical and chemical properties of seeds and kernels of *Jatropha curcas* L. *Ind. Crops Prod.* **2010**, *32*, 129–138. [CrossRef]
41. Sirisomboon, P.; Kitchaiya, P.; Pholpho, T.; Mahuttanyavanitch, W. Physical and mechanical properties of *Jatropha curcas* L. fruits, nuts and kernels. *Biosyst. Eng.* **2007**, *97*, 201–207. [CrossRef]
42. Mohsenin, N.N. *Physical Properties of Plant and Animal Materials*; Gordon & Breach Science Publishers Ltd.: Philadelphia, PA, USA, 1968.
43. ASAE. ASAE D272: Resistance to airflow of grains, seeds, other agricultural products, and perforated metal sheets. *Am. Soc. Agric. Biol. Eng.* **2003**, 3, 569–576.
44. Miller, R.W. *Flow Measurement Engineering Handbook*, 3rd ed.; McGraw-Hill: New York, NY, USA, 1983.
45. Mühlbauer, W. *Handbuch der Getreidetrocknung: Grundlagen und Verfahren*; Agrimedia: Clenze, Germany, 2009.
46. Tabatabaeefar, A. Moisture-dependent physical properties of wheat. *Int. Agrophys.* **2003**, *17*, 207–211.
47. Karimi, M.; Kheiralipour, K.; Tabatabaeefar, A.; Khoubakht, G.; Naderi, M.; Heidarbeigi, K. The effect of moisture content on physical properties of wheat. *Pak. J. Nutr.* **2009**, *8*, 90–95. [CrossRef]

48. Molenda, M.; Horabik, J. Mechanical Properties of Granular Agro-Materials and Food Powders for Industrial Practice. In *Part I: Characterization of Mechanical Properties of Particulate Solids for Storage and Handling*; Institute of Agrophysics Polish Academy of Sciences: Lublin, Poland, 2005; p. 145.
49. Wang, D.; Dowell, F.E.; Lacey, R.E. Single wheat kernel size effects on near-infrared reflectance spectra and color classification. *Cereal Chem.* **1999**, *76*, 34–37. [CrossRef]
50. Nelson, S. Moisture-Dependent Kernel- and Bulk-Density Relationships for Wheat and Corn. *Trans. ASAE* **1980**, *23*, 139–0143. [CrossRef]
51. Petingco, M.C.; Casada, M.E.; Maghirang, R.G.; Thompson, S.A.; McNeill, S.G.; Montross, M.D.; Turner, A.P. Influence of kernel shape and size on the packing ratio and compressibility of hard red winter wheat. *Trans. ASABE* **2018**, *61*, 1437–1448. [CrossRef]
52. Markowski, M.; Żuk-Gołaszewska, K.; Kwiatkowski, D. Influence of variety on selected physical and mechanical properties of wheat. *Ind. Crops Prod.* **2013**, *47*, 113–117. [CrossRef]
53. Gürsoy, S.; Güzel, E. Determination of physical properties of some agricultural grains. *Res. J. Appl. Sci. Eng. Technol.* **2010**, *2*, 492–498.
54. Jayas, D.S.; Cenkowski, S. Grain Property Values and Their Measurement. In *Handbook of Industrial Drying*, 3rd ed.; CRC Press: Boca Raton, FL, USA, 2006.
55. Muir, W.; Sinha, R. Physical properties of cereal and oilseed cultivars grown in western Canada. *Can. Agric. Eng.* **1988**, *30*, 51–55.
56. Kraszewski, A.W. Wheat moisture content and bulk density determination by microwave parameters measurement. *Can. Agric. Eng.* **1992**, *34*, 327–335.
57. Petingco, M.C.; Casada, M.E.; Maghirang, R.G.; Thompson, S.A.; Turner, A.P.; McNeill, S.G.; Montross, M. Discrete Element Method Simulation of Wheat Bulk Density as Affected by Grain Drop Height and Kernel Size Distribution. *J. ASABE* **2022**, *65*, 555–566. [CrossRef]
58. Stephens, L.E.; Foster, G.H. Grain Bulk Properties as Affected by Mechanical Grain Spreaders. *Trans. ASAE* **1976**, *19*, 354–0358. [CrossRef]
59. White, F.M.; Majdalani, J. *Viscous Fluid Flow*, 2nd ed.; McGraw-Hill: New York, NY, USA, 2006.
60. Bakker-Arkema, F.W.; Patterson, R.J.; Bickert, W.G. Static pressure-airflow relationships in packed beds of granular biological materials such as cherry pits. *Trans. ASAE* **1969**, *12*, 134–0136. [CrossRef]
61. Cheng, X.; Zhang, Q.; Shi, C.; Yan, X. Model for the prediction of grain density and pressure distribution in hopper-bottom silos. *Biosyst. Eng.* **2017**, *163*, 159–166. [CrossRef]
62. Guillard, F.; Golshan, P.; Shen, L.; Valdes, J.R.; Einav, I. Dynamic patterns of compaction in brittle porous media. *Nat. Phys.* **2015**, *11*, 835–838. [CrossRef]
63. Grundas, S.; Szot, B.; Wozniak, W. Variability of the porosity of cereal grain layer under the influence of static loading. *Zesz. Probl. Postępów Nauk. Rol.* **1978**, *203*, 33–40.
64. Jayas, D.S.; Sokhansanj, S.; Moysey, E.B.; Barber, E.M. Airflow resistance of canola (rapeseed). *Trans. ASAE* **1987**, *30*, 1484–1488. [CrossRef]
65. Kay, R.L.; Bern, C.J.; Hurburgh, C.R., Jr. Horizontal and Vertical Airflow Resistance of Shelled Corn at Various Bulk Densities. *Trans. ASAE* **1989**, *32*, 733–0736. [CrossRef]
66. Siebenmorgen, T.J.; Jindal, V.K. Airflow resistance of rough rice as affected by moisture content, fines concentration and bulk density. *Trans. ASAE* **1987**, *30*, 1138–1143. [CrossRef]
67. Haque, E. Void fraction as a function of depth and pressure drops of packed beds of porous media formed by granular materials. *Trans. ASABE* **2011**, *54*, 2239–2243. [CrossRef]
68. Khatchatourian, O.A.; Savicki, D.L. Mathematical Modelling of Airflow in an Aerated Soya Bean Store under Non-uniform Conditions. *Biosyst. Eng.* **2004**, *88*, 201–211. [CrossRef]
69. Gao, M.; Cheng, X.; Du, X. Simulation of bulk density distribution of wheat in silos by finite element analysis. *J. Stored Prod. Res.* **2018**, *77*, 1–8. [CrossRef]

applied sciences

MDPI

Article

Life Cycle Assessment for Environmental Impact Reduction and Evaluation of the Energy Indices in Lettuce Production

Ali Mousavi [1], Ebrahim Asadi Aghbolaghi [2], Ali Khorramifar [1], Marek Gancarz [3,4,*], Yousef Darvishi [5], Mateusz Stasiak [4], Anna Miernik [3] and Hamed Karami [1,6,*]

1 Department of Biosystems Engineering, University of Mohaghegh Ardabili, Ardabil 56199-11367, Iran
2 Department of Biosystems Engineering, Faculty of Agriculture, Ferdowsi University of Mashhad, Mashhad 91779-48974, Iran
3 Faculty of Production and Power Engineering, University of Agriculture in Kraków, Balicka 116B, 30-149 Krakow, Poland
4 Institute of Agrophysics, Polish Academy of Sciences, Doświadczalna 4, 20-290 Lublin, Poland
5 Department of Biosystems Engineering, University of Tehran, Tehran 11365-4117, Iran
6 Department of Petroleum Engineering, College of Engineering, Knowledge University, Erbil 44001, Iraq
* Correspondence: m.gancarz@urk.edu.pl (M.G.); hamedkarami@uma.ac.ir (H.K.); Tel.: +48-81-744-50-61 (M.G.); Fax: +48-81-744-50-67 (M.G.)

Abstract: Since the supply of basic needs, especially food, is among the strategic priorities of each country and conventional food production methods no longer suffice, food production methods are now transforming into industrial approaches. Industrialization, however, requires higher energy usage. Greater energy demand brings about the issue of energy sustainability. In particular, the depletion of fossil fuels results in serious challenges in food production processes. On the other hand, the utilization of energy carriers is accompanied by environmental contamination. In this regard, evaluating energy consumption and environmental pollution in the production systems can be a proper approach to finding the energy consumption and pollution centers for presenting applicable solutions to decrease pollution. In this study, energy indices of ER, EP, SE, and NEG were assessed to evaluate the energy consumption of lettuce production. The results showed values of 0.4, 17.28 kg/MJ, 0.06 MJ/kg, and 29,922 MG/ha for ER, EP, SE, and NEG, respectively. Among the consumption inputs, diesel fuel and nitrogen fertilizer had the highest consumption rate. Pollutants were also explored by the life cycle assessment method. Accordingly, chemicals and agricultural machinery led to the highest contaminating emissions. To reduce environmental contaminants, lowering the application of chemical pesticides, using biological approaches to combat pests, determining the proper amount of chemical fertilizers, using animal fertilizers, and using the proper agricultural machines should be considered.

Keywords: energy; environmental contamination; lettuce; life cycle assessment

Citation: Mousavi, A.; Aghbolaghi, E.A.; Khorramifar, A.; Gancarz, M.; Darvishi, Y.; Stasiak, M.; Miernik, A.; Karami, H. Life Cycle Assessment for Environmental Impact Reduction and Evaluation of the Energy Indices in Lettuce Production. *Appl. Sci.* **2022**, *12*, 10348. https://doi.org/10.3390/app122010348

Academic Editor: Stefano Loppi

Received: 13 September 2022
Accepted: 11 October 2022
Published: 14 October 2022

1. Introduction

Pressing problems have arisen in meeting basic human needs, including food, due to the daily increase in population, urban growth, and the reality of public welfare. Existing resources cannot meet the needs of people today [1,2]. Advanced agricultural approaches have been developed to meet that urgent need [1,2]. Food production processes are industrialized because traditional methods can no longer produce enough food [3,4]. Industrialization, however, requires higher levels of the limited resources of currently available energy [5,6]. This raises the issue of energy non-sustainability, especially fossil fuels, posing a serious challenge for food production processes [7,8]. Until renewable energy is better utilized, attention should be paid to energy efficiency in the production process [9,10]. In addition to higher energy consumption, reducing energy efficiency will lead to a higher energy loss from the production cycle, thereby contributing to global warming. An increase in energy efficiency can help to reduce the dire environmental consequences; it can promote the goals of sustainable agriculture. Therefore, proper use of energy in agriculture and breeding can result in sustainable production, cost-effectiveness, and slower depletion of fossil fuel sources while preventing air pollution [11]. Management of energy consumption is the only approach to this goal. One of the methods of energy management involves the analysis of energy consumption and the determination of the energy consumption indices [10,12]. Additionally, determining the emission of environmental pollutants per energy consumption can be an analytical tool to calculate the pollution induced by energy consumption, along with the energy consumption-related analysis [8,13]. There is a lack of literature on the management of energy consumption in lettuce production; thus, it will be discussed in this research.

Environmental indicators, as one of the sustainability criteria in various production and service activities, are of interest to researchers. One of the limitations of considering inputs in the agricultural sector is the release of various pollutants that have negative effects on the environment. Numerous studies on the determination of agricultural energy inputs and outputs and environmental pollution highlight the significance of this issue in agriculture and its subunits regarding agricultural strategies [14]. In the following paragraph, some of these studies are mentioned. Moraditochaee [15] evaluated the energy indices of cultivating tobacco under dryland farming in the north of Iran. They obtained a mean yield of 1112 kg/ha, corresponding to an energy input of 890 MJ/ha. The energy efficiency (the ratio of output energy to input energy) was determined as 0.03, indicating the inefficient use of energy in the tobacco production system. The share of nonrenewable energy in the total input energy was 94.09%. They finally concluded that tobacco production requires a modification of the energy consumption and application of renewable energies. Abeliotis et al. [16] addressed the life cycle assessment of different bean varieties in Greece. After evaluation based on product volume (kg), it was established that varieties with higher input requirements have higher yields and less environmental impact. Bartzas et al. [17] explored the life cycle assessment of lettuce production in Spain and Italy. They showed that the use of compost could be a proper strategy to maintain efficiency and increase sustainability in the agricultural sector. Additionally, compost production, irrigation systems, and greenhouse construction and maintenance were recognized as three steps with the highest energy consumption and environmental pollution. In this study, the energy and environmental indices of lettuce production will be assessed in Karaj, Iran.

The interest of communities in agriculture and sustainable foods is growing, which will increase the demand (from farmers, policymakers, agricultural jobs, media, public suppliers, and consumers) for information about the environmental performance of the agricultural systems, food chain, and food products [18,19].

Production of nutritious foods for humans and animals and economic development of the benefit-holders are among the main objectives of agriculture [20,21]. These goals can remarkably affect economic growth, such that the publications in bio-economics rose from 1000 papers in 2017 to about 3500 in 2021 [22].

Life cycle assessment (LCA) is a method to evaluate the environmental sustainability during the life cycle of a product which has been standardized by ISO 14040 and 14044 [23,24]. Energy and material flows, as well as environmental releases, are quantified and converted into environmental consequences. The LCA method includes four different steps (Figure 1) [25]:

- Goal and scope definition: This step defines the boundaries of the system and functional unit and establishes some of the assumptions made.
- Life cycle inventory (LCI): This step includes data collection (inputs, intermediate processes, and outputs).
- Life cycle impact assessment (LCIA): This step interprets the potential environmental effects such as acidification, global warming, ozone layer destruction, and ecotoxicity.

- Interpretation: LCI and LCIA results are summarized in this step. The critical points are identified and analyzed. Conclusions and recommendations are also presented for the future.

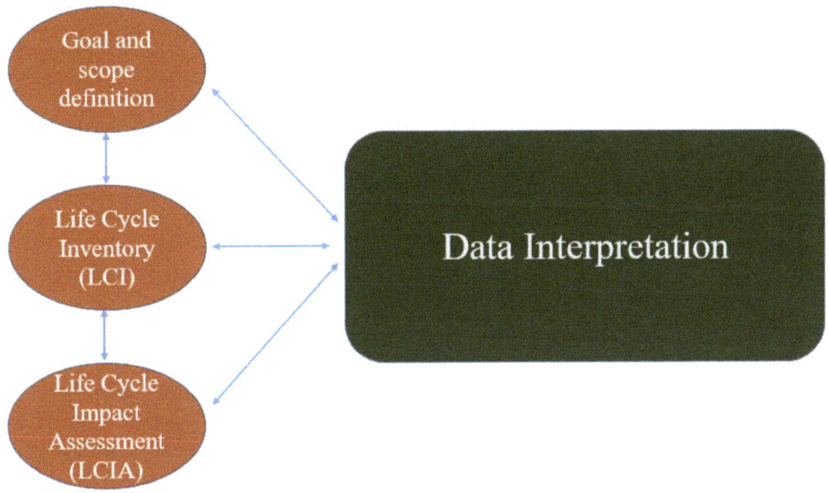

Figure 1. The framework of the life cycle assessment (LCA).

Today, LCA is the major method in the European Union's (EU's) development of a harmonized methodology for the determination of the environmental footprint of products (PEF), including several food classes [26].

Climate change, the consequences of the extraction of fossil fuels, the depletion of resources, and energy shortages are the most significant environmental concerns throughout the world. These concerns should be considered in addition to the current challenges in discovering renewable energy solutions [27].

The purpose of the present study was to evaluate the life cycle of lettuce production and estimate the extent of various environmental effects on the production process.

2. Materials and Methods

The required information for this study was collected by filling out the questionnaires at lettuce farms by the researcher. Local experts in the ministry of agriculture were also interviewed to obtain their opinion on the conditions and issues associated with lettuce cultivation. The equation proposed by Cochran (Equation (1)) was utilized to determine the sample size [28]:

$$n = N(S^*t)^2/((N-1)d^2 + (S^*t)^2) \tag{1}$$

2.1. Evaluation of Energy Indices

The energy ratio (ER), energy productivity (EP), specific energy (SE), and neat energy (NEG) were calculated by Equations (2)–(5) employing the energy-equivalent inputs and outputs (Table 1) [29].

$$ER = E_{out}\ E_{in} \tag{2}$$

$$EP = Y/E_{in} \tag{3}$$

$$SE = E_{in}/Y \tag{4}$$

$$NEG = E_{out} - E_{in} \tag{5}$$

In the above equations, E_{out} represents the output energy (MJ/ha), E_{in} (MJ/ha), and Y (kg/ha) represent the input energy and product yield, respectively. ER is a dimensionless value, while EP, SE, and NEG were reported in kg/ha, MJ/kg, and MJ/ha, respectively.

Table 1. All the energy value inputs and outputs in lettuce production.

Input and Output	Unit	Energy Equivalent (Megajoules per Unit)	Source
		Inputs	
Labor	H	1.96	Mobtaker, et al. [30]
Machines	Kg	142.7	Pimentel, et al. [31]
Diesel	L	56.31	Nabavi-Pelesaraei, et al. [32]
Electricity	kWh	12	Albright and de Villiers [33]
Nitrogen	Kg	66.14	Mousavi-Avval, et al. [34]
Phosphorous	Kg	12.44	Unakitan, et al. [35]
Potassium	Kg	11.15	Pahlavan, et al. [36]
Pesticide	Kg	120	Kitani [37]
Seeds	Kg	16.7	Albright and de Villiers [33]
		Output	
Lettuce	Kg	0.7	Razavinia, et al. [38]

The required energy for agricultural activities can be classified into direct and indirect groups. In this research, direct energy includes the labor force, electricity, and diesel fuel, while seeds, fertilizers, pesticides, and machines are classified as indirect energies. Energy can also be categorized into two types (renewable and nonrenewable). While diesel, fertilizers, pesticides, electricity, and agricultural machines are nonrenewable energies, the human workforce and seeds can be regarded as renewable energies [29].

2.2. Life Cycle Assessment

Life cycle assessment follows the trend presented in the ISO 14044 standard [24]. LCA includes four stages: goal and scope definition, life cycle inventory, evaluation of the consequences, and interpretation of the results. The goal and scope definition discusses how (scope) and why (goals) an LCA is carried out. The goal definition determines the benefits of the study; the scope determination copes with the description of the studied functional unit, production system and its boundaries, data collection and processing, and the environmental consequences. The inventory step examines the natural sources and other inputs of the system, as well as pollutant emissions and the other outputs of the process. The consequence evaluation step involves the presentation of the natural sources and environmental emission inputs in terms of their contribution to the intended sector. The results of the previous steps are finally interpreted in the last step [39].

The functional unit is a key concept in LCA studies; it enables the comparison of various products and services (ISO, 2006). The functional unit of the present research is based on mass and is determined as the production of one ton of lettuce within one agricultural year.

Selection of the system boundaries is one of the prominent and essential steps in goal and scope determination. Determination of the boundaries of the research is essential for a more precise calculation of the emission due to the consumption of the farm or after harvesting and exiting the farm during the processing steps [40]. LCA is an insight from start to finish, but it is possible to consider the system boundary as a part of the total system for more focus on the processes. In this way, the results are expressed based on their selected boundaries on a smaller scale [38]. In this study, the farm gate was taken as the system boundary. The boundary of the system is shown in Figure 2.

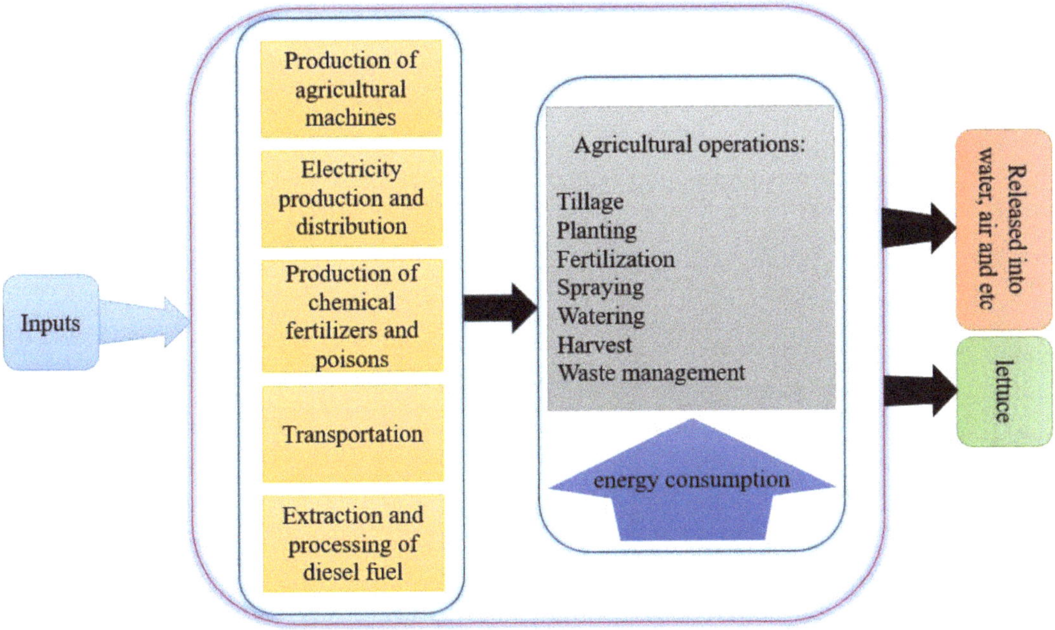

Figure 2. The boundary of the lettuce production system in the field.

Table 2 lists the impact categories of this study. The environmental consequences of lettuce production were analyzed based on the CML 2 baseline 2000 (Pere 2013).

Table 2. Impact categories and unit of measurement of each section.

Impact Categories	Symbol	Unit of Measurement
Reduction of inorganic substance	AD	kg Sb eq
Acidification	AC	kg SO_2 eq
Eutrophication	EU	kg PO_4^{3-} eq
Global warming	GW	kg CO_2 eq [a]
Ozone layer defect	OD	kg CFC-11 eq
Human toxicity	HT	kg 1,4-DCB eq [b]
Surface water toxicity	FAET	kg 1,4-DCB eq [b]
Free water toxicity	MAET	kg 1,4-DCB eq [b]
Soil toxicity	TE	kg 1,4-DCB eq [b]
Phytochemical oxidation	PhO	kg C_2H_4 eq

[a] Considering 100 years. [b] DCB = dichlorobenzene.

Data used to assess the environmental impact of a production process are divided into two categories: data related to farm operations and data related to the production of inputs used. The data relating to farm operations are the energy assessment data collected by completing the questionnaire. The data associated with producing consumable inputs were also obtained from the databases available in the SimaPro life cycle assessment software.

3. Results

Table 3 summarizes the average amount of energy consumed and produced on lettuce farms. As seen below, diesel fuel and nitrogen, with approximate values of 38 and 24%, respectively, had the highest rate of consumption, while the labor force and seeds possessed the lowest energy consumptions, with contributions below 1%.

Table 3. Amount and percentage of input and output energies in lettuce production.

Input and Output	Energy	Percentage
Labor force	452.76	0.9
Agricultural machines	5708	11.37
Diesel	19,145.4	38.14
Electricity	8400	16.74
Nitrogen	11,905.2	23.72
Phosphorous	1007.64	2.01
Potassium	903.15	1.8
Chemical pesticides	2664	5.31
Seeds	5.01	0.01
Lettuce	20,269.2	100

Table 4 presents the energy indices and their contribution to lettuce production. The ER was 0.4, leading to the negative value for the NEG, indicating the low energy consumption efficiency in lettuce production. The share of direct energy in lettuce production is equal to 55%, and the share of indirect energy is equal to 45%.

Table 4. Indicators and classification of energy types.

Index or Type of Energy	Unit	Average Value	Percent
Energy ratio	-	0.4	-
Energy productivity	kg/MJ	17.28	-
Specific energy	MJ/kg	0.06	-
NEG	MJ/ha	−29,922	-
Direct energy	MJ/ha	27,998.16	55.78
Indirect energy	MJ/ha	22,193	44.22
Renewable energy	MJ/ha	457.77	0.91
Nonrenewable energy	MJ/ha	49,733.39	99.09

The environmental consequences of lettuce production were also analyzed in the region, and the results were reported for each impact category for one ton of lettuce in Table 5. The contribution of each consumption input is depicted in Figure 3 for the impact categories. One of the most important parts of the effect investigated in this study is the global warming potential. Global warming potential is a way to express the share of gases released from agricultural systems. In this study, the amount of global warming potential per ton of produced product is estimated to equal 96.01 kg of carbon dioxide equivalent. As can be seen when comparing the sections of the eutrophication effect, acidification, soil toxicity, surface water toxicity, and the global warming potential of pollution caused by direct farm emissions (caused by burning diesel fuel, using chemical and organic fertilizers, atmospheric decomposition of fertilizers, mixing product residues with soil, using chemical poisons, and human breathing) the latter had the largest values of the mentioned indicators. With respect to the effect of reducing inorganic substances, the most impact is related to the process of diesel fuel and nitrogen fertilizer. In terms of the effect of the ozone layer defect, the biggest effect is related to the pesticides. In the areas of human toxicity and surface water toxicity, the production process of pesticides and the use of agricultural machinery in the farm has the greatest effect. In the field of photochemical oxidation, the greatest effect is related to pesticides, nitrogen fertilizer and agricultural machinery.

Table 5. Life cycle evaluation indicators in lettuce production.

Impact Categories	Unit	Estimated Value
Reduction of inorganic substances	kg Sb eq	0.64
Acidification	kg SO_2 eq	0.37
Eutrophication	kg PO_4^{3-} eq	0.07
Global warming	kg CO_2 eq	96.01
Ozone layer defect	kg CFC-11 eq	2.47×10^{-5}
Human toxicity	kg 1,4-DCB eq	31.51
Surface water toxicity	kg 1,4-DCB eq	5.6
Free water toxicity	kg 1,4-DCB eq	13,795.21
Soil toxicity	kg 1,4-DCB eq	0.08
Phytochemical oxidation	kg C_2H_4 eq	0.02

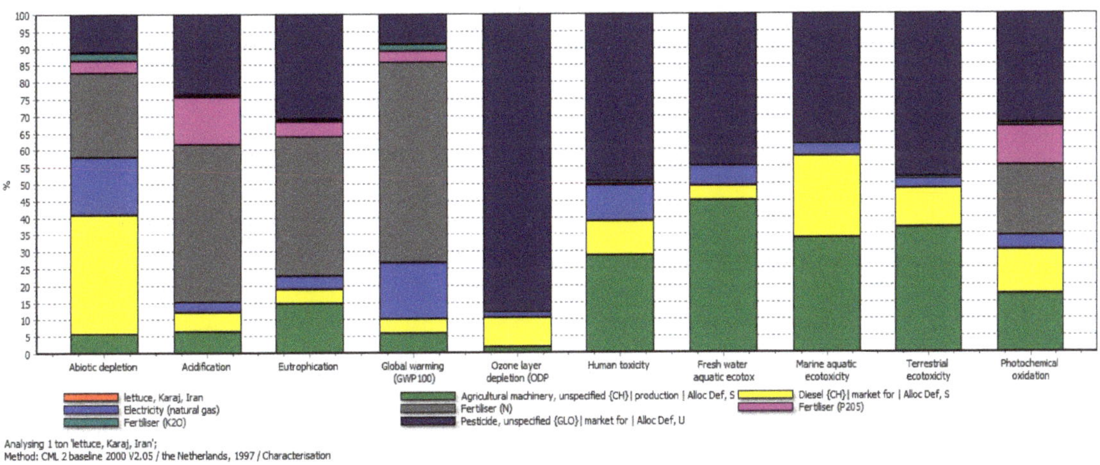

Figure 3. The share of consumption inputs in each of the sections of the studied work.

In terms of acidification, eutrophication, and global warming, nitrogen fertilizer had the highest impact. For the rest of the categories, the chemical pesticides had the highest pollution. After chemical pesticides, agricultural machines had the highest pollution for the impact categories of human toxicity, surface water toxicity, free water toxicity, soil toxicity, and phytochemical oxidation.

Each impact category had different units of measurement, making it impossible to compare their significance. Therefore, the impact categories were normalized to reach similar units of measurement, making their comparison possible, even for non-experts. The normalization results can be found in Figure 4. As seen, the environmental toxicity effect of marine aquatics with a normal value of 4.4×10^{-9} had the highest environmental load in lettuce production. For this category, the chemical pesticides had the highest pollution. After chemical pesticides, agricultural machines and diesel fuel had the next highest pollution contribution.

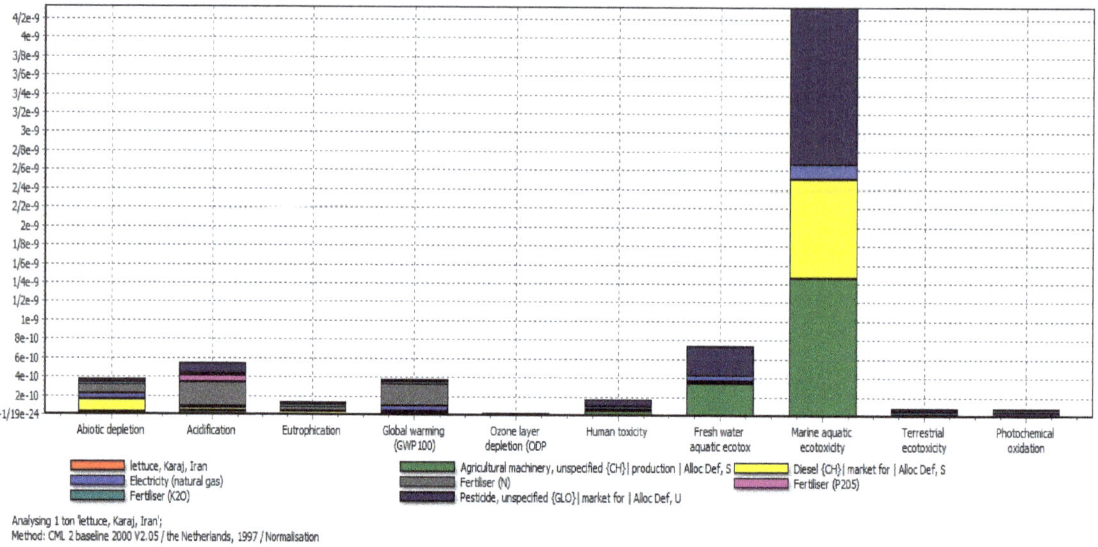

Figure 4. Parts of the normalized effect in lettuce production.

4. Discussion

This study was carried out in the Alborz province to evaluate the energy consumption and environmental pollution of lettuce production. The total energy consumption for lettuce production was 50,191.16 MJ/ha while the output energy of this system was 20,269.2 MJ/ha. The calculated energy indices indicated that lettuce production is not efficient. On the other hand, diesel fuel and chemical fertilizers had the highest energy consumption. Therefore, avoiding the use of old machines (to lower fuel consumption) and the use of the proper amount of chemical fertilizers can improve lettuce production. The share of nonrenewable energies in the production of lettuce was above 99%. The use of animal waste and green fertilizers can preserve soil fertility while decrementing environmental pollution. Razavinia, Fallah and Niknejad [38] explored the energy consumption for lettuce production in the Mazandaran province of Iran and declared that the highest energy consumption was for diesel fuel (47%). In one recent study (2018), researchers stated that the main processes are responsible for 51% of GW potential and almost entirely responsible for abiotic discharge (99%); both are related to the use of fossil fuels during a jam-production process [41]. For the research that Anthony Rouault et al. (2020) performed in the grape industry, they concluded that, for freshwater ecotoxicity and terrestrial ecotoxicity, an important contribution to the effects was related to the release of pesticides into the environment, and that was one of the reasons it was considered as fuel consumption. Finally, due to fuel consumption, mechanical operations also had a very important contribution to several impact categories, including: ozone depletion, climate change, particulate matter formation, and photochemical oxidant formation [42]. Grapes are one of the most important agricultural products around the world, which have been widely researched [43,44].

The ER in the production of processed products, such as bread and sugar, have higher values (0.52) [45]. The EP was 17.28 kg/MJ, suggesting the production of 17.28 kg product when consuming one MJ of energy. The SE (0.06) showed an inverse relationship with the EP, in the research results of Dekamin et al. [46], who reported energy efficiency in the production of coriander seeds as 0.06. Table 4 also indicates that about 56% of the consumed energy is direct, while 44% of that is indirect. The contribution of renewable energy is below 1%, so nonrenewable energy accounts for more than 99% of the consumed energy. Pourmehdi and Kheiralipour [47] reported the share of renewable energy in the production of wheat flour as 99.19%, and so the amount of indirect and renewable energy was far more than nonrenewable and direct energy.

It should be noted that in this study and other similar studies that are conducted with the approach of evaluating the life cycle of plant products, only the release of pollutants is examined, although plants can also absorb environmental pollutants. For example, plants absorb a great deal of carbon dioxide during their growth period, so the amount of carbon dioxide absorbed by the plant may be greater than its emission. The noteworthy point is that the agricultural sector is not the only emitter of this gas but it is the only absorber of this gas. In other words, plants have the ability to balance carbon in nature; therefore, due to the large volume of carbon dioxide emission from other sectors, the emission of pollutants in the agricultural sector should be minimized so that plants can establish a carbon balance.

5. Conclusions

The environmental consequences of lettuce production were explored by life cycle assessment; the results indicated that the highest contribution to emissions was from chemical pesticides and agricultural machines. Also, the toxicity of surface water, acidification, global warming, and abiotic depletion, respectively, have the largest contributions to the release of pollutants in lettuce production. Therefore, lower amounts of chemical fertilizers should be employed through the use of biological approaches in the combat against pests. Determining the correct amount of chemical fertilizers and the use of animal fertilizers and proper machinery should also be considered.

Author Contributions: Conceptualization, H.K. and A.K.; methodology, H.K.; software, A.M. (Ali Mousavi) and E.A.A.; validation, H.K., Y.D. and M.G.; formal analysis, A.K. and Y.D.; investigation, H.K.; resources, H.K.; data curation, H.K. and M.G.; writing—original draft preparation, A.K., A.M. (Ali Mousavi) and E.A.A.; writing—review and editing, H.K., M.S., A.M. (Anna Miernik) and Y.D.; visualization, M.G.; supervision, H.K.; project administration, H.K..; funding acquisition, M.G. All authors have read and agreed to the published version of the manuscript.

Funding: This research received no external funding.

Institutional Review Board Statement: Not applicable.

Informed Consent Statement: Not applicable.

Data Availability Statement: The datasets used and/or analyzed during the current study are available from the corresponding author upon reasonable request.

Conflicts of Interest: The authors declare no conflict of interest.

References

1. Khorramifar, A.; Rasekh, M.; Karami, H.; Malaga-Toboła, U.; Gancarz, M. A machine learning method for classification and identification of potato cultivars based on the reaction of MOS type sensor-array. *Sensors* **2021**, *21*, 5836. [CrossRef] [PubMed]
2. Khorramifar, A.; Rasekh, M.; Karami, H.; Mardani Korani, A. Classification of potato cultivars based on Toughness coupled with ANN and LDA methods. *J. Environ. Sci. Stud.* **2021**, *6*, 4230–4237.
3. Rasekh, M.; Karami, H.; Wilson, A.D.; Gancarz, M. Classification and identification of essential oils from herbs and fruits based on a MOS electronic-nose technology. *Chemosensors* **2021**, *9*, 142. [CrossRef]
4. Rasekh, M.; Karami, H.; Fuentes, S.; Kaveh, M.; Rusinek, R.; Gancarz, M. Preliminary study non-destructive sorting techniques for pepper (*Capsicum annuum* L.) using odor parameter. *LWT* **2022**, *164*, 113667. [CrossRef]
5. Karami, H.; Kaveh, M.; Mirzaee-Ghaleh, E.; Taghinezhad, E. Using PSO and GWO techniques for prediction some drying properties of tarragon (*Artemisia dracunculus* L.). *J. Food Process Eng.* **2018**, *41*, e12921. [CrossRef]
6. Karami, H.; Kaveh, M.; Golpour, I.; Khalife, E.; Rusinek, R.; Dobrzański, B.; Gancarz, M. Thermodynamic evaluation of the forced convective hybrid-solar dryer during drying process of rosemary (*Rosmarinus officinalis* L.) leaves. *Energies* **2021**, *14*, 5835. [CrossRef]
7. Parhizi, Z.; Karami, H.; Golpour, I.; Kaveh, M.; Szymanek, M.; Blanco-Marigorta, A.M.; Marcos, J.D.; Khalife, E.; Skowron, S.; Adnan Othman, N.; et al. Modeling and optimization of energy and exergy parameters of a hybrid-solar dryer for basil leaf drying using RSM. *Sustainability* **2022**, *14*, 8839. [CrossRef]

8. Kaveh, M.; Karami, H.; Jahanbakhshi, A. Investigation of mass transfer, thermodynamics, and greenhouse gases properties in pennyroyal drying. *J. Food Process Eng.* **2020**, *43*, e13446. [CrossRef]
9. Khorramifar, A. Using a single wheel tester in the soilbin to syudy the contact pressure between wheel and soil. *J. Environ. Sci. Stud.* **2021**, *6*, 4248–4255.
10. Karami, H.; Lorestani, A.N.; Tahvilian, R. Assessment of kinetics, effective moisture diffusivity, specific energy consumption, and percentage of thyme oil extracted in a hybrid solar-electric dryer. *J. Food Process Eng.* **2021**, *44*, e13588. [CrossRef]
11. Karami, H.; Rasekh, M.; Darvishi, Y. Effect of temperature and air velocity on drying kinetics and organo essential oil extraction efficiency in a hybrid dryer. *Innov. Food Technol.* **2017**, *5*, 65–75. [CrossRef]
12. Karami, H.; Rasekh, M.; Darvishi, Y.; Khaledi, R. Effect of drying temperature and air velocity on the essential oil content of *Mentha aquatica* L. *J. Essent. Oil Bear. Plants* **2017**, *20*, 1131–1136. [CrossRef]
13. Karami, H.; Rasekh, M. Kinetics mass transfer and modeling of tarragon drying (*Artemisia dracunculus* L.). *Iran. J. Med. Aromat. Plants Res.* **2018**, *34*, 734–747. [CrossRef]
14. Tatli, S.; Mirzaee-Ghaleh, E.; Rabbani, H.; Karami, H.; Wilson, A.D. Rapid detection of urea fertilizer effects on VOC emissions from cucumber fruits using a MOS E-nose sensor array. *Agronomy* **2022**, *12*, 35. [CrossRef]
15. Moraditochaee, M. Study energy indices of tobacco production in north of Iran. *J. Agric. Biol. Sci.* **2012**, *7*, 462–465.
16. Abeliotis, K.; Detsis, V.; Pappia, C. Life cycle assessment of bean production in the Prespa National Park, Greece. *J. Clean. Prod.* **2013**, *41*, 89–96. [CrossRef]
17. Bartzas, G.; Zaharaki, D.; Komnitsas, K. Life cycle assessment of open field and greenhouse cultivation of lettuce and barley. *Inf. Process. Agric.* **2015**, *2*, 191–207. [CrossRef]
18. Willett, W.; Rockström, J.; Loken, B.; Springmann, M.; Lang, T.; Vermeulen, S.; Garnett, T.; Tilman, D.; DeClerck, F.; Wood, A. Food in the Anthropocene: The EAT—Lancet Commission on healthy diets from sustainable food systems. *Lancet* **2019**, *393*, 447–492. [CrossRef]
19. Eyhorn, F.; Muller, A.; Reganold, J.P.; Frison, E.; Herren, H.R.; Luttikholt, L.; Mueller, A.; Sanders, J.; Scialabba, N.E.-H.; Seufert, V. Sustainability in global agriculture driven by organic farming. *Nat. Sustain.* **2019**, *2*, 253–255. [CrossRef]
20. Tsalidis, G.A. Human health and ecosystem quality benefits with life cycle assessment due to fungicides elimination in agriculture. *Sustainability* **2022**, *14*, 846. [CrossRef]
21. Gancarz, M.; Dobrzański, B.; Malaga-Toboła, U.; Tabor, S.; Combrzyński, M.; Ćwikła, D.; Strobel, W.R.; Oniszczuk, A.; Karami, H.; Darvishi, Y.; et al. Impact of coffee bean roasting on the content of pyridines determined by analysis of volatile organic compounds. *Molecules* **2022**, *27*, 1559. [CrossRef]
22. D'Adamo, I.; Gastaldi, M.; Morone, P.; Rosa, P.; Sassanelli, C.; Settembre-Blundo, D.; Shen, Y. Bioeconomy of sustainability: Drivers, opportunities and policy implications. *Sustainability* **2021**, *14*, 200. [CrossRef]
23. *ISO 14040:2006*; Environmental Management—Life Cycle Assessment—Principles and Framework. International Organization for Standardization: Geneva, Switzerland, 2006.
24. *ISO 14044:2006*; Environmental Management—Life Cycle Assessment—Requirements and Guidelines. International Organization for Standardization: Geneva, Switzerland, 2006.
25. Hanafiah, M.M. Quantifying Effects of Physical, Chemical and Biological Stressors in Life Cycle Assessment. Ph.D. Thesis, Radboud University, Nijmegen, The Netherlands, 2013.
26. van der Werf, H.M.; Knudsen, M.T.; Cederberg, C. Towards better representation of organic agriculture in life cycle assessment. *Nat. Sustain.* **2020**, *3*, 419–425. [CrossRef]
27. Ludin, N.A.; Mustafa, N.I.; Hanafiah, M.M.; Ibrahim, M.A.; Teridi, M.A.M.; Sepeai, S.; Zaharim, A.; Sopian, K. Prospects of life cycle assessment of renewable energy from solar photovoltaic technologies: A review. *Renew. Sustain. Energy Rev.* **2018**, *96*, 11–28. [CrossRef]
28. Romero-Gámez, M.; Suárez-Rey, E.; Antón, A.; Castilla, N.; Soriano, T. Environmental impact of screenhouse and open-field cultivation using a life cycle analysis: The case study of green bean production. *J. Clean. Prod.* **2012**, *28*, 63–69. [CrossRef]
29. Samavatean, N.; Rafiee, S.; Mobli, H.; Mohammadi, A. An analysis of energy use and relation between energy inputs and yield, costs and income of garlic production in Iran. *Renew. Energy* **2011**, *36*, 1808–1813. [CrossRef]
30. Mobtaker, H.G.; Akram, A.; Keyhani, A. Energy use and sensitivity analysis of energy inputs for alfalfa production in Iran. *Energy Sustain. Dev.* **2012**, *16*, 84–89. [CrossRef]
31. Pimentel, D.; Hurd, L.; Bellotti, A.; Forster, M.; Oka, I.; Sholes, O.; Whitman, R. Food production and the energy crisis. *Science* **1973**, *182*, 443–449. [CrossRef] [PubMed]
32. Nabavi-Pelesaraei, A.; Abdi, R.; Rafiee, S. Energy use pattern and sensitivity analysis of energy inputs and economical models for peanut production in Iran. *Int. J. Agric. Crop Sci.* **2013**, *5*, 2193.
33. Albright, L.; de Villiers, D. Energy investments and CO_2 emissions for fresh produce imported into New York State compared to the same crops grown locally. In *Final Report Prepared for the New York State Energy Research and Development Authority*; Cornell University: New York, NY, USA, 2008.
34. Mousavi-Avval, S.H.; Rafiee, S.; Jafari, A.; Mohammadi, A. Optimization of energy consumption for soybean production using Data Envelopment Analysis (DEA) approach. *Appl. Energy* **2011**, *88*, 3765–3772. [CrossRef]
35. Unakitan, G.; Hurma, H.; Yilmaz, F. An analysis of energy use efficiency of canola production in Turkey. *Energy* **2010**, *35*, 3623–3627. [CrossRef]

36. Pahlavan, R.; Omid, M.; Akram, A. Modeling and sensitivity analysis of energy inputs for greenhouse cucumber production. *J. Agric. Technol.* **2011**, *7*, 1509–1521.
37. Kitani, O. *CIGR Handbook of Agricultural Engineering*; Chapter 1 natural energy and biomass, part 1.3 biomass resources; CIGR: Liège, Belgium, 1999; Volume 5 Energy and Biomass Engineering.
38. Razavinia, B.; Fallah, H.; Niknejad, Y. Energy efficiency and economic analysis of winter cultivation (lettuce, bersim clover, broad bean) in Mazandaran province of Iran. *Biol. Forum* **2015**, *7*, 1452–1460.
39. Khoshnevisan, B.; Rafiee, S.; Mousazadeh, H. Environmental impact assessment of open field and greenhouse strawberry production. *Eur. J. Agron.* **2013**, *50*, 29–37. [CrossRef]
40. Suh, S.; Lenzen, M.; Treloar, G.J.; Hondo, H.; Horvath, A.; Huppes, G.; Jolliet, O.; Klann, U.; Krewitt, W.; Moriguchi, Y. System boundary selection in life-cycle inventories using hybrid approaches. *Environ. Sci. Technol.* **2004**, *38*, 657–664. [CrossRef]
41. Recanati, F.; Arrigoni, A.; Scaccabarozzi, G.; Marveggio, D.; Melià, P.; Dotelli, G. LCA towards sustainable agriculture: The case study of Cupuaçu Jam from agroforestry. *Procedia CIRP* **2018**, *69*, 557–561. [CrossRef]
42. Rouault, A.; Perrin, A.; Renaud-Gentié, C.; Julien, S.; Jourjon, F. Using LCA in a participatory eco-design approach in agriculture: The example of vineyard management. *Int. J. Life Cycle Assess.* **2020**, *25*, 1368–1383. [CrossRef]
43. Khorramifar, A.; Karami, H.; Wilson, A.D.; Sayyah, A.H.A.; Shuba, A.; Lozano, J. Grape cultivar identification and classification by machine olfaction analysis of leaf volatiles. *Chemosensors* **2022**, *10*, 125. [CrossRef]
44. Afkari-Sayyah, A.H.; Khorramifar, A.; Karami, H. Identification and classification of different grape cultivars using cultivar leaves by electroni nose. *J. Environ. Sci. Stud.* **2021**, *6*, 4382–4389.
45. Kheiralipour, K.; Sheikhi, N. Material and energy flow in different bread baking types. *Environ. Dev. Sustain.* **2021**, *23*, 10512–10527. [CrossRef]
46. Dekamin, M.; Kheiralipour, K.; Afshar, R.K. Energy, economic, and environmental assessment of coriander seed production using material flow cost accounting and life cycle assessment. *Environ. Sci. Pollut. Res.* **2022**. [CrossRef]
47. Pourmehdi, K.; Kheiralipour, K. Assessing the effects of wheat flour production on the environment. *Adv. Environ. Technol.* **2020**, *6*, 111–117.

applied sciences

MDPI

Article

Effect of Poultry and Goat Manures on the Nutrient Content of *Sesamum alatum* Leafy Vegetables

Khulekani Cyprian Mbatha [1], Charmaine Ntokozo Mchunu [2], Sydney Mavengahama [3] and Nontuthuko Rosemary Ntuli [1,*]

1 Department of Botany, University of Zululand, Private Bag X 1001, KwaDlangezwa 3886, South Africa; khulekanicyprian0@gmail.com
2 Soil Fertility and Analytical Services, KwaZulu-Natal Department of Agriculture and Rural Development, Pietermaritzburg 3200, South Africa; Charmaine.Mchunu@kzndard.gov.za
3 Food Security and Safety Niche Area Research Group, Crop Science Department, Faculty of Natural and Agricultural Sciences, North West University, Mmabatho 2735, South Africa; Sydney.Mavengahama@nwu.co.za
* Correspondence: NtuliR@unizulu.ac.za; Tel.: +27-35-902-6105

Abstract: *Sesamum alatum* Thonn. is one of the less-popular but nutritious leafy vegetables that is still collected from the wild or as weeds among crops in South Africa. The plant is also used in medicines and cosmetics in Africa and elsewhere. Despite its importance, the cultivation of *S. alatum* under different agronomic systems for improved harvestable yield and nutrient content is still lacking. The study aimed to determine the response of *S. alatum* nutrient content to the application of poultry and goat manures. Plants were grown in pots under rain-fed shade cloth conditions, with poultry and goat manures applied at 0, 1, 2, and 3 t ha^{-1} each, and they were laid in a completely randomized design. Shoot tips were harvested at 60 days after planting and analyzed for nutrient content. Shoots contained better nutrients in *S. alatum* plants grown during the first than the second season, with minor exceptions. Poultry and goat manure application led to an increase in Ca, Mg, K, P, and micro-nutrients. Goat manure had potential to increase the nutrient content in *S. alatum* than poultry manure, although differences were not substantial. Therefore, both manures could be equally used to improve nutrient content of *S. alatum*.

Keywords: calcium; magnesium; phosphorus; potassium; copper; iron; manganese; sodium; zinc; wild leafy vegetables

Citation: Mbatha, K.C.; Mchunu, C.N.; Mavengahama, S.; Ntuli, N.R. Effect of Poultry and Goat Manures on the Nutrient Content of *Sesamum alatum* Leafy Vegetables. *Appl. Sci.* **2021**, *11*, 11933. https://doi.org/10.3390/app112411933

Academic Editors: Tadeusz Juliszewski, Sławomir Kurpaska and Paweł Kiełbasa

Received: 23 November 2021
Accepted: 9 December 2021
Published: 15 December 2021

1. Introduction

Sesamum alatum Thonn., commonly known as winged-seed sesame, is a leafy vegetable that is collected from the wild in KwaZulu-Natal, South Africa [1] and other parts of the globe [2,3]. It is an erect, year-round, mucilaginous herbaceous plant [4]. In Africa and Asia, soft and immature leaves, shoots, pods, and seeds are cooked and eaten fresh or dried and added into soups, sauces, snacks and cakes [3–6]. *Sesamum alatum* has excellent amounts of proteins, carbohydrates, dietary fibre, fats, Ca, N, P, K, Mg, and other trace elements such as Cu, Fe, Mn, Na, and Zn, which compare well with *Amaranthus hydridus* [1,7]. It also contains essential antioxidant which can be linked to it use as an ayurvedic medicine against diseases such as, diarrhoea, intestinal disorders [4], and diabetes [8]. It is also used as an aphrodisiac for human and livestock fertility improvement [4]. Fresh leaves and seeds are valued for their oils and their mucilaginous substance that has qualities valued in hair shampoo, cloth washers, and hair conditioner [3].

Organic manures are organic residues of animal and plant by-products [9,10]. Small-scale farmers usually rely on organic manures due to high costs and less accessibility of inorganic manures [11–13], which includes animal manures, compost, and green manures [14]. Organic manures improve plant root rizoster conditions which enhance the assimilation of nutrients from the soil [15], as well as the nutrient content of the plant itself [12,16].

Applications of 2.5–6 t ha^{-1} poultry manure cause an increase in the Ca, Mg [17], ash, fats, N, P, and protein contents in *S. indicum* leaves [18]. The ash, moisture content, proteins, carbohydrates, fibre, fat, N, P, and K in *C. olitorious* leaves increase when poultry manure is applied at 1 and 2 t ha^{-1} [12]. The increase in the ash, fat, and protein content was also noted in *Amaranthus caudatus* and *A. cruentus* when poultry manure was applied [19]. Further, there is an increase in proximate and mineral contents with the application of 0.03 t ha^{-1} of poultry manure on *A. cruentus*, *A. hydridus*, *A. deflexus*, and *A. spinosus* [20]. Goat manure applied at rates ranging from 2.5–13 t ha^{-1} resulted in an increase in Ca, Mg, N, P, and K content in *Capsicum annuum* leaves [21]. Application of 8 t ha^{-1} goat manure increases Ca, Cu, Fe, K, Mg, Mn, Na, P, and Zn content in *Cleome gynandra* shoots [22].

The nutrient content of *S. alatum* is only known for plants collected from the wild [1]. *Sesamum alatum* is one of the wild-collected leafy vegetables with no agronomic studies on its improvements under different soil fertilities using animal manure. To improve nutrient richness in *S. alatum*, a soil with necessary nutrients that will facilitate production and maintain or improve the crop nutrient content is a prerequisite. Therefore, the objective of this study was to examine the effect of poultry and goat manures on the nutrient content of *S. alatum*.

2. Materials and Methods

2.1. Study Site and Seed Sourcing

The study was conducted at the University of Zululand, KwaDlangezwa campus (28° 85′ 24″ S, 31° 84′ 91″ E), South Africa. Pot trials were conducted under a shade cloth with 600 µmol/m^2/s light intensity (40% shade), over two summer seasons (September–December of 2018 and 2019). The agrometeorological data (temperature, rainfall, and relative humidity) for these seasons are presented in Table 1. Poultry and goat manure was obtained from the Agriculture Farm at the University of Zululand. Seeds were collected from Mseleni (27° 36′ 71″ S, 32° 53′ 93″ E), a rural village in the northern KwaZulu-Natal province.

Table 1. Meteorological data of the University of Zululand in 2018 and 2019 (South African Weather Services, Durban, 2020).

Month	Temperature (°C)				Rainfall (mm)		RH (%)	
	2018		2019		2018	2019	2018	2019
	Max.	Min.	Max.	Min.				
January	31.9	19.9	29.5	19.2	59.0	73.6	73	80
February	30.1	20.5	30.2	20.4	145.2	144.2	79	81
March	30.2	19.7	30.5	20.7	94.2	58.0	79	80
April	28.6	18.4	27.5	17.9	153.4	146.6	80	82
May	25.7	15.2	27.2	15.5	334.0	7.6	80	78
June	24.6	12.4	26.1	12.1	64.4	47.2	78	74
July	24.3	12.0	27.0	11.1	15.0	21.6	75	66
August	24.1	13.8	26.3	15.3	89.2	56.0	75	76
September	26.7	15.4	26.7	14.9	59.0	95.2	76	74
October	26.2	15.3	28.7	17.1	200.2	138.2	76	76
November	27.1	17.1	29.0	18.9	86.6	128.4	75	80
December	30.2	20.6	28.7	18.9	12.8	317.8	77	80
Mean	27.48	16.69	28.12	16.83	109.41	102.87	76.92	77.25

RH, relative humidity; max., maximum; min., minimum.

2.2. Experimental Design

River sand and loam soils collected at 30–60 cm soil depth were mixed at a ratio of 7:3 to make a sandy-loam mixture and used as a growth medium. Pots (20 L) were filled with 60 kg of growth medium, and manures were each applied at 0; 1; 2; and 3 t ha^{-1}, mixed with the soil, irrigated to pot capacity, and left to incubate for 10 days under the shade cloth. Manure in tons per hectare of application rates was converted from grams per 20 kg of soil, which was equivalent to 2,000,000 kg of soil per hectare for a furrow slice. Pots were arranged in a completely randomized design of 2 × 4 factorial combinations. Sowing was carried out immediately after the 10-day soil–manure incubation period. Seeds were broadcasted in each pot and covered with a layer (1–2 mm) of sprinkled sand. At 60 days after planting, soft shoot tips (4–8 cm long) were harvested from five plants of each treatment for nutrient analyses. Shoots were thoroughly washed with distilled water, cut into small pieces and then dried in an oven at 60 °C until a constant dry weight was reached. Dried shoots were ground into powder, packaged in airtight plastic containers and stored in a fridge (−4 °C) for further analysis.

Five replications of each of the samples were analysed for their nutritive value in human food. Proximate and mineral analyses were conducted according to the Association of Official Analytical Chemists (AOAC) [23] and Inductively Coupled Plasma Optical Emission Spectrometry (ICP-OES) methods [24]. Fertility analysis of soil and manure physicochemical properties (Table 2) and nutrient concentration in shoots was conducted at the Department of Agriculture and Rural Development, Soil Fertility and Analytical Services section in KwaZulu-Natal.

Table 2. Available nutrient content, pH, organic carbon and total nitrogen of soil and manures used in this study.

Available Nutrients	Season One			Season Two		
	Soil	Manures		Soil	Manures	
	Sandy-Loam	Poultry	Goat	Sandy-Loam	Poultry	Goat
Organic carbon (%)	0.5	>6	>6	0.6	>6	>6
Nitrogen (%)	<0.05	>0.6	>0.6	<0.05	>0.6	>0.6
Phosphorus (mg L^{-1})	5.00	890.00	350.00	15.00	940.00	500.00
Potassium (mg L^{-1})	32.00	4076.00	2216.00	58.00	2656.00	12,696.00
Calcium (mg L^{-1})	337.00	791.00	2966.00	340.00	1241.00	1773.00
Magnesium (mg L^{-1})	112.00	574.00	857.00	94.00	997.00	857.00
Zinc (mg L^{-1})	1.50	72.20	29.30	6.40	59.00	18.20
Manganese (mg L^{-1})	26.00	62.00	25.00	31.00	37.00	10.00
Copper (mg L^{-1})	0.80	4.90	0.90	3.70	4.20	0.80
pH (KCl)	4.68	7.70	7.19	4.63	7.06	7.54

2.3. Soil and Manure Analysis

Soil or manure samples were air-dried at room temperature, after which they were spread out in drying trays and air was forced over them. When dry, the samples were crushed between rubber belts on a soil or manure crusher and passed through a 1 mm sieve. Material coarser than 1 mm that cannot be crushed (such as stones, gravel, and concretions) were discarded. Samples were scooped into trays which each contained 11 PVC cups (capacity 70 mL); a tray was used for nine unknown samples, one standard sample (for quality control), and one blank. For operations such as dispensing and stirring and for quality control, batches of three trays (27 samples, three unknowns, and three blanks) were used. Multiple dispensers and diluter/dispensers were used to dispense aliquots of extractant or reagent to three samples at a time.

Soil or manure samples were analysed on a volume rather than a mass basis. To enable the conversion of the results to a mass basis, the mass of a 10 mL scoop of a dried and milled sample was measured and the calculated sample density was reported. For the determination of pH, 10 mL of soil or manure was scooped into sample cups. Then, 25 mL of 1 *M* KCl solution was added, and the suspension was stirred at 400 revolutions per minute (r.p.m.) for 5 min using a multiple stirrer. The suspension was allowed to stand for about 30 min, and the pH was measured using a gel-filled combination glass electrode while stirring. De-ionised water was substituted for the 1 *M* KCl solution if pH (water) was required.

Extractable (1 M KCl) calcium and magnesium: A total of 2.5 mL of soil or manure was scooped into sample cups. Then, 25 mL of 1 *M* KCl solution was added, and the suspension was stirred at 400 r.p.m. for 10 min using a multiple stirrer. The extracts were filtered using Whatman No.1 paper. A total of 5 mL of the filtrate was diluted with 20 mL of 0.0356 *M* $SrCl_2$, and Ca and Mg determined by atomic absorption. To determine extractable acidity, 10 mL of the filtrate was diluted with 10 mL of de-ionised water containing 2–4 drops of phenolphthalein and titrated with 0.005 *M* NaOH.

Extractable (Ambic-2) phosphorus, potassium, zinc, copper and manganese: The Ambic-2 extracting solution consisted of 0.25 *M* NH_4CO_3 + 0.01 *M* Na_2EDTA + 0.01 *M* NH_4F + 0.05 g L^{-1} Superfloc (N100), adjusted to pH 8 with a concentrated ammonia solution. An amount of 25 mL of this solution was added to 2.5 mL soil, and the suspension was stirred at 400 r.p.m. for 10 min using a multiple stirrer. The extracts were filtered using Whatman No.1 paper. Phosphorus was determined on a 2 mL aliquot of filtrate using a modification of the Murphy and Riley [25] molybdenum blue procedure [26]. Potassium was determined by atomic absorption on a 5 mL aliquot of the filtrate after dilution with 20 mL de-ionised water. Zinc, Cu, and Mn were determined by atomic absorption on the remaining undiluted filtrate.

Estimation of organic matter by the Walkley–Black method: Organic matter was based on the Walkley–Black procedure [27], which measured the readily oxidizable organic carbon. The organic matter was oxidized by potassium dichromate in a sulphuric acid medium. The excess dichromate was determined by titration with standard ferrous sulphate solution.

2.4. Proximate Analysis

Shoots were examined for the protein, fibre and fat content from each treatment.

Crude proteins: Two grams of dried sample powder were weighed and introduced into a 650 mL digestion flask along with 10 mL of distilled water. A digestion tablet along with 20 mL of concentrated H_2O_4 was added to the flask to act as the solution catalyst. Boiling chips were added to the mixture and the digestion was given time until a colourless solution was obtained. The digestion solution was allowed to cool before it was diluted with distilled ammonia-free water up to 100 mL. The Kjeldahl flask was pre-washed using distilled water for use. A total of 10 mL of the digest's 100 mL solution was pipetted into a distillation flask with 90 mL of distilled water. Twenty millimetres of 40% NaOH were pipetted into the distillation flask. A 250 mL conical flask containing 10 mL boric acid solution along with some drops of the indicator was used. Ammonia was then purified from the solution. About 100 mL of ammonia was extracted from boric acid. After extraction, the 100 mL solution was titrated against 0.1 N HCl until a pink colour appeared. The titrated volume was then subtracted from the original sample's titrated volume. Values obtained from the above process were used to calculate the N content percentage of the sample using the formula:

$$\%\ Nitrogen = \frac{millimetre\ acid \times normality\ standard\ acid}{weight\ of\ the\ sample\ in\ grams} \times 0.014 \times 100$$

The crude protein percentage was calculated by multiplying the nitrogen percentage by a conversion factor of 6.25.

Crude Fibre (Acid Detergent Fibre and Neutral Detergent Fibre): Crude fibre was estimated by 1.25% H_2SO_4 and 1.25% NaOH solutions. Ether extract residue was introduced into a digestion flask along with a 200 mL boiling H_2O_4 solution. Similarly, an anti-foaming agent was added and connected to the flask immediately with a condenser and heated for 30 min. After that time, the digestion flask was removed and filtered immediately for its content using linen cloth and then washed using boiling water till the acid was removed from the washings. After washing, the residue was again introduced into a flask containing 200 mL of boiling NaOH solution. The flask was then connected to a reflux condenser and heated for 30 min. After 30 min, the flask was removed, the precipitate content was filtered with a gooch crucible and then it was washed using boiling water. Filtered content was washed with 15 mL of 95% alcohol (ethanol) and it was dried along with the crucible at 110 °C till a constant weight was achieved. The content and crucible were cooled in a desiccator and weighed. The crucible content was incinerated for 30 min in a furnace at 550 °C until the carbonaceous matter was consumed. It was cooled again in a desiccator and the final weight measurements were taken. The following formula was used to calculate crude fibre as the loss in weight:

$$Crude\ fibre\ \% = \frac{wt.\ of\ dry\ crucible\ and\ sample - wt.\ of\ incinerated\ and\ ash}{sample\ weight} \times 100$$

Crude fats: The amount of fat contained was determined using petroleum ether as a reagent. Firstly, the weight of the bottle and lid to be used were ensured to be stable and sterilized. This was carried out by placing the bottle and lid in an incubator at 105 °C for 12 h. About 100 g of the sample was weighed and wrapped using a paper filter. The wrapped sample was extracted using a thimble and then transferred into a Soxhlet filled with about 250 mL of petroleum ether and then heated. The Soxhlet apparatus was connected and placed in water to cool down and then heated again for 14 h at a heating rate of 150 drops/min. The solvent was completely evaporated using a vacuum condenser and incubated at 80–90 °C until the bottle became dry. The lid of the bottle was transferred to the desiccator for cooling and the dried content of the bottle was reweighed. The results were expressed in percentages using the formula:

$$Fat\ (\%) = \frac{Weight\ of\ fat}{Weight\ of\ sample} \times 100$$

2.5. Mineral Analysis

The mineral elements analysis of Ca, Cu, Fe, K, Mg, Mn, Na, P, and Zn in the samples were quantified in triplicate using ICP–OES [24]. The plant material samples were analysed using the batch-handling procedure as described by Manson and Roberts [24]. The samples were dried at 75 °C and sieved using a 0.84 mm sieve. The product after sieving was ash dried overnight at 450 °C and was then taken up to 1 M of HCl. The following elements: Ca, Cu, Fe, K, Mg, Mn, Na, P, and Zn were determined using ICP–OES.

2.6. Data Analysis

Data were subjected to analysis of variance (ANOVA) using the General Linear Model procedure of the Statistical Analysis System (SAS) the General Linear Model procedure [28]. Tukey's Studentized Range test (HSD) was used to compare treatment means at $p < 0.05$.

3. Results

3.1. Effect of Seasonal Variation on Nutrient Content

Plants that were grown in the first season had better macro- and micro-nutrient content than in the second season (Table 3). However, the exception was recorded in acid detergent fibre, which was greater in plants grown during the second than in the first season, as well as neutral detergent fibre and fats, where both had insignificant differences during both seasons.

Table 3. Effect of season as well as poultry and goat manure on *S. alatum* nutrient content.

Season		Macronutrients (%)									Micronutrients (ppm)				
		ADF	NDF	Ca	Mg	K	K/Ca + Mg	Fat	P	Protein	Cu	Fe	Mn	Na	Zn
One		20.34 ^{b}	29.39	0.91 ^{a}	0.54 ^{a}	2.96 ^{a}	0.80 ^{a}	3.98	0.29 ^{a}	36.76 ^{a}	11.48 ^{a}	399.00 ^{a}	64.48 ^{a}	1500.00 ^{a}	49.67 ^{a}
Two		22.51 ^{a}	30.11	0.68 ^{b}	0.38 ^{b}	2.00 ^{b}	0.76 ^{b}	3.39	0.25 ^{b}	34.12 ^{b}	7.52 ^{b}	198.33 ^{b}	46.62 ^{b}	900.00 ^{b}	40.48 ^{b}
Significance		***	NS	***	***	***	***	NS	**	***	***	***	***	***	***
Manure (t ha^{-1})															
Control	0	22.57 ^{a}	30.75 ^{a}	0.35 ^{c}	0.20 ^{e}	1.06 ^{d}	0.80 ^{ab}	2.88 ^{ab}	0.08 ^{d}	36.09 ^{a}	5.00 ^{c}	253.00 ^{dc}	38.00 ^{c}	700.00 ^{d}	29.00 ^{d}
Poultry	1	20.96 ^{ab}	29.61 ^{ab}	1.01 ^{a}	0.58 ^{abc}	3.00 ^{ab}	0.78 ^{bc}	3.23 ^{ab}	0.33 ^{abc}	33.39 ^{c}	11.00 ^{ab}	307.67 ^{abc}	60.67 ^{b}	1300.00 ^{bc}	49.33 ^{ab}
	2	20.07 ^{b}	27.50 ^{b}	0.72 ^{b}	0.44 ^{bcd}	2.65 ^{abc}	0.76 ^{c}	2.03 ^{b}	0.27 ^{abc}	36.77 ^{a}	8.67 ^{bc}	269.17 ^{bcd}	52.00 ^{bc}	1100.00 ^{c}	45.83 ^{abc}
	3	21.11 ^{ab}	28.32 ^{ab}	0.67 ^{b}	0.36 ^{ed}	1.96 ^{c}	0.77 ^{bc}	3.56 ^{ab}	0.25 ^{bc}	35.91 ^{a}	8.50 ^{bc}	227.33 ^{d}	43.00 ^{cd}	1100.00 ^{c}	42.83 ^{bc}
Goat	1	21.68 ^{ab}	29.86 ^{ab}	1.08 ^{a}	0.59 ^{ab}	3.28 ^{a}	0.82 ^{a}	4.66 ^{a}	0.37 ^{a}	33.63 ^{bc}	13.67 ^{a}	352.50 ^{a}	91.50 ^{a}	1600.00 ^{a}	57.33 ^{a}
	2	21.20 ^{ab}	30.84 ^{a}	0.70 ^{b}	0.42 ^{cd}	2.23 ^{bc}	0.80 ^{ab}	4.78 ^{a}	0.22 ^{c}	36.69 ^{a}	5.67 ^{c}	349.00 ^{a}	48.17 ^{bcd}	1200.00 ^{bc}	36.00 ^{cd}
	3	22.37 ^{ab}	31.38 ^{a}	1.06 ^{a}	0.64 ^{a}	3.16 ^{a}	0.77 ^{bc}	4.65 ^{a}	0.35 ^{ab}	35.61 ^{ab}	14.00 ^{a}	332.00 ^{ab}	55.50 ^{bc}	1500.00 ^{ab}	55.17 ^{ab}
Significance		*	**	***	***	***	***	**	***	***	***	***	***	***	***

ADF, acid detergent fibre; NDF, neutral detergent fibre; NS, not significant. Different superscript letter(s) (a, b, c, d and e) within a column indicate differences between seasons as well as manure types and concentrations within each nutrient, in Tukey's Studentized Range (HSD) ($p < 0.05$). Significance level: NS, not significant; * $p < 0.05$; ** $p < 0.01$; *** $p < 0.001$.

3.2. *Macro-Nutrients*

3.2.1. Acid Detergent Fibre and Neutral Detergent Fibre

Sesamum alatum control plants had more acid detergent fibre (ADF) content compared with 2 t ha^{-1} of poultry manure-treated plants (Table 3). Again, the variation in manure types and quantities did not affect the ADF concentration in *S. alatum* (Table 3). Neutral detergent fibre (NDF) was higher in plants that were treated with goat manure at rates ≥2 t ha^{-1} than untreated plants (Table 3). The highest ADF content recorded in the control and in plants treated with 3 t ha^{-1} of poultry and goat manures in the first and second seasons, respectively, was similar to all other application rates except rates ≥ 2 t ha^{-1} of poultry manure (Table 4). The highest NDF contents obtained in plants without manure application in season one were comparable to the NDF contents achieved with 1 t ha^{-1} of poultry manure and all goat manure rates (Table 4). The lower ADF and NDF values of control and treated *S. alatum* from the current study (Tables 3 and 4) than values recorded in untreated species in Sudan (ADF = 34.9%, NDF = 39.2%) [7] were probably caused by differences in the harvest stages, plants were analysed at the seedling stage (60 DAP) where in this study, but in the flowering stage in Sudan. This generally means that these leafy vegetables are easily palatable and digestible prior to flowering [7].

3.2.2. Calcium, Magnesium, and Potassium

Generally, the calcium, magnesium and potassium contents were better in plants from soils treated with poultry and goat manure than the controls (Table 3). The highest calcium content was recorded in 1 t ha^{-1} of poultry as well as 1 and 3 t ha^{-1} of goat manure-treated plants. The highest magnesium content in 3 t ha^{-1} of goat manure-treated plants was similar to that obtained in plants exposed to 1 t ha^{-1} of poultry and goat manures. Again, the highest potassium content in 1 and 3 t ha^{-1} of goat manure-treated plants was similar to that of 1 and 2 t ha^{-1} of poultry manure-treated plants (Table 3). The interaction between seasons and manures shows that 3 t ha^{-1} goat manure-treated plants had a greater calcium content than the control in both seasons and poultry manure rates that were ≥2 t ha^{-1} as well as 2 t ha^{-1} of goat manure in season two (Table 4). Similarly, magnesium content was higher in plants that received 3 t ha^{-1} of goat manure but was only significantly different with rates of ≥2 t ha^{-1} of poultry manure, 2 t ha^{-1} of goat manure in the second season and the control in each season (Table 4). The potassium content was better with the application of 2 t ha^{-1} of poultry manure than the control in season one. No variations were noted between type and rate of manure application in *Sesamum alatum* potassium content (Table 4). In season two, the potassium content obtained with poultry manure (≥2 t ha^{-1}) and 2 t ha^{-1} of goat manure was comparable with the control. Further, the lower (<2 t ha^{-1}) rate of poultry manure was similar to 1 and 3 t ha^{-1} of goat manure in season two (Table 4).

The highest K/Ca + Mg content was recorded in 1 t ha^{-1} of goat manure-treated plants (Table 3). Manure and application rate variations showed that all levels of poultry manure-treated plants produced a lower K/Ca + Mg content than that of plants treated with goat manure, except in 3 1 t ha^{-1} (Table 3). Seasons and manure interaction show that the application of different manure types and quantities did not affect the K/Ca + Mg content in both seasons. Plants that received 2 t ha^{-1} of poultry manure in season two recorded the least K/Ca + Mg content, but levels were comparable with 3 t ha^{-1} of poultry manure and 2 t ha^{-1} of goat manure in the same season (Table 4).

Table 4. Interaction between season and manure on *Sesamum alatum* nutrient content.

Season × Manure ($t\ ha^{-1}$)		Macronutrients (%)									Micronutrients (ppm)				
		ADF	NDF	Ca	Mg	K	K/Ca + Mg	Fat	P	Protein	Cu	Fe	Mn	Na	Zn
S1—Control	0	23.59 ^{a}	32.31 ^{a}	0.35 ^{cd}	0.20 ^{c}	1.06 ^{c}	0.80 $^{a–d}$	2.86 ^{bc}	0.08 ^{d}	36.78 ^{abc}	5.00 ^{de}	253.00 ^{de}	38.00 ^{de}	700.00 ^{cd}	29.00 ^{de}
S1—Poultry	1	19.90 ^{abc}	29.73 ^{abc}	1.01 ^{ab}	0.58 ^{ab}	3.00 ^{a}	0.78 ^{bcd}	3.88 ^{abc}	0.33 ^{ab}	37.84 ^{ab}	11.00 $^{a–d}$	307.67 ^{cd}	60.67 ^{bc}	1500.00 ^{ab}	49.33 $^{a–d}$
	2	17.89 ^{c}	26.71 ^{bc}	0.82 ^{abc}	0.55 ^{ab}	3.76 ^{a}	0.83 ^{ab}	1.40 ^{c}	0.28 ^{abc}	39.89 ^{a}	10.00 ^{cd}	401.00 ^{bc}	65.00 ^{bc}	1500.00 ^{ab}	55.00 ^{abc}
	3	18.93 ^{bc}	26.07 ^{c}	1.17 ^{a}	0.63 ^{a}	3.45 ^{a}	0.80 $^{a–d}$	3.64 ^{abc}	0.43 ^{ab}	36.91 ^{abc}	16.33 ^{ab}	406.00 ^{bc}	74.00 ^{ab}	1900.00 ^{a}	63.00 ^{ab}
S1—Goat	1	20.69 ^{abc}	29.73 ^{abc}	1.07 ^{ab}	0.56 ^{ab}	3.14 ^{a}	0.81 ^{abc}	6.73 ^{a}	0.34 ^{ab}	33.63 ^{c}	16.33 ^{ab}	460.00 ^{ab}	92.00 ^{a}	1800.00 ^{a}	58.67 ^{ab}
	2	19.83 ^{abc}	30.53 ^{abc}	1.08 ^{ab}	0.62 ^{a}	3.48 ^{a}	0.85 ^{a}	4.82 ^{ab}	0.32 ^{ab}	36.69 ^{abc}	10.33 ^{bcd}	544.00 ^{a}	73.33 ^{ab}	1800.00 ^{a}	50.00 $^{a–d}$
	3	21.55 ^{abc}	31.13 ^{abc}	0.90 ^{ab}	0.62 ^{a}	2.84 ^{ab}	0.76 ^{cd}	4.51 ^{abc}	0.25 ^{bcd}	35.61 ^{bc}	11.33 ^{abc}	421.33 ^{b}	48.33 ^{cd}	1500.00 ^{ab}	42.67 $^{b–e}$
S2—Control	0	21.54 ^{abc}	29.18 ^{abc}	0.35 ^{cd}	0.20 ^{c}	1.06 ^{c}	0.80 $^{a–d}$	2.58 ^{bc}	0.08 ^{d}	35.61 ^{bc}	5.00 ^{de}	253.00 ^{de}	38.00 ^{de}	700.00 ^{cd}	29.00 ^{de}
S2—Poultry	1	22.03 ^{ab}	29.94 ^{abc}	1.01 ^{ab}	0.58 ^{ab}	3.00 ^{a}	0.78 ^{bcd}	2.66 ^{bc}	0.33 ^{ab}	28.94 ^{d}	11.00 $^{a–d}$	307.67 ^{cd}	60.67 ^{bc}	1000.00 ^{bc}	49.33 $^{a–d}$
	2	22.25 ^{ab}	28.29 ^{abc}	0.62 ^{bcd}	0.33 ^{bc}	1.55 ^{bc}	0.68 ^{e}	3.48 ^{abc}	0.25 ^{bcd}	33.65 ^{c}	7.33 ^{cd}	137.33 ^{fg}	39.00 ^{de}	800.00 ^{cd}	36.67 ^{cde}
	3	23.30 ^{a}	30.57 ^{abc}	0.17 ^{d}	0.10 ^{c}	0.48 ^{c}	0.74 ^{de}	4.51 ^{abc}	0.08 ^{d}	34.91 ^{bc}	0.67 ^{e}	48.67 ^{g}	12.00 ^{f}	400.00 ^{d}	22.67 ^{e}
S2—Goat	1	22.67 ^{ab}	29.98 ^{abc}	1.10 ^{a}	0.61 ^{a}	3.43 ^{a}	0.83 ^{ab}	2.59 ^{bc}	0.41 ^{ab}	33.63 ^{c}	11.00 $^{a–d}$	245.00 ^{def}	91.00 ^{a}	1400.00 ^{ab}	56.00 ^{abc}
	2	22.57 ^{ab}	31.16 ^{abc}	0.33 ^{d}	0.21 ^{c}	0.97 ^{c}	0.74 ^{de}	4.75 ^{ab}	0.12 ^{cd}	36.69 ^{abc}	1.00 ^{e}	154.00 ^{efg}	23.00 ^{ef}	500.00 ^{cd}	22.00 ^{e}
	3	23.20 ^{a}	31.64 ^{ab}	1.21 ^{a}	0.66 ^{a}	3.48 ^{a}	0.77 ^{bcd}	4.79 ^{ab}	0.46 ^{a}	35.61 ^{bc}	16.67 ^{a}	242.67 ^{def}	62.67 ^{bc}	1500.00 ^{ab}	67.67 ^{a}
Significance		**	*	***	***	***	***	**	***	***	***	***	***	***	***

ADF, acid detergent fibre; NDF, neutral detergent fibre. Different superscript letter(s) (a–g) within a column indicate differences between seasons as well as manure types and concentrations within each nutrient, in Tukey's Studentized Range (HSD) ($p < 0.05$). Significance level: * $p < 0.05$; ** $p < 0.01$; *** $p < 0.001$.

3.2.3. Fat

The greatest fat content was recorded in goat manure-treated plants (Table 3). Manure-treated plants recorded a slight increase in fat in both seasons, except for plants that were treated with 2 t ha^{-1} of poultry manure in season one. The fat content varied between 1.40 and 6.73% in the first season and 2.58 and 4.79% in the second season (Table 4). In season one, more (6.73%) fat content was recorded in plants that received 1 t ha^{-1} of goat manure, whereas the application of 3 t ha^{-1} of goat manure produced the highest (4.79%) fat content in season two.

3.2.4. Phosphorus

Poultry and goat manure increased the phosphorus content in *Sesamum alatum* plants when compared with the control (Table 3). The phosphorus content of plants that were exposed to 1 t ha^{-1} of goat manure was more than that plants that received 2 t ha^{-1} of goat manure and 3 t ha^{-1} of goat manure, but levels were similar with all other manure rates (Table 3). The interaction between season and manure shows that the application of poultry and goat manure at different levels increased the phosphorus content in both seasons. However, poultry manure rates $\geq$2 t ha^{-1} and 2 t ha^{-1} of goat manure had no effect on the phosphorus content in the second season. The highest (0.43%) P content in season one was recorded with the application of 3 t ha^{-1} of poultry manure but 3 t ha^{-1} of goat manure in season two (Table 4).

3.2.5. Protein

The protein amount recorded with 2 t ha^{-1} of poultry and goat manure was the greatest, but levels were similar to the protein contents achieved with 3 t ha^{-1} of poultry and goat manure, as well as no manure treated plants (Table 3). The smallest rates of manure application (1 t ha^{-1}) recorded the least protein content. Comparison in manure types and application rates show that poultry manure promotes better protein content in *S. alatum* than goat manure. In the first season, the highest protein content (39.89%) was recorded in plants that received 2 t ha^{-1} of poultry manure, but levels were comparable with other treatment rates except for some rates of goat manure (1 and 3 t ha^{-1}) (Table 4). The lowest protein content (28.94%) was 1 t ha^{-1} of poultry manure in season one.

3.3. Micro-Nutrients

Plants that were exposed to 1 and 3 t ha^{-1} of goat manure-amended soils produced the highest copper content but was similar to plants that received 1 t ha^{-1} of poultry manure (Table 3). The highest iron levels were obtained in both 1 and 2 t ha^{-1} goat manure-treated plants. Manganese content was highest in 1 t ha^{-1} goat manure-treated plants. The highest sodium content was also recorded in 1 t ha^{-1} goat manure-treated plants. Again, the zinc content was also recorded as highest in 1 t ha^{-1} goat manure-treated plants. Plants that had no manure applied produced the least micronutrient content.

The highest copper content was recorded in season two with 3 t ha^{-1} of goat manure, which was comparable to 1 and 3 t ha^{-1} of poultry and goat manure-treated plants in season one and 1 t ha^{-1} of poultry and goat manure. Manure application rates $\leq$2 t ha^{-1} of poultry manure and 2 t ha^{-1} of goat manure had no effect on the copper content (Table 4). Interaction between season and manure shows that the application of manure caused an increase in iron in season one when compared with the control. The highest iron content was recorded in season one with 2 t ha^{-1} of goat manure, but the content was similar to that in 1 t ha^{-1} of goat manure-treated plants. In season two, manure application had no effect on the iron content.

The highest manganese content was obtained with 1 t ha^{-1} of goat manure in both seasons (Table 4). The sodium content in season one was increased by the manure application. In season two, the increased sodium content was only obtained with 1 and 3 t ha^{-1} of goat manure (Table 4). The interaction between season and manure shows that high levels of poultry manure and lower rates of goat manure caused an increase in zinc content in season one, whereas a lower rate (1 t ha^{-1}) of poultry manure and goat manure ($\geq$2 t ha^{-1}) had no effect on the zinc content in season one. In season two, the zinc content was greatest with the application of 3 t ha^{-1} of goat manure but comparable with 1 t ha^{-1} of poultry and goat manure (Table 4).

4. Discussion

4.1. Macro-Nutrients

4.1.1. Acid and Neutral Detergent Fibres

Shoots of *Sesamum alatum* plants exposed to 2 t h^{-1} poultry manure had lesser acid detergent fibre (ADF) and neutral detergent fibre (NDF) contents than the control (Table 3). On the contrary, insignificant differences were recorded in the fibre percentage of *S. indicum* plants that were grown under 5 t h^{-1} poultry manure treatment when compared with the control [29]. Control plants of *S. alatum* recorded the highest fibre content (Tables 3 and 4), but *Corchorus olitorius* had a significant increase in fibre content under 1 and 2 t h^{-1} of poultry manure [12].

4.1.2. Calcium, Magnesium, and Potassium

The general increase in Ca, Mg, and K concentration of *Sesamum alatum* under manure-treated plants compared with control plants (Table 3) was probably due to nutrient release from poultry and goat manure (Table 2). However, Ca and Mg contents decreased but K content increased in *S. indicum* plants subjected to poultry manure [29]. Calcium is an important nutrient for healthy bones, contraction and relaxation of body muscles, and coagulation [21,30]. Magnesium is an enzyme co-factor in metabolic processes, protein synthesis, and like Ca, it also prevents bleeding disorders, degenerative diseases and immunological dysfunction [31]. Potassium also promotes good myosin–actin cycling and increases iron availability in the body, while it regulates the pH balance and neurotransmission [21].

The interactions between season and different manure application rates were significantly different for the Ca, Mg and K contents when compared with the control (Table 4). A similar significant interaction was also reported in *Abelmoschus esculentus* due to the effect of poultry manure application [16]. Goat manure applied at 2.5–12 t h^{-1} increased the concentration of Ca, Mg, and K in *Capsicum annum* leaves during both planting seasons [21]. Again, Ca, Mg, and K contents were better in *C. gynandra* [22] and *Amaranthus cruentus* [32] plants treated with goat manure compared with the control.

The highest Ca content recorded in 3 t ha^{-1} of poultry manure-treated plants in season one as well as 1 and 3 t ha^{-1} goat manure in season two (Table 4), which was possibly due to the high levels of calcium content recorded in poultry manure in the first season and goat manure in the second season (Table 2). Poultry manure with higher Mg content suppressed the Mg concentration in *S. alatum* edible shoots, but lower rates of application showed improvements when compared with the controls of each season. Organic manure is used in crop production for its ability to increase harvestable yields and also to replenish soil nutrients for improved plant quality [12].

4.1.3. Fat

Poultry manure-treated *S. alatum* plants had insignificant differences in fat content when compared with the control (Table 3), but *S. indicum* [29] and *Abelmoschus esculentus* [16] had a significant reduction in the fat content when poultry manure was applied. Further, 1 and 2 t ha^{-1} of poultry manure treatment resulted in an increase in the fat content of *Corchorus olitorius* [12]. Nitrogen applied at high rates limits fat availability in plants [16]. This plant growth nutrient plays a key role in plant metabolism [33]. The fat content recorded with 1 and 2 t ha^{-1} of poultry manure in the present study (Table 4) was greater than the fat content of *C. olitorius*, achieved with the same manure application rates [12].

4.1.4. Phosphorus

An increase in P content in *S. alatum* plants exposed to poultry and goat manures (Table 3) was similar to the higher *C. olitorius* P content under poultry manure [12] compared with control plants. Poultry manure application on *A. esculentus* significantly increased the fruit P concentration [16]. The increase in leaf P content was reported in *Amaranthus cruentus* [32], *Capsicum annuum* [21], and *C. gynandra* [22] when goat manure was applied. The increase in N accessibility to plants led to increased P absorption by plants [33].

4.1.5. Proteins

There were insignificant differences between manure treatment rates ≥2 t h^{-1} and controls on *Sesamum alatum* protein content; the highest was recorded in a 2 t h^{-1} poultry manure-treated plant (Table 3). A similar study on *S. indicum* reported an insignificant difference in the protein percentage when 5 t h^{-1} of poultry manure was applied [29]. The current findings are contrary to the significant increase in protein concentration in *Corchorus olitorius* [12] and *Abelmoschus esculentus* [16] plants treated with poultry manure. The highest protein content (39.89%) recorded with 2 t h^{-1} of poultry manure in season one (Table 4) was higher than the protein content recorded with same manure in *C. olitorius* [12] and *A. esculentus* [16]. Protein is a nitrogen-dependent nutrient [16]; this is also true for phosphorus [33]. The high rainfall seen in the second season may have altered the availability of soil nutrients, which promotes protein synthesis in *S. alatum* plants grown during this season, hence the variation (Table 1).

4.2. *Micro-Nutrients*

Plants that received 1 and 3 t ha^{-1} of goat manure had a significant increase in Cu content when compared with the control (Table 3). On the contrary, an insignificant response was reported in the Cu content of *S. indicum* under poultry manure treatment [29]. Goat manure application to *S. alatum* plants resulted in a significant increase in Fe content, whereas rates ≥2 t ha^{-1} of poultry manure had no significant effect when compared with the control (Table 3). The application of poultry manure at 5 t ha^{-1} caused a significant increase in the Fe content of *S. indicum* [29]. Plants that were treated with different rates of poultry and goat manure, except for 3 t ha^{-1} poultry and 2 t ha^{-1} goat manure, had a significant increase in Mn content when compared with the control (Table 3). The Mn content in *S. indicum* was increased with poultry manure application at 5 t ha^{-1} [29]. The manure application on *S. alatum* plants caused a significant increase in the sodium content. This was also true for the Zn content, except for plants that received 2 t ha^{-1} of goat manure. The Na and Zn contents were significantly increased with 1 t ha^{-1} goat manure when compared with the control (Table 3). Sodium and zinc concentrations in *S. indicum* plants grown under poultry manure soil amendments showed insignificant differences when compared with control plants [29]. Goat manure also increased the Cu, Fe, Mn, Na, and Zn content in *C. gynandra* plants [22].

5. Conclusions

Poultry and goat manure application led to increased levels of Ca, Mg, K, P, and micronutrients in *Sesamum alatum*. The increase in nutrient concentrations in *S. alatum* plants treated with manure over the control suggests that poultry and goat manures are useful in the improvement of crop nutrient content. However, ADF, fat, NDF and protein content in *S. alatum* needed no manure amendments to show its comparative potential in agronomy and nutrient concentration amongst food plants. The application of poultry and goat manure on *S. alatum* showed no significant differences in nutrient improvements except for the fat content recorded with 2 t ha^{-1} of poultry manure. Goat manure is best for the increase in nutrient content in *S. alatum*. However, poultry manure can also be used as an equal alternative in the absence of goat manure.

Author Contributions: Conceptualization, N.R.N. and S.M.; methodology, K.C.M., C.N.M., S.M. and N.R.N.; software, K.C.M., C.N.M., S.M. and N.R.N.; validation, C.N.M., S.M. and N.R.N.; formal analysis, K.C.M., C.N.M., S.M. and N.R.N.; investigation, K.C.M.; resources, C.N.M., S.M. and N.R.N.; data curation, K.C.M.; writing—original draft preparation, K.C.M.; writing—review and editing, K.C.M., C.N.M., S.M. and N.R.N.; visualization, K.C.M., C.N.M., S.M. and N.R.N.; supervision, S.M. and N.R.N.; project administration, S.M. and N.R.N.; funding acquisition, S.M. and N.R.N. All authors have read and agreed to the published version of the manuscript.

Funding: This research was funded by UNIVERSITY OF ZULULAND RESEARCH OFFICE, grant number UZREC 171110-030 PGM 2019/19.

Acknowledgments: Authors would like to thank the University of Zululand for funding, as well as North-West University and KwaZulu-Natal Department of Agriculture and Rural Development for their facilities.

Conflicts of Interest: The authors declare no conflict of interest.

References

1. Ntuli, N.R. Nutrient content of scarcely known wild leafy vegetables from northern KwaZulu-Natal, South Africa. *S. Afr. J. Botany* **2019**, *127*, 19–24. [CrossRef]
2. Jeyaray, S.; Suhara Beevy, S. Studies on anthecology and breeding systems of selected species of *Sesamum*, L. (Pedaliaceae) from Kerala. In *Plant Genetic Resources-Characterization and Utilization*; Swapna, T.S., Suhara Beevy, S., Radhamany, P.M., Eds.; Cambridge University Press: Cambridge, UK, 2021; pp. 18–25.
3. Bedigian, D. Feeding the forgotten: Wild and cultivated *Ceratotheca* and *Sesamum* (Pedaliaceae) that nourish and provide remedies in Africa. *Econ. Botany* **2019**, *72*, 496–542. [CrossRef]
4. Mariod, A.A.; Mirghani, M.E.S.; Hussein, I.H. Sesamum alatum (Thonn) winged-seed sesame. Principles of oil extraction, processing and oil composition. *Unconv. Oilseeds Oil Sources* **2017**, 39–43. [CrossRef]
5. Sarath Babu, B.; Dikshit, N.; Rameash, K.; Sivaraj, N. Maximum entropy modelling for predicting the potential distribution of wild sesame, *Sesamum alatum* Thonn. In India. *J. Oilseeds Res.* **2016**, *33*, 45–50.
6. Welcome, A.K.; Van Wyk, B.-E. An inventory and analysis of the food plants of southern Africa. *S. Afr. J. Bot.* **2019**, *122*, 136–179. [CrossRef]
7. Ali, H.A.M.; Ismail, A.B.O.; Fatur, M.; Ahmed, F.A.; Ahmed, E.H.O.; Ahmed, M.E.E. Nutritional evaluation and palatability of major range forbs from South Darfur, Sudan. *Open J. Anim. Sci.* **2016**, *5*, 42–48. [CrossRef]
8. Sundarakumar, M.; Karmegam, N. Antibacterial activity of ethanol extracts of *Sesamum alatum* Thonn. leaves. *Int. J. Curr. Res. Biosci. Plant Biol.* **2018**, *5*, 38–41. [CrossRef]
9. Hoover, N.L.; Law, J.Y.; Long, L.A.M.; Kanwar, R.S.; Soupir, M.L. Long-term impact of poultry manure on crop yield, soil and water quality, and crop revenue. *J. Environ. Manag.* **2019**, *252*, 109582. [CrossRef] [PubMed]
10. Ryne, N.; Aula, L. Livestock manure and the impacts on soil health: A review. *Soil Syst.* **2020**, *4*, 64. [CrossRef]
11. Sanyaolu, V.T.; Adepoju, I.H. Comparative evaluation of organic manure from some invasive plants, poultry manure and NPK on the growth of *Corchorus olitorius* L. *J. Appl. Sci. Environ. Manag.* **2018**, *22*, 1185–1192. [CrossRef]
12. Ayinla, A.; Alagbe, I.A.; Olayinka, B.U.; Lawal, A.R.; Aboyeji, O.O.; Etejere, E.O. Effects of organic, inorganic and organo-mineral fertilizer on the growth, yield and nutrient composition of *Corchorus olitorious* (L). *Ceylon J. Sci.* **2018**, *47*, 13–19. [CrossRef]
13. Yaw Debrah, P.; Afriyie-Debrah, C.; Nuhu, F.; Francisco Ribeiro, P.; Agyei Obeng, E. Response of cabbage (*Brassica oleracea* var *capitata* L.) to organic and inorganic fertilizer. *J. Exp. Agric. Int.* **2021**, *43*, 105–114. [CrossRef]
14. Ekwealor, K.U.; Anukwuorji, C.A.; Egboka, T.P.; Eze, H.N. Studies on the comparative effects of cow dung, goat dung and poultry manure in the restoration of gully eroded soil using *Amaranthus hybridus* as test plant. *Asian J. Soil Sci. Plant Nutr.* **2020**, *6*, 10–16. [CrossRef]
15. Sari, R.M.; Ali Akbar, S.; Astuti, T.; Afrini, D. The influence of some type of manure on the growth and production of elephant grass (Pennisetum purpureum) CV. Taiwan in acid soil. In Proceedings of the 7th International Conference on Sustainable Agriculture, Food and Energy, Phuket, Thailand, 18–21 October 2019; IOP Conference Series: Earth and Environmental Science: Vancouver BC, Canada, 2021; Volume 709. [CrossRef]
16. Adekiya, A.O.; Ejue, W.S.; Olayanju, A.; Dunsin, O.; Aboyeji, M.C.; Aremu, C.; Adegbite, K.; Akinpelu, O. Different organic manure sources and NPK fertilizer on soil chemical properties, growth, yield and quality of okra. *Sci. Rep.* **2020**, *10*, 16083. [CrossRef] [PubMed]
17. Akande, M.O.; Makinde, E.A.; Otuwe, M.O. Dry matter partitioning of sesame and nutrient dynamics with organic and inorganic fertilizers. *Trop. Subtrop. Agroecosyst.* **2011**, *14*, 1063–1069.
18. Anguria, P.; Chemining'wa, G.N.; Onwonga, R.N.; Ugen, M.A. Effect of organic manures on nutrient uptake and seed quality of sesame. *J. Agric. Sci.* **2017**, *9*, 135–144. [CrossRef]
19. Kahu, J.C. Effects of Different Types of Organic Fertilizers on Growth Performance, Nutrients and Toxicological Composition of Amaranthus caudatus and Amaranthus cruentus. Master's Thesis, Ahmadubello University, Zaria, Nigeria, 2017.

20. Oyedeji, S.; Animasaun, D.A.; Bello, A.A.; Agboola, O.O. Effect of NPK and poultry manure on growth, yield, and proximate composition of three Amaranths. *J. Bot.* **2014**, *2014*, 828750. [CrossRef]
21. Awodun, M.A.; Omonijo, L.I.; Ojeniyi, S.O. Effect of goat dung and NKP fertilizer on soil and leaf nutrient content, growth and yield of pepper. *Int. J. Soil Sci.* **2007**, *2*, 142–147.
22. Sowunmi, L.I. Nutritional value and cultivation requirements of Cleome gynandra L.: A wild vegetable growing in the Eastern Cape province, South Africa. Ph.D. Thesis, University of Fort Hare, Alice, South Africa, 2015.
23. Association of Official Analytical Chemists (AOAC). *Official Methods of Analysis of the Association of Official Analytical Chemists*, 15th ed.; Association of Official Analytical Chemists Inc: Washington, WA, USA, 1990.
24. Manson, A.D.; Roberts, V.G. *Analytical Methods Used by the Soil Fertility and Analytical Services Section*; KZN Agri-Report No. N/A/2001/4; KwaZulu-Natal Department of Agriculture and Rural Development: Pietermaritzburg, South Africa, 2000.
25. Murphy, J.; Riley, J.R. A modified single solution method for the determination of phosphate in natural waters. *Anal. Chim. Acta.* **1962**, *27*, 31–36. [CrossRef]
26. Hunter, A. *Tentative ISFEI Soil Extraction Procedure*; International Soil Fertility and Improvement Project; N.C. State University: Raleigh, NC, USA, 1974.
27. Allison, L.E. Organic carbon. In *Methods of Soil Analysis, Part 1*; Black, C.A., Ed.; John Wiley & Sons.: Madison, WI, USA, 1965; Volume 9, pp. 545–567.
28. SAS Institute. *Base SAS 9.4 Procedures Guide*; SAS Institute: Cary, NC, USA, 2015.
29. Alege, G.O.; Mustapha, O.T.; Ojo, S.; Awosemo, M.B. The morphological, proximate, and mineral responses of sesame to different nutrient sources. *Glob. J. Biosci. Biotechnol.* **2013**, *2*, 12–16.
30. Emebu, P.K.; Anyika, J.U. Proximate and mineral composition of Kale (*Brassica oleracea*) grown in Delta State, Nigeria. *Pak. J. Nutr.* **2011**, *10*, 190–194. [CrossRef]
31. Mohammed, M.I.; Sharif, N. Mineral composition of some leafy vegetables consumed in Kano, Nigeria. *Niger. J. Basic Appl. Sci.* **2011**, *19*, 208–211.
32. Ojeniyi, S.O.; Adejobi, K.B. Effect of ash and goat dung manure on leaf nutrients composition, growth and yield of *Amaranthus*. *Niger Agric. J.* **2002**, *33*, 46–57. [CrossRef]
33. Fageria, V.D. Nutrient interactions in crop plants. *J. Plant Nutr.* **2001**, *24*, 1269–1290. [CrossRef]

applied sciences

MDPI

Article

Compression and Fungal Heat Production in Maize Bulk Considering Kernel Breakage

Chaosai Liu [1], Yang Zhou [1,2,3,*], Guixiang Chen [1,2,3,*], Deqian Zheng [1,2,3] and Longfei Yue [1]

1 College of Civil Engineering, Henan University of Technology, Zhengzhou 450001, China; lcs@stu.haut.edu.cn (C.L.); deqianzheng@haut.edu.cn (D.Z.); yuelongfei1988@163.com (L.Y.)
2 Henan Key Laboratory of Grain Storage Facility and Safety, Zhengzhou 450001, China
3 Henan International Joint Laboratory of Modern Green Ecological Storage System, Zhengzhou 450001, China
* Correspondence: robertzhouy@haut.edu.cn (Y.Z.); cgx@haut.edu.cn (G.C.)

Abstract: Breakage in maize kernels and vertical pressure in grains lead to the uneven distribution of grain bulk density, which easily causes undesired problems in terms of grain storage. The objective of this study was, therefore, to determine the compression and heat production of the whole kernel (WK) and half kernel (HK) under two different loadings, i.e., 50 and 150 kPa, in maize bulk. An easy-to-use element testing system was developed by modification of an oedometer, and an empirical–analytical–numerical method was established to evaluate fungal heat production, considering kernel breakage and vertical pressure. Based on the experimental results, it was found that breakage induced larger compression; the compression of HK was 62% and 58% higher than that of WK at 50 kPa and 150 kPa, respectively. The creep model of the Hooke spring–Kelvin model in series can be used to accurately describe the creep behavior of maize bulk. Fungi and aerobic plate counting (APC) were affected significantly by the breakage and vertical pressure. APC in HK was 19 and 15 times that of WK under 150 and 50 kPa, respectively. The heat released by the development of fungi was found to be directly related to the APC results. The average temperatures of WK and HK under 150 and 50 kPa were 11.1%, 9.7%, 7.9%, and 7.6% higher than the room temperature, respectively. A numerical method was established to simulate the temperature increase due to fungi development. Based on the numerical results, heat production (Q) by fungi was estimated, and the results showed that the Q in HK was 1.29 and 1.32 times that of WK on average under 150 and 50 kPa. Additionally, the heat production results agreed very well with the APC results.

Keywords: maize bulk; kernel breakage; vertical pressure; deformation; heat production

Citation: Liu, C.; Zhou, Y.; Chen, G.; Zheng, D.; Yue, L. Compression and Fungal Heat Production in Maize Bulk Considering Kernel Breakage. *Appl. Sci.* **2022**, *12*, 4870. https://doi.org/10.3390/app12104870

Academic Editors: Paweł Kiełbasa, Tadeusz Juliszewski and Sławomir Kurpaska

Received: 24 March 2022
Accepted: 10 May 2022
Published: 11 May 2022

1. Introduction

Maize is among the major cereal crops, with a wide planting range and high harvest in China. Over 40 million hectares were sown, producing 260 million tons of maize in 2020 [1]. However, despite a massive increase in maize production, postharvest losses of maize during storage remain a significant challenge [2]. A considerable amount of maize kernel breakage is easily caused by postharvest processes such as transportation and grain loading. It reduces the quality grade of the grain and also increases the risk of grain security. The density change due to the self-weight of maize bulk further induces variation in grain properties, such as non-uniform airflow [3] and heat conductivity rate, and affects mildew on grain.

Broken kernels are more susceptible to the development of fungi, and they produce a considerable amount of heat because the physical integrity of kernels is compromised [4–7]. The distribution of fungi in broken kernels was preliminarily studied via milling and dehulling [5]. Pietri et al. [8] and Burger et al. [9] found that fungi are usually concentrated in the bran, and the damage to kernel brans often increases the risk of fungi infection. Oliveira et al. [7] analyzed the relationship between the physical properties of maize kernels and fungi infection and proposed that maize with a softer endosperm can present higher contamination by fungi. Coradi et al. [6] milled and separated breakage maize kernels into different particle sizes and concluded that aerobic plate counting (APC) was inversely proportional to the results of density and particle size.

Fungal growth can decompose organic matter in maize, thereby developing heat energy, especially in grain bulk with suitable temperature, moisture content, and oxygen concentration [10,11]. At a given height of grain bulk, the grain bulk will creep with storage time. The creep behavior will significantly influence the maize quality during and after storage [12]. However, the creep behavior of harvested maize grains during storage was rarely reported, especially the grain bulk deformation, considering broken kernels. In the food industry, creep behavior is mainly used to select high-quality raw materials and develop new products [13]. Sheng et al. [14] only established a creep model to describe the maize kernel, but there are few reports on the testing of maize bulk deformation. Thermal conductivity increased linearly with the bulk density of maize because the contact area between kernels influences heat transfer [15,16], resulting in the redistribution of moisture content and heat in the grain bulk [17,18]. The moisture–heat coupling model of grain bulk was established based on the principle of local heat and mass balance [19,20], but the effect of fungal heat production on the temperature of grain bulk was not considered. Recently, Wu et al. [21] proposed a solution for heat production in grain bulk, considering fungal activities induced rise in temperature, heat loss during conduction and convention, and water evaporation, but the effect of vertical pressure and kernel breakage on fungal heat production was not considered.

In this research, the authors developed an inexpensive and easy-to-use element testing system to study the effects of vertical pressure on compression and heat production. First, vertical pressure was applied to whole kernels (WKs) and half kernels (HKs) at 50 kPa and 150 kPa, respectively. The creep behavior was then investigated and analyzed using the model of Hooke spring–Kelvin model in series. Second, the temperature changes and APC were measured to study the fungi in the grain during compression. A newly empirical–analytical–numerical method was established to analyze the heat generated by fungi in grain bulk.

2. Materials and Methods

2.1. Test Materials

The maize used in this experiment was a normal, Zhengdan 958 hybrid maize with the conventional tillage, harvested in 2021, and stored in silos. Samples of maize kernels were collected directly from the silo. In total, 10 kg of the sample was taken at 20 random sampling points and sealed in a sterile sampling bag. The maize sample was filtered through a circular sieve to remove broken kernels and foreign impurities. Four maize samples were randomly selected from the filtered grains, and the mass of each sample was 400 g. Two samples were whole kernels, as shown in Figure 1a, and the kernels of the other two samples were cut into two parts along their middle, as shown in Figure 1b. The initial moisture content of maize was determined to be 9.8%, with samples of 30 g in three replications after drying at 103 °C for 72 h, and the moisture content after the test was also determined [22]. In this study, we focused on the numerical simulation and analytical investigation of the mildew process. Thus, we increased the moisture content to about 20% to shorten the test duration, which might not be the optimum moisture content for storage in reality. To obtain the desired target moisture content, maize was rewetted by adding distilled water mixed thoroughly and then hermetically sealed in polyethylene bags and stored at 4 °C for 48 h to allow moisture equilibrium [10]. Maize was mixed well, and about 400 g was randomly drawn for each treatment. The four actual levels of adjusted moisture content of maize were 21.02 ± 0.03%, 20.97 ± 0.29%, 21.04 ± 0.06% and 20.89 ± 0.13% wet basis.

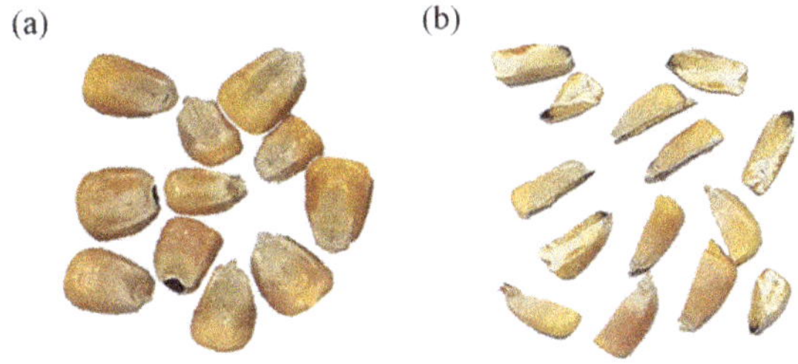

Figure 1. Maize kernels: (**a**) whole maize kernels and (**b**) half kernel cut along the middle.

2.2. Test Methods

To evaluate the effect of kernel breakage and compression on the mildew effect, a testing system was used in this study, which was modified from a conventional oedometer, normally used for soil testing, as shown in Figure 2. The testing system consists of a loading frame, a testing box, and a measuring system. The box is made of high-strength aluminum alloy and plexiglass, with a size of 120 by 120 by 55 mm. The metal loading plate is highly stiff, with a flexible rubber pad pasted below it [23]. A vertical loading screw is located in the groove so as to distribute the pressure applied to the maize bulk below. The extension length of the screw was adjusted according to the height of the maize bulk. The compression amount (s) of maize bulk under different vertical pressure levels and temperatures (T) were obtained by using a dial indicator and two T-type thermocouples (Applent Instruments Inc., Changzhou, China). The dial indicator was set on the loading screw. One thermocouple was placed in the geometric center of the maize sample, and another was attached to the external side of the box to observe the room temperature. The temperature results were recorded with an AT4508 temperature testing system (Applent Instruments Inc., China) and an automatic acquisition system.

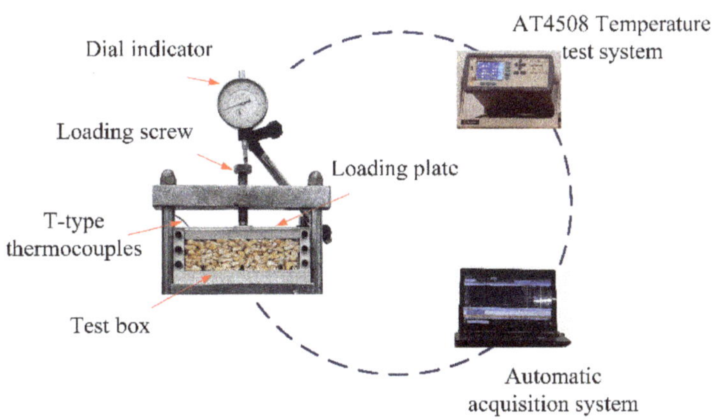

Figure 2. Testing system.

The walls of the testing box were coated with a thin layer of vaseline to eliminate side friction, and a 400 g sample of maize kernels was randomly taken and deposited in the testing box. As the sample area in this study was different from that of a conventional oedometer test, the vertical stress applied followed a sequence of 3, 8, 18, 28, 50, 75, 100, 125, and 150 kPa according to the calculation beforehand. In this study, two WK samples were loaded to 50 kPa and 150 kPa on multi-stages, respectively (i.e., WK_{50} and WK_{150}), and two HK samples were loaded in the same way (i.e., HK_{50} and HK_{150}). It is worth noting that each loading stage lasted for 20 min before reaching the final stress. Then, the four samples were continued for 7 d at constant stress. After the loading test, the APC test was performed to check mildew and grain quality.

2.3. *Inspection of APC*

Maize kernels at the center of the box, as shown in Figure 3, were selected to conduct the APC test. The sample was prepared using a rectangular plastic sampler with a size of 29 by 29 mm, which was used to take out about 30 g of maize.

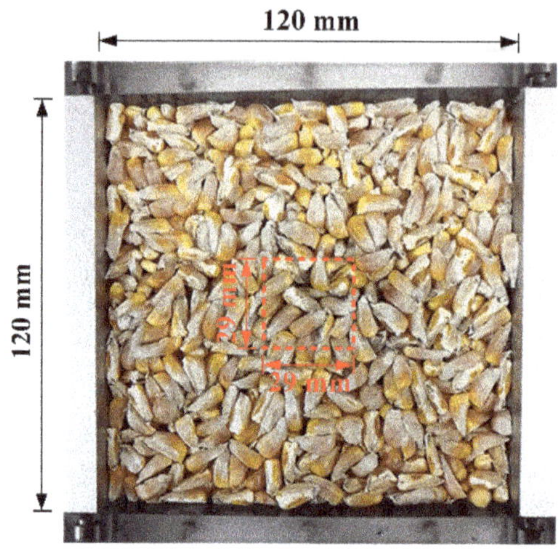

Figure 3. Sampling points for APC.

The specific preparation process for the APC test is as follows:

1. Preparation of the plate counting agar (PCA): First, the tryptone, yeast extract, glucose, and agar were mixed thoroughly and dissolved fully, and the pH was controlled at 7.0 ± 0.2. The mixed solution was autoclaved at 121 °C for 15 min in a conical flask;
2. Preparation of the bacterial suspension: Briefly, 25 g maize sample and 225 mL sterile water were put into a 500 mL conical flask. The conical flask was placed in an oscillator and vibrated for 30 min at a speed of 10,000 r·min^{-1}. The aim of this process was to fully diffuse the microorganisms on the surface of the maize kernels into the sterile water. By this process, a homogenate with a concentration of 1:10 was obtained;
3. Preparation of homogenates with different concentrations by serial dilution: For this step, 1 mL of 1:10 sample homogenate, drawn by a 1000 μL pipette, was mixed in a test tube containing 9 mL sterile water to obtain a sample homogenate, with 1:100 concentration. The above steps were repeated several times to obtain a series of solutions with different concentrations. Normally, two homogenates with different concentrations should be prepared for later fungus culture depending on the contamination conditions. In this study, sample homogenates with concentrations of 1:10 and 1:100 were selected for maize before testing, and 10^{-5} and 10^{-6} were selected after the test;
4. Fungus culture: Briefly, 1 mL of sample homogenate with desired concentration was taken and mixed with 15–20 mL PCA. Then, the mixture was cooled to 46 °C before injecting into 3 pre-sterilized Petri dishes. After the agar was coagulated, the dishes were turned over and cultured at 30 ± 1 °C for 72 h (MJPS-150, Shanghai Jing Hong Laboratory Instrument Co., Ltd., Shanghai, China);
5. Colony counting: For each sample, one from the 3 dishes with mold colony number ranging from 100 to 150 colony-forming units (cfus) were selected, and the type and number of colonies were recorded by observing under a microscope. The number of colonies was calculated as follows:

$$N = \frac{\Sigma C}{(n_1 + 0.1 n_2)d} \quad (1)$$

where N is the aerobic plate count; ΣC is the sum of bacterial colony number; n_1 is the number of colonies for 1:10 homogenate; n_2 is the number of colonies for 1:100 homogenate; d is the dilution. In this study, d is equal to 10^{-1} for maize before the test, and 10^{-5} after the test [24]. The procedure for the APC test is shown in Figure 4.

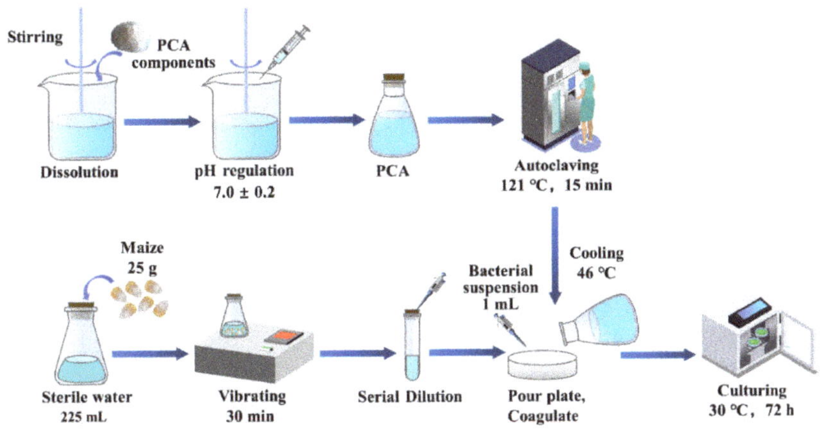

Figure 4. Inspection of APC.

3. Results and Discussion

3.1. Creep Behavior

Creep deformation analysis (i.e., creep) is considered a valuable method in the quality assessment of grain/food products [12]. The deformation of maize under vertical load (i.e., gravity or weight of machine) results in a reduction in void ratio and ultimately impacts the moisture content, temperature, and biochemical functions. The deformation curve of maize bulk is shown in Figure 5a. Kernel breakage and vertical pressure affected the vertical compression of maize. Compression deformations of WK_{50} and HK_{50} after compression were 3.73 and 6.04 cm when loading was 50 kPa, and compression deformations of WK_{150} and HK_{150} after compression were 6.15 and 9.72 cm when loading was 150 kPa. The vertical compression of maize bulk increased with time under constant load. The compression of WK was small under the same vertical pressure. The compression deformations of HK were 62% and 58% higher than those of WK at 50 kPa and 150 kPa, respectively. The reason is that the grain skeleton composed of WK was not easily compressed, whereas HK easily slid and filled the pores, resulting in large compression deformation.

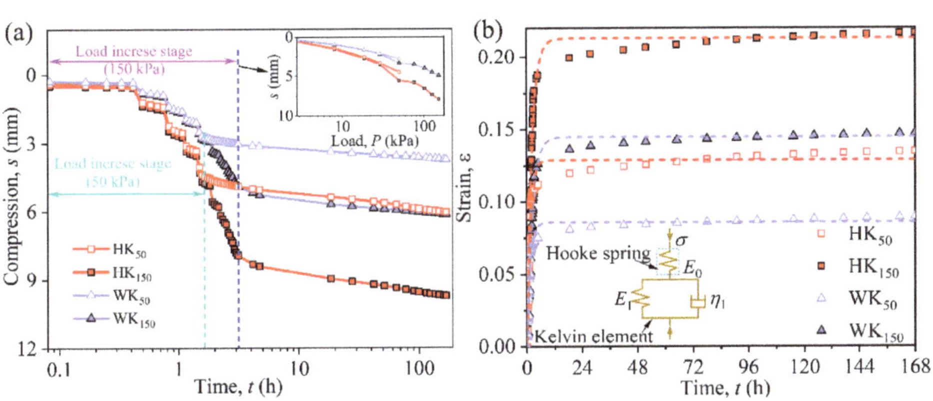

Figure 5. Creep behavior of maize bulk: (**a**) s~log (t) curve and (**b**) creep model.

The s~log (t) curve follows a reversed s-shape, and the deformation mainly underwent three stages: (1) Instantaneous deformation: the deformation was mainly affected by the magnitude of the load. The compression deformation occurred almost at the moment of load application, and the kernels slipped along the contact surface due to shear stress overcoming the friction at the contact surface. The kernels filled the large pores between kernels; (2) rapid deformation: the curve in this stage is mainly a section near the inflection point; the kernels were compressed, the pores between the kernels were filled, the resistance of the kernels increased, and the deformation rate decreased; (3) steady-state creep: the pores between kernels were basically filled, the resistance of kernels was greater, and the creep curve tended to be flat, roughly parallel to the abscissa axis.

Although s~log t can reflect the deformation trend of maize, it cannot clearly explain the internal characteristics and compression mechanism of grain, and it is not convenient to calculate the settlement of grain bulk in the process of long-term grain storage. To avoid this defect and consider the practicability of the model, a three-element model was used in this study, as seen in Figure 5b. The first region is represented by the Hooke spring unit, which shows the initiation of elastic deformation in a very short time frame, known as the instantaneous elastic deformation. The second region is represented by the Kelvin element, which indicates the retarded elastic deformation [14]. The Kelvin element is composed of Hooke spring and dashpot in parallel and can be expressed as

$$\sigma = E_1\varepsilon + \eta_1\dot{\varepsilon} \tag{2}$$

where σ is the applied stress (MPa); ε is the total strain (dimensionless); E_1 is the deformation modulus of creep stage (MPa); η_1 is the viscosity coefficient of Kelvin (MPa·h).

When the pressure is applied to the model, the spring will not be compressed immediately, due to the dashpot, and deformation gradually develops. Hooke spring is connected to the Kelvin element in series, and the creep model can be expressed as

$$\eta_1\dot{\varepsilon} + E_1\varepsilon = \frac{E_0 + E_1}{E_0}\sigma + \frac{\eta_1}{E_0}\dot{\sigma} \tag{3}$$

$$\varepsilon = \frac{\sigma}{E_0} + \frac{\sigma}{E_1}\left(1 - e^{-\frac{E_1}{\eta_1}t}\right) \tag{4}$$

where E_0 is the instantaneous elastic modulus (MPa); $\varepsilon_0 = \sigma/E_0$; t is the storage time (h).

The creep model of Hooke spring and Kelvin model in series can accurately describe the creep behavior of maize bulk under uniaxial compression. The fitting curve agreed very well with the test results, and all regression parameters had the coefficient of determination $R^2 > 0.97$, suggesting that the creep model can represent the creep behavior of maize bulk. Moreover, only three parameters—E_0, E_1, and η_1—need to be determined when the vertical stress σ, time t, and strain ε are known; the model parameters are relatively few, which is convenient to describe the creep behavior of grain bulk.

3.2. *Temperature and APC*

Due to respiration in wet maize kernels and microorganisms, a considerable amount of heat is released, thus raising the grain temperature [11,25]; therefore, the temperature difference is a sensitive indicator of maize mold activity [11]. The temperatures of maize bulk are shown in Figure 6. By observing the room temperature, we found that heating intensity was high during daytime and low during nighttime. The temperature of maize bulk was significantly influenced by the laboratory environment and was higher than the room temperature. There were considerable differences caused by vertical pressure and kernel breakage. The temperature was the highest in HK_{50} and lowest in WK_{150}. The average temperatures of HK_{50}, HK_{150}, WK_{50}, and WK_{150} were 11.1%, 9.7%, 7.9%, and 7.6%, respectively, higher than the room temperature. The reason i believed to be the higher respiration process in broken kernels and sufficient oxygen between kernels under low vertical pressure, thereby developing heat energy and leading to the rise in grain bulk temperature.

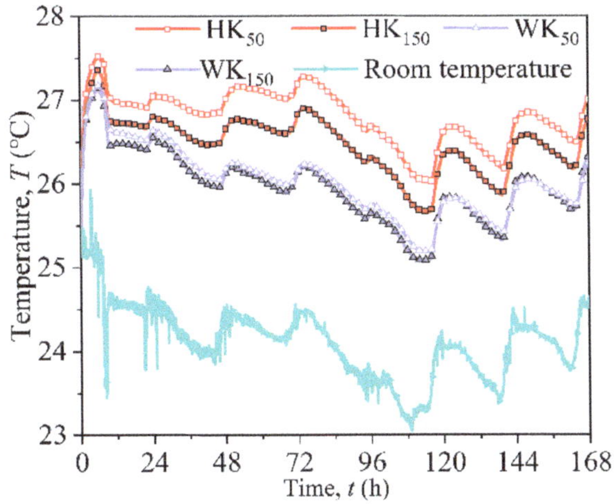

Figure 6. The change in maize bulk temperature.

APC of maize kernels is an important parameter to evaluate the bacterial reproductive dynamics and bacterial contamination during storage [6]. The APC values in the center of the maize sample were 3.0×10^6, 1.9×10^6, 0.2×10^6, and 0.1×10^6 cfu g^{-1} for HK_{50}, HK_{150}, WK_{50}, and WK_{150}, respectively, as seen in Figure 7. The endosperm of maize had celadon hyphae and strong musty off-odor in half kernels. In addition, the color of less moldy samples was relatively bright in WK_{150}. These results indicated that the broken kernels were more likely to be contaminated, compared with whole kernels. The larger vertical pressure compressed and deformed the maize bulk, reducing the pores between kernels; the oxygen concentration was also low, which inhibited the development of stored grain fungi. This result corresponded very well to the temperature change shown in Figure 6. Fewer molds produced a smaller amount of respiratory heat.

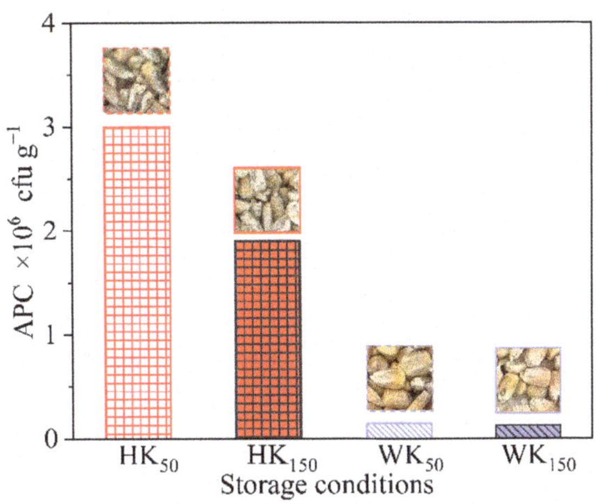

Figure 7. Aerobic plate count of maize stored for 7 d.

3.3. Physical Modeling of Heat Transfer and Verification

To investigate heat transfer due to physical factors during the testing process, finite element (FE) software COMSOL Multiphysics was used to establish numerical modeling. For simplicity, the maize kernels were assumed as porous media with isotropic properties and in local thermodynamic equilibrium with their surrounding air. Additionally, the laminar air in the void between kernels was incompressible under vertical load. However, the buoyant force and natural convection due to temperature gradient can be considered. The flow rate is assumed to be unchanged, and the 3D flow field can be expressed as

$$\frac{\partial u_j}{\partial x_j} = 0 \tag{5}$$

where u_j (j = 1,2,3) is the air flow rate in the direction x_j, $u_1 = u$, $u_2 = u_3 = v$. In rectangular Cartesian coordinate system, $x_1 = x$, $x_2 = y$, $x_3 = z$. The Brinkman–Darcy formulation was incorporated with the maize domain to represent the air flow. Additionally, the momentum satisfies the Boussinesiq's approximation, then

$$\rho_a \frac{\partial u_i}{\partial t} + \frac{\rho_a u_j}{\phi} \frac{\partial u_i}{\partial x_j} = -\frac{\partial p}{\partial x_i} + \frac{\partial}{\partial x_j}\left[\mu \frac{\partial u_i}{\partial x_j}\right] + \rho_0 g \beta (T - T_0) - \frac{\phi \mu u_i}{K} \tag{6}$$

where ρ_a is the density of air, u_i is the speed of airflow, t is the storage time, p is the air pressure, ϕ is the porosity of maize bulk, K is the permeability of maize bulk, T is the temperature of the air and maize kernels, ρ_0 is the density of air at the reference temperature (T_0), g is the gravity vector, β is the coefficient of volumetric expansion of the air, and μ is the viscosity of air.

As there is a local thermodynamic equilibrium between maize kernels and the surrounding air, the governing conservation equation of thermal energy is

$$\rho_b c_b \frac{\partial T}{\partial t} + (\rho_a c_a) u_j \frac{\partial T}{\partial x_j} = \frac{\partial}{\partial x_j}\left[k_b \frac{\partial T}{\partial x_j}\right] + \rho_b h_s \frac{\partial W}{\partial t} \tag{7}$$

where c_a is the specific heat of air; ρ_b, c_b, and k_b are the dry density, specific heat, and the effective thermal conductivity of maize bulk, respectively; h_s is the heat sorption of water on maize, and W is the moisture content of maize in dry basis. Maize is a living organism with water absorption and desorption feature. Thus, the moisture balance formula is written as

$$\frac{\partial}{\partial t}[(\varepsilon \rho_a w) + (\rho_b W)] + \rho_a u_j \frac{\partial w}{\partial x_j} = \frac{\partial}{\partial x_j}\left[\frac{D_v \varepsilon}{\tau} \frac{\partial}{\partial x_j}(\rho_a w)\right] \tag{8}$$

where τ is the tortuosity factor of maize bulk, D_v is a dimensionless rate coefficient for moisture exchange between air and maize kernels, and w is the moisture content of water vapor in the air on a dry basis.

To verify the effectiveness of the numerical model, physical heat transfer for maize was investigated with both testing and FE methods. The test was conducted by placing the sample within the model box (including the top plate) in a freezer. Before testing, the model box, top plate, and spread-open kernels were placed in the freezer, with the door open for hours to reach a uniform initial temperature. Then, the whole maize kernels were put into the box and covered by the top plate. The tests were begun after closing the door, and the temperature of the freezer was set at 6 °C. The test lasted for 350 min only. Therefore, biochemical heat production could be ignored, and only physical heat transfer occurred during the test. Two T-type thermocouples were set at the central point of the sample and in the freezer to monitor the temperature change. The material parameters of the model box and maize used in FE analysis are listed in Table 1, and the sources of the parameters are specified as well. As seen in Figure 2, the maize bulk was basically sealed by the box and top plate during the test. Thus, there was almost no air convection between the pores, and the maize moisture content slightly changed during the test. As a result, the flow rate of air was assumed to be zero, and water evaporation was not considered. A comparison of the tested and numerical results of T at the central point is shown in Figure 8. After placing in the freezer, the temperature of the freezer E_F gradually decreased from around 30 °C to 6 °C. It is worth noting that the maize tested T was slightly higher than the numerical value for in the late stage, e.g., after 300 min. The reason is that the biochemical heat production is not zero even under the very low freezer temperature of 6 °C. In conclusion, the heat transfer considering various physical factors, such as air convection, water migration and evaporation, and local thermodynamic equilibrium, could be well simulated using the established FE model.

Table 1. The material parameters.

Material	Property	Value
Aluminum alloy	Thickness (l_A)	0.01 m
	Thermal conductivity (k_A)	201 W·m^{-1}·°C^{-1}
	Density (ρ_A)	2720 kg·m^{-3}
	Specific heat (c_A)	90.64 J·kg^{-1}·°C^{-1}
PMMA	Thickness (l_P)	0.01 m
	Thermal conductivity (k_P)	0.19 W·m^{-1}·°C^{-1}
	Density (ρ_P)	1180 kg·m^{-3}
	Specific heat (c_P)	1424 J·kg^{-1}·°C^{-1}
Air	Air thermal conductivity (k_a)	0.025 W·m^{-1}·°C^{-1}
	Air density (ρ_a)	1.205 kg·m^{-3}
	Air specific heat (c_a)	1006 J·kg^{-1}·°C^{-1}
	Air tortuosity factor (τ)	1.2
	Air viscosity (μ)	1.79 × 10^{-5} Pa·s
Maize	Maize moisture content (M_{maize})	9.8%
	Maize density (ρ_{maize})	768 kg·m^{-3}
	Maize thermal conductivity (k_{maize})	0.12 W·m^{-1}·°C^{-1}
	Maize specific heat (c_{maize})	2223 J·kg^{-1}·°C^{-1}
	Maize permeability (K_{maize})	1.9 × 10^{-9} m^2

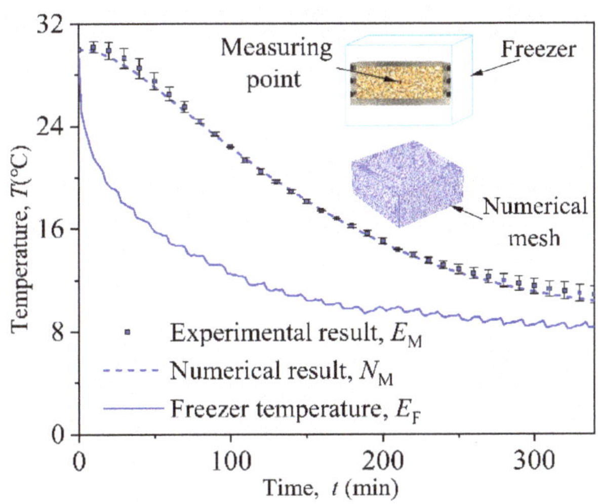

Figure 8. Experiment and validation of numerical model in low-temperature environment.

3.4. Simulation Results and Analysis

The current method normally assumed that maize is an isotropic porous medium, and the model for simulating heat and moisture transfer in grain bulk was also developed. However, the current model did not consider the effect of compression deformation on heat and moisture coupling transfer. Using the verified model mentioned in the previous section and considering vertical pressure, the variation in temperature (T) at the central point was simulated. The parameters involved in the FE simulation for maize are shown in Table 2. During the simulation, vertical-pressure-induced compression (Δh), considered by the input value of density (ρ_b), was calculated using Δh step by step, according to the loading procedure. Similarly, the variation in porosity and thermal conductivity during the test were considered step by step as well. The experimental result and simulated result are designated as T_E and T_N, respectively. Figure 9 presents the temperature variations in T_E and T_N for the four groups of maize bulk (HK_{50}, HK_{150}, WK_{50}, and WK_{150}). Clearly, it can be observed that T_E is higher than T_N. Compared with numerical T_N, which only includes physical effect, the test T_E resulted from physical heat transfer and biochemical effect (e.g., mildew). The temperature difference/rise actually is a consequence of heat accumulation due to a biochemical effect. Generally, half kernels under low vertical pressure, e.g., HK_{50}, yield larger ΔT, and WK_{150} yields smaller ΔT. The average values of ΔT were 2.7 °C, 2.4 °C, 1.9 °C, and 1.8 °C in HK_{50}, HK_{150}, WK_{50}, and WK_{150}, respectively. This result is proportional to the APC in Figure 7, which indicates that mildew is one of the main reasons for temperature rise in grain storage. Owing to the low thermal conductivity of air between the pores of the kernels, heat conduction is slow, and the heat generated by initial fungi inside the grain bulk cannot be quickly dissipated. Increased temperature in grain bulk further provides a favorable environment for subsequent fungal growth [10,16]. This cycle of "heat conduction–temperature rise–fungal growth" accounts for the initial condition significantly. The temperature differences ΔT of HK samples were larger than those in WK, and the result is consistent with the above APC test results.

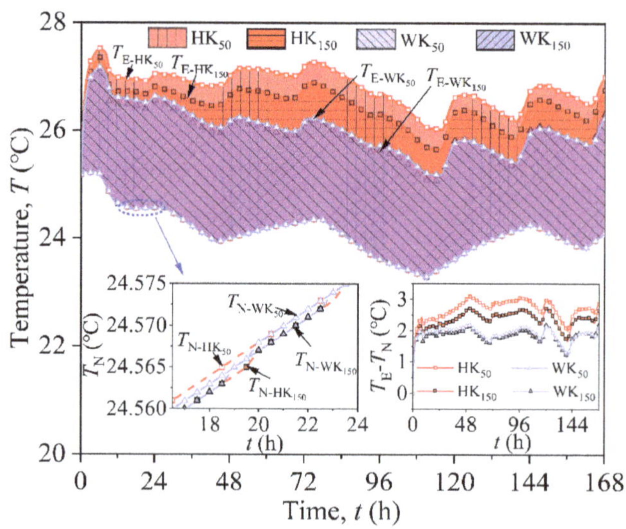

Figure 9. Comparison of experimental and numerical simulation results.

Table 2. The parameters involved in FE simulation for maize.

Material Property		Value
Moisture content (M)		21%
Mass (m)		0.4 kg
Initial height (h_0)	HK	0.042 m
	WK	0.045 m
Maize bulk density (ρ_b)		$m/0.0144 \times (h_0 - s)$
Particle Density (ρ_s)		1230.2 kg·m^{-3}
Porosity (ϕ)		$1 - \rho_b/\rho_s$
Thermal conductivity (k_b)		$0.0902 + 1.165 \times 10^{-4}\ \rho_b$ [16]

3.5. Estimation of Fungal Heat Production

Fungal growth decomposes organic matter in maize, thereby developing heat energy, and biochemical heat production is assumed to be heat production due to fungi development [21]. In the estimation of fungal heat production, some of the heat (E_T) is used to raise the temperature of the system, some heat is likely lost due to conduction (E_C) and convection (E_V), and some may be consumed during the water phase change (E_E), e.g., from liquid to vapor. In this study, E_C was not considered, as the sample was nearly sealed. E_E was also considered ignorable because the moisture content of the four samples lost only 1.1%~1.3% after being stored for 7 d. Therefore, the heat produced by fungi in the maize bulk can be determined as follows:

$$Q = f(\Delta T') = E_T + E_C \tag{9}$$

where $\Delta T'$ is the temperature difference in the maize bulk before and after fungus emergence. Heat absorbed by maize and the specific heat capacity of maize bulk were calculated using the following formula [21,26]:

$$E_T = \frac{c_b \rho_b \Delta V \Delta T'}{\Delta V} = c_b \rho_b \Delta T' \tag{10}$$

where ΔV is the spatial range affected by fungal activity (m^3).

The conduction heat loss (E_C) was determined by considering heat transfer from the geometric center of the maize sample to the wall of the testing box. The calculation space is $\Delta x = 0.12$ m,$\Delta y = 0.12$ m, and Δz is the height of the sample at the corresponding time ($\Delta V = \Delta x \Delta y \Delta z$). Taking into account the sum of the conduction heat loss on the six sides of the testing box, according to Fourier's law, E_C can be calculated as follows:

$$E_C = \frac{k_b \Delta t}{\Delta x \Delta y \Delta z} \sum_{i=1}^{6} \frac{A_i \Delta T_i}{l_i} \tag{11}$$

where Δt is the storage time, A_i is the area of one of the six faces of the testing box, and ΔT_i is the temperature difference between the center of the geometric center of the maize sample and the corresponding face A_i measured at a given time t. l_i is the distance between the center of the geometric center of the maize sample and the corresponding face A_i.

In this investigation, the value of temperature difference before and after fungi emergence ($\Delta T'$) can be equivalently regarded as the difference between the test and numerical results (ΔT). As mentioned above, the reason for ΔT between test and simulation is whether the biochemical effect or fungi was considered. The numerical result from COMSOL Multiphysics can only include the physical factors in the simulation. Using the test and numerical temperature difference, heat production (Q) by fungi was evaluated, as shown in Figure 10. Q was the least in WK_{150}, and in HK_{50}, HK_{150}, and WK_{50}, the values were 33.7%, 29.7%, and 1.4% higher than it on average. Additionally, it is clear that this variation in Q with time agreed very well with the APC and temperature results (Figures 7 and 9).

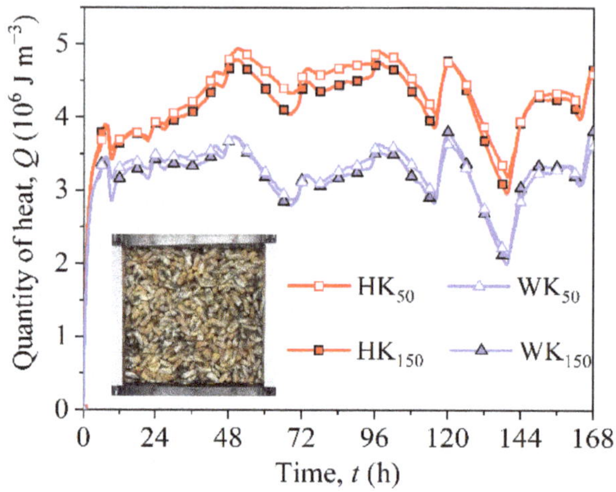

Figure 10. Variation in fungal heat production in maize bulk.

4. Conclusions

The deformation of maize under vertical load and kernel breakage of maize have an important impact on the biochemical functions of maize bulk. Maize bulk appeared to creep deformation under stable load, and the creep model of Hooke spring–Kelvin model in series can accurately describe the creep behavior of maize bulk. There was a significant difference between the compression of the whole kernel and half kernel—namely, the compression of the half kernel was 62% and 58% higher than that of the whole kernel at 50 kPa and 150 kPa, respectively. Broken kernels increase the risk of fungi infection, and deformation under vertical load results in a reduction in void ratio, thus affecting fungi development. The APC values in HK were 19 and 15 times those of WK under 150 and 50 kPa, respectively. The temperature and heat rates agreed very well with the APC results. The average temperatures of HK_{50}, HK_{150}, WK_{50}, and WK_{150} were 11.1%, 9.7%, 7.9%, and 7.6% higher than the room temperature, respectively. The heat production rate by fungi in HK was 1.29 and 1.32 times that of WK on average under 150 and 50 kPa.

Author Contributions: Conceptualization, Y.Z., G.C. and L.Y.; methodology, Y.Z. and D.Z.; software, C.L.; validation, C.L.; formal analysis, C.L. and Y.Z.; investigation, C.L. and L.Y.; resources, G.C. and D.Z.; data curation, C.L. and Y.Z.; writing—original draft preparation, C.L.; writing—review and editing, Y.Z., G.C. and D.Z.; supervision, G.C. and D.Z.; project administration, D.Z.; funding acquisition, G.C. All authors have read and agreed to the published version of the manuscript.

Funding: This research was funded by the National Grain Public Welfare Research Project of China, grant number 201513001-01; the Innovative Funds Plan of the Henan University of Technology, grant number 2020ZKCJ05; and the Open Project of Henan Key Laboratory of Grain and oil storage facility & safety, grant number 2020KF-A02.

Institutional Review Board Statement: Not applicable.

Informed Consent Statement: Not applicable.

Data Availability Statement: The data presented in this study are available in the article.

Conflicts of Interest: The authors declare no conflict of interest.

References

1. National Bureau of Statistics. *China Statistical Yearbook*; China Statistics Press: Beijing, China, 2020. Available online: http://www.stats.gov.cn/tjsj/ndsj/2021/indexch.htm (accessed on 1 November 2021).
2. Godfray, H.C.J.; Beddington, J.R.; Crute, I.R.; Haddad, L.; Lawrence, D.; Muir, J.F.; Pretty, J.; Robinson, S.; Thomas, S.M.; Toulmin, C. Food security: The challenge of feeding 9 billion people. *Science* **2010**, *327*, 812–818. [CrossRef] [PubMed]
3. Lawrence, J.; Maier, D.E. Three-dimensional airflow distribution in a maize silo with peaked, levelled and cored grain mass configurations. *Biosyst. Eng.* **2011**, *110*, 321–329. [CrossRef]
4. Ziegler, V.; Paraginski, R.T.; Ferreira, C.D. Grain storage systems and effects of moisture, temperature and time on grain quality—A review. *J. Stored Prod. Res.* **2021**, *91*, 101770. [CrossRef]
5. Schaarschmidt, S.; Fauhl-Hassek, C. The fate of mycotoxins during the primary food processing of maize. *Food Control* **2021**, *121*, 107651. [CrossRef]
6. Coradi, P.C.; Maier, D.E.; Channaiah, L.H.; Campabadal, C. Effects of the processing on the distribution of aflatoxin and fumonisin levels in corn fractions and feeds. *J. Food Process. Eng.* **2020**, *39*, 215–225. [CrossRef]
7. Oliveira, T.R.; Jaccoud, D.D.; Henneberg, L.; Michel, M.D.; Demiate, I.M.; Pinto, A.T.B.; Machinski, M.; Barana, A.C. Maize (*Zea Mays* L.) landraces from the southern region of Brazil: Contamination by fusarium sp, zearalenone, physical and mechanical characteristics of the kernels. *Braz. Arch. Biol. Technol.* **2009**, *52*, 11–16. [CrossRef]
8. Pietri, A.; Zanetti, M.; Bertuzzi, T. Distribution of aflatoxins and fumonisins in dry-milled maize fractions. *Food Addit. Contam. A* **2009**, *26*, 372–380. [CrossRef]
9. Burger, H.M.; Shephard, G.S.; Louw, W.; Rheeder, J.P.; Gelderblom, W.C.A. The mycotoxin distribution in maize milling fractions under experimental conditions. *Int. J. Food Microbiol.* **2013**, *165*, 57–64. [CrossRef]
10. Suleiman, R.; Bern, C.J.; Brumm, T.J.; Rosentrater, K.A. Impact of moisture content and maize weevils on maize quality during hermetic and non-hermetic storage. *J. Stored Prod. Res.* **2018**, *78*, 1–10. [CrossRef]
11. Mohapatra, D.; Kumar, S.; Kotwaliwale, N.; Singh, K.K. Critical factors responsible for fungi growth in stored food grains and non-chemical approaches for their control. *Ind. Crop. Prod.* **2017**, *108*, 162–182. [CrossRef]
12. Moreira, R.; Chenlo, F.; Arufe, S.; Rubinos, S.N. Physicochemical characterization of white, yellow and purple maize flours and rheological characterization of their doughs. *J. Food Sci. Technol. Mys.* **2015**, *52*, 7954–7963. [CrossRef] [PubMed]
13. Kamboj, U.; Guha, P.; Mishra, S. Changes in rheological properties of wheat due to storage. *J. Sci. Food Agric.* **2017**, *98*, 1374–1380. [CrossRef] [PubMed]
14. Sheng, S.Y.; Wang, L.J.; Li, D.; Mao, Z.H.; Adhikari, B. Viscoelastic behavior of maize kernel studied by dynamic mechanical analyzer. *Carbohyd. Polym.* **2014**, *112*, 350–358. [CrossRef] [PubMed]
15. Cheng, X.D.; Zhang, Q.; Shi, C.X.; Yan, X.J. Model for the prediction of grain density and pressure distribution in hopper-bottom silos. *Biosyst. Eng.* **2017**, *163*, 159–166. [CrossRef]
16. Chang, C.S. Thermal conductivity of wheat, corn, and grain sorghum as affected by bulk density and moisture content. *Trans. ASAE* **1986**, *29*, 1447–1450. [CrossRef]
17. Kara, M.; Ozturk, I.; Bastaban, S.; Kalkan, F. Thermal conductivity of safflower (*Carthamus tinctorius* L.) seeds. *Span. J. Agric. Res.* **2011**, *9*, 687–692. [CrossRef]
18. Jia, C.C.; Sun, D.W.; Cao, C.W. Mathematical simulation of temperature fields in a stored grain bin due to internal heat generation. *J. Food Eng.* **2000**, *43*, 227–233. [CrossRef]
19. Wang, Y.C.; Yang, K.M.; Zhang, Z.T.; Qi, W.; Yang, J. Natural convection heat and moisture transfer with thermal radiation in a cavity partially filled with hygroscopic porous medium. *Dry. Technol.* **2016**, *34*, 275–286. [CrossRef]
20. Hammami, F.; Ben Mabrouk, S.; Mami, A. Modelling and simulation of heat exchange and moisture content in a cereal storage silo. *Math. Comput. Model. Dyn. Syst.* **2016**, *22*, 207–220. [CrossRef]
21. Wu, Z.D.; Zhang, Q.; Yin, J.; Wang, X.M.; Zhang, Z.J.; Wu, W.F.; Li, F.J. Interactions of mutiple biological fields in stored grain ecosystems. *Sci. Rep.* **2020**, *10*, 9302. [CrossRef]
22. *ASAE S352.2 APR1988 (R2017)*; Moisture Measurement–Unground Grain and Seeds. ASAE: St. Joseph, MI, USA, 2017.
23. Talesnick, M. Measuring soil pressure within a soil mass. *Can. Geotechn. J.* **2013**, *50*, 716–722. [CrossRef]
24. *GB4789.2-2016*; National Standards of the People's Republic of China. Food Microbiological Analysis: Determination of Aerobic Plate Count. State Food and Drug Administration: Beijing, China, 2016.
25. Ubhi, G.S.; Sadaka, S. Temporal valuation of corn respiration rates using pressure sensors. *J. Stored Prod. Res.* **2015**, *61*, 39–47. [CrossRef]
26. *ASAE D243.4 MAY2003 (R2017)*; Thermal Properties of Grain and Grain Products. ASAE: St. Joseph, MI, USA, 2017.

applied sciences

Article

Efficacy of Four In Vitro Fungicides for Control of Wilting of Strawberry Crops in Puebla-Mexico

Alba Cruz Coronel [1], Conrado Parraguirre Lezama [1], Yesenia Pacheco Hernández [1], Olga Santiago Trinidad [2], Antonio Rivera Tapia [3] and Omar Romero-Arenas [1,*]

1 Centro de Agroecología, Instituto de Ciencias, Benemeérita Universidad Autoónoma de Puebla, Edificio VAL 1, Km 1, 7 Carretera a San Baltazar Tetela, San Pedro Zacachimalpa, Puebla 72960, Mexico; alba.cruzcor@alumno.buap.mx (A.C.C.); conrado.parraguirre@correo.buap.mx (C.P.L.); yesenia.pachecoh@gmail.com (Y.P.H.)
2 Campo Experimental El Palmar, CIR-Golfo Centro, Instituto Nacional de Invistigaciones Forestales Agrícolas y Pecuarias (INIFAP), Tezonapa, Veracruz 95083, Mexico; santiago.olga@inifap.gob.mx
3 Centro de Investigaciones en Ciencias Microbiológicas, Instituto de Ciencias, Benemérita Universidad Autónoma de Puebla, Ciudad Universitaria, Puebla 72570, Mexico; jose.riverat@correo.buap.mx
* Correspondence: biol.ora@hotmail.com; Tel.: +52-22-2229-5500 (ext. 3717)

Abstract: Strawberry wilt is an established disease of strawberry crops caused by fungus *Fusarium solani*. In Mexico, strawberry cultivation represents an important productive activity for several rural areas; however, wilt disease affects producers economically. The objectives of this research were: (a) to identify and morphologically characterize strain "MA-FC120" associated with root rot and wilting of strawberry crops in Santa Cruz Analco, municipality of San Salvador el Verde, Puebla-Mexico; (b) to evaluate the potential of single and multiple applications of four broad-spectrum fungicides used against *F. solani* in vitro. Plant tissue samples were collected from strawberry crops in Puebla-Mexico with presence of symptoms of desiccation and root rot. Strain "MA-FC120" was identified as *F. solani*, being the causal agent of wilt and root rot in strawberry plants from Santa Cruz Analco. Fungicide Benomyl 50® showed the highest percentage of inhibition on *F. solani* (100%) under in vitro conditions. The fungicide Mancosol 80® and Talonil 75® at low concentration (600 and 450 mg L^{-1}) showed no toxicity, being harmless to strain MA-FC120. However, fungicide Talonil 75® showed slight toxicity at the dose recommended by the manufacturer and moderate toxicity in high concentration (1350 mg L^{-1}). Likewise, Captan 50® in its three concentrations evaluated showed slight toxicity, obtaining around 50% on the classification scale established by International Organization for Biological Control (IOBC).

Keywords: fungicide resistance; PCR; broad-spectrum fungicides; *Fusarium solani*; toxicity

Citation: Coronel, A.C.; Parraguirre Lezama, C.; Pacheco Hernández, Y.; Santiago Trinidad, O.; Rivera Tapia, A.; Romero-Arenas, O. Efficacy of Four In Vitro Fungicides for Control of Wilting of Strawberry Crops in Puebla-Mexico. *Appl. Sci.* **2022**, *12*, 3213. https://doi.org/10.3390/app12073213

Academic Editors: Paweł Kiełbasa, Tadeusz Juliszewski, Sławomir Kurpaska and Gyungsoon Park

Received: 9 February 2022
Accepted: 10 March 2022
Published: 22 March 2022

1. Introduction

Strawberry (*Fragaria* × *ananassa*) is a widely distributed crop in the world due to its genotypic diversity, highly heterozygous nature, and wide range of environmental adaptations [1,2]. Strawberry is a favored food due to its nutraceutical properties and antioxidant effects. It is also a relevant source of bioactive compounds, due to its high level of vitamin C and phenolic compounds that provide beneficial effects on maintenance of consumers' health, especially by strengthening the immune system [3].

World strawberry production in 2021 was 12,106,585 tons, with China being the main producer with 3,213,000 tons, followed by USA (1,164,000 tons) and Mexico [4,5]. The cultivated area in Mexico is currently 12,913 ha, with a production of 557,514 tons in 2020 [6]. The State of Puebla ranks ninth in national production [7].

The strawberry crop has several fungal, bacterial, and viral diseases which cause large yield losses, with fungal diseases being of greatest concern and causing huge economic damage [8,9]. Soil-borne fungal pathogens such as *Fusarium oxysporum*, *F. solani*,

Macrophomina phaseolina, *Pythium* spp., *Phytophthora* spp., and *Rhizoctonia* spp., cause the death of feeder rootlets, plant deterioration, and blackening of the main root system, as well as a decrease in plant stand vigor and productivity, causing a decrease in crop yield [10,11].

Among the fungal diseases of the strawberry crop, wilting or drying is a disease caused by *F. solani* (Mart) Sacc, causing root rot in the crop, which causes stunting, wilting, and death of plants [12,13]. *F. solani* is a generalist plant pathogenic fungus with soil and seed origin present worldwide [14]. It is mainly characterized by abundant aerial mycelium, cottony texture, and sickle-shaped macroconidia with two to five septa [15]. *F. solani* has recently been reported in Spain in strawberry crops in intensive "nursery" systems, as well as in production fields [16]. *F. solani* has also been described as a pathogen of strawberry in Italy [17], Iran [18], and Pakistan [19]. However, there are few reports in Mexico [8].

Multiple fungicides to control various fungal diseases in strawberry crops, including demethylation inhibitors, quinone external inhibitors, and succinate dehydrogenase inhibitors (Fungicide Resistance Action Committee codes 3, 11, and 7, respectively), are available to producers. Benomyl and thiophanate-methyl have been consistently reported to be effective against various species of the genus *Fusarium* on several host plants [20,21]. Chemical management is an important tool for control of diseases, including soil-borne diseases. In addition, identification of effective fungicides would allow consolidation of different components needed to formulate integrated fungicides for disease management [22].

In Mexico, there is no publicly available information on the name, type, and quantity of each authorized pesticide applied nationally [23]. According to García et al. [24], the active ingredients (a.i.) Mancozeb®, Benomyl®, Cholothalonil®, and Captan® can be identified as most widely used in the country. Given this situation, chemical control is considered an efficient practice; however, extensive and repeated application of these chemicals has led to the frequent appearance of resistance to fungicides that could compromise their effectivity, as highlighted by incidences of resistance to carbendazim and tebuconazole [25].

In the summer of 2021, plant stunting, wilting, leaf chlorosis, and root rot were observed on strawberry var. "Camino Real" in Santa Cruz Analco, Puebla-Mexico, with symptoms indicative of *F. solani*. These symptoms were observed in 40% of the crops. Therefore, the objectives of this research were: (a) to identify the causal agent of this disease; (b) to evaluate the potential of single and multiple applications of four broad-spectrum fungicides used against *F. solani* in vitro.

2. Materials and Methods

2.1. Isolation Area

Samples of plant tissue with stem and root rot were collected from a 1000 m^2 plot of strawberry var. "Camino Real" during summer–autumn 2021 production. The agricultural plot corresponds to the community of Analco de Ponciano Arriaga (Santa Cruz Analco), which belongs to the municipality of San Salvador el Verde in the State of Puebla, Mexico. The study community has a temperate sub-humid climate (Cw) and average rainfall of 1000 mm [26]. Sampling was directed towards individuals with symptoms associated with genus *Fusarium*; all samples were kept in plastic bags in a cooler at 4 °C until they were transferred to the laboratory.

2.2. Isolation of *F. solani* from Plant Tissues

Samples were cut into small 5 mm pieces, surface-sterilized with 1% sodium hypochlorite for 3 min, and rinsed in sterile distilled water three times. Pieces were placed on potato dextrose agar medium (PDA, Bioxon, Becton Dickinson and Company, Queretaro, Mexico) and incubated at 28 °C for 10 days. The developed colonies were isolated and purified by monosporic cultures, which involved the transfer of a single conidium to potato dextrose agar (PDA, Bioxon, Becton Dickinson and Company, Queretaro, Mexico) medium. Subsequently, the plates were transferred to a microbiological incubator at 24 °C (Thermo Scientific, CA, USA) and subjected to a photoperiod of 12:12 L:D (12 h of light: 12 h of darkness) for 10 days [2].

2.3. Morphological Characterization

Identification of fungal colonies obtained was performed by analyzing morphological characteristics associated with genus *Fusarium*, using taxonomic identification keys from Barnett and Hunter [27], in a microculture system in an optical microscope (Carl Zeiss, Jena, Germany) at 1000× magnifications.

Once the isolates were characterized, the most representative was selected based on its ability to grow in three growth media tested: (a) potato dextrose agar medium (PDA, Bioxon, Becton Dickinson and Company, Queretaro, Mexico), (b) tryptone soybean agar (TSA, Sigma-Aldrich, Mexico [28]), and (c) water agar with 8 g/L carnation (CWA [29]). All culture media were incubated with a photoperiod of 12:12 L:D (12 h of light: 12 h of darkness) for 10 days [2]. Likewise, the diameter of the mycelium was measured every 12 h with a digital vernier caliper (CD-6 Mitutoyo) to estimate the growth rate (mm d^{-1}) [30].

2.4. DNA Extraction, PCR Amplification, and Sequencing

DNA was extracted from conidia, conidiophores, and mycelium of isolates obtained with morphological characteristics associated with *F. solani*. This procedure was performed with 2% cetyl trimethylammonium bromide (CTAB) [31]. Genomic DNA was suspended in 100 μL of sterile Milli-Q water and quantified by spectrophotometry in a Nano Drop 2000c (Thermo Scientific, Waltham, MA, USA). To determine the DNA quality, absorbance values between 1.8 and 2.2 at $A2_{80}/_{260}$ and $A_{230}/_{260}$ nm were considered acceptable. Finally, the DNA was diluted to 20 ng μL^{-1} and then stored at −20 °C for until further processing.

From the most representative colonies present in diseased strawberry plants, isolate 20 from plot 1 was chosen and named strain MA-FC120. Molecular identification of strain MA-FC120 was carried out based on the analysis of internal transcribed spacer (ITS) region sequences using primer pairs ITS5/ITS4 [32], and by sequencing a part of translation elongation factor-1 alpha (EF-1α) using primers EF-1 and EF-2 [16]. Amplified PCR products were verified by electrophoresis on a 1.5% agarose gel (Seakem, CA, USA). They were then purified and ligated into pGEM T-Easy Vector (Promega) and bidirectionally sequenced by Macrogen, Seoul, Korea. Sequences were assembled and edited using SeqMan (DNAStar, Madison, WI, USA) and compared to established sequences in GenBank™ using the Blast algorithm.

Phylogenetic analyses were performed with concatenated sequence alignment of genes ITS/TEF-1α using MEGA X software [33]. The analysis involved 11 nucleotide sequences obtained from the GenBank™ database (Table 1). Sequences from the ITS and TEF-1α regions of *Fusarium oxysporum* were used as an outgroup. Evolutionary history was inferred using the maximum likelihood method and the Tamura–Nei model [34], applying the Neighbor-Join and BioNJ algorithms to a distance matrix using the all-sites option. There was a total of 17,513 pb in the final dataset per 1000 bootstrap replicates to assess relative branch stability.

2.5. Pathogenicity Tests

Fifty Camino Real strawberry plants provided by local farmers were used. Each two-month-old plant was planted individually in a 1 L plastic pot containing a sterilized mixture of Peatmoss and Agrellite (1:1 *v*/*v*). Inoculation of strain MA-FC120 was performed by direct spray to runoff with 1×10^6 conidia/mL conidia suspension from pure cultures, where they had already developed macroconidia; these were taken using a 0.5–1 μL micropipette in a laminar flow hood, added to 10 mL of sterile physiological saline solution and incubated for 7 days at 25 °C [30]. Plant development was carried out under greenhouse conditions (25 °C and 70% relative humidity) until the appearance of disease symptoms. For the case of the control group, 50 healthy plants were used that were inoculated by direct spray to runoff with sterile water and kept under the same greenhouse conditions.

Table 1. *Fusarium solani* species complex gene (FSSC) bank sequences to corroborate the identity of strain MA-FC120.

Species (FSSC)	Strain	Isolation Source	Country	No. Access	
				ITS	EF-1α
Fusarium solani	MA-FC120	Strawberry	México	OM473287	OM616884
Fusarium solani	SVY-402-1	Pea	EU	KJ437436	KM044428
Fusarium solani	GuangX17	Bitter grand	China	KY785014	KY785024
Fusarium solani	CB24-4	Ginseng	China	MN637861	MN650097
Fusarium solani	MJ	Strawberry	Spain	MH300474	MH300509
Fusarium solani	MY	Strawberry	Spain	MH300482	MH300517
Fusarium solani	MS	Strawberry	Spain	MH300452	MH300523
Fusarium solani	MRC 2635	Wheat	India	MH582403	MH582423
Fusarium solani	MRC 2805	Wheat	India	MH582404	MH582424
Fusarium solani	NRRL 52778	Chinch	Syria	JF740931	JF740846
Fusarium oxysporum	Esm3	Golden berry	Colombia	KJ936621	JX465113

2.6. In Vitro Sensitivity Bioassay

The efficiency of a systemic fungicide Benomyl 50® and three protectors, Talonil 75®, Mancosol 80®, and Captan 50® with three concentrations per product was evaluated: (a) commercial dose authorized by the manufacturer; (b) half the recommended dose, and; (c) double the commercial dose expressed in mg L^{-1} (Table 2).

Table 2. Fungicides at different concentrations evaluated.

Fungicide (Commercial Name)	Active Ingredient	Molecular Formula	Concentration (mg L^{-1})		
			Low	Recommended	High
Control	Water	H_2O	-	-	-
Captan 50®	Captan	$C_9H_8Cl_3NO_2S$	450	900	1350
Mancosol 80®	Mancozeb	$C_4H_6MnN_2S_4$	600	1200	1800
Talonil 75®	Chlorothalonil	$C_8Cl_4N_2$	450	900	1350
Benomyl 50®	Benomyl	$C_{14}H_{18}N_4O_3$	250	500	750

2.7. Mycelial-Growth Inhibition

The controlled-poisoning technique [35] was used, which consisted of placing 5 mm-diameter discs with mycelium of pathogens in the center of the Petri dish with potato dextrose agar medium (PDA, Bioxon, Becton Dickinson and Company, Queretaro, Mexico) plus fungicide at different concentrations (Table 2), which were incubated in dark conditions at 28 °C for 10 days. Mycelial diameter was measured every 12 h with a digital vernier (CD-6 Mitutoyo) to estimate growth rate (mm d^{-1}), which was calculated with the linear growth function [30] Equation (1), obtaining the average of four measurements of the longitudinal diameter per experimental unit.

$$Y = MX + B \quad (1)$$

where:

Y = distance
M = pending
X = time
B = the constant factor.

As a control, pathogen discs were used on PDA medium without fungicides. Evaluation was concluded when the mycelium of the pathogen completely covered the plate of control treatment. Each treatment consisted of three replicates and was carried out in duplicate in a completely randomized experimental design [36]. The mycelial-growth inhibition percentage (I%) was calculated with the Abbot formula [37] (2):

$$(I\%) = (C-T)/C \times 100 \quad (2)$$

where:

I% = percent growth reduction in *F. solani*
C = colony growth (cm) in the control
T = colony growth (cm) in the treatment.

To determine the toxicity according to inhibition percentages obtained, the results were classified based on the classification scale established by the IOBC [38] (Table 3).

Table 3. Toxicity classification scale [38] (Reproduced with permission from Viñuela et al., Phytoma; published by PHYTOMA-España, 1993).

Growth Inhibition (%)	Classification
<30	Harmless
30–75	Slightly toxic
75–90	Moderately toxic
>90	Toxic

2.8. Statistical Analysis

Statistical analyses were performed with IBM SPSS Statistics version 25 software, using a completely randomized factorial design. The variable (I%) was expressed as a percentage and transformed with angular arccosine $\sqrt{x} + 1$. Data were analyzed with analysis of variance (MULTIVARIANT ANOVA) using a quadratic model of response as a function of mycelial-growth inhibition percentage (I%), development rate (mm h^{-1}), and growth rate (mm d^{-1}) to determine significant differences among treatments, under the following mathematical model: (3):

$$I\gamma j = \mu + ti + \varepsilon ij \quad (3)$$

where:

$I\gamma j$ = value of response variable of experimental unit associated with the γ-th treatment, and to the j-th repetition
μ = corresponds to general average of response variable in the experiment
ti = effect of γ-th treatment
$\varepsilon\gamma j$ = error of experimental unit associated with the γ-th treatment
j-th = repetition
$\gamma = 1, 2, 3, 4 \ldots 13$
j = 1, 2, 3
I = variable under study (I%).

3. Results

3.1. Isolation, Characterization, and Identification of Causal Agent of Wilting on Strawberry Crops

Ten representative isolates from 50 different plants developed pinkish-white colonies with granular and powdery textures after 10 days of incubation; no aerial mycelium was observed. Soma consisted of septate hyphae 2.579–9.58 (5.789) µm wide (n = 50), and some false heads 18.497–72.566 (29.906) µm in diameter were visualized. Kidney-shaped oval microconidia 9.178–14.791 (12.78) µm long × 3.649–4.473 (4.232) µm wide were observed (n = 50). Spindle-shaped, slightly curved macroconidia 15.054–33.729 (21.33) µm long × 4.482–7.336 (5.256) µm wide (n = 50) were also observed (Figure 1A–F).

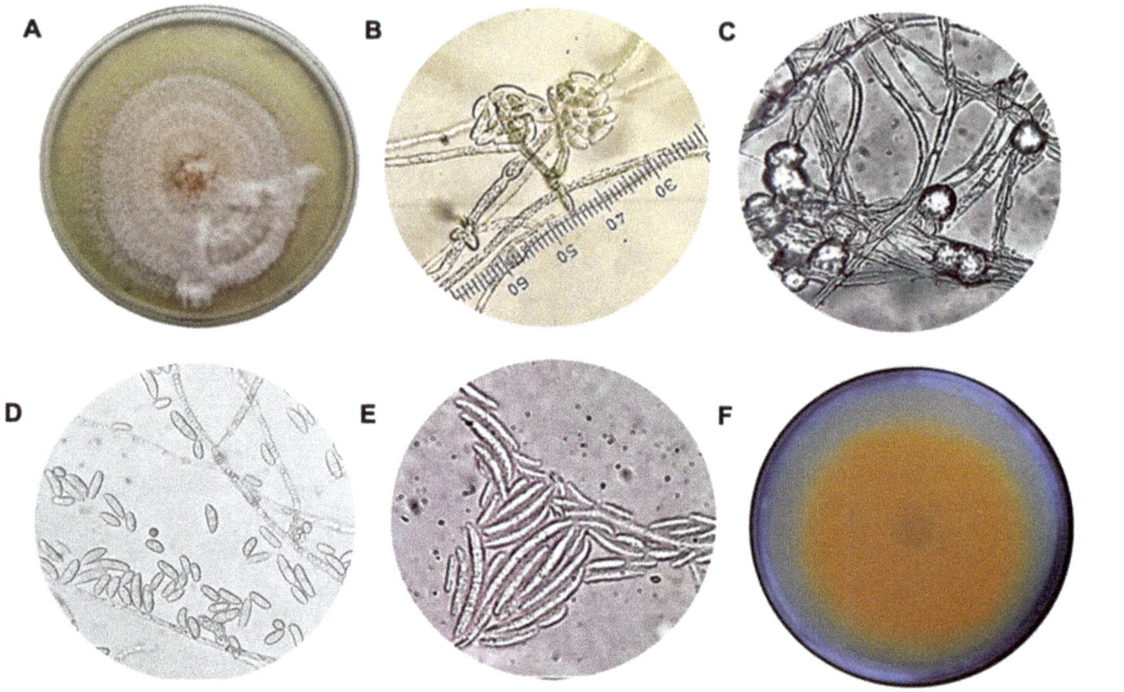

Figure 1. Morphology of the MA-FC120 strain: (**A**) Appearance of colony on PDA culture medium; (**B**) False heads of microconidia held in monophialides seen at 40× and 1000×; (**C**) Chlamydospora seen at 1000×; (**D**,**E**) microconidia and macroconidia seen at 400× and 100×, respectively; (**F**) Colony reverse staining on PDA.

Of the 10 characteristic isolates of *F. solani* from diseased strawberry plants, isolate MA-FC120 was chosen based on its high rate of development and speed of growth in three culture media tested (PDA, TSA, and CWA). The results obtained from the alignment of ITS gene sequences (551 bp) presented 100% identity with *F. solani*, as well as the elongation factor EF-1α (686 bp) confirmed the identity with *F. solani*, obtaining between 100 and 99, 85% identity, from data obtained from the nucleotide base of the GenBank™ of the National Center for Biotechnology Information.

The phylogenetic analysis generated from concatenated ITS/EF-1α sequences (Figure 2) clearly distinguishes at the species level, showing that strain MA-FC120 corresponds to the *Fusarium solani* species complex (FSSC).

3.2. Pathogenicity Tests

Koch's postulates confirmed that the MA-FC120 strain belonging to the *F. solani* species complex (FSSC) produced typical symptoms of desiccation, wilting, and root rot 20 days after inoculation. In addition, ascending wilt (Figure 3C), loss of turgor, and generalized vascular wilt (Figure 3B) were observed, finally leading to death of the entire plant 45 days after inoculation. No disease symptoms were observed in the control group (Figure 3A).

3.3. In Vitro Sensitivity Bioassay and Mycelial-Growth Inhibition

The treatments under study showed highly significant differences (p = 0.0001) in colony diameter, growth rate, and percentage inhibition, as well as interaction of the active ingredient of the different fungicide concentrations, at a 95% confidence level (Table 4).

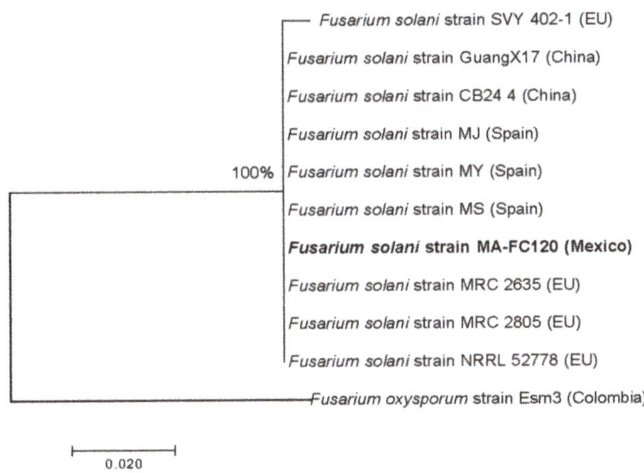

Figure 2. The optimal tree with the sum of branch length = 0.1818 is shown. The tree is drawn to scale, with branch lengths in the same units as those of the evolutionary distances used to infer the phylogenetic tree generated from concatenated ITS/EF-1α sequences. *Fusarium oxysporum* was used as the outgroup. The MA-FC120 strains obtained in this study are shown in bold black.

Figure 3. Pathogenicity tests on strawberry plants (var. Camino Real): (**A**) healthy control plant without symptoms of drying, (**B**) plant with wilting and death of foliage, (**C**) cross-section of strawberry root showing ascending root rot.

Table 5 shows the summary of the MULTIVARIATE ANOVA analysis for the quadratic response surface model, finding highly significant statistical differences. Model value for F = 523.500 (Treatments), 0.310 (Rate of Development), and 330.807 (Growth Rate), which implies a highly significant model for the percentage inhibition of mycelial growth (I%) in vitro.

All fungicides inhibited the growth rate of strain MA-FC120 compared to the control. However, systemic fungicide Benomyl 50® was the most effective product, presenting highly significant statistical differences with the protectant fungicides, inhibiting 100% of fungal development from the first days of incubation in the three doses applied (Figure 3).

The growth rate in the negative control group was 6.7104 mm $d^{-1.}$ It was observed that, with the different concentrations of fungicides, it decreased considerably; however, the systemic fungicide Benomyl 50® did not show growth after 10 days of incubation in the three doses applied (Table 4). The protective fungicide Talonil 75® at double the commercial dose (1350 mg L^{-1}) showed an inhibition percentage of 75.70% (Table 4) on mycelial growth and development. However, the protective fungicides Captan 50® and Mancosol 80® showed an inhibition percentage of 50% on the mycelial growth and development of strain MA-FC120 under in vitro conditions after ten days of incubation. It is worth mentioning that the fungicide Mancosol 80® at the lowest dose of 600 mg L^{-1} showed a harmless effect on mycelial growth and development, obtaining 10.73%. Inhibition percentage (Table 4). In general, as the dose of the fungicides increased, an increase in the percentage of inhibition was observed.

Table 4. In vitro efficacy of four fungicides on the rate of development, speed of growth, and percentage of inhibition of mycelial growth of the MA-FC120 strain.

Fungicides (Commercial name)	Concentration (mg L^{-1})	Development Rate * (mm h^{-1})	Growth Rate * (mm d^{-1})	I%
Control	-	0.2873 ± 0.0082 ^{d}	6.7104 ± 0.0888 ^{g}	0.00 ^{h}
Captan 50®	450	0.1557 ± 0.0008 ^{bc}	3.5296 ± 0.0833 ^{d}	47.54 ± 3.12167 ^{e}
	900	0.1426 ± 0.0077 ^{bc}	3.2432 ± 0.1408 ^{d}	51.71 ± 1.31683 ^{d}
	1350	0.1366 ± 0.0030 ^{bc}	3.1600 ± 0.0806 ^{d}	52.90 ± 1.57778 ^{d}
Mancosol 80®	600	0.2632 ± 0.0076 ^{d}	5.9992 ± 0.0847 ^{f}	10.73 ± 4.3486 ^{g}
	1200	0.1658 ± 0.0114 ^{c}	4.4512 ± 0.3004 ^{e}	33.68 ± 5.1496 ^{f}
	1800	0.1648 ± 0.0011 ^{c}	4.3376 ± 0.0237 ^{e}	35.46 ± 1.64257 ^{f}
Talonil 75®	450	0.1427 ± 0.0279 ^{bc}	4.7025 ± 0.7000 ^{e}	29.95 ± 7.43654 ^{f}
	900	0.0961 ± 0.0223 ^{b}	2.4399 ± 0.5936 ^{c}	63.48 ± 6.91699 ^{c}
	1350	0.0923 ± 0.0244 ^{b}	1.6305 ± 0.5031 ^{b}	75.70 ± 9.65249 ^{b}
Benomyl 50®	250	0.00 ± 0.00 ^{a}	0.00 ± 0.00 ^{a}	100.00 ^{a}
	500	0.00 ± 0.00 ^{a}	0.00 ± 0.00 ^{a}	100.00 ^{a}
	750	0.00 ± 0.00 ^{a}	0.00 ± 0.00 ^{a}	100.00 ^{a}

* Means followed by the same letter are not significantly different for $p \leq 0.05$ according to Tukey test.

Table 5. Quadratic model of response surface for different concentrations of fungicides on rate of development, speed of growth, and percentage of inhibition of mycelial growth (I%) of the MA-FC120 strain in vitro.

Origin		Sum of Squares (Type III)	gL	Mean Square	F	Sig.
Corrected model	X_1 = Treatments	523.500 ^{a}	26	20.135	10.738	<0.001
	X_2 = Rate of Development	0.310 ^{b}	26	0.012	202.898	<0.001
	X_3 = Growth Rate	330.807 ^{c}	26	12.723	133.673	<0.001
Intersection	X_1	782.759	1	782.759	417.471	<0.001
	X_2	0.621	1	0.621	10580.715	<0.001
	X_3	477.343	1	477.343	5015.036	<0.001
I (%)	X_1	523.500	26	20.135	10.738	<0.001
	X_2	0.310	26	0.012	202.898	<0.001
	X_3	330.807	26	12.723	133.673	<0.001
Error	X_1	22.500	12	1.875		
	X_2	0.001	12	5.874×10^{-5}		
	X_3	1.142	12	0.095		
Total	X_1	1950.000	39			
	X_2	0.937	39			
	X_3	782.328	39			
Total corrected	X_1	546.000	38			
	X_2	0.311	38			
	X_3	331.949	38			

(a) R-squared = 0.959 (R-squared adjusted = 0.870); (b) R-squared = 0.998 (R-squared adjusted = 0.993); (c) R-squared = 0.997 (R-squared adjusted = 0.989).

The effect of fungicides on mycelial development and morphology of fungal colonies of strain MA-FC120 was corroborated after 10 days of incubation with respect to control, showing a marked inhibition of mycelial development and pronounced radial growth deformation compared to the control group (Figure 3). However, strain MA-FC120 showed higher resistance to Mancosol 80® and Talonil 75® at lower doses (Figure 3). At the 600 mg L^{-1} concentration, the strain reached a 5.9992 mm d^{-1} development rate, while at the 450 mg L^{-1} concentration, strain MA-FC120 reached a 4.7025 mm d^{-1} development rate, in both cases presenting a harmless effect on the classification scale established by IOBC [38] (Figure 4).

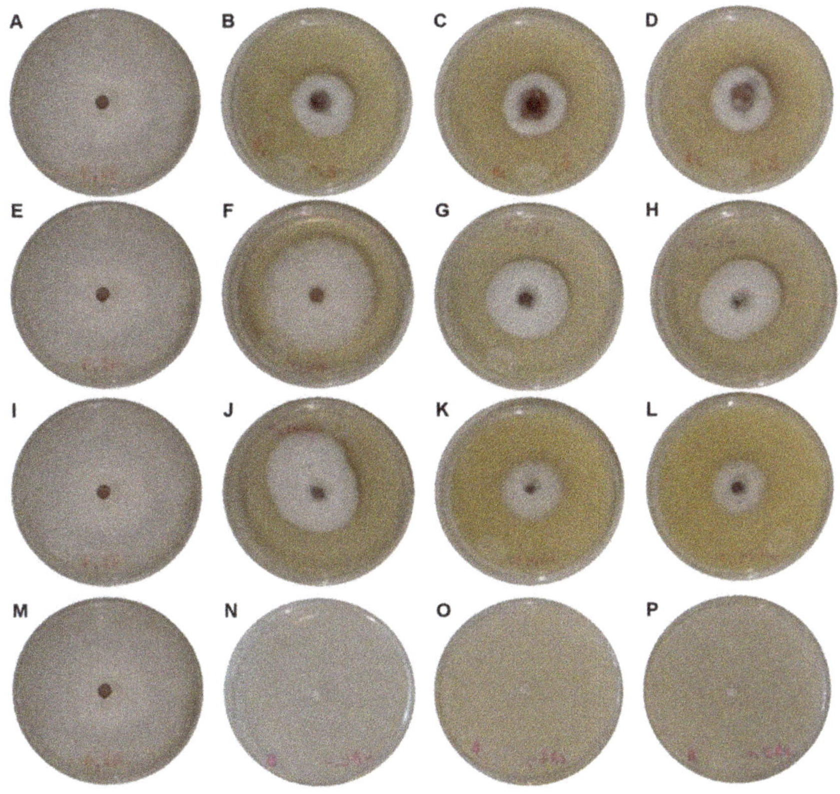

Figure 4. Strain MA-FC120 (FSSC) in culture medium with fungicide. (**A**,**E**,**I**,**M**) Controls; Captan 50®: (**B**) 450 mg L^{-1}, (**C**) 900 mg L^{-1}, (**D**) 1350 mg L^{-1}; Mancosol 80®: (**F**) 600 mg L^{-1}, (**G**) 1200 mg L^{-1}, (**H**) 1800 mg L^{-1}; Talonil 75®: (**J**) 450 mg L^{-1}, (**K**) 900 mg L^{-1}, (**L**) 1350 mg L^{-1}; Benomyl 50®: (**N**) 250 mg L^{-1}, (**O**) 500 mg L^{-1}, and (**P**) 750 mg L^{-1}.

At the concentrations evaluated and according to the IOBC [38] classification scale, the protective fungicides Mancosol 80® and Captan 50® were slightly toxic; however, fungicide Talonil 75® at a dose of 1350 mg L^{-1} showed moderate toxicity (Figure 4).

4. Discussion

According to the morphological and taxonomic characteristics corresponding to *F. solani* [19,39,40], it was confirmed that the strain MA-FC120 belongs to this fungal genus. Likewise, the phylogenetic analysis of the concatenated sequences (ITS-EF-1α) shows 100% identity with the species *F. solani*. In this regard, Kurt et al. [41] used ITS and EF-1α molecular markers to confirm the identity of CFs4 (ITS: MF972071; EF-1α: MF972074) and CFs8 (ITS: MF972072; EF-1α: MF972075) strains, with both strains belonging to the *F. solani* species complex (FSSC), associated with citrus dry root rot in the eastern Mediterranean region of Turkey; the same molecular markers were used in the present study.

The phytopathogen *F. solani* can affect more than 100 plant species of agricultural interest, including strawberry crops [42]. It has been reported as a pathogen that causes drying, stunting, wilting, necrosis, and death in strawberry plants [12]. It was detected for the first time in 2012, in strawberry nursery plants with symptoms of crown rot similar to those found in the present study.

Pathogenicity tests revealed similar results to those already reported by Pastrana et al. [12] on strawberry (*Fragaria* × *ananassa*) plants. Symptoms consisted of wilting of foliage, stunting, drying, and death of plants. Villarino et al. [16] reported vascular wilt, chlorosis, and root rot in strawberry crops. They observed necrotic roots, gray-green wilted leaves, and plant death, symptoms similar to those observed in this study. Fungi of genus *Fusarium* can invade the vascular tissues of plants, impede water transport through the Xylem by inducing vessel blockage and causing wilting of foliage [43]. In addition, *F. solani* concentrates in the collar area, causing rotting and subsequently plant death [44].

The lack of research limits the ability of growers to choose effective products and increases the risk of generating strains resistant to the active ingredients used in fungicides. However, the traditional method of control against pathogenic fungi is still chemical pesticides [45]. Likewise, management of *Fusarium* wilting infection is mainly performed through chemical soil fumigation for strawberry cultivation [46].

Fungicides are commonly used in conventional agriculture to protect strawberries from field rot and improve yield [47]. Da Silva et al. [48] studied in vitro the effect of Mancozeb® on *F. solani* and reported that this pesticide inhibited 42.5% of mycelial growth of pathogens. Another study by Kumar-Gupta et al. [49] reported the effect of Mancozeb® on isolates of *F. solani*, the causal agent of root rot in papaya obtaining 61.46% mycelial-growth inhibition. Both studies presented higher values than those reported in the present research. According to FRAC (2018) [50], Mancozeb® has a multisite mode of action; therefore, it is broad-spectrum and protective, it acts by forming compounds that bind with sulfhydryl groups of amino acids and causes denaturation of proteins and enzymes, and, consequently, suppresses pathogen growth. However, a study by Andrabi et al. [51] reported the effect of Mancozeb® on *F. solani*, the causal agent of root rot in chickpea, where they obtained 54.82% inhibition of mycelial growth, with a concentration of 500 ppm, values similar to those reported in present investigation and considered slightly toxic on the IOBC [38] classification scale. Dithiocarbamate fungicides such as Mancozeb® are among the most widely used pesticides today in the control of a wide variety of diseases in seeds, fruits, and vegetables. However, there are several studies where the fungicide Mancozeb® is related to health diseases such as Parkinson's, teratogenesis, and carcinogenesis, for which reason it is prohibited in several countries [52]. Due to its high toxicity, the minimum purity proposed by the EU NMS is 850 g/kg, while the minimum purity proposed by Agria is 915 g/kg [53].

In the case of the fungicide Captan®, Ayvar-Serna et al. [54] observed that, under in vitro conditions, it inhibited 100% of mycelial growth of *F. solani*, the causal agent of root rot in tomato, at a concentration of 0.044 g. Another study by Shah et al. [55] reported that this pesticide inhibited 75.90% of mycelial growth of pathogens at a concentration of 1000 $\mu g\ mL^{-1}$ in vitro. Both studies presented higher values than those reported in the present investigation. Fungicide Captan® presents a fungitoxic compound because it affects energy transport in mitochondria and the physiological functioning of membranes in fungal cells, obstructing the activity of sulfhydryl enzymes and resulting in the release of thiophosgene ($CSCl_2$), a highly toxic compound that kills pathogens [50].

The present study investigated the fungistatic ability of Chlorothalonil®, which is a foliar fungicide with widespread use worldwide [56]. It is a non-systemic organochlorine fungicide. Specifically it is a broad-spectrum polychlorinated aromatic that causes mycelial-growth retardation and inhibits spore germination. It acts on respiration of fungal cells;

that is, it binds to sulfhydryl groups of some amino acids that affect the Krebs cycle by reducing ATP synthesis, causing cell death [57].

Studies by Dwivedi et al. [58] reported that fungicide Chlorothalonil® reduced mycelial growth of *F. solani* by 82.34%, in pathogens isolated from roots in eggplant crops, and these values are similar to those reported in the present investigation, considered moderately toxic on the IOBC [38] rating scale for the highest concentration (1350 mg L^{-1}). Bhaliya et al. [59] reported that coriander root rot caused by *F. solani* was controlled by fungicide Chlorothalonil® by 69.96%. A similar finding was observed by Madhusndhan et al. [60] for *F. solani*, the causal agent of root rot in sweet orange, obtaining 62.82% mycelial-growth inhibition, with a concentration of 250 ppm, with values similar to those reported in the present investigation and considered slightly toxic on the IOBC [38] classification scale for the recommended concentration (900 mg L^{-1}). For this reason, Salazar et al. [61] report its extensive use in strawberry cultivation under greenhouse conditions in Mexico.

The disappearance of honey bees is only the most visible part of a widespread phenomenon due to massive loss of biodiversity that negatively impacts agroecosystems [62]. The fungicide Chlorothalonil has been shown to alter the intestinal bacterial communities of honey bees, in addition to decreasing larval survival in the hive [63]. However, growers continue to rely on the chemical, without considering the negative effects for the future.

According to the literature, Benomyl® is a systemic fungicide and is banned in developed countries., However, its use is very frequent in developing countries [64], as in the case of Mexico. It is a member of the benzimidazole class, which penetrates the fungal cell and reacts with proteins forming chelates; it also alters intracellular redox balance, inducing oxidative stress, and is a potent inhibitor of DNA synthesis in fungi [65,66].

Recently, Asadboland et al. [67] reported 76% inhibition of mycelial growth of *F. solani* (IRAN 11C) by 0.1 g/mL Benomyl® under in vitro conditions on PDA medium. In this sense, Kordali et al. [68] also reported a reduction in mycelial growth on a strain of *F. solani* from Atatürk University, obtaining 79.8% inhibition with a concentration of 1 mg/mL. However, Dorugade et al. [69] observed that, under in vitro conditions, the fungicide Benomyl® (100 ppm) inhibited 100% mycelial growth of *F. solani* (Fs-3), the causal agent of root rot on elephant-foot sweet potato. These values are like those reported in the present investigation, considered toxic on the IOBC [38] rating scale for the three concentrations used (Figures 4 and 5).

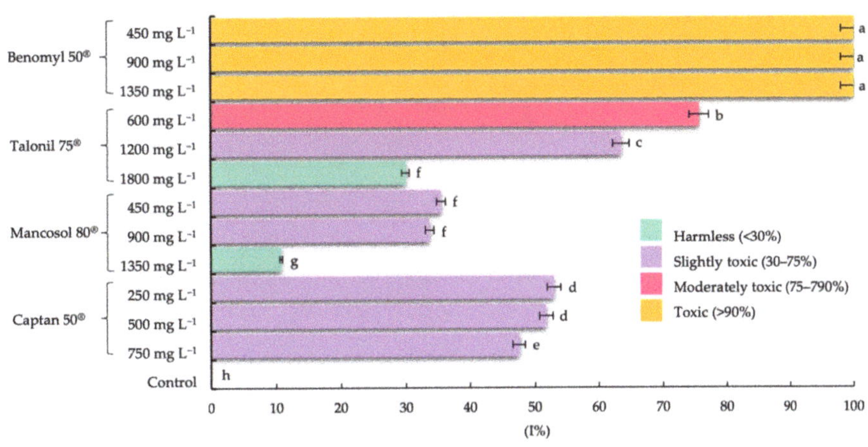

Figure 5. The effect of four fungicides at different concentrations according to the classification scale established by IOBC. Means followed by the same letter are not significantly different for $p \leq 0.05$ according to Tukey test. [Reproduced with permission from Viñuela et al., Phytoma; published by PHYTOMA-España, 1993].

Despite advances in international regulation, considerable information gaps and uncertainties remain about the risks of pesticides, particularly for low- and middle-income countries compared to high-income countries [70].

5. Conclusions

The MA-FC120 strain belongs to the *Fusarium solani* species complex (FSSC) and is associated with the wilting and root rot of the Camino Real variety of the strawberry crop, located in the rural area of Santa Cruz Analco, belonging to the municipality of San Salvador el Green Puebla-Mexico. The fungicides Mancosol 80® and Talonil 75® at low concentration (600 and 450 mg L^{-1}) showed no toxicity, being harmless to strain MA-FC120. However, Talonil 75® showed slight toxicity at the dose recommended by the manufacturer and moderate toxicity at high concentration (1350 mg L^{-1}). Likewise, Captan 50® fungicide, in its three concentrations evaluated, showed slight toxicity, obtaining around 50% on the classification scale established by IOBC. Unlike the protective fungicides, systemic fungicide Benomyl 50® showed 100% inhibition of mycelial growth of *F. solani* at the three concentrations evaluated. This research represents the first study of the most widely used fungicides for strawberry production in Mexico and their effect on *F. solani*, obtaining results that can lead to reduction in chemical products and generate less resistance to persistent fungal strains of the strawberry crop variety "Camino Real", from the Santa Cruz Analco community, Puebla-Mexico.

Author Contributions: Conceptualization, O.R.-A., C.P.L., and A.C.C.; methodology, A.R.T., A.C.C., and O.R.-A.; software, O.S.T., and O.R.-A.; validation, Y.P.H., A.R.T., and O.R.-A.; formal analysis, A.C.C., C.P.L., and O.R.-A.; resources, O.R.-A. and C.P.L.; original—draft preparation, Y.P.H., O.S.T., and O.R.-A.; writing—review and editing, A.R.T. and O.R.-A.; visualization, O.R.-A. and A.C.C.; supervision, O.S.T.; project administration, O.R.-A.; funding acquisition, O.R.-A. All authors have read and agreed to the published version of the manuscript.

Funding: This research was supported by the program PRODEP 2020 of the Secretaría de educación Pública of Mexico (SEP); the Consejo Nacional de Ciencia y Tecnología (CONACyT) number 72760 and Benémerita Universidad Autónoma of Puebla, number 100420500.

Institutional Review Board Statement: Not applicable.

Informed Consent Statement: Not applicable.

Data Availability Statement: Informed consent was obtained from all subjects involved in the study.

Acknowledgments: The authors are grateful to Consejo Nacional de Ciencia y Tecnología (CONACyT) and laboratory 204 of the Center for Agroecology at Benémerita Universidad Autónoma of Puebla.

Conflicts of Interest: The authors declare no conflict of interest.

References

1. Sharma, V.K.; Godara, A.K. Growth responses of strawberry (Fragaria × ananassa duch.) plants grown at different planting density using pvc pipe under protected cultivation. *Bangladesh J. Bot.* **2019**, *48*, 1–7. [CrossRef]
2. Morales-Mora, L.A.; Andrade-Hoyos, P.; Valencia-de Ita, M.; Romero-Arenas, O.; Silva-Rojas, H.V.; Contreras-Paredes, C.A. Caracterización de hongos asociados al cultivo de fresa y efecto antagonista in vitro de Trichoderma harzianum. *Rev. Mex. Fitopatol.* **2020**, *38*, 434–449.
3. Yadav, R.B. Potential benefits of berries and their bioactive compounds as functional food component and immune boosting food. In *Immunity Boosting Functional Foods to Combat COVID-19*, 1st ed.; Giri, A., Ed.; CRC Press: London, UK, 2021; pp. 75–90. [CrossRef]
4. U.S. Department of Agriculture (USDA). National Agricultural Statistics Service. Available online: https://downloads.usda.library.cornell.edu/usda-esmis/files/zs25x846c/sf269213r/6t054c23t/ncit0521.pdf (accessed on 21 January 2022).
5. Antunes, L.E.C.; Reisser Juniory, C.; Bonow, S. Morango produção aumenta ano a ano. *Anuário HF* **2021**, *1*, 87–90.
6. Organización de las Naciones Unidas para la Alimentación y la Agricultura (FAO). Available online: https://www.fao.org/faostat/es/#data/QCL (accessed on 21 January 2022).
7. Servicio de Información Agroalimentaria y Pesquera (SIAP). Available online: https://nube.siap.gob.mx/gobmx_publicaciones_siap/pag/2020/Atlas-Agroalimentario-2020 (accessed on 21 January 2022).

8. Ceja-Torres, L.F.; Mora-Aguilera, G.; Téliz, D.; Mora-Aguilera, A.; Sánchez-García, P.; Muñoz-Ruíz, C.; Tlapal-Bolaños, B.; De La Torre-Almaraz, R. Ocurrencia de hongos y etiología de la secadera de la fresa con diferentes sistemas de manejo agronómico. *Agrociencia* **2008**, *42*, 451–461.
9. Martínez, F.; Castillo, S.; Carmona, E.; Avilés, M. Dissemination of *Phytophthora cactorum*, cause of crown rot in strawberry, in open and closed soilless growing systems and the potential for control using slow sand filtration. *Sci. Hortic.* **2010**, *1254*, 756–760. [CrossRef]
10. Abd-Elgawad, M.M.M.; Elshahawy, I.E.; Abd-El-Kareem, F. Eficacia de la solarización del suelo en la enfermedad de la podredumbre de la raíz negra y especulación sobre su influencia en nematodos y malezas de fresa en Egipto. *Bull. Natl. Res. Cent.* **2019**, *43*, 175. [CrossRef]
11. Bárcenas-Santana, D.; Guillén-Sánchez, D.; Yazmín-Basaldua, C.; Ramos-García, M.L.; Valle-de la Paz, M. Etiology Strawberry dry wilt (Fragaria spp.) in Morelos, Mexico. *Rev. Mex. Fitopatol.* **2019**, *37*, 454–463. [CrossRef]
12. Pastrana, A.M.; Capote, N.; De los Santos, B.; Romero, F.; Basallote-Ureba, M.J. First report of *Fusarium solani* causing crown and root rot on strawberry crops in southwestern Spain. *Plant Dis.* **2014**, *98*, 161. [CrossRef]
13. De la Lastra, E.; Villarino, M.; Astacio, J.D.; Larena, I.; De Cal, A.; Capote, N. Genetic diversity and vegetative compatibility of *Fusarium solani* species complex of strawberry in Spain. *Phytopathology* **2019**, *109*, 2142–2151. [CrossRef]
14. Summerell, B.A.; Leslie, J.F. Fifty years of Fusarium: How could nine species have ever been enough? *Fungal Divers.* **2011**, *50*, 135–144. [CrossRef]
15. Chitrampalam, P.; Abraham, N.; Nelson, B.D.A., Jr. Culture-independent PCR-based assay to detect the root rot pathogen *Fusarium solani* species complex 11 from soybean roots and soil. *Plant Dis.* **2018**, *102*, 327–333. [CrossRef] [PubMed]
16. Villarino, M.; De la Lastra, E.; Basallote-Ureba, M.J.; Capote, N.; Larena, I.; Melgarejo, P.; De Cal, A. Characterization of *Fusarium solani* populations associated with Spanish strawberry crops. *Plant Dis.* **2019**, *103*, 1974–1982. [CrossRef] [PubMed]
17. Manici, L.M.; Caputo, F.; Baruzzi, G. Additional experiences to elucidate the microbial component of soil suppressiveness towards strawberry black root rot complex. *Ann. App. Biol.* **2005**, *146*, 421–431. [CrossRef]
18. Ayoubi, N.; Soleimani, M.J. Morphological and molecular identification of pathogenic *Fusarium* spp. on strawberry in Iran. *Sydowia* **2016**, *68*, 163–171.
19. Mehmood, N.; Riaz, A.; Jabeen, N.; Anwaar, S.; Hussain, S.Z.; Abbas, M.F.; Zhang, X.; Rosli, H.; Gleason, M.L. First report of *Fusarium solani* causing fruit rot of strawberry in Pakistan. *Plant Dis.* **2017**, *101*, 1681. [CrossRef]
20. Nel, B.; Steinberg, C.; Labuschagne, N.; Viljoen, A. Evaluation of fungicides and sterilants for potential application in the management of Fusarium wilt of banana. *Crop Prot.* **2007**, *26*, 697–705. [CrossRef]
21. Tarekegn, G.; Sakhuja, P.K.; Swart, W.J.; Tamado, T. Integrated management of groundnut root rot using seed quality and fungicide seed treatment. *Int. J. Pest Manag.* **2007**, *53*, 53–57. [CrossRef]
22. Madhavi, G.B.; Bhattiprolu, S.L. Evaluation of Fungicides, Soil Amendment Practices and Bioagents against *Fusarium solani*—Causal Agent of Wilt Disease in Chilli. *J. Hortic. Sci.* **2011**, *6*, 141–144.
23. Bejarano, G.F. (Ed.) *Los Plaguicidas Altamente Peligrosos en México*; Red de Acción sobre Plaguicidas y Alternativas en México, A.C. (RAPAM): Texcoco, México, 2017; pp. 59–110.
24. García, H.J.; Leyva, M.J.B.; Martínez, R.I.E.; Hernández, O.M.I.; Aldana, M.M.L.; Rojas, G.A.E.; Betancourt, L.M.; Perez, H.N.E.; Perera, R.J.H. Estado actual de la investigación sobre plaguicidas en México. *Rev. Int. De Contam. Ambient.* **2018**, *34*, 29–60. [CrossRef]
25. Qian, H.; Du, J.; Chi, M.; Sun, X.; Liang, W.; Huang, J.; Li, B. The Y137H mutation in the cytochrome P450 FgCYP51B protein confers reduced sensitivity to tebuconazole in Fusarium graminearum. *Pest Manag. Sci.* **2018**, *74*, 1472–1477. [CrossRef]
26. García, E. *Modificaciones al Sistema de Clasificación Climática de Köppen*, 5th ed.; Instituto de Geografía, Universidad Autónoma de México: México City, México, 2004; pp. 11–90.
27. Barnett, H.L.; Hunter, B.B. *Illustrated Genera of Imperfect Fungi*, 4th ed.; The American Phytopatological Society: St. Paul, MN, USA, 2006; pp. 1–197.
28. Báez-Vallejo, N.; Camarena-Pozos, D.A.; Monribot-Villanueva, J.L.; Ramírez-Vázquez, M.; Carrión-Villarnovo, G.J.; Guerrero-Analco, J.A.; Partida-Martínez, L.P.; Reverchon, F. Forest tree associated bacteria for potential biological control of *Fusarium solani* and of *Fusarium kuroshium*, causal agent of *Fusarium* dieback. *Microbiol. Res.* **2020**, *235*, 126440. [CrossRef] [PubMed]
29. Fisher, N.L.; Marasas, W.F.O.; Toussoun, T.A. Taxonomic importance of microconidial chains in Fusarium section Liseola and effects of water potential on their formation. *Mycología* **1983**, *74*, 693–698. [CrossRef]
30. Morales-Mora, L.A.; Martínez-Salgado, S.J.; Andrade-Hoyos, P.; Valencia de Ita, M.A.; Silva-Rojas, H.V.; Romero-Arenas, O. First report of leaf spot and anthracnosis caused by *Pestalotiopsis* sp., on strawberry in Puebla, Mexico. *Plant Dis.* **2019**, *103*, 2668. [CrossRef]
31. Rivera-Jiménez, M.N.; Zavaleta-Mancera, H.A.; Rebollar-Alviter, A.; Aguilar-Rincón, V.H.; García-de-los-Santos, G.; Vaquera-Huerta, H.; Silva-Rojas, H.V. Phylogenetics and histology provide insight into damping-off infections of 'Poblano' pepper seedlings caused by Fusarium wilt in greenhouses. *Mycol. Prog.* **2018**, *17*, 1237–1249. [CrossRef]
32. White, T.J.; Bruns, T.; Lee, S.; Taylor, J. Amplification and direct sequencing of fungal ribosomal RNA genes for phylogenetics. In *PCR Protocols, A Guide to Methods and Applications*; Innis, M.A., Gelfand, D.H., Sninsky, J.J., White, T.J., Eds.; Academic Press: New York, NY, USA, 1990; Volume 18, pp. 315–322. [CrossRef]

33. Kumar, S.; Stecher, G.; Li, M.; Knyaz, C.; Tamura, K. MEGA X: Molecular evolutionary genetics analysis across computing platforms. *Mol. Biol. Evol.* **2018**, *35*, 1547. [CrossRef]
34. Stecher, G.; Tamura, K.; Kumar, S. Molecular evolutionary genetics analysis (MEGA) for macOS. *Mol. Biol. Evol.* **2020**, *37*, 1237–1239. [CrossRef]
35. Azza, R.E.; Hala, M.I.; Sanaa, A.M. The role of storage on Mancozeb fungicide formulations and their antifungal activity against *Fusarium oxysporium* and *Rhizoctonia solani*. *Arab. J. Chem.* **2021**, *14*, 103322. [CrossRef]
36. Ortiz-Martínez, L.E.; Robles-Yerena, L.; Leyva-Mir, S.G.; Camacho-Tapia, M.; Juárez-Rodríguez, L. *Fusarium* sp., causal agent of vascular wilt in citrus and its sensitivity to fungicides. *Rev. Mex. Fitopatol.* **2021**, *40*, 1–17. [CrossRef]
37. Miranda-Granados, J.; Chacón, C.; Ruiz-Lau, N.; Vargas-Díaz, M.E.; Zepeda, L.G.; Álvarez-Gutiérrez, P.; Meza-Gordillo, R.; Lagunas-Rivera, S. Alternative use of extracts of chipilín leaves (*Crotalaria longirostrata* Hook. & Arn) as antimicrobial. *Sustainability* **2018**, *10*, 883. [CrossRef]
38. Viñuela, E.; Jacas, J.A.; Marco, V.; Adrón, A.; Badia, F. Los efectos de los plaguicidas sobre los organismos beneficiosos en la agricultura y el grupo de trabajo de la OILB. Plaguicidas y organismos beneficiosos I. Insecticidas y acaricidas. *Phytoma* **1993**, *45*, 18–25.
39. Leslie, J.F.; Summerell, B.A. *The Fusarium Laboratory Manual*, 1st ed.; Blackwell Publishing Professional: Ames, IA, USA, 2006; pp. 3–278.
40. Šišić, A.; Baćanović-Šišić, J.; Al-Hatmi, A.M.S.; Karlovsky, P.; Ahmed, S.A.; Maier, W.; de Hoog, G.S.; Finckh, M.R. The forma specialis issue in *Fusarium*: A case study in *Fusarium solani* f. sp. *pisi*. *Sci. Rep.* **2018**, *8*, 1252. [CrossRef]
41. Kurt, Ş.; Uysal, A.; Soylu, E.M.; Kara, M.; Soylu, S. Characterization and pathogenicity of *Fusarium solani* associated with dry root rot of citrus in the eastern Mediterranean region of Turkey. *J. Gen. Plant Pathol.* **2020**, *86*, 326–332. [CrossRef]
42. Mohamed, A.A.; Behiry, S.I.; Ali, H.M.; EL-Hefny, M.; Salem, M.Z.M.; Ashmawy, N.A. Phytochemical Compounds of Branches from *P. halepensis* Oily Liquid Extract and *S. terebinthifolius* Essential Oil and Their Potential Antifungal Activity. *Processes* **2020**, *8*, 330. [CrossRef]
43. Yan, K.; Han, G.; Ren, C.; Zhao, S.; Wu, X.; Bian, T. *Fusarium solani* infection depressed photosystem performance by inducing foliage wilting in apple seedlings. *Front. Plant Sci.* **2018**, *9*, 479. [CrossRef] [PubMed]
44. Robledo-Buriticá, J.; Ángel-García, C.; Castaño-Zapata, J. Microscopía electrónica de barrido ambiental del proceso de infección de Fusarium solani f. sp. passiflorae en plántulas de maracuyá (*Passiflora edulis* f. flavicarpa). *Rev. Acad. Colomb. Cienc. Exactas* **2017**, *41*, 213–221. [CrossRef]
45. Gu, K.X.; Song, X.S.; Xiao, X.M.; Duan, X.X.; Wang, J.X.; Duan, Y.B.; Hou, Y.P.; Zhou, M.G. A *β2-tubulin* dsRNA derived from *Fusarium asiaticum* confers plant resistance to multiple phytopathogens and reduces fungicide resistance. *Pestic. Biochem. Physiol.* **2019**, *153*, 36–46. [CrossRef]
46. Essa, T. Response of some commercial strawberry cultivars to infection by wilt diseases in Egypt and their control with fungicides. *Egypt. J. Pathol.* **2015**, *43*, 113–127. [CrossRef]
47. Kelly, K.; Yagiz, Y.; Li, Z.; Mahnken, G.; Borejsza-Wysocki, W.; Marshall, M.; Sims, C.A.; Peres, N.; do Nascimento, N.M.C. Sensory and Physicochemical Quality, Residual Fungicide Levels and Microbial Load in 'Florida Radiance' Strawberries from Different Disease Control Treatments Exposed to Simulated Supply Chain Conditions. *Foods* **2021**, *10*, 1442. [CrossRef]
48. Da Silva, A.N.; de Azevedo, G.B.; Sobrinho, G.G.R.; de Novaes, Q.S. Efecto de los productos químicos y de *Trichoderma* spp. En control del *Fusarium solani* de la fruta de pasión. *Interciencia* **2014**, *39*, 398–403.
49. Kumar, G.P.K.; Singh, S.K.; Shikha, S. In vitro efficacy of different fungicides against *Fusarium solani* isolate causing root rot of papaya (*Carica papaya* L). *Int. J. Chem. Stud.* **2020**, *8*, 221–224. [CrossRef]
50. FRAC, Fungicide Resistance Action Committee. FRAC Code List 2018: Fungicides Sorted by Mode of Action (Including FRAC Code Numbering). Available online: http://www.phi-base.org/images/fracCodeList.pdf (accessed on 21 January 2022).
51. Andrabi, M.; Vaid, A.; Razdan, V.K. Evaluation of different measures to control wilt causing pathogens in chickpea. *J. Plant Prot. Res.* **2011**, *51*, 55–59. [CrossRef]
52. Ibañez, D.; Izquierdo-Bote, D.; González-García, M.B.; Hernández-Santos, D.; Fanjul-Bolado, P. Development of a New Screen-Printed Transducer for the Electrochemical Detection of Thiram. *Chemosensors* **2021**, *9*, 303. [CrossRef]
53. European Parliament. *Precision Agriculture: An Opportunity for EU Farmers–Potential Support with the CAP 2014–2020*; European Parliament: Brussels, Belgium, 2014; pp. 11–45.
54. Ayvar-Serna, S.; Díaz-Nájera, J.F.; Vargas-Hernández, M.; Enciso-Maldonado, G.A.; Alvarado-Gómez, O.G.; Ortíz-Martínez, A.I. Actividad antifúngica de pesticidas biológicos, botánicos y químicos sobre el agente causal de la marchitez vascular del jitomate. *Rev. Fitotec. Mex.* **2021**, *44*, 617–624. [CrossRef]
55. Shah, M.U.D.; Ahmad, M.; Sagar, V.; Padder, F.A.; Ahanger, F.A.; Sofi, T.A.; Nabi, A.; Mir, A.A.; Kausar, S. In vitro evaluation of bioagents and fungitoxicants against Fusarium oxysporum and Fusarium solani causing corm rot of saffron (Crocus sativus) in Kashmir, India. *Acta Hortic.* **2018**, *1200*, 125–132. [CrossRef]
56. Monadjemi, S.; El Roz, M.; Richard, C.; Ter Halle, A. Photoreduction of Chlorothalonil fungicide on plant leaf models. *Environ. Sci. Technol.* **2011**, *45*, 9582–9589. [CrossRef]
57. National Center for Biotechnology Information. Available online: https://pubchem.ncbi.nlm.nih.gov/compound/Chlorothalonil (accessed on 12 November 2021).

58. Dwivedi, S.K.; Singh, E. In vitro cellulase activity of two wilt causing soil fusaria (*Fusarium solani* and *F. oxysporum* f. sp. *lycopersici*) and efficacy of some pesticides against the said fusaria. *J. Appl. Hortic.* **2015**, *17*, 58–65.
59. Bhaliya, C.M.; Jadeja, K.B. Efficacy of different fungicides against *Fusarium solani* causing coriander root rot. *BioScan* **2014**, *9*, 1225–1227.
60. Madhusndhan, P.; Gopal, K.; Haritha, V.; Sangale, U.R.; Rao, S.V.R.K. Compatibility of *Trichoderma viride* with fungicides and efficiency against *Fusarium solani*. *J. Plant Dis. Sci.* **2019**, *5*, 23–26.
61. Salazar, M.J.A.; Somoza, V.C.E.; Pérez, A.B.; Velásquez, S.M.; Torres, G.G.; Huerta de la Pea, A.; Ortega, M.L.D. Uso y manejo de plaguicidas en diferentes sistemas de producción de fresa en México. *PAyDS* **2017**, *6*, 27–42. [CrossRef]
62. Van Lexmond, M.B.; Bonmatin, J.M.; Goulson, D.; Noome, D.A. Worldwide integrated assessment on systemic pesticides. *Environ. Sci. Pollut. Res.* **2015**, *22*, 1–4. [CrossRef]
63. Wu, T.; Han, B.; Wang, X.; Tong, Y.; Liu, F.; Diao, Q.; Dai, P. Chlorothalonil alters the gut microbiota and reduces the survival of immature honey bees reared in vitro. *Pest Manag. Sci.* **2022**. [CrossRef] [PubMed]
64. Kara, M.; Öztaş, E.; Boran, T.; Karaman, E.F.; Veskoukis, A.S.; Tsatsakis, A.M. Ameliorative Effects of the Sesquiterpenoid Valerenic Acid on Oxidative Stress Induced in HepG2 Cells after Exposure to the Fungicide Benomyl. *Antioxidants* **2021**, *10*, 746. [CrossRef] [PubMed]
65. Pemán, J.; Canton, E.; Espinel-Ingroff, A. Antifungal drug resistance mechanisms. *Expert Rev. Anti Infect. Ther.* **2009**, *7*, 453–460. [CrossRef] [PubMed]
66. National Center for Biotechnology Information. Available online: https://pubchem.ncbi.nlm.nih.gov/source/hsdb/1655 (accessed on 12 November 2021).
67. Asadboland, R.; Eslahi, M.R.; Iranbakhsh, A.R.; Shirzadian, S. Growth inhibition effects of extracts of eight mosses on the phytopathogenic fungus *Fusarium solani*. *J. Crop Prot.* **2021**, *10*, 615–622.
68. Kordali, S.; Usanmaz, A.; Cakir, A.; Komaki, A.; Ercisli, S. Antifungal and herbicidal effects of fruit essential oils of four *Myrtus communis* genotypes. *Chem. Biodivers.* **2016**, *13*, 77–84. [CrossRef]
69. Dorugade, S.P.; Walawade, M.N.; Kamble, S.S. Toxicity of fungicides on *Fusarium solani* causing dry rot of elephant foot yam. *Int. J. Adv. Res.* **2015**, *3*, 1501–1504.
70. Valbuena, D.; Cely-Santos, M.; Obregón, D. Agrochemical pesticide production, trade, and hazard: Narrowing the information gap in Colombia. *J. Environ. Manag.* **2021**, *286*, 112141. [CrossRef]

applied sciences

Article

Effect of Commercial Microbial Preparations Containing *Paenibacillus azotofixans*, *Bacillus megaterium* and *Bacillus subtilis* on the Yield and Photosynthesis of Winter Wheat and the Nitrogen and Phosphorus Content in the Soil

Arkadiusz Stępień [1,*], Katarzyna Wojtkowiak [2] and Ewelina Kolankowska [2]

1 Department of Agroecosystems and Horticulture, University of Warmia and Mazury in Olsztyn, pl. Łódzki 3, 10-719 Olsztyn, Poland
2 Department of Heavy-Duty Machines and Research Methodology, University of Warmia and Mazury in Olsztyn, ul. Oczapowskiego 11, 10-719 Olsztyn, Poland
* Correspondence: arkadiusz.stepien@uwm.edu.pl; Tel.: +48-895233266

Abstract: The present state of knowledge and biotechnological advances have allowed the potential of microorganisms to be used effectively in crop cultivation. A field study on the use of commercial bacterial preparations in the cultivation of winter wheat (*Triticum aestivum* L.) was carried out in the years 2017–2019 at the Educational and Experimental Station in Tomaszkowo (53°71′ N, 20°43′ E), Poland. This study analysed the effect of commercial microbial preparations containing *Paenibacillus azotofixans*, *Bacillus megaterium* and *Bacillus subtilis*, applied during the winter wheat growing season, on the grain yield, protein content, leaf greenness index (SPAD), the course of photosynthesis and the N-NO_3, N-NH_4 and P contents in the soil. The highest grain yield was noted following the application of mineral fertilisation and the three microbial preparations in combination (*Paenibacillus azotofixans*, *Bacillus megaterium* and *Bacillus subtilis*), as well as NPK with *Paenibacillus azotofixans*, in relation to mineral fertilisation alone (by 19.6% and 18.4%, respectively). The microbial preparations had a significant effect on the leaf greenness index (SPAD) at both test dates. No interaction was recorded between the years of study and the preparations applied on the SPAD values. The highest leaf photosynthetic index at both observation dates was noted for the application of NPK + *P. azotofixans*, as well as for NPK and all the preparations combined (*P. azotofixans*, *B. megaterium*, *B. subtilis*). The highest N-NO_3, N-NH_4 and P contents in the soil were obtained using NPK and all microbial preparations combined. Strong correlations were found between the SPAD index and the photosynthetic index value and the protein content in wheat grains and between the N-NO_3, N-NH_4 and P contents in the soil and the wheat grain yield.

Keywords: winter wheat; leaf greenness index (SPAD); protein; photosynthesis; nitrogen; phosphorus

Citation: Stępień, A.; Wojtkowiak, K.; Kolankowska, E. Effect of Commercial Microbial Preparations Containing *Paenibacillus azotofixans*, *Bacillus megaterium* and *Bacillus subtilis* on the Yield and Photosynthesis of Winter Wheat and the Nitrogen and Phosphorus Content in the Soil. *Appl. Sci.* **2022**, *12*, 12541. https://doi.org/10.3390/app122412541

Academic Editor: Joachim Müller

Received: 19 October 2022
Accepted: 3 December 2022
Published: 7 December 2022

1. Introduction

The design of stable winter wheat yields is determined by the use of proven cultivation methods and technologies under specific soil and climate conditions [1–3]. Intensive farming systems contribute to a reduction in soil fertility, which can result in environmental degradation as well as a drop in the quantity and quality of crop yields [4–6]. An alternative to conventional production resources used to support crop production may be biological preparations serving as biofertilisers and biopesticides. These are intended to protect plants against pathogens, influence soil fertility and shape the growth and development of plants [7–11]. An important group of biological preparations are those containing plant-growth-promoting rhizobacteria (PGPR), which comprise bacterial strains of *Azoarcus*, *Azospirillum*, *Azotobacter*, *Arthrobacter*, *Bacillus*, *Clostridium*, *Enterobacter*, *Gluconacetobacter*, *Pseudomonas* and *Serratia* [12]. The rhizosphere bacteria are supportive of the recovery of nutrients from the soil and are important for soil fertility [9,13–16].

Bacteria of the genus *Bacillus* are widely distributed in nature. Most of them are species that are safe for humans, animals and plants. Due to the above-mentioned characteristics of the bacteria *Bacillus* sp., they are used for the production of commercial preparations in the form of insecticides or biostimulants and are involved in supporting plant production [17–20]. *Bacillus* sp. *bacteria* (*B. subtilis*, *B. cereus*, *B. thuringiensis*, *B. pumilus*, *B. megaterium*, etc.) have evolved mechanisms to stimulate plant growth by increasing the availability of the nutrients: N, P, potassium (K) and iron (Fe). Moreover, *Bacillus* strains are capable of fixing molecular nitrogen [19,21–23].

Bacillus subtilis is a saprophyte which decomposes organic compounds of plant origin. This bacterium produces peptide antibiotics, e.g., polymyxin B and subtilin, amino acids, polysaccharide inulin and enzymes, e.g., amylase and protease [24–28]. It also forms siderophores (bacillobactin) which are capable of binding iron ions by binding all available forms of iron into chelates and sharing them with plants. The protein of these bacteria also contains hydrophobin BslA, which reduces surface tension, thus increasing the wetting of the surface on which the bacteria are located, leading to an improvement in moisture within the root system and covering it with an additional protective film (which is particularly important during dry periods) [29–31]. It also stabilises soil colloids [32]. The bacterium *Bacillus subtilis* reduces the number of pathogenic fungi and bacteria in the soil, thus contributing to improved phytosanitary status in crops. [33,34]. *Bacillus subtilis* can also solubilise P in the soil, contribute to nitrogen fixation and produce siderophores which inhibit the growth of pathogens. *Bacillus subtilis* increases stress tolerance in its plant hosts by inducing the expression of stress response genes, phytohormones and stress-related metabolites [20,35]. *Bacillus megaterium* bacteria are described as soil microorganisms with a natural ability to produce acids or enzymes as metabolites, making them capable of dissolving phosphorus [36,37]. *Paenibacillus azotofixans* was initially classified as *Bacillus azotofixans* and then reclassified as belonging to the genus *Paenibacillus* [38]. *Paenibacillus* presents the nifH gene encoding the Fe nitrogenase protein, an enzyme responsible for nitrogen fixation [39]; it has been suggested as a potential biofertiliser for certain crops, e.g., maize or wheat [40–43]. What is more, *Paenibacillus* are well known as effective plant growth promoters in many crops, e.g., maize, wheat, or sorghum [39,44]. *Paenibacillus azotofixans* is a nitrogen-fixing bacterium often found in the soil and rhizospheres of various plants [45–49]. The use of biological preparations in the form of biofertilisers may be a way to ensure environmental stability in intensive agricultural production. To date, no studies have focused on the comprehensive effect of biological preparations on the photosynthesis processes, the yield of wheat grain and the content of nitrogen and phosphorus in soil.

The aim of the study was to assess the effect of commercial microbial preparations containing *Paenibacillus azotofixans*, *Bacillus megaterium* and *Bacillus subtilis*, applied during the winter wheat growing season, on the grain yield, protein content, leaf greenness index, the course of photosynthesis and the N-NO_3, N-NH_4 and P contents in the soil.

2. Materials and Methods

2.1. Experimental Design, Growing Conditions and Treatments

UPTOHERE The research experiment was carried out at the Educational and Experimental Station in Tomaszkowo (53°71′ N, 20°43′ E), Poland. The experiments were carried out in the years 2017–2019 using commercial bacterial preparations that are currently used in the cultivation of winter wheat (*Triticum aestivum* L.). The method employed was that of a static field experiment carried out in four replicates in a randomised block design. The area of the seeding plot was 9.90 m^2 and that of the harvest plot was 8.00 m^2. Winter wheat of the KWS OZON cultivar was sown annually in succession, with winter triticale as the nurse crop in 2016. The wheat was sown at a density of 500 seeds·m^{-2}, with a row spacing of 12.5 cm.

Wheat was cultivated on lessive soil with a granulometric composition of a medium dusty loam. The particle composition of the mineral surface soil horizon included strong loamy sand. The soil was slightly acidic (in KCl solution with pH 5.7), the carbon content was 10.0 g kg^{-1}, and the total nitrogen content was 0.99 g kg^{-1}. The soil abundance in available nutrients was high for P (85.1 mg kg^{-1}) and medium for K (155.0 mg kg^{-1}). Weather conditions registered during field trials are presented in Table 1. The factor under research was the application of commercial bacterial preparations in combination with NPK mineral fertilisers (Table 2).

Table 1. Monthly air temperature and monthly rainfall in the 2016–2019 season. Meteorological data against the years 1981–2010 (Data obtained from the Meteorological Station at Tomaszkowo (53°71′ N, 20°43′ E), Poland).

Growing Season	Mean Temperature (°C)								
	IX *	X	XI-III	IV	V	VI	VII	VIII	Av.
2016/2017	13.6	6.1	2.4	5.7	12.1	15.7	16.8	17.4	7.5
2017/2018	12.8	8.7	3.9	10.8	15.7	17.2	19.7	19.2	8.6
2018/2019	14.5	8.7	3.3	8.0	11.6	20.2	17.1	18.5	8.8
1981–2010	12.8	8.0	2.9	7.7	13.5	16.1	18.7	17.9	7.9
	Rainfall (mm)								Sum
2016/2017	21.1	104.3	84.8	59.1	25.1	74.5	107.6	63.1	693.8
2017/2018	168.1	114.9	42.4	33.5	25.0	53.7	141.0	44.6	713.4
2018/2019	20.3	84.7	16.0	0.0	142.8	120.6	56.3	55.9	677.6
1981–2010	56.9	42.6	44.8	33.3	58.5	80.4	74.2	59.4	581.8

* Month/Phenological growth stages (BBCH scale): IX/germination–leaf development (BBCH 00–19); X/tillering (BBCH 20 …); XI-III/winter dormancy; IV/starting vegetation; V/stem elongation–booting (BBCH 30–49); VI/Inflorescence emergence, heading–flowering, anthesis (BBCH 51–69); VII/development of fruit–senescence (BBCH 71–99); VIII/harvesting.

Table 2. Design of the field experiment. Dose and date of application microbial preparations used in the field experiment.

Treatment (Shortcut)	Component	Application Date/Dose
NPK (NPK)	N (ammonium sulphate 34%; N (ammonium sulphate 34%; P (triple superphosphate 20%); K (potash salt, 49.8%)	$BBCH_{23-24}$/90.0 kg ha^{-1} $BBCH_{31-32}$/60.0 kg ha^{-1} Pre-sowing/70.0 kg ha^{-1} Pre-sowing/100.0 kg ha^{-1}
NPK * + *Paenibacillus azotofixans* (NPK + PA)	*Paenibacillus azotofixans* 1 × 10^9 CFU ** in 1 g of the product (maltodextrin)	$BBCH_{23-24}$/1.0 L ha^{-1} $BBCH_{31-32}$/1.0 L ha^{-1}
NPK * + *Bacillus megaterium* (NPK + BM)	*Bacillus megaterium* 1 × 10^9 CFU in 1 g of the product (maltodextrin)	$BBCH_{23-24}$/1.0 L ha^{-1} $BBCH_{31-32}$/1.0 L ha^{-1}
NPK * + *Bacillus subtilis* (NPK + BS)	*Bacillus subtilis* 5 × 10^9 CFU in 1 g of the product (maltodextrin)	$BBCH_{23-24}$/1.0 L ha^{-1} $BBCH_{31-32}$/1.0 L ha^{-1}
NPK * + *Paenibacillus azotofixans* + *Bacillus megaterium* + *Bacillus subtilis* (NPK + PA + BM + BS)	*Paenibacillus azotofixans* *Bacillus megaterium* *Bacillus subtilis*	$BBCH_{23-24}$/1.0 L ha^{-1} $BBCH_{31-32}$/1.0 L ha^{-1} $BBCH_{23-24}$/1.0 L ha^{-1} $BBCH_{31-32}$/1.0 L ha^{-1} $BBCH_{23-24}$/1.0 L ha^{-1} $BBCH_{31-32}$/1.0 L ha^{-1}

* NPK—mineral fertilisers were applied on all plots at the same doses and dates as on the NPK plot (control), ** CFU—colony-forming unit.

The sowing, cultivation treatment and harvesting of the wheat were carried out in accordance with the agrotechnical requirements specific to the plant species. No protection against pests or diseases was performed. Weeds were controlled using herbicides: BBCH 31–32–Axial 50 EC 0.8 L/ha (pinoxaden)–50 g/L (5.05%), Mustang Forte 195 SE 1.0 L/ha (Florasulam–5 g/L Aminopyralid–10 g/L, 2,4-D–180 g/L).

2.2. Yield and Quality Analysis Samplings

The wheat grain was harvested during the first ten-day period of August using a plot harvester (Wintersteiger Classic 1540, Austria). The wheat grain yield was determined at a moisture of 15%. The protein content was determined on 1.0 kg of samples using a NIR System Infratec 1241 Analyzer camera (Foss, Hillerod, Denmark).

2.3. Leaf Gas Exchange, SPAD Index Measurement

During the winter wheat growing season, the net photosynthetic intensity [mmol $CO_2 \cdot m^{-2} \cdot s^{-1}$] and the leaf greenness index (SPAD) were assessed at two different times:

(1) stem elongation–first node at least 1 cm above tillering node ($BBCH_{31}$)
(2) inflorescence emergence, heading–beginning of heading ($BBCH_{51}$).

The measurements of photosynthesis were carried out on a fully developed, youngest leaf on five randomly selected plants. Ten records were made on each leaf at 5-s intervals. The measurement was performed on a sunny day, in the forenoon hours (09:00 a.m.–11.00 a.m.). Photosynthesis was assessed using an LCi compact camera (ADC BioScientific LCi Analyser Serial No. 32568) manufactured by Eijkelkamp.

The measurements of the leaf greenness index (SPAD) were taken on a fully developed, youngest leaf on ten randomly selected plants. The leaf greenness index was assessed using a SPAD–502 Plus camera by Konica Minolta.

2.4. Physicochemical Soil Analyses

Soil pH was measured potentiometrically in 1M KCl in soil suspension to a 1:5 solution. Total N was determined by mineralisation of a sample with sulphonic acid with an addition of a catalyst (Se mixture), the distillation of products of sodium hydroxide reaction, followed by titration with 0.1 M hydrochloric acid solution against the Tashiro indicator. NH_4^+-N and NO_3^--N was determined calorimetrically, with Nessler's reagent and with phenyldisulphophenolic acid as colouring agents, respectively (UV-1201 V spectrophotometer, Shimadzu Corporation Kyoto, Japan). Available phosphorus (P) and potassium (K) in the soil (mg/kg) was measured by the Egner-Riehm method in an aqueous solution containing calcium lactate ($(CH_3CHOHCOO)_2Ca$) acidified with hydrochloric acid to pH 3.6. Organic C content was determined by oxidation with potassium dichromate in sulphonic acid solution and measurement of the absorbance on a spectrophotometer UV-1201 V (Shimadzu Corporation, Kyoto, Japan).

2.5. Statistical Analysis

Results were statistically analysed using the Statistica 13.1 PL statistical software package. The differences between the experimental plots were determined using a one-way analysis of variance (ANOVA), and Tukey's test was applied to identify homogeneous groups. The calculations were made at a significance level $\alpha = 0.05$ [50]. Using a regression analysis, equations describing the photosynthetic process as determined by the temperature and rainfall occurring at the $BBCH_{31}$ and $BBCH_{51}$ stages were determined [51]. The correlation of the influence of the applied biological preparations on the leaf greenness index, photosynthesis, grain yield, protein content in grains and the N-NO_3, N-NH_4 and P contents in the soil in wheat cultivation was also determined. To this end, the principal component analysis (PCA) method was used to determine the strength and direction of the correlation between the measurement variables [52].

3. Results and Discussion

3.1. Grain Yield and Protein Content

The lowest yield obtained in 2018 results from the unfavourable rainfall distribution during this growing season (Figure 1). Excessive rainfall during the sowing and emergence of wheat in 2017 (September–October) might have contributed to the poor rooting of the plants, which could have resulted in lower yields in 2018. Furthermore, low rainfall during winter (February–March 2018), as well as water shortages during the intensive growth (May–June 2018), adversely affected the growth and development of the plants. During this period, the unfavourable water balance was exacerbated by the occurrence of higher temperatures. A reduced yield is a typical response to water stress, as both the photosynthetic intensity and the plant growth processes are reduced [53–57]. High temperatures combined with low rainfall increase the loss of nutrients (N, P, K), particularly in poor soils [58]. Adequate weather conditions occurred in the growing seasons of 2016/2017 and 2018/2019, which resulted in higher yields (by 2.64 t and 1.72 t ha^{-1}, respectively) as compared to the season of 2017/2018. A higher yield can primarily result from a relatively normal temperature and the appropriate distribution of the optimum rainfall amount during the May-June period, which is critical for cereal growth in Poland [59]. Irrespective of the years, as compared to the control plot (NPK), the highest grain yield was noted following the application of the three combined microbial preparations (NPK + *P. azotofixans* + *B. megaterium* + *B. subtilis*), as well as NPK + *P. azotofixans* by 19.6% and 18.4%, respectively. In the years of the study, the effect of the application of the microbial preparations on the grain yield was not uniform (Table 3). In a study by Turan et al. [60], the introduction of *Bacillus subtilis*, *Bacillus megaterium* and *Azospirillum brasilense* into the soil increased the grain yield by 24%, 19% and 19%, respectively, while their application in combination resulted in an increase in the grain yield by 33% as compared to the control (with no microbial preparations). In the authors' studies in 2017 and 2019, the highest grain yields were obtained following the application of NPK + *P. azotofixans* and NPK and all the preparations combined (*P. azotofixans* + *B. megaterium* + *B. subtilis*), while in 2018, optimal yields were observed only following the application of all the three preparations in combination. The preparation with *P. azotofixans* was applied under optimal conditions for wheat growth and development, and all the preparations applied in combination performed well under extreme weather conditions thanks to their ability to supplement each other. In general, the effects of microbial preparations are more noticeable under unfavourable weather conditions [61] and less noticeable under optimal conditions prevailing during the crop-growing season [62]. Bacterial endophytes found in plants employ direct or indirect mechanisms to improve the growth and development of plants and increase their tolerance to biotic and abiotic stresses [63–65]. According to de Lima et al. [66], Ali and Khan [67], Maslennikova and Lastochkina [68], plants inoculated with *B. subtilis* can improve their growth under water stress conditions, mainly due to increased water use efficiency.

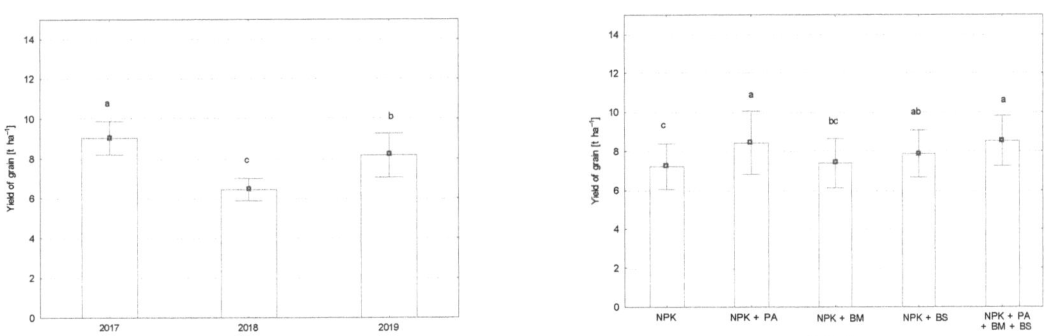

(a) (b)

Figure 1. The winter wheat grain yield after application of microbial preparations (mean value and standard deviation), (**a**) average for the years, (**b**) average for treatment. Values followed by the same letters do not differ significantly in Tukey's (HSD) test ($p < 0.05$). PA—*Paenibacillus azotofixans*; BM—*Bacillus megaterium*; BS—*Bacillus subtilis*.

Table 3. The grain yield after application of microbial preparations (mean value and standard deviation), interaction between the years and treatments, t ha^{-1}.

Year	NPK	NPK + PA *	NPK + BM **	NPK + BS ***	NPK + PA + BM + BS
2017	8.72 ± 0.56 ab	9.81 ± 0.23 a	8.38 ± 0.69 abc	8.60 ± 0.58 abc	9.76 ± 0.89 a
2018	6.46 ± 0.32 e	6.41 ± 0.84 e	5.81 ± 0.29 e	6.45 ± 0.17 e	7.13 ± 0.07 cd
2019	6.51 ± 0.38 de	9.10 ± 0.60 ab	7.98 ± 0.45 bcd	8.57 ± 0.91 abc	8.71 ± 0.75 ab

* PA—*Paenibacillus azotofixans*; ** BM—*Bacillus megaterium*; *** BS—*Bacillus subtilis*. Values followed by the same letters do not differ significantly in Tukey's (HSD) test ($p < 0.05$).

The quality characteristics of wheat grains are determined by the genetic conditions and modified by the cultivation technology type and the weather conditions [1–3]. Certain microorganism species (plant growth-promoting rhizobacteria) can extract nutrients from the soil or the atmosphere and, consequently, contribute to improved plant nutrition [20,21,28,69]; however, these may not be sufficient quantities to satisfy all plant nutrient needs [70–72]. In the authors' own study, the highest protein contents in grains were noted in 2018; lower contents were observed in other years (Figure 2, Table 4). Jie et al. [73] showed that weather conditions, including the temperature, insolation duration and rainfall during the plant growth period, are the main factors that determine the wheat protein content and quality. Protein accumulation under high temperatures and no rainfall conditions is related to metabolism, transport, photosynthesis, responses to stress, detoxication and protein synthesis [74]. According to Yang et al. [75], Liu et al. [76] and Sehgal et al. [55], mild stress due to water shortage can promote the remobilisation of carbon assimilates into the grains, thus speeding up the grain filling and ultimately improving the yield quality. The influence of microbial preparations on the protein content in wheat grain was not demonstrated (Figure 2).

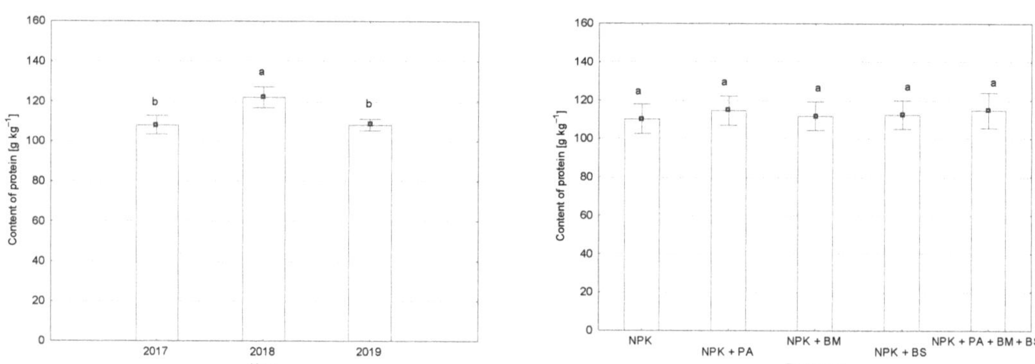

(a) (b)

Figure 2. Protein content in wheat grains after application of microbial preparations (mean value and standard deviation), (**a**) average for the years, (**b**) average for treatment, g kg^{-1}. Values followed by the same letters do not differ significantly in Tukey's (HSD) test ($p < 0.05$). PA—*Paenibacillus azotofixans*; BM—*Bacillus megaterium*; BS—*Bacillus subtilis*.

Table 4. Protein content in wheat grains after application of microbial preparations (mean value and standard deviation), interaction between the years and treatments, g kg^{-1}.

Year	NPK	NPK + PA *	NPK + BM **	NPK + BS ***	NPK + PA + BM + BS
2017	103 ± 2.36 ^{b}	108 ± 4.43 ^{b}	108 ± 2.44 ^{b}	109 ± 5.57 ^{b}	109 ± 3.30 ^{b}
2018	121 ± 4.78 ^{a}	121 ± 2.21 ^{a}	121 ± 3.30 ^{a}	121 ± 6.18 ^{a}	124 ± 6.39 ^{a}
2019	109 ± 2.50 ^{b}	108 ± 3.56 ^{b}	106 ± 3.77 ^{b}	108 ± 1.50 ^{b}	110 ± 2.89 ^{b}

* PA—*Paenibacillus azotofixans*; ** BM—*Bacillus megaterium*; *** BS—*Bacillus subtilis*. Values followed by the same letters do not differ significantly in Tukey's (HSD) test ($p < 0.05$).

3.2. The Leaf Greenness Index (SPAD) and the Photosynthetic Rate

Weather conditions during the research had a significant effect on the SPAD index value only at the 1 node stage ($BBCH_{31}$), with its highest value noted in 2019 (Figure 3). The microbial preparations had a significant effect on the SPAD index value at both test dates ($BBCH_{31}$ and $BBCH_{51}$). Following the application of NPK and the preparation containing *P. azotofixans* and NPK with all the biopreparations (*P. azotofixans* + *B. megaterium* + *B. subtilis*), significantly higher values of the SPAD index were noted at both growing stages.

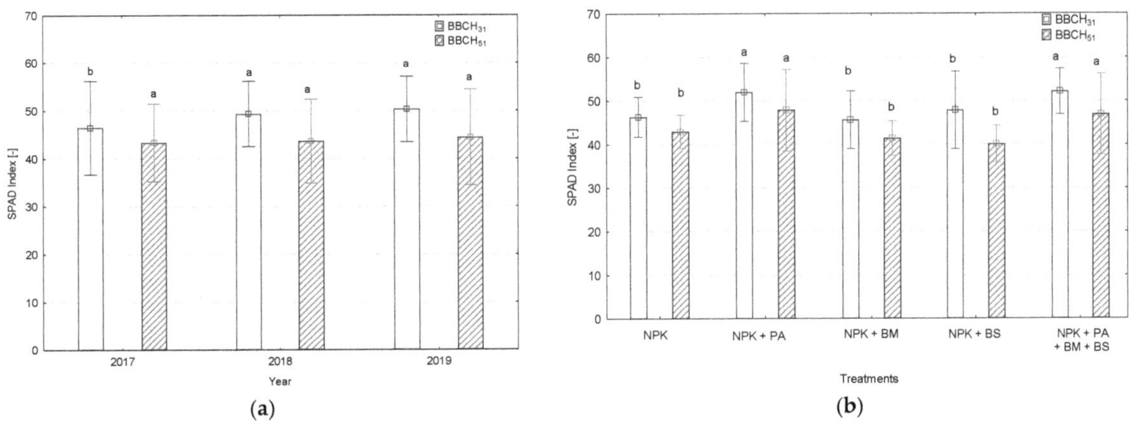

Figure 3. The leaf greenness index (SPAD) after application of microbial preparations (mean value and standard deviation), (**a**) average for the years, (**b**) average for treatment. Values followed by the same letters do not differ significantly in Tukey's (HSD) test ($p < 0.05$). PA—*Paenibacillus azotofixans*; BM—*Bacillus megaterium*; BS—*Bacillus subtilis*.

According to Govindasamy et al. [28], a higher SPAD index value results from the provision of a sufficient amount of N, supported by microbial preparations, which enables the improvement of the condition of plants. An increase in nitrogen fertilisation is accompanied by an increase in both the leaf greenness index value and the leaf area index (LAI) [77]. Islam et al. [78] and Monostori et al. [79] also noted that nitrogen fertilisation significantly contributes to an increase in the flag leaf SPAD index value, which is positively correlated with the wheat grain yield. In the present study, no interaction was noted between the years of study and the preparations applied to the SPAD index values (data not shown).

The photosynthetic process rate positively correlates with the mineral content in the soil [80,81]. The photosynthetic intensity, however, can be impaired by almost any adverse environmental factor [81,82]. One such factor is a soil water deficit, which results from the weather condition pattern. Abiotic stresses primarily reduce the photosynthetic performance of plants due to their adverse effect on chlorophyll biosynthesis, photosystem performance, electron transport mechanisms, gas exchange parameters and many more [57].

On both measurement dates, the years of the study unevenly affected the photosynthetic rate shaping (Figure 4), and on both observation dates, irrespective of the years, the highest photosynthetic rate was noted following the application of NPK + *P. azotofixans* and NPK and all the preparations combined (*P. azotofixans* + *B. megaterium* + *B. subtilis*). The weather conditions throughout the years of the study differentiated the effect of biological preparations on the photosynthesis index. (Table 5). Such an example, as compared to the control plot (NPK), is obtaining lower photosynthetic rate values at the $BBCH_{31}$ stage following the application of NPK + *B. megaterium* in 2018 and at the $BBCH_{51}$ stage in 2018 and 2019. This was confirmed by the strong positive correlation between the grain yield and the net photosynthetic intensity, as noted in a study by Olszewski et al. [82], Carmo-Silva et al. [83], Murchie et al. [84] and Sanchez-Bragado et al. [85]. Such a relationship is indicative of the correct use of photosynthetic products in yield formation [86]. In the present study, the regression equations describing the relationship of the photosynthetic rate at the wheat development stages of $BBCH_{31}$ and $BBCH_{51}$ are characterised by a determination coefficient of 0.40–0.99 (Table 6).

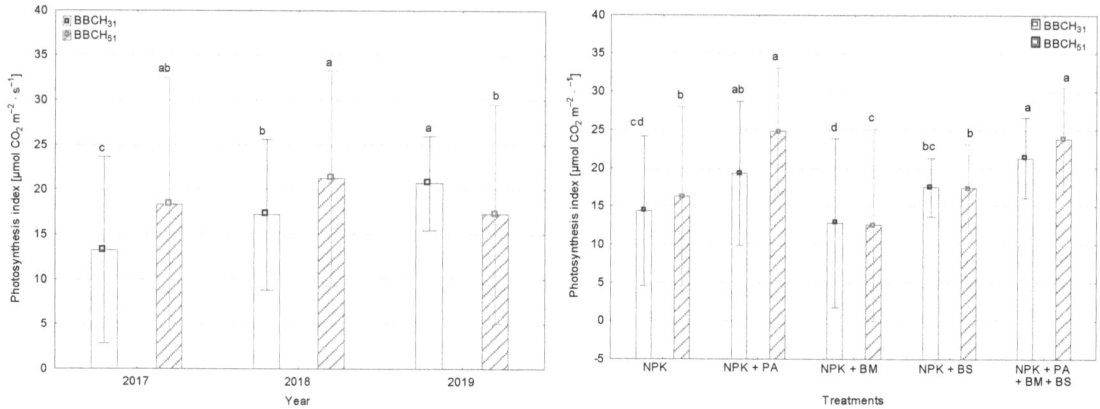

(a) (b)

Figure 4. The photosynthetic index after application of microbial preparations (mean value and standard deviation), (**a**) average for the years, (**b**) average for treatment. Values followed by the same letters do not differ significantly in Tukey's (HSD) test ($p < 0.05$). PA—*Paenibacillus azotofixans;* BM—*Bacillus megaterium;* BS—*Bacillus subtilis.*

Table 5. The photosynthetic index after application of microbial preparations (mean value and standard deviation), interaction between the years and treatments (µmol CO_2 m^{-2} s^{-1}).

Year	NPK	NPK + PA *	NPK + BM **	NPK + BS ***	NPK + PA + BM + BS
			$BBCH_{31}$		
2017	8.18 ± 3.04 ^{e}	13.14 ± 0.46 ^{cde}	9.33 ± 5.35 ^{e}	15.58 ± 1.14 ^{bcd}	20.02 ± 3.03 ^{ab}
2018	17.23 ± 1.00 ^{bc}	20.99 ± 0.29 ^{ab}	10.57 ± 4.03 ^{de}	17.17 ± 0.38 ^{bc}	20.25 ± 2.11 ^{ab}
2019	17.75 ± 0.07 ^{bc}	23.77 ± 0.06 ^{a}	18.58 ± 0.33 ^{abc}	19.76 ± 0.16 ^{ab}	23.77 ± 0.06 ^{a}
			$BBCH_{51}$		
2017	10.22 ± 0.91 ^{de}	27.07 ± 5.15 ^{a}	16.85 ± 4.10 ^{bcd}	13.75 ± 0.65 ^{cde}	23.74 ± 3.81 ^{ab}
2018	19.53 ± 7.13 ^{abc}	25.80 ± 0.52 ^{a}	14.93 ± 6.07 ^{cd}	19.12 ± 1.92 $^{a–d}$	27.08 ± 0.21 ^{a}
2019	19.12 ± 0.71 $^{a–d}$	21.55a ± 3.75 ^{bc}	5.89 ± 0.46 ^{e}	19.33 ± 0.12 ^{abc}	20.67 ± 0.06 ^{abc}

* PA—*Paenibacillus azotofixans;* ** BM—*Bacillus megaterium;* *** BS—*Bacillus subtilis.* Values followed by the same letters do not differ significantly in Tukey's (HSD) test ($p < 0.05$).

At the wheat growth phase of BBCH31, when fertilising wheat with NPK alone, NPK + *P. azotofixans*, NPK + *B. megaterium* and NPK + *B. subtilis*, an increase in the photosynthetic rate value was obtained with the maximum amount of rainfall and at the maximum temperature (Figure 5). As regards wheat fertilisation with NPK + *P. azotofixans* + *B. megaterium* + *B. subtilis*, an increase in the photosynthetic rate value occurred with an increase in the amount of rainfall, irrespective of the temperatures occurring during the period of measurements (May). At the $BBCH_{51}$ stage, the maximum increase in the photosynthetic rate was obtained at the minimum amount of rainfall and at the maximum temperature in the fertilisation variants of NPK, NPK + *B. subtilis*, NPK + *B. megaterium* and NPK + *P. azotofixans* + *B. megaterium* + *B. subtilis*. As regards the fertilisation with NPK + *P. azotofixans* at the BBCH51 stage, an increase in the photosynthetic rate was obtained with a minimum amount of rainfall and at a minimum temperature.

Table 6. Regression equations describing the photosynthetic rate (A) depending on the rainfall (R) and the temperature (T) at the winter wheat development stages ($BBCH_{31}$ and $BBCH_{51}$) after application of microbial preparations.

Treatments	$BBCH_{31}$		$BBCH_{51}$	
	Regression Equation	Coefficient of Determination R^2	Regression Equation	Coefficient of Determination R^2
NPK	$A = 0.0921{\cdot}R + 2.5164{\cdot}T - 24.5848$	0.883	$A = -0.1753{\cdot}R + 3.7715{\cdot}T - 35.9334$	0.585
NPK + PA *	$A = 0.0996{\cdot}R + 2.1819{\cdot}T - 15.7609$	0.996	$A = -0.0160{\cdot}R - 1.0631{\cdot}T + 44.9450$	0.354
NPK + BM **	$A = 0.0801{\cdot}R + 0.3463{\cdot}T + 3.1292$	0.599	$A = -0.0479{\cdot}R - 1.9456{\cdot}T + 50.9656$	0.629
NPK + BS ***	$A = 0.0374{\cdot}R + 0.4448{\cdot}T + 9.2531$	0.889	$A = -0.0972{\cdot}R + 2.2360{\cdot}T - 14.1208$	0.867
NPK + PA + BM + BS	$A = 0.0321{\cdot}R + 0.0641{\cdot}T + 18.4380$	0.464	$A = -0.1206{\cdot}R + 0.5517{\cdot}T + 24.0601$	0.653

* PA—Paenibacillus azotofixans; ** BM—Bacillus megaterium; *** BS—Bacillus subtilis.

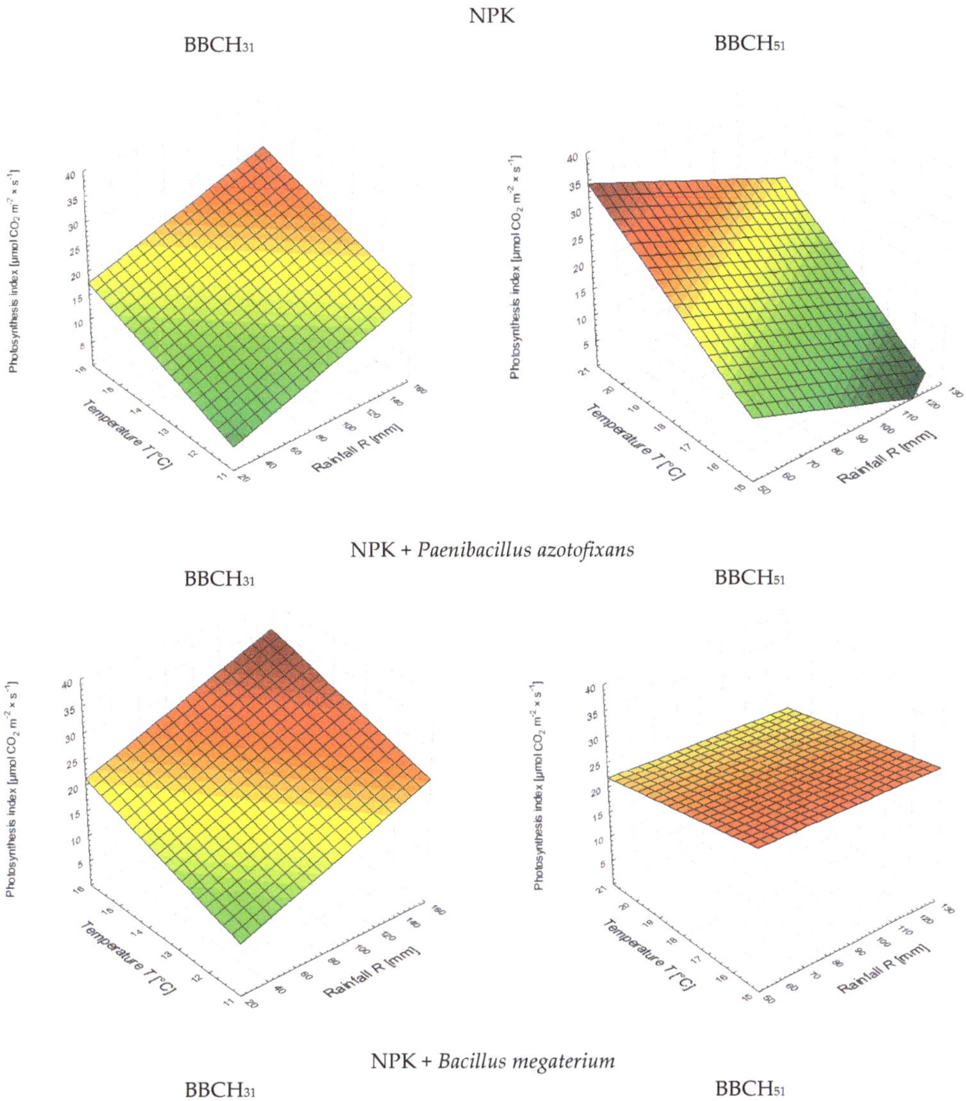

Figure 5. *Cont.*

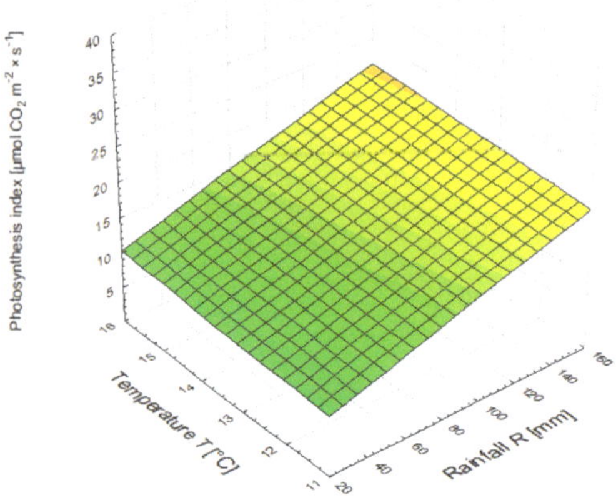

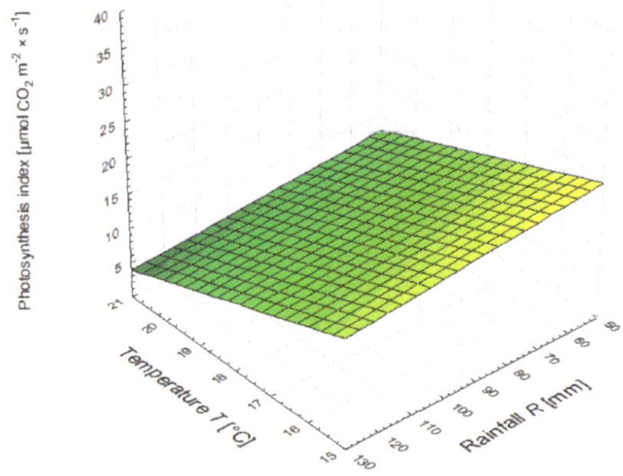

NPK + *Bacillus subtilis*

BBCH31 BBCH51

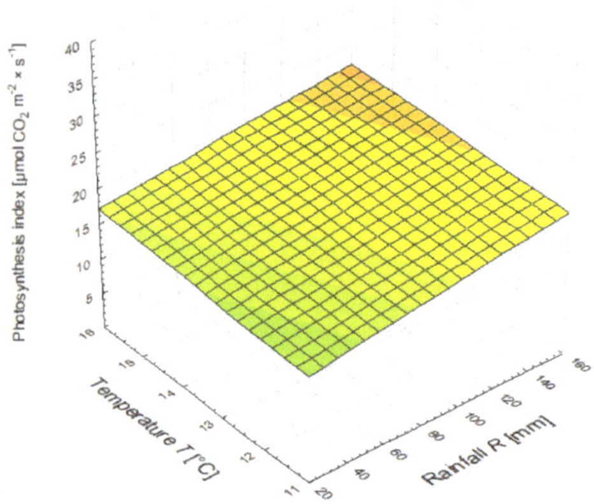

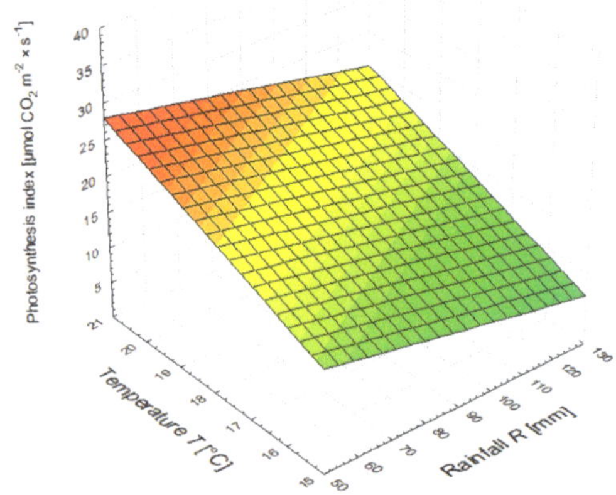

NPK + *Paenibacillus azotofixans* + *Bacillus megaterium* + *Bacillus subtilis*

BBCH31 BBCH51

Figure 5. *Cont.*

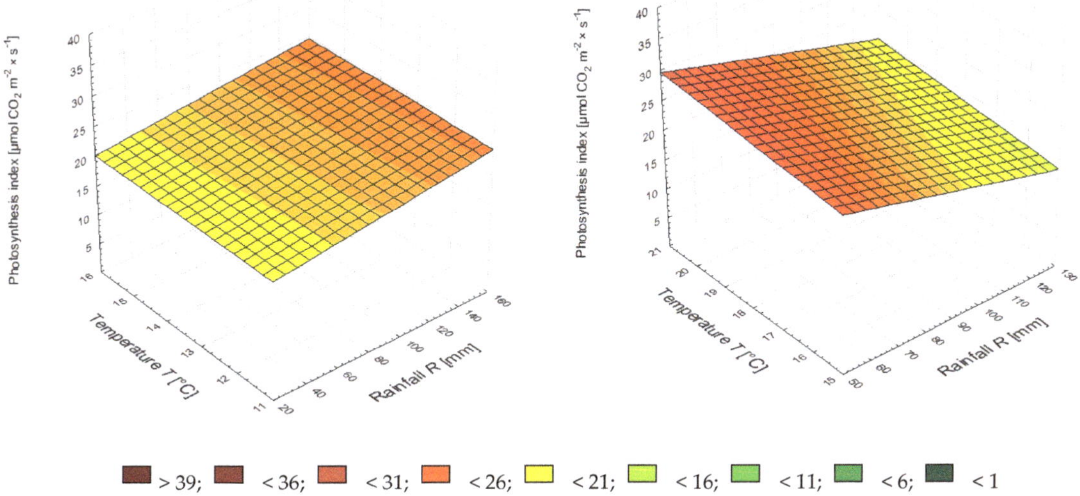

Figure 5. The photosynthetic index depending on the temperature and rainfall at the stages $BBCH_{31}$ and $BBCH_{51}$ after application of microbial preparations.

3.3. N-NO_3,N-NH_4, and P Contents in the Soil

The highest mineral nitrogen (N-NO_3 and N-NH_4) contents in the soil were noted in 2017 and were, on average, higher as compared to the subsequent years of the study by 43.3% and 45.5%, respectively (Figure 6). Irrespective of the years, an increase in the N-NH_4 content in relation to the control plot (NPK) was noted following the introduction of each plot. An increase in the N-NO_3 content in relation to the control plot was noted following the application of NPK + *P. azotofixans* and NPK and all the preparations combined (*P. azotofixans* + *B. megaterium* + *B. subtilis*). In a study by Kołodziejczyk [87], microbial preparations Proplantan AM and Effective Microorganisms significantly reduced the N-NO_3 and N-NH_4 contents in the soil. In the authors' own study, the principle of N-NO_3 content following the application of NPK + *P. azotofixans* being higher than that on the control plot (NPK) was not reflected in all the years of the study (Table 7), with an increase in the N-NH_4 content in relation to the control plot (NPK) noted in 2017 following the application of NPK + *P. azotofixans* and NPK + *B. subtilis*. The N-NH_4 contents were also significantly higher in 2018 and 2019 than those on the control plot (NPK) and were only found following the application of NPK and all of the microbial preparations in combination (*P. azotofixans* + *B. megaterium* + *B. subtilis*).

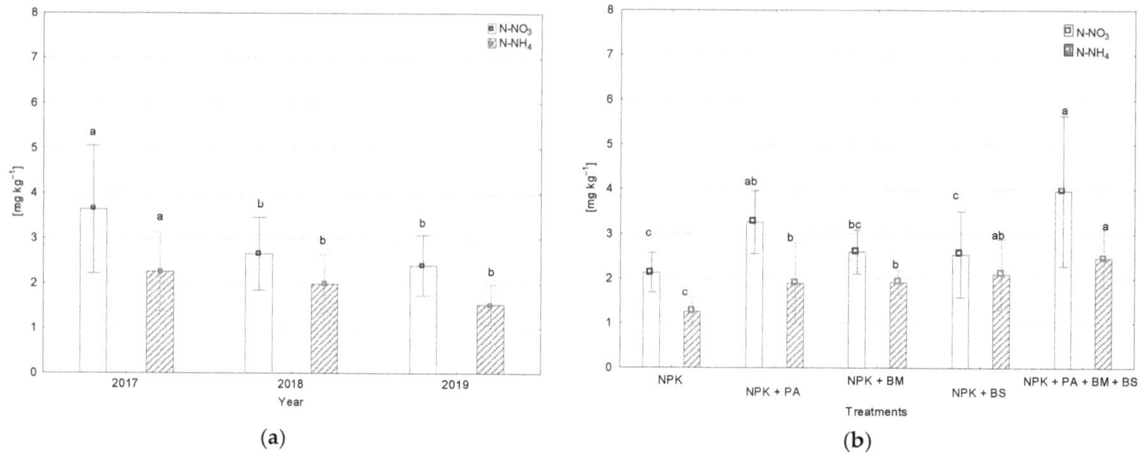

Figure 6. The $N\text{-}NO_3$ and $N\text{-}NH_4$ contents in the soil after application of microbial preparations (mean value and standard deviation), (**a**) average for the years, (**b**) average for treatment. Values followed by the same letters do not differ significantly in Tukey's (HSD) test ($p < 0.05$). PA—*Paenibacillus azotofixans*; BM—*Bacillus megaterium*; BS—*Bacillus subtilis*. ^{a,b,c}—statistically homogenous groups, $p \leq 0.05$.

Table 7. $N\text{-}NO_3$ and $N\text{-}NH_4$ contents in the soil after application of microbial preparations (mean value and standard deviation), interaction between the years and treatments.

Year	NPK	NPK + PA *	NPK + BM **	NPK + BS ***	NPK + PA + BM + BS
			$N\text{-}NO_3$		
2017	2.32 ± 0.52 ^{bc}	3.79 ± 0.45 ^{b}	2.77 ± 0.43 ^{bc}	3.30 ± 0.57 ^{bc}	6.05 ± 0.73 ^{a}
2018	1.82 ± 0.48 ^{c}	3.25 ± 0.81 ^{bc}	2.82 ± 0.45 ^{bc}	2.60 ± 1.17 ^{bc}	2.80 ± 0.45 ^{bc}
2019	2.25 ± 0.18 ^{bc}	2.75 ± 0.51 ^{bc}	2.22 ± 0.42 ^{c}	1.75 ± 0.33 ^{c}	3.05 ± 0.96 ^{bc}
			$N\text{-}NH_4$		
2017	1.12 ± 0.11 ^{d}	2.97 ± 0.81 ^{a}	2.04 ± 0.14 $^{a\text{–}d}$	3.03 ± 0.52 ^{a}	2.13 ± 0.78 $^{a\text{–}d}$
2018	1.48b ± 0.50 ^{cd}	1.46 ± 0.07 ^{bcd}	2.15 ± 0.01 ^{abc}	1.88 ± 0.02 ^{bcd}	2.99 ± 0.58 ^{a}
2019	1.15 ± 0.07 ^{cd}	1.29 ± 0.21 ^{bcd}	1.56 ± 0.08 ^{bcd}	1.38 ± 0.33 ^{bcd}	2.27 ± 0.32 ^{ab}

* PA—*Paenibacillus azotofixans*; ** BM—*Bacillus megaterium*; *** BS—*Bacillus subtilis*. Values followed by the same letters do not differ significantly in Tukey's (HSD) test ($p < 0.05$).

In 2018, the P content in the soil was lower by an average of 16.8% than that noted in 2017 and 2019 (Figure 7). Irrespective of the year, the highest P content, similar to $N\text{-}NO_3$ and $N\text{-}NH_4$, was noted following the application of NPK and all the microbial preparations in combination (*P. azotofixans* + *B. megaterium* + *B. subtilis*). This effect of the applied NPK and all the biopreparations in combination was confirmed in all years of the study (Table 8). Furthermore, in 2018, a higher P content was noted after the application of NPK + *B. megaterium* than after the application of NPK alone. According to Kocoń and Jadczyszyn [88], favourable weather conditions (a high amount of evenly distributed rainfall and high temperatures) had a positive effect on the performance of microbial preparations (EM, EM-Farming, UGmax) on the available phosphorus content in the soil. *Bacillus* spp., which fixes atmospheric nitrogen and dissolves soil phosphorus resources, stimulates plant growth [23,37,89]. In a study by Turan et al. [60], preparations containing *Bacillus subtilis*, *Bacillus megaterium*, or *Azospirillum brasilense*, applied individually or in combination, increased the concentration of labile and moderately labile P fraction in the rhizosphere soil. Tahir et al. [90] found that mineral fertilisation in combination with a phosphorus-solubilising bacterial strain (*Bacillus strain* MWT-14) increased the P content by 3.3% and the N content by 3.7%, on average.

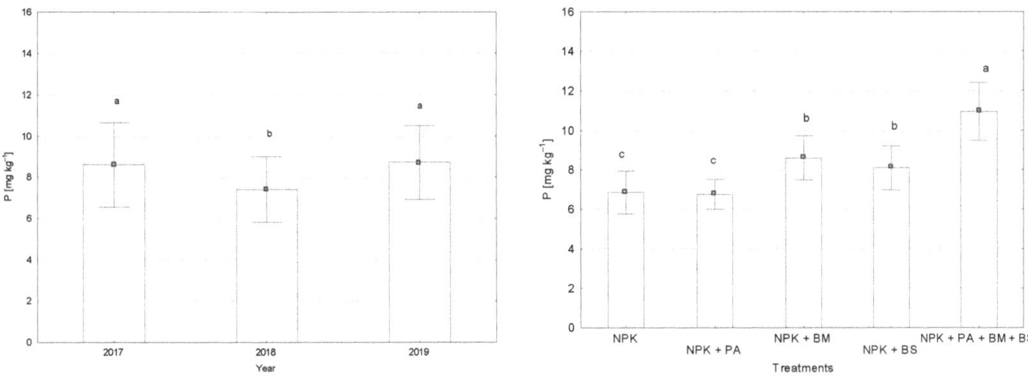

(a) (b)

Figure 7. P content in the soil after application of microbial preparations (mean value and standard deviation), (**a**) average for the years, (**b**) average for treatment. Values followed by the same letters do not differ significantly in Tukey's (HSD) test ($p < 0.05$). PA—*Paenibacillus azotofixans*; BM—*Bacillus megaterium*; BS—*Bacillus subtilis*.

Table 8. P content in the soil after application of microbial preparations (mean value and standard deviation), interaction between the years and treatments.

Year	NPK	NPK + PA *	NPK + BM **	NPK + BS ***	NPK + PA + BM + BS
2017	$7.78 \pm 1.34^{b\text{–}e}$	6.74 ± 0.84^{de}	8.73 ± 1.24^{bcd}	$7.99 \pm 1.52^{b\text{–}e}$	11.84 ± 0.24^{a}
2018	5.91 ± 0.18^{e}	6.14 ± 0.35^{e}	8.54 ± 1.62^{bcd}	$7.26 \pm 0.18^{b\text{–}e}$	9.24 ± 1.28^{b}
2019	6.89 ± 0.09^{cde}	$7.39 \pm 0.46^{b\text{–}e}$	8.56 ± 0.65^{bcd}	9.03 ± 0.38^{bc}	11.77 ± 0.48^{a}

* PA—*Paenibacillus azotofixans*; ** BM—*Bacillus megaterium*; *** BS—*Bacillus subtilis*. Values followed by the same letters do not differ significantly in Tukey's (HSD) test ($p < 0.05$).

3.4. Principal Component Analysis–PCA

The analyses concerning the effect of the application of microbial preparations in winter wheat cultivation on the SPAD index, photosynthesis, grain yield, protein content in grains and the N-NO_3; N-NH_4; and P contents in the soil were supplemented by the determination of correlations between the above-mentioned factors (Figure 8). To this end, the principal component analysis (PCA) method was used to determine the links (the strength and direction of the correlation) between the measurement variables. The analysis showed a strong correlation between the leaf greenness index (SPAD) as well as photosynthesis and the protein content in wheat grains during both development stages ($BBCH_{31}$ and $BBCH_{51}$). The N-NO_3, N-NH_4 and P contents in the soil were also strongly correlated with the grain yield.

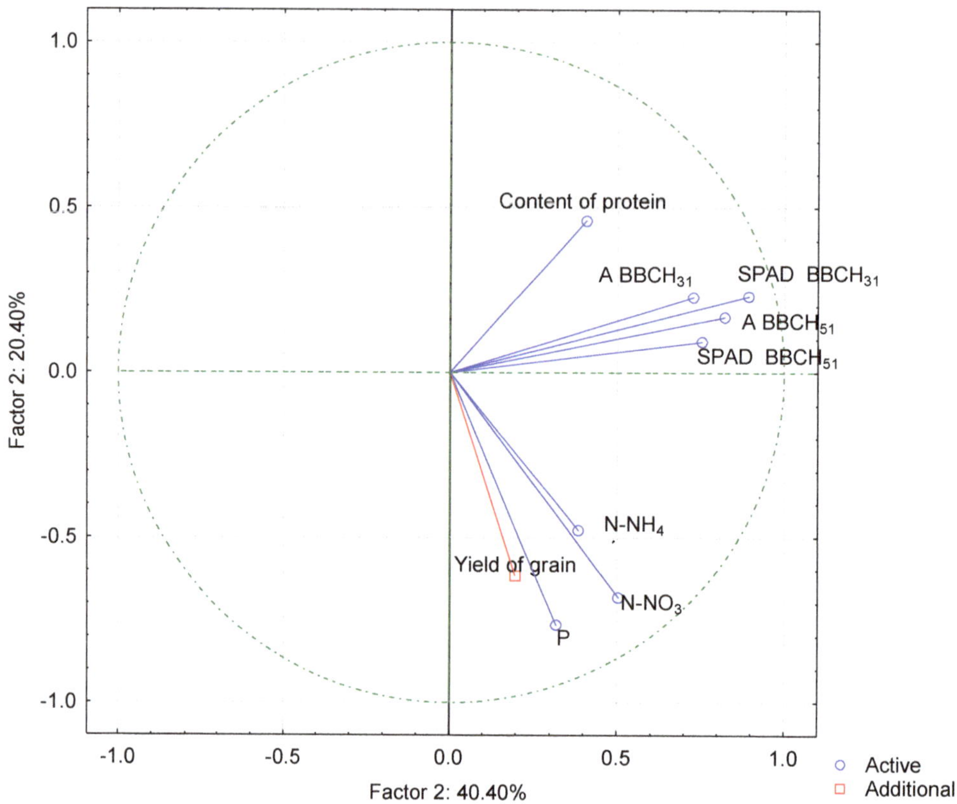

Figure 8. Diagram illustrating the correlation between the photosynthetic index (A), SPAD, protein content in grains, $N\text{-}NO_3$, $N\text{-}NH_4$ and P contents in the soil and the winter wheat grain yield.

4. Conclusions

The highest grain yield was noted following the application of mineral fertilisation and the three microbial preparations combined (*Paenibacillus azotofixans*, *Bacillus megaterium* and *Bacillus subtilis*), as well as NPK with *Paenibacillus azotofixans*. The preparation containing *Paenibacillus azotofixans* performed well under optimal conditions for wheat growth and development, while all the preparations applied in combination (*P. azotofixans*, *B. megaterium*, *B. subtilis*) performed well under extreme weather conditions by supplementing each other. The microbial preparations had a significant effect on the leaf greenness index (SPAD) at both test dates ($BBCH_{31}$ and $BBCH_{51}$). No interaction was noted between the years of study and the preparations applied on the SPAD values. The highest leaf photosynthetic index at both observation dates was noted following the application of NPK + *P. azotofixans* and NPK and all the preparations combined (*P. azotofixans*, *B. megaterium* and *B. subtilis*). At the development phase of $BBCH_{31}$, when fertilising wheat with NPK alone and applying NPK and the microbial preparations individually, an increase in the photosynthetic rate value was obtained at the maximum amount of rainfall and at the maximum temperature. At the $BBCH_{51}$ stage, the maximum increase in the photosynthetic rate was obtained with the minimum amount of rainfall and at the maximum temperature in the fertilisation variants of NPK, NPK + *B. subtilis*, NPK + *B. megaterium* and NPK + all the preparations in combination. The highest $N\text{-}NO_3$, $N\text{-}NH_4$ and P contents in the soil were noted following the application of NPK and all microbial preparations combined. Strong correlations were noted between the leaf greenness index (SPAD) and the photosynthetic index value and the protein content in wheat grains and between the $N\text{-}NO_3$, $N\text{-}NH_4$ and P contents in the soil and the wheat grain yield.

Author Contributions: Conceptualisation, A.S., K.W. and E.K.; methodology, A.S., K.W. and E.K.; validation, A.S., K.W. and E.K.; formal analysis, A.S., K.W. and E.K.; investigation, A.S.; resources, A.S., K.W. and E.K.; data curation, A.S., K.W. and E.K.; writing—original draft preparation, A.S., K.W. and E.K.; writing—review and editing, A.S., K.W. and E.K.; visualisation, A.S. and E.K.; supervision, A.S., K.W. and E.K.; project administration, A.S.; funding acquisition, A.S., K.W. and E.K. All authors have read and agreed to the published version of the manuscript.

Funding: Project financially supported by the Minister of Education and Science under the program entitled "Regional Initiative of Excellence" for the years 2019–2023, Project No. 010/RID/2018/19, amount of funding 12.000.000 PLN. The results presented in this paper were obtained as part of a comprehensive study financed by the University of Warmia and Mazury in Olsztyn, Faculty of Agriculture and Forestry, Department of Agroecosystems and Horticulture (grant No. 30.610.015-110).

Institutional Review Board Statement: Not applicable.

Informed Consent Statement: Not applicable.

Data Availability Statement: Not applicable.

Conflicts of Interest: The authors declare no conflict of interest.

References

1. Jaskulska, I.; Jaskulski, D.; Gałęzewski, L.; Knapowski, T.; Kozera, W.; Wacławowicz, R. Mineral Composition and Baking Value of the Winter Wheat Grain under Varied Environmental and Agronomic Conditions. *J. Chem.* **2018**, *2018*, 5013825. [CrossRef]
2. Woźniak, A.; Rachoń, L. Effect of Tillage Systems on the Yield and Quality of Winter Wheat Grain and Soil Properties. *Agriculture* **2020**, *10*, 405. [CrossRef]
3. Buczek, J.; Migut, D.; Jańczak-Pieniążek, M. Effect of Soil Tillage Practice on Photosynthesis, Grain Yield and Quality of Hybrid Winter Wheat. *Agriculture* **2021**, *11*, 479. [CrossRef]
4. Ju, X.-T.; Xing, G.-X.; Chen, X.-P.; Zhang, S.-L.; Zhang, L.-J.; Liu, X.-J.; Cui, Z.-L.; Yin, B.; Christie, P.; Zhu, Z.-L.; et al. Reducing environmental risk by improving N management in intensive Chinese agricultural systems. *Proc. Natl. Acad. Sci. USA* **2009**, *106*, 3041–3046. [CrossRef] [PubMed]
5. Kosicka, D.; Wolna-Maruwka, A.; Trzeciak, M. The influence of microbial inoculates on morphological traits in plants. *Kosmos* **2015**, *64*, 327–335. (In Polish)
6. Bonanomi, G.; De Filippis, F.; Zotti, M.; Idbella, M.; Cesarano, G.; Al-Rowaily, S.; Abd-ElGawad, A. Repeated applications of organic amendments promote beneficial microbiota, improve soil fertility and increase crop yield. *Appl. Soil Ecol.* **2020**, *156*, 103714. [CrossRef]
7. Itelima, J.U.; Bang, W.J.; Onyimba, I.A.; Sila, M.D.; Egbere, O.J. Bio-fertilizers as key player in enhancing soil fertility and crop productivity. *Direct Res. J. Agric. Food Sci.* **2018**, *6*, 73–83. [CrossRef]
8. Mącik, M.; Gryta, A.; Frąc, M. Biofertilizers in agriculture: An overview on concepts, strategies and effects on soil microorganisms. *Adv. Agron.* **2020**, *162*, 31–87. [CrossRef]
9. Basu, A.; Prasad, P.; Das, S.N.; Kalam, S.; Sayyed, R.Z.; Reddy, M.S.; El Enshasy, H. Plant Growth Promoting Rhizobacteria (PGPR) as Green Bioinoculants: Recent Developments, Constraints, and Prospects. *Sustainability* **2021**, *13*, 1140. [CrossRef]
10. Mahmud, A.A.; Upadhyay, S.K.; Srivastava, A.K.; Bhojiya, A.A. Biofertilizers: A Nexus between soil fertility and crop productivity under abiotic stress. *Curr. Res. Environ. Sustain.* **2021**, *3*, 100063. [CrossRef]
11. Sagar, A.; Yadav, S.S.; Sayyed, R.Z.; Sharma, S.; Ramteke, P.W. *Bacillus subtilis*: A Multifarious Plant Growth Promoter, Biocontrol Agent, and Bioalleviator of Abiotic Stress. In *Bacilli in Agrobiotechnology*; Springer: Cham, Switzerland, 2022; pp. 561–580. [CrossRef]
12. Hurek, T.; Reinhold-Hurek, B. *Azoarcus* sp. strain BH72 as a model for nitrogen-fixing grass endophytes. *J. Biotechnol.* **2003**, *106*, 169–178. [CrossRef]
13. Ahemad, M.; Kibret, M. Mechanisms and applications of plant growth promoting rhizobacteria: Current perspective. *J. King Saud Univ. Sci.* **2014**, *26*, 1–20. [CrossRef]
14. Pii, Y.; Mimmo, T.; Tomasi, N.; Terzano, R.; Cesco, S.; Crecchio, C. Microbial interactions in the rhizosphere: Beneficial influences of plant growth-promoting rhizobacteria on nutrient acquisition process. A review. *Biol. Fertil. Soils* **2015**, *51*, 403–415. [CrossRef]
15. Ramakrishna, W.; Yadav, R.; Li, K. Plant growth promoting bacteria in agriculture: Two sides of a coin. *Appl. Soil Ecol.* **2019**, *138*, 10–18. [CrossRef]
16. Kumar, S.; Diksha; Sindhu, S.S.; Kumar, R. Biofertilizers: An ecofriendly technology for nutrient recycling and environmental sustainability. *Curr. Res. Microb. Sci.* **2022**, *3*, 100094. [CrossRef] [PubMed]

17. Schallmey, M.; Singh, A.; Ward, O.P. Developments in the use of *Bacillus* species for industrial production. *Can. J. Microbiol.* **2004**, *50*, 1–17. [CrossRef]
18. Alabouvette, C.; Olivain, C.; Steinberg, C. Biological Control of Plant Diseases: The European Situation. *Eur. J. Plant Pathol.* **2006**, *114*, 329–341. [CrossRef]
19. Borriss, R. Use of plant-associated *Bacillus* strains as biofertilizers and biocontrol agents in agriculture. In *Bacteria in Agrobiology: Plant Growth Responses*; Springer: Berlin/Heidelberg, Germany, 2011; pp. 41–76. [CrossRef]
20. Hashem, A.; Tabassum, B.; Fathi Abd Allah, E. *Bacillus subtilis*: A plant-growth promoting rhizobacterium that also impacts biotic stress. *Saudi J. Biol. Sci.* **2019**, *26*, 1291–1297. [CrossRef]
21. Goswami, D.; Thakker, J.N.; Dhandhukia, P.C. Portraying mechanics of plant growth promoting rhizobacteria (PGPR): A review. *Cogent Food Agric.* **2016**, *2*, 1127500. [CrossRef]
22. Di Benedetto, N.A.; Corbo, M.R.; Campaniello, D.; Cataldi, M.P.; Bevilacqua, A.; Sinigaglia, M.; Flagella, Z. The role of Plant Growth Promoting Bacteria in improving nitrogen use efficiency for sustainable crop production: A focus on wheat. *AIMS Microbiol.* **2017**, *3*, 413–434. [CrossRef]
23. Miljaković, D.; Marinković, J.; Balešević-Tubić, S. The Significance of *Bacillus* spp. in Disease Suppression and Growth Promotion of Field and Vegetable Crops. *Microorganisms* **2020**, *8*, 1037. [CrossRef]
24. Fujita, Y.; Yasuda, Y.; Tochikubo, K. Permeability of Gentamicin and Polymyxin B into the Inside of *Bacillus subtilis* Spores. *Microbiol. Immunol.* **1990**, *34*, 1013–1023. [CrossRef] [PubMed]
25. Nurmatov, S.; Akhmedova, Z.R.; Rakhimov, D.A. Isolation, Purification, and Properties of α-Amylase from *Bacillus* subtilis-7A. *Chem. Nat. Compd.* **2001**, *37*, 364–368. [CrossRef]
26. Rey, M.W.; Ramaiya, P.; Nelson, B.A.; Brody-Karpin, S.D.; Zaretsky, E.J.; Tang, M.; De Leon, A.L.; Xiang, H.; Gusti, V.; Clausen, I.G.; et al. Complete genome sequence of the industrial bacterium *Bacillus licheniformis* and comparisons with closely related *Bacillus* species. *Genome Biol.* **2004**, *5*, r77. [CrossRef]
27. Stein, T. *Bacillus subtilis* antibiotics: Structures, syntheses and specific functions. *Mol. Microbiol.* **2005**, *56*, 845–857. [CrossRef]
28. Govindasamy, V.; Senthilkumar, M.; Magheshwaran, V.; Kumar, U.; Bose, P.; Sharma, V.; Annapurna, K. *Bacillus* and *Paenibacillus* spp.: Potential PGPR for sustainable agriculture. In *Plant Growth and Health Promoting Bacteria*; Springer: Berlin/Heidelberg, Germany, 2010; pp. 333–364.
29. Vlamakis, H.; Chai, Y.; Beauregard, P.; Losick, R.; Kolter, R. Sticking together: Building a biofilm the *Bacillus subtilis* way. *Nat. Rev. Microbiol.* **2013**, *11*, 157–168. [CrossRef] [PubMed]
30. Hobley, L.; Ostrowski, A.; Rao, F.V.; Bromley, K.M.; Porter, M.; Prescott, A.R.; MacPhee, C.E.; van Aalten, D.M.F.; Stanley-Wall, N.R. BslA is a self-assembling bacterial hydrophobin that coats the *Bacillus subtilis* biofilm. *Proc. Natl. Acad. Sci. USA* **2013**, *110*, 13600–13605. [CrossRef] [PubMed]
31. Kobayashi, K.; Iwano, M. BslA(YuaB) forms a hydrophobic layer on the surface of *Bacillus subtilis* biofilms. *Mol. Microbiol.* **2012**, *85*, 51–66. [CrossRef]
32. Hong, Z.-N.; Yan, J.; Lu, H.-L.; Jiang, J.; Li, J.-Y.; Xu, R.-K. Inhibition of phosphate sorptions on four soil colloids by two bacteria. *Environ. Pollut.* **2021**, *290*, 118001. [CrossRef]
33. Bardin, M.; Ajouz, S.; Comby, M.; Lopez-Ferber, M.; Graillot, B.; Siegwart, M.; Nicot, P.C. Is the efficacy of biological control against plant diseases likely to be more durable than that of chemical pesticides? *Front. Plant Sci.* **2015**, *6*, 566. [CrossRef]
34. Blake, C.; Christensen, M.N.; Kovács, Á.T. Molecular Aspects of Plant Growth Promotion and Protection by *Bacillus subtilis*. *Mol. Plant-Microbe Interact.* **2021**, *34*, 15–25. [CrossRef] [PubMed]
35. Bhattacharyya, P.N.; Jha, D.K. Plant growth-promoting rhizobacteria (PGPR): Emergence in agriculture. *World J. Microbiol. Biotechnol.* **2011**, *28*, 1327–1350. [CrossRef] [PubMed]
36. Labuda, M.; Saeid, A.; Chojnacka, K.; Górecki, H. Use of *Bacillus megaterium* in solubilization of phosphorus. *Przem. Chem.* **2012**, *91*, 837–840. (In Polish)
37. Saeid, A.; Prochownik, E.; Dobrowolska-Iwanek, J. Phosphorus Solubilization by *Bacillus* Species. *Molecules* **2018**, *23*, 2897. [CrossRef]
38. Heyndrickx, M.; Vandemeulebroecke, K.; Hoste, B.; Janssen, P.; Kersters, K.; DE Vos, P.; Logan, N.A.; Ali, N.; Berkeley, R.C.W. Reclassification of *Paenibacillus* (formerly *Bacillus*) pulvifaciens (Nakamura 1984) Ash et al. 1994, a Later Subjective Synonym of *Paenibacillus* (formerly *Bacillus*) larvae (White 1906) Ash et al. 1994, as a Subspecies of *P. larvae*, with Emended Descriptions of *P. larvae* as *P. larvae* subsp. larvae and *P. larvae* subsp. pulvifaciens. *Int. J. Syst. Evol. Microbiol.* **1996**, *46*, 270–279. [CrossRef]
39. Liu, X.; Li, Q.; Li, Y.; Guan, G.; Chen, S. *Paenibacillus* strains with nitrogen fixation and multiple beneficial properties for promoting plant growth. *PeerJ* **2019**, *7*, e7445. [CrossRef] [PubMed]
40. Hao, T.; Chen, S. Colonization of Wheat, Maize and Cucumber by Paenibacillus polymyxa WLY78. *PLoS ONE* **2017**, *12*, e0169980. [CrossRef] [PubMed]
41. Hussain, A.; Ahmad, M.; Nafees, M.; Iqbal, Z.; Luqman, M.; Jamil, M.; Maqsood, A.; Mora-Poblete, F.; Ahmar, S.; Chen, J.-T.; et al. Plant-growth-promoting *Bacillus* and *Paenibacillus* species improve the nutritional status of *Triticum aestivum* L. *PLoS ONE* **2020**, *15*, e0241130. [CrossRef]
42. Aquino, J.P.A.; Antunes, J.E.L.; Bonifácio, A.; Rocha, S.M.B.; Amorim, M.R.; Neto, F.A.; Araujo, A.S.F. Plant growth-promoting bacteria improve growth and nitrogen metabolism in maize and sorghum. *Theor. Exp. Plant Physiol.* **2021**, *33*, 249–260. [CrossRef]

43. Kaziūnienė, J.; Mažylytė, R.; Krasauskas, A.; Toleikienė, M.; Gegeckas, A. Optimizing the Growth Conditions of the Selected Plant-Growth-Promoting Rhizobacteria *Paenibacillus* sp. MVY-024 for Industrial Scale Production. *Biology* **2022**, *11*, 745. [CrossRef]
44. Dos Santos, R.M.; Diaz, P.A.E.; Lobo, L.L.B.; Rigobelo, E.C. Use of Plant Growth-Promoting Rhizobacteria in Maize and Sugarcane: Characteristics and Applications. *Front. Sustain. Food Syst.* **2020**, *4*, 136. [CrossRef]
45. Seldin, L.; van Elsas, J.D.; Penido, E.G.C. *Bacillus* polymyxa bacteriophages from Brazilian soils. *Antonie Leeuwenhoek* **1984**, *50*, 39–51. [CrossRef] [PubMed]
46. De Lima, B.D.; Penido, E.G.C. Inoculation in vitro of wheat seedlings and wheat straw cultures with *Bacillus azotofixans*. *Plant Soil* **1989**, *113*, 133–136. [CrossRef]
47. Rosado, A.S.; Duarte, G.F.; Seldin, L.; Van Elsas, J.D. Genetic Diversity of *nifH* Gene Sequences in *Paenibacillus azotofixans* Strains and Soil Samples Analyzed by Denaturing Gradient Gel Electrophoresis of PCR-Amplified Gene Fragments. *Appl. Environ. Microbiol.* **1998**, *64*, 2770–2779. [CrossRef] [PubMed]
48. Padda, K.P.; Puri, A.; Chanway, C.P. Paenibacillus polymyxa: A prominent biofertilizer in the whole organism and biocontrol agent for sustainable agriculture. In *Agriculturally Important Microbes for Sustainable Agriculture*; Springer: Singapore, 2017; pp. 165–191.
49. Seldin, L. Paenibacillus, nitrogen fixation and soil fertility. In *Endospore-Forming Soil Bacteria*; Springer: Berlin/Heidelberg, Germany, 2011; pp. 287–307.
50. Stanisz, A. Accessible Course in Statistics Based on the STATISTICA PL Software on Examples from Medicine. In *Basic Statistics*; StatSoft Polska: Kraków, Poland, 2007; Volume 1. (In Polish)
51. Rabiej, M. *Statistics with the Program Statistica*; Wydawnictwo Helion: Gliwice, Poland, 2012. (In Polish)
52. Wątroba, J. *Statystyka I Data Mining W Badaniach Naukowych*; Wydawnictwo StatSoft Polska: Kraków, Poland, 2006. (In Polish)
53. Starck, A. Integration of biomass partitioning includes total export from photosynthate donors and partitioning of assimilates between various sinks. *Zesz. Probl. Post. Nauk Rol.* **2002**, *481*, 113–123.
54. Anjum, S.A.; Xie, X.Y.; Wang, L.C.; Saleem, M.F.; Man, C.; Lei, W. Morphological, physiological and biochemical responses of plants to drought stress. *Afr. J. Agric. Res.* **2011**, *6*, 2026–2032. [CrossRef]
55. Sehgal, A.; Sita, K.; Siddique, K.H.M.; Kumar, R.; Bhogireddy, S.; Varshney, R.K.; HanumanthaRao, B.; Nair, R.M.; Prasad, P.V.V.; Nayyar, H. Drought or/and Heat-Stress Effects on Seed Filling in Food Crops: Impacts on Functional Biochemistry, Seed Yields, and Nutritional Quality. *Front. Plant Sci.* **2018**, *9*, 1705. [CrossRef]
56. Hlaváčová, M.; Klem, K.; Rapantová, B.; Novotná, K.; Urban, O.; Hlavinka, P.; Smutná, P.; Horáková, V.; Škarpa, P.; Pohanková, E.; et al. Interactive effects of high temperature and drought stress during stem elongation, anthesis and early grain filling on the yield formation and photosynthesis of winter wheat. *Field Crop. Res.* **2018**, *221*, 182–195. [CrossRef]
57. Sharma, A.; Kumar, V.; Shahzad, B.; Ramakrishnan, M.; Sidhu, G.P.S.; Bali, A.S.; Handa, N.; Kapoor, D.; Yadav, P.; Khanna, K.; et al. Photosynthetic Response of Plants Under Different Abiotic Stresses: A Review. *J. Plant Growth Regul.* **2020**, *39*, 509–531. [CrossRef]
58. Jeppesen, E.; Kronvang, B.; Olesen, J.E.; Audet, J.; Søndergaard, M.; Hoffmann, C.C.; Andersen, H.E.; Lauridsen, T.L.; Liboriussen, L.; Larsen, S.E. Climate Change effect on nitrogen loading from cultivated catchments in Europe: Implications for nitrogen retention, ecological stat of lakes and adaptation. *Hydrobiologia* **2011**, *663*, 1–21. [CrossRef]
59. Skowera, B.; Kopcinska, J.; Kolodziejczyk, M.; Kopec, B. Precipitation deficiencies and excesses during the growing season of winter rape and winter wheat in Poland (1971–2010). *Acta Agrophys.* **2015**, *22*, 193–207.
60. Turan, M.; Gulluce, M.; von Wirén, N.; Sahin, F. Yield promotion and phosphorus solubilization by plant growth–promoting rhizobacteria in extensive wheat production in Turkey. *J. Plant Nutr. Soil Sci.* **2012**, *175*, 818–826. [CrossRef]
61. Schütz, L.; Gattinger, A.; Meier, M.; Müller, A.; Boller, T.; Mäder, P.; Mathimaran, N. Improving Crop Yield and Nutrient Use Efficiency via Biofertilization—A Global Meta-analysis. *Front. Plant Sci.* **2018**, *8*, 2204. [CrossRef] [PubMed]
62. Mayer, J.; Scheid, S.; Widmer, F.; Fliessbach, A.; Oberholzer, H.-R. How effective are 'Effective microorganisms® (EM)'? Results from a field study in temperate climate. *Appl. Soil Ecol.* **2010**, *46*, 230–239. [CrossRef]
63. Santoyo, G.; Moreno-Hagelsieb, G.; del Carmen Orozco-Mosqueda, M.; Glick, B.R. Plant growth-promoting bacterial endophytes. *Microbiol. Res.* **2016**, *183*, 92–99. [CrossRef]
64. Shahzad, R.; Khan, A.L.; Bilal, S.; Waqas, M.; Kang, S.-M.; Lee, I.-J. Inoculation of abscisic acid-producing endophytic bacteria enhances salinity stress tolerance in Oryza sativa. *Environ. Exp. Bot.* **2017**, *136*, 68–77. [CrossRef]
65. Lastochkina, O.; Pusenkova, L.; Yuldashev, R.; Babaev, M.; Garipova, S.; Blagova, D.; Khairullin, R.; Aliniaeifard, S. Effects of *Bacillus subtilis* on some physiological and biochemical parameters of *Triticum aestivum* L. (wheat) under salinity. *Plant Physiol. Biochem.* **2017**, *121*, 80–88. [CrossRef]
66. de Lima, B.C.; Moro, A.L.; Santos, A.C.P.; Bonifacio, A.; Araujo, A.S.F.; de Araujo, F.F. *Bacillus subtilis* ameliorates water stress tolerance in maize and common bean. *J. Plant Interact.* **2019**, *14*, 432–439. [CrossRef]
67. Ali, S.; Khan, N. Delineation of mechanistic approaches employed by plant growth promoting microorganisms for improving drought stress tolerance in plants. *Microbiol. Res.* **2021**, *249*, 126771. [CrossRef]
68. Maslennikova, D.; Lastochkina, O. Contribution of Ascorbate and Glutathione in Endobacteria *Bacillus subtilis*-Mediated Drought Tolerance in Two *Triticum aestivum* L. Genotypes Contrasting in Drought Sensitivity. *Plants* **2021**, *10*, 2557. [CrossRef]
69. Khan, N.; Mehmood, A.; Ali, S.; Shahid, M.A. Plant growth-promoting rhizobacteria and their role as bio-inoculants for sustainable agriculture under stressful environments. In *Biofertilizers*; Elsevier: Amsterdam, The Netherlands, 2021; pp. 313–321. [CrossRef]

70. Masclaux-Daubresse, C.; Daniel-Vedele, F.; Dechorgnat, J.; Chardon, F.; Gaufichon, L.; Suzuki, A. Nitrogen uptake, assimilation and remobilization in plants: Challenges for sustainable and productive agriculture. *Ann. Bot.* **2010**, *105*, 1141–1157. [CrossRef]
71. Kuan, K.B.; Othman, R.; Rahim, K.A.; Shamsuddin, Z.H. Plant Growth-Promoting Rhizobacteria Inoculation to Enhance Vegetative Growth, Nitrogen Fixation and Nitrogen Remobilisation of Maize under Greenhouse Conditions. *PLoS ONE* **2016**, *11*, e0152478. [CrossRef] [PubMed]
72. Gupta, K.; Dubey, N.K.; Singh, S.P.; Kheni, J.K.; Gupta, S.; Varshney, A. Plant Growth-Promoting Rhizobacteria (PGPR): Current and Future Prospects for Crop Improvement. In *Current Trends in Microbial Biotechnology for Sustainable Agriculture*; Springer: Singapore, 2021; pp. 203–226. [CrossRef]
73. Pan, J.; Zhu, Y.; Cao, W.; Dai, T.; Jiang, D. Predicting the Protein Content of Grain in Winter Wheat with Meteorological and Genotypic Factors. *Plant Prod. Sci.* **2006**, *9*, 323–333. [CrossRef]
74. Rollins, J.A.; Habte, E.; Templer, S.E.; Colby, T.; Schmidt, J.; von Korff, M. Leaf proteome alterations in the context of physiological and morphological responses to drought and heat stress in barley (*Hordeum vulgare* L.). *J. Exp. Bot.* **2013**, *64*, 3201–3212. [CrossRef] [PubMed]
75. Yang, J.; Zhang, J.; Wang, Z.; Zhu, Q.; Liu, L. Abscisic acid and cytokinins in the root exudates and leaves and their relationship to senescence and remobilization of carbon reserves in rice subjected to water stress during grain filling. *Planta* **2002**, *215*, 645–652. [CrossRef] [PubMed]
76. Liu, E.; Mei, X.; Yan, C.; Gong, D.; Zhang, Y. Effects of water stress on photosynthetic characteristics, dry matter translocation and WUE in two winter wheat genotypes. *Agric. Water Manag.* **2016**, *167*, 75–85. [CrossRef]
77. Buczek, J.; Jańczak-Pieniążek, M. Hybrid wheat response to high nitrogen application rates and foliar fertilization. *Biul. Inst. Hod. I Aklim. Roślin* **2021**, *296*, 17–24. [CrossRef]
78. Islam, M.R.; Haque, K.S.; Akter, N.; Karim, M.A. Leaf chlorophyll dynamics in wheat based on SPAD meter reading and its relationship with grain yield. *Sci. Agric.* **2014**, *4*, 13–18. [CrossRef]
79. Monostori, I.; Árendás, T.; Hoffman, B.; Galiba, G.; Gierczik, K.; Szira, F.; Vágújfalvi, A. Relationship between SPAD value and grain yield can be affected by cultivar, environment and soil nitrogen content in wheat. *Euphytica* **2016**, *211*, 103–112. [CrossRef]
80. Wojcieska, U. Fizjologiczna rola azotu w kszlatowaniu plonu roslin. Czesc II. Zywienie roslin azotem a fotosynteza, fotorespiracja i oddychanie ciemniowe. *Postępy Nauk Rol.* **1994**, *41*, 127–143. (In Polish)
81. Dong, Z.; Zhang, X.; Li, J.; Zhang, C.; Wei, T.; Yang, Z.; Cai, T.; Zhang, P.; Ding, R.; Jia, Z. Photosynthetic characteristics and grain yield of winter wheat (*Triticum aestivum* L.) in response to fertilizer, precipitation, and soil water storage before sowing under the ridge and furrow system: A path analysis. *Agric. For. Meteorol.* **2019**, *272–273*, 12–19. [CrossRef]
82. Olszewski, J.; Makowska, M.; Pszczółkowska, A.; Okorski, A.; Bieniaszewski, T. The effect of nitrogen fertilization on flag leaf and ear photosynthesis and grain yield of spring wheat. *Plant Soil Environ.* **2014**, *60*, 531–536. [CrossRef]
83. Carmo-Silva, E.; Andralojc, P.J.; Scales, J.C.; Driever, S.; Mead, A.; Lawson, T.; A Raines, C.; Parry, M. Phenotyping of field-grown wheat in the UK highlights contribution of light response of photosynthesis and flag leaf longevity to grain yield. *J. Exp. Bot.* **2017**, *68*, 3473–3486. [CrossRef] [PubMed]
84. Murchie, E.H.; Kefauver, S.; Araus, J.L.; Muller, O.; Rascher, U.; Flood, P.J.; Lawson, T. Measuring the dynamic photosynthome. *Ann. Bot.* **2018**, *122*, 207–220. [CrossRef] [PubMed]
85. Sanchez-Bragado, R.; Vicente, R.; Molero, G.; Serret, M.D.; Maydup, M.L.; Araus, J.L. New avenues for increasing yield and stability in C3 cereals: Exploring ear photosynthesis. *Curr. Opin. Plant Biol.* **2020**, *56*, 223–234. [CrossRef] [PubMed]
86. Ashraf, M. Effect of salicylic acid applied through rooting medium on drought tolerance of wheat. *Pak. J. Bot.* **2006**, *38*, 1127–1136.
87. Kołodziejczyk, M. Effect of nitrogen fertilization and application of soil properties improving microbial preparations on the content of mineral nitrogen in soil after spring wheat harvesting. *J. Cent. Eur. Agric.* **2013**, *14*, 306–318. [CrossRef]
88. Kocoń, A.; Jadczyszyn, T. The effect of microbiological preparations, method of their application and doses of nitrogen fertilization on the content of available phosphorus in the soil, and other selected chemical indicators of soil fertility. *Pol. J. Agron.* **2015**, *21*, 11–18. (In Polish)
89. Saxena, A.; Kumar, M.; Chakdar, H.; Anuroopa, N.; Bagyaraj, D.; Murugan, K. *Bacillus* species in soil as a natural resource for plant health and nutrition. *J. Appl. Microbiol.* **2020**, *128*, 1583–1594. [CrossRef]
90. Tahir, M.; Khalid, U.; Ijaz, M.; Shah, G.M.; Naeem, M.A.; Shahid, M.; Mahmood, K.; Ahmad, N.; Kareem, F. Combined application of bio-organic phosphate and phosphorus solubilizing bacteria (*Bacillus* strain MWT 14) improve the performance of bread wheat with low fertilizer input under an arid climate. *Braz. J. Microbiol.* **2018**, *49*, 15–24. [CrossRef]

applied sciences

MDPI

Article

Detection and Factors That Induce *Stenocarpella* spp. Survival in Maize Stubble and Soil Suppressiveness under Tropical Conditions

Felipe Augusto Moretti Ferreira Pinto [1,*], Victor Biazzotto Correia Porto [2], Rafaela Araújo Guimarães [3], Carolina da Silva Siqueira [3], Mirian Rabelo de Faria [4], José da Cruz Machado [3], Henrique Novaes Medeiros [3], Dagma Dionísia da Silva [5], Helon Santos Neto [3], Edson Ampelio Pozza [3] and Flávio Henrique Vasconcelos de Medeiros [3,*]

[1] Empresa de Pesquisa e Extensão Rural-EPAGRI, Estação Experimental de São Joaquim, São Joaquim 88600-000, SC, Brazil

[2] Fundação de Apoio à Pequisa e Desenvolvimento do Oeste Baiano—Fundação BA, Luiz Eduado Magalhães 47850-000, BA, Brazil; victorbcporto@gmail.com

[3] Department of Plant Pathology, Universidade Federal de Lavras, P.O. Box 3037, Lavras 37200-900, MG, Brazil; rafaela.guimaraes3@ufla.br (R.A.G.); kerolpet@gmail.com (C.d.S.S.); machado@ufla.br (J.d.C.M.); medeirosxd@gmail.com (H.N.M.); helonareado@hotmail.com (H.S.N.); eapozza@ufla.br (E.A.P.)

[4] FCA/Campus Botucatu, Universidade Estadual Paulista Júlio Mesquita Filho, Botucatu 186108-034, SP, Brazil; mirianrabelofaria@yahoo.com.br

[5] Embrapa Milho e Sorgo, P.O. Box 285, Sete Lagoas 35701-970, MG, Brazil; dagma.silva@embrap.br

* Correspondence: felipepinto@epagri.sc.gov.br (F.A.M.F.P.); flaviomedeiros@ufla.br (F.H.V.d.M.)

Citation: Pinto, F.A.M.F.; Porto, V.B.C.; Guimarães, R.A.; Siqueira, C.d.S.; Faria, M.R.d.; Machado, J.d.C.; Medeiros, H.N.; Silva, D.D.d.; Santos Neto, H.; Pozza, E.A.; et al. Detection and Factors That Induce *Stenocarpella* spp. Survival in Maize Stubble and Soil Suppressiveness under Tropical Conditions. *Appl. Sci.* **2022**, *12*, 4974. https://doi.org/10.3390/app12104974

Academic Editors: Angeles Sanroman Braga and Vicente Mariscal

Received: 16 March 2022

Accepted: 12 May 2022

Published: 14 May 2022

Abstract: *Stenocarpella* spp. causes stalk and ear rot in maize and overwinters in stubble during the off-season. Understanding the factors that guide saprophytic colonization is a crucial strategy for management. In this study, we analyzed the abiotic factors and crop management practices in relation to the inoculum of *Stenocarpella* spp. in stubble by qPCR. Soil samples were used for suppressiveness tests against *Fusarium verticillioides*, *Fusarium graminearum*, and *Stenocarpella maydis*. In the 29 fields, different levels of *Stenocarpella* spp. were detected. Only three fields were considered suppressive for the three pathogens. Heat maps showed that soil suppressiveness was inversely related to the pathogen concentration, and the suppressiveness of one pathogen was correlated with the suppressiveness of other pathogens. Under no-tillage systems in which rotation with soybeans was adopted, *Stenocarpella* spp. were detected at lower concentrations than in areas that adopted no-tillage systems with maize monocultures. While in tillage systems, the maize–maize monocropping increases the inoculum level of *Stenocarpella* spp. Crop rotation is a factor related to the observed reduction in the pathogen concentration and increases in the broad-spectrum antagonistic microbial communities. These communities guide the suppressiveness of soil-borne diseases in maize fields cultivated under tropical conditions.

Keywords: *Zea mays*; stalk diseases; crop rotation; stubble; suppressive soils

1. Introduction

Maize (*Zea mays* L.) stalk and ear rot diseases result in significant post-harvest economic losses and reduce the quality and quantity of grains. Several fungi can cause stalk and ear rot, the most important being *Stenocarpella* spp. and *Fusarium* spp. [1]. These pathogens colonize the aerial parts of maize plants due to conidial dispersal by wind, water splashing, and contaminated seeds. Indeed, seeds disperse the inoculum of the pathogen over long distances, but there is no report of *Stenocarpella*-caused disease outbreaks in the first season in which maize is cultivated. In order for an outbreak to occur, the inoculum levels in the maize stubble need to reach sufficient levels [2].

Stenocarpella macrospora and *S. maydis* are the predominant species within the genus and can be collectively detected using a genus-specific primer [3]. A molecular biology approach based on qPCR (quantitative PCR) is an easy, fast, and sensitive test for specific targets than conventional methods. When more than one pathogen causes infection in a plant and requires accurate detection, qPCR is a suitable method for detection and quantification [4].

S. macrospora and *S. maydis* are seed-transmitted, cause stalk rot and ear rot, and *S. macroscopora* is associated with macrospora leaf spot. These diseases may not only compromise photosynthesis by the reduction in the green leaf area but also serve as a reservoir for the pathogen inoculum build up and infect the kernels causing ear rot or at least transmit the pathogen. These species can survive on seed and/or colonize the maize stubble. In this environment, the spores can increase or remain dormant until the following raining and planting season [5]. Such stubble also serves as a reservoir for other maize pathogens that compete for the same nutrients (stalk and kernels), and the most important are *F. verticillioides* and *F. graminearum* [6].

Conversely, maize stubble does not last long in tropical agriculture systems due to the speed with which such organic matter is decomposed. Therefore, there until now is no evidence of the role of the different maize stubble (in relation to size, crop system, and climatic conditions) in the survival of *Stenocarpella* app. on maize. Furthermore, *Stenocarpella* spp. is host-specific; therefore, the presence of a maize monoculture likely plays a detrimental role in the overwintering of the pathogen [7], especially in tropical agriculture systems, where the weather is favorable for volunteer maize plants to not only harbor but also build up inoculum of *Stenocarpella* spp. [8].

Within the integrated management of diseases, soil biological properties play an important role. It can influence the selection of antagonistic microbial communities associated with the pathogen reduction and elimination, resulting in the reduction in the disease. These soil conditions, the use of fungicides (chemical and biological), and the resistance of cultivars do not necessarily result in satisfactory disease control [9].

In Brazil, no-tillage cropping is the default practice, but it relies on the sequence of soybean and maize planted in a succession, not in rotation [10]. This system reduces the soil preparation time and maintains soil humidity, thus reducing erosion [11]. However, crop stubble on the soil surface supports the survival of many plant-stubble-borne pathogens, which are saprophytes and can cause disease outbreaks [12].

There is considerable knowledge on the epidemiology of other species that cause ear rot on maize plants, such as *F. graminearum* and *F. verticillioides* [13]. In contrast, knowledge of the occurrence and epidemiology of the *Stenocarpella* complex is scarce, particularly with regard to tropical agriculture. This knowledge is essential for the development of preventive measures to reduce the risk of fungal infections during the growing season. The objectives of this study were to identify the source of inoculum for *Stenocarpella*, the contribution of maize growing areas and crop rotation on the *Stenocarpella* spp., as well as to understand the suppressiveness of soil towards *S. maydis*, *F. verticillioides*, and *F. graminearum*.

2. Materials and Methods

2.1. Identification of the Inoculum Source for Stenocarpella spp. within Maize Stubble

To determine the major sources of inoculum, samples were taken in a maize field under a no-tillage management system in Lavras (Minas Gerais, Brazil) during the season of 2015/2016. The evaluated samples consisted of stalks, grains, cobs, decaying maize leaves, and dead weeds (species of weed were not identified). Each type of the different plant-derived material (e.g., stalks together) was used separately as a sample for the study of *Stenocarpella* spp. inoculum sources in the field sampling. For each type of sample, the DNA was extracted, and relative quantification of *Stenocarpella* spp. was conducted.

2.2. Classification of Decomposition on Maize Stubble

Once the maize plant part within the stubble that harbored the highest inoculum source was determined, a second evaluation was performed on the inoculum level according to the decomposition rate of the corresponding plant part. Maize stubbles were classified into three levels of decomposition: low, medium, and high decomposition of stubble. The low decomposition classified the stubble regarding fragmental size as entire. The medium decomposition was partially disintegrated, and the high decomposition was disintegrated.

2.3. Sampling in Maize Fields

After determining the maize plant part within the stubble that harbored the highest concentration of *Stenocarpella* spp. inocula, 29 samples of this plant part were collected in 15 different locations (Figure 1). The sampling system took into account the rotation system (whether the grower adopted crop rotation) and soil tillage (no-tillage or conventional tillage). The number of replicates and the sampling strategy were similar to those described above.

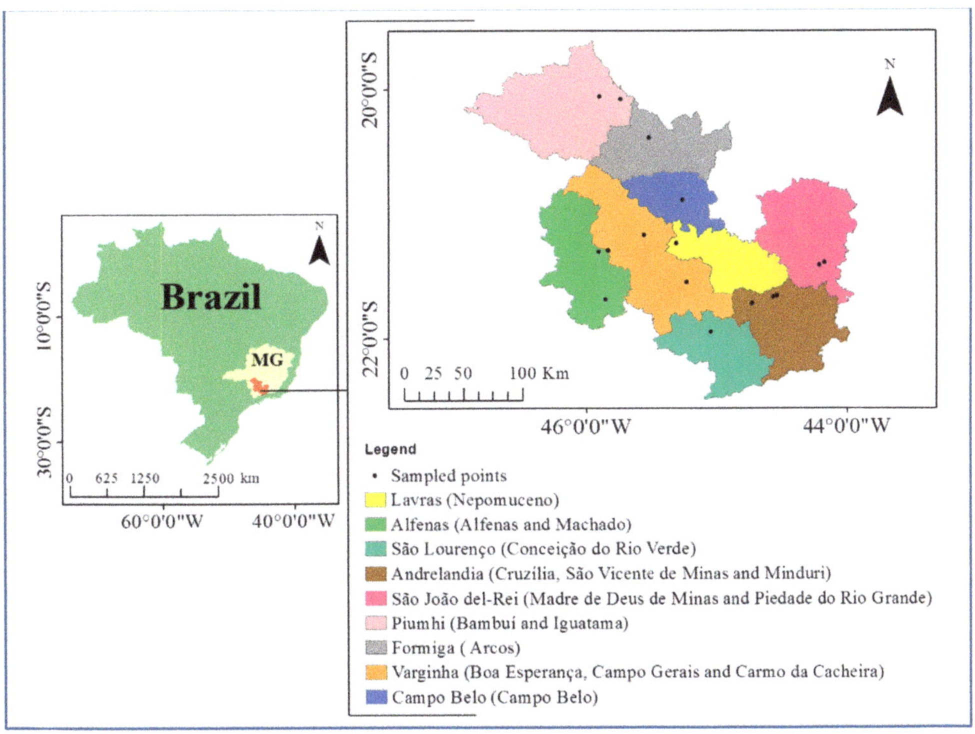

Figure 1. Location map of sampled areas in Minas Gerais state (MG), Brazil.

2.4. Sample Processing, DNA Extraction, and Quantitative Real-Time PCR (qPCR)

The samples were ground with a 1-mm mesh sieve, freeze-dried, and stored at −18 °C until DNA extraction was conducted. Approximately 40 mg of lyophilized sample were used for DNA extraction, employing a Wizard® Genomic DNA Purification Kit (Promega, Madison, WI, USA), following the protocol recommended by the manufacturer [14].

All DNA samples from the same area were pooled to determine the presence of *Stenocarpella* spp. in a given field. The qPCR analysis was conducted using an SYBR Green PCR Master Mix on a Rotor-Gene 6500 (Corbett Research, Mortlake, Australia). For each reaction, a 2.0-µL sample was mixed with 23 µL of reaction mix containing 12.5 µL SYBR Green PCR Kit (Qiagen, Hilden, Germany) and 0.75 µM of each forward and reverse primer. The primers used were P1 (GTTGGGGGTTTAACGGCACG) and P2 (GTTGCCTCGGCACAGGCCGG), sequenced and designed from the ITS1 and ITS2 regions of rDNA as described by [3], which are specifically for the detection of the levels of DNA copies from *Stenocarpella* spp. in biological samples, according to [5]. A 5-fold serial dilution (20 ng to 0.002 ng of DNA) of *S. maydis* (isolate LAPS 698) was included in each run as a positive control and to calculate the number of DNA copies generated in the cycles.

The qPCR conditions were as follows: 95 °C for 3 min; 40 cycles of 94 °C for 30 s, 60 °C for 1 min, and 72 °C for 1 min; and, finally, 72 °C for 10 min to melt the double-stranded DNA. The specificity of the amplicons was confirmed with melting-curve analysis of the qPCR products at the different DNA concentrations. The threshold curve was calculated with the Ct value, which was determined as the number of cycles in which the fluorescence generated within a reaction crossed the threshold. The comparative Ct method was also used. Samples showing the lowest expression of each gene were used as calibration samples, and relative expression was measured using the relative standard curve method. The values obtained corresponding to the sample DNA levels were compared to the control DNA level. To calculate the gene expression levels, the following were considered: the Ct values (exponential increase in PCR product) of the target gene and endogenous control, ΔCt = Ct (sample) − Ct (endogenous control), and $\Delta\Delta$Ct = ΔCt (sample) − ΔCt (calibrator). The gene expression levels were then calculated using the formula RQ = 2 − $\Delta\Delta$Ct, where RQ means relative quantifications [14].

2.5. Suppressiveness against Soil-Borne Diseases

To identify the suppressiveness of soils from the different sample sites, we collected soils (0–5 cm depth) from the same sites where maize stalks of bulk soil were sampled (Supplementary Table S1). The suppressiveness was evaluated for the three pathogens that frequently overwinter on maize stubble and cause similar stalk and ear rot damage to maize: *F. verticillioides*, *F. graminearum*, and *S. maydis*. The methodology proposed by [15] was used with modifications. The soil samples were individually placed into Petri plates (9 mm), watered to 60% field capacity, and inoculated with 5 mL of 10^5 conidia mL^{-1} of each pathogen (*F. verticillioides*, *F. graminearum*, and *S. maydis*). Finally, the sampled soils were classified as suppressive (considerable bait colonization reduction), intermediate (some effect in bait colonization reduction), and conducive (considerable bait colonization increase), as proposed by [16].

2.6. Data Processing and Statistical Analysis

The experiments were carried out in a randomized block design (RBD). For statistical analyses, the Shapiro–Wilk test was performed for normalization. For data regarding the different types of stalks used in qPCR, the Ct values and the classification of decomposition on maize stubble were submitted to one-way ANOVA and Tukey's multiple comparisons of means at a 95% family-wise confidence level. The incidence data gathered in each replicate from the soil suppressiveness assay were submitted to variance analysis (ANOVA). For the evaluation of significant effects, the means were compared according to Scott–Knott's test ($p \leq 0.05$). Both analyses were conducted using the software R.

The relative quantifications (Ct values) and locations of the collected samples were submitted to an agglomerative hierarchical cluster analysis based on Ward's grouping method. A correlation matrix [17] was constructed to group the 16 different municipalities. The clustering was based on a dissimilarity matrix of Euclidean distances between individual municipalities. The same methods were also used for the sampled sites, and clusters were used to define categories for the creation of a map of the quantification of *Stenocarpella* spp. inocula. Thus, heat maps were created for the classes of the relative quantifications of *Stenocarpella* spp. inoculum through Pearson's correlation with the annual average temperature, annual average precipitation, altitude, and average maize yield. The heat maps were created using the inverse distance weighting interpolation method. The software used to build the maps was ArcGIS 10.3.

3. Results

Among the different maize parts found in the stubble, stalks were the only substrate encountered in all sampled areas that harbored *Stenocarpella* spp. However, *Stenocarpella* spp. was observed in higher numbers on grains (Ct = 26.86) and cobs (Ct = 25.08) when they were found in the main crop stubble. The decaying maize leaves (Ct = 32.21) and dead weeds (Ct not detected) did not harbor inoculum at high enough levels to be detected (Figure 2). Since stalks were the only maize found at all sampled sites, we decided to consider this plant part with which to make comparisons and inferences.

A

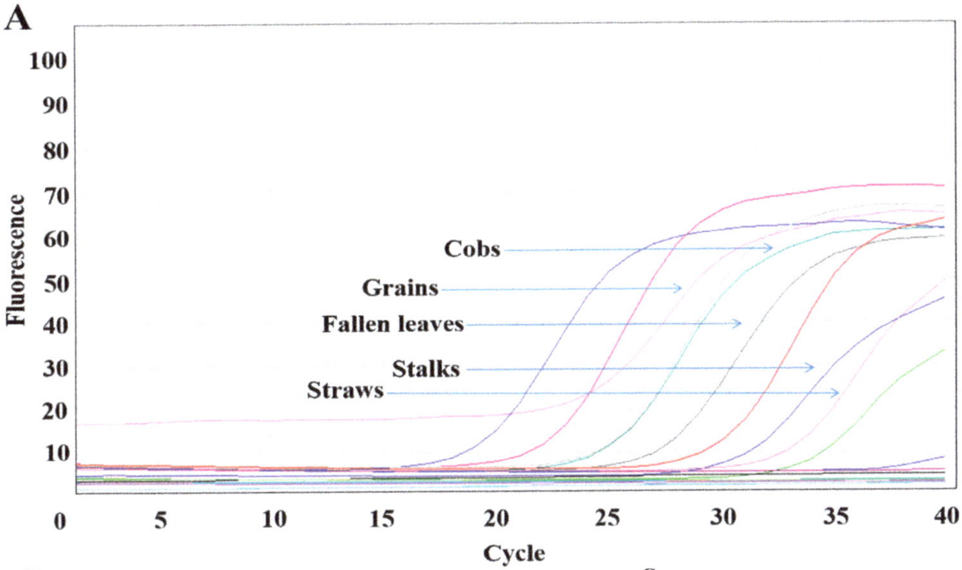

B

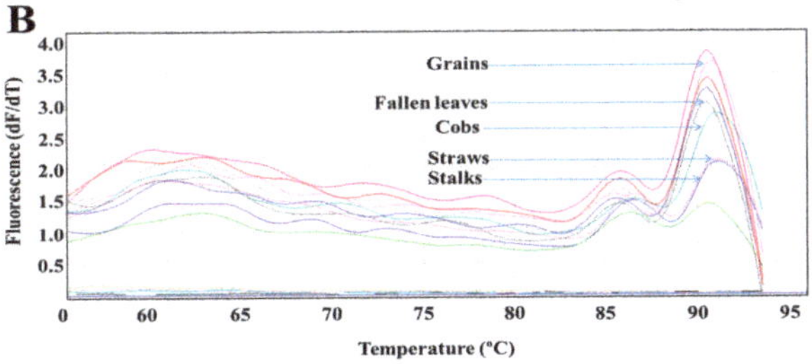

C

Sample	CT	Temperature
Grains	26.86	90.7
Cobs	25.08	91.0
Stalks	30.17	91.0
Fallen maize leaves	32.21	91.3
Straws	33.95	90.5
Positive control 1	19.49	90.5
Positive control 2	22.41	90.5
Positive control 3	25.22	90.5
Positive control 4	27.40	90.5
Positive control 5	30.00	90.5

Figure 2. Melting curve analysis for *Stenocarpella* spp. using real-time polymerase reaction (PCR) assay. (**A**) The amplification curve of different substrates and controls obtained with primers P1/P2 [3]. (**B**) Melting curve analysis of the same samples shows the presence of the specific PCR product. (**C**) Comparison chart with cycle threshold (Ct) values and melting temperatures. Ct means the cycle threshold values. P1 = Primer used (GTTGGGGGTTTAACGGCACG) as reference for *Stenocarpella* spp. P2 = Primer used (GTTGCCTCGGCACAGGCCGG) as reference for *Stenocarpella* spp.

RQ was higher in fields (fields 15 and 24) that were cultivated with the adoption of conventional tillage practices. On the other hand, in sites managed under no-tillage systems, *Stenocarpella* spp. was always detected, almost always in smaller concentrations (20 of the total 27 fields that adopted no-tillage). However, in 7 fields (1, 16, 18, 20, 21, 25, and 29), RQ was higher. This value represents 25% of no-tillage fields. In the monocropping fields (fields 10, 11, 14, 17, 19, 21, 23, and 24), RQ levels were different, and only 2 fields (21 and 24) increased RQ levels (Table 1).

Table 1. Relative quantification (RQ) of *Stenocarpella* spp. (DNA copies) determined by qPCR in stubble and colonization of baits (%) by soil-borne pathogens (*Stenocarpella maydis, Fusarium graminearum,* and *Fusarium verticillioides*) in collected soils.

Field [1]	Crop in Sequence [2]	Tillage [3]	Maize Stubble	Baits Colonization in Soils (%)		
			RQ of *Stenocarpella* spp.	*Stenocarpella maydis*	*Fusarium graminearum*	*Fusarium verticillioides*
1	M-S	No	1	95 a *	85 b	33 c
2	M-S	No	8.49×10^{-11}	100 a	100 a	100 a
3	M-B-S	No	6.06×10^{-11}	10 c	14 c	10 c
4	M-S	No	7.10×10^{-10}	5 c	24 c	5 c
5	M-B	No	1.61×10^{-10}	71 b	62 b	38 c
6	M-S	No	1.31×10^{-10}	95 a	95 a	100 a
7	S-M-S	No	3.93×10^{-9}	71 b	100 a	90 b
8	S-M-S	No	8.59×10^{-9}	33 c	67 b	29 c
9	M-W-S	No	4.87×10^{-11}	71 b	62 b	71 b
10	M-M	No	6.34×10^{-11}	76 b	95 a	86 b
11	M-M	No	1.58×10^{-10}	100 a	100 a	86 b
12	M-W-M	No	3.90×10^{-10}	29 c	29 c	43 c
13	S-W-M	No	9.80×10^{-7}	95 a	95 a	100 a
14	M-M	No	7.89×10^{-11}	86 b	71 b	76 b
15	M-B	Yes	1	38 c	76 b	33 c
16	M-S	No	1	100 a	100 a	100 a
17	M-M	No	1.70×10^{-11}	33 c	100 a	100 a
18	M-S	No	1	95 a	100 a	95 a
19	M-M	No	2.15×10^{-12}	100 a	100 a	100 a
20	M-S	No	1	76 b	100 a	81 b
21	M-M	No	1	86 b	90 a	95 a
22	M-S	No	5.86×10^{-12}	100 a	95 a	100 a
23	M-M	No	1.23×10^{-11}	100 a	100 a	95 a
24	M-M	Yes	1	95 a	100 a	100 a
25	M-S	No	1	100 a	100 a	100 a
26	M-M	No	7.71×10^{-12}	100 a	100 a	100 a
27	M-S	No	8.52×10^{-11}	81 b	71 b	71 a
28	M-S	No	4.18×10^{-8}	95 a	100 a	100 a
29	M-S	No	1	81 b	86 b	86 b

[1] Field locations are available in Supplementary Table S1. [2] Crop in sequence: M-S (Maize-Soybean), M-B-S (Maize-Bean-Soybean), M-B (Maize-Bean), S-M-S (Soybean-Maize-Soybean), M-W-S (Maize-Wheat-Soybean), M-M (Maize-Maize), S-W-M (Soybean-Wheat-Maize), and M-B (Maize-Bean). [3] Tillage system: No (No-tillage) and Yes (Tillage). * Lowercase letters in columns indicate no statistical differences by Scott–Knott cluster analysis ($p \leq 0.05$).

According to one-way ANOVA and Tukey's multiple comparisons of means, stubble particle size contributed to pathogen survival, as measured by the DNA quantification of *Stenocarpella* spp. The relative quantification of stalks classified as entire was different from those classified as disintegrated and was equal to those classified as partially disintegrated. Areas with disintegrated stalks were different from areas with partially disintegrated stalks.

The rate of the relative quantification (RQ) of *Stenocarpella* spp. did not necessarily group according to the site location, i.e., neighboring municipalities. For example, Piedade do Rio Grande and Madre de Deus de Minas were grouped in different clades, and Campo Belo and Boa Esperança, which are neighboring sites, were not grouped together; therefore, it was not the location of the site but rather other factors that likely governed such inoculum levels (Figure 3).

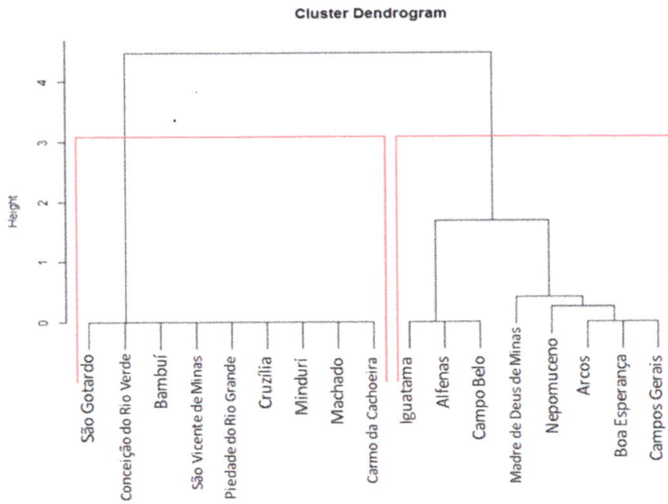

Figure 3. Cluster dendrogram of the DNA qPCR-obtained relative quantifications of *Stenocarpella maydis* by sampled location. The dendrogram was generated by data submitted to an agglomerative hierarchical cluster analysis based on Ward's grouping method and correlation matrix [17].

Based on the obtained results (Table 1), the sampled sites were classified into three groups for each pathogen, following these characteristics: suppressive (considerable bait colonization reduction), intermediate (some effect in bait colonization reduction), and conducive (considerable bait colonization increase). From the 29 sampled fields, only 3 fields (3, 4, and 12) were classified as suppressive (all were classified as 'c' in Scott–Knott cluster analysis) to all tested pathogens (Table 1). In addition, fields 3, 4, 8, 12, 15, and 17 were suppressive against *S. maydis* and fields 3, 4, and 12 were suppressive against *F. graminearum*. For *F. verticillioides*, fields 3, 4, 8, 12, 15, and 17 were considered suppressive. Whereas only 3 fields (5, 9, and 14) were classified as intermediate (all were classified as 'b' in Scott–Knott cluster analysis) to all tested pathogens. Thus, fields 5, 7, 9, 10, 14, 20, and 21 were intermediate against *S. maydis*, and fields 1, 5, 8, 9, 14, 15, 27, and 29 were intermediate against *F. graminearum*. For *F. verticillioides*, the fields 5, 7, 9, 10, 14, 20, 21, 27, and 29 were considered intermediate. While 13 fields (fields 2, 6, 11, 13, 16, 18, 19, 22, 23, 24, 25, 26, and 28) were classified as conducive (all were classified as 'a' in Scott–Knott cluster analysis) to the 3 pathogens tested. The fields 1, 2, 6, 11, 13, 16, 18, 19, 22, 23, 24, 25, 26, and 27 were conducive to *S. maydis*, fields 2, 6, 7, 10, 11, 13, 16, 17, 18, 19, 20, 21, 22, 23, 24, 25, 26, and 28 were conducive to *F. graminearum*, and fields 1, 2, 6, 11, 13, 16, 18, 19, 22, 23, 24, 25, 26, and 28 were conducive to *F. verticillioides* (Table 1).

Among the sampled soils, 45% were conducive to all 3 tested pathogens. Whereas 10% of these soils were suppressive, and another 10% were intermediate in reducing the inoculum of all evaluated pathogens. By considering the soil suppressiveness levels of the different sampled sites for *F. verticillioides*, *F. graminearum*, and *S. maydis*, a significant correlation was found (Table 2).

Finally, heat maps were generated based on the data of the *Stenocarpella* inoculum quantifications: numbers of maize stalk baits colonized by each of the pathogens (*S. maydis*, *F. verticillioides*, and *F. graminearum*), altitudes, rainfall levels, maize grain yields, and temperatures (Figure 4). The *Stenocarpella* inoculum map was designed based on the relative quantification of the pathogen in the stalks on soil and ranged from 0 to 3 (the closer to 0, the more inoculum and the closer to 3, the less inoculum of *Stenocarpella* ssp., as also suggested in Table 1). The lowest figures represent the highest concentrations of the pathogen. A pattern of lower inoculum levels (green color) was observed on the edges of the evaluated area, and higher inoculum levels were observed in the middle of the study area.

Table 2. Pearson's correlation regarding of colonization baits in soils (%) and soil-borne diseases: *Fusarium verticillioides, Fusarium graminearum*, and *Stenocarpella maydis*.

Colonization Baits (%)	Pearson's Correlation		
	Fusarium graminearum	*Fusarium verticillioides*	*Stenocarpella maydis*
Fusarium graminearum	1 *	0.86	0.80
Fusarium verticillioides	0.86	1	0.78
Stenocarpella maydis	0.80	0.78	1

* The level of significance of evaluation is $p \leq 0.05$.

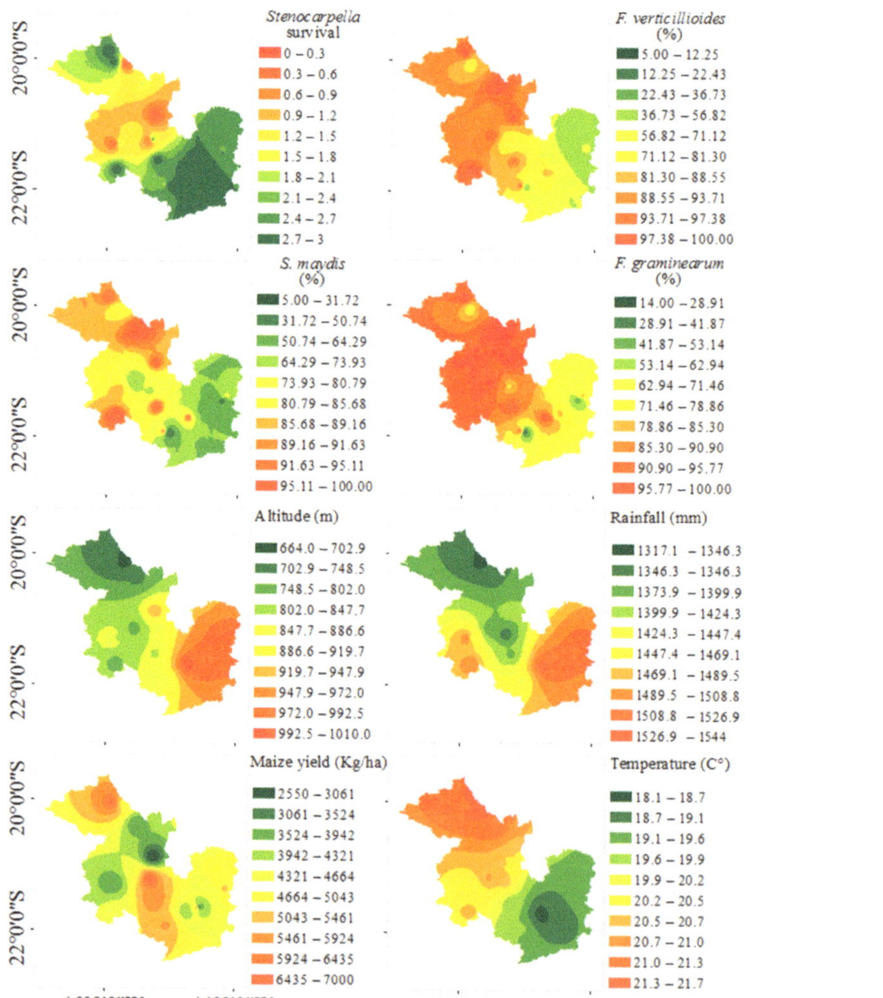

Figure 4. Heat maps of the classes of *Stenocarpella* spp. survival (classes were grouped by cluster analysis of *Stenocarpella* spp. DNA relative quantifications), soil suppressiveness to *Stenocarpella maydis, Fusarium graminearum*, and *Fusarium verticillioides* (% of maize stubble baits colonized), altitude, annual average rainfall, maize yield, and annual average temperature (°C).

The altitude, rainfall, and temperature maps were very similar to each other, i.e., the higher the altitude, the higher the rainfall and the lower the temperature. Such conditions were not exclusive drivers of the importance of *Stenocarpella*-caused diseases since there were regions where the higher the temperature (>20 °C) was, the higher the inoculum level was, as observed for the Varginha sampled sites (Figures 1 and 4). However, under the same temperature condition, such as at the Piumhi location (Figures 1 and 4), a lower inoculum level was found, and this was not related to the suppressiveness of the soil to the disease.

4. Discussion

The rate of survival of *Stenocarpella* spp. on maize stubble during the off-season [18] has been the focus of previous studies on grains and stalks [19] or on seeds [5]. Although other maize plant parts such as cobs or grains may serve as the pathogen reservoir of the pathogen within the stubble, they were not always encountered. Indeed, some of the sampled fields had been harvested for forage, and, as such, the cob, along with the grains and most of the shoot, was taken out from the field and only the basal part of the stalks were left. In this way, we set out to study these dynamics in stalks, which were the substrate for the pathogen present in all sampled areas.

Although such estimates of the pathogen inoculum concentrations in maize stubble have been previously carried out, the exclusive observation of the fungal structures of these pathogens, such as their pycnidia and conidia, may underestimate the pathogen inoculum concentration, since inoculum may also be associated with the plant tissue as dormant mycelia. The contribution of crop management systems to *Stenocarpella* spp. survival using DNA-based quantification was addressed in maize stubble in maize fields for the first time, and, as such, not only conidia but also mycelia and any other fungal structures were recovered from the substrate [4].

In addition, the persistence of maize stubble as stalks has a direct impact on the inoculum importance, and the lifespan of the stubble determines the survival of pathogens [20]. In turn, the factors that govern the decomposition of stubble are the C/N ratio, particle size, lignin content, polyphenols, lignin/N ratio, lignin +N/polyphenol ratio, presence of toxic elements, physical and chemical conditions of the soil, and type of microorganisms present in the soil; therefore, these factors govern the survival of *Stenocarpella* spp. [21], which matches our findings. The stalks are representing substrates on which the substantial pathogen concentrations were encountered. These substrates are rich in lignin [22] and therefore last longer than other plant parts [23]. On the other hand, lignin-rich substrates do not offer an environment that is attractive to antagonistic microbial colonization; as a result, *Stenocarpella* spp. should be encountered at higher concentrations in less decomposed stalks, and, in turn, crop management practices would have an impact on such stubble decomposition.

In this regard, the tillage system and crop rotation are major drivers of stubble decomposition. While a no-till system implies a longer lifespan of stubble, crop rotation with legumes (soybean or common bean) results in the amendment of nitrogen to stubble and accelerates the decomposition rate of straw [8]. Furthermore, the disintegration of stubble increases the surface area of its contact with the soil microbiome; whenever the microbiome encompasses antagonistic communities [24], the stubble offers an environment favorable to biological control with a sustained humidity level, stable temperature, and low incidence of ultraviolet radiation [20]. Other approaches may result in the breakdown of straw, favoring a reduction in the survival of pathogens [25].

Cluster analysis for relative quantification by municipalities showed that nearby municipalities do not necessarily group with each other. Although the pathogens are exposed to similar environmental conditions under different crop management practices, different practices lead to different patterns of pathogen occurrence and soil suppressiveness. The development of suppressive soils is a more cost-effective and efficient alternative to microbe application [26]. If soils are evaluated for their suppressiveness prior to planting, the results will foster decisions regarding the implementation of *Stenocarpella* spp. management practices, such as plowing, fallowing, or not planting maize for the length of time necessary for the maize stubble to decompose.

The observed suppressiveness differed according to the considered pathogen, i.e., the microbiome of the soil acts in a certain way towards a given pathogen and does not necessarily suppress another pathogen, although some of the studied soils could suppress all tested pathogens [24]. The heat maps generated showing the suppressiveness between *Fusarium* species are more similar to each other than to the map generated for the suppressiveness of *Stenocarpella*.

Additionally, the map showing the suppressiveness of *Stenocarpella* spp. and other pathogens does not always match with the concentration of *Stenocarpella* spp. inoculum (Figure 4). The suppressive soils to *Stenocarpella maydis* is an important indication of the lower pathogen inoculum level of *Stenocarpella* spp., but other disease management practices that do not result in suppressive soils are also resulting in the lower pathogen inoculum level (Table 1), and this may be related to the frequency into which the corn is planted within the season and considering maize is the only host of the pathogen not having maize and/or lower maize stubble would result in lower pathogen inoculum. Furthermore, if no pathogen is present, there is a lower chance of selecting for an antagonistic microbial community [27].

The maize yield was also not necessarily explained by the *Stenocarpella* spp. inoculum concentration, suppressiveness of the soil to the different stubble-borne pathogens, or environmental conditions. Although all these factors are reported to be related to maize yield, since each grower undertakes a different combination of the factors, different yields may be obtained. In the maps, it is not possible to clearly determine the relationships between the survival of *Stenocarpella* spp. and maize yield, maize production, or temperature (Figure 4). However, it seems that there are trends among the survival of *Stenocarpella* spp. and the soil suppressiveness, altitude, and precipitation. In areas where more *Stenocarpella* spp. DNA was found, the soil suppressiveness levels were low or absent; this result corroborates the hypothesis that soil suppressiveness can reduce the survival of *Stenocarpella* spp. Areas with high altitudes had higher occurrences of *Stenocarpella* spp. than areas with low altitudes (Figure 4). In [28], when comparing the map classes of *Stenocarpella* spp. survival and precipitation, they concluded that high precipitation favored the survival of *Stenocarpella* spp. [12].

For the integrated management of *Stenocarpella* ssp. in areas where stubble with possibly different inoculum levels are present, farmers use a specific tool to apply fungicide. However, [29] found that the application of fungicide did not consistently reduce white ear rot or improve yield when working with different fungicide spray programs at three different maize growth stages. This failure of fungicides in the management of the disease implies the importance of reducing the inoculum levels of the pathogen in maize stubble [2]. Additionally, glyphosate-resistant, overwintering volunteer maize seedlings [30], which are frequently encountered throughout the year in tropical agricultural fields, represent *Stenocarpella* spp. reservoirs that add to the one already found in the stubble with the advantage of lower competition since many are microorganisms able to colonize the dead plant debris, though fewer have the ability to colonize the live plant.

Therefore, we propose that the distribution of soils that are suppressive to stubble-borne pathogens in tropical soils and *Stenocarpella* inoculum are not necessarily distributed according to the maize growing region or environmental conditions but are more likely to be related to the adopted crop management practices. Although crop rotation and no-till systems were most frequently associated with suppressive soils, these cannot be taken as defaults since there were exceptions and other factors governing both pathogen survival and build-up in the offseason that need to be further dissected.

The pathogen concentration in the stalks was directly associated with the decomposition rate. The crop rotation under no-till systems was associated with soil suppressiveness and reduced pathogen concentrations. Such suppressiveness, when encountered for one pathogen causing stalk-rot, is not necessarily widespread, but the suppressiveness to *Stenocarpella* spp. is a strong indication of the lower pathogen concentration in the field. Factors other than the antagonistic dominant microbial communities govern the lower concentration of *Stenocarpella* spp.

Supplementary Materials: The following supporting information can be downloaded at: https://www.mdpi.com/article/10.3390/app12104974/s1, Table S1: Areas sampled to assess *Stenocarpella* spp. inoculum.

Author Contributions: Conceptual Idea: F.H.V.d.M., F.A.M.F.P. and D.D.d.S.; Methodology design: F.H.V.d.M., F.A.M.F.P., J.d.C.M. and D.D.d.S.; Data collection: F.A.M.F.P., V.B.C.P., H.N.M., M.R.d.F. and F.H.V.d.M.; Data analysis and interpretation: F.A.M.F.P., F.H.V.d.M., R.A.G., E.A.P., H.S.N. and C.d.S.S.; Writing and editing: F.A.M.F.P., F.H.V.d.M., R.A.G., V.B.C.P., J.d.C.M. and E.A.P. All authors have read and agreed to the published version of the manuscript.

Funding: We thank FAPEMIG (Research Support Foundation of Minas Gerais) grants APQ 02059-13 and APQ01578-15, CNPq (National Council for Research and Development) for the EAP, JCM, and FHVM productivity fellowship (grant 309307/2017-1), CAPES (Coordenação de apoio pessoal de nível superior) and CNPq for the FAMFP, VBCP, RAG, MRF, and HSN scholarships.

Institutional Review Board Statement: Not applicable.

Informed Consent Statement: Not applicable.

Data Availability Statement: Not applicable.

Acknowledgments: The authors would like to thank Jürgen Köhl, Pierter Kastelein, and Helen Goossen van de Geijn from Wageningen University and Research Center for technical support on the DNA extraction and qPCR protocol for the *Stenocarpella* quantification, as well as all growers for allowing samples to be collected in their fields.

Conflicts of Interest: The authors declare no conflict of interest. The funders had no role in the design of the study; in the collection, analyses, or interpretation of data; in the writing of the manuscript, or in the decision to publish the results.

References

1. Matiello, R.R.; dos Santos, D.D.P.M.; Coelho, C.J.; Pria, M.D.; Gardingo, J.R. Damage in maize ears associated with methods of inoculation of *Stenocarpella maydis*. *Afr. J. Agric. Res.* **2015**, *10*, 2711–2716. [CrossRef]
2. Flett, B.C.; McLaren, N.W.; Wehner, F.C. Incidence of *Stenocarpella maydis* ear rot of corn under crop rotation systems. *Plant Dis.* **2001**, *85*, 92–94. [CrossRef] [PubMed]
3. Xia, Z.; Achar, P.N. Random amplified polymorphic DNA and polymerase chain reaction markers for the differentiation and detection of *Stenocarpella maydis* in maize seeds. *J. Phytopathol.* **2001**, *149*, 35–44. [CrossRef]
4. Barros, E.; Crampton, M.; Marais, G.; Lezar, S. A DNA-based method to quantify *Stenocarpella maydis* in maize. *Maydica* **2008**, *53*, 125.
5. Siqueira, C.D.S.; Barrocas, E.N.; Machado, J.D.C.; Silva, U.A.D.; Dias, I.E. Effects of *Stenocarpella maydis* in seeds and in the initial development of corn. *J. Seed Sci.* **2014**, *36*, 79–86. [CrossRef]
6. Li, L.; Qu, Q.; Cao, Z.; Guo, Z.; Jia, H.; Liu, N.; Wang, Y.; Dong, J. The Relationship Analysis on Corn Stalk Rot and Ear Rot According to *Fusarium* Species and Fumonisin Contamination in Kernels. *Toxins* **2019**, *11*, 320. [CrossRef]
7. Flett, B.C.; Wehner, F.C. Incidence of *Stenocarpella* and *Fusarium* cob rots in monoculture maize under different tillage systems. *J. Phytopathol.* **1991**, *133*, 327–333. [CrossRef]
8. Berghetti, J.; Casa, R.T.; Ferreira, E.Z.; Zanella, E.J.; Scheidt, B.T.; Sangoi, L. Incidence of stalk rots in corn hybrids influenced by sowing time and nitrogen rates. *Bragantia* **2019**, *78*, 371–378. [CrossRef]
9. van Rensburg, B.J.; Mc Laren, N.W.; Schoeman, A.; Flett, B.C. The effects of cultivar and prophylactic fungicide spray for leaf diseases on colonisation of maize ears by fumonisin producing *Fusarium* spp. and fumonisin synthesis in South Africa. *Crop Prot.* **2016**, *79*, 56–63. [CrossRef]
10. Barbosa, M.; Sousa-Ferraz, R.L.; Coutinho, E.L.M.; Neto, A.M.C.; da Silva, M.S.; Fernandes, C.; Rigobelo, E.C. Multivariate analysis and modeling of soil quality indicators in long-term management systems. *Sci. Total Environ.* **2019**, *657*, 457–465. [CrossRef]
11. Derpsch, R.; Friedrich, T.; Kassam, A.; Li, H. Current status of adoption of no-till farming in the world and some of its main benefits. *Int. J. Agric. Biol. Eng.* **2010**, *3*, 1–25. [CrossRef]
12. Casa, R.T.; Reis, E.M.; Zambolim, L. Doenças do milho causadas por fungos do gênero *Stenocarpella*. *Fitopatol. Bras.* **2006**, *31*, 427–439. [CrossRef]

13. Munkvold, G.P. Cultural and genetic approaches to managing mycotoxins in maize. *Annu. Rev. Phytopathol.* **2003**, *41*, 99–116. [CrossRef]
14. Livak, K.J.; Schmittgen, T.D. Analysis of relative gene expression data using real-time quantitative PCR and the $2^{-\Delta\Delta CT}$ method. *Methods* **2001**, *25*, 402–408. [CrossRef]
15. Ghini, R.; Kimati, H. *Método de Iscas Para Obtenção de Isolados de Trichoderma Antagônicos a Botrytis cinerea*; EMBRAPA-CNPDA: Jaguariúna, Brazil, 1989.
16. Bonanomi, G.; Antignani, V.; Pane, C.; Scala, F. Suppression of soilborne fungal diseases with organic amendments. *J. Plant Pathol.* **2007**, *89*, 311–324.
17. Webster, R.; Oliver, M.A. *Statistical Methods in Soil and Land Resource Survey*; Oxford University Press: New York, NY, USA, 1990.
18. Shurtleff, M.C. *Compendium of Maize Diseases*; American Phytopathological Society: St. Paul, MI, USA, 1992.
19. Reis, E.M.; Mario, J.L. Quantificação do inóculo de Diplodia macrospora e de D. maydis em restos culturais, no ar, e sua relação com a infecção em grãos de milho. *Fitopatol. Bras.* **2003**, *28*, 143–147. [CrossRef]
20. Casa, R.T.; Reis, E.M.; Zambolim, L. Decomposição dos restos culturais do milho e sobrevivência saprofítica de Stenocarpella macrospora e S. maydis. *Fitopatol. Bras.* **2003**, *28*, 355–361. [CrossRef]
21. Handayanto, E.; Giller, K.E.; Cadisch, G. Regulating N release from legume tree prunings by mixing residues of different quality. *Soil Biol. Biochem.* **1997**, *29*, 1417–1426. [CrossRef]
22. Santiago, R.; Barros-Rios, J.; Malvar, R.A. Impact of cell wall composition on maize resistance to pests and diseases. *Int. J. Mol. Sci.* **2013**, *14*, 6960–6980. [CrossRef]
23. Jung, H.G.; Casler, M.D. Maize stem tissues: Impact of development on cell wall degradability. *Crop Sci.* **2006**, *46*, 1801–1809. [CrossRef]
24. Köhl, J.; Lombaers, C.; Moretti, A.; Bandyopadhyay, R.; Somma, S.; Kastelein, P. Analysis of microbial taxonomical groups present in maize stalks suppressive to colonization by toxigenic *Fusarium* spp.: A strategy for the identification of potential antagonists. *Biol. Control* **2015**, *83*, 20–28. [CrossRef]
25. Kerdraon, L.; Laval, V.; Suffert, F. Microbiomes and pathogen survival in crop residues, an ecotone between plant and soil. *Phytobiomes J.* **2019**, *3*, 246–255. [CrossRef]
26. Lapsansky, E.R.; Milroy, A.M.; Andales, M.J.; Vivanco, J.M. Soil memory as a potential mechanism for encouraging sustainable plant health and productivity. *Curr. Opin. Biotechnol.* **2016**, *38*, 137–142. [CrossRef] [PubMed]
27. Faria, M.R.D.; Guimarães, R.A.; Pinto, F.A.M.F.; Siqueira, C.D.S.; Silva, C.A.; Medeiros, F.H.V.; Bettiol, W. Contribution of organic amendments to soil properties and survival of Stenocarpella on maize stalk. *Sci. Agric.* **2020**, *77*, e20180289. [CrossRef]
28. Malusha, J.M.; Karama, M.; Makokha, A.O. Household maize storage practices and aflatoxin contamination in Makueni County, Kenya. *Afr. J. Health Sci.* **2016**, *29*, 89–105.
29. Romero-Luna, M.P.; Wise, K.A. Development of molecular assays for detection of *Stenocarpella maydis* and *Stenocarpella macrospora* in corn. *Plant Dis.* **2015**, *99*, 761–769. [CrossRef]
30. Soltani, N.; Nurse, R.E.; Sikkema, P.H. Response of glyphosate-resistant soybean to dicamba spray tank contamination during vegetative and reproductive growth stages. *Can. J. Plant Sci.* **2016**, *96*, 160–164. [CrossRef]

applied sciences

MDPI

Article

Method for Prolonging the Shelf Life of Apples after Storage

Bogdan Saletnik *, Grzegorz Zaguła, Aneta Saletnik, Marcin Bajcar, Ewelina Słysz and Czesław Puchalski

Department of Bioenergetics, Food Analysis and Microbiology, Institute of Food Technology and Nutrition, College of Natural Science, Rzeszow University, Ćwiklińskiej 2D, 35-601 Rzeszow, Poland; gzagula@ur.edu.pl (G.Z.); asaletnik@ur.edu.pl (A.S.); mbajcar@ur.edu.pl (M.B.); eslysz@ur.edu.pl (E.S.); cpuchalski@ur.edu.pl (C.P.)
* Correspondence: bsaletnik@ur.edu.pl

Abstract: This study investigated the effects of the use of low magnetic fields as a potential method for improving the quality of apples after storage. The fruit were exposed to 100 μT magnetic fields for 8 h per day and kept for a period of two weeks in room conditions. The results showed that the samples that were treated with a magnetic field generally had a higher value ratio of total soluble solid and titratable acidity compared to the untreated samples, which indicated their higher quality. Continuous treatment with a magnetic field influenced the mechanical properties of apples, as demonstrated by the greater firmness, lower weight loss and suppressed CO_2 production of the apples that were stored in room conditions. After the treatment of the apples, a new product was produced with greater firmness, higher quality potential (the ratio of total soluble solid and titratable acidity) and an extended shelf life/lower respiration rate. Therefore, treatment with a magnetic field can be used to extend the shelf life of apples and needs to be demonstrated by further investigations.

Keywords: magnetic fields; quality parameters; firmness; total soluble solid; titratable acidity; dry matter; respiration rate; apple fruit

Citation: Saletnik, B.; Zaguła, G.; Saletnik, A.; Bajcar, M.; Słysz, E.; Puchalski, C. Method for Prolonging the Shelf Life of Apples after Storage. *Appl. Sci.* **2022**, *12*, 3975. https://doi.org/10.3390/app12083975

Academic Editors: Paweł Kiełbasa, Tadeusz Juliszewski and Sławomir Kurpaska

Received: 16 March 2022
Accepted: 12 April 2022
Published: 14 April 2022

1. Introduction

Apples are an important source of sugars, amino acids, proteins, organic acids, vitamins and minerals in the human diet. This is important because it is recommended that they are eaten fresh. However, they are perishable products with a high metabolism and respiration rate, which are responsible for their limited shelf life [1]. Improvements in the postharvest treatment of fresh fruit are very important for increasing food availability [2] because losses during postharvest handling can reach 25–28% [3].

Apples have varied nutritional qualities that are lost during postharvest processing. Losses that occur during storage depend on external and internal conditions. The temperature and relative humidity during postharvest handling activities have the greatest impact on the storage efficiency of apples and the quality characteristics of fruit and thus, affect their shelf life [4,5]. They affect, for example, fruit firmness, weight loss, titratable acidity and the content of soluble solids [6]. During ripening, there is a process of pectin degradation that plays an important role in cell wall degradation, which also has an effect on the shelf life of fruit [7].

Proper postharvest handling involves altering the natural conditions of the product in order to prolong their postharvest life [8]. Therefore, there is a need to find a new method for reducing the softening and senescence and prolonging the shelf life of fruit that are kept at room temperature after storage. This would provide a much-needed improvement in postharvest technology.

This need can be met using physical methods (high voltage electric field, UV-C light, IR radiation, microwave radiation, etc.).

The use of an alternating magnetic field (AMF) also has the potential to meet this need. AMFs are naturally occurring, easy to use, low on energy consumption and safe for the environment [9]. Magnetic fields have been reported to exert a positive effect on living organisms by influencing metabolic pathways [10–12]. Magnetic fields can change the selective permeability of a membrane, the position of the main membrane components (protein and lipid domains) and the molecular position of the electric charge. This can affect turgor pressure variations. The above changes may affect the rate of the physical, biochemical and physiological processes of the tissue of biological material. They also influence the activation of protein synthesis, thus increasing the viability of organisms [13–16]. Alternating magnetic fields (AMFs) have been used by researchers to treat seedlings, plants, fruit and irrigated water. Numerous scientists have reported the positive effects of AMFs on plant growth. The treatment of plant crops with a magnetic field increases the germination rate, size and dry weight of the plant. After the application of a magnetic field, there can also be improvement in the parameters of fruit growth, such as an increase in the number of leaves, the leaf surface, the macroelement content in the leaves and the number of flowers and fruit per plant. A variable magnetic field can also increase the total fruit yield as well as the average weight of the fruit [17].

High voltage electric fields (HVEFs) can be used to extend the shelf life and improve the quality of fruit and vegetables. The electric field reduces the rate of cellular respiration and ethylene production, thus extending the postharvest shelf life. The electric field increases the action of the SOD, APX and CAT enzymes and increases the concentrations of reduced glutathione, ascorbic acid and polyphenols [17].

UV-C light has direct and indirect inhibitory and damaging effects on living cells; therefore, it is widely used for disinfection purposes. UV-C light stimulates the synthesis and accumulation of health-promoting phytochemicals before and after harvest, extends the shelf life of fruit and vegetables and stimulates the mechanisms of adaptation to biotic and abiotic ranges. This could be related to the stimulation of the production of reactive oxygen species (ROS) and the stimulation of molecules and antioxidant mechanisms. UV-C can also trigger and regulate signalling pathways due to its effect on ROS production [18].

Microwave radiation is a component of the electromagnetic spectrum that ranges from 300 MHz to 300 GHz. Microwave radiation can cause different biological effects depending on the field strength, wave form, modulation and duration of the exposure. The positive effects of microwave radiation on the germination of seeds and tubers have been demonstrated. Microwave radiation modifies the germination process and reduces crop infections. The exposure of seeds to microwave radiation appears to be beneficial for increasing the germination, growth and vigour of emerging seedlings and the accumulation of biomass in various plant species. The susceptibility of plants to microwave radiation depends on environmental conditions. MW radiation in living tissues causes ion movement, dipole rotation and the distortion of the electronic orbit, which results in rapid and selective heating. Non-destructive MW treatments are widely used to disinfect seeds before sowing or storage [18].

IR radiation is a part of the electromagnetic spectrum within the wavelength range of 0.5 to 1000 mm and is mainly used in food processing. Infrared (IR) radiation is used as an effective method for the disinfection and drying of seeds and the postharvest disinfection of fresh fruit and vegetables. Gamma (γ) radiation is a type of infrared radiation that can penetrate and interact with living tissues. Gamma radiation is used to remove microbial contamination or control insect pests and pathogens, thereby preventing plant diseases. Γ radiation is also used to delay fruit ripening and vegetable germination by inhibiting the activation of key enzyme activities, which results in the extended shelf life of crops. Γ rays are used by plant breeders to obtain the desired characteristics of plants or develop new varieties. The biological effects of γ rays are strongly dependent on the intensity, dose and duration of the exposure. Irradiation with lower doses of γ rays has a positive effect on the morphological features of plants and improves seed germination efficiency. Low-dose irradiation also has a beneficial effect on the production characteristics of crops. Low doses of γ radiation have a stimulating effect on enzymatic activity and the synthesis of nucleic acids and proteins in treated seeds [18].

The use of the above-mentioned physical methods can contribute to satisfying consumer demand for healthy fruit and vegetables with extended storage and storage times.

The objective of this research was to study the effects of treatment with a 100 µT MF on the quality and shelf life of apples after storage that were kept at room temperature.

2. Materials and Methods

2.1. Materials

Idared (Idaredest) and Legal (Red Sour) apples that were harvested at commercial maturity were procured from local producers and stored in a cold store at 3 °C until the experiment began. Experiments on apples were performed in February and March, after storage. The basic postharvest characteristics of the fruit that were tested are presented in Table 1.

Table 1. Characteristics of the studied fruit.

		Parameters			
Variety	**Harvest**	**Firmness (N)**	**TSS (°Bx)**	**TA (%)**	**DM (%)**
Idared	20 October	88.3	11.3	8.37	14.2
Ligol	5 October	78.5	11.5	5.84	14.0

Prior to the trial, fruit were manually selected on the basis of their color, size, uniformity and the lack of physical damage to their surface. Then, they were divided into 2 groups. They were kept for two weeks. The control group was kept in room conditions (at a temperature of 22 °C and relative humidity of 50%). The study group was placed on the base of the central point between the coils as shown in Figure 1 and treated with a magnetic field of 100 µT for 8 h per day in room conditions. Fruit were arranged horizontally with each of their calyxins facing towards the south magnetic field. After connecting the power supply, the passing current produced a magnetic field with homogeneous distribution inside the coil, which had a diameter of 10 cm (Figure 2).

Figure 1. Coil designed for fruit storage in room conditions.

The quality parameters of the MF and control groups were evaluated after cold storage and after 14 days of storage in room conditions.

The experiment was performed 3–4 times after 120 days of cold storage and at 14-day intervals.

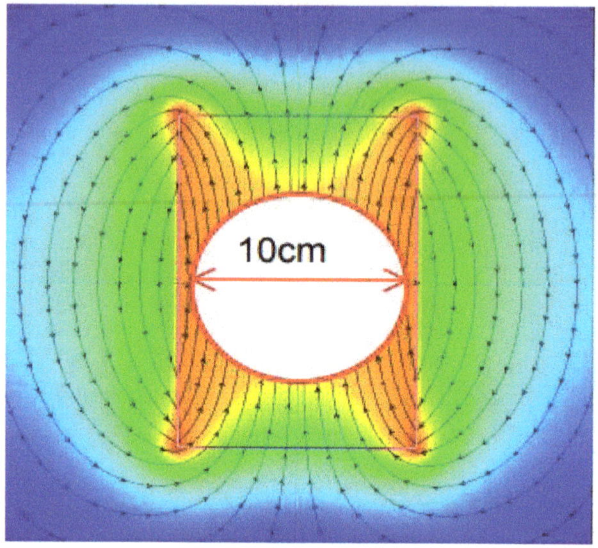

Figure 2. Distribution of the magnetic field in the coil.

2.2. Magnetic Field Treatment (MF)

The parameters of the above-mentioned magnetic treatment were outlined in a previous study [16].

The magnetic field was generated using cylindrical magnets (length, 100 mm; width, 10 mm). The fruit were exposed to a 100 μT magnetic field during storage at room temperature. The control fruit were not exposed to a magnetic field.

The resulting MFD density was measured using the Kanetec TM801 Tesla. Slow changing fields were produced by means of a 12-layer induction coil with the dimensions of 40 cm × 3.5 cm, which was powered by a 50 Hz AC autotransformer [16].

It consisted of two parallel couples of Helmholtz coils, with the radius of each coil being the same as the distance between them. Two pairs of coils were connected in series and connected to a potential transformer that was capable of converting direct current into alternating current power. The fruit were placed on a substrate at the central point between the coils in an area with a uniform magnetic field. After the treatment, the quality of the apples was assessed on the basis of firmness, TSS, TA, DM and respiration rate.

2.3. Firmness (N)

Tests were performed using an Effigi penetrometer with a diameter of 11.1 mm and a cylindrical tip with a convex face for the penetration of the apple flesh. Penetration tests to a depth of 8 mm were conducted on fruit flesh without a layer of skin. The skin puncture tests were carried out on the two opposite sides of each apple, near the centre, and the average values were reported. The penetration tests were carried out with a 2 mm/s speed of approach of the probe. The peak force (N) indicated the maximum force that was registered during the penetration, which was related to the firmness of the fruit.

2.4. Total Soluble Solid (TSS, °Bx)

The total soluble carbohydrate concentration and acidity were determined using a near-infrared (NIR) fruit selector (F-750 PRODUCE Quality METER, Felix Instruments Inc., Camas, WA, USA), which was calibrated for apples.

To measure the total soluble solids, apple juice was extracted using a home juicer and then filtered through a cotton muslin cloth. The TSS of the filtered juice was measured using a manual refractometer with automatic temperature compensation (DIGITAL REFRRACTOMETER PAL-3, Atago Co.,Ltd., Bellevue, WA, USA), which had at least 0.2% Brix. The following formula was used for the conversion: total soluble solids TSS/acid = titratable acidity.

2.5. Titratable Acidity (TA, %)

The acidity (malic acid mg/5 g fr.wt) was obtained using the Association of Official Analytical Chemists [19] method with slight modifications and the results were presented as percentage of malic acid. Three repetitions were carried out for the determination of TSS and acidity values in the apple juice and the obtained results were recorded. The apple juice was diluted with distilled water (25 mL) and titrated with a 0.1 mol/L NaOH solution with phenolphthalein used as the indicator. The calculation was made using the following formula:

Titratable acidity (malic acid mg/5 g fr.wt) = d $\times$ 0.006 (when the NaoH solution is 0.1 N) a $\times$ b $\times$ c, where a is the weight of the sample, b is the volume of the aliquot being examined, c is the total volume with the distilled water and d is the average burette reading for the sample.

2.6. Dry Matter (DM, %)

The DMC was calculated as the ratio of dry to fresh mass and expressed as a percentage. In the test material, the dry weight was determined in accordance with the methodology [20]. The DMC was determined using samples with a flesh thickness of 10 mm and skin tissues from the cross section that was removed from the equatorial area of each apple. Next, they were oven-dried at 65 °C for 48 h. The dry mass content of each fruit was determined in relation to the fresh mass and the results were expressed as a percentage.

2.7. Respiration Rate (CO_2, mg/g fr.wt)

The respiration rate of the apples was calculated according to the method of Zagała et al., 2020, with some modifications [21]. The respiration rate was expressed as the milligrams of CO_2 that was produced per g of fruit per 24 h (mg/g). The increase in carbon dioxide was measured by monitoring the CO_2 production of the treated and untreated fruit, which were enclosed in a chamber with a volume of 10 dm^3 for 24 h at a room temperature of 20 °C and RH of 55%.

The fruit's production of CO_2 was measured using ULTRAMAT 23 apparatus (Siemens, London, England). The following tests were replicated three times. After 72 h, the measuring sensors of the Siemens apparatus were connected to each container in order to measure the level of oxides (the intensity of respiration).

The sensors were connected to each container, then the stopper at the bottom of the container was removed and the apparatus was able to detect the CO_2 that was discharged over 72 h from the cellular respiration process of the apples. The apparatus collected a gas sample at a rate of 1 dcm^3/min using a probe. The gas concentrations were measured for 10 min until the chamber was completely empty.

2.8. Statistical Analysis

The results were expressed as mean values and significant differences were marked with different Arabic letters. The data that were obtained from four independent measurements of Idared and three independent measurements of Ligol apples after storage in a cold store and being treated with MF or kept as a control in room conditions were subjected to a one-way analysis of variance (ANOVA) to detect significant differences between the mean parameters by means of Statistica 10, using Tukey's post-hoc test, at $\alpha = 0.05$ and the number of repetitions n = 3.

Linear regression analysis was used to determine the trends of change in the total soluble solids of the apples during storage maturity and between the dry matter DM and the TSS/TA ratio.

3. Results and Discussion

3.1. Fruit Firmness

Firmness is an important parameter that is associated with the ripeness, harvest time and quality grade of fruit [22]. The results of the measurements of the firmness parameter for the control and magnetic field-treated fruit of the Idared and Ligol varieties are shown in Figures 3 and 4, respectively. Fruit firmness in both groups decreased continuously over the duration of cold storage and after the 14 days in room conditions for each experiment. After 14 days in room conditions, fruit firmness was higher in the apples that were treated with the magnetic field compared to the control in all cases that were tested. It could be seen that using a magnetic field of 100 μT with a frequency of 50 Hz for both of the studied varieties produced an observed increase in the firmness parameter of up to 10%, depending on the characteristics of the fruit. There were significant differences between the control fruit and those treated with a magnetic field. These could have been influenced by the metabolism of the fruit, which is responsible for the rapid conversion of protopectin into pectin and is related to fruit firmness [23]. The higher firmness values that were found in this study after magnetic field treatment were also likely due to the electroporation that was induced by the electrical impulses, which increases the permeability of cell membranes [24]. In this way, the cell membranes can recover their integrity and structure, which results in firmer tissues. These results showed that the MF treatment retarded the softening of the fruit tissue. Therefore, the application of MF treatment at 100 μT for 8 h per day is the proposed method for maintaining the firmness of apples. The same observations were confirmed by the study of Zaguła et al. [25]. Their results also showed an increase in the postharvest firmness of strawberries of up to 30% using an AMF of 50–150 μT in comparison to a control. MFs penetrate deeper into tissues, which may cause changes in cellular metabolic pathways [11]. Another team of scientists analysed the effects of an electric field on tomatoes and apples. They reported that the electric field permeabilised the cell membrane, which resulted in tissue softening. The pulsating electric field reduced the loss of turgor and the fracture of apple cell membranes [26] and the firmness value of tomatoes [27].

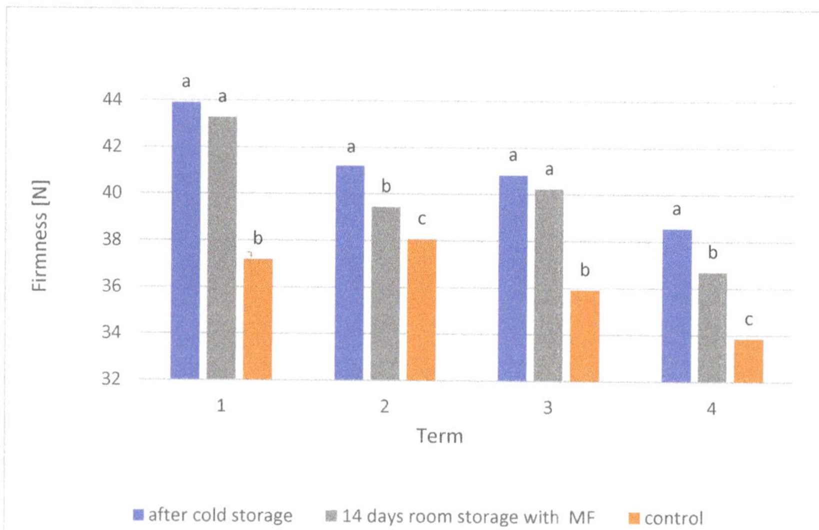

Figure 3. Effects of MF on the firmness of Idared apples after 120 days of cold storage. For each experiment, apples were stored in room conditions for 14 days. Statistically significant differences are marked by different letters. Differences between average values that are marked with the same letters (a,b) are not statistically significant. The data were analysed separately for each experiment.

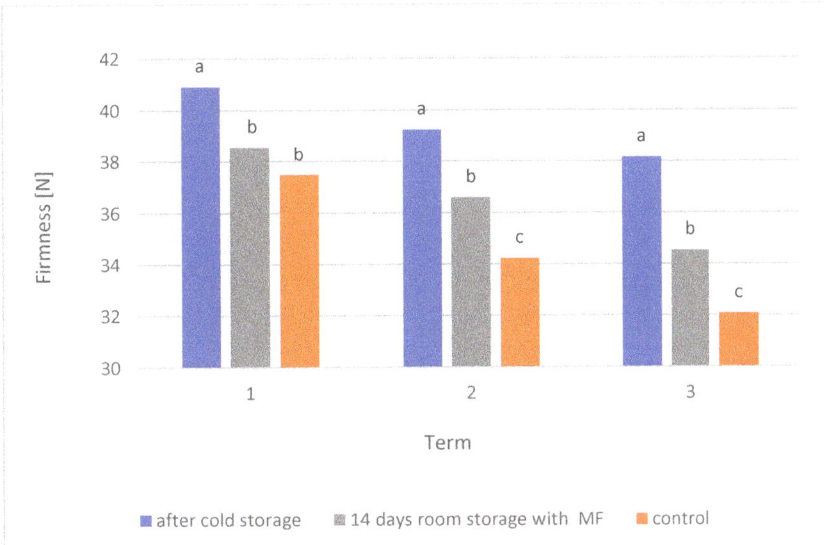

Figure 4. Effects of MF on the firmness of Ligol apples after 120 days of cold storage. For each experiment, apples were stored in room conditions for 14 days. Statistically significant differences are marked by different letters. Differences between average values that are marked with the same letters (a,b) are not statistically significant. The data were analysed separately for each experiment.

3.2. Total Soluble Solids

The soluble solids percentage values of the fruit that were treated with a magnetic field compared to those of the control samples are shown in Figures 5 and 6. The samples that were treated with a magnetic field had a lower percentage change in soluble sugar concentration than the control samples. After 14 days in room conditions, the TSS values were higher in the apples that were treated with a magnetic field than in the control fruit for each experiment. However, significant differences were only found in the Idared variety between experiments two and three. The MF treatment induced changes in the total soluble solids (TSS) content of the apples. The first TSS measurement for the control group averaged 11.8% and was significantly higher after MF treatment. In this way, TSS values increased by as much as 5% and reached maximum concentrations in the apples that were subjected to MF treatment. Scientists have proved that soluble carbohydrates are cell components of structural and metabolic importance [28]. An increase in TSS may be in response to osmotic stress [29,30]. The application of MF treatment may also the cause of the disorganisation of cell wall polysaccharides and molecular bonds [31], which could lead to a change in the TSS content.

The TSS of the stored apples varied significantly ($p < 0.05$) in relation to temperature and duration of storage. The total soluble solid content of apples in cold storage (12.2%, on average) was higher than that of apples in room conditions with MF treatment (12.0%, on average) and the control samples (11.7%, on average). The study showed that the storage of apples at room temperature (22 °C) or in cold storage (3 °C) resulted in a decreased TSS content. The TSS value of the Idared variety generally decreased over the course of the four experiments of the study from 12.6 to 11.4% and from 12.2 to 11.3% for apples in room conditions with MF treatment and the control group, respectively. This could have been caused by the use of monosaccharides (glucose, sucrose and fructose) in the cellular respiration process during storage [32]. Similarly, storing apples indoors showed an adverse effect on their TSS content. Extending the shelf life resulted in a significant decrease in TSS content. The variability in TTS content during the storage of apples is illustrated by a linear regression with the coefficient of determination R^2 of 0.94–0.97. The slope coefficient with a value of 0.371 was smaller for the MF-treated apples compared to those in cold storage and the control samples, which indicated the less variable dynamic of this parameter during storage. The regression equations are presented in Table 2. The same trend for TSS has been reported in the literature, although described with a third-order polynomial [33,34].

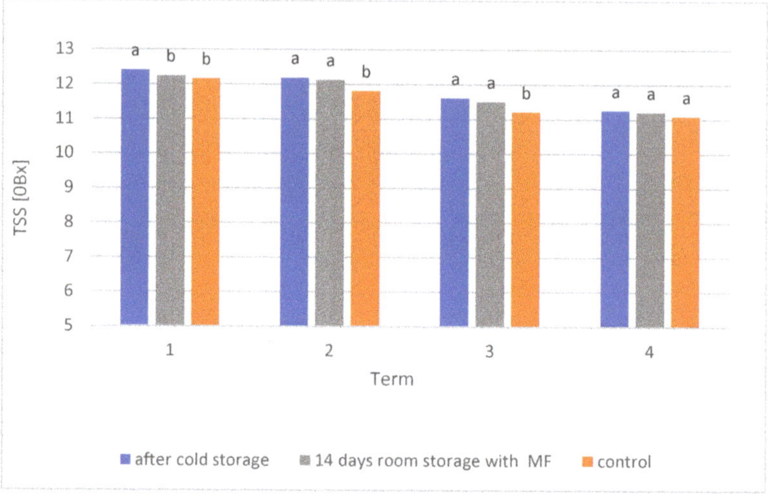

Figure 5. Effects of MF on the TSS of Idared apples after 120 days of cold storage. For each experiment, apples were stored in room conditions for 14 days. Statistically significant differences are marked by different letters. Differences between average values that are marked with the same letters (a,b) are not statistically significant. The data were analysed separately for each experiment.

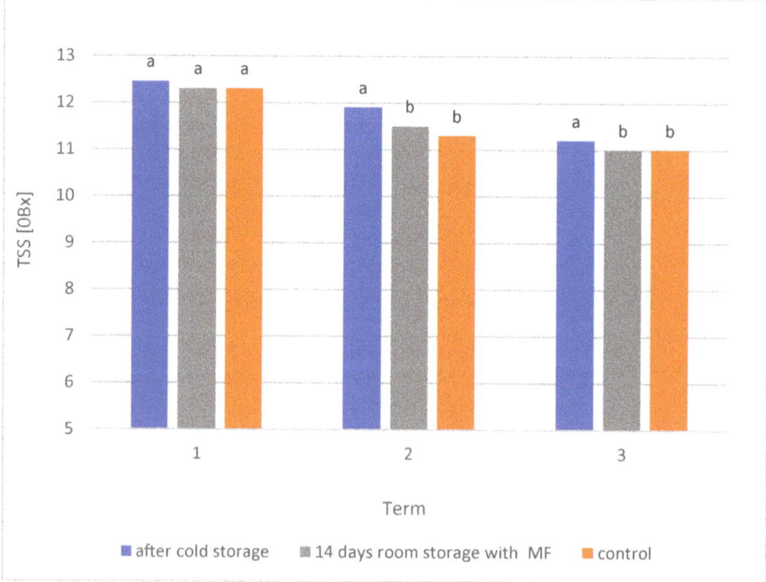

Figure 6. Effects of MF on the TSS of Ligol apples after 120 days of cold storage. For each experiment, apples were stored in room conditions for 14 days. Statistically significant differences are marked by different letters. Differences between average values that are marked with the same letters (a,b) are not statistically significant. The data were analysed separately for each experiment.

Table 2. Regression equations describing the variations in TSS between the different storage conditions of the Idared variety.

Storage Condition	Equation (y)	R
Cold Storage	−0.403x + 12.865	0.98
Room Storage MF	−0.371x + 12.69	0.97
Control	−0.381x + 12.51	0.96

3.3. Titratable Acidity

The test apples were ripened in a cold store and then stored in room conditions. The titratable acidity values of the fruit that were treated with a magnetic field compared to that of the control samples are shown in Figures 7 and 8. In this study, differences in titratable acidity were noted between the test (MF-treated) and control samples. The content of titratable acid affects the taste of apples [35]. Less acidity with minor changes may delay changes in the texture of the apple flavour. In the presented study, the apples that were treated with MF had up to 4.4% less titratable acid than untreated samples. A study by Martiñon et al. [36] reported that titratable acid changed slightly in freshly cut cantaloupes and successfully delayed the postharvest maturation process [36]. These data indicated that treating apples with a magnetic field of 2 mT for 15 min can prolong the life of the proper flavour of apples. The differences between the titratable acidity of the treated and untreated samples were significant and could have been caused by a different metabolic rate, which consumes organic acids and leads to a decrease in acidity [37–40].

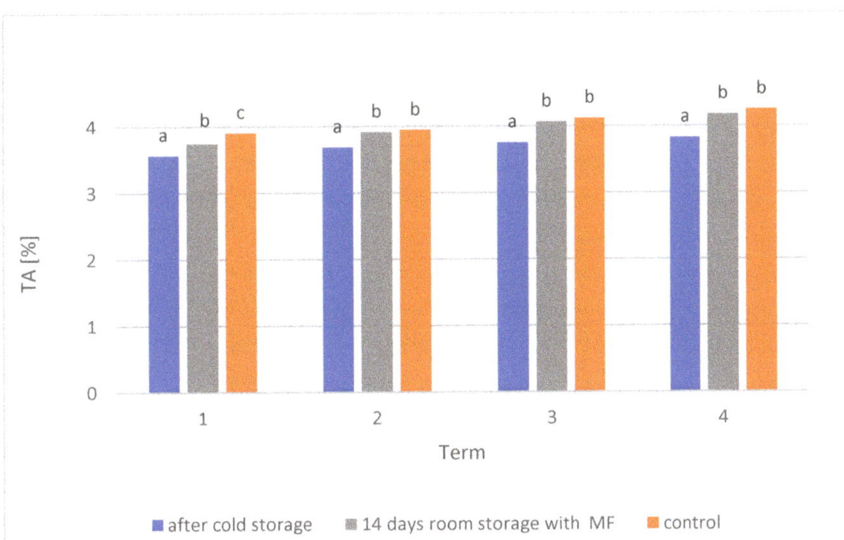

Figure 7. Effects of MF on the TA of Idared apples after 120 days of cold storage. For each experiment, apples were stored in room conditions for 14 days. Statistically significant differences are marked by different letters. Differences between average values that are marked with the same letters (a,b) are not statistically significant. The data were analysed separately for each experiment.

The difference between the titratable acidity of the studied varieties was not significant. The titratable acidity of the apples did not change much after four months of storage. However, research [41–43] has shown a marked difference in the acidity of apples after storage and a decline in the malic acid content of apples after six months of storage.

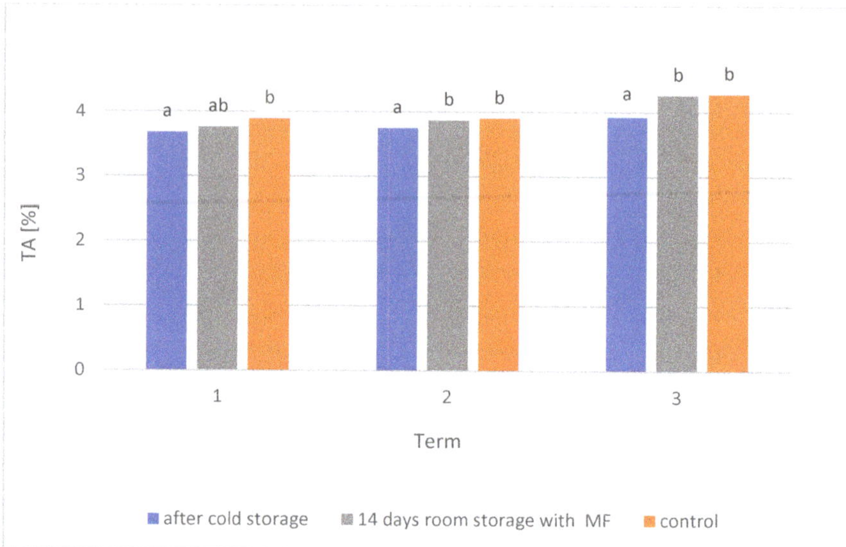

Figure 8. Effects of MF on the TA of Ligol apples after 120 days of cold storage. For each experiment, apples were stored in room conditions for 14 days. Statistically significant differences are marked by different letters. Differences between average values that are marked with the same letters (a,b) are not statistically significant. The data were analysed separately for each experiment.

Apple cultivars have been shown to have significant differences in TSS and acidity [33]. The values of the TSS/TA ratio in the studied fruit are presented in Table 3. The TSS/acid ratio is a major quality parameter and is also essential for predicting the maximum shelf life of fruit [44,45]. The average TSS/acid ratio for three experiments was recorded as being 3.06 for the Idared variety and 2.94 for Ligol. The MF treatment increased the value of this ratio by up to 5% compared to the control group. The duration of storage had also a significant effect on the TSS/acid ratio of the apples and caused the value of this parameter to gradually decrease by up to 18%. Ligol apples experienced a decrease in the value of this ratio by 2.58.

Table 3. Values of the TSS/TA ratio of MF-treated and control apples for each experiment.

		TSS/TA ratio		
Term	**Idared**		**Ligol**	
	MF	Control	MF	Control
1	3.27	3.11	3.27	3.16
2	3.10	2.99	2.97	2.89
3	2.83	2.72	2.58	2.57
4	2.68	2.61	-	-

3.4. Dry Mass

Dry matter can be used to determine the potential quality of fruit. It is the result of the accumulation of carbohydrates, starch, sugars, proteins, cell walls, minerals, etc. The DM of the fruit that were subjected to the magnetic field treatment relative to that of the control samples is shown in Figures 9 and 10. The results showed the slower loss of DM for the test samples (those treated with MF) versus the control samples. On the basis of the obtained results, a decrease in DM by up to 6% was observed in the test sample compared to the control samples. The magnetic field treatment reduced the loss of moisture and hardness in the fruit, thus maintaining their structural integrity and fresh appearance for longer during storage. From the conducted research, it could be concluded that the magnetic field treatment can inhibit the action of tissue enzymes [46].

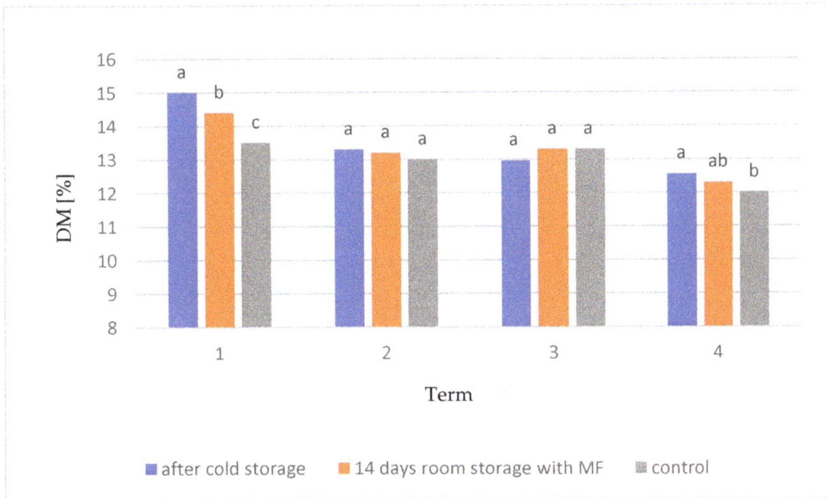

Figure 9. Effects of MF on the DM of Idared apples after 120 days of cold storage. For each experiment, apples were stored in room conditions for 14 days. Statistically significant differences are marked by different letters. Differences between average values that are marked with the same letters (a,b) are not statistically significant. The data were analysed separately for each experiment.

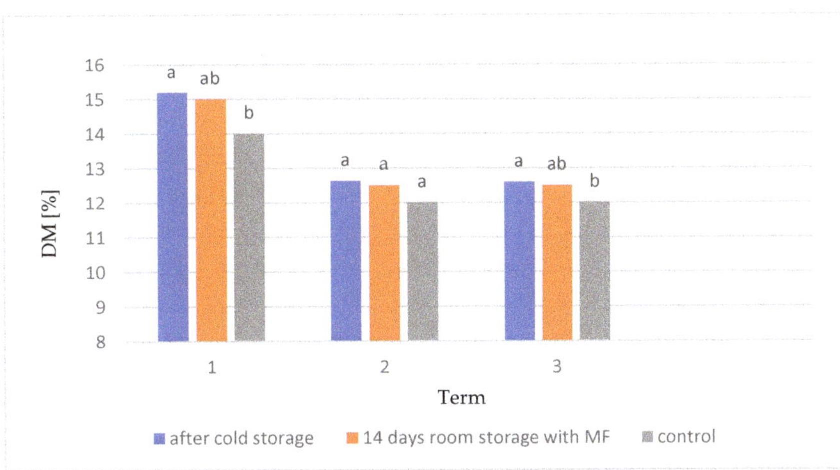

Figure 10. Effects of MF on the DM of Ligol apples after 120 days of cold storage. For each experiment, apples were stored in room conditions for 14 days. Statistically significant differences are marked by different letters. Differences between average values that are marked with the same letters (a,b) are not statistically significant. The data were analysed separately for each experiment.

Significant differences were observed between the MF-treated samples and the control group of the tested varieties ($p < 0.05$) during the first two experiments after cold storage. The DM loss for the samples that were treated with an MF of 100 µT was, on average, 1.5% during experiments 3–4 of the study, while it was 4.6% for the control samples. Some research has proved the efficacy of strong electric fields in extending the shelf life of fruit at 20 °C [47–49]. Their results showed that the physiological loss of mass during shelf life was lower in the electrical field-treated samples than in the control samples.

During the first experiment, a DM loss of about 10% was recorded in the Idared variety and about 8% in the Ligol variety. However, the effects of the cultivar on DM loss are not stated. The DM loss percentage of the test apples increased significantly with an incremental increase in storage duration for specimens in cold storage, room conditions with MF treatment and control conditions. The DM content in the fruit decreased linearly with the increase in storage duration due to an increase in the metabolic rate of the fruit [50]. The study showed a linear regression between the dry matter (DM) and TSS/TA ratio of the test apples with an R^2 coefficient of 0.99 (Figure 11).

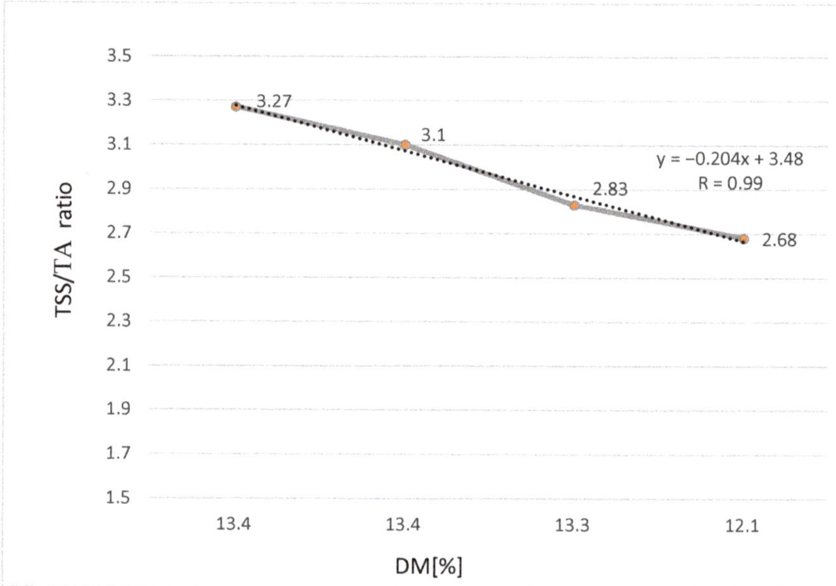

Figure 11. The dry matter versus TSS/TA ratio of Idared apples.

3.5. Respiration Rate

The respiration rate values of the fruit that were exposed to the magnetic field and the fruit in the control group are shown in Figure 12. The respiration rate was expressed as the production of CO_2 in the fruit. The production of CO_2 was lower in the MF-treated fruit compared to the control apples. The magnetic field that affected the fruit for eight hours a day reduced the production of CO_2 compared to the apples that were kept in room conditions. The magnetic and electric fields are elements of plant physiology [51]. The internal electric fields of plants are dependent on those coming from outside impulses, which may cause changes in fruit metabolism that are related to fruit respiration. Many studies [10,21,52] have reported that magnetic and electric fields reduce the respiration rate of various products. Respiration plays an important role in adapting plant metabolism to changing conditions [30,53–56].

Generally, the higher the yield of carbon dioxide, the shorter the shelf life of the product [57–59]. The level of carbon dioxide in the apples that were exposed to MF treatment was determined using samples that were stored for 14 days at 22 °C and 50% relative humidity. A study by Ko et al. [46] showed that electric fields reduce the activity of enzymes, which is the same feature that inhibits the increase in carbon dioxide yield [46].

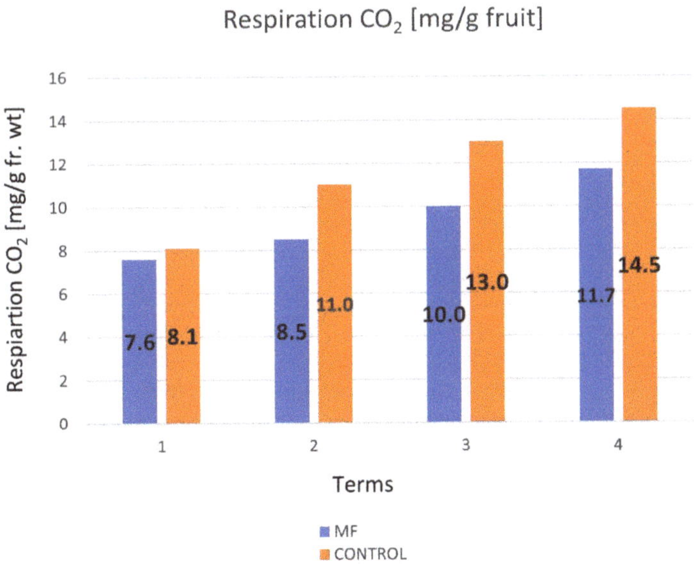

Figure 12. Effects of MF on the respiration rate of Idared apples.

The carbon dioxide yields were 7.6 and 8.1 mg of CO_2 per gram of fruit and increased to 11.7 and 14.5 mg of CO_2 per gram of fruit during storage for treated and untreated samples, respectively. In fact, the MF treatment of fruit reduced the rate of carbon dioxide yield from 6.2% to 23%, depending on the time after storage. This same treatment may have the ability to inhibit metabolism. Generally, the application of MF treatment to apples had an impact on their respiration rate, which led to a decrease in carbon dioxide production. Thus, it contributed to the extension of the shelf life of the apples.

3.6. *Summarisation of the Effects of MF on Apples*

After the MF treatment of the apples, a new product was produced with a greater firmness and quality potential (ratio of TSS/TA) and an extended shelf life, which was expressed by a lower respiration rate (Figure 13).

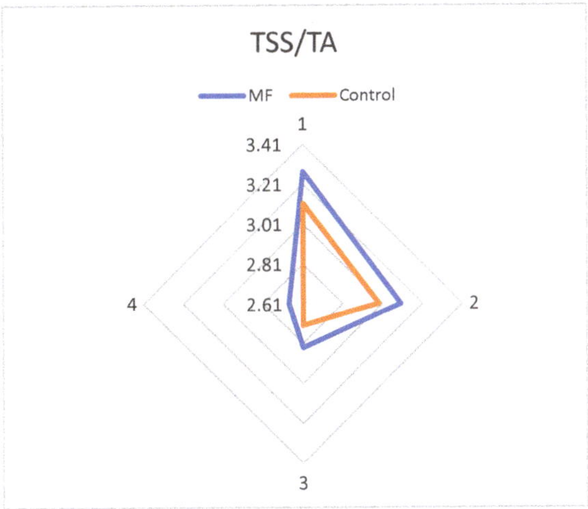

Figure 13. Comparison of quality parameters depending on the development stage of the fruit (1, 2, 3, 4) after storage (parameter values increase from the centre outwards).

4. Conclusions

The authors of the study aimed to investigate the influence of a magnetic field on the shelf life of apples after storage. The parameters of firmness, total soluble solids, titratable acidity, dry weight and CO_2 production were used as the indicators of fruit quality. Since the magnetic field method has a low energy consumption, its use would be a cheap and accessible way to improve the shelf life and quality of fruit. Based on this research, it was shown that the magnetic field-treated samples generally had a higher value of SSC/TA ratio compared to untreated samples, which indicated their higher quality and predicted the longer shelf life of the apples. Short periods (8 h per day) of exposure to the magnetic field changed the mechanical properties of the apples, as demonstrated by a greater firmness and lower dry mass loss. It was noticed that apples in the test sample were harder than those in the control group, depending on variety and development stage. Short periods (8 h per day) of exposure to the magnetic field inhibited CO_2 production in fruit that were kept in room conditions. The MF treatment of the fruit reduced the rate of carbon dioxide yield from 6.2% to 23%, depending on the time after storage. Magnetic field treatment with 100 μT can be used as a method to extend the freshness and shelf life of apples that are stored in room conditions, but it requires further investigation.

Author Contributions: Conceptualisation, C.P. and B.S.; formal analysis, C.P.; investigation, B.S. and G.Z.; resources, C.P.; data curation, C.P. and E.S.; writing—original draft preparation, C.P. and B.S.; writing—review and editing, A.S. and E.S.; visualisation, B.S. and M.B.; supervision, B.S., G.Z. and M.B.; project administration, B.S. and A.S.; funding acquisition, C.P. All authors have read and agreed to the published version of the manuscript.

Funding: The project was financed by the Minister of Science and Higher Education programme named "Regional Initiative of Excellence" for the years 2019–2022 (project number 026/RID/2018/19; the amount of financing PLN 9 542 500.00).

Institutional Review Board Statement: Not applicable.

Informed Consent Statement: Not applicable.

Data Availability Statement: Not applicable.

Conflicts of Interest: The authors declare no conflict of interest.

References

1. Seymur, G.B.; Taylor, J.E.; Tucker, G.A. *Biochemistry of Fruit Ripening*, 1st ed.; Chapman & Hall: London, UK, 1993; pp. 3–51.
2. Kader, A.A. Fruit maturity, ripening, and quality relationships. *Acta Hort.* **1999**, *485*, 203–208. [CrossRef]
3. Ilyas, M.B.; Ghazanfar, M.U.; Khan, M.A.; Khan, C.A.; Bhatti, M.A.R. Post harvest losses in apple and banana during transport and storage. *Pakistan J. Agri. Sci.* **2007**, *44*, 534–539.
4. LeBlanc, D.; Stark, R.; MacNeil, B.; Goguen, B.; Beraulieu, C. Perishable food temperature in retail stores. In *New Development in Refrigeration for Food Safety and Quality*; Refrigeration Science and Technology Proceedings of the Meeting of Commission C2, with Commissions B2, D1 and D2-3; International Institute of Refrigeration: Lexington, KY, USA, 1996; Volume 6, pp. 42–57.
5. Tano, K.; Oulé, M.K.; Doyon, G.; Lencki, R.W.; Arul, J. Comparative evaluation of the effect of storage temperature fluctuation on modified atmosphere packages of selected fruit and vegetables. *Postharvest Biol. Technol.* **2007**, *46*, 212–221. [CrossRef]
6. Tu, K.; Nicolaï, B.; De Baerdemaeker, J. Effects of relative humidity on apple quality under simulated shelf temperature storage. *Sci. Hortic.* **2000**, *85*, 217–229. [CrossRef]
7. Gwanpua, S.G.; Mellidou, I.; Boeckx, J.; Kyomugasho, C.; Bessemans, N.; Verlinden, B.E.; Hertog, M.; Hendrickx, M.E.; Nicolai, B.M.; Geeraerd, A.H. Expression analysis of candidate cell wall-related genes associated with changes in pectin biochemistry during postharvest apple softening. *Postharvest Biol. Technol.* **2016**, *112*, 176–185. [CrossRef]
8. Gonza'lez-Aguilar, G.A.J.F.; Ayala-Zavala, G.I.; Olivas, L.A.; De la Rosa, E. Preserving quality of fresh-cut products using safe technologies. *J. Verbr. Lebensm.* **2010**, *5*, 65–72. [CrossRef]
9. Jia, J.; Wang, X.; Lv, J.; Gao, S.; Wang, G. Alternating Magnetic Field Prior to Cutting Reduces Wound Responses and Maintains Fruit Quality of Cut *Cucumis melo* L. cv Hetao. *Open Biotechnol. J.* **2015**, *9*, 230–235. [CrossRef]
10. Boe, A.A.; Salunkhe, D.K. Effects of Magnetic Fields on Tomato Ripening. *Nature* **1963**, *199*, 91–92. [CrossRef]
11. Funk, R.H.; Monsees, T.; Özkucur, N. Electromagnetic effects–From cell biology to medicine. *Prog. Histochem. Cytochem.* **2009**, *43*, 177–264. [CrossRef]
12. Markov, M.S. Magnetic Field Therapy: A Review. *Electromagn. Biol. Med.* **2007**, *26*, 1–23. [CrossRef]

13. Hunt, R.W.; Zavalin, A.; Bhatnagar, A.; Chinnasamy, S.; Das, K.C. Electromagnetic Biostimulation of Living Cultures for Biotechnology, Biofuel and Bioenergy Applications. *Int J. Mol. Sci.* **2009**, *10*, 4515–4558. [CrossRef] [PubMed]
14. Nakasono, S.; Saiki, H. Effect of ELF Magnetic Fields on Protein Synthesis inEscherichia coliK12. *Radiat. Res.* **2000**, *154*, 208–216. [CrossRef]
15. Valiron, O.; Peris, L.; Rikken, G.; Bs, A.S.; Bs, Y.S.; Remy, C.; Job, D. Cellular disorders induced by high magnetic fields. *J. Magn. Reson. Imaging* **2005**, *22*, 334–340. [CrossRef] [PubMed]
16. Zaguła, G.; Puchalski, C. Glucose and fructose changes in apple subject to permanent and slow-ranging action magnetic fields. *Żywność Nauka Technologia Jakość* **2013**, *2*, 162–172. (In Polish)
17. Saletnik, B.; Zaguła, G.; Saletnik, A.; Bajcar, M.; Słysz, E.; Puchalski, C. Effect of Magnetic and Electrical Fields on Yield, Shelf Life and Quality of Fruits. *Appl. Sci.* **2022**, *12*, 3183. [CrossRef]
18. De Sousa Araújo, S.; Paparella, S.; Dondi, D.; Bentivoglio, A.; Carbonera, D.; Balestrazzi, A. Physical Methods for Seed Invigoration: Advantages and Challenges in Seed Technology. *Front. Plant. Sci.* **2016**, *7*, 646. [CrossRef]
19. AOAC. *Official Methods of Analysis*, 15th ed.; Helrich, K., Ed.; Association of Official Analytical Chemists, Inc.: Arlington, VA, USA, 1990.
20. Regulation of the Minister of Agriculture and Rural Development of 2 December 2004 on the methodology of analytical proceedings in the scope of determining the content of nutrients and feed additives in feed materials, premixes, compound feed and medicated feeds. *J. Laws Repub. Poland* **2004**, *271*, 2688.
21. Zaguła, G.; Tarapatskyy, M.; Bajcar, M.; Saletnik, B.; Puchalski, C.; Marczuk, A.; Andrejko, D.; Oszmiański, J. Near-Null Geomagnetic Field as an Innovative Method of Fruit Storage. *Processes* **2020**, *8*, 262. [CrossRef]
22. Peng, Y.; Lu, R. Prediction of apple fruit firmness and soluble solids content using characteristics of multispectral scattering images. *J. Food Eng.* **2007**, *82*, 142–152. [CrossRef]
23. Paniagua, C.; Posé, S.; Morris, V.J.; Kirby, A.R.; Quesada, M.A.; Mercado, J.A. Fruit softening and pectin disassembly: An overview of nanostructural pectin modifications assessed by atomic force microscopy. *Ann. Bot.* **2014**, *114*, 1375–1383. [CrossRef]
24. Korohoda, W.; Grys, M.; Madeja, Z. Reversible and irreversible electroporation of cell supensions flowing throught a localized dc electric field. *Cell. Mol. Biol. Lett.* **2013**, *18*, 102–119. [CrossRef] [PubMed]
25. Zaguła, G.; Gorzelany, J.; Puchalski, C. Using a computer video system to examine the impact of magnetic and electromagnetic fields on quality of strawberries. *Inż. Rol.* **2010**, *2*, 293–300.
26. Lebovka, N.I.; Praporscic, I.; Vorobiev, E. Combined treatment of apples by pulsed electric fields and by heating at moderate temperature. *J. Food Eng.* **2004**, *65*, 211–221. [CrossRef]
27. González-Casado, S.; Martín-Belloso, O.; Elez-Martinez, P.; Soliva-Fortuny, R. Enhancing the carotenoid content of tomato fruit with pulsed electric field treatments: Effects on respiratory activity and quality attributes. *Postharvest Biol. Technol.* **2018**, *137*, 113–118. [CrossRef]
28. Rosa, M.; Prado, C.; Podazza, G.; Interdonato, R.; González, J.; Hilal, M.; Prado, F.E. Soluble sugars: Metabolism, sensing and abiotic stress: A complex network in the life of plants. *Plant Signal Behav.* **2009**, *4*, 388–393. [CrossRef] [PubMed]
29. Atkinson, N.J.; Dew, T.P.; Orfila, C.; Urwin, P.E. Influence of Combined Biotic and Abiotic Stress on Nutritional Quality Parameters in Tomato (*Solanum lycopersicum*). *J. Agric. Food Chem.* **2011**, *59*, 9673–9682. [CrossRef]
30. Fraire-Velazquez, S.; Balderas-Hernández, V.E. Abiotic stress in plants and metabolic responses. In *Abiotic Stress. Plant Responses and Applications in Agriculture*, 1st ed.; Vahdati, K., Leslie, C., Eds.; InTech: Rijeka, Croaita, 2013; Volume 2, pp. 25–48.
31. Cholet, C.; Delsart, C.; Petrel, M.; Gontier, E.; Grimi, N.; L'Hyvernay, A.; Ghidossi, R.; Vorobiev, E.; Mietton-Peuchot, M.; Gény, L. Structural and Biochemical Changes Induced by Pulsed Electric Field Treatments on Cabernet Sauvignon Grape Berry Skins: Impact on Cell Wall Total Tannins and Polysaccharides. *J. Agric. Food Chem.* **2014**, *62*, 2925–2934. [CrossRef]
32. Eskin, N.; Henderson, H.; Townsend, R. *Biochemistry of Foods*; Elsevier: Amsterdam, The Netherlands, 1971. [CrossRef]
33. Ali, M.A.; Raza, H.; Khan, M.A.; Hussain, M. Effect of Different Periods of Ambient Storage on Chemical Composition of Apple Fruit. *Int. J. Agri. Biol.* **2004**, *6*, 568–571.
34. Ahmad, F.; Zaidi, S.; Arshad, M. Postharvest quality assessment of apple during storage at ambient temperature. *Heliyon* **2021**, *7*, e07714. [CrossRef]
35. Vallone, S.; Sivertsen, H.; Anthon, G.E.; Barrett, D.M.; Mitcham, E.J.; Ebeler, S.E.; Zakharov, F. An integrated approach for flavour quality evaluation in muskmelon (*Cucumis melo* L. reticulatus group) during ripening. *Food Chem.* **2013**, *139*, 171–183. [CrossRef]
36. Martiñon, M.E.; Moreira, R.G.; Castell-Perez, M.E.; Gomes, C. Development of a multilayered antimicrobial edible coating for shelf-life extension of fresh-cut cantaloupe (*Cucumis melo* L.) stored at 4 °C. *LWT* **2014**, *56*, 341–350. [CrossRef]
37. Mahajan, B.V.C. Biochemical and enzymatic changes in apple during cold storage. *India. J. Food Sci. Technol.* **1994**, *31*, 142–152.
38. Clark, C.J.; McGlone, V.A.; Jordan, R.B. Detection of brownheart in 'Braeburn' apple by transmission NIR spectroscopy. *Postharvest Biol. Technol.* **2001**, *28*, 87–96. [CrossRef]
39. Riveria, J. Cutting shape and storage temperature affect overall quality of fresh cut papaya cv. Maradol. *J. Food Sci.* **2005**, *70*, 488–489.
40. Ghafir, S.A.M.; Gadalla, S.O.; Murajei, B.N.; El-Nady, M.F. Physiological and anatomical comparison between four different apple cultivars under cold-storage conditions. *Afr. J. Plant Sci.* **2009**, *3*, 133–138.
41. Radenkovs, V.; Juhnevica-Radenkova, K. Effect of storage technology on the chemical composition of apples of the cultivar 'Auksis'. *Zemdirb.-Agric.* **2017**, *104*, 359–368. [CrossRef]

42. Tahir, I.I.; Ericsson, N.A. Effect of postharvest heating and ca– storage on storability and quality of apple cv. 'Aroma'. *Acta Hort.* **2003**, *600*, 410–415. [CrossRef]
43. Bilisili, A.A.; Ayanoglu, A.; Baykent, N. Quality changes in stored apples. *Teknol. Lab. Yalova* **1970**, *3*, 50–55.
44. Weibel, F.P.; Treutter, D.; Graf, U.; Haseli, A. Sensory and health-related fruit quality of organic apples: A comparative field study over three years using conventionaland holistic methods to assess fruit quality. In Proceedings of the 11th International Conference on Cultivation Technique and Phytopathological Problems in Organic Fruit Growing, Weinsberg, Germany, 3–5 February 2004.
45. Peck, G.; Andrews, P.K.; Reganold, J.P.; Fellman, J.K. Apple Orchard Productivity and Fruit Quality under Organic, Conventional, and Integrated Management. *HortScience* **2006**, *41*, 99–107. [CrossRef]
46. Ko, W.C.; Shi, H.Z.; Chang, C.K.; Huang, Y.H.; Chen, Y.A.; Hsieh, C.W. Effect of adjustable parallel high voltage on biochemical indicators and actomyosin Ca2þ-ATPase from tilapia (*Orechromis niloticus*). *LWT-Food Sci. Technol.* **2016**, *69*, 417–423. [CrossRef]
47. Kharel, G.P.; Hashinaga, F. Effect of High Electric Field on Shelf Life of Strawberries. *Food Sci. Technol. Int. Tokyo* **1996**, *2*, 198–202. [CrossRef]
48. Bajgai, T.R.; Hashinaga, F.; Isobec, S.; Raghavan, G.S.V.; Ngadi, M.O. Application of high electric field (HEF) on the shelf-life extensionof emblic fruit (*Phyllanthus emblica* L.). *J. Food Eng.* **2006**, *4*, 308–313. [CrossRef]
49. Palanimuthu, V.; Rajkumar, P.; Orsat, V.; Gariepy, Y.; Raghavan, G.S.V. Improving cran-berry shelf life using highvoltage electric field treatment. *J. Food Eng.* **2009**, *90*, 365–371. [CrossRef]
50. Atungulu, G.; Nishiyama, Y.; Koide, S. Respiration and climacteric patterns of apples treated with continuous and intermittent direct current electric field. *J. Food Eng.* **2004**, *63*, 1–8. [CrossRef]
51. Scott, B.I.H. Electric fields in plants. *Annu. Rev. Plant Physiol.* **1967**, *18*, 409–418. [CrossRef]
52. Toda, S. Preservation of foods and vegetables by application of electric field. *Shokuhin Rhytsu Gijitsu* **1990**, *19*, 62–64.
53. Łukaszuk, E.; Ciereszko, I. Plant responses to wounding stress. In *Biological Diversity—From Cell to Ecosystem*, 1st ed.; Łaska, G., Ed.; Polish Botanical Society: Białystok, Poland, 2012; pp. 73–85.
54. Rakhmankulova, Z.F.; Fedyaev, V.V.; Podashevka, O.A.; Usmanov, I.Y. Alternative Respiration Pathways and Secondary Metabolism in Plants with Different Adaptive Strategies under Mineral Deficiency. *Russ. J. Plant Physiol.* **2003**, *50*, 206–212. [CrossRef]
55. Sabbagh, E.; Lakzayi, M.; Keshtehgar, A.; Rigi, K. The effect of salt stress on respiration, PSII function, chlorophyll, carbohydrate and nitrogen content in crop plants. *Int. J. Farming* **2014**, *3*, 988–993.
56. Yuan, X.K.; Yang, Z.Q.; Li, Y.X.; Liu, Q.; Han, W. Effects of different levels of water stress on leaf photosynthetic characteristics and antioxidant enzyme activities of greenhouse tomato. *Photosynthetica* **2016**, *54*, 28–39. [CrossRef]
57. Bhande, S.D.; Ravindra, M.R.; Goswami, T.K. Respiration rate of banana fruit under aerobic conditions at different storage temperatures. *J. Food Eng.* **2008**, *87*, 116–123. [CrossRef]
58. Famiani, F.; Farinelli, D.; Palliotti, A.; Moscatello, S.; Battistelli, A.; Walker, R.P. Is stored malate the quantitatively most important substrate utilised by respiration and ethanolic fermentation in grape berry pericarp during ripening? *Plant Physiol. Biochem.* **2014**, *76*, 52–57. [CrossRef] [PubMed]
59. Kan, J.; Wang, H.-M.; Jin, C.-H. Changes of Reactive Oxygen Species and Related Enzymes in Mitochondrial Respiration during Storage of Harvested Peach Fruits. *Agric. Sci. China* **2011**, *10*, 149–158. [CrossRef]

applied sciences

MDPI

Review

Effect of Magnetic and Electrical Fields on Yield, Shelf Life and Quality of Fruits

Bogdan Saletnik *, Grzegorz Zaguła, Aneta Saletnik, Marcin Bajcar, Ewelina Słysz and Czesław Puchalski

Department of Bioenergetics, Food Analysis and Microbiology, Institute of Food Technology and Nutrition, College of Natural Science, Rzeszow University, Ćwiklińskiej 2D, 35-601 Rzeszow, Poland; gzagula@ur.edu.pl (G.Z.); asaletnik@ur.edu.pl (A.S.); mbajcar@ur.edu.pl (M.B.); eslysz@ur.edu.pl (E.S.); cpuchalski@ur.edu.pl (C.P.)

* Correspondence: bsaletnik@ur.edu.pl

Abstract: The presented article is a review of the literature reports on the influence of magnetic and electric fields on the growth, yield, ripening, and durability of fruits and their quality. The article shows the potential application of MF and EF in agricultural production. Magnetic and electrical fields increase the shelf life of the fruit and improve its quality. Alternating magnetic fields (AMF) with a value of 0.1–200 mT and a power frequency of 50 Hz or 60 Hz improve plant growth parameters. MF cause an increase in firmness, the rate of maturation, the content of beta-carotene, lycopene, and fructose, sugar concentration, and a reduction in acidity and respiration. The most common is a high-voltage electric field (HVEF) of 2–3.61 kV/cm. These fields extend the shelf life and improve the quality of fruit by decreasing respiration rate and ethylene production. The presented methods seem to be a promising way to increase the quantity and quality of crops in agricultural and fruit production. They are suitable for extending the shelf life of fruit and vegetables during their storage. Further research is needed to develop an accessible and uncomplicated way of applying MF and AEF in agricultural and fruit production.

Keywords: electrical fields; magnetic fields; high-voltage electric field; growth of fruits; ripening of fruits; shelf life of fruits

Citation: Saletnik, B.; Zaguła, G.; Saletnik, A.; Bajcar, M.; Słysz, E.; Puchalski, C. Effect of Magnetic and Electrical Fields on Yield, Shelf Life and Quality of Fruits. *Appl. Sci.* **2022**, *12*, 3183. https://doi.org/10.3390/app12063183

Academic Editors: Paweł Kiełbasa, Tadeusz Juliszewski and Sławomir Kurpaska

Received: 24 February 2022
Accepted: 18 March 2022
Published: 21 March 2022

1. Introduction

Magnetic and electrical fields occur in the environment and constantly affect living organisms and affect the course of many biological processes [1]. All plants live and develop in the presence of the Earth's magnetic field and the electrical field that occurs between clouds and the Earth [2]. The Earth's magnetic field is mainly produced by the dynamo action of turbulent flows in the fluid metallic outer core of the planet [3]. MF is related to the movement of electrical charges. There are differences on the ground in the strength and direction of the Earth's magnetic (geomagnetic) field. The vertical component of this field is at a maximum at the magnetic pole (about 67 μT) and is zero at the magnetic equator, while the horizontal component is at a maximum at the magnetic equator (about 33 μT) and is zero at the magnetic poles [4]. An electric field is produced by stationary charges [5]. Shawanroy (2012) noted that the combination of an electric field and a magnetic field can be viewed as an electromagnetic field (EMF) [6]. One of the sources producing an electromagnetic field is high-voltage power transmission lines, which may affect the growth of plants and selected crops. A magnetic field (MF) is a source of energy, thereby it has effects on the metabolism of cells [7] and has an impact on meristem cell division [8]. In addition, an MF affects water absorption, preservation, and ionisation [9]. It may cause magnetophoresis in macromolecules [10] and an increase in the chemical reactions of plants, causing a positive effect on photochemical activity, respiration ratio, and enzyme activity [11,12]. This same stimulation of plants with a magnetic field is a way to increase the quantity and quality of seed germination, seedling development, and yields of different species, such as field, fodder, and industrial crops, herbs, and medicinal plants [5].

In addition, yield improvement can be achieved with a pre-sowing treatment of seeds with electrical and electromagnetic fields [12–14]. Research studies have shown that a high-voltage electric field (HVEF) can be used to maintain the freshness of food by extending its shelf life, improving product quality and inhibiting microbial growth [15–17]. It is classified as a non-thermal and low-energy treatment. The HVEF process is based on the production of an electrical wind by corona discharge between at least two electrodes. Different electrode configurations, such as point-ring, needle-ring, needle-plate, and wire-plate, are commonly used for preparing the electrical field. The ions produced in a small area around the needle electrodes are then accelerated by an electrical field. Corona discharge involves the partial electrical breakdown of the gaseous medium [18,19]. Research has illustrated that outer electrical and magnetic fields impact on the actuation of ions, polarisation of electric doublet, water absorption, preservation, and ionisation [9,20]. Many studies have shown that magnetically activated water used in agriculture helps improve germination, plant growth, flowers, fruit, and crop yield. Magnetic treatment of water enhances the overall physical characteristics of the fruit and water productivity [21–23].

Fruit is essential in human nutrition and health and it contains a very high percentage of water compared with other plant-derived foods, such as seeds. Therefore, they exhibit relatively high metabolic activity after harvest and, in consequence, this leads to the rapid breakdown of the structure of the fruit and the loss of its properties [24]. During fruit ripening, many physiological and biochemical processes take place in their cells. These processes result in changes in the colour, taste, and texture of the fruit [25]. These features determine the acceptance of the fruit by the consumer [26]. During ripening and storage, the fruit softens and loses its hard texture [27]. The rate of change is genetically programmed, but also influenced by environmental factors [28]. The strength of fruit tissues, and thus the stability of the whole fruit, is influenced by the mechanical properties of the cell walls, the presence of intercellular fluid, and intercellular interactions in the middle lamina [29]. These factors are subject to changes and constant modifications during the ripening of the fruit. The cell wall is responsible for the physical endurance and shape of the cell [30]. The properties of the cell wall are determined by cell size, inter-cell adhesion, packing and turgor, wall thickness, wall composition, and cell response to shear stress [29]. Most often, plant cell walls on a dry weight basis contain 15–40% cellulose, 30–50% pectin polysaccharides, 20–30% xyloglucan, and smaller amounts of other hemicelluloses and structural proteins [27]. The structure of the cell walls is constantly changing during the growth and storage of the fruit. The main factor responsible for the properties of the cell wall, i.e., stiffness, porosity, and interactions between cells, are pectins [31]. Fruit texture involves a series of alterations in the cell wall, which are genetically programmed, but is also influenced by environmental factors. Pectinase (PE) is involved in the degradation of pectins [32], and thus leads to the degradation of the cell wall. This has a negative effect on fruit firmness [33,34]. Fruits are classified by their respiration profile and ethylene production as climacteric and non-climacteric ones. Climacteric fruits include tomato (*Solanum lycopersicum* L.), banana (*Musa acuminata Colla*), apple (*Malus domestica*), mango (*Mangifera indica*), and pear (*Pyrus amygdaliformis*), while grape (*Vitis vinifera*), citrus (*Citrus medica*), and watermelon (*Citrullus lanatus*) belong to the non-climacteric class [35]. Ethylene is closely associated with these changes in climacteric fruits [36,37], but its role cannot be excluded in the softening of non-climacteric fruit [38–41].

The purpose of this paper is to review scientific results and summarise the emerging topic of the effects of magnetic and electrical fields on the growth, yield, ripening, and shelf life of fruits and on fruit quality, as well as the potential application of MF and EF in agricultural production.

The characteristics of the data presented take into account the following variable factors: variety, method used, field intensity, and time of exposure. Treatment with magnetic and electrical fields has been applied in the whole plant production system, namely on the pre-treatment of seeds and the treatment of seedlings, plants, and fruit and of irrigation water.

2. Effect of Magnetic Fields and Electrical Fields on the Growth, Development, and Yield of Fruit

Research work on the effect of MF on the growth, development, and yield of fruit is reviewed and summarised in Table 1. The results obtained are divided into the following areas depending on the intensity of the magnetic field and the duration of exposure to it: seed and seedling treatments, plant treatment, water irrigation, and water treatment.

Table 1. Summary of the total effect of magnetic fields on the yield and growth of fruit.

Variety	Method	MF Parameter	Effect	References
Tomato *Campbell 28*	Seed treatment	90 mT for 10 min 154 mT for 3 min MF	Increase in leaf area, leaf dry weight, SLA in growth rate of stem and roots, enhanced mean weight, diameter and yield of fruit	De Souza et al., 2006 [42]
Tomato *Vyta*	Seed treatment	120 mT for 10 min 80 mT for 8 min LFMF	Increase in root and stem length, fresh dry root, and dry stem weight, mean fruit weight and fruit yield	De Souza et al., 2005 [43]
Tomato *Lycopersicon Esculentum* L.	Seed treatment	Static MF 50, 100, 150 mT for 1 h	Increase in plant height, shoot and root weight, and dry weight of plant, also improved and enhanced tomato growth parameters: number of leaves/plant, leaf area, number of flowers/plant, number of fruits/plant	Kutby et al., 2020 [44]
Tomato *MST/32*	Seed treatment	MF 332.1; 108.7; 50.6 mT for 1, 2, 24 h	Increase in speed of germination, stem weight ratio, the longer the exposure to MF, the higher the accumulation of biomass in the leaves and stem	Poinapen et al., 2013 [41]
Tomato *Pusa Rohini*	Seed treatment	Static MF 100 mT for 30 min	Increase in speed of germination, reactive oxygen species (superoxide and hydrogen peroxide), activities of antioxidant enzymes, relative expression of various genes in germinating tomato seeds and decrease in total antioxidant capacity	Anand et al., 2019 [45]
Tomato Onion (*Allium cepa* L.)	Seed treatment	Static MF generated by a coil	Increase in weight and yield	Kireva, Mihov 2018 [46]
Tomato	Seed are harvested from plant and are induced by MF and infected by *Fusarium* sp. monospore suspension	MF 0.2 mT for 7 min 48 s, 11 min 44 s, 15 min 36 s	Increase in germination % and rate, dry weight	Agustrina et al., 2018 [47]

Table 1. *Cont.*

Variety	Method	MF Parameter	Effect	References
Melon (*Cucumis melo*): *Ravi*	Seed treatment	MF 100, 200 mT for 5–20 min	Increase in germ germination, root and shoot extents, vigour indices, plantlets fresh and dry mass, leaf region, alpha amylase, protease, catalase, chlorophyll	Iqbal et al., 2016 [48]
Passion fruit (*Passiflora edulis Sims*)	Seed treatment	Static MF 200 mT during germination test 14 days	Increase in germination speed index, germination %, emergence speed index	Menegatti et al., 2019 [49]
Sesame (*Sesamum indicum* L.): *Winas*	Seed treatment	MF from 2 parallel coils 0.1–0.5 mT for 20 min every day for 5 days	Increase in stem height, chlorophyll content, fruit weight and resistance to *F. oxysporum*	Tirono et al., 2021 [50]
Tomato (*Solanum lycopersicum* L.)	Seedling treatment	MF 1–3 mT for daily exposure	Increase in chlorophyll level with low MF and exposure time	Răcuciu 2020 [51]
Strawberry (*Fragaria × ananasa*) *Camarosa*	Plant treatment with electric wire	MF 96, 192, 384 mT	Increase in fruit yield, average fruit weight, number of leaves, fresh and dry root weight Expansion in macroelements and Zn content of plant leaves	Eşitken, Turan 2003 [52]
Strawberry *Camarosa*	Plant treatment with electric wire	MF 96, 192, 384 mT	Increase in fruit yield, average fruit weight, quantity of leaves, fresh and dry root scale Increase in N, K, Ca, Mg, Fe, Mn, and Zn content of plant leaves	Eşitken, Turan 2003 [52]
Strawberry *Camarosa* Tomato *Micro-Tom*	Treatment of culture medium	Magnet NdFeB	Increase in the number of leaves, shoot and root fresh weight, root and shoot length and chlorophyll content and Mg, Ca, Fe, K, P, and Na uptake	Taimourya 2017 [53]
Banana *Williams*	Magnetically treated irrigation water	Device with magnets up to 136 mT	Increase in fruit length, weight and yield	El-Kholy et al., 2015 [54] Patil 2014 [29]
Tomato	Irrigation water magnetised	MF 12.4 31.9 71.9 mT	Increase in stem diameter, height of tomato plant and yield of tomato	Yusuf, Ogunlela 2015–18 [55,56]
Mandarin (*Citrus reticulata*) *Balady Fremont*	Treatment of trees with magnetic water	Two magnets 3.5–93 mT	Increase in pulp and peel weight of fruits and yield	Taimourya et al., 2018 [57]
Strawberry *Camarosa*	Magnetically treated irrigation water	MF in the range of 3.5–136 mT	Increase in number of flowers, fruits, yield and quality of fruits	Taimourya et al., 2018 [57]

Table 1. *Cont.*

Variety	Method	MF Parameter	Effect	References
Tomato *Pavlina*	Treatment of seeds and plants	MF for 50 Hz 20, 40 and 60 mT for 20 min seeds and plants for 48 days	Increase in seed germination, growth of young plant, size of fruit, stem length, weight of tomatoes and earlier fruit setting	Jedlička et al., 2015 [58]
Tomato *Castlerock*	Seed treatment with distilled water and irrigation with magnetised water	MF 0.1; 0.15 and 0.2 T for 1, 5, 10 and 15 min	Optimal magnetic treatment 0.1 T for 15 min, increase in stem length and diameter, Leaf area and fresh and dry weight	El-Yazied et al., 2011 [59]
Tomato *Rocco* and *Monza*	Seeds Seedbed plots, Irrigation with magnetised water	MF 4–6 mT for 2.5 s—seeds 3 times -seedbeds 3 movement—plots	Yield increased on *Monza* 28–51% and bloomed 3–4 days earlier	Danilov et al., 1994 [60]
Tomato *Castlerock*	Seed treatment Irrigation with magnetised water	MF 10 mT for 10 min Magnetic field treatment of water 80 mT	Increase in height, yield and fresh weight, P mineral content in plant leaves	El-Yazied et al., 2012 [61]

2.1. *Effect on Germination*

Many studies show that a magnetic field has a positive effect on seed [27,40,41].

The strength of the MF and exposure time are among the most significant factors influencing seed germination, emergence rate, and seed yield [62]. There were significant interaction effects of strength of magnetic field and the duration of exposure on germination percentage [59]. The highest value was noted when seeds were treated with a strength of 0.1 Tesla for 15 min. By contrast, the lowest germination percentage was at an exposure of 0.2 Tesla for 15 min. Similar results were obtained by Souza et al. (1999) with an exposure time of 10 min [63]. Magnetic treatment can accelerate plant emergence by 2–3 days, compared with control plants [63]. El-Yazied et al. (2011) and Aladjadjiyan (2012) showed that the MF dose and the duration of exposure can affect the germination traits of different seeds, including tomato and broad bean. They demonstrated that the strength of MF plays a significant role in germination percentage [59,64]. Poinapena et al. (2013) observed a germination that was approximately 11.0% stronger in magnetically exposed seeds than in unexposed seeds, although the seedlings emerging from the SMF treatments did not show a constant increase in biomass accumulation [41]. They confirm that longer seed exposure (24 h) resulted in better germination, and that shorter exposure (1 and 2 h) resulted in faster germination [40].

The germination rate is an indicator of the metabolic advance of germination events in the treated seeds. In magnetoprimed seeds, the germination rate was 30.3% higher than in unprimed seeds [45]. The results of Agustrina et al. (2018) show that 0.2 mT MF induced parental tomato seeds to yield F1 seeds that have a higher percentage germination rate and dry weight than control tomato parental seeds. A magnetic field of 0.2 mT was applied for 7 min 48 s. [47].

Scientists report that exposure of tomato seeds to magnetic fields may lead to a significant improvement in the growth and development of plants produced by them [41]. This improvement induced by the magnetic treatment was also consistent with the results of other studies [65] that also report enhanced root and stem growth and fresh weight in tomato plants. Socorro et al. (1999) also reported a positive effect of magnetic treatment on leaf thickness in crop tomatoes, leading to a noticeable increase in the thickness of the spongy tissue, and in the length and width of chlorophyll-containing cells and the upper and lower epidermal cells [66]. Research has reported that magnetic treatment of tomato seeds resulted in a significant increase in leaf area, leaf dry weight, average weight of the fruits, as well as an increase in the harvest of tomatoes per unit area [42,43,58]. Iqbal et al. (2016) showed that MF pre-sowing treatment of melon seed enhanced germination by 14.6%, root and shoot lengths by 36.4 and 22.8%, vigour indices I and II by 40.6 and 28.8%, seedling fresh weight by 9.6%, and dry mass by 12.9%, leaf area by 50.0%, alpha amylase by 80.0%, protease by 92.5%, catalase by 36.5%, and chlorophyll "a" and "b" content by 50.4 and 80.9% when compared to control treatments [48]. Fruit yield per plant was also significantly influenced by magnetic treatment, with increases of 19.4–28.5% when compared with controls [42,66,67]. Moon and Chung (2000) confirmed higher mean fruit weight of tomatoes, as well as overall higher yields, in experiments with magnetically treated seeds of tomatoes [27]. Similar effects have been reported on strawberry yield parameters by Esitken and Turan (2004) [68].

Treatment with a magnetic flux density (MFD) from 0.1 mT to 0.3 mT made sprouts grow faster, increased stem height and chlorophyll content, shortened flowering time, and resulted in heavier fruit weight than control treatments [50]. A study by Fu (2012) reported that treatment with an MFD of 0.33 T increased plant height from 4.18 cm to 5.25 cm at four weeks of age [69]. Chlorophyll content is closely related to plant health, with higher chlorophyll content being associated with better plant health [70].

Răcuciu (2020) and Taimourya et al. (2017) showed that the exposure of plants to a magnetic field increased the chlorophyll levels with low magnetic flux density and short duration of exposure, respectively [51,57]. Given that chloroplasts have paramagnetic properties, the influence of magnetic field on plants increases its inner energy, which is distributed among the atoms causing accelerated metabolism [71].

The MF affected various plant characteristics, such as germination of seeds, root growth, rate of seedling growth, reproduction growth rate of the meristem cells, and quantity of chlorophyll [72]. The researchers have shown that the magnetic field changed the characteristics of the cell membrane, causing some changes in cell metabolism. The magnetic field affected various plant properties, such as gene expression, protein biosynthesis, and enzyme activities, and affected various plant functions in either plant organs or tissues [73,74]. The mechanisms of action are represented in Figure 1.

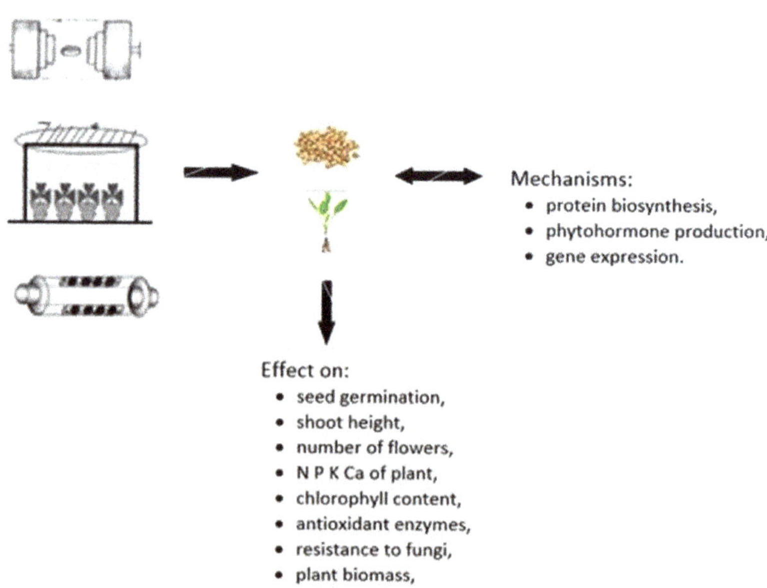

Figure 1. Magnetic field affecting the morphophysiological properties on a fruit plant with possible mechanisms.

2.2. Effect on Plants

Esitken (2003) treated whole strawberry plants with an MF. He showed that increasing MF strength to 0.096 T increased fruit yield per plant by about 18%; however, higher field values reduced fruit yield and had a negative effect on plant growth [52]. All the MF strengths expanded average fruit mass compared with the control, even though the largest fruits of about 8.92.g were set on at 0.096 T. On the other hand, MF had beneficial effect in terms of number of leaves, and fresh and dried root heaviness. Tomato plants cultivated in magnetic treatment by Taimourya et al. (2017) also showed significant increases in the number of leaves, shoot length, and root length, with an increment of 152.9% in number of leaves, 14.4% in shoot length, and 23% in root length [53]. These results are in line with those of De Souza et al. (2006) who observed that pre-treatment of seeds with magnetic field or irrigation with magnetic treatment increased leaf, stem, and root fresh and dry weight of tomato [42]. The same results were confirmed by Eşitken and Turan (2004) [68]. Additionally, they stated that an increase in strength of the MF up to 0.384 T increased the concentration of macroelements and Zn of flora leaves. However, it decreased contents of P as well as S compared with the control. It was similarly reported by Matsuda et al. (1993) that MF had a beneficial effect on crop in strawberry [75]. In shoots of strawberry plants cultivated in a culture medium subject to magnetic treatment, Taimourya et al. (2017) observed an increase in magnesium content by 26.8%; calcium content by 30.2%; iron content by 62.4%; potassium content by 25.3%; phosphorous content by 12.8%; and sodium content by 26.9% [57]. In roots, mineral uptake was increased by 36.6% for magnesium; 23.1% for calcium; 184.8% for iron; 19.3% for potassium; 32.2% for phosphorous; and 58.9% for sodium. Jedlička et al. (2015) stated that there was a positive effect from treatment of young plants of the tomato variety "Pavlina" with magnetic fields of 40 mT. The mean weights of the ripe fruit and the aboveground parts were higher on average by 55.2% and 11.2% than the control variant, respectively [58].

Many studies report the positive effects of treating young plants with a magnetic field. The use of appropriate magnetic field values increases plant growth and crop yield. However, the use of high field values reduced fruit yield and had a negative effect on plant growth.

2.3. Effect of Magnetised Irrigation Water on Plant Growth and Crop Yield

Magnetised aqua is obtained by volatile water across a magnetic field (constant magnets or else electro magnets) in a supply pipeline. A literature overview relates to the beneficial effects of irrigation with magnetised water [61]. Grewal and Maheshwari (2011) reported that some transformation was performed in the physical and chemical features of water because of magnetic treatment [76]. These are mainly related to hydrogen bonding, polarity, surficial tension, conductivity, and pH, as well as dissolubility of salts. Magnetic water has a lower surface tension; therefore, nutrients are more readily absorbed in the water [77,78]. The mechanism in which the magnetic field operated was linked by Phirke et al. (1996), Turker et al. (2007), Maheshwari and Grewal (2009), and Hozayn and Abdul Qados (2010) to the activation of phytohormones, such as gibberellic acid equivalents indole-3-acetic acid and trans-zeatin, and the activation of the bio-enzyme systems leading to improved cell activity and plant growth [30,79,80]. This observation is consistent with the work of De Souza et al. (2005) and Dhawi (2014) that magnetic water accelerates the reactions of enzymes related to auxin (plant hormone), which could increase the growth rate and promote fruit ripening [43,81]. Stange et al. (2002) stated that electromagnetic fields improve the rate of ion shipping through the plasma membrane and influence the structure of the cellule membrane lipid protein and may result in changes in the permeability of the plasma membrane [82]. The results of Yusuf and Ogunlela (2015) revealed that the use of magnetic flux densities of 124, 319, and 719 G for treating the irrigation water caused tomato plants to grow faster and influenced their vegetative growth and stem diameter [55]. This confirms the work of El-Sayed and Sayed (2014) [83]. Taimourya et al. (2018) found that, when using magnetic irrigation water on strawberry plants, there was a boost in the amount of fruits and fruit crop by over 30% [57]. Fernandez et al. (1996) reported that seedlings developed with magnetically treated irrigation were over 100% higher than the control [84]. In a nursery experiment, El-Yazied et al. (2011) applied the optimal magnetic seed treatment of tomato (0.1 T for 15 min) and/or irrigated with magnetised water. This produced significant increases in transplant stem length, stem diameter, leaf area, and fresh and dry weight than those in the control treatment [59]. These results are consistent with those of Grewal and Maheshwari (2011) who reported that tomato grain (cultivar Castle Rock) treatment with a magnetic field of 100 gauss for 15 min, as well as with magnetically treated hydrated water, boost their vegetable rise. Furthermore, treatment increased the total phosphorus component of tomato leaves, as well as total yield, while reducing the pH valuation in soil extraction [76]. In addition, this study revealed that irrigation with magnetically treated water increased nutrient mobility in soil and enhanced extraction and uptake of P, K, N, and Fe by plants [28,53,77,85]. The interaction between aqua treatment and grain treatment, as well as NPK levels, was also tested. The highest valuation of P and the lowest concentrations of Na in tomato leaves were observed with 100 or 75% treatments of NPK fertiliser combined with magnetised seeds and irrigating water [76]. Mohamed (2013) and Ajitkumar (2014) found that magnetic aqua significantly boosts fruit dimensions, the fresh and dry weights of fruits, as well as crop yield compared with a control variant [29,78]. Hamdy et al. (2015) irrigated Balady and Fremont mandarin trees with magnetic water during the growth season and demonstrated a significant increase in yield of nearly 35% in comparison with the control [86]. Yusuf et al. (2018) confirmed an increase in tomato yield with magnetic irrigation water. The percentage increments in tomato yields were 39.9 and 68.7% with magnetic flux densities of 124 and 719 G, respectively. They also showed increased concentrations of calcium, potassium, sodium, nitrogen, phosphorus, and organic carbon when water was treated [56]. Changes in water properties by magnetic fields can alter the characteristics of plant growth and production. Many researchers have concluded that magnetic treatment of irrigation water increases plant growth and crop yield [16,30,61,80,87–90].

2.4. *Effect of Magnetic Fields and Electrical Fields on the Ripening and Shelf Life of Fruits*

A summary of the effect of MF and EF on the ripening and shelf life of fruits is presented in Table 2.

Table 2. Summary of the total effect of magnetic fields and electrical fields on the ripening and shelf life of fruits.

Variety	Method	MF Parameter	Effect	References
Apples *Melrose, Champion, Cortland*	Fruit treatment	SPMF 200 mT within storage exposure time 5 × 30	Change in firmness depending on variety, increase in *Melrose* decrease in *Champion,* no change in *Cortland*	Puchalski 2001 [91]
Tomato *V.R Moscow*	Fruit treatment	MF 60 mT within ripening period (15 days)	Increase in ripening rate, beta carotene and lycopene, sugar concentration and decrease in acidity and respiration rate	Boe et al., 1968 [92]
Tomato *Starbuck*	Fruit treatment	MF 2.5 mT for storage period	Low MF did not influence ripening, no significant differences in colour, firmness, lycopene concentration	Bourget et al., 2012 [93]

Table 2. *Cont.*

Variety	Method	MF Parameter	Effect	References
Strawberries *Bogota, Elkat, Ventana, Honeyoe, Salut*	Fruit treatment during vegetative period on plants	SPMF 5–100 mT, AMF 50–150 uT and 5–100 mT for 5 min with 5 replicates each 5 days	Increase in firmness with AMF for 50–150 uT up to 30%	Zaguła et al., 2010 [94]
Apple *Jonagold*	Fruit treatment with compensated vertical component of Geographic MF	Near null GMF after 4, 5 and 6 weeks of storage	Increase in storage time by slowing down the process of enzymatic decomposition of starch and intensity of respiration. Lower content of simple sugars, ash and greater calorific value, volatile compounds and water content were recorded	Zaguła et al., 2020 [95]
Apples *Fuji, Rome Beauty, Golden Delicious* and *Starking Delicious*	Fruit treatment using needles as the cathode and an aluminium plate	EF 1.25–2.5 kV/cm for 1 and 2 h period	Increase in soluble sugar concentration, and decrease in CO_2 production, it becomes tougher and stiffer material	Atungulu et al., 2003 [96]
Apples *Fuji, Sansa, Starking Delicious* and *Golden Delicious*	Fruit treatment with two parallel aluminium plates or needle and plate	EF 3; 3.75; 6 kV/cm EF continuous for week and intermittent 3 h every 5 days	Decrease in CO_2 evolution during storage, and suppressing CO_2 evolution at the peak of the climacteric. Direction of electric field with parallel plate electrode influenced apple respiration	Atungulu et al., 2004 [97]
Apple *Fuji* Pear *culta Nakai* Plum *Karari* Banana *Cavendish*	Fruit treatment with Cu wire and plate (post and pre-climacteric period depending on variety)	EF 4.3 kV/cm for 5 and 20 min	Decrease in respiration rate of fruit	Kharel et al., 1996 [98]
Cherry tomato	Fruit treatment before storage with parallel plates	EF 1, 2, 3 kV/cm for 1 or 2 h	Decrease in fruit weight loss, fruit softening percentage, peak of climacteric delayed by 3 days	Kusuma et al., 2018 [99]
Emblic fruit (*Phyllanthus emblica* L.)	Fruit treatment with 2 copper plates	(AC) and (DC) current ((HVEF) of field strength 4.3 kV/cm for 2 h	AC HVEF can be used to extend the shelf life of emblic fruits, rotting significantly decreased after 25 days storage	Bajgai et al., 2006 [23]
Peach fruit (*Prunus persica*)	Fruit treatment with corona discharge producing ozone within the chamber	EF with voltage 7 kV Within storage	Increase in fruit hardness, decrease in soluble solid content	Shiina et al., 2009 [100]
Persimmon fruits (*Diospyros kaki*).	Fruit treatment between two plates	EF field of 6 kV/cm for 30, 60, 90 or 120 min, 6 duplicates, before storage	Decrease in rate of weight loss, rate of decrease in hardness, rate of carbon dioxide production	Liu et al., 2017 [101]
Tomato *Chaoyan-219*	Fruit treatment with two stainless steel plates used as cathode or anode	EF +/−1, +/−2, +/−3 kV/cm for 2 h within 20 days of storage	Decrease in respiration rate and ethylene production and extending postharvest life by about 7–10 days. HVEF may delay the decline in firmness and the change in colour, total soluble sugar and titratable acidity of fruits	Wang et al., 2008 [102]

Table 2. *Cont.*

Variety	Method	MF Parameter	Effect	References
Tomato *Chaoyan-219*	Fruit treatment with 2 stainless steel plates	EF +/−2 kV/cm for 2 h	Lower production rate of O_2—and the content of H_2O_2, efficient in hold up ripening, as well as the aging process	Zhao et al., 2011 [103]
Mini-tomato (*Lycopersicon esculentum*) *Chika*	Fruit treatment using needles or an aluminium plate as the cathode	EF 1, 2, 3, 4 kV/cm 2 h per day within storage temperature (0–17 C)	Decrease in respiration rate and increase in moisture loss using needle cathode	Atungulu et al., 2005 [104]
Strawberries	Fruit treatment using multiple pin-to-plate electrodes	EF 3.61, 4.56, and 5.13 kV/cm for 1 h	Decrease in mass losses in fruit and extended cold storage and preventing growth of *Botrytis cinerea*	Esehaghbeygi et al., 2021 [105]
	Fruit treatment with glass plate and Cu wire or Cu plate	EF 0.71–4.3 kV/cm for 0.5–1.5 h and continuous	Increase in shelf life of fruits without affecting major chemical constituents. In the 6 days of storage, reduction of fruit rotting by about 80%	Kharel and Hashina 1996 [106]

The effect of magnetic and electrical fields on the postharvest physiology of fruits is a new and a not yet exhausted area of research. Metabolic activity continues after harvest of fruits, thus making most types of fruit highly perishable commodities. As the fruit ripens, respiration increases and their flesh begins to soften and will reach senescence [24,107–109].

Boe and Salunkhe (1963, 1968) reported that a magnetic field exerted a positive effect on the ripening of fruits [92]. MF brought forward the ripening rate of tomato fruit. This acceleration was manifested in increased colour development. The increase in colour development was due to accelerated synthesis of carotenoids, beta carotene, and lycopene. The fruit ripened in the magnetic field also showed characteristic changes in acids and sugars that accompany ripening [92]. However, Bourget et al. (2012) showed that MF, at the low doses used in these experiments, did not influence the ripening of tomato fruit during storage at 16 °C. No significant difference was observed in colour change and weight loss between control and magnetic-field-exposed fruit. Lycopene concentration, firmness, and membrane permeability were also no different between the control and exposed fruit. It was concluded that a higher intensity of magnetic field may be necessary to have an impact on the ripening behaviour of tomatoes [93]. Puchalski (2001) stated that a strong magnetic field of 200 mT boosts the firmness of Melrose apples and, as a result, the storage period was extended by 20 days. However, in the case of the Champion variety, a field of this value was destructive [91]. Valentinuzzi (1964) stated that magnetic fields could be inhibitory in nature because of the reduced incidence of molecular collisions due to the orientation of paramagnetic molecules [107].

Treatment with a strong electrical field has been, in theory, used as a method of foodstuff preservation because of its skill to sterilise microbes, as well as inhibit the activity of enzymes engaged in the loss of food property [22,109–111].

Literature reports that high voltage electrostatic field (HVEF) treatments affect both the physical and biochemical properties of food. Murr (1963) proved that field action connected with the polarisation of organic radicals in the flora biosystems and molecular fatigue can affect the reactions of plants [112]. This results in the maintenance of the postharvest quality of some fruits and the extension of the shelf life of the product [23,96–98,102,106,113–116]. Studies indicate that electric fields cause gas ionisation in room conditions, which then moved towards the opposite electrode at a high velocity. It is a nonthermal treatment useful in extending the shelf life of fruits [23], which impacts the cell membrane permeability and influences enzyme activity and the inhibition of microbial action [103,109,111,113,117]. A corona electrical field might help reduce decay and spore production in stored materials [105]. Kharel and Hashinaga (1996) applied an external electrical field to observe a reduction in the decay rate of strawberries [106]. HVEF treatment of cranberries enhanced the fruit's shelf life [114]. A study by Esehaghbeygi et al. (2021) showed that HVEF at 4.56 kV·cm^{-1} was capable of controlling fungal infections, preventing the growth of *B. cinerea* in treated strawberries, conserving the fruit's moisture content and freshness, and extending the strawberries' storage time [105]. Yu et al. (1995) found that electron beam irradiation suppressed fungal growth on fresh strawberries, with an extension of their shelf life [118].

Atungulu et al. (2005) applied different methods relating to high-voltage treatment to examine their effect on tomato storage [104]. They noted that weight loss in fruit depended on the electrode configuration used. Using needle electrodes as the cathode in the high-voltage treatment (10 kV, 20 kV, and 30 kV) resulted in greater moisture loss than in the untreated samples [104]. Another study compared the use of different methods with electric field direction on the postharvest treatment of apples [97]. Using parallel plates, the arrangement in the "reversed" electric field treatment gave higher weight loss than in the "forward" electric field treatment (apples on an anode plate). However, in corona discharge this process was reversed [97].

Electrostatic treatment is a low energy, nonheated treatment technology and may be influenced to maintain a specific respiration rate for specific fruits and vegetables. Expanding the shelf life of the fruit is possible through curtailment and lag of the respiration of fruits resulting off the containment of ethylene generation [102]. Wang et al. (2008) found that a negative high-level treatment electrostatic field (−2 kV/cm) treating, with its capability to upgrade respiration and ethylene out-turn, can daff tomato fruit maturation and prolong postharvest life during storage for about 7–10 days [102]. Kharel and Hashinaga (1996) stated that a high-voltage electric field (430 kV/m) treatment of pears, plums, and bananas in the preclimacteric period suppressed the respiration rate during the climacteric period. Similarly, little effect on the respiration rate was also observed in apples treated by HVEF in a postclimacteric period [98]. The physicochemical properties (i.e., Brix, pH, hardness, and Hunter "Lab" values) of bananas 17 days after HVEF treatment indicated that HVEF treatment retarded ripening (1.5–2 days) as compared to the control. However, the ethylene production rate in apples and bananas was not affected by a HVEF. The ripening of mature green bananas and sweet peppers was delayed by the exposure of HVEF [98]. Kusuma et al., 2018 found that the climacteric peak for tomatoes was delayed for 3 days in all HVEF pre-treatments at 2 kV/cm applied for 2 h [99]. Wang et al. (2008) assumed that treatment of a −2 kV/cm electrostatic field can efficiently avoid the decrease in firmness and colour progression in tomato fruit, while similar effects were not monitored in fruit treated with a positive electrostatic field. The pitch of respiration as well as ethylene generation of tomato fruit during storage were lagged by 6 days [102]. Wang at al. (2008) and Zhao et al. (2011) confirmed the earlier observations that ethylene output of tomato fruits treated from HVEF was lower before that of the control fruits and that a negative HVEF treatment can inhibit the production of ethylene more efficiently than a positive HVEF treatment [102,103]. However, Shivashankara et al. (2004) demonstrated that a positive high-voltage electrical field (1.5 kV/cm) pre-treatment had an effect on the respiration and antioxidant capability of mango fruit, though did not have some prominent effect on firmness and rind colour of mango fruit. HVEF can also affect the inhibition rate of carbon dioxide, indicating that it has the ability to inhibit metabolism and delay organisational deterioration [119]. Atungulu et al. (2004) found that, after 21 and 39 days in storage, apples treated with 36 and 48 kV showed a lower amount of CO_2 than samples in a non-reversed electric field [97]. In another study, it was shown that 1- or 2-h treatment periods at 10 and 20 kV reduced respiration in apples [96]. Other research showed that an uncompensated geomagnetic field may have an effect on the inhibition rate of carbon dioxide during storage of Jonagold apples. The difference between treated and control samples was 20 mg per kg of fruit in storage [97].

The literature reports that a magnetic field of appropriate intensity accelerates the rate of fruit ripening. The magnetic field accelerates the synthesis of carotenoids, beta-carotene and lycopene. The ripening of the fruit in a magnetic field accelerates the characteristic changes in acids and sugars. PM increases the firmness of the fruit, extending its shelf life [91–93,107]. Many studies confirm that the electric field extends the shelf life of the fruit. HVEF slows down the respiration rate, affects the production of ethylene, and slows down the ripening of the fruit. The electric field prolongs the shelf life and freshness of the fruit by inhibiting the activity of enzymes and the inhibition of microbial action [97,102,109,119].

2.5. Effect of Magnetic Fields and Electrical Fields on Fruit Quality

Magnetic and electrical fields affect various plant characteristics and stimulate the growth of the plant. Strengthening the anabolic processes taking place in the plant by applying an electric and magnetic field may lead to an improvement in the quality of the fruit [5,9,15,18,65,106,112,113,120,121]. A summary of the effect of MF and EF on fruit quality is presented in Table 3.

Table 3. Summary of the total effect of magnetic fields and electrical fields on fruit quality.

Variety	Method	MF Parameter	Effect	References
Apple *Gloster* *Jonagold* *Ligol* *Rubin*	Fruit treatment during growth and ripening on the tree	PMF 5–100 mT AMF 50–150 uT and 5–100 mT for 5 min with 6 replicates each for 7 days	Increase in extract content for 100 uT, 8% more fructose and 25% more glucose were produced in the fruits stimulated during ripening	Zaguła, Puchalski 2013 [122]
Cherry fruits (*Prunus avium*)	Fruit treatment	PMF 0–20 mT AMF 0–2 mT During freezing process	Decrease in the phase change, time with PMF consumes less energy, the average ice crystal area achieved a reduction of 67% and 78% with PMF and AMF, respectively	Tang et al., 2021 [123]
Mandarin *Balady* and *Fremont*	Irrigating trees with magnetic water	MF in the range of 3.5–136 mT	Increase in firmness, total soluble solid, vitamin C, fruit weight and yield, but decrease in total acidity	Hamdy et al., 2015 [86]
Melon *Hetao*	Treatment of intact fruit using 2 parallel couples of coils, then cut in pieces and stored	AMF 2 mT for 5, 10, 15, 20 and 25 min	Increase in firmness, soluble solids (after 15 min), decrease in decomposition rate and titratable acid (after 2 days)	Jia et al., 2015 [124]
Strawberries *Elkat, Ventana, Honeyoe*	Fruit treatment during vegetation time on plants	SMF 5–100 mT, AMF 50–150 uT and 5–100 mT for 5 min with 5 replicates each 5 days	Increase in fructose content for 100 uT and with a frequency of 50–100 Hz	Zaguła et al., 2017 [125]
Tomato	Irrigation of trees with magnetic water	MF 12.4, 31.9, 71.9 mT	Increase in heavy metal content (Pb). No differences in vitamin content (A and C)	Yusuf and Ogunlela 2016 [126]
Apples *Fuji* and *Golden Delicious*	Fruit treatment with needle and plate electrode (CD) or 2 parallel plates (PP), reversed EF and forward EF	EF +/−1.25 kV/cm within storage	Using (CD), increase in weight loss, soluble sugar concentration, respiration and decrease in degree of change of hue. Using (PP), decrease in weight loss, soluble sugar concentration and respiration	Atungulu et al., 2003 [96]

Table 3. *Cont.*

Variety	Method	MF Parameter	Effect	References
Mango fruits *Irwin*	Fruit treatment with titanium plate and needle	EF 1.5 kV/cm of electric field for 45 min before storage	Decrease in antioxidative capacity of ripe fruit, no changes in ascorbic acid, carotene, quercetin, total phenols soluble solids, titratable acidity, firmness	Shivashankara et al., 2004 [119]
Persimmon fruits (*Diospyros kaki*)	Fruit treatment between two plates	EF field of 6 kV/cm for 30, 60, 90 or 120 min, 6 duplicates, before storage	Decrease in malondialdehyde content and pectinesterase activity, no change in total phenols (but some increase in MDA after treatment)	Liu et al. 2017 [101]
Persimmon fruits (*Diospyros kaki*)	Fruit treatment between the two plates	EF 7 kV/cm for 3, 6, 9 days within storage	Increase in total content of phenolic compounds, antioxidant activity and firmness	Jaisue et al., 2020 [127]
Seagrape (*Coccoloba uvifera*)	Fruit treatment with two-tier-parallel board before 9 days storage	EF 7.5 kV/cm for 60 min	Decrease in water loss, malondialdehyde (MDA), increase in total phenolic content (TPC), total chlorophyll content	Sulaimana et al., 2021 [128]
Strawberries *Selva*	Fruit treatment using a pin-to-plate	EF intensity of 3.61, 4.56, and 5.13 kV/cm for 60 min	Decrease in loss of mass of fruit, without changes in soluble solid content, pH, titratable acidity, softness and colour but preventing *Botrytis cinerea* growth	Esehaghbeygi et al., 2021 [105]
Tomato *Chaoyan-219*	Fruit treatment with 2 stainless steel plates	EF +/−2 kV/cm For 2 h	Growth in content of GSH, ascorbic acid as well as polyphenols. The activity of enzymes of SOD, APX and CAT was meaningfully higher than that of the control while storage	Zhao et al., 2011 [103]
Tomato *Pannovy*	Fruit treatment at harvest with steel plates as electrodes	Direct electric current 100, 300, 500 mA for 15, 30, 60 min	Increase in content of lycopene, B-carotene, total phenol content and antioxidant activity	Dannehl et al., 2011 [129]

Researchers irrigated the plants with magnetic water to enhance the quality of the fruit. Hamdy et al. (2015) indicated an increase in the total soluble solids of mandarin fruits when irrigation aqua was subject to a magnetic field in comparison with the control [86]. This may be due to increasing ion mobility and ion uptake and in the activities of antioxidant enzymes. They also stated that, in comparison with the control, there is an increase in fruit firmness of one variety and of the vitamin C content of fruits expressed as ascorbic acid/100 mL of fruit in both varieties when treated with magnetic water [86]. This same trend in comparison with that of the control was found by Yusuf and Ogunlela (2016) for the content of vitamins A and C when magnetic water was used for the irrigation of tomatoes. The highest values of these parameters were recorded at magnetic field values of 124 G and 719 G, respectively [126]. The treatment of whole fruit with a field improved their structural properties. The study of Zaguła et al. (2010) also showed an increase in firmness of strawberry fruits of up to 30% using an AMF of 50–150 uT in comparison with the control [94]. MFs penetrate deeper into cells and tissues; therefore, they can influence metabolic pathways at the cellular level [130]. On the other hand, Bourget et al. (2012) stated that a magnetic field, at low doses of 2.5 mT, did not influence the firmness of tomato fruit during storage at 16 °C [93]. However, treatment with AMF at 2.5 mT before the cutting of melo fruits influenced firmness, which was 54.9% higher in comparison with the control. These results showed that AMF retarded the softening of fruit tissue [124].

Other studies showed that the selected magnetic field of 100 μT at 50 Hz that was applied significantly influenced the quality parameters of fruits. Apple fruits treated with an MF had 8 and 25% more fructose and glucose, respectively, when compared to the control. There was also a general tendency to improve the content of the total extract of strawberries from 6–9% depending on variety [122,125,131]. A similar tendency was noted by Boe et al. (1968), who showed that tomato fruits treated with an MF of 6 mT had higher concentrations of glucose, beta carotene, and lycopene relative to control fruit [92]. However, Tang et al. (2020) used magnetic field pre-treatment of 10 mT for PMF and 1.26 mT for AMF to improve the quality of frozen cherry fruit. The application of a magnetic field could reduce drip loss to some extent. Ice crystals formed homogeneously in the cells when a magnetic field was applied. Compared to the control group, this achieved a reduction in the average area of ice crystals of 67% and 78% with a PMF and AMF, respectively [123].

Many studies have proved that an electric field may also affect the quality of fruits. The mass loss at the end of storage for those samples treated with a HVEF at 4.56 $kV \cdot cm^{-1}$ was 1.36%, while it was 3.98% for the control [105]. Another study revealed the effect of ACEF in reducing water loss after nine days of storage of sea grapes [128]. Wang et al. (2008) showed that firmness in tomato fruit treated with a −2 kV/cm electric field was 102.9% higher than that of control fruit on day 20 of storage [102]. The same results were obtained by Jaisue et al. (2020) with an electric field (7 kV/cm), showing that, after 15 d of storage, firmness of treated persimmons was higher than the control [127]. Other researchers pointed out that the application of HVEF can effectively reduce deterioration of apple colour postharvest and weight loss during storage [96,97,102,131].

Use of high-voltage electrical field treatment may also maintain the content of ascorbic acid, carotene, antioxidant capacity, taste, fruit colour, and a high level of soluble solids in mango fruits (*Mangifera indica*) for up to 20 days in storage at 5 °C when compared with a control [119]. Zhao et al. (2011) also showed that the content of non-enzyme antioxidants, such as GSH, ascorbic acid, and polyphenols, and the activities of antioxidant enzymes containing catalase, superoxide dismutase, ascorbate peroxidase, as well as peroxidase, in −2 kV/cm HVEF-treated tomato fruit were pointedly higher before those in the control during storage at 13 °C [103]. Dannehl et al. (2011) confirmed that direct current (DC 500 mA for 15 min) treatment of harvested tomatoes affected the secondary metabolism of fruits, which showed a high content of total phenol, lycopene, ß-carotene, and antioxidant activity [129]. However, Wang et al. (2008) indicated that the lycopene content in tomato fruit treated with an electrical field was invariably lower before that in control fruit amid the entire storage period [102]. The study of Liu et al. (2017) also showed the benefits of the application of an electrical field in maintaining the physicochemical properties of persimmon fruits. However, they are different according to structure, physiological metabolism, and electric properties from other products [101]. Persimmons treated with 600 kV/m for different lengths of time (0, 30, 60, 90, or 120 min) in samples stored at 25 °C showed almost 2.5-fold lower pectinesterase (PE) activity after 12 days of storage in comparison with the control [101]. These changes involved the action of cell wall hydrolysis enzymes, such as pectinesterase (PE) and polygalacturonase (PG) [132,133]. Pectin degradation plays an important role in the ripening of fruit via the disassembly of cellulose that contributes to fruit firmness. Hsieh et al. (2020) explained the effect of phenolic content on cell membrane integrity [134].

Sulaimana et al. (2021) reported that alternating current electric field (ACEF) with 125 kV/m of intensity for 60 min reduced malondialdehyde (MDA) production and chlorophyll degradation and increased total phenolic content (TPC) of sea grapes during nine days of storage in comparison with the control [128]. The observation by Han and Fumio (1997) that alternant high-voltage electric field treatment could hold up the comedown of chlorophyll and flavonoids in mandarin fruit was confirmed [115]. Delaying the degradation of chlorophyll and malondialdehyde (MDA) production under the effect of an electrical field was also demonstrated by Wang et al. (2008) and Liu et al. (2017) by comparison with the control [101,102]. Jaisue et al. (2020) showed increased total phenolic content and enhanced antioxidant activity in persimmon fruits by application of electric field treatment and comparison with a control [127]. This is consistent with the results obtained by Jeya et al. (2018) for grape extract and Rodríguez-Roque et al. (2015) for blueberry, showing the response of plants to induced stress and the inactivation of the enzymes involved in phenolic compound oxidation by an electric field [135–137]. Jiang et al. (2010) stated that the presence of phenolic compounds helps to contribute to the sensory attributes (taste, colour, texture, and aroma) [138]. The mechanism of HVEF is to interfere with the fruit's physiological biochemical metabolism by a high-voltage electrostatic field and, thus, affect the molecules' interaction and dynamics in cells. From the study of Wagner et al. (2006) it might be assumed that the electrical field increases cell membrane permeability by influencing the voltage-gated ion channels [139]. In addition, calcium-dependent protein kinases (CDPKs or CPKs) induced by stress are involved in stress signalling, hormone response, and regulation of a given metabolic pathway [140–142]. Consequently, it may result in an increase in freshness in fruit and vegetables [143]. On the other hand, a high-voltage electrical field of −35 kV/cm leads to irreversible damage to the cell membrane, resulting in an increase in the lycopene and ß-carotene content [144].

The literature reports numerous benefits of using magnetic and electric fields on fruit quality [5,145,146]. The treatment with an electric or magnetic field improves the structural properties of the fruit and inhibits the loss of water in the tissues. Magnetic and electric fields significantly delay the softening of the fruit tissue and increase the firmness of the fruit. Treatment with an electric or magnetic field significantly affects the quality parameters of the fruit and increases the concentration of glucose, beta carotene, lycopene, and vitamins A and C. The use of MF and EF has a positive effect on the content of phenolic compounds and antioxidant enzymes [5,108]. Effects of magnetic and electric field on the morphophysiological properties of fruits, with possible mechanism of action, are represented in Figure 2.

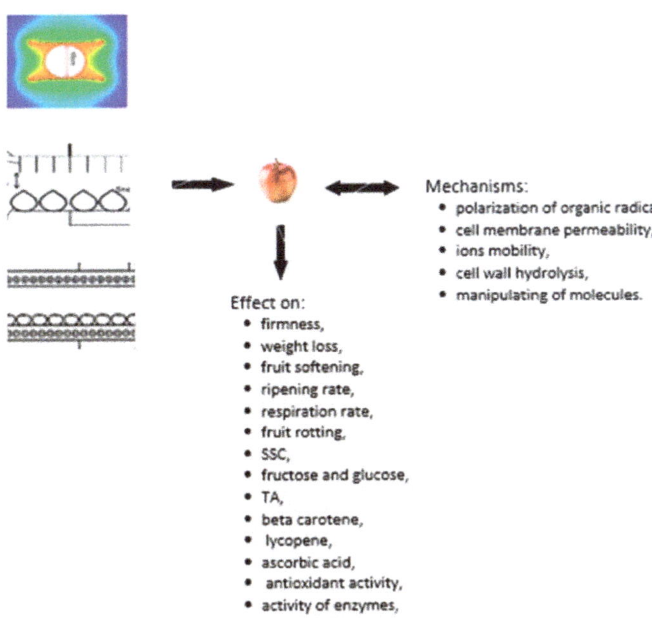

Figure 2. Magnetic and electric field affecting the morphophysiological properties of fruit, with possible mechanisms.

3. Conclusions

The objective of this paper was to review scientific results and summarise the emerging field of the effects of magnetic and electrical fields on the growth, yield, ripening, and shelf life of fruits and fruit quality, as well as the potential application of MF and EF in agricultural production. The results varied depending on the plant system, namely on the pre-treatment of seedlings, seed germination, plant growth, development and ripening, and the different categories of magnetic and electrical field intensity and exposure time applied to plants [49,105,145,147].

Alternating magnetic fields (AMF) with a value of 0.1–200 mT were the most commonly used with 50 Hz or 60 Hz power frequency for the treatment of seedlings, plants, fruits, and irrigated water. This magnetic field mainly caused the following effects on plant growth regardless of the treatment method: increase in speed of germination, plant height, shoot and root weight, and dry weight of the plant. It has contributed to improving fruit growth parameters, number of plant leaves, leaf area, number of flowers/plant, number of fruits/plant, fruit yield, average fruit weight, and the N, K, Ca, Mg, Fe, Mn, and Zn in foliage. These matters also influenced the shelf life of the fruit and its quality, mainly through an increase in firmness, ripening rate, beta carotene and lycopene, sugar concentration, fructose content, and a decrease in acidity and respiration rate.

High-voltage electrical fields (HVEF) with values of 2–3.61 kV/cm were the most commonly used treatments of fruits. Electrical fields were used to extend the shelf life and improve the quality of the fruit mainly through: a decrease in respiration rate and ethylene generation, and extension of post-harvest life. This is also related to delaying the decline in firmness, as well as the transformation in colour, entire soluble sugar, and also titratable acidity of fruit and an increase in the content of reduced glutathione, ascorbic acid, and polyphenols. The action of enzymes of SOD, APX, and CAT was suggestively higher than that of the control when in storage.

The use of electric and magnetic fields is a potential method to increase crops, improve crop quality, prevent disease and pest damage, and inhibit fruit aging. Exposing plants and fruits to an electric and magnetic field is an economic, safe, and environmentally friendly method. Despite numerous attempts, there are still difficulties in applying magnetic and electric fields in field and greenhouse conditions. Further experimentation is needed to develop magnetic- and electric-field-generating devices suitable for use in field and greenhouse crops. Moreover, the inconsistencies of the research results suggest that the effects of magnetic and electric fields on plants and fruit are species-specific and depend on the field strength and exposure time. This has promising research potential for the coming years. The application of electric and magnetic fields in field and greenhouse conditions also causes difficulties. This is another topic for further research.

Author Contributions: Conceptualisation, C.P. and B.S.; formal analysis, C.P.; investigation, B.S. and G.Z.; resources, C.P.; data curation, C.P. and E.S.; writing—original draft preparation, C.P. and B.S.; writing—review and editing, A.S. and E.S.; visualisation, B.S. and M.B.; supervision, B.S., G.Z. and M.B.; project administration, B.S. and A.S.; funding acquisition, C.P. All authors have read and agreed to the published version of the manuscript.

Funding: This research was funded by the Minister of Science and Higher Education in the range of the program entitled "Regional Initiative of Excellence" for the years 2019–2022, Project No. 026/RID/2018/19, amount of funding 9 542 500,00 PL.

Institutional Review Board Statement: Not applicable.

Informed Consent Statement: Not applicable.

Data Availability Statement: Not applicable.

Conflicts of Interest: The authors declare no conflict of interest.

Abbreviations

AMF	alternating magnetic fields
MF	magnetic fields
EF	electrical fields
HVEF	high voltage electrical field
PE	pectinase
MFD	magnetic flux density
SOD	superoxide dismutase
APX	ascorbate peroxidase
CAT	catalase
LFMF	low frequency magnetic field
SLA	specific leaf area
N	nitrogen
K	potassium
Ca	calcium
Mg	magnesium
Fe	iron
Mn	manganese
Zn	zinc
P	phosphorus
Na	sodium
Cu	copper
Pb	lead
PP	parallel plate
NdFeb	neodymium magnet
SMF	static magnetic field
NPK	nitrogen phosphorus potassium
SPMF	Sequentially programmed magnetic field
GMF	geomagnetic field
DC	direct current
AC	alternating current
PMF	pulsed magnetic field
MDA	malondialdehyde
TPC	total phenolic content
GSH	glutathione
ACEF	alternating current electric field
PG	polygalacturonase
CDPK	the calcium-dependent protein kinase
CPK	the calcium-dependent protein kinase

References

1. Nilsen, E.; Orcutt, D.M. *The Physiology of Plants under Stress–Abiotic Factors*, 2nd ed.; John Wiley and Sons: New York, NY, USA, 1996.
2. Oomori, U. *Bioelectromagnetics and Its Applications*; Fuji Technosystem Ltd.: Fuji, Japan, 1992; Volume 2, pp. 340–346.
3. Qamili, E.; De Santis, A.; Isac, A.; Mandea, M.; Duka, B.; Simonyan, A. Geomagnetic jerks as chaotic fluctuations of the Earth's magnetic field. *Geochem. Geophys. Geosystems* **2013**, *14*, 839–850. [CrossRef]
4. Kobayashi, M.; Soda, N.; Miyo, T.; Ueda, Y. Effects of combined DC and AC magnetic fields on germination of hornwort seeds. *Bioelectromagnetics* **2004**, *25*, 552–559. [CrossRef] [PubMed]
5. Nyakane, N.E.; Markus, E.D.; Sedibe, M.M. The effects of magnetic fields on plants growth: A comprehensive review. *ETP Int. J. Food Eng.* **2019**, *5*, 79–87. [CrossRef]
6. Shawan, R. *Electromagnetic Field*; BUBT University: Dhaka, Bangladesh, 2012.
7. Belyavskaya, N.A.; Fomicheva, V.M.; Govorun, R.D.; Danilov, V. Structural-functional organisation of the meristem cells of pea, lentil and flax roots in conditions of screening the geomagnetic field. *Biophysics* **1992**, *37*, 657–666.
8. Aladjadjiyan, A. The use of physical methods for plant growing stimulation in Bulgaria. *J. Cent. Eur. Agric.* **2007**, *8*, 369–380.
9. Taia, W.; Al-Zahrani, H.; Kotbi, A. The effect of static magnetic forces on water contents and photosynthetic pigments in sweet basil *Ocimum basilicum* L. (Lamiaceae). *Saudi J. Biol. Sci.* **2007**, *14*, 103–107.
10. Paul, A.; Robert, F.; Meisel, M. High magnetic field induced changes of gene expression in arabidopsis. *Biomagn. Res. Technol.* **2006**, *4*, 7. [CrossRef]

11. Martinez, E.; Carbonell, M.V.; Amaya, J.M. A static magnetic field of 125 mT stimulates the initial growth stages of barley (*Hordeum vulgare* L.). *Electro- Magn.* **2000**, *19*, 271–277.
12. Carbonell, M.V.; Martinez, E.; Amaya, J.M. Stimulation of germination in rice (*Oryza sativa* L.) by a static magnetic field. *Electro-Magnetobiol.* **2000**, *19*, 121–128. [CrossRef]
13. Aladjadjian, A. Study of influence of Magnetic field on some Biological Characteristics of Zea Mais. *J. Cent. Eur. Agric.* **2002**, *3*, 89–94.
14. Cho, E.G.; Kweon, S.J.; Suh, D.Y.; Suh, H.S.; Lee, S.K.; Sohn, J.K.; Oh, J.F. Studies of utilization of magnetic force in agricultural genetic engineering. *Res. Rep. Rural. Dev. Adm. Biotechnol.* **1992**, *34*, 10–14.
15. Sale, A.; Hamilton, W. Effects of high electric fields on microorganisms: I. Killing of bacteria and yeasts. *Biochim. Biophys. Acta (BBA) Gen. Subj.* **1967**, *148*, 781–788. [CrossRef]
16. Bajgai, T.R.; Hashinaga, F.; Isobe, S.; Raghavan, G.S.V.; Ngadi, M.O. Application of high electric field (HEF) on the shelf-life extension of embic fruit (*Phyllanthus emblica* L.). *J. Food Eng.* **2006**, *74*, 308–313. [CrossRef]
17. Singh, A.; Orsat, V.; Raghavan, G.S.V. A comprehensive review on electro hydrodynamic drying and high-voltage electric field in the context of food and bioprocessing. *Dry. Technol.* **2012**, *30*, 1812–1820. [CrossRef]
18. Baumgarten, B.E. EHD-Enhanced Heat Transfer in a Metallic and a Ceramic Compact Heat Exchanger. Master's Thesis, Department of Mechanical Engineering, Master of Science, University Maryland, College Park, MD, USA, 2003.
19. Rahbari, M.; Hamdami, N.; Mirzaei, H.; Jafari, S.M.; Kashaninejad, M.; Khomeiri, M. Effects of high voltage electric field thawing on the characteristics of chicken breast protein. *J. Food Eng.* **2018**, *216*, 98–106. [CrossRef]
20. Moon, J.-D.; Chung, H.-S. Acceleration of germination of tomato seed by applying AC electric and magnetic fields. *J. Electrost.* **2000**, *48*, 103–114. [CrossRef]
21. Diaz, D.C.; Riquenes, J.A.E.; Sotolongo, B.; Portuondo, M.A.; Quintana, E.O.; Perez, R. Effects of magnetic treatment of irrigation water on the tomato crop. *Hortic. Abstr.* **1997**, *69*, 494.
22. Patil, A.G. Device for magnetic treatment of irrigation water and its effects on quality and yield of banana plants. *Int. J. Biol. Sci. Appl.* **2014**, *1*, 152–156.
23. Maheshwari, L.; Basant, L.; Grewal, H.S. Magnetic treatment of irrigation water: Its effects on vegetable crop yield and water productivity. *Agric. Water Manag.* **2009**, *96*, 1229–1236. [CrossRef]
24. Taylor, J.E.; Tucker, G.A. *Biochemistry of Fruit Ripening*, 1st ed.; Springer: Dordrecht, The Netherlands, 1993; pp. 3–51.
25. Li, Z.; Yang, H.; Li, P.; Liu, J.; Wang, J.; Xu, Y. Fruit biomechanics based on anatomy: A review. *Int. Agrophysics* **2013**, *27*, 97–106. [CrossRef]
26. Billy, L.; Mehinagic, E.; Renard, C.M.G.C.; Prost, C. Relationship between texture and pectin composition of two apple cultivars during storage. *Postharvest Biol. Technol.* **2008**, *47*, 315–324. [CrossRef]
27. Szymańska-Chargot, M.; Chylińska, M.; Pieczywek, P.M.; Rösch, P.; Schmitt, M.; Popp, J.; Zdunek, A. Raman imaging of changes in the polysaccharides distribution in the cell wall during apple fruit development and senescence. *Planta* **2016**, *243*, 935–945. [CrossRef] [PubMed]
28. Ng, J.K.; Schröder, R.; Sutherland, P.W.; Hallett, I.C.; Hall, M.I.; Prakash, R.; Smith, B.G.; Melton, L.D.; Johnston, J.W. Cell wall structures leading to cultivar differences in softening rates develop early during apple (*Malus × domestica*) fruit growth. *BMC Plant Biol.* **2013**, *13*, 183. [CrossRef]
29. Volz, R.K.; Harker, F.R.; Lang, S. Firmness decline in gala apple during fruit development. *J. Am. Soc. Hortic. Sci.* **2003**, *128*, 797–802. [CrossRef]
30. Xia, Y.; Petti, C.; Williams, M.A.; De Bolt, S. Experimental approaches to study plant cell walls during plant-microbe interactions. *Front. Plant Sci.* **2014**, *5*, 1–7. [CrossRef]
31. Cybulska, J.; Zdunek, A.; Kozioł, A. The self-assembled network and physiological degradation of pectins in carrot cell walls. *Food Hydrocoll.* **2015**, *43*, 41–50. [CrossRef]
32. Nakamura, Y.; Wakabayashi, K.; Hoson, T. Temperature modulates the cell wall mechanical properties of rice coleoptiles by altering the molecular mass of hemicellulosic polysaccharides. *Physiol. Plant.* **2003**, *118*, 597–604. [CrossRef] [PubMed]
33. Gwanpua, S.G.; Mellidou, I.; Boeckx, J.; Kyomugasho, C.; Bessemans, N.; Verlinden, B.E.; Geeraerd, A.H. Expression analysis of candidate cell wall-related genes associated with changes in pectin biochemistry during postharvest apple softening. *Postharvest Biol. Technol.* **2016**, *112*, 176–185. [CrossRef]
34. Cherian, S.; Figueroa, C.; Nair, H. 'Movers and shakers' in the regulation of fruit ripening: A cross-dissection of climacteric versus non-climacteric fruit. *J. Exp. Bot.* **2014**, *65*, 4705–4722. [CrossRef]
35. Bapat, V.A.; Trivedi, P.K.; Ghosh, A.; Sane, V.A.; Ganapathi, T.R.; Nath, P. Ripening of fleshy fruit: Molecular insight and the role of ethylene. *Biotechnol. Adv.* **2010**, *28*, 94–107. [CrossRef]
36. Vicente, A.; Saladi, M.; Rose, J.K.C.; Labavitch, J.M. The linkage between cell wall metabolismand fruit softening: Looking to the future. *J. Sci. Food Agric.* **2007**, *87*, 1435–1448. [CrossRef]
37. Chen, Y.; Grimplet, J.; David, K.; Castellarin, S.D.; Terol, J.; Darren, C.; Wong, D.C.J.; Luo, Z.; Schaffer, R.; Celton, J.M.; et al. Ethylene receptors and related proteins in climacteric and non-climacteric fruits. *Plant Sci.* **2018**, *276*, 63–72. [CrossRef] [PubMed]
38. Villarreal, N.M.; Marina, M.; Nardi, C.F.; Civello, P.M.; Martínez, G.A. Novel insights of ethylene role in strawberry cell wall metabolism. *Plant Sci.* **2016**, *252*, 1–11. [CrossRef] [PubMed]

39. Kafkaletou, M.; Tsantili, E. The paradox of oleuropein increase in harvested olives (*Olea europea* L.). *J. Plant Physiol.* **2018**, *224–225*, 132–136. [CrossRef]
40. Martinez, E.; Carbonell Padrino, M.V.; Florez, M.; Amaya, J. Germination of tomato seeds (*Lycopersicon esculentum* L.) under magnetic field. *Int. Agrophysics* **2009**, *23*, 45–49.
41. Poinapen, D.; Beeharry, G.K.; Brown, D.C. Seed orientation and magnetic field strength have more influence on tomato seed performance than relative humidity and duration of exposure to non-uniform static magnetic fields. *J. Plant Physiol.* **2013**, *170*, 1251–1258. [CrossRef]
42. De Souza, A.; Garcia, D.; Sueiro, L.; Licea, L.; Porras, E. Pre-sowing magnetic treatment of tomato seeds increase the growth and yield of plants. *Bioelectromagnetics* **2006**, *27*, 247–257. [CrossRef]
43. De Souza, A.; Garcia, D.; Sueiro, L.; Licea, L.; Porras, E. Pre-sowing magnetic treatment of tomato seeds: Effects on the growth and yield of plants cultivated late in season. *Span. J. Agric. Res.* **2005**, *3*, 113–122. [CrossRef]
44. Kutby, A.M.; Al-Zahrani, H.S.; Hakeem, K.R. Role of Magnetic Field and Brassinosteroids in Mitigating Salinity Stress in Tomato (*Lycopersicon esculentum* L.). *Int. J. Eng. Res. Technol.* **2020**, *9*, 306–319.
45. Anand, A.; Kumari, A.; Thakur, M.; Koul, A. Hydrogen peroxide signaling integrates with phytohormones during the germination of magnetoprimed tomato seeds T. *Sci. Rep.* **2019**, *9*, 1–11.
46. Kireva, R.; Mihov, M. Impact of magnetic treatment of tomato and onion seeds on their productivity. *Int. Sci. J. Mech. Agric.* **2018**, *2*, 68–71.
47. Agustrina, R.; Nurcahyani, L.N.; Irawan, B. The Germination and Growth of Induced F1 Tomato Seeds by Exposure to 0.2mt of Magnetic Field and Fusarium Sp. Infection. *J. Agric. Vet. Sci.* **2018**, *11*, 84–88.
48. Iqbal, M.; Haq, Z.U.; Jamil, Y.; Nisar, J. Pre-sowing seed magnetic field treatment influence on germination, seedling growth and enzymatic activities of melon (*Cucumis melo* L.). *Biocatal. Agric. Biotechnol.* **2016**, *6*, 176–183. [CrossRef]
49. Menegatti, R.D.; Oliveira de Oliveira, L.; Costa, A.; Braga, E.J.B.; Bianchi, V.J. Magnetic field and gibberlic acid as pre-germination treatment of passion fruit seeds. *Ciência Agrícola* **2019**, *17*, 5–22.
50. Tirono, M.; Hananto, F.S.; Suhariningsih, V.Q.A. An Effective Dose of Magnetic Field to Increase Sesame Plant Growth and Its Resistance to Fusarium oxysporum Wilt. *Int. J. Des. Nat. Ecodynamics* **2021**, *16*, 285–291. [CrossRef]
51. Răcuciu, M. Development of tomato (*Solanum lycopersicum* L.) seedlings under the action of extremely low frequency magnetic field in a controlled environment conditions. *AIP Conf. Proc.* **2020**, *2206*, 030003.
52. Estiken, A. Effects of magnetic fields on yield and growth in strawberry Camarosa. *J. Hortic. Sci. Biotechnol.* **2003**, *78*, 145–147. [CrossRef]
53. Taimourya, H.; Oussible, M.; Baamal, L.; Harif, A.E.; Zaid, H.; Guedira, A.; Smouni, A. Magnetic treatment of culture medium enhance growth and minerals uptake of strawberry (*Fragaria × ananassa* Duch.) and tomato (*Solanum lycopersicum*) in Fe deficiency conditions. *Int. J. Sci. Eng. Res.* **2017**, *8*, 1414–1436.
54. El-Kholy, M.F.; Samia, F.; Hosny, S.; Farag, A.A. Effect of Magnetic Water and Different Levels of NPK on Growth, Yield and Fruit Quality of Williams Banana Plant. *Nat. Sci.* **2015**, *13*, 94–101.
55. Yusuf, K.O.; Ogunlela, A.O. Impact of Magnetic Treatment of Irrigation Water on the Growth and Yield of Tomato. *Not. Sci. Biol.* **2015**, *7*, 345–348. [CrossRef]
56. Yusuf, K.O.; Ogunlela, A.O. Effect of Magnetically Treated Water on Precipitation of some Macro Elements in the Soil for Tomato Growth. *J. Eng. Technol.* **2018**, 3, 108–112. [CrossRef]
57. Taimourya, H.; Oussible, M.; Baamal, L.; Bourarach, H.; Hassanain, N.; Masmoudi, L.; El Harif, A. Magnetically Treated Irrigation Water Improves the Production and the Fruit Quality of Strawberry Plants (*Fragaria × ananassa* Duch.) in the Northwest of Morocco. *J. Agric. Sci. Technol.* **2018**, *8*, 145–156. [CrossRef]
58. Jedlicka, J.; Paulen, O.; Ailer, S. Research of effect of low frequency magnetic field on germination, growth and fruiting of field tomatoes. *Acta Hortic. Regiotect.* **2015**, *1*, 1–4. [CrossRef]
59. El-Yazied, A.; Shalaby, A.O.A.; El-Gizawy, A.M.; Khalf, S.M.; El-Satar, M. Effect of Magnetic Field on Seed Germination and Transplant Growth of Tomato. *J. Am. Sci.* **2011**, *7*, 306–312.
60. Danilov, V.; Bas, T.; Eltez, M.; Rizakulyeva, A. Artifical magnetic field effect on yield and quality of tomatoes. *Acta Hortic.* **1994**, *366*, 279–285. [CrossRef]
61. El-Yazied, A.; El-Gizawy, A.M.; Khalf, S.M.; El-Satar, A.; Shalaby, O.A. Effect of Magnetic Field Treatments for Seeds and Irrigation Water as Well as N, P and K Levels on Productivity of Tomato Plants. *J. Appl. Sci. Res.* **2012**, *8*, 2088–2099.
62. Amaya, J.M.; Carbonell, M.V.; Martinez, E.; Raya, A. Effects of stationary magnetic fields on germination and growth of seeds. *Horticulturae* **1996**, *68*, 1363.
63. Souza, A.D.; Porras, L.E.; Casate, F.R. Effect of magnetic treatment of tomato (*Lycopersicon esculentum* Mill) seeds on germination and seedling growth. Invest. *Agric. Prod. Prot. Veg.* **1999**, *14*, 437–444.
64. Aladjadjiyan, A. Physical Factors for Plant Growth Stimulation Improve Food Quality. In *Food Production Approaches, Challenges and Tasks*, 1st ed.; Anna Aladjadjiyan; InTech: Rijeka, Croatia, 2012; Volume 9, pp. 145–168.
65. Amaya, J.M.; Carbonell, M.V.; Martinez, E.; Raya, A. Incidence of static magnetic fields on seed germination and growth. *Agricultura* **1999**, 1049–1052. (In Spanish)

66. Socorro, A.; Gil, M.; Labrada, A.; Díaz, C.; Lago, E. Cell model of seed tissue treated with magnetic field. In Proceedings of the II International Symposium on Applied Nuclear and Related Techniques in Agricultura, Industry and Environment, La Habana, Cuba, 26–29 October 1999.
67. Mitrov, P.P.; Kroumova, Z.; Baidanova, V.D. Auxin content of corn and tomato plants following magnetic treatments. *Fiziol. No Rastenyata* **1988**, *14*, 18–23.
68. Eşitken, A.; Turan, M. Alternating magnetic field effects on yield and plant nutrient element composition of strawberry (*Fragaria x ananassa* cv. *camarosa). Acta Agric. Scand. Sect. B Soil Plant Sci.* **2004**, *54*, 135–139. [CrossRef]
69. Fu, E. The effects of magnetic fields on plant growth and health. *Young Sci. J.* **2012**, *11*, 38–43. [CrossRef]
70. Pavlovic, D.; Nikolic, B.; Djurovic, S.; Waisi, H.; Andjelkovic, A.; Marisavljevic, D. Chlorophyll as a measure of plant health: Agroecological aspects. *J. Pestic. Phytomed.* **2014**, *29*, 21–34. [CrossRef]
71. Campbell, G.S.; Norman, J.M. *An Introduction to Environmental Biophysics*, 2nd ed.; Springer Science + Business Media: New York, NY, USA, 2012.
72. Reina, F.G.; Pascual, L.A.; Fundora, I.A. Influence of a Stationary Magnetic Field on water relations in lettuce Seeds. Part II: Experimental Results. *Bioelectromagnetics* **2001**, *22*, 596–602. [CrossRef] [PubMed]
73. Goodman, E.M.; Greenabaum, B.; Morron, T.M. Effects of electromagnetic fields onmolecules and cells. *Int. Rev. Cytol.* **1995**, *158*, 279–325. [PubMed]
74. Atak, C.; Emiroglu, O.; Aklimanoglu, S.; Rzakoulieva, A. Stimulation of regeneration by magnetic field in soybean (Glycine max L. Merrill) tissue cultures. *J. Cell Mol. Biol.* **2003**, *2*, 113–119.
75. Matsuda, T.; Asou, H.; Kobayashi, M.; Yonekura, M. Influences of magnetic fields on growth and fruit production of strawberry. *Acta Hortic.* **1993**, *348*, 378–380. [CrossRef]
76. Grewal, H.S.; Maheshwari, B.L. Magnetic treatment of irrigation water and snow pea and chickpea seeds enhances early growth and nutrient contents of seedlings. *Bioelectromagnetics* **2011**, *32*, 58–65. [CrossRef]
77. Ali, Y.; Samaneh, R.; Kavakebian, F. Applications of Magnetic Water Technology in Farming and Agriculture Development: A Review of Recent Advances. *Curr. World Environ.* **2014**, *9*, 695–703. [CrossRef]
78. Mohamed, A. Effects of Magnetized Low Quality Water on Some Soil Properties and Plant Growth. *Int. J. Res. Chem. Environ.* **2013**, *3*, 140–147.
79. Turker, M.; Temirci, C.; Battal, P.; Erez, M.E. The effects of an artificial and static magnetic field on plant growth, chlorophyll and phytohormone levels in maize and sunflower plants. *Phyton Ann. Rei Bot. Horn* **2007**, *46*, 271–284.
80. Hozayn, M.; Qados, A.M.S.A. Irrigation with magnetized water enhances growth, chemical constituent and yield of chickpea (*Circer arietinum* L.). *Agric. Biol. J. N. Am.* **2010**, *1*, 671–676.
81. Dhawi, F. Why magnetic fields are used to enhance a plant's growth and productivity? *J. Annu. Res. Rev. Biol.* **2014**, *4*, 886–896. [CrossRef]
82. Stange, B.C.; Rowland, R.E.; Rapley, B.I.; Podd, J.V. ELF magnetic fields increase amino acid uptake into *Vicia faba* L. roots and alter ion movement across the plasma membrane. *Bioelectromagnetics* **2002**, *23*, 347–354. [CrossRef]
83. El-Sayed, H.; Sayed, A. Impact of magnetic water irrigation for improve the growth, chemical composition and yield production of broad bean (*Vicia faba* L.) plant. *Am. J. Exp. Agric.* **2014**, *4*, 476–496.
84. Fernandez, L.; Teran, Z.; Leon, M. The effect of magnetically treated irrigation water on quality of onion seedlings grown in zeoponics. *Cultiv. Trop. (INCA)* **1996**, *17*, 55–59.
85. Hilal, M.H.; Shata, S.M.; Abdel-Dayem, A.A.; Hillal, M.M. Application of magnetic technologies in desert agriculture. III-Effect of Magnetized Water on yield and uptake of certain elements by citrus in relation to nutrients mobilization in soil. *Egypt. J. Soil Sci.* **2020**, *42*, 43–55.
86. Hamdy, A.E.; Khalifa, S.M.; Abdeen, S.A. Effect of magnetic water on yield and fruit quality of some mandarin varieties. *Ann. Agric. Sci.* **2015**, *53*, 657–666.
87. Lin, I.; Yotvat, J. Exposure of irrigation and drinking water to a magnetic field with controlled power and direction. *J. Magn. Magn. Mater.* **1990**, *83*, 525–526. [CrossRef]
88. Moussa, H.R. The impact of magnetic water application for improving common bean (*Phaseolus vulgaris* L.) production. *N. Y. Sci. J.* **2011**, *4*, 15–20.
89. Anand, A.; Nagarajan, S.; Verma, A.P.S.; Joshi, D.K.; Pathak, P.C.; Bhardwaj, J. Pre-treatment of seeds with static magnetic field ameliorates soil water stress in seedling of maize (*Zea mays* L.). *Indian J. Biochem. Biophys.* **2012**, *49*, 63–70. [PubMed]
90. Chern, C.C. Application of Magnetic Water to Stimulate the Lady's Finger (*Abelmosculentus* L.) Moench Plan Growth. Ph.D. Thesis, University of Technology, Johor Bahru, Malaysia, 2012.
91. Puchalski, C. Methodological aspects of testing apple friction and firmness in terms of assessing their quality. *Sci. J. Agric. Univ. Krakow* **2001**, *275*, 1233–4189.
92. Boe, A.A.; Do, J.Y.; Salunke, D.K. Tomato Ripening: Effects of Light Frequency, Magnetic Field, and Chemical Treatments. *Econ. Bot.* **1968**, *22*, 124–134. [CrossRef]
93. Bourget, S.; Corcuff, R.; Angers, P.; Arul, J. Effect of the Exposure to Static Magnetic Field on the Ripening and Senescence of Tomato Fruits. *Acta Hortic.* **2012**, *945*, 129–134. [CrossRef]
94. Zaguła, G.; Gorzelany, J.; Puchalski, C. Using a computer video system to examine the imoact of magnetic and electromagnetic fields on quality of strawberries. *Inżynieria Rol.* **2010**, *2*, 293–300.

95. Zaguła, G.; Tarapatskyy, M.; Bajcar, M.; Saletnik, B.; Puchalski, C.; Marczuk, A.; Andrejko, D.; Oszmiański, J. Near-Null Geomagnetic Field as an Innovative Method of Fruit Storage. *Processes* **2020**, *8*, 262. [CrossRef]
96. Atungulu, G.; Nishiyama, Y.; Koide, S. Use of an Electric Field to extend the Shelf Life of Apples. *Biosyst. Eng.* **2003**, *85*, 41–49. [CrossRef]
97. Atungulu, G.; Nishiyama, Y.; Koide, S. Respiration and climacteric patterns of apples treated with continuous and intermittent direct current electric field. *J. Food Eng.* **2004**, *63*, 1–8. [CrossRef]
98. Kharel, G.P.; Hasinaga, F.; Shintani, R. Effect of High Electric Fields on Some Fruits and Vegetables. *J. Japan. Soc. Cold Preserv. Food* **1996**, *22*, 17–22. [CrossRef]
99. Kusuma, R.A.; Pujantoro, L.; Wulandani, D. Effect of High Electrostatic Field Pre-treatment on Quality of Cherry Tomato during Storage. *J. Keteknikan Pertan.* **2018**, *6*, 31–38. [CrossRef]
100. Shiinaa, T.; Nei, D.; Nakamura, N.; Thammawong, M. Evaluation of High Electric Field Chamber for Shelf Life Extension of Food and Agricultural Commodities. *Acta Hortic.* **2010**, *880*, 517–524. [CrossRef]
101. Liu, C.E.; Chen, W.; Chang, C.; Li, P.; Lu, P.; Hsieh, C. Effect of a high voltage electrostatic field (HVEF) on the shelf life of persimmons (Diospyros kaki). *LWT* **2017**, *75*, 236–242. [CrossRef]
102. Wang, Y.; Wang, B.; Li, L. Keeping quality of tomato fruit by high electrostatic field pretreatment during storage. *J. Sci Food Agric.* **2008**, *88*, 464–470. [CrossRef]
103. Zhao, R.; Hao, J.; Xuea, J.; Liua, H.; Li, L. Effect of high-voltage electrostatic field pretreatment on the antioxidant system in stored green mature tomatoes. *J. Sci. Food Agric.* **2011**, *91*, 1680–1686. [CrossRef] [PubMed]
104. Atungulu, G.; Atungulu, E.; Okada, R.; Nishiyama, Y. Eficcacy of High Voltage Treatment on Tomato Storage. *J. Food Technol.* **2005**, *3*, 209–215.
105. Esehaghbeygi, A.; Hajisadeghian, A.; Nasrabad, M.N. Role of a corona field application in the physicochemical properties of stored strawberries. *Res. Agric. Eng.* **2021**, *67*, 58–64. [CrossRef]
106. Kharel, G.P.; Hashin, F. Effect of High Electric Field on Shelf Life of Strawberries. *Food Sci. Technol.* **1996**, *2*, 198–202. [CrossRef]
107. Valentinuzzi, M. Rotational diffusion in a magnetic field and its possible magnetobiological implications. In *Biological Effects of Magnetic Fields*; Barnothy, M.F., Ed.; Springer: Boston, MA, USA, 1964.
108. Sarraf, M.; Kataria, S.; Taimourya, H.; Santos, L.O.; Menegatti, R.D.; Jain, M.; Ihtisham, M.; Liu, S. Magnetic Field (MF) Applications in Plants: An Overview. *Plants* **2020**, *9*, 1139. [CrossRef]
109. Van Loey, A.; Verachtert, B.; Hendrickx, M. Effects of high electric field pulses on enzymes. *Trends Food Sci. Technol.* **2002**, *12*, 94–102. [CrossRef]
110. Hulseger, H.; Potel, J.; Niemann, E.G. Electric field effect on bacteria and yeast cells. *Radiat. Environ. Biophys.* **1986**, *22*, 149–162. [CrossRef]
111. Sakurauchi, Y.; Kondo, E. Lethal effect of high electric fields on microorganism. *Nippon. NouGeikagaku Kaishi* **1980**, *54*, 837–844. [CrossRef]
112. Murr, L.E. Plant growth response in a simulated electric field environment. *Nature* **1963**, *200*, 490–491. [CrossRef]
113. Toda, S. Preservation of foods and vegetables by application of electric field. *Shokuhin Rhytsu Gijitsu* **1990**, *19*, 62–64.
114. Palanimuthu, V.; Rajkumar, P.; Orsat, V.; Gariepy, Y.; Raghavan, G.S.V. Improving cran-berry shelf life using high voltage electric field treatment. *J. Food Eng.* **2009**, *90*, 365–371. [CrossRef]
115. Zhang, H.; Hashinaga, F. Effect of high electric field on quality off satsuma mandarin fruits. *J. Soc. High Technol. Agric.* **1997**, *9*, 107–113. [CrossRef]
116. Karaca, H.; Velioglu, Y.S. Ozone applications in fruit and vegetable processing. *Food Rev. Int.* **2007**, *23*, 91–106. [CrossRef]
117. Samaranayake, C.P.; Sastry, S.K. Effects of controlled-frequency moderate electric fields on pectin methylesterase and polygalacturonase activities in tomato homogenate. *Food Chem.* **2016**, *199*, 265–272. [CrossRef]
118. Yu, L.; Reitmeier, C.A.; Gleason, M.L.; Nonnecke, G.R.; Olson, D.G.; Gladon, R.J. Quality of electron beam irradiated strawberries. *J. Food Sci.* **1995**, *60*, 1084–1087. [CrossRef]
119. Shivashankara, K.S.; Isobe, S.; Al-Haq, M.M.; Takenaka, M.; Shiina, T. Fruit antioxidant activity, ascorbic acid, total phenol, quercetin, and carotene of Irwin mango fruits stored at low temperature after high electric field pretreatment. *J. Agric. Food Chem.* **2004**, *52*, 1281–1286. [CrossRef]
120. Ruzic, R.; Berden, M.; Jerman, I. The effects of oscillating electromagnetic fields on plants. In *Summary Report, Proceedings of the First World Congress on the Bioeffects of Electricity and Magnetism on the Natural World, Madeira, UK, 1–6 October 1998*; Coghill Research Laboratories: Pontypool, UK, 2011.
121. Panda, D.; Mondal, S. Seed enhancement for sustainable agriculture: An overview of recent trends. *Plant Arch.* **2020**, *20*, 2320–2332.
122. Zaguła, G.; Puchalski, C. Glucose-fructose changes in apples exposed to constant and slowly changing magnetic fields. *Food Sci. Technol. Qual.* **2013**, *2*, 162–172. [CrossRef]
123. Tang, J.; Zhang, H.; Tian, C.; Shao, S. Effects of different magnetic fields on the freezing parameters of cherry. *J. Food Eng.* **2020**, *278*, 109949. [CrossRef]
124. Jia, J.; Wang, X.; Lv, J.; Gao, S.; Wang, G. Alternating Magnetic Field Prior to Cutting Reduces Wound Response and Maintains Fruit Quality of Cut *cucumis melo* L. cv Hetao. *Open Biotechnol. J.* **2015**, *9*, 230–235. [CrossRef]

125. Zaguła, G.; Puchalski, C.; Czernicka, M.; Bajcar, M.; Saletnik, B.; Woźny, M.; Szeregii, E. The magnetic field stimulation system applied on strawberry fruits. *Econtechmod. Int. Q.* **2017**, *6*, 117–122.
126. Yusuf, K.O.; Ogunlela, A.O. Effect of magnetically treated water on the quality of tomato. *J. Sci. Eng. Technol. Katmandu* **2016**, *12*, 29–33. [CrossRef]
127. Jaisue, N.; Setha, S.; Hamanaka, D.; Naradisorn, M. Impact of Electric Field on Physicochemical Properties and Antioxidant Activity of Persimmon (*Diospyros kaki* L.). *EAEF* **2020**, *13*, 98–104. [CrossRef]
128. Sulaimana, A.S.; Chang, C.; Hou, C.; Yudhistira, B.; Punthi, F.; Lung, C.; Cheng, K.; Santoso, S.P.; Hsieh, C. Effect of Oxidative Stress on Physicochemical Quality of Taiwanese Seagrape (*Caulerpa lentillifera*) with the Application of Alternating Current Electric Field (ACEF) during Post-Harvest Storage. *Processes* **2021**, *9*, 1011. [CrossRef]
129. Dannehl, D.; Huyskens-Keil, S.; Eichholz, I.; Ulrichs, C.; Schmidt, U. Effects of direct-electric-current on secondary plant compounds and antioxidant activity in harvested tomato fruits (*Solanum lycopersicon* L.). *Food Chem.* **2011**, *126*, 157–165. [CrossRef]
130. Zaguła, G.; Puchalski, C.; Gorzelany, J. Spectroscopy method of evaluation of the influence of permanent and low-frequency magnetic fields during the increase and ripening on the balance of glucose and fructose of selected apple varieties. *Inżynieria Rol.* **2011**, *9*, 269–276.
131. Funk, H.W.R.; Monsees, T.; Özkucur, N. Electromagnetic effects–From cell biology to medicine. *Prog. Histochem. Cytochem.* **2009**, *43*, 177–264. [CrossRef]
132. Fischer, R.L.; Bennett, A.B. Role of cell wall hydrolases in fruit ripening. *Annu. Rev. Plant Physiol. Plant Mol. Biol.* **1991**, *42*, 675–703. [CrossRef]
133. King, G.A.; O'Donoghue, E.M. Unravelling senescence: New opportunities for delaying the inevitable in harvested fruit and vegetables. *Trends Food Sci. Technol.* **1995**, *6*, 385e389. [CrossRef]
134. Hsieh, C.C.; Chang, C.K.; Wong, L.W.; Hu, C.C.; Lin, J.A.; Hsieh, C.W. Alternating current electric field inhibits browning of Pleurotus ostreatus via inactivation of oxidative enzymes during postharvest storage. *LWT-Food Sci. Technol.* **2020**, *134*, 110212. [CrossRef]
135. Jeya, S.T.; Gowri, S.V.; Priyanka, A.; Sundararajan, R. Antioxidant and anticancer activity of pulsed electric field treated grape extract. *J. Nanomed. Biother. Discov.* **2018**, *8*, 1000159.
136. Rodríguez-Roque, M.J.; Ancos, B.D.; Sánchez-Moreno, C.; Cano, M.P.; Elez-Martínez, P.; Martín-Belloso, O. Impact of food matrix and processing on the in vitro bioaccessibility of vitamin C, phenolic compounds, and hydrophilic antioxidant activity from fruit juice-based beverages. *J. Funct. Foods* **2015**, *14*, 33–43. [CrossRef]
137. Dziadek, K.; Kopec, A.; Drozdz, T.; Kiełbasa, P.; Ostafin, M.; Bulski, K.; Oziembłowski, M. Effect of pulsed electric field treatment on shelf life and nutritional value of apple juice. *J. Food Sci. Technol.* **2019**, *56*, 1184–1191. [CrossRef]
138. Jiang, T.; Jahangir, M.M.; Jiang, Z.; Lu, X.; Ying, T. Influence of UV-C treatment on antioxidant capacity, antioxidant enzyme activity and texture of postharvest shiitake (*Lentinus edodes*) mushrooms during storage. *Postharvest Biol. Technol.* **2010**, *56*, 209–215. [CrossRef]
139. Wagner, E.; Lehner, L.; Normann, J.; Veit, L.; Albrechtová, J. Hydroelectrochemical integration of the higher plant–Basis for electrogenic flower induction. In *Communication in Plants*; Baluska, F., Mancuso, S., Volkmann, D., Eds.; Springer: Berlin, Germany, 2006; pp. 369–387.
140. Xing, T.; Wang, X.J.; Malik, K.; Miki, B.L. Ectopic expression of an Arabidopsis calmodulin-like domain protein kinase-enhanced NADPH oxidase activity and oxidative burst in tomato protoplasts. *Mol. Plant-Microbe Interact.* **2001**, *14*, 1261–1264. [CrossRef]
141. Valmonte, G.R.; Arthur, K.; Higgins, C.M.; MacDiarmid, R.M. Calcium-dependent protein kinases in plants: Evolution, expression and function. *Plant Cell Physiol.* **2014**, *55*, 551–569. [CrossRef]
142. Crizel, R.L.; Perin, E.C.; Vighi, I.L.; Woloski, R.; Seixas, A.; da Silva Pinto, L.; Rombaldi, C.V.; Galli, V. Genome-wide identifcation, and characterization of the CDPK gene family reveal their involvement in abiotic stress response in *Fragaria × ananassa*. *Sci. Rep.* **2020**, *10*, 11040. [CrossRef]
143. Robertson, D.; Miller, M.W. Inhibition and recovery of growth process in roots of *Pisum sativum* L. exposed to 60-Hz electric fields. *Bioelectromagnetics* **1981**, *2*, 329–340. [CrossRef] [PubMed]
144. Odriozola-Serrano, I.; Aguilo-Aguayo, I.; Soliva-Fortuny, R.; Gimeno-Ano, V.; Martiin-Belloso, O. Lycopene, vitamin C, and antioxidant capacity of tomato juice as affected by high-intensity pulsed electric fields critical parameters. *J. Agric. Food Chem.* **2007**, *55*, 9036–9042. [CrossRef] [PubMed]
145. González-Casado, S.; Martín-Belloso, O.; Elez-Martínez, P.; Soliva-Fortuny, R. Application of pulsed electric fields to tomato fruit for enhancing the bioaccessibility of carotenoids in derived products. *Food Funct.* **2018**, *9*, 2282–2289. [CrossRef] [PubMed]
146. El Kantar, S.; Boussetta, N.; Lebovka, N.; Foucart, F.; Rajha, H.N.; Maroun, R.G.; Louka, N.; Vorobiev, E. Pulsed electric field treatment of citrus fruits: Improvement of juice and polyphenols extraction. *Innov. Food Sci. Emerg. Technol.* **2018**, *46*, 153–161. [CrossRef]
147. Abobatta, W.F. Overview of Role of Magnetizing Treated Water in Agricultural Sector Development. *Adv. Agric. Technol. Plant Sci.* **2019**, *2*, 180023.

applied sciences

MDPI

Review

The Effect of UV-C Irradiation on the Mechanical and Physiological Properties of Potato Tuber and Different Products

Addis Lemessa, Ernest Popardowski *, Tomasz Hebda and Tomasz Jakubowski

Faculty of Production and Power Engineering, University of Agriculture in Krakow, Balicka 116B, 30-149 Krakow, Poland; addislemessa@gmail.com (A.L.); tomasz.hebda@urk.edu.pl (T.H.); tomasz.jakubowski@urk.edu.pl (T.J.)
* Correspondence: ernest.popardowski@urk.edu.pl

Abstract: Amongst the surface treatment technologies to emerge in the last few decades, UV-C radiation surface treatment is widely used in food process industries for the purpose of shelf life elongation, bacterial inactivation, and stimulation. However, the short wave application is highly dose-dependent and induces different properties of the product during exposure. Mechanical properties of the agricultural products and their derivatives represent the key indicator of acceptability by the end-user. This paper surveys the recent findings of the influence of UV-C on the stress response and physiological change concerning the mechanical and textural properties of miscellaneous agricultural products with a specific focus on a potato tuber. This paper also reviewed the hormetic effect of UV-C triggered at a different classification of doses studied so far on the amount of phenolic content, antioxidants, and other chemicals responsible for the stimulation process. The combined technologies with UV-C for product quality improvement are also highlighted. The review work draws the current challenges as well as future perspectives. Moreover, a way forward in the key areas of improvement of UV-C treatment technologies is suggested that can induce a favorable stress, enabling the product to achieve self-defense mechanisms against wound, impact, and mechanical damage.

Keywords: UV-C radiation; stress response; mechanical properties; stimulation; potato tuber

Citation: Lemessa, A.; Popardowski, E.; Hebda, T.; Jakubowski, T. The Effect of UV-C Irradiation on the Mechanical and Physio-Logical Properties of Potato Tuber and Different Products. *Appl. Sci.* **2022**, *12*, 5907. https://doi.org/10.3390/app12125907

Academic Editor: Claudio De Pasquale

Received: 12 May 2022
Accepted: 7 June 2022
Published: 10 June 2022

1. Introduction

The non-thermal technique is relatively fruitful for elongating fresh foods' shelf life. Nevertheless, its potentiality of adversely influencing sensory attributes has been highlighted [1–7]. Several non-thermal technologies for the treatment of food crops have come to emerge over the last few decades. This technology included irradiations (pulsed ultraviolet, gamma, and X-ray) pulsed electric field, pulsed electromagnetic field, cold plasma, ultra-sonication, microwave, supercritical technology, high-pressure processing, etc. [8–11]. In the classification of surface treatment, microwave surface treatment can be considered as either a thermal or non-thermal surface treatment depending on the intensity used [12,13]. It is reported that the irradiation technology is of high energy. Therefore, alongside its desirable effects, it is expected to induce undesirable alterations by interacting with different structures as well as chemical constituents of the potato tubers [14]. Among other different crop treatment techniques, such as gamma irradiation and fumigation [15], the UV-C radiation technique is one of the non-thermal technologies that is dominantly used for the surface disinfection and decontamination of crops and fruits and is by far regarded as effective [16]. UV-C radiation, emitted at wavelengths of 200–280 nm along with UV-A (320–400 nm) and UV-B (280–320 nm), is reported to be retained by the ozone layer. The different regions of UV radiation are shown in Figure 1 [17–19]. UV-C radiation, because of its high absorption level by the ozone layer, does not penetrate the earth in any appreciable amount. The shorter the wavelength, the higher the energy of the photon as depicted from the Planck relation. This short wavelength carries a higher pocket of energy which is capable of destroying microorganisms by damaging the microbial DNA, causing cross-linking between neighboring thiamine and cytosine in the same DNA strand [20,21].

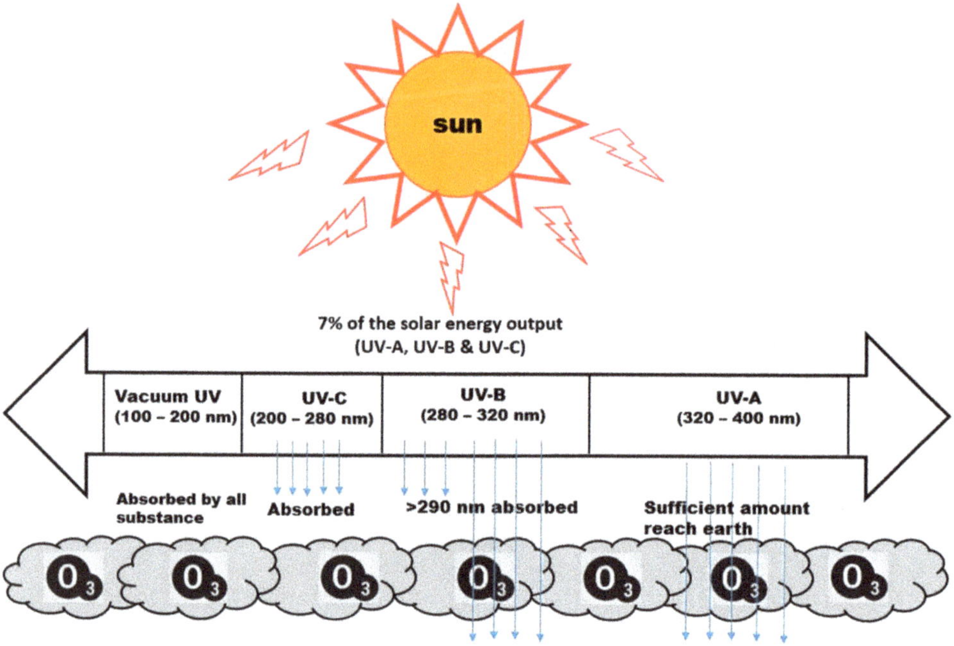

Figure 1. Different regions of UV radiation.

Even if this surface treatment has consent from Food and Drug Administration and other institutes as recognized disinfectant technology [22,23], it is important to study its impact on the characteristics of the crop. Numerous studies have already been focused on the bactericidal and fungicidal capacity of UV-C treatment in plant products, but information on product quality attributes is scanty. Limited studies have been reporting that short wave UV-C has a pronounced effect on plants' physiology and structure. It was reported that there induces a significant structural damaging effect on the plant cell, most specifically on the chloroplasts, mitochondria, and membranes [24,25]. The change in cellular structure is directly related to the overall alteration of the mechanical properties of the plant. This correlation is well described by a previous study conducted on the potato tuber cell structural parameters having a significant effect on the overall mechanical characteristics such as strength and modulus of elasticity [26]. Moreover, in the other study on UV-C exposed strawberries, cellular structural changes in relation to the physical properties of the fruit were studied stating that delayed softening and higher firmness are caused by changes in the activity of enzymes and proteins responsible for cell wall disassembly [27]. It was clearly stated that the change in mechanical characteristics of food derived from plants arises from three structural factors at the cellular level are turgor, cell wall rigidity, and cell-cell adhesion [28,29].

The mechanical characteristics of crops play a prominent role in deciding the acceptability in the chains of target users, such as food processing industries and people. Since plants are biological materials with complex structures that have high exposure to mechanical change, these changes can lead to irreversible changes in structure and other crop characteristics where these changes are reported to be invisible, having inner deformations [26]. Potatoes are the fourth-ranking food crop in the world in terms of calories and are extensively produced throughout the world [30–32]. The mechanical properties of the potato or derived products following surface treatment by irradiation during either growth or storage represent an imperative factor for meeting consumers' needs, as they are required for determining other properties when characterizing food qualities. Most importantly, one of the most important features here is the texture that potentially undergoes change, since external factors are concerned [33–37].

Studies have shown the influence of different physical treatment methods inducing a change in potato mechanical characteristics. Microwave treatment methods have found extensive applications in various processing and operations [38–41]. Studies have demonstrated their influential effects on the texture, compression test, weight loss, morphological, and microstructural changes of potatoes [42–44]. In the process of extending the shelf life of potatoes, gamma radiation is also an important facility nowadays. Several studies on the effect of gamma irradiation on physicomechanical properties of potatoes have been carried out on different potato cultivars for the evaluation of sprout weight, appearance quality, texture, specific gravity, morphology, puncture test, shear test, and compressive tests at varied dosses ranging from 0.04 to 100 kGy under different storage conditions [14,45–48]. From the result, Gamma irradiated resulted in an intact, rigid cell wall, more cavities, and bigger lamellar structure. At lower dose, mechanical properties were maintained, and sprouting was inhibited, while increased specific gravity at higher dose was also indicated. Texture was reduced with increasing dose, and the appearance of potato tuber was firm and slightly shrivelled. One of the non-destructive evaluation techniques is an ultrasound that applies a low range of intensities (0.1–20 kHz), but higher intensities (20–100 kHz) are reported to alter the physical, mechanical, and structural characteristics of starch granules derived from potato [49,50]. The results indicated that the sonicated potato tubers showed a change in molecular order in the crystalline region. A transparent appearance of the potato starch was also found. The effect of pulsed magnetic field treatment on the firmness, energy for cutting, and smoothening of the surface of potato was investigated at lower pulses ranging from 1 to 2.5 kV. It was concluded that firmness and energy for cutting were reduced while smoothening of the surface increased [8,51].

A study conducted on the impact of UV-C on mechanical characteristics of certain plant types reports that the mechanical properties are the result of the cellular change and rheology [52]. Despite the popular demand for potato tuber, there are only limited research outputs, making this area untapped for research. However, it was possible to survey some of the most up-to-date research that recognized the effect of UV-C on different characteristics, including the acrylamide content [53], color [42], density [54,55], and chemical constituents [56] of potato tuber and its derivatives. The objective of this review work is to critically review the current state of the art in the damaging effect, physiological change, and related scenarios of UV-C on the potato. Of course, this review also surveys the mechanically damaging effect of UV-C on different postharvest crops and fruits.

2. Influence of UV-C on the Physical and Mechanical Properties of Different Crops

UV-C surface treatment technology is a set of techniques by which a specific wavelength of 254 nm is delivered from a source for the purpose of microbial inactivation, disinfection, and stimulation. The short wave carries a pocket of energy necessary to inhibit microorganisms from the surface of the product by developing defense mechanisms. However, the application is highly dose dependent and results a positive and negative impact on the mechanical properties of the given product. In the quest of the current state of the art, a number of findings were reported in different types of products and varieties as well. It is apparently clear that the UV-C effect on the plant products is highly dependent on UV-C dose.

2.1. Physical Properties

With a specific focus in food characterization, textural profile analysis (TPA) offers comprehensive properties important to the acceptability of a product by the end user. A wide set of parameters traced from TPA analysis includes hardness, cohesiveness, viscosity, elasticity (springiness), adhesiveness, resilience, brittleness (fracturability), chewiness, and gumminess. This is one of the most important parameters as far as the UV-C stimulation is concerned with the cell physiology. The properties of firmness, hardness, and change in mass are critically assessed in this review as they are greatly influenced by the anatomy of the plant tissue, the strength of the cell wall, and ultimately other mechanical properties. It is one of the most important parameters when it comes to UV-C stimulation and cell physiology [57].

The firmness of bell pepper fruit treated at 0.25 kJ·m^{-2} and its control sample significantly dropped at the initial storage period, but after seven days of storage the UV-C treated sample was found to be firmer than the control. A weight loss experiment in the same fruit resulted in lower weight loss in the treated sample [58]. An experiment on the UV-C treated in green asparagus confirmed an appreciable increment in tissue toughness and a significant increase in cutting energy (sheer force) even at a lower UV-C dose of 1 kJ·m^{-2} and 8 min exposure time [59]. Conversely, a low UV-C dose up to 3.8 kJ·m^{-2} is indicated to have an insignificant change in textural properties [60]. Observations on the fresh-cut melon after UV-C exposure confirms no significant difference in the firmness as a result of UV-C application at 1.2 kJ·m^{-2} followed by storage at 6 °C [61]. No significant difference was observed among treated and control samples in delaying softening at a lower UV-C dose. The research work on the firmness and weight loss characteristics of UV-C exposed fresh-cut apples were analyzed by different authors and resulted in no significant change in the firmness [62,63], while a decrease in weight loss by 6% was noted at an exposure time of 1 min and intensity of 1.2 kJ·m^{-2} UV-C [62]. For fruit crops, weight loss is an important parameter as it is associated with dehydration, which significantly determines the commercial value of the fruit. The weight loss of amaranth vegetables treated at a moderate UV-C dose of 1.7 kJ·m^{-2} reduced the weight loss significantly better than the control [64]. The weight loss experiment on different UV-C exposed tomato cultivars, namely "cherry tomato" and *Elpida*, showed no effect and reduced weight loss when treated at 3.7 kJ·m^{-2} and 4 kJ·m^{-2}, respectively [65,66]. It was depicted that a lower UV-C dose of up to 4 kJ·m^{-2} has a minor effect on the firmness of tomato [67] and blueberry fruits [68]. The firmness value of different tomato cultivars was enhanced when exposed to UV-C treated in the range of 3 kJ·m^{-2} to 4.2 kJ·m^{-2} [65,66,69–72]. An experiment conducted on the quality parameters of UV-C exposed pineapple at the dose of 4.5 kJ·m^{-2} over the range of 0–90 s and followed by a 10 °C storage condition was able to retain the firmness properties better than the control samples [73]. Exposure to 7.11 kJ·m^{2} induced higher tissue softening in lettuce and caused the development of abnormal color in cauliflower [74,75].

Research work on a comparative study of the effect of UV-A and UV-C over the temperature range of 25–100 °C on oyster mushrooms reported UV-C exposed mushroom samples had a higher increase in loss modulus and loss factor than that of UV-A exposed ones. From the outcome, it was concluded UV-C light had a greater impact on the mechanical properties of oyster mushrooms compared to UV-A light [76]. Storage modulus was lower for the samples exposed to both UV-A and UV-C, indicating the samples had a viscoelastic characteristic.

In the quest for the previous state of the art, the most decisive factor was found to be firmness, which was studied dominantly following the UV-C irradiation and storage. It was also noted from recent research that firmness is an important sensory characteristic for consumers and is the factor that is highly affected by mechanical processes that potentially induce the ejection of intracellular fluid due to tissue rupture [77].

The textural aspect of mechanical property is dominantly used in the processing, as well as with raw vegetables and fruits, which is connected with the rheological characteristics of biological materials called firmness or hardness. It expresses the maximum force required to attain a specific strain in compression, puncture, and cut tests [57,78].

2.2. Mechanical Properties

Mechanical properties of harvested vegetables and fruits are important characterization throughout processing, storage, and consumption. Mechanical properties may be defined by cell structure and are dependent on the physical state, flow properties, and porosity. In the light of measurements, force-deformation methods are commonly used in textural or mechanical properties of solid foods, fresh vegetable/fruit, and their derivatives in their solid state. The basic mechanical properties that are determined from the force-deformation includes rupture force, toughness, cutting force, shear force, and strength [79], which are effective for different purposes, such as product standardization, transportation, handling, and design purposes as well. If we take account of the dimension of a product in the mechanical testing, stress–strain characterization, for example, can be used and some important properties can be obtained from the result of this test, such as yield strength, Young's modulus, ultimate strength, and modulus of elasticity.

Recent research conducted intensive work on mechanical changes of UV-C irradiated strawberries at a varied irradiation dose of 0.8, 2, and 4 kJ·m^{-2} by changing the storage duration from 0 to eight days at 0 °C. Resistance to compression, crush resistance, and distance to tissue failure attributes dropped significantly during storage. According to the author, the resistance to compression was higher at a lower UV-C dose. At 2 kJ·m^{-2} and 4 kJ·m^{-2}, tissue deformation reported a higher value [80]. In the latest work on the effect of UV-C irradiation on some of the mechanical characteristics of blueberries *'O'Neal'* a puncture test was performed at UV-C intensities of 5.3, 8.3, and 11.4 kJ·m^{-2}, exposure time of 7, 11, and 15 min, and storage time from 0 to 15 days. According to the report, mechanical parameters were not affected by UV-C treatments until 15 days of storage time when irradiated samples showed higher values of rupture force, deformation, and weight loss [81]. In a related study, a compression test (stress –strain) was performed on UV-C exposed fresh-cut apples at an exposure time of 10, 15, and 25 min and UV-C intensity at 5.6, 8.4, and 14.1 kJ·m^{-2}. True stress and deformability modulus were noticeably reduced at lower exposure time and UV-C dose [63]. The effect of UV-C on the mechanical characterization as well as the dimensions (length, width, and height) of *Faba* bean during storage was studied by considering various UV-C exposure times (0, 30, 60, and 90 min) and during storage periods of 0, three, six, and nine months. The result indicated that main dimensions, mass and bulk volume, and true volume were decreased by increasing the storage period and decreasing ultraviolet irradiation time. UV-C irradiation time increased with reducing the storage time, although, length and thickness decreased slightly. From this research, very important mechanical tests, such as shear force and shear penetration, were also investigated on the UV exposed *Faba* bean. The authors described the decrement of both the sheer force and penetration force of seed as UV-C exposure time increased during the given storage time [82]. The mechanical properties of different UV-C irradiated plant commodities are briefly presented in Table 1.

Table 1. Mechanical and related characteristics of UV-C exposed plant products.

Commodity	Operational Condition	Key Finding	Reference
Sweet corn kernels	UV-C dose at 0. 1.94, and 4.01 kJ·m^{-2}, controlled atmosphere of with %oxygen: %carbondioxide: %nitrogen ratios of 21:0.03:78, 3:10:87, and 3:15:82 at 6 °C for 20 h.	Hardness remains unchanged	[83]
Fresh-cut green onion	UV-C exposure time at 3, 5, 10, and 15 min and storage days of 5, 10, 15 days and storage temperature of 5 °C.	Higher UV-C exposure results in higher weight loss (%).	[84]

Table 1. *Cont.*

Commodity	Operational Condition	Key Finding	Reference
Tomato (*Lycopersicon esculentum* L.)	UV-C dose 3.7 kJ·m^{-2} from 0 to 25 days of storage duration time at 16 °C and relative humidity of 95%.	Firmness (Newton) decreased with storage duration. Higher resistance penetration compared to the control sample.	[85]
Cucumber (*Cucumis sativus* L.)	UV-C dose of 4.5 kJ·m^{-2} stored for 15 days at 4 °C, a combination of UV-C with Nano-coating Nanocapsules.	The UV-C control sample brought better firmness, as the loss was delayed to day 9 of storage.	[86]
Peeled garlic (*Allium sativum* L.)	UV-C dose of 2 kJ·m^{-2} stored for 15 days at room temperature.	High firmness value with the UV-C treated sample.	[87]
Cherry tomato	UV-C dose of 3.7 kJ·m^{-2}	UV-C treated and control sample both mass loss and firmness were unaffected.	[65]
Tomato (*Lycopersicon esculentum* L.)	UV-C dose of 4.2 kJ·m^{-2} for 8 min	The firmness decreased gradually during storage in both the control and UV-C treated tomatoes.	[72]
Common dandelion and purple coneflower	UV-C dose of 3.8 J·m^{-2}, 1 m of distance from light source, 10 to 120 exposure time, and 21 days of storage period.	Fresh and dry weight loss recorded for both dandelion and purple coneflower is higher than the control sample.	[88]
Strawberry	UV-C dose of 1.70 kJ·m^{-2} for 4.8 min and Storage duration of 0, 2, and 4 days at 21 °C.	No difference in firmness between fruit from control and UV-C-treated samples.	[89]
Mango (*Kensington Pride*)	UV-C dose at 4.0, 8.3, and 11.7 kJ·m^{-2}), room temperature (20 ± 1 °C), relative humidity at 80%, and ethylene storage duration from 3 to 12 days.	At a higher UV-C dose, the firmness is significantly higher than untreated fruits after 6 days of storage. No significant difference in weight loss with the control sample.	[90]
Tahitian lime (*Citrus latifolia*)	The doses were 3.4, 7.2, and 10.5 kJ·m^{-2}. Fruits were located 20 cm from the UV-C light source.	Higher dose (10.5 kJ·m^{-2}) reduced weight loss.	[91]
Mango (*Tommy Atkins*)	UV-C dose of 8220 mW·m^{-2} and exposure time of 10 and 20 min. and storage temperature 5 and 20 °C	Lower weight loss and high firmness for Samples exposed to 10 min and 5 °C.	[92]

3. Effect of UV-C on the Mechanical Properties of Potato

Potatoes belong to the family of *Solanum tuberosum* L. (Solanaceae), which is an important crop plant produced globally. There are a number of products derived from potato and consumed in the form of processed and dehydrated products, such as chips, French fries, granules, snacks, etc. The mechanical properties of the potato tuber and its derivatives represent an important quality parameter for the food industries and customers. Hence, due to the growing demand and massive production, the applicability of surface treatment is intensive. A limited number of studies have reported on the effect of UV-C on the mechanical properties of potato tuber.

Potato tuber treated with UV-C at operational conditions of 20 cm distance away from the source, exposure time of 1 h, followed by a storage period of 0–28 days at 4 °C was studied against a weight loss experiment. The author indicated that there is no significant difference observed between the control and the UV-C treated potatoes during the entire duration of the storage [93]. However, another author [94] stated reduced weight loss in different varieties of potato tuber after exposing the samples to UV-C at a power density from 80 to 100 μW·cm^{-2} and at different stages of the storage period [66]. It was reported that reduced weight could occur at the epicuticular wax morphology of the pretreated tuber, while transpiration and respiration also contributed to the enhanced weight loss. Preliminary work on the mechanical property of pre-injured potato was conducted on different varieties of potato tuber at a wavelength of 253.7 nm, a power density of 80–100 $\mu W \cdot cm^{-1}$, and exposure time of 5, 10, and 15 min and five months of storage. The result led to a smaller weight loss of samples exposed to UV-C for 10 and 15 min compared to control samples [95]. The author suggested the use of UV-C in combination with conveyors and a multi-spaced irradiator to potentially reduce weight loss caused as a result of mechanical injury during post-harvest handling. Conversely, in other research work, the effect of UV-C irradiation in the presence of fluorescence light and darkness on the weight loss of two potato species (*Agata* and *Monalisa*) was investigated at 0, 2.3, 6.9, 11.5, or 34.5 $kJ \cdot m^{-2}$ dose, 3.83 $mW \cdot cm^{-2}$ flow density and 254 nm. the author found higher weight loss than the control sample for tubers stored under fluorescent light and UV-C in darkness and they related this result with the early-stage development of sprouting as this requires dry matter from the tuber [96]. It has been found that the UV-C dose in the range of 5–20 $kJ \cdot m^{-2}$ is able to suppress the sprout propagation [97]. Research on a comparative study between sodium acid sulfate and UV-C treatment on the storage quality of fresh-cut potato was conducted. Firmness as part of the textural analysis was tested after exposing the potato for 3 min from 0 to 25 days of storage duration at 4 °C. Firmness was found to be stable in the entire duration of storage for UV-C treatment. However, it was indicated that, in the combined treated sample with UV-C and Sodium acid sulfate, the firmness was higher than the control sample in the later storage duration [98].

A study reporting on the effect of UV-C light on a sprout length depicted that those potatoes treated with 13.6 $kJ \cdot m^{-2}$ resulted in a shorter length of sprouts compared to other UV-C doses and untreated tubers [99]. The dimensional growth of potato sprouts caused because of UV-C exposure was reported to be unknown. However, it was speculated to be a result of physical change or related to its tubular biochemistry [97]. A similar study was conducted on the impact of UV-C on the deformation of potato tuber (variety *Vineta*, *Lord* and *Owacja*) at varied UV-C doses (69.4, 86.3, and 171.9 $\mu W \cdot cm^{-2}$), exposure time (1, 10, and 30 min.), and irradiator height (40, 70 and 100 cm). Consequently, it was confirmed that UV-C induced a minor percentage of deformation of tubers, as compared to the control combination [100].

4. Stimulation Process of UV-C

4.1. Effect of UV-C on Cell Physiology and Mechanical Properties

Several studies suggested that UV-C radiation is effective not only because of its disinfecting effect, but because it may also stimulate plant defenses against mechanical damages. Exposing UV-C (or hormetin) in food crops has an important effect, called the hormetic effect [101,102], that is responsible for the production of phenylalanine ammonia-lyase (PAL), which induces the formation of a phenolic compound referred to as phytoalexins. The phytochemicals released from PAL having UV absorbing characteristics are found to be chlorogenic acid, gallic acid, epicatechin, and quercetin. This phenolic compound is capable of improving the resistance of fruits and vegetables to microorganisms and defense mechanisms. Phytoalexins give rise to the accumulation of other inducible defenses, such as cell wall modifications [103]. The phenolic content of different crops is discussed as it is the secondary metabolites that are directly connected to plant defense responses. Despite the sophisticated hormetic response process, the role of the hormetic effect in the fruits and their cellular mechanisms by the respective enzymes was demonstrated by a few researchers. The enzymes and genes can be seen in three types based on their response to the UV-C irradiation [104].

Peroxidases and reductases: responsible for oxidative burst and formation of lignin polymers creating boundaries against microorganisms.

Glucanases and chitinases: responsible for lytic activities towards major fungal cell wall components.

L-Phenylalanine ammonia-lyase (PAL): responsible for the biosynthesis of phenolic that led to cell wall modification.

A multitude of studies have shown that UV-C irradiation, depending on its dose, induces the production of phenolic components and other related defense responsive chemicals. The amount of UV-C energy per unit time (dose) released from the light source is an important factor when dealing with surface treatment [105].

4.2. Low UV-C Dose Stimulation

A number of studies investigated different properties of postharvest products when exposed to UV-C irradiation by considering a range of UV-C doses. In the subsequent sections of the review, we discuss the stimulation process of UV-C by classifying low, moderate, and high doses. The classification is mostly based on the relative comparison among the treatment levels and control samples taken in the particular work of art. There is no clear demarcation set for the UV-C dose irradiation from low to high. This is due to the fact that the change induced by UV-C is varied in different products. The effect might even be different among different varieties of the same product. However, the classification takes account of the proposition and findings of previous research. A work by a researcher depicted that a fruit could resist the effect of UV-C dose up to 4.1 kJ·m^{-2} [106]. Furthermore, the latest research carried out an experiment on the effect of low UV-C dose in the range below 4 kJ·m^{-2} [107].

Recent work on the sprouting suppression and quality attributes of potato tuber exposed to low UV-C dose indicated that the role of UV-C light is not only limited to the anti-germicidal effect, but also enables resistance to damage repair and sprouting [108]. A low UV-C dose is reported to induce a hormetic effect on vegetative crops without damaging the crop, which is termed as UV-C hormesis, making use of a harmful low dose of irradiation to induce a favorable stress response on crops [109]. In addition to the hormetic effect, the plant crop is able to offer health-promoting advantages when used by the customer which is referred to as xenohormesis [110]. This is due to the phenolic compound phytoalexins having antioxidant properties further inducing a secondary resistance in the end user's body [111]. Basically, the purpose of using non-thermal technology on harvested crops is primarily to maintain the quality without undergoing damage, which in turn delays senescence. The influence of UV-C offers a pronounced effect over UV-A and UV-B in the process of delaying senescence, and over the other non-thermal processes UV-C is considered as releasing optimized oxidative stress for defense against germ and wound response [112–115]. These effects were well demonstrated by the experiment conducted on lettuce, tomatoes, peaches, and strawberries [116–119]. From a molecular perspective, the plant develops the response by producing ROS (reactive oxygen species) that signals the physiological modification in the chloroplast, mitochondria, DNA and produces secondary metabolites [120,121]. An experimental study performed on tomato fruit in response to UV-C treatment elucidated that the response of the plant involves much greater expression of a number genes responsible for cell wall disassembly, which may be linked to physiological changes, damage resistance, and most importantly delaying softening or maintaining the firmness of the final product [122]. A recent investigation of broccoli florets exposed to a hormetic dose of 1.2 kJ·m^{-2} of UV-C radiation was able to maintain the lowest weight loss as compared to untreated samples [123]. Higher firmness and delayed softening at a lower UV-C dose on strawberry fruit were confirmed and it was explained that a set of genes was able to overcome cell wall degradation through the encoding process. In a previous study, UV-C radiation in strawberry fruit also maintained higher firmness and delayed softening due to the presence of a set of genes encoding for proteins and enzymes involved in cell wall degradation [27,124]. UV-C exposed fresh-cut melon at 1.18 $\times$ 103 mW·cm^{-2} caused a greater firmness during storage that is related to the accumulation of peroxidase enzyme [125]. This enzyme, as discussed in the previous section, is responsible for creating a barrier between the cell wall and the pathogen, reducing softening and enhancing firmness.

4.3. Moderate and High UV-C Dose Stimulation

The application of a high dose during treatment is indicated to have a severe effect on the physiological constituent of plant crops and potentially affect the mechanical attributes during storage. Firmness is an important quality indicating parameter that indicates the degree of softness as the crops undergo a given storage duration. Studies have not yet sufficiently shown the effect of the mechanical changes as a result of high UV-C dose. Nevertheless, some research works depicted the negative impact of high UV-C doses on the physiological constituents that are discussed in this section. The correlation between the physiological change and firmness demonstrated by a previous study indicates that softening of the fruit during storage is a result of cell wall damage that is caused by the generation of two major degrading enzymes in a cell wall, namely polygalacturonase (PG) and pectin methylesterase (PME) [126]. An experiment conducted on the physiological damage of cells of certain vegetable crops at a high UV-C dose was reported to cause cell membrane damage. A research work conducted on spinach exposed to a high UV-C dose from 7.94 to 11.35 kJ·m^{-2} 13 days of storage duration and 8 °C of storage temperature depicted the reduction of antioxidant chemicals up to 75% [127]. Likewise, in other experimental work on pepper exposed to the UV-C dose of 7 kJ·m^{2}, 18 days of storage duration, and 10 °C storage temperature resulted in lower antioxidant chemicals but with fewer changes [128]. The latest research work on satsuma mandarin fruit resulted in a significant decrease of DPPH (antioxidant) and phenolic compounds when exposed to a high UV-C dose of 10 kJ·m^{-2} [129]. An experimental study on Mangosteen treated over the range of 6–40 kJ·m^{-2} at 25 °C for seven days resulted in a higher defense response and decreased weight loss at 13 kJ·m^{-2} than the lower and the higher UV-C dose, which showed no significant difference from the control sample [130]. However, UV-C irradiated persimmon and cucumber at 12.9 kJ·m^{-2} at varied exposure times and storage periods were found to be ineffective for the enhancement of phytochemicals [131]. Fresh-cut lotus treated with UV-C from 0.3 to 12 kJ·m^{-2} indicated the moderate UV-C intensity from 1.5 to 3 kJ·m^{-2} was significant in inducing phenolic content [132]. A comparable result was noted in a tomato treated at a UV-C dose of 4 kJ·m^{-2} and storage temperature of 13 °C, elucidating an increase in phenolic content throughout the storage period significantly higher than the control sample [133]. In other research, using a similar dose of UV-C, the total phenolic content of fresh-cut strawberries was enhanced significantly [134]. The UV-C stimulation at a mild dose is predominantly reported to induce phytochemicals and antioxidants for the role of defense response. However, the reduction in these defense-triggering chemicals when treated at a high dose is due to the depleting activity of UV-C, resulting in membrane damage that gives rise to the change in the chemical compositions, hence lowering the antioxidant activities of the crop and lowering the firmness. The physiological change in terms of weight loss and firmness in relation to the phenolic content was excellently indicated by a recent research article conducted on UV-C exposed bitter gourd sample, maintaining two-fold lower weight loss than the control sample at prolonged days of storage [135]. The softening of tissues during the prolonged storage period is attributed to the gradual alteration in cell wall composition and cell separation [85]. Besides, a moderate dose could also result in higher softening. For instance, a tomato fruit treated at a moderate UV-C dose of 3.7 kJ·m^{-2} resulted in higher weight loss compared to untreated ones [136]. This is due to the lower production of defense response compounds. The change is induced not only by the effect of UV-C operational factors, but the inherent characteristic of the fruit is also important. In general, high-dose UV-C conditions offer unfavorable stress resulting from low quality and defense mechanisms. Some investigations concluded that low dose irradiation led to hormonal effects improvement and microbial load reduction. On the other hand, high doses or continuous exposure could lead to a significant quality loss, decrease in shelf life, and lower antimicrobial activity [137,138]. The disadvantages of inducing high UV-C dose were reported by previous research stating that accelerated ripening, senescence, lower stress response for bacterial wound, lower shelf life, and economical loss were associated with high dose exposure in postharvest fresh cuts [139]. It is highly important to ensure the optimal UV-C dose in order to avoid compromising the textural quality, shelf-life elongation, and feasibility. The linked relationship between the UV-C dose with cellular level modification and the final quality attribute of the vegetable crop is depicted diagrammatically in Figure 2 [127–129,137,138].

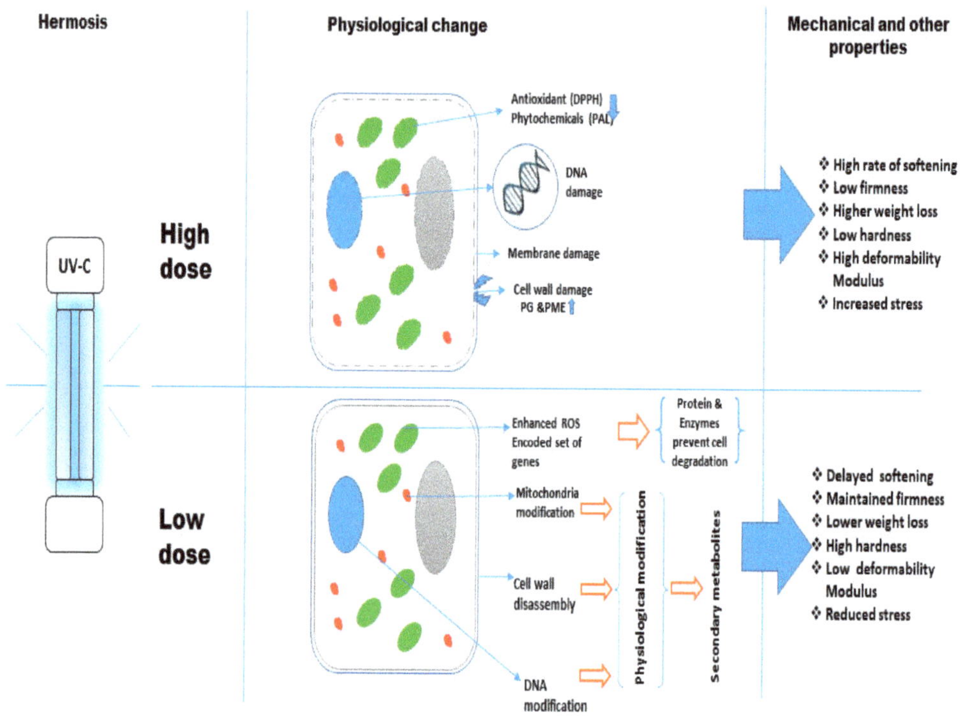

Figure 2. Diagrammatic representation for stress response as a result of low and high UV-C hermosis effect. The physiological modification encountered gives rise to changes in mechanical and other attributes of the vegetable.

Table 2 shows miscellaneous horticulture products exposed at different UV-C doses and their resulting changes in quality texture, antioxidant, enzymes, and phenolic compounds.

Table 2. Different horticultural product's metabolic activities when subjected to a varied UV-C dose and other operational conditions.

Commodity	Operating Condition	Key Finding	Reference
Tomato (*Solanum lycopersicum*)	UV-C dose of 4 kJ·m^{-2}, exposure time of 6 min.	Increase in phenolic compounds content.	[133]
Blueberries	UV-C dose of 0.43, 2.15, 4.30 and 6.45 kJ·m^{-2}.	Higher antioxidant capacity in fruit treated with 2.15, 4.30, and 6.45 kJ·m^{-2} compared to the control fruit. Increased phenolic components in a lesser amount at 0.43 kJ·m^{-2}.	[140]
Fragrant pear (*Korla*)	UV-C irradiation of 0.12, 0.24, 0.36, 0.48, 0.72 and 1.08 kJ·m^{-2}.	low-dose UV-C irradiation (0.36 kJ·m^{-2}) enhanced the phenolic compound.	[141]

Table 2. *Cont.*

Commodity	Operating Condition	Key Finding	Reference
Mandarin (*Satsuma*)	UV-C dose of 0.75, 1.5 and 3.0 kJ·m^{-2}.	Phenolic acids and antioxidant capacity were not significantly affected by UV-C treatments, while 1.5 and 3.0 kJ·m^{-2} significantly increased flavonoids and total phenolics. 0.75 kJ·m^{-2} was infective to induce any change.	[142]
Garlic (*Danyang*)	UV-C of 25 kJ·m^{-2}, exposure time of 380 s.	The UV-C treatment reduces microorganisms present and no significant differences in quality attributes, phenolic compounds, and antioxidants.	[143]
Fresh-cut Rocket Leaves	UV-C dose at 1, 3 and 5 kJ·m^{-2}.	The optimum dose of UV-C for enhancing total anthocyanin content was 3.0 kJ·m^{-2}.	[144]
Mango (*Haden*)	UV-C dose of 2.46 and 4.93 kJ·m^{-2}.	the highest accumulation of phytochemicals in mangoes exposed to 4.93 kJ·m^{-2}.	[145]

5. Factors Affecting UV-C Process in Stimulation

The efficiency of UV stimulation on crops is affected by factors, such as light source, product composition, flow profile, and geometric configuration. UV light has a great impact on the quality attribute of food in relation to wavelength, intensity, and exposure. Total dosage (energy per unit area) is the main factor determining fruit and vegetable responses to UV-C, but the intensity of the radiation (dose per unit time) may also determine treatment outcome. The degree to which the stimulation occurs by UV radiation is directly related to the UV dose. The UV-C dose (J·cm^{-2}) is expressed in terms of the UVC intensity flux (W·cm^{-2}) and exposure time (s), as calculated from the correlation presented in Equation (1).

$$\text{Dose} = \text{UVC Intensity}\cdot\text{Exposure time} \tag{1}$$

The UV-C intensity rate is the total amount of radiant flux passing from all angles through a unit area that is determined based on the location of the sample under exposure. However, the UV-C intensity is not equal to the amount of energy absorbed by the exposed sample. The effect of UV-C exposure time (20–40 min at 15 cm distance) on the defense response chemical content is well experimented with in recent research on bitter gourd resulting in a significant change in the quality attributes [135]. Distance from the source to the sample is an important factor that affects intensity flux and stimulation. The shorter the distance, the higher the radiant flux to effectively stimulate or disinfect the sample of the crop. This correlation is well described by the Equation (2) shown below [146,147].

$$\text{UVC intensity } (r) = \left(\frac{P}{2\pi r}\right)\cdot e^{-\alpha . r} \tag{2}$$

The UVC fluence rate at radial distance r from the lamp is proportional to P, which is UV-C power emitted per unit arc length of the lamp, and α is the absorption coefficient (cm^{-1}). During UV-C irradiation into the sample, the performance of absorption to induce a change is also dependent on the composition of the food product. The surface of the food item determines the penetration ability of UV-C light. A higher value of α leads to a decrease in the penetration ability of UV-C, resulting in the minimal effectiveness of the stimulation of UV-C dose [148]. Complete information and characterization of the absorptive characteristics of a sample are very important when estimating the required UVC dose. For example, in fruits, there is a presence of epicuticular waxes, which contain a microcrystalline structure. Exposure of fruit to UV-C changes the wax morphology, composition, and structure, eventually resulting in a change in surface permeability. This phenomenon can lead to weight loss (water vapor, volatile compounds, etc.), which is a decisive parameter in the quality attribute of fruit [149]. In some vegetable plants, such as Brussels sprout, exposure to a hormetic dose of UV-C resulted in reduced weight loss due to the modification from shorter crystalline structures leading to a dense wax structure, lowering permeability because of a better protective layer [150]. The purity of the sample also affects the stimulation process. The existence of suspended solids and soluble components in the food matrix weakens the application of UVC radiation by inducing light scattering, absorption, and reflection [151]. Due to these factors influencing the efficiency of the UV-C irradiation, the UV-C station needs a proper and operational dose followed by validation using a computer program to confirm the distribution of the irradiation and its stress.

6. Innovative Technologies Used in Combination with UV-C

The investigation of the combined effects of different methods and UV-C on the textural attributes of fruits and vegetables is recently come to emerge showing a glimpse of light in the improvement of the surface treatment. However, it is very limited and difficult to examine the advantageous facets as compared to the conventional single UV-C irradiation. Few studies have reported on the combined effects on the textural attributes.

The quality of strawberries subjected to different patterns of cyclic and repetitive low dose resulted in an improvement in some of the quality parameters. In their study, they exposed the sample at 4 $kJ \cdot m^{-2}$ prior to storage for single time, two step at 2 $kJ \cdot m^{-2}$ consecutively at harvest and after four days of storage, and multistep 0.8 $kJ \cdot m^{-2}$ after zero, two, four, six, and eight days of storage, respectively. From their result, a single step UV-C exposure resulted in a tendency of sustaining higher firmness and delayed softening than the control sample. Two- and multi-step treatments resulted in a higher textural property [107]. The quality of sweet cherry was studied by exposing the fruit to a single and combined operation of UV-B (21.6 $kJ \cdot m^{-2}$), UV-C (21.6 $kJ \cdot m^{-2}$), and combination with a coating of the sample with 1% chitosan for 24 days and 8 °C. Their result indicated that the combination of UV-C, UV-B, and chitosan coating was able to slow the weight loss and maintain firmness better than the control sample in comparison to the singular effects [152]. In a similar study, grapes treated at 6 kJ/m^2 UV-C and 0.5% chitosan reduced weightlessness. A combined treatment of UV-C and Chlorine dioxide (ClO_2) on spinach leaves and tomato surface resulted in no significant effect on the texture of the sample after seven days of storage [153]. The combined effect of UV-C irradiation at 2 $kJ \cdot m^{-2}$ and modified atmosphere packaging in cold storage was studied on the quality of cherry tomato, resulting in lower weight loss and maintained firmness [154]. Asparagus exposed to a combined treatment technique of UV-C at 1 $kJ \cdot m^{-2}$ and ozonized water at 3 ppm for 3 s resulted in improved cutting energy compared to the control over a four-day storage period [70]. The quality of date palm was studied by exposing it to alkaline electrolyte water, neutral electrolyzed water, ozonated water, and UV-C [155]. They found that the combination of neutral electrolyzed water and UV-C treated at 6.22 $kJ \cdot m^{-2}$ resulted in a lower weight loss while a triple combination between neutral, alkaline, and UV-C at 6.22 $kJ \cdot m^{-2}$ brought a higher firmness value as compared to the control and other combinations.

7. Conclusions

The research works on the UV-C surface treatment of fruits and vegetables is slowly increasing due to the growing industrialization, the concern for quality, health, and safety, and customer demand for fresh harvested fruit and vegetable crops. This multidimensional factor makes the area untapped for research.

The determination of the appropriate dose to encounter a favorable stress response and to achieve required mechanical properties is highly varied with different plants exhibiting their own physicochemical properties. Furthermore, the standard textural quality requirements in processing industries for different products are various. The combined technologies are sophisticated and less feasible in terms of energy and mass requirement. The need to focus on the optimization of the process parameter of the UV-C chamber is by far promising. In addition to this, the physicochemical characteristics in the capacity of perceiving light require careful investigation prior to determining the irradiation dose. From the review, low and moderate doses are reported to induce a better favorable physiological stress response, while the high dose reduces the enzymatic role (PAL) to produce phenolic and antioxidants responsible for cell wall modification. From the review, it was possible to understand the strong link between the cellular physiological changes that give rise to the changes in mechanical characteristics of the product. Little is known about the study of UV-C exposed potato tuber mechanical properties, as it is the largest food crop in the world and the key economic driving agricultural product in the frozen market. More studies are needed to address this hurdle process with the priority for stable crops that are produced extensively throughout the globe, such as potato tuber, from which a number of derivative products are produced. Optimization and intensification of the process will offer opportunities to improve the quality and ultimately scale-up towards effective commercialization.

Author Contributions: Conceptualization, T.J. and T.H.; methodology, T.J.; software, E.P.; validation, T.H. and T.J.; formal analysis, T.H.; investigation, A.L.; resources, E.P.; data curation, E.P.; writing—original draft preparation, A.L.; writing—review and editing, A.L.; visualization, A.L.; supervision, T.J and T.H. All authors have read and agreed to the published version of the manuscript.

Funding: This research received no external funding.

Institutional Review Board Statement: Not applicable.

Informed Consent Statement: Not applicable.

Data Availability Statement: The data presented in this study is available upon request of the respective author. The data is not publicly available due to the possibility of their commercial use by the unit in which the authors are employed.

Conflicts of Interest: The authors declare no conflict of interest.

References

1. Soares, I.G.; Silva, E.B.; Amaral, A.J.; Machado, E.C.; Silva, J.M. Physico-chemical and sensory evaluation of potato (*Solanum tuberosum* L.) after irradiation. *An. Acad. Bras. Ciências* **2016**, *88*, 941–950. [CrossRef] [PubMed]
2. Nawara, P.; Gliniak, M.; Popardowski, E.; Szczuka, M.; Trzyniec, K. Control system of a prototype measurement system for the identification of ultra-low photonic emission of organic materials. In Proceedings of the 2018 Progress in Applied Electrical Engineering, Koscielisko, Poland, 18–22 June 2018. [CrossRef]
3. Jakubowski, T. Transfer of microwave irradiation effects of seed potatoes (*Solanum tuberosum* L.) to the plants of next generations. *Bulg. J. Agric. Sci.* **2015**, *21*, 1185–1193.
4. Kielbasa, P.; Drozdz, T.; Nawara, P.; Drozdz, M.; Trzyniec, K. Assessment of the potential of using photon emission to identify selected qualitative features of organic matter. In Proceedings of the 2018 Applications of Electromagnetics in Modern Techniques and Medicine, Racławice, Poland, 9–12 September 2018; pp. 117–120. [CrossRef]
5. Nawara, P.; Trzyniec, K.; Dróżdż, T.; Popardowski, E.; Juliszewski, T.; Zagórda, M.; Miernik, A. Analysis of the possibility of identifying the quality parameters of the oil using ultra-weak secondary luminescence. *Prz. Elektrotech.* **2020**, *96*, 117–120. (In Polish) [CrossRef]
6. Jakubowski, T. The influence of selected physical methods on the content of starch and simple sugars in stored potato tubers. In Proceedings of the 2019 Applications of Electromagnetics in Modern Engineering and Medicine, Janow Podlaski, Poland, 9–12 June 2019; pp. 63–66. [CrossRef]
7. Jakubowski, T. The reaction of garden cress (*Lepidium sativum* L. to microwave radiation. In Proceedings of the 2018 Applications of Electromagnetics in Modern Techniques and Medicine, Racławice, Poland, 9–12 September 2018; pp. 81–84. [CrossRef]
8. Fauster, T.; Schlossnikl, D.; Rath, F.; Ostermeier, R.; Teufel, F.; Toepfl, S.; Jaeger, H. Impact of pulsed electric field (PEF) pretreatment on process performance of industrial French fries production. *J. Food Eng.* **2018**, *235*, 16–22. [CrossRef]
9. Jadhav, H.B.; Annapure, U.S.; Deshmukh, R.R. Non-thermal technologies for food processing. *Front. Nutr.* **2021**, *8*, 657090. [CrossRef]
10. Huang, H.W.; Wu, S.J.; Lu, J.K.; Shyu, Y.T.; Wang, C.Y. Current status and future trends of high-pressure processing in food industry. *Food Control* **2017**, *72*, 1–8. [CrossRef]

11. Jakubowski, T. The effect of stimulation of seed potatoes (*Solanum tuberosum* L.) in the magnetic field on selected vegetation parameters of potato plants. *Prz. Elektrotech.* **2020**, *96*, 166–169.
12. Jakubowski, T. The influence of microwave radiation at the frequency 2.45 GHz on the germination. *Prz. Elektrotech.* **2018**, *94*, 254–325. [CrossRef]
13. Jakubowski, T. Effect of microwave radiation on the germination of *Solanum tuberosum* L. tubers. *Bangladesh J. Bot.* **2016**, *45*, 1255–1257.
14. Mahto, R.; Das, M. Effect of gamma irradiation on the physico-mechanical and chemical properties of potato (*Solanum tuberosum* L.), cv. 'Kufri Sindhuri', in non-refrigerated storage conditions. *Postharvest Biol. Technol.* **2014**, *92*, 37–45. [CrossRef]
15. Jamieson, L.E.; Meier, X.; Page, B.; Zulhendri, F.; Page-Weir, N.; Brash, D.; McDonald, R.M.; Stanley, J.; Woolf, A.B. A review of postharvest disinfestation technologies for selected fruits and vegetables. *Plant Food Res. Client Rep.* **2009**, *19*, 36072.
16. Lu, C.; Ding, J.; Park, H.K.; Feng, H. High intensity ultrasound as a physical elicitor affects secondary metabolites and antioxidant capacity of tomato fruits. *Food Control* **2020**, *113*, 107176. [CrossRef]
17. Hockberger, P.E. A History of Ultraviolet Photobiology for Humans, Animals and Microorganisms. *Photochem. Photobiol.* **2002**, *76*, 561–579. [CrossRef]
18. Bintsis, T.; Litopoulou-Tzanetaki, E.; Robinson, R.K. Existing and potential applications of ultraviolet light in the food industry–a critical review. *J. Sci. Food Agric.* **2000**, *80*, 637–645. [CrossRef]
19. Darras, A.I.; Tsikaloudakis, G.; Lycoskoufis, I.; Dimitriadis, C.; Karamousantas, D. Low doses of UV-C irradiation affects growth, fruit yield and photosynthetic activity of tomato plants. *Sci. Hortic.* **2020**, *267*, 109357. [CrossRef]
20. Vanhaelewyn, L.; Van Der Straeten, D.; De Coninck, B.; Vandenbussche, F. Ultraviolet radiation from a plant perspective: The plant-microorganism context. *Front. Plant Sci.* **2020**, *11*, 597642. [CrossRef]
21. Rameš, J.; Chaloupecký, V.; Sojková, N.; Bencko, V. An attempt to demonstrate the increased resistance of selected bacterial strains during repeated exposure to UV radiation at 254 nm. *Cent. Eur. J. Public Health* **1997**, *5*, 30–31.
22. United States Food and Drug Administration—FDA. Ultraviolet radiation for the processing and treatment of food. In *Code of Federal Regulations*; 21 Part, Section 179.39; FDA: Silver Spring, MD, USA, 2002.
23. National Archives and Records Administration, Code of Federal Regulations. 21 CFR 179.39—Ultraviolet Radiation for the Processing and Treatment of Food. Available online: https://www.govinfo.gov/app/details/CFR-2011-title21-vol3/CFR-2011-title21-vol3-sec179-39/context (accessed on 4 May 2022).
24. Frohnmeyer, H.; Staiger, D. Ultraviolet-B radiation-mediated responses in plants. Balancing damage and protection. *Plant Physiol.* **2003**, *133*, 1420–1428. [CrossRef]
25. Bassham, J.A.; Calvin, M. The path of carbon in photosynthesis. In *Die CO_2-Assimilation. The Assimilation of Carbon Dioxide. Handbuch der Pflanzenphysiologie*; Pirson, A., Ed.; Encyclopedia of Plant Physiology; Springer: Berlin/Heidelberg, Germany, 1960; p. 5. [CrossRef]
26. Konstankiewicz, K.; Pawlak, K.; Zdunek, A. Influence_of_structural_parameters. *Int. Agrophys.* **2000**, *15*, 243–246.
27. Pombo, M.A.; Dotto, M.C.; Martínez, G.A.; Civello, P.M. UV-C irradiation delays strawberry fruit softening and modifies the expression of genes involved in cell wall degradation. *Postharvest Biol. Technol.* **2009**, *51*, 141–148. [CrossRef]
28. Jackman, R.L.; Stanley, D.W. Perspectives in the textural evaluation of plant foods. *Trends Food Sci. Technol.* **1995**, *6*, 187–194. [CrossRef]
29. Waldron, K.W.; Smith, A.C.; Parr, A.J.; Ng, A.; Parker, M.L. New approaches to understanding and controlling cell separation in relation to fruit and vegetable texture. *Trends Food Sci. Technol.* **1997**, *8*, 213–221. [CrossRef]
30. Ichiki, H.; Van, N.N.; Yoshinaga, K. Stone-clod Separation and Its Application to Potato Cultivation in Hokkaido. *Eng. Agric. Environ. Food* **2013**, *6*, 77–85. [CrossRef]
31. Zhang, D.Q.; Mu, T.H.; Sun, H.N.; Chen, J.W.; Zhang, M. Comparative study of potato protein concentrates extracted using ammonium sulfate and isoelectric precipitation. *Int. J. Food Prop.* **2017**, *20*, 2113–2127. [CrossRef]
32. Ezekiel, R.; Singh, N.; Sharma, S.; Kaur, A. Beneficial phytochemicals in potato—A review. *Food Res. Int.* **2013**, *50*, 487–496. [CrossRef]
33. Flegg, P.B.; Spencer, D.M.; Wood, D.A. *The Biology and Technology of the Cultivated Mushroom*; John Wiley & Sons Ltd.: Chichester, UK, 1985; pp. i–xii+347.
34. Miernik, A.; Jakubowski, T. Selected methods for starch content determination in plant materials. *J. Phys. Conf. Ser.* **2021**, *1782*, 012019. [CrossRef]
35. Nawara, P.; Jakubowski, T.; Sobol, Z. Application of the CIE L*a*b* method for the evaluation of the colour of fried products from potato tubers exposed to C band ultraviolet light. *E3S Web Conf.* **2019**, *132*, 02004. [CrossRef]
36. Błaszczak, W.; Sadowska, J.; Fornal, J.; Vacek, J.; Flis, B.; Zagórski-Ostoja, W. Influence of cooking and microwave heating on microstructure and mechanical properties of transgenic potatoes. *Food/Nahrung* **2004**, *48*, 169–176. [CrossRef]
37. Jakubowski, T. A system for the control and recording of physical parameters inside a chamber for UV-C irradiating of biological material. *E3S Web Conf.* **2019**, *132*, 01006. [CrossRef]
38. Gliniak, M.; Tomasik, M.; Popardowski, E.; Knaga, J.; Lis, A.; Gliniak, M. Application of natural luminescence for analysis of the radionuclide migration path during hard coal combustion. In Proceedings of the 2018 Applications of Electromagnetics in Modern Techniques and Medicine, Racławice, Poland, 9–12 September 2018; pp. 61–64. [CrossRef]

39. Kharchenko, S.; Borshch, Y.; Kovalyshyn, S.; Piven, M.; Abduev, M.; Miernik, A.; Popardowski, E.; Kiełbasa, P. Modeling of aerodynamic separation of preliminarily stratified grain mixture in vertical pneumatic separation duct. *Appl. Sci.* **2021**, *11*, 4383. [CrossRef]
40. Jakubowski, T.; Syrotyuk, S.; Yankovska, K. The use of microwave radiation with a frequency of 2.45 GHz as a factor reducing the storage losses of potato tubers. *J. Phys. Conf. Ser.* **2021**, *1782*, 012011. [CrossRef]
41. Gliniak, M.; Tomasik, M.; Popardowski, E.; Knaga, J.; Lis, A.; Gliniak, M. Application of natural luminescence for assessment of hard coal quality. In Proceedings of the 2018 Applications of Electromagnetics in Modern Techniques and Medicine, Racławice, Poland, 9–12 September 2018; pp. 73–76. [CrossRef]
42. Sobol, Z.; Jakubowski, T.; Nawara, P. The effect of UV-C stimulation of potato tubers and soaking of potato strips in water on color and analyzed color by CIE l*a*b*. *Sustainability* **2020**, *12*, 3487, Correction in *Sustainability* **2020**, *12*, 7473. [CrossRef]
43. Shen, H.; Fan, D.; Huang, L.; Gao, Y.; Lian, H.; Zhao, J.; Zhang, H. Effects of microwaves on molecular arrangements in potato starch. *RSC Adv.* **2017**, *7*, 14348–14353. [CrossRef]
44. Xie, Y.; Yan, M.; Yuan, S.; Sun, S.; Huo, Q. Effect of microwave treatment on the physicochemical properties of potato starch granules. *Chem. Cent. J.* **2013**, *7*, 113. [CrossRef]
45. Teixeira, B.S.; Inamura, P.Y.; Mastro, N.L. The influence of gamma irradiation on texture, color and viscosity properties of potato starch. In Proceedings of the 2015 International Nuclear Atlantic Conference—INAC 2015, São Paulo, Brazil, 4–9 October 2015.
46. Wang, J.; Chao, Y. Effect of gamma irradiation on quality of dried potato. *Radiat. Phys. Chem.* **2003**, *66*, 293–297. [CrossRef]
47. Chung, H.J.; Liu, Q. Molecular structure and physicochemical properties of potato and bean starches as affected by gamma-irradiation. *Int. J. Biol. Macromol.* **2010**, *47*, 214–222. [CrossRef]
48. Rezaee, M.; Almassi, M.; Minaei, S.; Paknejad, F. Impact of post-harvest radiation treatment timing on shelf life and quality characteristics of potatoes. *J. Food Sci. Technol.* **2013**, *50*, 339–345. [CrossRef]
49. Zhu, J.; Li, L.; Chen, L.; Li, X. Study on supramolecular structural changes of ultrasonic treated potato starch granules. *Food Hydrocoll.* **2012**, *29*, 116–122. [CrossRef]
50. Chung, K.M.; Moon, T.W.; Kim, H.; Chun, J.K. Physicochemical properties of sonicated mung bean, potato, and rice starches. *Cereal Chem.* **2002**, *79*, 631–633. [CrossRef]
51. Liu, C.; Grimi, N.; Lebovka, N.; Vorobiev, E. Effects of preliminary treatment by pulsed electric fields and convective air-drying on characteristics of fried potato. *Innov. Food Sci. Emerg. Technol.* **2018**, *47*, 454–460. [CrossRef]
52. Gormley, T.R. Texture studies on mushrooms. *Int. J. Food Sci. Technol.* **1969**, *4*, 161–169. [CrossRef]
53. Sobol, Z.; Jakubowski, T.; Surma, M. Effect of Potato Tuber Exposure to UV-C Radiation and Semi-Product Soaking in Water on Acrylamide Content in French Fries Dry Matter. *Sustainability* **2020**, *12*, 3426. [CrossRef]
54. Sobol, Z.; Jakubowski, T. The effect of storage duration and UV-C stimulation of potato tubers, and soaking of potato strips in water on the density of intermediates of French fries production. *Prz. Elektrotech.* **2020**, *96*, 242–245. [CrossRef]
55. Sobol, Z.; Jakubowski, T.; Wrona, P. The effect of UV-C stimulation of potato tubers and soaking of potato strips in water on density differences of intermediates for French-fry production. In Proceedings of the Contemporary Research Trends in Agricultural Engineering, Proceedings of the BIO Web Conferences, Kraków, Poland, 25–27 September 2017; EDP Sciences: Les Ulis, France, 2018; p. 02031. [CrossRef]
56. Pelai, Z.; Pedisi, S.; Repaji, M.; Zori, Z.; Levaj, B. Effect of UV-C Irradiation, Storage and Subsequent Cooking on Chemical Constituents of Fresh-Cut Potatoes. *Foods* **2021**, *10*, 1698. [CrossRef] [PubMed]
57. Peleg, M. On fundamental issues in texture evaluation and texturization—A view. *Food Hydrocoll.* **2006**, *20*, 405–414. [CrossRef]
58. Ma, L.; Wang, Q.; Li, L.; Grierson, D.; Yuan, S.; Zheng, S.; Wang, Y.; Wang, B.; Bai, C.; Fu, A.; et al. UV-C irradiation delays the physiological changes of bell pepper fruit during storage. *Postharvest Biol. Technol.* **2021**, *180*, 111506. [CrossRef]
59. Huyskens-Keil, S.; Hassenberg, K.; Herppich, W.B. Impact of postharvest UV-C and ozone treatment on textural properties of white asparagus (*Asparagus officinalis* L.). *J. Appl. Bot. Food Qual.* **2012**, *84*, 229–234.
60. Poubol, J.; Lichanporn, I.; Puthmee, T.; Kanlayanarat, S. Effect of ultraviolet-C irradiation on quality and natural microflora of asparagus spears. *Acta Hortic.* **2010**, *875*, 257–262. [CrossRef]
61. Manzocco, L.; da Pieve, S.; Maifreni, M. Impact of UV-C light on safety and quality of fresh-cut melon. *Innov. Food Sci. Emerg. Technol.* **2011**, *12*, 13–17. [CrossRef]
62. Manzocco, L.; Da Pieve, S.; Bertolini, A.; Bartolomeoli, I.; Maifreni, M.; Vianello, A.; Nicoli, M.C. Surface decontamination of fresh-cut apple by UV-C light exposure: Effects on structure, colour and sensory properties. *Postharvest Biol. Technol.* **2011**, *61*, 165–171. [CrossRef]
63. Gómez, P.L.; Alzamora, S.A.; Castro, M.A.; Salvatori, D.M. Effect of ultraviolet-C light dose on quality of cut-apple: Microorganism, color and compression behavior. *J. Food Eng.* **2010**, *98*, 60–70. [CrossRef]
64. Gogo, E.O.; Opiyo, A.M.; Hassenberg, K.; Ulrichs, C.; Huyskens-Keil, S. Postharvest UV-C treatment for extending shelf life and improving nutritional quality of African indigenous leafy vegetables. *Postharvest Biol. Technol.* **2017**, *129*, 107–117. [CrossRef]
65. Vunnam, R.; Hussain, A.; Nair, G.; Bandla, R.; Gariepy, Y.; Donnelly, D.J.; Kubow, S.; Raghavan, G.S. Physico-chemical changes in tomato with modified atmosphere storage and UV treatment. *J. Food Sci. Technol.* **2014**, *51*, 2106–2112. [CrossRef] [PubMed]
66. Cote, S.; Rodoni, L.; Miceli, E.; Concellón, A.; Civello, P.M.; Vicente, A.R. Effect of radiation intensity on the outcome of postharvest UV-C treatments. *Postharvest Biol. Technol.* **2013**, *83*, 83–89. [CrossRef]

67. Tzortzakis, N.; Borland, A.; Singleton, I.; Barnes, J. Impact of atmospheric ozone-enrichment on quality-related attributes of tomato fruit. *Postharvest Biol. Technol.* **2007**, *45*, 317–325. [CrossRef]
68. Perkins-Veazie, P.; Collins, J.K.; Howard, L. Blueberry fruit response to postharvest application of ultraviolet radiation. *Postharvest Biol. Technol.* **2008**, *47*, 280–285. [CrossRef]
69. Stevens, C.; Liu, J.; Khan, V.A.; Lu, J.Y.; Kabwe, M.K.; Wilson, C.L.; Igwegbe, E.C.; Chalutz, E.; Droby, S. The effects of low-dose ultraviolet light-C treatment on polygalacturonase activity, delay ripening and Rhizopus soft rot development of tomatoes. *Crop Prot.* **2004**, *23*, 551–554. [CrossRef]
70. Tiecher, A.; de Paula, L.A.; Chaves, F.C.; Rombaldi, C.V. UV-C effect on ethylene, polyamines and the regulation of tomato fruit ripening. *Postharvest Biol. Technol.* **2013**, *86*, 230–239. [CrossRef]
71. Obande, M.A.; Tucker, G.A.; Shama, G. Effect of preharvest UV-C treatment of tomatoes (*Solanum lycopersicon* Mill.) on ripening and pathogen resistance. *Postharvest Biol. Technol.* **2011**, *62*, 188–192. [CrossRef]
72. Bu, J.; Yu, Y.; Aisikaer, G.; Ying, T. Postharvest UV-C irradiation inhibits the production of ethylene and the activity of cell wall-degrading enzymes during softening of tomato (*Lycopersicon esculentum* L.) fruit. *Postharvest Biol. Technol.* **2013**, *86*, 337–345. [CrossRef]
73. Pan, Y.G.; Zu, H. Effect of UV-C radiation on the quality of fresh-cut pineapples. *Procedia Eng.* **2012**, *37*, 113–119. [CrossRef]
74. Allende, A.; McEvoy, J.L.; Luo, Y.; Artes, F.; Wang, C.Y. Effectiveness of two-sided UV-C treatments in inhibiting natural microflora and extending the shelf-life of minimally processed 'Red Oak Leaf' lettuce. *Food Microbiol.* **2006**, *23*, 241–249. [CrossRef] [PubMed]
75. Artés, F.; Gómez, P.; Aguayo, E.; Escalona, V.; Artés-Hernández, F. Sustainable sanitation techniques for keeping quality and safety of fresh-cut plant commodities. *Postharvest Biol. Technol.* **2009**, *51*, 287–296. [CrossRef]
76. Edward, T.L.; Kirui, M.S.K.; Omolo, J.O.; Ngumbu, R.G.; Odhiambo, P.M. Effect of Ultraviolet-A (UV-A) and Ultraviolet-C (UV-C) Light on Mechanical Properties of Oyster Mushrooms during Growth. *J. Biophys.* **2014**, *2014*, 687028. [CrossRef] [PubMed]
77. Meng, X.; Zhang, M.; Zhan, Z.; Adhikari, B. Changes in Quality Characteristics of Fresh-cut Cucumbers as Affected by Pressurized Argon Treatment. *Food Bioprocess Technol.* **2014**, *7*, 693–701. [CrossRef]
78. Tabilo-Munizaga, G.; Barbosa-Cánovas, G.V. Rheology for the food industry. *J. Food Eng.* **2005**, *67*, 147–156. [CrossRef]
79. Ohwovoriole, E.N.; Oboli, S.; Mgbeke, A.C. Studies and preliminary design for a cassava tuber peeling machine. *Trans. ASAE* **1988**, *31*, 380–0385. [CrossRef]
80. Araque, L.C.O.; Ortiz, C.M.; Darré, M.; Rodoni, L.M.; Civello, P.M.; Vicente, A.R. Role of UV-C irradiation scheme on cell wall disassembly and surface mechanical properties in strawberry fruit. *Postharvest Biol. Technol.* **2019**, *150*, 122–128. [CrossRef]
81. Jaramillo Sánchez, G.; Contigiani, E.V.; Coronel, M.B.; Alzamora, S.M.; García-Loredo, A.; Nieto, A.B. Study of UV-C treatments on postharvest life of blueberries 'O'Neal' and correlation between structure and quality parameters. *Heliyon* **2021**, *7*, e07190. [CrossRef]
82. Mohammed, H.E.S.H.; Suliman, A.E.R.E.; Ahmed, A.E.R.; Ebrahim, M.A. Ultraviolet effect on faba bean seed quality during storage. *Asian J. Plant Sci.* **2020**, *19*, 26–34. [CrossRef]
83. Chudhangkura, A.; Teangpook, C.; Sikkhamondhol, C.; Jariyavattanavijit, C. Effects of ultraviolet C, controlled atmosphere, and ultrasound pretreatment on free ferulic acid in canned sweet corn kernels. *J. Food Sci. Technol.* **2018**, *55*, 4167–4173. [CrossRef] [PubMed]
84. Kasim, M.U.; Kasim, R.; Erkal, S. UV-C treatments on fresh-cut green onions enhanced antioxidant activity, maintained green color and controlled 'telescoping'. *J. Food Agric. Environ.* **2008**, *6*, 63–67.
85. Ait Barka, E.; Kalantari, S.; Makhlouf, J.; Arul, J. Impact of UV-C irradiation on the cell wall-degrading enzymes during ripening of tomato (*Lycopersicon esculentum* L.) fruit. *J. Agric. Food Chem.* **2000**, *48*, 667–671. [CrossRef]
86. Zambrano-Zaragoza, M.L.; Quintanar-Guerrero, D.; González-Reza, R.M.; Cornejo-Villegas, M.A.; Leyva-Gómez, G.; Urbán-Morlán, Z. Effects of uv-c and edible nano-coating as a combined strategy to preserve fresh-cut cucumber. *Polymer* **2021**, *13*, 3705. [CrossRef] [PubMed]
87. Park, M.H.; Kim, J.G. Low-dose UV-C irradiation reduces the microbial population and preserves antioxidant levels in peeled garlic (*Allium sativum* L.) during storage. *Postharvest Biol. Technol.* **2015**, *100*, 109–112. [CrossRef]
88. Castronuovo, D.; Sofo, A.; Lovelli, S.; Candido, V.; Scopa, A. Effects of UV-C radiation on common dandelion and purple coneflower: First results. *Int. J. Plant Biol.* **2017**, *8*, 61–64. [CrossRef]
89. Forges, M.; Bardin, M.; Urban, L.; Aarrouf, J.; Charles, F. Impact of UV-C radiation applied during plant growth on pre-and postharvest disease sensitivity and fruit quality of strawberry. *Plant Dis.* **2020**, *104*, 3239–3247. [CrossRef]
90. Pristijono, P.; Golding, J.B.; Bowyer, M.C. Postharvest UV-C treatment, followed by storage in a continuous low-level ethylene atmosphere, maintains the quality of 'Kensington pride' mango fruit stored at 20 °C. *Horticulturae* **2019**, *5*, 1. [CrossRef]
91. Pristijono, P.; Bowyer, M.C.; Papoutsis, K.; Scarlett, C.J.; Vuong, Q.V.; Stathopoulos, C.E.; Golding, J.B. Improving the storage quality of Tahitian limes (Citrus latifolia) by pre-storage UV-C irradiation. *J. Food Sci. Technol.* **2019**, *56*, 1438–1444. [CrossRef]
92. González-Aguilar, G.A.; Wang, C.Y.; Buta, J.G.; Krizek, D.T. Use of UV-C irradiation to prevent decay and maintain postharvest quality of ripe 'Tommy Atkins' mangoes. *Int. J. Food Sci. Technol.* **2008**, *36*, 767–773. [CrossRef]
93. Lin, Q.; Xie, Y.; Liu, W.; Zhang, J.; Cheng, S.; Xie, X.; Guan, W.; Wang, Z. UV-C treatment on physiological response of potato (*Solanum tuberosum* L.) during low temperature storage. *J. Food Sci. Technol.* **2017**, *54*, 55–61. [CrossRef] [PubMed]
94. Jakubowski, T.; Królczyk, J.B. Method for the reduction of natural losses of potato tubers during their long-term storage. *Sustainability* **2020**, *12*, 1048. [CrossRef]

95. Jakubowski, T. Use of UV-C radiation for reducing storage losses of potato tubers. *Bangladesh J. Bot.* **2018**, *47*, 533–537. [CrossRef]
96. Rocha, A.B.O.; Honório, S.L.; Messias, C.L.; Otón, M.; Gómez, P.A. Effect of UV-C radiation and fluorescent light to control postharvest soft rot in potato seed tubers. *Sci. Hortic.* **2015**, *181*, 174–181. [CrossRef]
97. Cools, K.; del Carmen Alamar, M.; Terry, L.A. Controlling sprouting in potato tubers using ultraviolet-C irradiance. *Postharvest Biol. Technol.* **2014**, *98*, 106–114. [CrossRef]
98. Xie, Y.; Lin, Q.; Guan, W.; Cheng, S.; Wang, Z.; Sun, C. Comparison of Sodium Acid Sulfate and UV-C Treatment on Browning and Storage Quality of Fresh-Cut Potatoes. *J. Food Qual.* **2017**, *2017*, 5980964. [CrossRef]
99. Pristijono, P.; Bowyer, M.C.; Scarlett, C.J.; Vuong, Q.V.; Stathopoulos, C.E.; Golding, J.B. Effect of UV-C irradiation on sprouting of potatoes in storage. In Proceedings of the VIII International Postharvest Symposium: Enhancing Supply Chain and Consumer Benefits-Ethical and Technological Issues 1194, Cartagena, Spain, 21–24 June 2016; pp. 475–478. [CrossRef]
100. Jakubowski, T. Impact of UV-C Irradiation of Potato Seed Tubers on the Defects in Potato Plant Crops. *Agric. Eng.* **2019**, *23*, 71–77. [CrossRef]
101. Stevens, C.; Khan, V.A.; Lu, J.Y.; Wilson, C.L.; Chalutz, E.; Droby, S.; Kabwe, M.K.; Haung, Z.; Adeyeye, O.; Pusey, L.P.; et al. Induced resistance of sweetpotato to Fusarium root rot by UV-C hormesis. *Crop Prot.* **1999**, *18*, 463–470. [CrossRef]
102. Stevens, C.; Khan, V.A.; Lu, J.Y.; Wilson, C.L.; Pusey, P.L.; Igwegbe, E.C.; Kabwe, K.; Mafolo, Y.; Liu, J.; Chalutz, E.; et al. Integration of ultraviolet (UV-C) light with yeast treatment for control of postharvest storage rots of fruits and vegetables. *Biol. Control* **1997**, *10*, 98–103. [CrossRef]
103. Mercier, J. Role of phytoalexins and other antimicrobial compounds from fruits and vegetables in postharvest disease resistance. In *Proceedings-Phytochemical Society of Europe*; Oxford University Press Inc.: Oxford, UK, 1996; Volume 41, pp. 221–242.
104. Gonzalez-Aguilar, G.A.; Villa-Rodriguez, J.A.; Ayala-Zavala, J.F.; Yahia, E.M. Improvement of the antioxidant status of tropical fruits as a secondary response to some postharvest treatments. *Trends Food Sci. Technol.* **2010**, *219100*, 475–482. [CrossRef]
105. Shama, G. Process challenges in applying low doses of ultraviolet light to fresh produce for eliciting beneficial hormetic responses. *Postharvest Biol. Technol.* **2007**, *44*, 1–8. [CrossRef]
106. Pan, J.; Vicente, A.R.; Martínez, G.A.; Chaves, A.R.; Civello, P.M. Combined use of UV-C irradiation and heat treatment to improve postharvest life of strawberry fruit. *J. Sci. Food Agric.* **2004**, *84*, 1831–1838. [CrossRef]
107. Ortiz Araque, L.C.; Rodoni, L.M.; Darré, M.; Ortiz, C.M.; Civello, P.M.; Vicente, A.R. Cyclic low dose UV-C treatments retain strawberry fruit quality more effectively than conventional pre-storage single high fluence applications. *LWT-Food Sci. Technol.* **2018**, *92*, 304–311. [CrossRef]
108. Hassan, H.; Abd El-Rahman, A.; Liela, A. Sprouting suppression and quality attributes of potato tubers as affected by post-harvest UV-C treatment under cold storage. *Int. J. Adv. Res* **2016**, *4*, 241–253. [CrossRef]
109. Shama, G.; Alderson, P. UV hormesis in fruits: A concept ripe for commercialization. *Trends Food Sci. Technol.* **2005**, *16*, 128–136. [CrossRef]
110. Terry, L.A.; Joyce, D.C. Elicitors of induced disease resistance in postharvest horticultural crops: A brief review. *Postharvest Biol. Technol.* **2004**, *32*, 1–13. [CrossRef]
111. Xu, Y.; Charles, M.T.; Luo, Z.; Mimee, B.; Tong, Z.; Véronneau, P.Y.; Rolland, D.; Roussel, D. Preharvest ultraviolet C treatment affected senescence of stored strawberry fruit with a potential role of microRNAs in the activation of the antioxidant system. *J. Agric. Food Chem.* **2018**, *66*, 12188–12197. [CrossRef]
112. Zhang, W.; Jiang, W. UV treatment improved the quality of postharvest fruits and vegetables by inducing resistance. *Trends Food Sci. Technol.* **2019**, *92*, 71–80. [CrossRef]
113. Xu, Y.; Charles, M.T.; Luo, Z.; Mimee, B.; Tong, Z.; Roussel, D.; Rolland, D.; Véronneau, P.Y. Preharvest UV-C treatment affected postharvest senescence and phytochemicals alternation of strawberry fruit with the possible involvement of abscisic acid regulation. *Food Chem.* **2019**, *299*, 125138. [CrossRef]
114. Vàsquez, H.; Ouhibi, C.; Lizzi, Y.; Azzouz, N.; Forges, M.; Bardin, M.; Nicot, P.; Urban, L.; Aarrouf, J. Pre-harvest hormetic doses of UV-C radiation can decrease susceptibility of lettuce leaves (*Lactuca sativa* L.) to *Botrytis cinerea* L. *Sci. Hortic.* **2017**, *222*, 32–39. [CrossRef]
115. Ouhibi, C.; Attia, H.; Nicot, P.; Lecompte, F.; Vidal, V.; Lachaâl, M.; Urban, L.; Aarrouf, J. Effects of nitrogen supply and of UV-C irradiation on the susceptibility of *Lactuca sativa* L. to *Botrytis cinerea* and *Sclerotinia minor*. *Plant Soil* **2015**, *393*, 35–46. [CrossRef]
116. Yang, Z.; Cao, S.; Su, X.; Jiang, Y. Respiratory activity and mitochondrial membrane associated with fruit senescence in postharvest peaches in response to UV-C treatment. *Food Chem.* **2014**, *161*, 16–21. [CrossRef]
117. Pombo, M.A.; Rosli, H.G.; Martínez, G.A.; Civello, P.M. UV-C treatment affects the expression and activity of defense genes in strawberry fruit (*Fragaria* × *ananassa*, Duch.). *Postharvest Biol. Technol.* **2011**, *59*, 94–102. [CrossRef]
118. Ouhibi, C.; Attia, H.; Rebah, F.; Msilini, N.; Chebbi, M.; Aarrouf, J.; Urban, L.; Lachaal, M. Salt stress mitigation by seed priming with UV-C in lettuce plants: Growth, antioxidant activity and phenolic compounds. *Plant Physiol. Biochem.* **2014**, *83*, 126–133. [CrossRef]
119. Charles, M.T.; Tano, K.; Asselin, A.; Arul, J. Physiological basis of UV-C induced resistance to *Botrytis cinerea* in tomato fruit. V. Constitutive defence enzymes and inducible pathogenesis-related proteins. *Postharvest Biol. Technol.* **2009**, *51*, 414–424. [CrossRef]
120. Urban, L.; Charles, F.; de Miranda, M.R.A.; Aarrouf, J. Understanding the physiological effects of UV-C light and exploiting its agronomic potential before and after harvest. *Plant Physiol. Biochem.* **2016**, *105*, 1–11. [CrossRef]

121. Urban, L.; Chabane Sari, D.; Orsal, B.; Lopes, M.; Miranda, R.; Aarrouf, J. UV-C light and pulsed light as alternatives to chemical and biological elicitors for stimulating plant natural defenses against fungal diseases. *Sci. Hortic.* **2018**, *235*, 452–459. [CrossRef]
122. Liu, C.; Cai, L.; Han, X.; Ying Tiejin, T. Temporary effect of postharvest UV-C irradiation on gene expression profile in tomato fruit. *Gene* **2011**, *486*, 56–64. [CrossRef]
123. Duarte-Sierra, A.; Nadeau, F.; Angers, P.; Michaud, D.; Arul, J. UV-C hormesis in broccoli florets: Preservation, phyto-compounds and gene expression. *Postharvest Biol. Technol.* **2019**, *157*, 110965. [CrossRef]
124. Li, D.; Luo, Z.; Mou, W.; Wang, Y.; Ying, T.; Mao, L. ABA and UV-C effects on quality, antioxidant capacity and anthocyanin contents of strawberry fruit (*Fragaria ananassa* Duch.). *Postharvest Biol. Technol.* **2014**, *90*, 56–62. [CrossRef]
125. Lamikanra, O.; Kueneman, D.; Ukuku, D.; Bett-Garber, K.L. Effect of Processing Under Ultraviolet Light on the Shelf Life of Fresh-Cut Cantaloupe Melon. *J. Food Sci.* **2005**, *70*, C534–C539. [CrossRef]
126. Prasanna, V.; Prabha, T.N.; Tharanathan, R.N. Fruit ripening phenomena—An overview. *Crit. Rev. Food Sci. Nutr.* **2007**, *47*, 1–19. [CrossRef]
127. Artés-Hernández, F.; Escalona, V.H.; Robles, P.A.; Martínez-Hernández, G.B.; Artés, F. Effect of UV-C radiation on quality of minimally processed spinach leaves. *J. Sci. Food Agric.* **2009**, *89*, 414–421. [CrossRef]
128. Vicente, A.R.; Pineda, C.; Lemoine, L.; Civello, P.M.; Martinez, G.A.; Chaves, A.R. UV-C treatments reduce decay, retain quality and alleviate chilling injury in pepper. *Postharvest Biol. Technol.* **2005**, *35*, 69–78. [CrossRef]
129. Phonyiam, O.; Ohara, H.; Kondo, S.; Naradisorn, M.; Setha, S. Postharvest UV-C Irradiation Influenced Cellular Structure, Jasmonic Acid Accumulation, and Resistance against Green Mold Decay in Satsuma Mandarin Fruit (*Citrus unshiu*). *Front. Sustain. Food Syst.* **2021**, *5*, 336. [CrossRef]
130. Sripong, K.; Jitareerat, P.; Uthairatanakij, A. UV irradiation induces resistance against fruit rot disease and improves the quality of harvested mangosteen. *Postharvest Biol. Technol.* **2019**, *149*, 187–194. [CrossRef]
131. Imaizumi, T.; Yamauchi, M.; Sekiya, M.; Shimonishi, Y.; Tanaka, F. Responses of phytonutrients and tissue condition in persimmon and cucumber to postharvest UV-C irradiation. *Postharvest Biol. Technol.* **2018**, *145*, 33–40. [CrossRef]
132. Wang, D.; Chen, L.; Ma, Y.; Zhang, M.; Zhao, Y.; Zhao, X. Effect of UV-C treatment on the quality of fresh-cut lotus (*Nelumbo nucifera* Gaertn.) root. *Food Chem.* **2019**, *278*, 659–664. [CrossRef]
133. Liu, C.; Zheng, H.; Sheng, K.; Liu, W.; Zheng, L. Effects of postharvest UV-C irradiation on phenolic acids, flavonoids, and key phenylpropanoid pathway genes in tomato fruit. *Sci. Hortic.* **2018**, *241*, 107–114. [CrossRef]
134. Li, M.; Li, X.; Han, C.; Ji, N.; Jin, P.; Zheng, Y. UV-C treatment maintains quality and enhances antioxidant capacity of fresh-cut strawberries. *Postharvest Biol. Technol.* **2019**, *156*, 110945. [CrossRef]
135. Prajapati, U.; Asrey, R.; Varghese, E.; Singh, A.K.; Pal Singh, M. Effects of postharvest ultraviolet-C treatment on shelf-life and quality of bitter gourd fruit during storage. *Food Packag. Shelf Life* **2021**, *28*, 100665. [CrossRef]
136. Charles, M.T.; Benhamou, N.; Arul, J. Physiological basis of UV-C induced resistance to *Botrytis cinerea* in tomato fruit: III. Ultrastructural modifications and their impact on fungal colonization. *Postharvest Biol. Technol.* **2008**, *47*, 27–40. [CrossRef]
137. Quintero-Cerón, J.P.; Bohorquez-Pérez, Y.; Valenzuela-Rea, C.; Solanilla-Duque, J.F. Advancements in the application of short-wave ultraviolet light (UVC) in whole and fresh-cut fruit and vegetables: A review. *Rev. Tumbaga* **2013**, *8*, 29–60.
138. Rivera-Pastrana, D.M.; Gardea Béjar, A.A.; Martínez-Téllez, M.A.; Rivera-Domínguez, M.; González-Aguilar, G.A. Postharvest biochemical effects of UV-C irradiation on fruit and vegetables. *Artículo Revisión Rev. Fitotec. Mex* **2007**, *30*, 361–372.
139. Nigro, F.; Ippolito, A.; Lima, G. Use of UV-C light to reduce Botrytis storage rot of table grapes. *Postharvest Biol. Technol.* **1998**, *13*, 171–181. [CrossRef]
140. Wang, C.Y.; Chen, C.T.; Wang, S.Y. Changes of flavonoid content and antioxidant capacity in blueberries after illumination with UV-C. *Food Chem.* **2009**, *117*, 426–431. [CrossRef]
141. Sun, T.; Ouyang, H.; Sun, P.; Zhang, W.; Wang, Y.; Cheng, S.; Chen, G. Postharvest UV-C irradiation inhibits blackhead disease by inducing disease resistance and reducing mycotoxin production in 'Korla'fragrant pear (*Pyrus sinkiangensis*). *Int. J. Food Microbiol.* **2022**, *362*, 109485. [CrossRef]
142. Shen, Y.; Sun, S.; Qiao, L.; Chen, J.; Liu, D.; Ye, X. Effect of UV-C treatments on phenolic compounds and antioxidant capacity of minimally processed Satsuma mandarin during refrigerated storage. *Postharvest Biol. Technol.* **2013**, *76*, 50–57. [CrossRef]
143. Gutiérrez, D.R.; Rodríguez, S.D.C. Combined Effect of UV-C and Ozone on Bioactive Compounds and Microbiological Quality of Fresh-Cut Rocket Leaves. *Am. J. Food Sci. Technol.* **2019**, *7*, 71–78. [CrossRef]
144. Wu, J.; Liu, W.; Yuan, L.; Guan, W.Q.; Brennan, C.S.; Zhang, Y.Y.; Zhang, J.; Wang, Z.D. The influence of postharvest UV-C treatment on anthocyanin biosynthesis in fresh-cut red cabbage. *Sci. Rep.* **2017**, *7*, 5232. [CrossRef]
145. González-Aguilar, G.A.; Zavaleta-Gatica, R.; Tiznado-Hernández, M.E. Improving postharvest quality of mango 'Haden' by UV-C treatment. *Postharvest Biol. Technol.* **2007**, *45*, 108–116. [CrossRef]
146. Watada, A.E.; Ko, N.P.; Minott, D.A. Factors affecting quality of fresh-cut horticultural products. *Postharvest Biol. Technol.* **1996**, *9*, 115–125. [CrossRef]
147. Brecht, J.K. Physiology of lightly processed fruits and vegetables. *HortScience* **1995**, *30*, 18–22. [CrossRef]
148. Abe, F.; Saito, K.; Miura, K.; Toriyama, K. A single nucleotide polymorphism in the alternative oxidase gene among rice varieties differing in low temperature tolerance. *FEBS Lett.* **2002**, *527*, 181–185. [CrossRef]
149. Ribeiro, C.; Canada, J.; Alvarenga, B. Prospects of UV radiation for application in postharvest technology. *Emir. J. Food Agric.* **2012**, *24*, 586–597. [CrossRef]

150. Reed, D.W.; Tukey, H.B.J. Light intensity and temperature effects on epicuticular wax morphology and internal cuticle ultrastructure of carnation and Brussels sprouts leaf cuticles. *J. Am. Soc. Hortic. Sci.* **1982**, *107*, 417–420.
151. Delorme, M.M.; Guimarães, J.T.; Coutinho, N.M.; Balthazar, C.F.; Rocha, R.S.; Silva, R.; Margalho, L.P.; Pimentel, T.C.; Silva, M.C.; Freitas, M.Q.; et al. Ultraviolet radiation: An interesting technology to preserve quality and safety of milk and dairy foods. *Trends Food Sci. Technol.* **2020**, *102*, 146–154. [CrossRef]
152. Abdipour, M.; Sadat Malekhossini, P.; Hosseinifarahi, M.; Radi, M. Integration of UV irradiation and chitosan coating: A powerful treatment for maintaining the postharvest quality of sweet cherry fruit. *Sci. Hortic.* **2020**, *264*, 109197. [CrossRef]
153. Park, S.H.; Kang, J.W.; Kang, D.H. Inactivation of foodborne pathogens on fresh produce by combined treatment with UV-C radiation and chlorine dioxide gas, and mechanisms of synergistic inactivation. *Food Control* **2018**, *92*, 331–340. [CrossRef]
154. Choi, D.S.; Park, S.H.; Choi, S.R.; Kim, J.S.; Chun, H.H. The combined effects of ultraviolet-C irradiation and modified atmosphere packaging for inactivating Salmonella enterica serovar Typhimurium and extending the shelf life of cherry tomatoes during cold storage. *Food Packag. Shelf Life* **2015**, *3*, 19–30. [CrossRef]
155. Jemni, M.; Gómez, P.A.; Souza, M.; Chaira, N.; Ferchichi, A.; Otón, M.; Artés, F. Combined effect of UV-C, ozone and electrolyzed water for keeping overall quality of date palm. *LWT-Food Sci. Technol.* **2014**, *59*, 649–655. [CrossRef]

applied sciences

MDPI

Article

Preliminary Research on the Influence of a Pulsed Magnetic Field on the Cationic Profile of Sunflower, Cress, and Radish Sprouts and on Their Germination Rate

Grzegorz Zaguła *, Bogdan Saletnik, Marcin Bajcar, Aneta Saletnik and Czesław Puchalski

Department of Bioenergetics, Food Analysis and Microbiology, Institute of Food Technology and Nutrition, College of Natural Sciences, University of Rzeszów, Ćwiklińskiej 2D, 35-601 Rzeszów, Poland; bogdan.saletnik@urz.pl (B.S.); mbajcar@ur.edu.pl (M.B.); a.saletnik@ur.edu.pl (A.S.); cpuchal@ur.edu.pl (C.P.)
* Correspondence: g_zagula@ur.edu.pl

Abstract: Magnetic stimulation of seeds before sowing can have a significant impact on the speed of their germination. Sprouts are sought after by consumers for their high nutrient content. The purpose of the study was to investigate the influence of a pulsed magnetic field on the dynamics of seed germination and on the content of ions in sunflower, cress, and radish sprouts. The research material in the experiment was provided by seeds of sunflower (*Helianthus annuus* L.), garden cress (*Lepidium sativum* L.), and garden radish (*Raphanus sativus* L.) intended for sprouting, which were supplied by PNOS Ożarów Mazowiecki. The research methods involved germinating seeds under strictly defined conditions for 14 days. Then, the mineral composition of the previously mineralised sprout material was determined using emission spectrometry on a ICP-OES iCAP Duo 6500 Termo spectrometer. Greater dynamics of germination were noted in the first half of the growth period in seeds stimulated with a pulsed magnetic field with the parameters 100 μT and 100 Hz. However, the application of the magnetic field produced no increase in the capacity of the seeds to germinate. The research showed an increase in the content of macronutrients in sprouts, such as calcium, magnesium, phosphorus, and sulphur. In the case of the field with parameters of 100 μT and 200 Hz, the effect was similar for both the germination percentage and the accumulation of macronutrients. However, in the case of both frequencies of magnetic field applied, the effect on individual plant seed species was different. Pre-sowing stimulation of seeds with a pulsed magnetic field may affect the rate of seed germination and the content of ions in the sprouts; however, these effects vary in individual plant matrices.

Keywords: sprouts; stimulation with a pulsed magnetic field; micro and macro components; ICP-OES

Citation: Zaguła, G.; Saletnik, B.; Bajcar, M.; Saletnik, A.; Puchalski, C. Preliminary Research on the Influence of a Pulsed Magnetic Field on the Cationic Profile of Sunflower, Cress, and Radish Sprouts and on Their Germination Rate. *Appl. Sci.* **2021**, *11*, 9678. https://doi.org/10.3390/app11209678

Academic Editor: José Miguel Molina Martínez

Received: 29 September 2021
Accepted: 15 October 2021
Published: 17 October 2021

1. Introduction

Seed germination is a process consisting of three phases, referred to as the imbibition phase, in which there is intensive cellular respiration and water uptake, the catabolic phase, where hydrolysis and metabolic processes again take place, and the third phase, referred to as anabolic, in which further growth and development are observed [1]. In Phase I, rapid water uptake takes place, which initiates the metabolism in which DNA and mitochondrial repair occurs along with protein synthesis using existing mRNA. During Phase II, further water uptake is limited because the water potential of the grain is almost in balance with that of the surrounding environment. This phase is also referred to as the activation or retardation phase. Major metabolic changes take place in this phase, such as the synthesis of hydrolytic enzymes (e.g., α-amylase, endoxylanase, and phytase) and other processes necessary for the development of the embryo. In phase III, there is a second rapid water intake. The rootlets appear, i.e., the so-called visible germination, hereinafter referred to as germination [2,3]. Sprouts are formed from seeds during germination and are an excellent source of protein, vitamins, and minerals, containing nutrients as important to health as glucosinolates, plus phenolic and selenium components in cruciferous plants or isoflavones in soybeans. Since sprouts are consumed early in the growth phase, their nutrient concentration remains very high. In addition to nutrients, sprouts contain phytochemicals, vitamins, minerals, enzymes, and amino acids, which are of the greatest importance as they are most beneficial to human health [4]. Sprouts have been consumed in China and eastern countries for many years, and, recently, as a result of a changing lifestyle, sprouts are becoming more and more popular among people around the globe due to their nutritional value and health benefits. A large number of epidemiological studies consistently show that the daily consumption of plant-based foods is associated with a reduction in risk factors for chronic diseases such as cardiovascular diseases, diabetes, and obesity. The health-promoting effects of plant-based foods may be related to the presence of several bioactive compounds in the edible parts, such as phenolic compounds, carotenoids, glucosinolates, vitamin C, and tocopherols, which show various biological properties. Numerous studies have proven that sprouting is an inexpensive and effective method of accumulation of bioactive compounds in the seeds of legumes, cereals, vegetables, fruit, flowers, and medicinal plant seeds [5].

A significant problem in the production of sprouts is their growth rate. Innovative and novel methods of accelerating plant germination are currently sought, including for sprouts [6,7]. The known alternative methods of accelerating the germination process include, inter alia, laser stimulation [8–10], the use of a permanent magnetic field [11,12], and seeds exposed to a static magnetic field [13,14]. Now, a pulsating magnetic field [15], in which stimulating effects acting at least in two planes in space give an opportunity to significantly accelerate the impact of this alternative method on the speed with which a valuable product with increased nutritional potential is obtained, is now also worthy of attention. This is because plant organisms, including seeds, can be treated as specific antennae that receive electrical and magnetic signals from the external environment [16]. This phenomenon is stronger in the case of resonance, and, often, the energy causes biological effects that are many times stronger [17]. The research carried out so far has shown the beneficial effect of the magnetic field on seeds and the plants grown therein [18]. The results obtained are measured by the effect of accelerated germination, the beneficial effect on growth, and the chemical composition of the grown plants. However, the results show that the changes of these parameters depend on many physical factors, such as the exposure dose, the type of magnetic field, the design of the device, or the plant species and cultivar. The issue of the influence of the magnetic field on the development and yield of plants has been largely documented by scientific results; however, a comprehensive explanation of this phenomenon requires further research.

A pulsating magnetic field is a new and very contemporary topic. Authors dealing with the issue of the impact of a PMF direct their research towards, inter alia, stimulation to improve the healing of bone fractures [19–21] and the impact of such a pulsating field at the cellular level [22–26].

The manuscript presents the results of research on the influence of a pulsed magnetic field with induction of 100 µT and two non-ionising frequencies 100 Hz and 200 Hz on the germination potential and mineral composition of three plant species intended for sprouts, i.e., sunflower (*Helianthus annuus* L.), garden cress (*Lepidium sativum* L.), and garden radish (*Raphanus sativus* L.).

2. Materials and Methods

The research material in the experiment was provided by seeds of sunflower (*Helianthus annuus* L.), garden cress (*Lepidium sativum* L.), and garden radish (*Raphanus sativus* L.) intended for sprouting and supplied by PNOS Ożarów Mazowiecki.

The study material consisted of 450 seeds for each species, divided into 3 groups. The first group was stimulated with a pulsed magnetic field with the parameters of 100 µT and 100 Hz, the second group was stimulated with a pulsed magnetic field with the parameters of 100 μT and 200 Hz. The third group consisted of seeds intended for the control sample, which had not been subjected to magnetic stimulation.

2.1. Magnetic Stimulation Site

The seeds were subjected to pre-sowing magnetic stimulation with the use of a set of coils generating alternating magnetic fields pulsed in the XY plane. In this system, it is possible to set the field rotation frequency in the range of 20–100 Hz, the field generation time from 1 s to 100 h, and the field magnetic induction value in the range of 0–500 μT. All the aforementioned operating parameters of the device are determined using a PC and an application in the LabVIEW environment. The seeds were subjected to magnetic stimulation with the induction and frequency parameters of 100 μT and 100 Hz, as well as 100 μT and 200 Hz, and the exposure time to the magnetic field was 30 min (Figures 1 and 2).

Figure 1. A set of coils for generating alternating magnetic fields (own study).

The system used consists of the following elements:

- A 2-channel function generator, DDS-JDS6600, which generates 2 sinusoidal voltage waveforms for channel X(CH1) and Y(CH2) with fixed amplitudes, and the same frequency for both channels.
- The MONACOR SA-200 voltage amplifier, to which the signals from the generator are transferred.
- A signal splitter/current intensity sensor (BOX)—the amplified signal from the amplifier passes to the signal splitter BOX, from where it is transferred to the individual windings of 2 pairs of Helmholtz coils. For channels X and Y, the current intensity is measured by measuring the voltage drop across precision resistors installed in the distributor. Subsequently, the voltage signal from the resistors is transferred from the splitter to the myDAQ digital oscilloscope.
- A group of Helmholtz coils—the current flowing through the coils creates a variable pulsed magnetic field Bxy in the XY plane within them. The X and Y coils consist of two sections, X1 and X2, and Y1 and Y2, attached to the base. These sections, in turn, are connected in series in the BOX distributor so that the current flowing through each

pair of coils has the same value. In the working area, there is a table on which the sample is placed.

- A NI myDAQ device used as a digital oscilloscope, recording voltage waveforms from measuring resistors
- Control software that determines the parameters for the current flowing through the coils and the magnetic field generated, and also controls the work of the JDS6600 generator by determining the frequency and phase shift of the signals and directs the amplitude of the waveforms in real time.

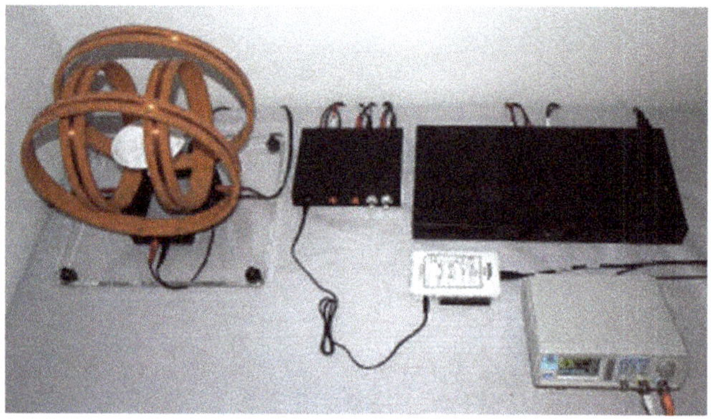

Figure 2. View of the overall set-up (own study).

2.2. *The Dynamics of Seed Germination*

Each of the 3 groups of seeds was divided into three samples, with 50 seeds in each, sown on plates. Initially, 100 mL of demineralised water was added to each batch of seeds. The seeds were hydrated for 4 h at 20 °C and then transferred to airtight containers and incubated at 20 °C for 2 days, with daily airing. Incubation was carried out at 20 °C on moist filter paper in Petri dishes and the number of germinated seeds was counted daily. Seeds, the germinal roots of which had pierced the seed coat and had a length of at least 1 mm, were considered to be germinated.

2.3. *Assessment of the Mineral Composition*

After 14 days, all sprouts were cut in order to determine the content of micro and macro elements therein, using the researcher's own procedure.

The samples of sprouts were mineralised under high pressure in super pure 65% HNO3. Samples of 0.5 g were weighed and placed in Teflon vessels, which were then filled with 8 mL of nitric acid and sealed tightly. For each group of nine samples, the rotor of a digestion system was also filled with a blank sample. The samples were digested at an algorithm of temperature increasing as specified for biological samples, never exceeding 200 °C. This procedure was carried out in an Ethos One microwave digestion system from Milestone. The vessels were opened after the mineralisation process had been completed and the samples with acid were brought to room temperature. Afterwards, they were replenished with water to a volume of 50 mL. The detection threshold obtained for each element was not lower than 0.01 mg kg^{-1} (with an assumed detection capacity of the measuring apparatus at a level exceeding 1 ppb). The measurements were performed on a Thermo iCAP Dual 6500 ICP-OES spectrometer with horizontal plasma and with a capacity of detection being determined both along and across the plasma flame (Radial and Axial). Before measuring each batch, the equipment was calibrated with the use of certified Merck models, with concentrations of 10,000 ppm for Ca, Fe, K, Mg, and P, and 1000 ppm for Al, Ba, Cd, Cu, Na, and Pb. The measurement result for each element was adjusted to account for the measurement of elements in the blank sample.

In each case, a 3-point calibration curve was used for each element, with optical correction applying the method of internal models, in the form of ytterbium and ytterbium ions, at concentrations of 2 mg L^{-1} and 5 mg L^{-1}, respectively. The analytical methods were validated using two independent tests. Certified Reference Material (NIST—1515) was used and the recovery obtained for specific elements is shown in Table 1. The method of adding a model with a known concentration was applied in order to identify the relevant measurement lines and to avoid possible interferences (Table 1).

Table 1. The lengths of measurement lines and the recovery obtained for the specific elements examined.

Element	Slotted Line	Recovery According to CRM	Recovery According to Method of Standard Addition
	nm	%	%
Al	167.079	98	99
Ca	317.933	103	98
Cd	228.802	99	97
Fe	259.940	99	97
K	766.490	101	97
Mg	279.533	99	98
Mn	257.610	98	97
Mo	203.844	101	99
Na	588.995	98	100
Ni	231.604	99	100
P	177.495	100	101
Pb	220.353	101	98
S	180.731	103	99
Se	196.090	101	98
Zn	213.856	102	101

2.4. Statistical Analysis

All parts of the experiment were independently repeated three times. The results obtained were subjected to statistical analyses using Statistica ver. 10.0. The results were statistically analysed with multiway ANOVA and the differences between the means were assessed using the Tuckey test.

3. Results and Discussion

The first parameter analysed throughout the growth period of sunflower, cress, and radish seed sprouts is the germination parameter. The results obtained for radish sprouts clearly show the tendency to accelerate the seed germination process during the first period of germination. After this period has elapsed, the number of germinated seeds in the control samples is equal to the number in the samples stimulated by two selected exposure doses of pulsating magnetic fields (Figure 3). The improvement of the germination process has been of interest to many authors. The factor improving this parameter was described by Menegatti et al. [27]. They indicated that the exposure of passion fruit seeds to a sublime magnetic field in an isolated manner stimulates seed germination, emergence, and vigour. In another text of the authors [28], they demonstrated that chickpea seeds (*Cicer arietinum* L.) subjected to magnetic treatment showed an improvement in seed germination efficiency in terms of the speed and length of the sprouts and the dry matter content of the seedling, and that the reaction varied depending on the intensity of the field and exposure time. The same positive effect on the rate of seed germination and vigour index was found in cucumber seeds (Cucumis sativus L. Var. Barsati) [29], lettuce (*Lactuca sativa* L.) [30], maize seeds (Zea mays L. Var. HQPM-1) [31], tomato seeds (*Solanum lycopersicum* L. Var. MST/32) [32], and radish seeds [33,34].

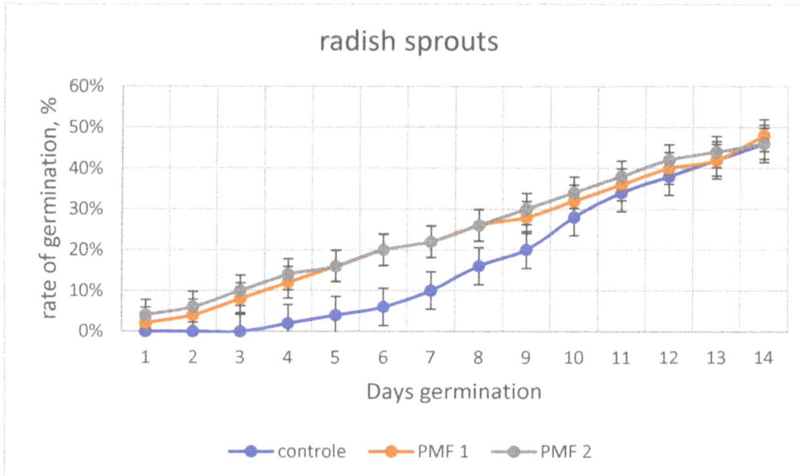

Figure 3. Radish sprout germination dynamics for the control sample and for magnetic field 1 (PMF 1–100 µT and 100 Hz) and magnetic field 2 (PMF 2–100 µT and 200 Hz), mean standard error $\alpha = 0.05$.

Similar results of improvement in germination capacity are seen in trials of growth enhancement effects caused by plasma irradiation of atmospheric air dielectric discharge and thermal shock with radish sprout seeds (*Raphanus sativus* L.). Interactions between radicals and seeds over a short time period of 3 min lead to an enhancement of the growth of radish sprouts over the long period of 7 days, and the maximum average length is 3.7 times larger than in the case of the control. The growth enhancement effects gradually weaken over time, therefore the ratio of mean duration of plasma irradiation to control irradiation decreases from 3.7 on the first day to 1.3 within seven days. The average length of the thermal shock of 60 °C for 10 min and 100 °C for 3 min is longer than in the case of the control, and the maximum average length is 1.3 times longer than in the case of the control. Thermal shock contributes little to the increase due to plasma irradiation because the maximum temperature due to plasma irradiation is less than 60 °C [35].

The results obtained for sunflower sprouts (Figure 4) clearly show the tendency to accelerate the seed germination process for almost the entire germination period up to day 12. After this point, the control samples have the same number of germinated seeds as the samples stimulated by the two selected exposure doses of pulsating magnetic fields. The study [36] shows the results obtained in terms of the effect of treating sunflower seeds subjected to accelerated ageing and cold test on seed vigour. The seeds were treated with distilled water, a solution of potassium nitrate (0.2%), and a solution of gibberellic acid (0.04%). The following parameters were examined: germination energy, germination, the share of abnormal seedlings, length of roots, and shoots of normal seedlings. Accelerated ageing for three and five days resulted in a statistically significant reduction in germination and germination energy; this adversely affected the length of the roots and shoots, and increased the share of abnormal seedlings. Treating the seeds with all three solutions mitigated the adverse effect of three-day accelerated ageing on germination energy. Moreover, treatment of seeds with gibberellic acid prior to the three-day accelerated ageing procedure had a positive effect on seed germination and neutralised the negative effect of accelerated ageing on the number of abnormal seedlings and the length of shoots and roots of normal seedlings. The cold test (at 5 °C for seven days) had a negative effect on the germination energy and root length, increased the share of abnormal seedlings, and did not affect seed germination. Treating seeds with distilled water prior to the cold test completely neutralised the negative influence of low temperatures on germination energy. Eventually, the treatment of the seeds with all three solutions completely neutralised the adverse effect of the cold test on root length.

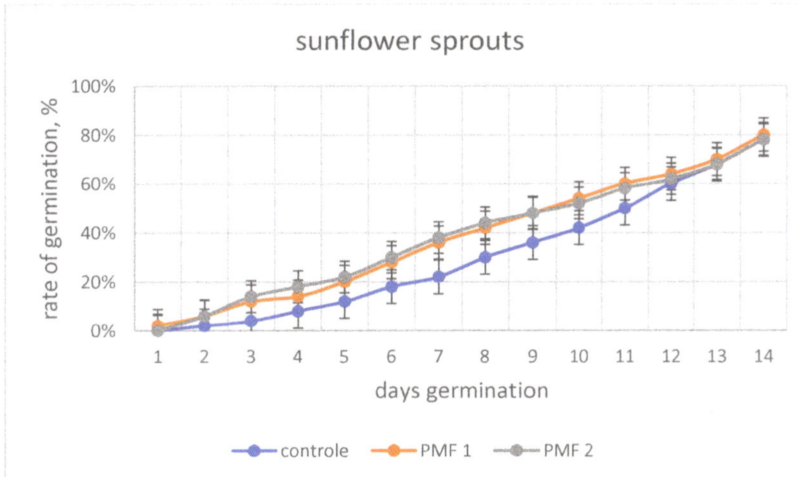

Figure 4. Sunflower sprout germination dynamics for the control sample and for magnetic field 1 (PMF 1–100 μT and 100 Hz) and magnetic field 2 (PMF 2–100 μT and 200 Hz), mean standard error $\alpha = 0.05$.

The results obtained for cress sprouts (Figure 5) show a tendency to accelerate the seed germination process by day seven. After this time, the control samples have the same number of germinated seeds with the samples stimulated by the two selected exposure doses of pulsating magnetic fields. Similar results for improved germination capacity are described by the authors using laser stimulation for cress seeds [37–40].

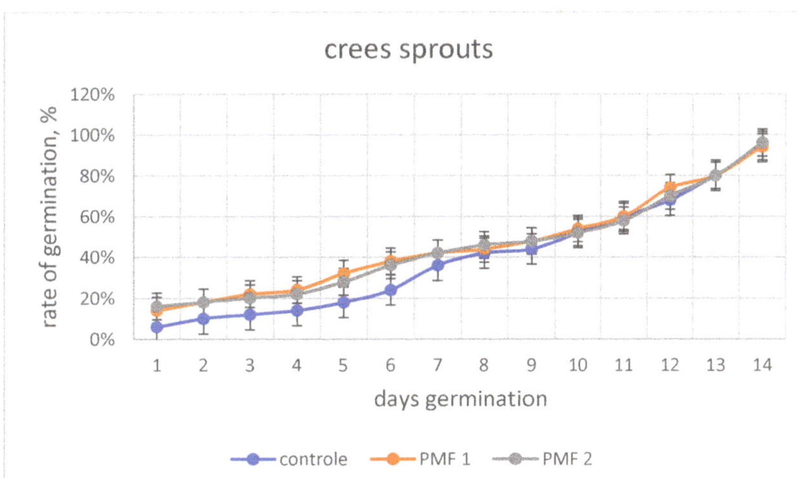

Figure 5. Cress sprout germination dynamics for the control sample and for magnetic field 1 (PMF 1–100 μT and 100 Hz) and magnetic field 2 (PMF 2–100 μT and 200 Hz), mean standard error $\alpha = 0.05$.

The use of an increased frequency of magnetic field seems to be irrelevant for the acceleration of the germination processes. This is because the dynamics of germination improvement are maintained for both 100 Hz and 200 Hz. The determining factor here, therefore, is the magnetic field induction of 100 μT.

Another parameter examined was the total content of micro and macro components in seed sprouts (Table 2). As a reservoir of valuable bio-nutrients, sprouts are especially valuable due to the content of basic micro and macro components. In recent years, sprouts and micro-herbs have been of great interest in many disciplines [41–44].

Table 2. Changes in the content of basic micro and macro components and heavy metals under the influence of applied pulsed magnetic fields for cress sprouts.

Ions	Control	PMF 1	PMF 2
	mg/100 g	mg/100 g	mg/100 g
Al	2.07 ± 0.86 [a]	2.20 ± 1.27 [a]	2.23 ± 1.28 [a]
Ca	492.51 ± 13.91 [c]	540.92 ± 5.12 [b]	699.75 ± 5.65 [a]
Cd	<LOQ	<LOQ	<LOQ
Fe	2.50 ± 0.25 [a]	2.99 ± 0.84 [a]	2.60 ± 1.21 [a]
K	877.92 ± 52.04 [b]	981.83 ± 7.91 [a]	858.08 ± 5.43 [b]
Mg	324.25 ± 25.00 [c]	352.08 ± 2.69 [b]	379.91 ± 2.98 [a]
Mn	0.72 ± 0.08 [b]	1.01 ± 0.04 [a]	1.05 ± 0.12 [a]
Mo	0.05 ± 0.01 [a]	0.07 ± 0.01 [a]	0.10 ± 0.02 [a]
Na	588.75 ± 6.14 [a]	577.51 ± 4.86 [a]	560.75 ± 3.07 [a]
Ni	<LOQ	<LOQ	<LOQ
P	333.31 ± 14.43 [b]	373.66 ± 2.25 [a]	371.00 ± 0.66 [a]
Pb	<LOQ	0.02 ± 0.01 [a]	0.02 ± 0.01 [a]
S	631.56 ± 14.43 [c]	704.73 ± 0.66 [b]	830.90 ± 0.14 [a]
Se	<LOQ	0.06 ± 0.03 [a]	0.07 ± 0.05 [a]
Zn	5.67 ± 1.15 [b]	4.91 ± 0.12 [b]	8.36 ± 0.01 [a]

Mean values with ±SD and statistical analysis using Tukey's test at the significance level of $\alpha = 0.05$; the results indicated with the same letters do not demonstrate a statistically significant difference (test performed as part of a comparison of the concentrations of individual ions with each other after the application of slowly changing magnetic fields and with a control group); <LOQ—result below the Limit of Quantification.

By analysing the results obtained, it is possible to identify the influence of the pulsed magnetic fields on the ionic composition of the cress sprouts. The highest statistically significant increases in the ion content were recorded for calcium with a magnetic field with a frequency of 200 Hz. A similar trend was observed for magnesium and sulphur. On the other hand, the magnetic field with a frequency of 100 Hz was more favourable and more effective in increasing the content of such ions as potassium and phosphorus. However, it should be clearly stated that the magnetic fields applied have a positive effect on the content of macronutrients in particular. Numerous authors have been involved in the comparison of changes in the content of the basic macro- and micro elements during the growth process [45,46].

By analysing the results obtained, it is possible to identify the influence of the pulsed magnetic fields on the ionic composition of the radish sprouts (Table 3). The highest statistically significant increases in ion content were recorded for calcium with a magnetic field with a frequency of 100 Hz. A similar trend was observed for potassium, magnesium, and sulphur. On the other hand, the magnetic field with a frequency of 200 Hz was more favourable and resulted in an increase in the content of such ions as iron and was less beneficial in relation to sodium from the nutritional point of view. Radish sprouts contain particularly large amounts of these minerals, the amounts of which increased after the application of a magnetic field with a frequency of 100 Hz. These amounts are also confirmed by research [47] and valuable bioactive ingredients as in the following studies [48–50].

Table 3. Changes in the content of basic micro and macro components and heavy metals under the influence of applied pulsed magnetic fields for radish sprouts.

Ions	Control	PMF 1	PMF 2
	mg/100 g	mg/100 g	mg/100 g
Al	1.47 ± 0.66 [a]	2.53 ± 1.12 [a]	1.59 ± 0.54 [a]
Ca	626.33 ± 6.97 [b]	802.58 ± 3.26 [a]	658.50 ± 5.07 [b]
Cd	0.02 ± 0.01 [a]	0.03 ± 0.01 [a]	0.01 ± 0.01 [a]
Fe	4.77 ± 1.05 [b]	7.25 ± 1.24 [a]	4.41 ± 0.88 [b]
K	659.08 ± 4.93 [c]	1001.25 ± 2.17 [a]	846.83 ± 6.08 [b]
Mg	344.58 ± 3.46 [c]	489.41 ± 1.60 [a]	390.83 ± 2.74 [b]
Mn	1.61 ± 0.15 [c]	2.37 ± 0.18 [a]	2.00 ± 0.08 [b]
Mo	0.05 ± 0.01 [a]	0.06 ± 0.01 [a]	0.05 ± 0.01 [a]
Na	315.08 ± 3.59 [c]	543.91 ± 2.02 [a]	485.16 ± 2.67 [b]
Ni	<LOQ	<LOQ	<LOQ
P	353.35 ± 15.48 [c]	762.75 ± 1.28 [a]	661.58 ± 3.02 [b]
Pb	<LOQ	<LOQ	<LOQ
S	738.40 ± 5.62 [c]	1137.48 ± 6.50 [a]	914.48 ± 5.87 [b]
Se	<LOQ	0.15 ± 0.02 [a]	0.11 ± 0.01 [b]
Zn	7.03 ± 0.06 [c]	11.39 ± 0.03 [a]	8.69 ± 0.04 [b]

Mean values with ±SD and statistical analysis using Tukey's test at the significance level of $\alpha = 0.05$; the results indicated with the same letters do not demonstrate a statistically significant difference (test performed as part of a comparison of the concentrations of individual ions with each other after the application of slowly changing magnetic fields and with a control group); <LOQ—result below the Limit of Quantification.

By analysing the results obtained, it is possible to identify the influence of the pulsed magnetic fields on the ionic composition of the sunflower sprouts (Table 4). The highest statistically significant increases in the ion content were recorded for calcium with a magnetic field with a frequency of 200 Hz. A similar trend was observed for magnesium, potassium, phosphorus, and sulphur. On the other hand, the magnetic field with a frequency of 100 Hz was more favourable and effective in increasing the content of such ions as sodium and iron.

The dominant trend seems to be a beneficial effect of the use of a pulsed magnetic field on the growth of bound components that act as electrolytes in the human body. These are extremely important ingredients for the body's nutrition. In the clinical setting, the measurement of electrolytes or substances dissolved in the blood is one of the most frequently ordered laboratory tests. The test is known by many names, including electrolyte, "solid", basal metabolic profile (BMP), serum chemistry, CHEM-7, CHEM-10, and sequential multiple analysis-7 (SMA-7). Interpretation of the results provides information on the volume status, acid-base status, and basal renal function. Assessment of any electrolyte imbalance can be difficult and requires an understanding of the pathophysiology of diseases that may cause electrolyte imbalance, the body's counter-regulatory pathways to correct imbalances, and various approaches to increasing the amount of a specific electrolyte through the use of drugs, replacement of a specific electrolyte, or electrolyte removal. Understanding these disorders can improve patient care, be cost-effective, prevent complications from the primary disease, and ultimately reduce mortality and morbidity [51].

The summary of the sum of minerals, including the division into micro–macro elements and heavy metals, including their fluctuations after stimulation with a pulsating magnetic field, turned out to be particularly interesting (Figures 6–8). The division into two groups of sprouts was clearly visible. A magnetic field with a frequency of 200 Hz had a better effect on cress sprouts and their saturation with macronutrients and microelements, while a pulsed field with a frequency of 100 Hz was more favourable for sunflower and radish sprouts. These results suggest the advisability of research into the use of a pulsed magnetic field and the matching of the magnetic field parameters to the subject of research.

Table 4. Changes in the content of basic micro and macro components and heavy metals under the influence of applied pulsed magnetic fields for sunflower sprouts.

Ions	Control	PMF 1	PMF 2
	mg/100 g	mg/100 g	mg/100 g
Al	0.32 ± 0.06 [a]	0.28 ± 0.12 [a]	0.32 ± 0.04 [a]
Ca	429.00 ± 1.88 [c]	543.85 ± 3.74 [b]	613.25 ± 2.00 [a]
Cd	0.02 ± 0.01 [a]	0.02 ± 0.01 [a]	0.03 ± 0.01 [a]
Fe	4.98 ± 0.25 [a]	5.66 ± 0.38 [a]	5.45 ± 0.58 [a]
K	751.12 ± 1.32 [c]	1380.25 ± 4.76 [a]	1074.08 ± 6.12 [b]
Mg	560.50 ± 1.80 [b]	649.33 ± 4.50 [a]	544.67 ± 2.18 [b]
Mn	1.61 ± 0.07 [b]	1.86 ± 0.11 [a]	1.85 ± 0.02 [a]
Mo	0.07 ± 0.01 [a]	0.06 ± 0.01 [a]	0.06 ± 0.01 [a]
Na	80.38 ± 0.48 [c]	197.03 ± 7.01 [a]	151.35 ± 1.34 [b]
Ni	<LOQ	<LOQ	<LOQ
P	746.83 ± 0.87 [c]	1064.08 ± 6.44 [a]	816.97 ± 8.81 [b]
Pb	<LOQ	<LOQ	<LOQ
S	291.23 ± 2.17 [c]	349.90 ± 2.02 [a]	307.90 ± 1.02 [b]
Se	0.02 ± 0.01 [a]	0.02 ± 0.01 [a]	0.01 ± 0.01 [a]
Zn	8.17 ± 0.02 [c]	9.79 ± 0.03 [a]	9.03 ± 0.05 [b]

Mean values with ±SD values and statistical analysis using Tukey's test at the significance level of $\alpha = 0.05$; the results indicated with the same letters do not demonstrate a statistically significant difference (test performed as part of a comparison of the concentrations of individual ions with each other after the application of slowly changing magnetic fields and with a control group); <LOQ—result below the Limit of Quantification.

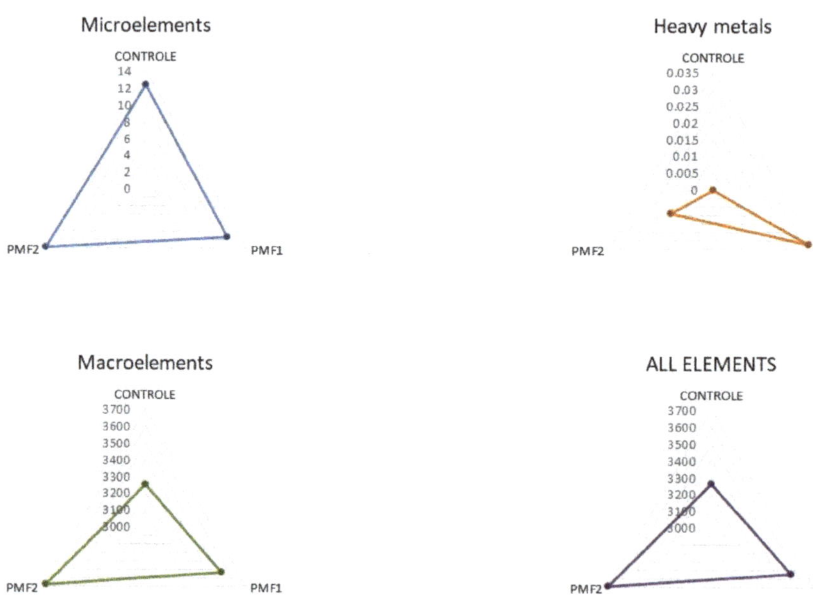

Figure 6. Changes in micro–macro elements and heavy metals and the sum thereof under the influence of the application of doses of pulsating magnetic field for cress sprouts.

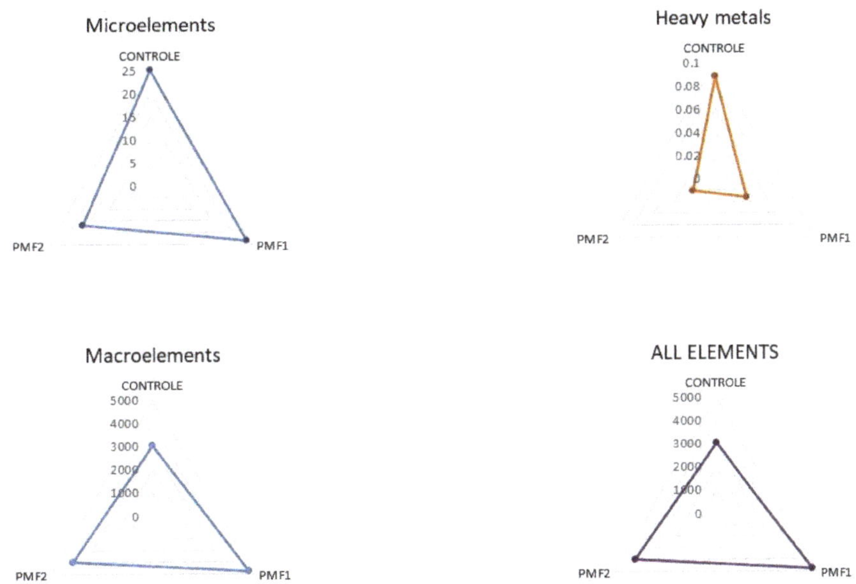

Figure 7. Changes in micro–macro elements and heavy metals and the sum thereof under the influence of the application of doses of pulsating magnetic field for radish sprouts.

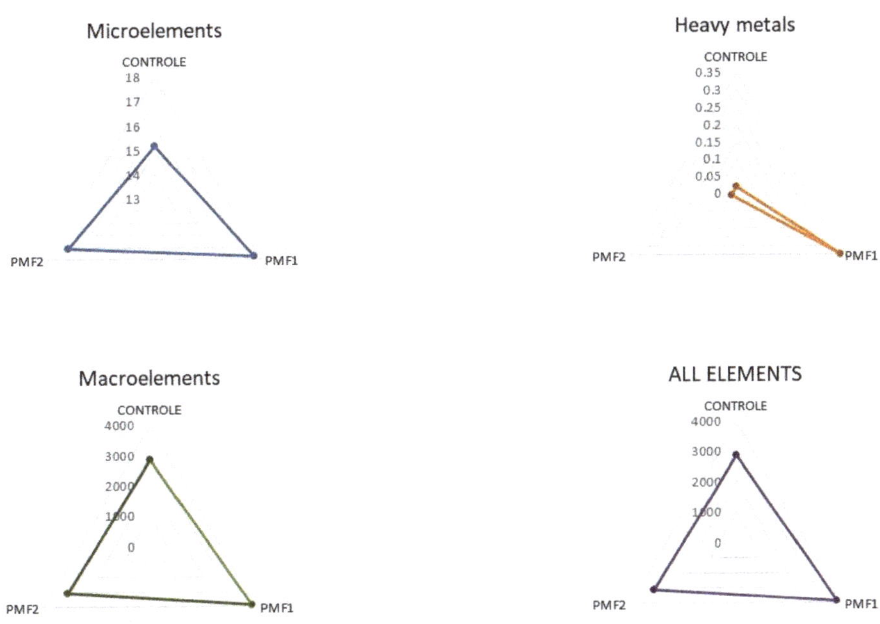

Figure 8. Changes in micro–macro elements and heavy metals and the sum thereof under the influence of the application of doses of the pulsating magnetic field for sunflower sprouts.

4. Conclusions

A magnetic field, especially a pulsating magnetic field, the essence of which is a fluctuating change in the direction of the magnetic field induction vector interaction over time, can be an excellent physical tool when used to accelerate seed germination. The authors proved that, in this case, a pulsating magnetic field with low induction reaching

100 μT and preferably frequencies of 100 Hz and 200 Hz, depending on the matrix of the subject of research, is particularly useful. The germination process starts much earlier when a pre-sowing pulsed magnetic field stimulant is applied for 30 min when compared to the control. This process allows for a significant improvement in the number of germinated seeds, especially by the 7–10th day of the seed germination process. The improvement is also noticeable in the quality of the sprouts, i.e., the content of basic macronutrients. This improvement depends on the species, with sunflower and radish sprouts storing the most nutritiously valuable macroelements under the influence of a pulsating field with a frequency of 100 Hz, while, for cress sprouts, it was with a pulsating field with a frequency of 200 Hz. These increases totalled 12% for cress sprouts, 57% for radish sprouts, and 38% for sunflower sprouts. These results are very significant, especially considering the nutritional value of plant sprouts. The above research should be continued and developed with new species of plant seeds and other parameters of the factor itself, i.e., the pulsating magnetic field.

Author Contributions: Conceptualization, G.Z.; formal analysis, G.Z., B.S.; investigation, G.Z., B.S., M.B.; resources, G.Z.; data curation, A.S., C.P.; writing—original draft preparation, G.Z.; writing—review and editing, G.Z., B.S., M.B., A.S.; visualization, A.S., B.S.; supervision, C.P.; project administration, B.S.; funding acquisition, C.P. All authors have read and agreed to the published version of the manuscript.

Funding: This research was funded by the Minister of Science and Higher Education in the range ofthe program entitled "Regional Initiative of Excellence" for the years 2019–2022, Project No.026/RID/2018/19, amount of funding 9 542 500.00 PL.

Institutional Review Board Statement: Not applicable.

Informed Consent Statement: Not applicable.

Data Availability Statement: Not applicable.

Conflicts of Interest: The authors declare no conflict of interest.

Abbreviations

PMF	pulsed magnetic field
Al	aluminum
Ca	calcium
Cd	cadmium
Fe	iron
K	potassium
Mg	magnesium
Mn	manganese
Mo	molybdenum
Na	sodium
Ni	nickel
P	phosphorus
Pb	lead
S	sulfur
Se	selenium
Zn	zinc

References

1. Rajjou, L.; Duval, M.; Gallardo, K.; Catusse, J.; Bally, J.; Job, C.; Job, D. Seed Germination and Vigor. *Annu. Rev. Plant Biol.* **2012**, *63*, 507–533. [CrossRef] [PubMed]
2. Lemmens, E.; Deleu, L.J.; De Brier, N.; De Man, W.L.; De Proft, M.; Prinsen, E.; Delcour, J.A. The Impact of Hydro-Priming and Osmo-Priming on Seedling Characteristics, Plant Hormone Concentrations, Activity of Selected Hydrolytic Enzymes, and Cell Wall and Phytate Hydrolysis in Sprouted Wheat (*Triticum aestivum* L.). *ACS Omega* **2019**, *4*, 22089–22100. [CrossRef] [PubMed]
3. Tavili, A.; Zare, S.; Moosavi, S.A.; Enayati, A. Effects of Priming Techniques on Seed Germination and Early Growth Characteristics of *Bromus tomentellus* L. and *Bromus inermis* L. *Not. Sci. Biol.* **2010**, *2*, 104–108. [CrossRef]

4. Márton, M.; Mándoki, Z.; Csapó, J. Evaluation of biological value of sprouts. Fat content, fatty acid composition. *Acta Univ. Sapientiae Aliment.* **2010**, *3*, 53–65.
5. Liu, H.; Kang, Y.; Zhao, X.; Liu, Y.; Zhang, X.; Zhang, S. Effects of elicitation on bioactive compounds and biological activities of sprouts. *J. Funct. Foods* **2018**, *53*, 136–145. [CrossRef]
6. Cakmak, T.; Dumlupinar, R.; Erdal, S. Acceleration of germination and early growth of wheat and bean seedlings grown under various magnetic field and osmotic conditions. *Bioelectromagnetics* **2009**, *31*, 120–129. [CrossRef]
7. Dembélé, S.; Zougmoré, R.B.; Coulibaly, A.; Lamers, J.P.A.; Tetteh, J.P. Accelerating Seed Germination and Juvenile Growth of Sorghum (Sorghum bicolor L. Moench) to Manage Climate Variability through Hydro-Priming. *Atmosphere* **2021**, *12*, 419. [CrossRef]
8. Soliman, A.; Harith, M. EFFECTS OF LASER BIOSTIMULATION ON GERMINATION OF ACACIA FARNESIANA (L.) WILLD. *Acta Hortic.* **2010**, *854*, 41–50. [CrossRef]
9. Klimek-Kopyra, A.; Dłużniewska, J.; Ślizowska, A.; Dobrowolski, J. Impact of Coherent Laser Irradiation on Germination and Mycoflora of Soybean Seeds—Innovative and Prospective Seed Quality Management. *Agriculture* **2020**, *10*, 314. [CrossRef]
10. Hasan, M.; Hanafiah, M.M.; Taha, Z.A.; Alhilfy, I.H.H.; Said, M.N.M. Laser Irradiation Effects at Different Wavelengths on Phenology and Yield Components of Pretreated Maize Seed. *Appl. Sci.* **2020**, *10*, 1189. [CrossRef]
11. Pietruszewski, S.; Martínez, E. Magnetic field as a method of improving the quality of sowing material: A review. *Int. Agrophy.* **2015**, *29*, 377–389. [CrossRef]
12. Athari Nia, M.; Noori, M.; Ghanati, F.E. Effect of static magnetic field on certain physiological and biochemical features of Cicer arietinum in vegetative growth phase. *Pajouhesh Sazandegi* **2008**, *21*, 62–68.
13. Vashisth, A.; Nagarajan, S. Effect on germination and early growth characteristics in sunflower (Helianthus annuus) seeds exposed to static magnetic field. *J. Plant Physiol.* **2010**, *167*, 149–156. [CrossRef] [PubMed]
14. Efthimiadou, A.; Katsenios, N.; Karkanis, A.; Papastylianou, P.; Triantafyllidis, V.; Travlos, I.; Bilalis, D.J. Effects of Presowing Pulsed Electromagnetic Treatment of Tomato Seed on Growth, Yield, and Lycopene Content. *Sci. World J.* **2014**, *2014*, 1–6. [CrossRef] [PubMed]
15. Rakoczy, R.; Przybył, A.; Kordas, M.; Konopacki, M.; Drozd, R.; Fijałkowski, K. The study of influence of a rotating magnetic field on mixing efficiency. *Chem. Eng. Process. Process. Intensif.* **2017**, *112*, 1–8. [CrossRef]
16. Aladjadjiyan, A. Study of the influence of magnetic field on some biological characteristics of Zea mais. *J. Cent. Eur. Agric.* **2002**, *3*, 89–94.
17. Van As, H.; Windt, C.W. Magnetic Resonance Imaging of Plants: Water Balance and Water Transport in Relation to Photosynthetic Activity. *Biophys. Tech. Photosynth.* **2008**, *26*, 55–75. [CrossRef]
18. Sarraf, M.; Kataria, S.; Taimourya, H.; Santos, L.; Menegatti, R.; Jain, M.; Ihtisham, M.; Liu, S. Magnetic Field (MF) Applications in Plants: An Overview. *Plants* **2020**, *9*, 1139. [CrossRef] [PubMed]
19. Varani, K.; Vincenzi, F.; Pasquini, S.; Blo, I.; Salati, S.; Cadossi, M.; De Mattei, M. Pulsed Electromagnetic Field Stimulation in Osteogenesis and Chondrogenesis: Signaling Pathways and Therapeutic Implications. *Int. J. Mol. Sci.* **2021**, *22*, 809. [CrossRef]
20. Shi, H.-F.; Xiong, J.; Chen, Y.-X.; Wang, J.-F.; Qiu, X.-S.; Wang, Y.-H.; Qiu, Y. Early application of pulsed electromagnetic field in the treatment of postoperative delayed union of long-bone fractures: A prospective randomized controlled study. *BMC Musculoskelet. Disord.* **2013**, *14*, 35. [CrossRef] [PubMed]
21. Varani, K.; De Mattei, M.; Vincenzi, F.; Gessi, S.; Merighi, S.; Pellati, A.; Ongaro, A.; Caruso, A.; Cadossi, R.; Borea, P. Characterization of adenosine receptors in bovine chondrocytes and fibroblast-like synoviocytes exposed to low frequency low energy pulsed electromagnetic fields. *Osteoarthr. Cartil.* **2008**, *16*, 292–304. [CrossRef]
22. Vincenzi, F.; Targa, M.; Corciulo, C.; Gessi, S.; Merighi, S.; Setti, S.; Cadossi, R.; Goldring, M.B.; Borea, P.A.; Varani, K. Pulsed Electromagnetic Fields Increased the Anti-Inflammatory Effect of A2A and A3 Adenosine Receptors in Human T/C-28a2 Chondrocytes and hFOB 1.19 Osteoblasts. *PLoS ONE* **2013**, *8*, e65561. [CrossRef]
23. Chen, X.; Qin, Z.; Zhao, J.; Yan, X.; Ye, J.; Ren, E.; Wang, J.; Yang, X.; Heng, S.; Zheng, L.; et al. Pulsed Magnetic Field Stimuli Can Promote Chondrogenic Differentiation of Superparamagnetic Iron Oxide Nanoparticles-Labeled Mesenchymal Stem Cells in Rats. *J. Biomed. Nanotechnol.* **2018**, *14*, 2135–2145. [CrossRef]
24. Wang, J.; Tang, N.; Xiao, Q.; Zhang, L.; Li, Y.; Li, J.; Wang, J.; Zhao, Z.; Tan, L. Pulsed Electromagnetic Field May Accelerate in Vitro Endochondral Ossification: PEMF Stimulation of Endochondral Ossification. *Bioelectromagnetics* **2015**, *36*, 35–44. [CrossRef] [PubMed]
25. Parate, D.; Franco-Obregón, A.; Fröhlich, J.; Beyer, C.; Abbas, A.A.; Kamarul, T.; Hui, J.H.P.; Yang, Z. Enhancement of Mesenchymal Stem Cell Chondrogenesis with Short-Term Low Intensity Pulsed Electromagnetic Fields. *Sci. Rep.* **2017**, *7*, 9421. [CrossRef] [PubMed]
26. Parate, D.; Kadir, N.D.; Celik, C.; Lee, E.H.; Hui, J.H.P.; Franco-Obregón, A.; Yang, Z. Pulsed electromagnetic fields potentiate the paracrine function of mesenchymal stem cells for cartilage regeneration. *Stem Cell Res. Ther.* **2020**, *11*, 1–16. [CrossRef]
27. Menegatti, R.D.; De Oliveira, L.O.; Da Costa, V.L.; Braga, E.J.B.; Bianchi, V.J. MAGNETIC FIELD AND GIBBERELIC ACID AS PRE-GERMINATION TREATMENTS OF PASSION FRUIT SEEDS. *Ciência Agrícola Rio Largo* **2019**, *17*, 15–22. [CrossRef]
28. Vashisth, A.; Nagarajan, S. Exposure of seeds to static magnetic field enhances germination and early growth characteristics in chickpea (*Cicer arietinum* L.). *Bioelectromagnetics* **2008**, *29*, 571–578. [CrossRef]

29. Bhardwaj, J.; Anand, A.; Nagarajan, S. Biochemical and biophysical changes associated with magnetopriming in germinating cucumber seeds. *Plant Physiol. Biochem.* **2012**, *57*, 67–73. [CrossRef]
30. Reina, F.G.; Pascual, L.A. Influence of a stationary magnetic field on water relations in lettuce seeds. Part I: Theoretical considerations. *Bioelectromagnetics* **2001**, *22*, 589–595. [CrossRef]
31. Vashisth, A.; Joshi, D.K. Growth characteristics of maize seeds exposed to magnetic field. *Bioelectromagnetics* **2016**, *38*, 151–157. [CrossRef]
32. Poinapen, D.; Brown, D.C.; Beeharry, G.K. Seed orientation and magnetic field strength have more influence on tomato seed performance than relative humidity and duration of exposure to non-uniform static magnetic fields. *J. Plant Physiol.* **2013**, *170*, 1251–1258. [CrossRef] [PubMed]
33. Konefał-Janocha, M.; Banaś-Ząbczyk, A.; Bester, M.; Bocak, D.; Budzik, S.; Górny, S.; Larsen, S.; Majchrowski, K.; Cholewa, M. The Effect of Stationary and Variable Electromagnetic Fields on the Germination and Early Growth of Radish (Raphanus sa-tivus. *Pol. J. Environ. Stud.* **2019**, *28*, 709–715.
34. Muszynski, S.; Gadyszewska, B. Representation of He-Ne laser irradiation effect on radish seeds with selected germination indices. *Int. Agroph.* **2008**, *22*, 151–157.
35. Sarinont, T.; Amano, S.; Kitazaki, K.; Koga, G.; Uchida, M.; Shiratani, N. Growth enhancement effects of radish sprouts: Atmospheric pressure plasma irradiation vs. *heat shock, J. Phys. Conf. Ser.* **2014**, *518*, 012017. [CrossRef]
36. Lekić, S.; Draganić, I.; Milivojević, M.; Todorović, G. Germination and Seedling Growth Response on Sunflower Seeds to Priming and Temperature Stress. *Helia* **2015**, *38*, 241–252. [CrossRef]
37. Aladjadjiyan, A. The use of physical methods for plant growing stimulation in Bulgaria. *J. Cent. Eur. Agric.* **2007**, *8*, 369–380.
38. Hernandez, A.C.; Dominguez, P.A.; Cruz, O.A.; Ivanov, R.; Carballo, C.A.; Zepeda, B.R. Laser in agriculture. *Int. Agrophys.* **2010**, *24*, 407–422.
39. Koper, R. Pre-sowing laser biostimulation of seeds of cultivated plants and its results in agrotechnics. *Int. Agrophys.* **1994**, *8*, 593–596.
40. Dziwulska-Hunek, A.; Kornarzynski, K.; Matwijczuk, A.; Pietruszewski, S.; Szot, B. Effect of laser and variable magnetic field stimulation on amaranth seed germination. *Int. Agroph.* **2009**, *23*, 229–235.
41. Galieni, A.; Falcinelli, B.; Stagnari, F.; Datti, A.; Benincasa, P. Sprouts and Microgreens: Trends, Opportunities, and Horizons for Novel Research. *Agronomy* **2020**, *10*, 1424. [CrossRef]
42. Benincasa, P.; Falcinelli, B.; Lutts, S.; Stagnari, F.; Galieni, A. Sprouted Grains: A Comprehensive Review. *Nutrients* **2019**, *11*, 421. [CrossRef]
43. Bianchi, G.; Falcinelli, B.; Tosti, G.; Bocci, L.; Benincasa, P. Taste quality traits and volatile profiles of sprouts and wheatgrass from hulled and non-hulled Triticum species. *J. Food Biochem.* **2019**, *43*, e12869. [CrossRef]
44. Frassinetti, S.; Moccia, E.; Caltavuturo, L.; Gabriele, M.; Longo, V.; Bellani, L.; Giorgi, G.; Giorgetti, L. Nutraceutical potential of hemp (*Cannabis sativa* L.) seeds and sprouts. *Food Chem.* **2018**, *262*, 56–66. [CrossRef]
45. Chiriac, E.R.; Chiţescu, C.L.; Sandru, C.; Geană, E.-I.; Lupoae, M.; Dobre, M.; Borda, D.; Gird, C.E.; Boscencu, R. Comparative Study of the Bioactive Properties and Elemental Composition of Red Clover (*Trifolium pratense*) and Alfalfa (*Medicago sativa*) Sprouts during Germination. *Appl. Sci.* **2020**, *10*, 7249. [CrossRef]
46. Pongrac, P.; Potisek, M.; Fraś, A.; Likar, M.; Budič, B.; Myszka, K.; Boros, D.; Necemer, M.; Kelemen, M.; Vavpetič, P. Com-position of mineral elements and bioactive compounds in tartary buckwheat and wheat sprouts as affected by natural miner-al-rich water. *J. Cereal Sci.* **2016**, *69*, 9–16. [CrossRef]
47. Baenas, N.; Gómez-Jodar, I.; Moreno-Fernández, D.; Garcia-Viguera, C.; Periago, P.M. Broccoli and radish sprouts are safe and rich in bioactive phytochemicals. *Postharvest Biol. Technol.* **2017**, *127*, 60–67. [CrossRef]
48. Li, R.; Zhu, Y. The primary active components, antioxidant properties, and differential metabolite profiles of radish sprouts (*Raphanus sativus* L.) upon domestic storage: Analysis of nutritional quality. *J. Sci. Food Agric.* **2018**, *98*, 5853–5860. [CrossRef]
49. Baenas, N.; Villaño, D.; García-Viguera, C.; Moreno, D.A. Optimizing elicitation and seed priming to enrich broccoli and radish sprouts in glucosinolates. *Food Chem.* **2016**, *204*, 314–319. [CrossRef] [PubMed]
50. Yuan, G.; Wang, X.; Guo, R.; Wang, Q. Effect of salt stress on phenolic compounds, glucosinolates, myrosinase and antioxidant activity in radish sprouts. *Food Chem.* **2010**, *121*, 1014–1019. [CrossRef]
51. Maday, K.R. Understanding electrolytes: Important diagnostic clues to patient status. *J. Am. Acad. Physician Assist.* **2013**, *26*, 26–31. [CrossRef] [PubMed]

applied sciences

MDPI

Article

Influence of Broadleaved Wood Conditioning by Pulsed Electric Field on Its Combustion Heat Characteristics

Ernest Popardowski * and Paweł Kiełbasa

Department of Machinery Exploitation, Ergonomics and Production Processes, University of Agriculture in Krakow, Balicka 116B, 30-149 Krakow, Poland; pawel.kielbasa@urk.edu.pl
* Correspondence: ernest.popardowski@urk.edu.pl

Abstract: This publication presents changes in sawdust of selected deciduous trees as a consequence of impulse electric field (PEF) stimulation. The analyzed changes concerned the time–temperature characteristics created during the measurement of the heat of combustion of the audited material. Based on experience from previous studies, two alternatives of electric field strength and one variant of capacitor discharges (pulses) were adopted. The results were compared with the sample not treated with PEF. The selected parameters were the result of previous studies, in which the applied variants seemed to be the most promising, i.e., they gave the most diverse results. The research presented in this work has shown that the pulsed electric field affects the time and temperature characteristics of biological material. The changes are most pronounced for the last period of the combustion process, from the moment the maximum temperature was reached to the end of the process. The obtained results indicate that birch and ash react to PEF conditioning in a similar manner. The second group, due to the similarity of the obtained results, is oak and linden. It seems that, apart from the electric field strength, the obtained results are also influenced by the cellulose content in the tested wood. The described process has a very low energy-efficiency, but the reduction of the energy needed to generate the impulse could lead to the possibility of applying the obtained results in industry.

Keywords: biomass combustion; broadleaved tree; pulsed electric field; calorific value

Citation: Popardowski, E.; Kiełbasa, P. Influence of Broadleaved Wood Conditioning by Pulsed Electric Field on Its Combustion Heat Characteristics. *Appl. Sci.* **2022**, *12*, 5048. https://doi.org/10.3390/app12105048

Academic Editors: Roberto Romaniello and José Miguel Molina Martínez

Received: 3 April 2022
Accepted: 16 May 2022
Published: 17 May 2022

1. Introduction

Among the multitude of energy sources used to generate both electricity and heat, biomass is considered the most promising fuel in many countries [1–4]. This opinion is confirmed by the fact that statistically, biomass is one of the most-used energy sources in the world [2]. An even greater increase in the share of energy derived from the combustion of wood biomass was seen in energy plants such as willow (*Salix viminalis*) or poplar (*Populus tremula*). However, in retrospect, these systems turned out to be economically unprofitable. The reason for this is that growth is not fast enough to achieve profitable biofuel production [5]. Therefore, various, often highly advanced technologies began to be used to transform the energy contained in the wood of trees [6–9]. From the point of view of energy use of fuel (including wood), some of the most important physicochemical parameters are the heat of combustion and the calorific value [10,11]. The aforementioned parameters may have different—often very different—values, even for the same fuel. An example is birch wood, which in an air-dry state can reach a calorific value from 17.90 MJ·kg^{-1} [12] to 18.72 MJ·kg^{-1} [13]. The heat of combustion is influenced by many factors, including the content of nonflammable parts (often due to contamination of the wood) and the water content. Water in wood may be present in the form of transient moisture, which is the result of the material and the surrounding air reaching the moisture balance, as well as permanent moisture, i.e., that stored in the cell walls and in the spaces between dead cells [14,15]. Usually, the determination of the heat of combustion is carried out with the use of a calorimeter. The combustion process in the calorimetric bomb is automatically registered. Due to this, it is possible to determine the duration of individual combustion stages and the temperature increase, which is an indirect measurement of the temperature of the medium receiving heat from the bomb body [16]. As a result of this registration, a time–temperature characteristic is created. To obtain reliable data, it is also necessary to designate the so-called corrected temperature-rise and corrections for heat exchange between the calorimeter and the thermostat [17].

Some researchers suggest that using appropriate conditioning of biological material, it is possible to modify the heat of combustion of this material [18–22]. Thermal methods constitute a large group among the methods of pretreatment of biological material. The action of the temperature factor contributes to the improvement of the mass exchange conditions, but at the same time, irreversibly changes some properties of the conditioned products, such as their color or the amount of biologically active compounds [23–25]. However, from the perspective of wood used for combustion in energy-generating installations, these negative effects are of minor importance. Among the techniques of nonthermal processing, the use of high voltage pulses, which generate a time-varying electric field—PEF (Pulsed Electric Field), is becoming interesting and gaining popularity [26–28]. PEF causes the effect of electroporation in biological material, i.e., changes in the structure of the cell membrane by enlarging the existing or creating new pores of the membranes. The resulting perforation of the natural structure of the cell membrane causes the diffusion of water. Thus, the resistance to mass movement is reduced, which contributes to osmotic dehydration, acceleration of the drying process, and improvement of the extraction conditions [29,30].

The aim of the research presented in this study was to determine the effect of pulsed electric field on the time–temperature characteristics of the heat of combustion of selected broadleaved wood species by determining the combustion curve for each of the analyzed variants.

2. Materials and Methods

The research material selected for analysis in this study was sawdust from four broad-leaved trees, i.e., birch, oak, linden, and ash tree. The sawdust was subjected to a pulsed electric field with an electric field intensity of 25 $kV \cdot cm^{-1}$ and the number of 300 capacitor discharges (pulses) and an electric field intensity of 30 $kV \cdot cm^{-1}$, where the number of 300 capacitor discharges (pulses) was also used. The results were compared with the sample not treated with PEF. The selected parameters were the result of previous studies [31], in which the applied variants seemed to be the most promising, i.e., they gave the most diverse results. A constant time between successive capacitor discharges (pulses) was assumed, regardless of the test variant used. The entire research process, i.e., preparation of the material for research, its conditioning, and measurement of the calorific value was carried out at the Laboratory of Experimental Research Techniques for Biological Products and Raw Materials, which is part of the University of Agriculture in Krakow.

In addition to comparing the results of studies of individual broadleaved tree species with a different number of discharges (pulses), a so-called blank sample was also introduced. It was made of a material that had not been treated with PEF.

Conditioning of sawdust with pulsed electric field was performed on the Ertec Su-1 stand (Figure 1a) used daily as a tool for low-temperature inactivation of microorganisms in food products. For the purposes of this research, minor adjustments were made to the stand. The PEF interaction device consists of a high-voltage pulse generator, a working chamber, and a control system. In the apparatus used in these tests, it is possible to generate a voltage up to 31 kV, which—with a distance between bases of the cylinder of c—gives a maximum electric field strength of 31 $kV \cdot cm^{-1}$. The station is equipped with control programming that enables process conditions to be set for several hundred PEF pulses, with widths from 1 to 400 μs and spacing between them in the range of 1–10 s. The generated pulses have an exponential shape. The average energy consumption when generating a single impulse is 65.6 mWh, with a variation coefficient of 6.9% [31].

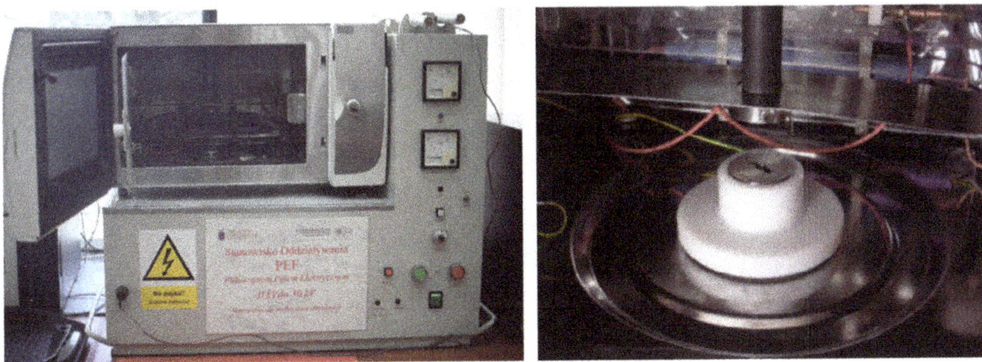

(**a**) (**b**)

Figure 1. Ertec Su-1 test stand: (**a**) discharge chamber; (**b**) stationary sterilization cell.

A detailed description of the devices used in this study was described in the previous article by the authors [32], where analogous studies were carried out for sawdust from coniferous trees.

Properly prepared sawdust, constituting the test material, was introduced into a stationary sterilization cell (Figure 1b). The material tightly filled the 1 cm wide space between the steel cylinders. The remainder of the cell is made of Teflon dielectric. The cells were introduced into the stand chamber, where the electrodes were attached to the bases of the cylinder. In the presented research, stimulation with a pulsed electric field was used with a pulse time width of 160 μs and a time between successive discharges equal to 10 s.

In accordance with the requirements set by the accredited laboratories, before starting the experiments, the material was homogenized and dried to an air-dry state in accordance with the EN 14780:2017-07 standard [33]. In accordance with the assumption presented earlier, a control sample, also known as a blank sample, was separated from the stabilized material (within each species of wood). In each of the variants, the moisture content in the analytical state was tested in accordance with the EN ISO 18134-3:2015 standard [34]. For samples of the material subjected to the pulsed electric field, the measurement of analytical moisture was carried out both before and after stimulation. This was to exclude any discrepancies in the obtained results of the heat of combustion, which, if the control was not carried out, could be caused not by the action of PEF itself, but by a reduction of the water content in the tested material. The value of 0.1% was adopted as the limit of reproducibility of the moisture results. The combustion heat of the tested biological material was determined by the method of combustion in a calorimetric bomb, based on the EN ISO 18125:2017 standard [17].

The measurement is based on the combustion of a fuel sample in the amount of 1 ± 0.1 g in a vessel called a calorimetric bomb. A special nickel–chromium wire, pressed in its central part into the material, is connected to the electrodes that extend outside the vessel. The bomb is tightly closed and filled with oxygen. The excess gas guarantees the correct combustion stoichiometry. The thus-prepared element of the station is introduced into a vessel of known capacity, previously filled with water. The water jacket surrounding the calorimetric vessel is the thermal insulation from the environment. The only elements protruding above the water level are the electrode leads, which are connected to the central unit of the station. After stabilizing the water temperature with the use of a stirrer, current flows through the electrodes, and the resistance wire in direct contact with the fuel sample causes it to ignite. The calorimeter measures and then automatically records three characteristic temperature values, three basic time-temperature periods of the combustion process and, most importantly, to determine the heat of combustion, the temperature and time at which ignition occurred and the combustion period ended. All the listed elements of the characteristics are shown in Figure 2 [17]. Due to the methodology used and the low density of sawdust, the material had to be compressed with a manual press before taking measurements. Only such a prepared portion could be used to make measurements.

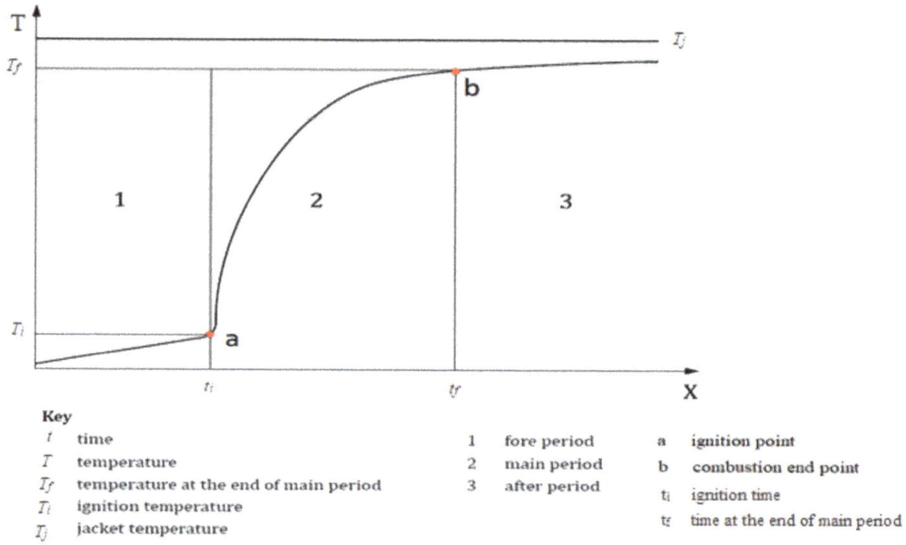

Figure 2. Example time-temperature curve from EN ISO 18125:2017-07.

Based on the designated characteristics and knowing the sample quantity of the tested material, as well as the content of nitrogen and sulfur (determined by titration in accordance with the method presented in the above-mentioned standard [17]), it is possible to determine the value of the heat of combustion.

The hypothesis about the existence of a normal distribution among the obtained results was verified based on the Kolmogorov–Smirnov test, at the significance level of $\alpha = 0.05$. This is a nonparametric significance test that gives you the opportunity to check whether two populations have the same distribution most commonly used for exponential distributions. Afterwards, selected basic descriptive statistics significant due to the described phenomena were calculated in accordance with ISO standards [35].

3. Results

Each of the tested materials (birch, oak, linden, and ash tree) was subjected to a separate analysis.

3.1. Birch Sawdust

Birch sawdust samples, both those that were not conditioned with PEF and those that were the control sample, had analytical moisture at the level of 7.7 ± 0.1%. By analyzing the temperature characteristics during combustion of the biological material, which was birch (Figure 3), it was found that the highest temperature was obtained for the sample not exposed to a pulsed electric field. PEF-conditioned sawdust with an electric field strength of 30 kV·cm^{-1} and a number of pulses of 300 showed the lowest temperature during material combustion. For birch sawdust, in all three variants used, the temperatures obtained at individual points of the characteristic were characterized by low variability. These results indicate that for a given biological material, no significant influence of PEF on the temperature characteristics of the combustion process was observed. It was found that the obtained differences are not statistically significant and are not of great importance for the entire analyzed process.

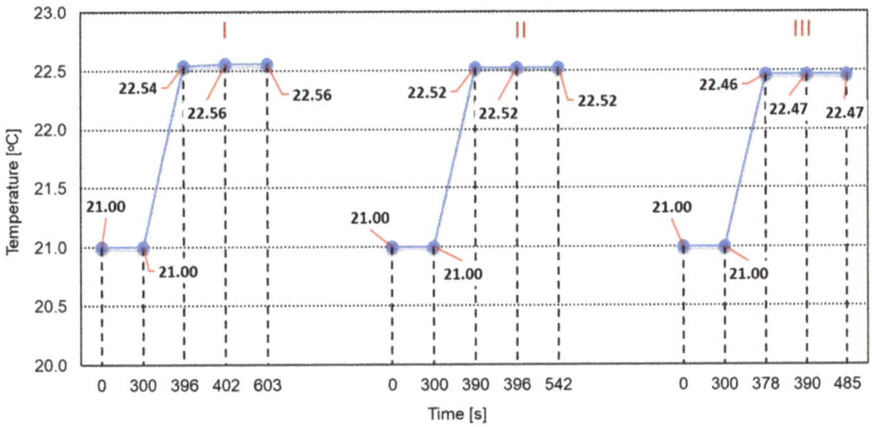

Figure 3. Temperature characteristics of individual variants of birch three combustion: I (nonconditioned with a pulsed electric field) and conditioned with a pulsed electric field: II (electric field intensity 20 kV·cm^{-1} and number of pulses 300); III (electric field intensity 30 kV·cm^{-1} and number of pulses 300).

The second of the analyzed elements of the heat of combustion characteristics is the duration of individual stages of the combustion process. The main period seems to be particularly important when there is a rapid increase in the measured temperature as a result of heat released by the combustion sample. Nevertheless, the total duration of the process seems to be important, taking into account not only the period of temperature rise, but also the interval, including the time when the high temperature is kept constant. Figure 4 shows the time characteristics of individual stages of calorimetric tests on birch, previously subjected to stimulation with an impulse electric field, and compared with the control sample.

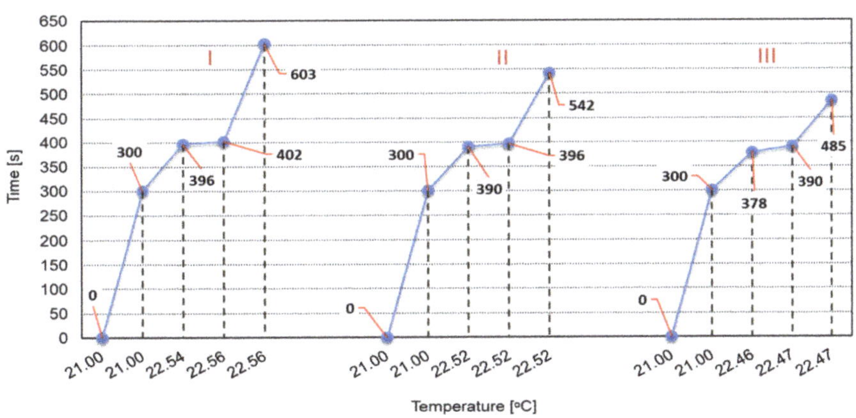

Figure 4. Time characteristics of individual combustion stages of a of birch tree: I (nonconditioned with a pulsed electric field) and conditioned with a pulsed electric field: II (electric field intensity 20 kV·cm^{-1} and number of pulses 300); III (electric field intensity 30 kV·cm^{-1} and number of pulses 300).

It was found that both the entire combustion process and the main period (from ignition to reaching the maximum temperature) lasted the longest for the sample nonconditioned with a pulsed electric field. These times were 603 s and 96 s, respectively. As the electric field intensity increased, the periods mentioned were decreasing and in the case of material under the influence of PEF with an electric field intensity of 30 kV·cm^{-1}, where the number of pulses was 300, they were shorter by 118 s and 18 s, respectively. These differences are significant in terms of absolute values. Much less within the range of interest for the setting phenomenon, it represents a significant proportion of the value. It proves the measurable effect of the pulsed electric field on the biological material

3.2. Oak Sawdust

Oak sawdust samples, both those that were not conditioned with PEF and those that were the control sample, had analytical moisture at the level of 7.2% ± 0.1%. Analyzing the temperature characteristics during combustion of biological material, i.e., oak sawdust, previously subjected to the impulse electric field (Figure 5) and without the influence of PEF, it was found that the temperature differences are insignificant—the temperature changes reached a maximum of three hundredths of a degree Celsius. The highest temperature was obtained for PEF with an electric field strength of 20 kV·cm^{-1}, where the number of pulses was 300. The lowest combustion temperature concerned sawdust that was not subjected to the pulsating electric field. As in the case of sawdust from birch wood, the obtained differences are not statistically significant and are not of great importance for the entire analyzed process.

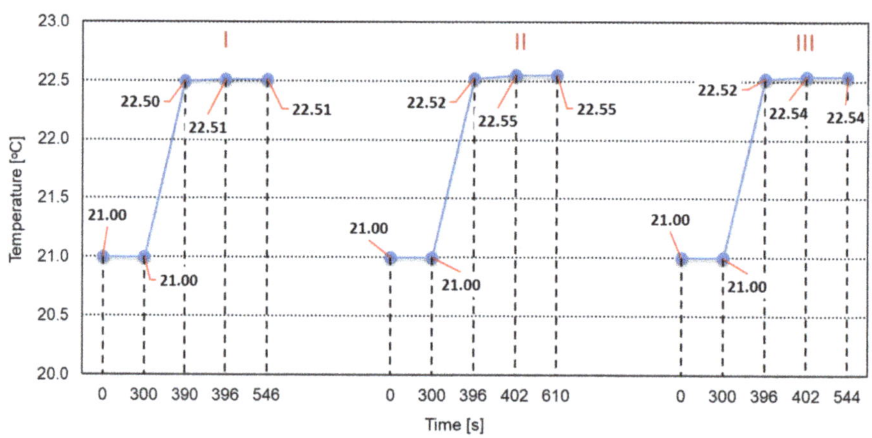

Figure 5. Temperature characteristics of individual variants of oak three combustion: I (nonconditioned with a pulsed electric field) and conditioned with a pulsed electric field: II (electric field intensity 20 kV·cm^{-1} and number of pulses 300); III (electric field intensity 30 kV·cm^{-1} and number of pulses 300).

In the case of the time from ignition to reaching the maximum temperature, no significant differentiation of the obtained results was observed, but the total burning time of the materials was clearly different for individual variants. Figure 6 shows the time characteristics of individual stages of calorimetric tests on oak sawdust previously exposed to a pulsating electric field, compared to the control sample. It was found that the longest duration of the discussed process was obtained for the interaction of PEF with the parameters of the second combination (electric field strength 20 kV·cm^{-1} and number of pulses 300), while the shortest time was obtained for the third combination of the experiment.

Referring to the characteristics representing the time of individual stages of the combustion process, it can be seen that in the second phase, from the recording of the second temperature to the recording of the third temperature, the values for the material treated with PEF were the same. At the same time, this stage was extended by 6 s in relation to the analogous stage of burning the unconditioned sample. In the case of the last combustion phase (the difference between the time values recorded for the fifth and fourth temperatures), much greater differences in their duration were noted. It should be emphasized that there was a difference of over a minute in the duration of the phase between the combinations of experiments II and II, where the differentiation of the stimulation resulted only from the applied intensity of the electric field of the discharge with the same number of pulses. The difference between combinations I and III is insignificant and amounts to 8 s. The trend noted for birch wood was not repeated in the case of beech wood.

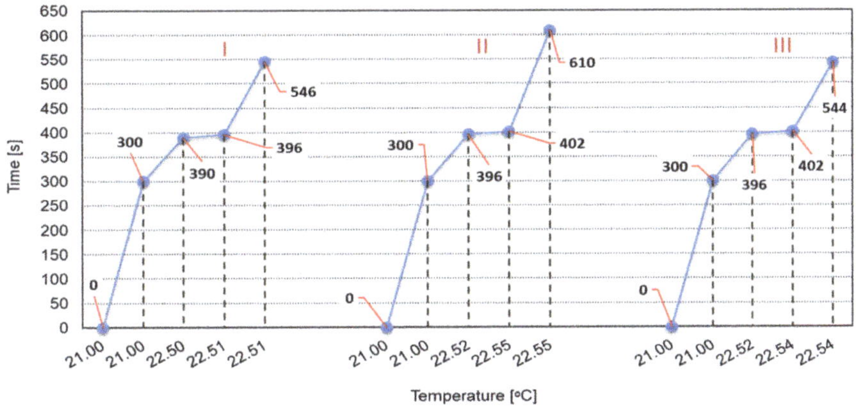

Figure 6. Time characteristics of individual combustion stages of oak tree: I (nonconditioned with a pulsed electric field) and conditioned with a pulsed electric field: II (electric field intensity 20 kV·cm^{-1} and number of pulses 300); III (electric field intensity 30 kV·cm^{-1} and number of pulses 300).

3.3. Linden Sawdust

The linden sawdust samples were characterized by analytical moisture at the level of 8.1%. As a result of exposure to a pulsating electric field, their humidity decreased by 0.1%. This difference was within the assumed error limit. Analyzing the temperature characteristics during combustion of biological material, which was linden sawdust previously subjected to the impulse electric field (Figure 7), a tendency was found that was almost identical to that for beech wood. The differences within the recorded temperatures were insignificant for all three variants, and the maximum difference in absolute values was recorded for variants II and III and amounted to two hundredths of a degree Celsius. The main period lasted the same for the material not subjected to the pulsed electric field as for the interaction with the field with the discharge intensity equal to 20 kV·cm^{-1} and the number of pulses equal to 300. From a statistical point of view, the obtained results do not differ significantly.

Similarly to oak sawdust, the main period (time from ignition to reaching the maximum temperature) for burning lime sawdust did not differ between materials stimulated with a field intensity of 20 kV·cm^{-1} and 30 kV·cm^{-1}. Figure 8 shows the time characteristics of individual stages of calorimetric tests on linden sawdust previously exposed to a pulsating electric field, compared to the control sample. The duration of the discussed process was longer by 6 s in the case of variants II and III of the experiment than for the sample not stimulated with the pulsating electric field. In terms of absolute values, the observed time differences are relatively small. However, when analyzing the phenomenon as a whole and taking into account the specificity of the measurement of the calorific value, including the mass of the test sample and the mass of the medium receiving thermal energy, the obtained discrepancies constitute a significant value.

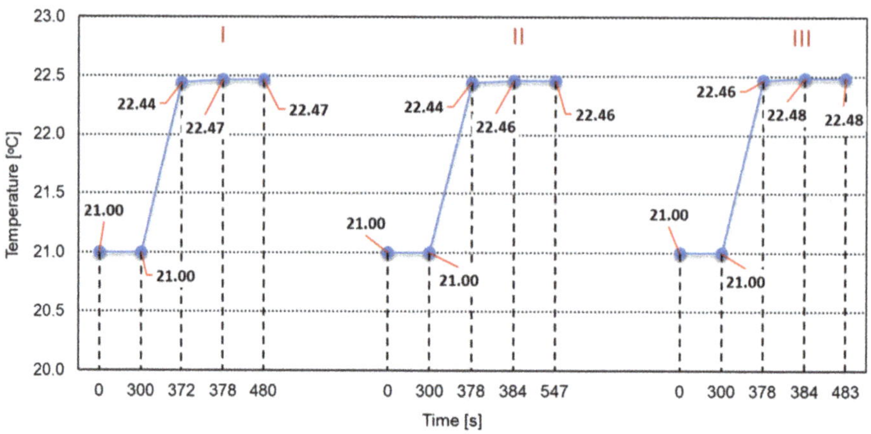

Figure 7. Temperature characteristics of individual variants of linden three combustion: I (nonconditioned with a pulsed electric field) and conditioned with a pulsed electric field: II (electric field intensity 20 kV·cm^{-1} and number of pulses 300); III (electric field intensity 30 kV·cm^{-1} and number of pulses 300).

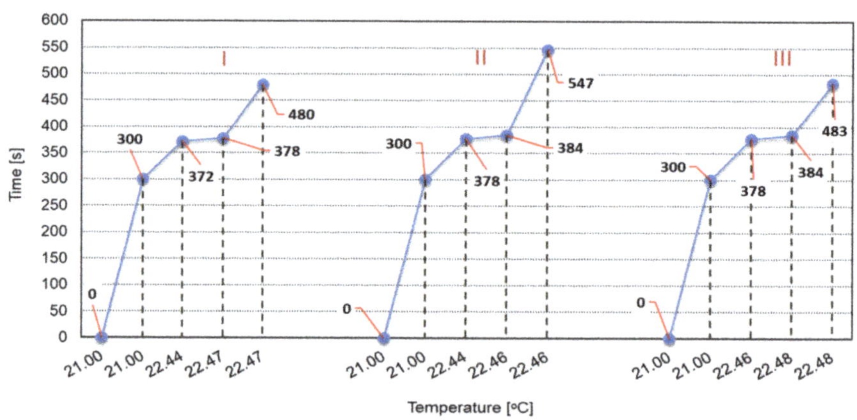

Figure 8. Time characteristics of individual combustion stages of a fir tree: I (nonconditioned with a pulsed electric field) and conditioned with a pulsed electric field: II (electric field intensity 20 kV·cm^{-1} and number of pulses 300); III (electric field intensity 30 kV·cm^{-1} and number of pulses 300).

The analysis of the total burning time of lime wood sawdust indicates a trend similar to that observed in the case of oak sawdust. The longest time, 547 s, was obtained for PEF with an electric field strength of 20 kV·cm^{-1}, where the number of pulses was 300. This is an increase of nearly 14% compared to the PEF-untreated sample. Again, these differences are a result of the lengthening of the last analyzed period; that is, between the fourth and fifth temperature recorded. This relationship is not in line with the trend obtained for birch sawdust, where increasing the field intensity decreased the duration of the entire combustion process. There was no significant difference between variant I and variant III. Only the discussed variant II differs significantly from the others.

3.4. Ash Sawdust

Ash sawdust samples, both those that were not conditioned with PEF and those that were the control sample, had analytical moisture at the level of 8.6% ± 0.1%. By analyzing the temperature characteristics during combustion of the biological material, which was ash tree (Figure 9), as in the case of other tree species discussed in this study, no statistically significant differences were found in the characteristic temperatures of the combustion process in the calorimetric bomb. As for birch, a slight decrease in the maximum and final temperature was observed for the subsequent variants of the experiment. The differences were small and amounted to a maximum of seven hundredths of a unit.

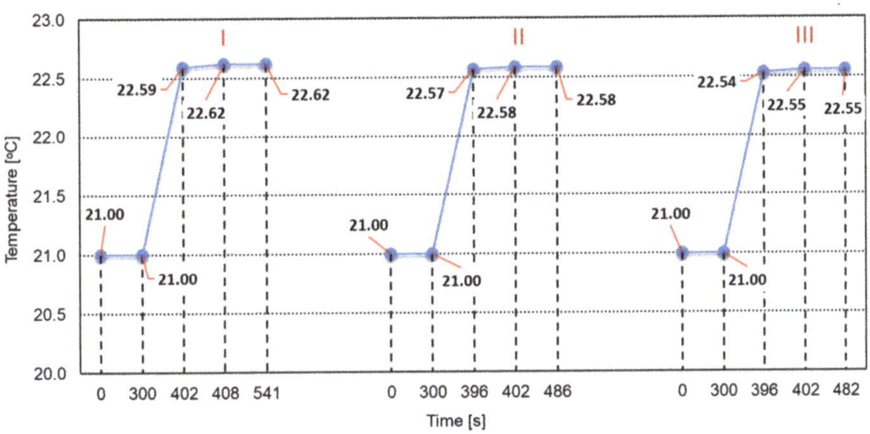

Figure 9. Temperature characteristics of individual variants of ash three combustion: I (nonconditioned with a pulsed electric field) and conditioned with a pulsed electric field: II (electric field intensity 20 kV·cm^{-1} and number of pulses 300); III (electric field intensity 30 kV·cm^{-1} and number of pulses 300).

Figure 10 shows the time characteristics of individual stages of calorimetric tests on ash sawdust previously exposed to a pulsating electric field, compared to the control sample. As for oak and linden, there were no significant differences in the time between ignition and reaching the maximum temperature. This statement applies to all three analyzed variants. These differences did not exceed the value of 6 s, and the longest time was recorded for the sample not subjected to stimulation by pulsed electric field.

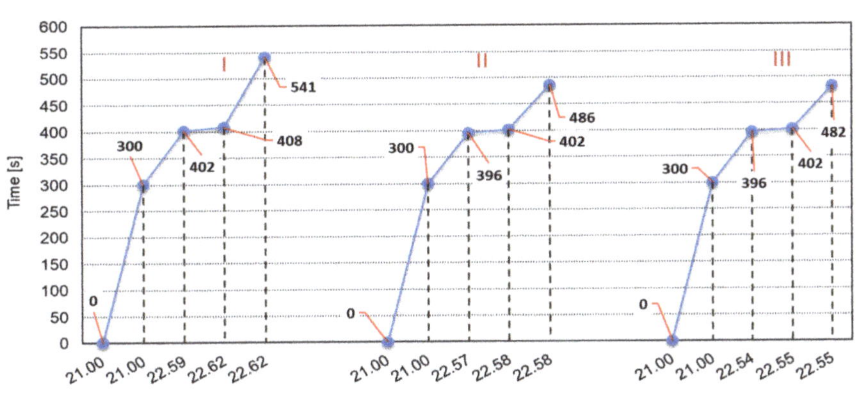

Figure 10. Time characteristics of individual combustion stages of ash tree: I (nonconditioned with a pulsed electric field) and conditioned with a pulsed electric field: II (electric field intensity 20 kV·cm^{-1} and number of pulses 300); III (electric field intensity 30 kV·cm^{-1} and number of pulses 300).

Similar to birch, and also in the case of ash sawdust, a trend was noticed according to which the duration of the combustion process, and above all the last of the analyzed periods, decreased with increasing field intensity. For the control sample (variant I), the difference between the fifth and fourth recorded temperature was 133 s, while for the material stimulated with a field with a intensity of 20 kV·cm^{-1} (variant II) and 300 pulses was 84 s, and for 30 kV·cm^{-1} (variant III) and the same number of pulses—80 s. It was stated that the resulting differences are statistically significant and are of great importance for the entire analyzed process.

4. Discussion and Conclusions

The use of pulsed electric field (PEF) to stimulate, condition, or change some properties of biological materials is the subject of numerous studies [36–51]. However, the scientific literature lacks the use of PEF for wood, especially in the context of changes that could affect its energy use.

The heat of combustion of firewood is the basic parameter for determining its energy potential. During calorimetric tests, modern calorimeters automatically determine the time–temperature characteristics. They have not been the subject of interest so far, and their role has been limited to only one of the stages of the process determining the value of the heat of combustion. It seems, however, that in addition to the amount of energy produced, the time in which it was released is also important, especially during commercial use of wood as a raw energy material.

Rhen et al. [52] presented the results of research on the combustion time of coal from the pyrolysis of spruce sawdust pellets, suggesting at the same time that knowledge of the fuel combustion time is necessary for the proper control of this process. The authors obtained results ranging from 4.5 to 6.5 min, and in their opinion, these differences mainly result from the type of fuel used. There is few research on combustion time. The situation is different with the combustion temperature. The authors of such publications [53–55] believe that the most important role is played by the ignition point and the peak temperature. The aforementioned researchers point out that the first of them, regardless of the type of fuel, has similar values, while the second depends mainly on the content of volatile substances. The time–temperature characteristics of fuel combustion, which is an element of the combustion heat determination process, allow for the analysis of both these parameters, i.e., the duration and temperature obtained at individual stages of the process. The research on the influence of PEF on lignocellulosic material carried out so far indicates the formation of the phenomenon of electroporation causing changes in the structure of the cell membrane. The effect of this is the loss of semi-permeability, which leads to the passage of enzymes hydrolyzing polysaccharides and lignin [29,30]. Such an effect was already obtained at an electric field of 1 to 10 kV·cm^{-1}, with which wood chips were treated [56]. Still, the described mechanisms have not been precisely parameterized [57]. It seems, however, that the same mechanisms may cause differences in the temperature and duration of individual stages of the combustion process.

Both the studies presented in this paper and the authors' previous studies [32] have shown that the pulsating electric field affects the temperature–time characteristics of the wood combustion process. The previously published results allowed us to conclude that the dominant factor of changes in the combustion characteristics of biological material treated with PEF is the electric field intensity. However, extending the research to new types of wood and limiting to two combinations of stimulation parameters partially changes the proposed thesis. The key issue to note is the fact that in each of the analyzed variants, the value of the heat of combustion (expressed in energy units per unit of mass) is the same. Slight temperature differences in the subsequent periods of the combustion process between the variants used do not cause changes exceeding the standardized repeatability limit. The situation is different in the case of time measurements. The observed changes are most pronounced in the last period of the combustion process, i.e., from the moment of reaching the maximum temperature to the end of the process. These results seem to be interesting because, as can be concluded from the research, it is possible to modify the time of obtaining thermal energy depending on one's needs. Thus, it becomes possible to accelerate certain technological processes based on the production of energy from fuel combustion, or, on the contrary, to extend the time of energy release, which is important, for example, from the point of view of central heating installations in buildings.

Therefore, we propose introducing the term of rate of heat of combustion, which will cover both the amount of heat produced and the time of its release. Its basic unit should be the expression $kW \cdot kg^{-1}$, which is also used in some fields to express the so-called specific power.

The presented results indicate that birch and ash react to PEF conditioning in a similar manner. The second group, due to the similarity of the obtained results, comprises oak and linden. Within both groups, the trees were compared in terms of their botany, especially the content of elementary chemical elements, the composition and structure of cell membranes, physical properties (weight density, porosity, contractility), as well as anatomical features (size and arrangement of vessels in the growth ring). It would seem that the main factor of changes should be porosity, which affects electroporation, and this in turn is the basis for modifications caused in wood during treatment with a pulsed electric field. However, the only demonstrated correlation between the obtained results and the biological, physical, and chemical parameters of the tested samples concerned the average content of pentosan-free cellulose in wood. Its values are 45.50% for birch, 45.50% for ash, 42.80% for oak, and 43.16% for linden, respectively [58]. Therefore, it seems that it is the cellulose content (in addition to the electric field strength shown in previous studies) that will be the factor determining the effect caused by PEF in wood.

Analyzing the obtained data in terms of their utilitarian use, an important parameter is the energy efficiency of the described process. The highest recorded consumption of electricity needed to generate an electric impulse in the used station is equal to 71.3 mWh, as described in the introduction [31]. Taking into account the amount of material that is introduced into the cell and the number of pulses used in each variant, the energy needed to stimulate 1 kg of sawdust is about 17.1 MJ. Referring again to the introduction to this paper—where, as an example, the amount of energy obtained from the combustion of 1 kg of birch sawdust, which is in the range from 17.90 to 18.72 $MJ \cdot kg^{-1}$ [12,13], is possible without any analysis—this process is unprofitable from an energy point of view. It should be emphasized, however, that the devices used by the authors were made in 2017, and its subsequent modifications concerned the control elements, and not the components responsible for generating the pulse. It seems that the use of the most modern methods would be able to reduce the amount of energy needed during stimulation, but research is still required to reduce the energy consumption of this process. Likewise, further experimentation is needed to cover a wider range of plants and other electric field parameters, leading to the parameterization of the described phenomenon.

Author Contributions: Conceptualization, P.K. and E.P.; methodology, P.K.; software, E.P.; validation, P.K. and E.P.; formal analysis, E.P.; investigation, E.P.; resources, P.K.; data curation, P.K.; writing—original draft preparation, E.P.; writing—review and editing, P.K.; visualization, E.P.; supervision, P.K. All authors have read and agreed to the published version of the manuscript.

Funding: This research received no external funding.

Institutional Review Board Statement: Not applicable.

Informed Consent Statement: Not applicable.

Data Availability Statement: The data presented in this study are available upon request of the respective author. The data are not publicly available due to the possibility of their commercial use by the unit in which the authors are employed and the use of source data for further research and development works of an accredited laboratory (Polish accreditation number AB 1698), whose staff are the authors of this publication.

Conflicts of Interest: The authors declare no conflict of interest.

References

1. Balat, M.; Acici, N.; Ersoy, G. Trends in the Use of Biomass as an Energy Source. *Energy Sources Part B Econ. Plan. Policy* **2006**, *1*, 367–378. [CrossRef]
2. Uddin, M.; Techato, K.; Taweekun, J.; Rahman, M.; Rasul, M.; Mahlia, T.M.I.; Rahman, S.M.A. An Overview of Recent Developments in Biomass Pyrolysis Technologies. *Energies* **2018**, *11*, 3115. [CrossRef]
3. Papoutsidakis, M.; Symeonaki, E.; Tseles, D.; Drosos, C. The biomass as an energy source and its application benefits. *Int. J. Eng. Appl. Sci. Technol.* **2018**, *2*, 1–5.
4. Spîrchez, C.; Lunguleasa, A.; Croitoru, C. The importance of the wood biomass in environment protection. *AIP Conf. Proc.* **2017**, *1918*, 20007.
5. Ko, J.H.; Kim, W.C.; Cho, J.S.; Choi, Y.I.; Park, E.J.; Im, J.H.; Han, O.; Keathley, D.; Han, K.H. EliteTree TM: An advanced biomass tree crop technology that features greater wood density and accelerated stem growth. *Biofuels Bioprod. Biorefin.* **2017**, *11*, 521–533. [CrossRef]
6. Karlsson, O.; Sidorova, E.; Moren, T. Influence of Heat Transferring Media on Durability of Thermally Modified Wood. *Bioresources* **2011**, *6*, 356–372. [CrossRef]
7. Nanda, S.; Reddy, S.N.; Vo, D.V.N.; Sahoo, B.N.; Kozinski, J.A. Catalytic gasification of wheat straw in hot compressed (subcritical and supercritical) water for hydrogen production. *Energy Sci. Eng.* **2018**, *6*, 448–459. [CrossRef]
8. Norouzi, O.; Safari, F.; Jafarian, S.; Tavasoli, A.; Karimi, A. Hydrothermal gasification performance of Enteromorpha intestinalis as an algal biomass for hydrogen-rich gas production using Ru promoted Fe–Ni/γ-Al2O3 nanocatalysts. *Energy Convers. Manag.* **2017**, *141*, 63–71. [CrossRef]
9. De Araujo Guilherme, A.; Dantas, P.V.F.; Padilha, C.E.D.A.; dos Santos, E.S.; de Macedo, G.R. Ethanol production from sugarcane bagasse: Use of different fermentation strategies to enhance an environmental-friendly process. *J. Environ. Manag.* **2019**, *234*, 44–51. [CrossRef]
10. Toscano, G.; Foppa, P.E. Calorific value determination of solid biomass fuel by simplified method. *J. Agric. Eng.* **2009**, *3*, 1–6. [CrossRef]
11. Bell, D.A.; Towler, B.F.; Fan, M. (Eds.) The Nature of Coal. In *Coal Gasification and Its Applications*, 1st ed.; William Andrew Publishing: Norwich, NY, USA, 2011; Volume 1, pp. 1–15.
12. Günther, B.; Gebauer, K.; Barkowski, R.; Rosenthal, M.; Bues, C.T. Calorific value of selected wood species and wood products. *Eur. J. Wood Wood Prod.* **2012**, *70*, 755–757. [CrossRef]
13. Dibdiakova, J.; Wang, L.; Li, H. Heating value and ash content of downy birch forest biomass. *Energy Procedia* **2017**, *105*, 1302–1308. [CrossRef]
14. Demirbaş, A. Effects of moisture and hydrogen content on the heating value of fuels. *Energy Sources Part A Recovery Util. Environ. Eff.* **2007**, *29*, 649–655. [CrossRef]
15. Charbel, A.T.; Trinchero, B.D.; Morais, D.D.; Mesquita, H.; Birchal, V.S. Evaluation of the Potential of Fruit Peel Biomass after Conventional and Microwave Drying for Use as Solid Fuel. *Appl. Mech. Mater.* **2015**, *798*, 480–485. [CrossRef]
16. Aniszewska, M.; Gendek, A. Comparison of heat of combustion and calorific value of the cones and wood of selected forest trees species. *For. Res. Pap.* **2014**, *75*, 231–236. [CrossRef]
17. *European Standard EN ISO 18125:2017*; Solid Biofuels—Determination of Calorific Value. ISO: Geneva, Switzerland, 2007.
18. Jayakumar, E.; Chittibabu, S.; Shanmugasundaram, S.; Lope, G.T. Synergetic effect of microwave heated alkali pre-treatment on densification of rice (*Oryza sativa*) husk biomass grinds. *Energy Sources Part A Recovery Util. Environ. Eff.* **2019**. [CrossRef]
19. Phanphanich, M.; Mani, S. Impact of torrefaction on the grindability and fuel characteristics of forest biomass. *Bioresour. Technol.* **2011**, *102*, 1246–1253. [CrossRef]
20. Yu, N.; Cai, Y.; Li, X.; Fan, Y.; Yin, H.; Zhang, R. Catalytic pyrolysis of rape straw for upgraded bio-oil production using HZSM-5 zeolite. *Nongye Gongcheng Xuebao Trans. Chin. Soc. Agric. Eng.* **2014**, *30*, 264–271.
21. Hassan, S.S.; Williams, G.A.; Jaiswal, A.K. Emerging technologies for the pretreatment of lignocellulosic biomass. *Bioresour. Technol.* **2018**, *262*, 310–318. [CrossRef]
22. Zhao, X.; Zhang, L.; Liu, D. Pretreatment of Siam weed stem by several chemical methods for increasing the enzymatic digestibility. *Biotechnol. J.* **2010**, *5*, 493–504. [CrossRef]
23. Aguilera, J.M.; Chiralt, A.; Fito, P. Food dehydration and product structure. *Trends Food Sci. Technol.* **2003**, *14*, 432–437. [CrossRef]
24. Welti-Chanes, J.; Guerrero, J.; Bárcenas, M.E.; Aguilera, J.; Vergara, F.; Barbosa-Cánovas, G. Glass transition temperature (Tg) and water activity (aw) of dehydrated apple products. *J. Food Process Eng.* **2007**, *22*, 91–101. [CrossRef]
25. Donsì, F.; Ferrari, G.; Pataro, G. Applications of Pulsed Electric Field Treatments for the Enhancement of Mass Transfer from Vegetable Tissue. *Food Eng. Rev.* **2010**, *2*, 109–130. [CrossRef]
26. Toepfl, S. Pulsed Electric Fields (PEF) for Permeabilization of Cell Membranes in Food- and Bioprocessing—Applications, Process and Equipment Design and Cost Analysis. Ph.D. Thesis, Technischen Universität, Berlin, Germany, 2006.
27. Tylewicz, U.; Tappi, S.; Mannozzi, C.; Romani, S.; Dellarosa, N.; Laghi, L.; Ragni, L.; Rocculi, P.; Dalla Rosa, M. Effect of pulsed electric field (PEF) pre-treatment coupled with osmotic dehydration on physico-chemical characteristics of organic strawberries. *J. Food Eng.* **2017**, *4*, 66–78. [CrossRef]
28. Wiktor, A.; Witrowa-Rajchert, D. Applying Pulsed Electric Field to enhance plant tissue dehydration process. *Żywność Nauka Technol. Jakość* **2012**, *2*, 22–32. [CrossRef]

29. Kumar, A.K.; Sharma, S. Recent updates on different methods of pretreatment of lignocellulosic feedstocks: A review. *Bioresour. Bioprocess* **2017**, *4*, 7. [CrossRef]
30. Barba, F.J.; Parniakov, O.; Pereira, S.A.; Wiktor, A.; Grimi, N.; Boussetta, N.; Saraiva, J.A.; Raso, J.; Martin-Belloso, O.; Witrowa-Rajchert, D.; et al. Current applications and new opportunities for the use of pulsed electric fields in food science and industry. *Food Res. Int.* **2015**, *77*, 773–798. [CrossRef]
31. Wesolowski, M.; Necka, K.; Drozdz, T.; Kiełbasa, P. The concept of modeling a pulsed electric field discharge (PEF) in products of the agri-food industry. *Prz. Elektrotech.* **2018**, *94*, 119–123. (In Polish)
32. Kiełbasa, P.; Dróżdż, T.; Popardowski, E. Influence of Coniferous Wood Conditioning by Pulsed Electric Field on Its Combustion Heat Characteristics. *Appl. Sci.* **2021**, *11*, 983. [CrossRef]
33. *European Standard EN ISO 14780:2017-07*; Solid Biofuels—Sample Preparation. ISO: Geneva, Switzerland, 2017.
34. *European Standard EN ISO 18134-3:2015*; Solid Biofuels—Determination of Moisture Content—Oven Dry Method—Part: Moisture in General Analysis Sample. ISO: Geneva, Switzerland, 2015.
35. *Standard ISO 3534-1:2006*; Statistics—Vocabulary and Symbols—Part: General Statistical Terms and Terms Used in Probability. ISO: Geneva, Switzerland, 2006.
36. Sepulveda, D.R.; Gongora-Nieto, M.M.; Guerrero, J.A.; Barbosa Canovas, G.V. Shelf life of whole milk processed by pulsed electric fields in combination with PEF-generated heat. *Food Sci. Technol.* **2009**, *42*, 735–739. [CrossRef]
37. Monfort, S.; Gayán, E.; Condón, S.; Raso, J.; Álvarez, I. Design of a combined process for the inactivation of *Salmonella enteritidis* in liquid whole egg at 55 °C. *Int. J. Food Microbiol.* **2011**, *145*, 476–482. [CrossRef] [PubMed]
38. Monfort, S.; Saldaña, G.; Condón, S.; Raso, J.; Álvarez, I. Inactivation of *Salmonella* spp. in liquid whole egg using pulsed electric fields, heat, and additives. *Food Microbiol.* **2012**, *30*, 393–399. [CrossRef] [PubMed]
39. Nguyen, P.; Mittal, G.S. Inactivation of natturally occuring microorganisms in tomato juice using pulsed electric field (PEF) with and without antimicrobials. *Chem. Eng. Process.* **2007**, *46*, 360–365. [CrossRef]
40. Mosqueda-Melgar, J.; Raybaudi-Massilia, R.M.; Martin-Belloso, O. Non-thermal pasteurization of fruit juices by combining high-intensity pulsed electric fields with natural antimicrobials. *Innov. Food Sci. Emerg. Technol.* **2008**, *9*, 328–340. [CrossRef]
41. Grimi, N.; Mamouni, F.; Lebovka, N.; Vorobiev, E.; Vaxelaire, J. Impact of apple processing modes on extracted juice quality: Pressing assisted by pulsed electric fields. *J. Food Eng.* **2011**, *103*, 52–61. [CrossRef]
42. Liang, Z.; Cheng, Z.; Mittal, G.S. Inactivation of spoilage microorganisms in apple cider using a continuous flow pulsed electric field system. *Food Sci. Technol.* **2006**, *39*, 351–357. [CrossRef]
43. Iu, J.; Mittal, G.S.; Griffiths, M.W. Reduction in levels of *Escherichia coli* O157:H7 in apple cider by pulsed electric fields. *J. Food Prot.* **2001**, *64*, 964–969. [CrossRef]
44. Abenoza, M.; Benito, M.; Saldaña, G.; Álvarez, I.; Raso, J.; Sánchez-Gimeno, A.C. Effects of pulsed electric field on yield extraction and quality of olive oil. *Food Bioprocess Technol.* **2013**, *6*, 1367–1373. [CrossRef]
45. Lamaunauskas, N.; Pataro, G.; Bobinas, C.; Satkauskas, S.; Viskelis, P.; Bobonaite, R.; Ferarri, G. Impact of pulsed electric field treatment on juice yield and recovery of bioactive compounds from raspberries and their by-products. *Zemdirb. Agric.* **2016**, *103*, 83–90. [CrossRef]
46. Wu, Y.; Guo, Y.; Zhang, D. Study of the Effect of High-Pulsed Electric Field treatment on Vacuum Freeze-Drying of Apples. *Dry. Technol.* **2011**, *29*, 1714–1720. [CrossRef]
47. Wu, Y.; Zhang, D. Effect of Pulsed Electric Field on Freeze-Drying of Potato Tissue. *Int. J. Food Eng.* **2014**, *10*, 857–862. [CrossRef]
48. Dalvi-Isfahan, M.; Hamdami, N.; Le-Bail, A.; Xanthakis, E. The principles of high voltage electric field and its application in food processing: A review. *Food Res. Int.* **2016**, *89*, 48–62. [CrossRef] [PubMed]
49. Raschke, D. Pulsed Electric Fields—Influence on Physiology, Structure and Extraction Processes of the Oleaginous Yeast *Waltomyces lipofer*. Ph.D. Thesis, Technischen Universität, Berlin, Germany, 2010.
50. Bobinaitė, R.; Pataro, G.; Raudonis, R.; Vškelis, P.; Bobinas, C.; Šatkauskas, S.; Ferrari, G. Improving the extraction of juice and anthocyanin compounds from blueberry fruits and their by-products by pulsed electric fields. *IFMBE Proc.* **2015**, *53*, 363–366.
51. Mahnič-Kalamiza, S.; Vorobiev, E.; Miklavčič, D. Electroporation in food processing and biorefinery. *J. Membr. Biol.* **2014**, *247*, 1279–1304. [CrossRef] [PubMed]
52. Rhen, C.; Ohman, M.; Gref, R.; Wasterlund, I. Effect of raw material composition in woody biomass pellets on combustion characteristics. *Biomass Bioenergy* **2007**, *31*, 66–72. [CrossRef]
53. Haykırı-Acma, H. Combustion characteristics of different biomass materials. *Energy Convers. Manag.* **2003**, *44*, 155–162. [CrossRef]
54. Kucukbayrak, S.; Haykırı-Acma, H.; Ersoy-Mericboyu, A.; Yaman, S. Effect of lignite properties on reactivity of lignite. *Energy Convers. Manag.* **2001**, *42*, 613–626. [CrossRef]
55. Haykırı-Acma, H.; Ersoy-Mericboyu, A.; Kucukbayrak, S. Effect of mineral matter on the reactivity of lignite chars. *Energy Convers. Manag.* **2001**, *42*, 11–20.
56. Kumar, P.; Barrett, D.M.; Delwiche, M.J.; Stroeve, P. Pulsed electric field pretreatment of switchgrass and wood chip species for biofuel production. *Ind. Eng. Chem. Res.* **2011**, *50*, 10996–11001. [CrossRef]
57. Qin, J.; Pasko, V.P. On the propagation of streamers in electrical discharges. *J. Phys. Part D Appl. Phys.* **2014**, *47*, 435202. [CrossRef]
58. Galewski, W.; Korzeniowski, A. *Atlas of the Most Important Types of Wood*, 1st ed.; Panstwowe Wydawnictwo Rolnicze i Lesne: Warsaw, Poland, 1958; pp. 52–55, 71–75, 169–172, 206–208.

MDPI
St. Alban-Anlage 66
4052 Basel
Switzerland
www.mdpi.com

Applied Sciences Editorial Office
E-mail: applsci@mdpi.com
www.mdpi.com/journal/applsci

www.ingramcontent.com/pod-product-compliance
Lightning Source LLC
Chambersburg PA
CBHW061503220926
51850CB00004B/156
* 9 7 8 3 0 3 6 5 9 4 9 6 5 *